The Voices of Nature

The Voices of Nature

HOW AND WHY ANIMALS COMMUNICATE

NICOLAS MATHEVON

WITH A FOREWORD BY
BERNIE KRAUSE

ILLUSTRATIONS BY
BERNARD MATHEVON

PRINCETON UNIVERSITY PRESS

PRINCETON & OXFORD

Originally published in French under the title *Les animaux parlent—Sachons les écouter* © 2021 by Humensciences/Humensis

Published in English by Princeton University Press
41 William Street, Princeton, New Jersey 08540
99 Banbury Road, Oxford OX2 6JX

press.princeton.edu

Library of Congress Cataloging-in-Publication Data
Names: Mathevon, Nicolas, author. | Krause, Bernie, 1938- writer of foreword.
Title: The voices of nature : how and why animals communicate / Nicolas Mathevon; with a foreword by Bernie Krause.
Other titles: Animaux parlent. English
Description: Princeton : Princeton University Press, [2023] | "Originally published in French under the title Les animaux parlent—Sachons les écouter © 2021 by Humensciences/Humensis"—title page verso. | Includes bibliographical references and index.
Identifiers: LCCN 2022036324 (print) | LCCN 2022036325 (ebook) | ISBN 9780691236759 (hardcover) | ISBN 9780691236766 (ebook)
Subjects: LCSH: Animal communication. | Animal sounds. | BISAC: SCIENCE / Life Sciences / Zoology / Ethology (Animal Behavior) | SCIENCE / Life Sciences / Neuroscience
Classification: LCC QL776 .M37313 2023 (print) | LCC QL776 (ebook) | DDC 591.59/4—dc23/eng/20220729
LC record available at https://lccn.loc.gov/2022036324
LC ebook record available at https://lccn.loc.gov/2022036325

British Library Cataloging-in-Publication Data is available

Editorial: Alison Kalett & Hallie Schaeffer
Production Editorial: Ali Parrington
Jacket Design: Katie Osborne
Production: Danielle Amatucci
Publicity: Matthew Taylor & Kate Farquhar-Thomson
Copyeditor: Jennifer McClain

Jacket art and book illustrations by Bernard Mathevon

This book has been composed in Arno and Helvetica Neue

Printed on acid-free paper. ∞

Printed in the United States of America

10 9 8 7 6 5 4 3 2

To my beloved wife and children, Patricia, Elise,
Pauline, and Baptiste.

To my family and friends.

To my partners in adventure.

To my students and colleagues at the ENES
Bioacoustics Research lab.

To all those who have allowed my research.

To the boys and girls curious about the world around them.

To the naturalists.

To the next generation of bioacousticians.

To Paul Géroudet and Christian Zuber.

To all those who made me dream.

And to Thierry Aubin. Of course.

What is this faculty we observe in them, of complaining, rejoicing, calling to one another for succour, and inviting each other to love, which they do with the voice, other than speech? And why should they not speak to one another?

THE WORKS OF MONTAIGNE (W. HAZLITT, ED.; JOHN TEMPLEMAN, LONDON, 1369)

CONTENTS

Foreword by Bernie Krause   xi

Note to the reader   xv

1   Animal chatters: Tinbergen's four questions   1

2   Making circles in water: A short vade
mecum of physical acoustics   8

3   The warbler's eyebrows: Why do birds sing?   15

4   Cocktails between birds:
Noise and communication theory   32

5   Family dinner: Parent-offspring communication   46

6   Submarine ears: Underwater bioacoustics   65

7   The tango of the elephant seals: Vocal signals
and conflict ritualization   78

8   The caiman's tears: Acoustic communication
in crocodiles   88

9   Hear, at all costs: Mechanisms of audition   99

10   Tell me what you look like:
Production of sound signals   113

11 Networking addiction: Acoustic communication
    networks   139

12 Learning to talk: Vocal learning in birds
    and mammals   158

13 Inaudible speech: Ultrasounds, infrasounds,
    and vibrations   183

14 The laughing hyena: Communications
    and complex social systems   203

15 Ancestral fears: The acoustic expression
    of emotions   229

16 The booby's foot: Acoustic communications
    and sex roles   244

17 Listening to the living: Ecoacoustics
    and biodiversity   260

18 Words . . . words: Do animals have a language? 280

Acknowledgments   297
Glossary   305
Notes   309
Index   355

FOREWORD

If we ever hope to understand how a living system operates, we'd better learn to listen to its voice. On the other hand, the marvels of natural sound have been made clear on every page of Professor Nicolas Mathevon's book *The Voices of Nature: How and Why Animals Communicate*. With his new publication, Mathevon explores with us one of the more elusive portals, the voice of the natural world in all its varied parts. Professor Mathevon's narrative reminded me of the time in the mid-1970s when I served as a consultant on a doctoral committee headed by the late naturalist Gregory Bateson. The young candidate had called a meeting to discuss his frustration at having failed to answer one question central to the subject of his dissertation. After a short appeal for help by the student, everyone in the room grew quiet. And then Bateson spoke up. "Here's a story that might help," he said, directing his comments to the anxious young man. "At the end of World War I, the great English philosopher, Alfred North Whitehead, was invited to join Harvard's faculty. As a condition of his acceptance, Whitehead insisted that his good friend and colleague, Bertrand Russell, with whom he'd collaborated on the publication of the *Principia Mathematica*, be included as part of the package. During that period, Harvard faculty members were required to give a lecture introducing themselves and their work. Russell chose as his topic a clarification of Max Planck's *Quantum Theory*. Given on a hot August night, the chapel filled with Boston society and the school's faculty, Russell labored for ninety minutes in the sweltering heat. When he finished, exhausted and drenched in sweat, he turned away from the podium and walked slowly to his seat. After a moment of polite applause, Alfred Whitehead stood up, glanced kindly at his colleague, and moved to the podium. In his high-pitched,

English-accented voice, he declared, 'I'd like to thank Professor Russell for his brilliant exposition. And, especially for leaving unobscured the vast darkness of the subject.'"

To some extent the sonic world is not that difficult to understand. Many of us operate intuitively at various levels of success within its borders. But it is difficult to explain because we lack the precise words to express acoustic experience. Most of us have the capacity to hear. And some of us are even gifted with the ability to listen. What makes grasping the biophonic world even more tricky and obscure is that, in the West, we rely primarily on what we see. What we know, what we talk about, the objects we tend to enjoy, are mostly understood through a graphic lens. Yet, our vocabulary is shy of terms that define the elements of acoustic environments which we are all a part of, whether they be urban, rural, or wild.

In *The Voices of Nature*, Nicolas Mathevon meets these issues head-on, leading us eloquently along evolutionary paths of communication that have informed us from the time when at least half a dozen of our hominid species appeared on this planet. This is a terrific exposition into the world of natural sounds and the organisms that produce them. Filled in equal measure with accessible narratives of discovery, revelations, and the characters who have observed the patterns that connect, we are transfixed by Mathevon's passionate stories of The Others and the researchers who identified the behaviors as they unfolded. Like a fine novel, this notable work is impossible to put down once you begin.

The sense of wonder that emanates from those dedicated to this topic issues from every page of Mathevon's chronicle. From Peter Marler's studies of white-crowned sparrow dialects to Thierry Aubin's studies of parent-offspring vocal recognition in penguins, the sense of devotion to unraveling the secrets of biophony is inspirational. As the naturalist and professor from Pitzer College, Paul Shepard, once suggested, "We may never fully understand the permutations of the natural world. And perhaps we weren't meant to." I would add to that a quip from the late Professor Kenneth Norris, Emeritus, University of California, Santa Cruz, the marine biologist who discovered how dolphins echolocate. When we were trying to solve the mysteries of singing sand dunes in the American Southwest, Norris responded to my puerile complaint about the incredible amount of time we had spent without a clear result, "Bernie, we're having such a grand time searching for the answers, I hope we *never* find out!" *The Voices of Nature* brings new proximity to that immense journey and our reconnection to the route.

Bernie Krause
Author of *The Great Animal Orchestra:*
Finding the Origins of Music in the World's Wild Places
Wild Sanctuary, Sonoma, California
April 2021

NOTE TO THE READER

This book is accompanied by a website where you can listen to some examples of songs and other sound signals produced by animals: https://mathevon0.wixsite.com/soundjourneybook or https://tinyurl .com/228ns885 or scan this QR code:

Although intended for a wide audience, this book is also written for students and researchers in bioacoustics. If you are one of them, I encourage you to pay attention to the notes (numbered in the text and listed at the end of the book). You will find more details and, above all, a substantial and up-to-date list of papers published in international peer-reviewed journals (around 700 references). An excellent way to start a bibliography on a new subject.

One last message before you dive into the book: Feel free to send me (mathevon@univ-st-etienne.fr) the publications that would have raised your interest and that are not referenced in this book, including yours, of course!

1

Animal chatters

TINBERGEN'S FOUR QUESTIONS

Calanques National Park, near Marseille. It is noontime under the sun of Provence. The heat is intense, the light bright. The garigue smells of thyme, rosemary, and lavender. The background music ("tchik-tchik-tchik") is provided by the cicadas, pressing against the bark of the pine trees, their rostra stuck into the trunk to pump out the beneficial sap. In the deep blue sky, swifts glide like arrows: "Weer!! . . . Weer!!!" A large locust spreads its colorful wings and flies in front of me, then lands a few meters away. Its long hind legs oscillate rapidly as they rub against its wings, producing a strange chirping sound, like rustling sheets of paper. When the legs freeze, the sound stops. On its branch, a subalpine warbler emits its cheerful ritornello, briskly playing sometimes fluted, sometimes squeaky notes. Suddenly, the warbler falls silent and dives into a bush. It soon comes out to sing again. In the background, far away, some sheep bleat. A dog barks. Later, when the sun has waned and it is getting cooler, the cicadas will stop their relentless concert. Others will take their place, and the entire night will rustle with the song of locusts, grasshoppers, toads, and other nocturnal creatures, in an apparent cacophony. Just before daybreak, the dawn chorus of birds will come alive. The cicadas will remain silent, waiting for the heat to start vibrating their cymbals, those small membranes hidden under their wings, which rattle several hundred times a second. "Tchik-tchik-tchik . . . tchik-tchik-tchik . . . tchik-tchik-tchik. . . ." This is the great concert of life!

The soundscape of the Mediterranean scrubland is unique. There is this multitude of sounds produced by animals, to which are sometimes added the breath of wind in the trees and the sound of waves crashing on the shore. For me, they are associated with all the times I've spent in this region of France. Have you ever wondered why animals make these sounds? Not simply to charm our senses, of course: They are not intended for us. In fact, their purpose is to communicate. To *communicate*— that's a big word! And yet . . . these songs, cries, and other shrill sounds are *signals* that, like our human words, allow them to converse with other animals. What do they say to each other, you may ask? This is what I propose to discover in this book. You are about to enter worlds of sound, some of which are familiar but most of which are completely unknown to you and which you never even knew existed. How could you, since some of them are not even accessible to our ears?

Many animals exploit the sound-transmitting properties of water or air to communicate: to find a partner, to defend a territory, to signal the presence of a predator or food source, to collaborate in hunting, to recognize and interact with members of the group. These communications are essential for many species, including our own. We know this well— we whose articulated language demonstrates an incredible complexity, commensurate with that of our social interactions; we whose simple cries, from the moment we are born, signal our emotions and needs to other humans. The fact that animals are comparable to humans has been demonstrated by a great deal of scientific work over the last forty years. We can no longer set our species apart from other animals: each species has its own biological, ecological, social, and sometimes cultural characteristics that define its own world. Acoustic communication systems are therefore diverse, but all are worthy of interest. They are evidence of the diversity of life.[1]

How are animal vocalizations produced? What information do they contain? Can we understand animal languages? For a long time, the diversity of these sound worlds was difficult to access, but technical advances—such as the tape recorder and then the computer—have changed this. In recent decades, scientists have begun to read the scores of animal concerts and decipher their meaning.

I'm involved in the science that studies animal acoustic communications, called *bioacoustics*. Bioacousticians are working to decipher how animals make and hear sounds, what information is encoded in their sound signals, what this information is used for in their daily lives, and also how their acoustic communication systems have developed over the history of life. We will see that studying the richness and complexity of animal acoustic communication can help us understand how our own communication system works—our words, our laughter, our cries. Bioacousticians are a bit like Champollion, the French historian who deciphered Egyptian hieroglyphics using the Rosetta Stone, a fragment of a stele where the same text is written in several languages. To decode animal languages, many other methods must be used, but the goal remains the same: to decipher their meaning. The sounds produced by animals are signals carrying information whose meaning we are trying to decipher.

Bioacoustics is a discipline rooted in ethology, the science of animal and human behavior. The development of this branch of biology is relatively recent, dating back to the 1960s. In 1973, the Nobel Prize in Physiology and Medicine was awarded to three ethologists. The first was Konrad Lorenz. You may have already heard of this Austrian researcher, who became famous for his experiments on imprinting, in which memory of certain events or individuals is built up very quickly and very early in life. Lorenz discovered imprinting in his observations of geese. If goslings hatch in the presence of a human being, they consider that person to be their mother and follow him wherever he goes.[2] The second Nobel laureate was Karl von Frisch. He discovered the dance of the bees, this unique communication system through which the worker bee, on her return to the hive, can inform her sisters of the whereabouts of new flowers.[3] It is impressively precise: the angle formed between the axis of the bee's walk along one of the honeycombs and the vertical axis corresponds to the angle formed between the direction of the sun and the direction of the flowers when exiting the hive. Simply incredible! But there is more: the frequency of the vibrations of the insect's body and wings contains information about the amount of food provided by the flowers. It is by vibrating that the bee signals it is worthwhile to go on a shopping spree. The third researcher was Nikolaas Tinbergen. Of the three, he is my favorite. Tinbergen spent most of his career studying animal behavior using the experimental method.[4] He invented ways of questioning animals in order to understand the causes and consequences of their behavior. For example, in order to test whether it was the red spot on the herring gull's beak that caused the chicks to beg for food, he offered them various objects (sometimes simple sticks) more or less faithfully reproducing an adult's head and bearing a bigger or smaller spot, and in different colors. He then measured the intensity of the chick's behavioral response—its speed in beating the lure with its beak. Tinbergen thus highlighted the importance of the "red spot" signal in the parent-chick relationship in this seabird species. In addition to being a remarkable experimenter, he sought to formalize scientific research in ethology. He explained that in order to fully understand animal behavior, four questions had to be answered. This method is still valid today,[5] and every bioacoustician keeps Tinbergen's four questions in mind when studying sound communication:

(1) What are the *mechanisms* of the behavior I observe?
(2) What are the *evolutionary causes* that explain the existence of this behavior?
(3) How did this behavior *develop over the course of the individual's life*?
(4) What has the *evolutionary history* of this behavior been over geological time?

Let's look at these four fundamental points in more detail. Let's imagine, for example, that you want to understand why American robins, *Turdus migratorius*, sing in the spring and what the drivers of this communication are.

You first need to understand the *mechanisms* of both the production and reception of signals, i.e., the processes that lead an animal to produce a sound and those that explain a behavioral response to what it perceives—for example, to understand why, when a male robin hears another male robin singing, it responds by singing in turn and sometimes by attacking the intruder. What is it about the song that causes this reaction? First and foremost, there must be particular acoustic characteristics identifying the American robin, which ensure that its song is not confused with that of another animal species, especially another bird species. Second, why is the reaction aggressive? Is it, for example, because the robin is ready to reproduce, and the high level of sex hormones circulating in its blood increases its reactivity? If we want to study these proximal causes of the behavior, we need to describe the properties of the stimuli that provoke the robin's reaction, both external (the intruder's song) and internal (hormonal balances). We also want to understand all the physiological processes, from the reception of the stimulus (How does hearing work?) to the expression of the behavioral response (Why all this agitation? To defend one's territory?). To explore these questions, you can set up experiments in acoustic playback with a loudspeaker placed near where the robin sings, and question it directly: "Is this song a territorial signal for you?"

Once you have addressed the first of Tinbergen's questions, you can turn your interest to the second question: the *evolutionary causes* of the communication. Why has this singing behavior rather than another been favored during the evolution of the species? In other words, how does singing confer advantages that might explain why, once it appeared, it has been retained over time? Does singing increase a male robin's likelihood of being noticed by a female? Will an aggressive individual, singing louder, more often, and for longer than others, be more effective in defending its territory and food resources? These two aspects would increase his reproductive success, i.e., the number of young he fathers and who survives into adulthood. Singing behavior would then be favored by the two facets of sexual selection: intersexual selection—females prefer some singers to others—and intrasexual selection—males drive off insufficiently aggressive colleagues more easily. But beware of the other side of the coin: Doesn't singing like a madman increase the probability of being spotted and captured by a predator such as a hawk or any other bird of prey? You could hypothesize that natural selection may have limited this behavior and favored individuals inclined to sing less loudly. Thus, sexual selection, like natural selection—the two major evolutionary mechanisms identified by Charles Darwin—probably participates in the evolution of communication behavior. You can see that things are complicated and that establishing the evolutionary causes of sound communication is not easy: all behavior is the result of a balance between constraints that sometimes have the opposite effect. You should not forget that evolution is also very much subject to chance (so-called stochastic processes). Your task as an evolutionary biologist will certainly be very difficult.

Let's see if it's easier to answer the third question formulated by Tinbergen: How was this communication behavior *acquired during the life* of our American robin? At birth, the robin chick cannot sing. It simply makes short calls to beg its parents for food. In the weeks following hatching, its brain develops and the chick gradually acquires the ability to produce more complex vocalizations. It is then essential that the young chick be able to hear adult songs, which it will learn by imitating. How are the two types of processes articulated? There are the innate processes (a robin will never sing like a wren; it has a genetic predisposition to sing "American robin") and the acquired processes (the young robin learns to sing by imitating an adult). This is a vast field of investigation. We discuss it in detail in chapter 12.

The fourth question remains, which is by far the most difficult to address: What is the *evolutionary history* of the communication that you are studying? To put it plainly, what are the stages that gradually led from the ancestor of birds—a kind of dinosaur, perhaps emitting dinosaur vocalizations[6]—to a robin singing a song? Quite a story, isn't it? All animal species, including humans, are rooted in the depths of time and share common ancestors. While we are beginning to understand the evolutionary mechanisms of diversification of living species rather well, particularly with regard to their genetic heritage, anatomy, physiology, and morphology, reconstructing the evolution of behavior remains a challenge. How and when did birdsong emerge over the course of evolution? Why in some species is it only the males that sing, whereas in many other species females also vocalize? Is song an ancestral trait in both sexes? Were dinosaurs, the ancestors of today's birds, capable of producing sounds? Did they use them to communicate? Can we imagine a tyrannosaurus "singing" to call his or her partner? Did young tyrannosauruses learn their vocalizations by imitating an adult? When and how did this learning happen? Answering all these questions is difficult, if not impossible, because behavior leaves few fossil traces.[7] My great frustration as a bioacoustician is not being able to listen to and record extinct species. I dream of being able to record baby tyrannosauruses and then have their parents listen to these vocalizations. And to see their reactions! It's obviously unlikely that we'll ever be able to achieve this kind of thing—but who knows? Maybe one day we'll be able to reconstruct "real" dinosaurs from fossil genomes, like in *Jurassic Park*. A Japanese team is trying to revive the woolly mammoth in this way. However, there are scientific methods that make it possible to establish solid hypotheses about the evolution of communications. We'll talk about that too.

Most bioacousticians focus their research on only one of Tinbergen's four questions and therefore do not aim to understand all aspects of the sound communication being studied. But keeping all of these questions in mind provides a fertile framework for thinking. Even when one is interested in relatively simple mechanisms, such as how a sound stimulates the robin's eardrum and is then transformed into nerve potentials that can be interpreted by the bird's brain, it is useful to consider that these mechanisms have a history.

Therefore, to conduct research in bioacoustics, a solid knowledge of biology is required. Most of my PhD students and postdocs have years of study in zoology, anatomy, physiology, neurobiology, ecology, ethology, and evolution. But bioacoustics requires, like many other disciplines in the life sciences, proficiency in scientific fields other than biology. In bioacoustics, there is certainly the prefix *bio* (living), but there is above all the root *acoustic*. Studying animal sound communications requires an understanding of what a sound is. Follow me and hang on tight: we're going on a detour through the physics of acoustic waves. Don't worry; it's not that complicated, and it's important for our journey into the world of animal sound. If, however, physics gives you a serious headache, and after trying to read the following chapter (science does require some effort, after all) you have trouble understanding what I have written, I'll take you straight to chapter 3. There, we start our journey by venturing into the Brazilian rainforest. For now, let's try bravely—a little courage! I first explain what a sound is, how it propagates, and how it can be described.

2

Making circles in water

A SHORT VADE MECUM OF PHYSICAL ACOUSTICS

When I started looking for a place to prepare my PhD thesis—I was fascinated by birds at the time—I was lucky enough to enter what was probably the last laboratory to be housed in a private home. Jean-Claude Brémond was waiting for me there. Mr. Brémond—I never called him by his first name—was director of research at the famous French *Centre national de la recherche scientifique* (CNRS). He told me straightaway: "If you want to work with us, that's OK, but there won't be any research position for you after you've completed your thesis." His words had the merit of being clear, and I was grateful to him for that. As a high school teacher at the time, I was not looking for a paid research position. On the contrary, I was ready to accept anything to satisfy my passion. The walls of the small room where we met were covered with posters of birds, and the atmosphere enthused me. I jumped in with both feet, thinking I would soon be wandering through the woods to listen to the birds singing. It was the beginning of the computer age, and Thierry Aubin—then junior researcher at the CNRS and later my companion in so many expeditions—emerged from a jumble of electronic machines to introduce himself and begin my training. But instead of discussing the latest scientific discoveries about birds and their sound communications, the first lesson Thierry gave me was an acoustics course. I'm not sure I followed all his explanations well: I had a solid background as a naturalist, but was seriously out of my league in physics and mathematics. Even today, after so many years spent observing oscillograms and spectrograms, I am still learning. "A person who wants to work in bioacoustics must have a feeling for sound," Thierry once told me. Never forgetting this maxim, I always ask students who want to work on my research team if they have a feeling for sound. When the students are musicians or music lovers (this is common among bioacousticians), there is no doubt. Otherwise, I am always attentive to their answers. Studying animal acoustic communications requires an understanding of the physics of sound waves and their methods of analysis. To hope to decipher an animal language, one must first understand such notions as frequency, amplitude, modulation, and many others. We discover the essentials in this chapter. If you have some knowledge of physics, the following lines may make you smile. You will find my metaphors approximate, and I hope you will forgive me. But if this is the first time you have dealt with these concepts, don't worry; we'll walk through them together.

To understand the physical nature of a sound, imagine that you are having fun throwing a pebble into a pond and looking at the circles in the water that form and become larger and larger as they move away from the point where the pebble landed. These ripples, which start when the pebble hits the water and then move away from the point of impact in concentric circles, are waves. As a wavelet passes, the water level rises and then falls for a short distance. The water molecules do not move with the wave: they return to their initial position once the wave has passed. It is the waves that propagate through the water, not the water itself. In fact, if you put a small cork on the water, you can see that it moves up and down as the wave passes, but remains at the same distance from the point hit by the pebble. While our circles in the water form and propagate only on the surface of the pond in as many circles, sound waves propagate in three dimensions. To picture this, imagine a loudspeaker—or, better yet, look at one. Its centerpiece is a membrane that vibrates and makes the air in contact with it vibrate. This membrane is the equivalent of a pebble hitting the surface of the pond. The sound waves created by the speaker's membrane are small variations in the pressure of the air (or water in the case of an underwater speaker) that move apart, forming larger and larger spheres, like the ever-larger circles in the water created by our pebble.

The analogy of circles in water, of course, has its limits. While the pebble only disturbs the water's surface for a brief moment, creating a very small number of circles, the vibrating membrane of the loud-speaker produces successions of sound waves. Moreover, the speed of our sound waves is much greater than that of the circles in the water: about 340 meters per second if the sound propagates through the air and much faster if it propagates in water, where sound waves travel at 1500 meters (1.5 km) per second. The speed of sound in air or water, however, is infinitely slower than the speed of light, which is close to 300 million meters per second! This discrepancy between the speed of light and the speed of sound is particularly striking during a thunder-storm. When you see a flash of lightning streaking across the sky, the sound of thunder comes later, with a greater or lesser time lag. In 0.00001 seconds, light travels 3 kilometers, while sound waves need almost 9 seconds to travel the same distance. It is therefore possible to estimate the distance from the impact of lightning by counting the sec-onds between the lightning and the thunder, then multiplying the number of seconds by 340 meters: 9 seconds multiplied by 340 equals 3060 meters, or about 3 kilometers.

A sound is therefore a wave that is transmitted in a propagation me-dium. To exist, sound needs particles: air molecules or water molecules (or solid molecules). There can be no sound in a vacuum. The micro-phones of the Perseverance rover have thus succeeded in recording sound on the planet Mars, since this planet has an atmosphere.[8] How-ever, it is impossible for sound waves to travel from Mars to Earth (or vice versa), since they would have to travel through a vacuum. Intersidereal space, where air is absent, is totally silent. If the propagation medium is air, the sound wave corresponds to a variation in atmospheric pressure. We hear about atmospheric pressure (also referred to as barometric pressure) in weather reports: if the pressure rises, the weather is going to be fine; if it falls, a low-pressure system—i.e., rain—is coming. At-mospheric pressure is the result of gravity acting on the layer of air (the atmosphere) that surrounds our planet. Gravity, the force of attraction exerted by the Earth, gives this layer of air a weight, which exerts pressure on any object on Earth. We are not aware of this because atmospheric pressure surrounds us constantly. Yet it weighs on our shoulders. The only time we perceive it is when we suddenly climb to high altitude, in a cable car, for example, or in a poorly pressurized plane: we then feel this pressure in our eardrums, and our ears can become blocked. This is because the internal ducts of our ears have not had time to adapt to the significant variation in atmospheric pressure due to the change in altitude: the higher the altitude, the thinner the layer of air above us, the lighter its weight, and the lower the atmospheric pressure. While at sea level (altitude = 0 meters) it is slightly above 1000 hectopascals (hPa), it barely reaches 900 hPa at 1000 meters and 700 hPa at 3000 meters. At the peak of Mont Blanc in the Alps (4810 meters), the atmospheric pressure is around 550 hPa. In short, let's just remember that the air around us is under pressure and that this pressure varies by a few hundred hectopascals depending on the altitude. It also varies with the temperature and humidity of the air: the difference in atmospheric pressure between a rainy day and a sunny day is in the tens of hectopascals.

In comparison, the order of magnitude of air pressure variations due to the passage of sound waves is ridiculously low. Thus, the human ear only begins to perceive them when they reach 20 millionths of a pascal (Pa). If we take the pascal as a unit of measurement (100 times less than 1 hectopascal), the variation in atmospheric pressure due to a change in altitude of 1000 meters is on the order of 10,000 Pa; the variation caused by the passage from bad to good weather is around 40 Pa; and the variation caused by the passage of sound waves begins to be audible when it is equal to 0.000002 Pa! On the other hand, when the pressure differential caused by the passage of sound reaches 20 Pa, hearing becomes unbearable and even painful. If we take up the analogy of the pebble in the pond, the variations in atmospheric pressure due to variations in weather conditions would correspond to variations in the average water level in the pond on the order of a meter, while the height of our waves spreading out from the impact of the pebble and representing sound waves would be on the order of a millimeter. Our ears—and the ears of all animals, including insects—are therefore ultrasensitive sensors of variations in atmospheric pressure. Moreover, when caused by atmospheric conditions, these changes occur slowly—in contrast to pressure variations due to the passage of sound waves, which are very fast. The ears are also therefore ultrafast sensors: we are able to detect oscillations in atmospheric pressure due to the passage of sound waves between 20 and 20,000 times per second! Our senses are particularly acute when these variations are repeated between 200 and 10,000 times per second.

If you're feeling like this is a little over your head, don't panic. The important thing to remember is that a sound is a wave of pressure variation that is transmitted by vibrating the air, and that these weak vibrations are perceived by our ears. Hence the two essential notions that define a sound produced by an animal: its intensity and its frequency. The *intensity* of a sound—also known as its amplitude—is the power at which we hear it: it can be of low intensity (we have difficulty hearing it; when someone whispers, for example) or high intensity (when someone shouts). This intensity corresponds to the variation in atmospheric pressure we have just been talking about. Again, let's use the analogy of the pebble in the pond: a small pebble will create small ripples while a larger one will create larger ones. The height of the wavelets is the amplitude of the pressure variation. We have seen that the human ear perceives pressure variations between 0.000002 and 20 Pa. Below that, we hear nothing; above that, the sensation is painful and we are unable to tell if the sound becomes louder or not. In order to take measurements, a logarithmic scale was invented, which spreads these variations between 0 and 120 decibels (dB). A differential of 0.000002 Pa compared with the local atmospheric pressure corresponds to 0 dB and therefore defines the threshold of minimum intensity audible to the human ear. At the other end of the scale, 120 dB corresponds to a pressure variation of 20 Pa. Beyond that, the intensity of sounds is such that it destroys the hair cells of our inner ear, at the risk of leaving us irremediably deaf. A remarkable property of the decibel scale is that an increase of 6 dB corresponds to a doubling of the value of the pressure differential. Thus, when we go from 0.000002 to 0.000004 Pa, the intensity of sound increases by 6 dB. Unfortunately, this dB scale was constructed for the human species. Not all animals have a sound detection threshold at 0.000002 Pa (i.e., 0 dB): some are certainly more sensitive and capable of detecting sounds at lower intensities (negative decibel values), others are less sensitive and only hear sounds above 0 dB. This does not prevent decibels from being very useful for reporting the intensity of a sound, however, and they are widely used in bioacoustics.[9]

Now let's take a little detour to the aquatic environment. We already know that sound waves propagate faster in water than in air. This is due to a difference in resistance to the passage of the sound wave between the two propagation media. Because water molecules are closer to each other than air molecules, they collide with each other more quickly, facilitating the transmission of pressure variations and thus increasing the speed of propagation of the sound pressure wave. In fact, the denser the propagation medium, the faster mechanical vibrations, such as sound, propagate. It is not for nothing that, in Western movie scenes, the gangsters listen for the arrival of the train they plan to attack by pressing their ears against the railroad track. They can hear it coming long before an airborne sound reaches them, because the vibrations propagate much faster in the metal than in the air, at more than 5000 meters per second (compared to the 340 meters per second of sound waves moving through the air). Water as a propagation medium lies in between. It is less dense than metal but denser than air: sounds move through it at about 1500 meters per second. In short, the decibel scale is different from one medium to another. When we work on the aquatic environment, we cannot use the minimum threshold of sound intensity for the human species that has been measured in the air (the 0.000002 Pa discussed earlier). This is another reference point that we use, independent of the human species: 0.000001 Pa, which therefore corresponds to 0 decibels underwater. Yes, all this is very complicated and not very rational. Let's just remember that a measurement in decibels is very beneficial for getting an idea of the intensity of sounds, and that measurements in decibels taken in air and in water are not easily comparable.

We now know that a sound wave is a variation in pressure, which propagates at a speed of 340 meters per second in air and 1500 meters per second in water, and whose wave amplitude defines the intensity of the sound. We also know from experience that a sound can be more or less high-pitched or low-pitched. This characteristic defines the *frequency* of the sound, which in turn corresponds to the rhythm of pressure variations. If this frequency is rapid, in other words, if the pressure alternately decreases and increases at a rapid rate, the frequency of the sound will be high and the sound will be high-pitched. If this rhythm is slower, it will be low and the sound will be lower. Let's take a last look at our analogy of the pebble in the pond. If we throw not just one but several pebbles one after the other, we will create a succession of circles in the water. If our pebbles are thrown very close together, the distance between the circles in the water—the wavelength—will be short; the little cork floating on the water will bob up and down very often. A short wavelength is synonymous with a high frequency. Conversely, if the pebbles are thrown in a more nonchalant way, the distance between the rings will be greater and the small cork will wobble less often. The wavelength is long, which is synonymous with a low frequency. I have already said above that our ear is particularly capable of detecting variations in air pressure when they are repeated between 200 and 10,000 times per second and that the human ear can perceive sound waves with a frequency between 20 and 20,000 cycles per second. Each animal species that communicates through sound has a different range of frequency sensitivity, measured in hertz (Hz). For example, elephants hear very low sounds: they perceive frequencies below 20 Hz, which are inaccessible to us (they are called infrasounds for this reason). Bats, on the other hand, produce extremely high-pitched sounds (up to 100,000 Hz for some species), which are also inaudible to humans (they are called ultrasounds). There is no difference in physical nature between infrasounds, sounds, and ultrasounds; these designations just indicate a difference in human hearing ability.

If you've followed me this far, bravo! Now you know what a sound is: a pressure wave that propagates, in water or in air (or in metal or other solid media), with a certain intensity and frequency.[10] You are now perfectly armed to enter the world of animal acoustic communications. Follow me. We're about to begin our journey in a South American rainforest, the *Mata Atlântica*, where a small bird, sweetly named the white-browed warbler, awaits us.

3

The warbler's eyebrows

WHY DO BIRDS SING?

Morro Grande Reserve, Atlantic Forest, Brazil. A first song reaches our ears: notes that are at first high-pitched, then gradually fall in a regular modulation. A few seconds later: the same song, muffled and indistinct this time. It is a reply. The second bird must be a good 50 meters away and the dense vegetation of the tropical forest greatly reduces the range of its vocalizations. This November morning, two white-browed male warblers engage in a territorial dispute. Thierry and I are the attentive witnesses. The second individual approaches, his more intense song further riling up his opponent. The latter is singing at the top of his lungs, perched a few meters high, then he flies resolutely toward the intruder. The battle is won, the newcomer flees. In birds, as in many other species, the owner of a territory usually wins, because his motivation to keep his property is unparalleled: I'm here and I'm staying put! In the following days, white-browed warblers will have to deal with a new intruder, and by no means the least formidable. We'll arrive armed with a tape recorder and loudspeaker and begin aural jousts with them . . . we'll learn a lot.

Why study this greenish warbler from the Atlantic Forest, the *Mata Atlântica*, on the east coast of Brazil? The idea had been given to us by Jacques Vielliard, then professor at the University of Campinas in Brazil, whose knowledge of the birds of South America and their songs was unequaled. We were interested in the problems that the forest environment poses to the sound communication of birds—dense vegetation being an obstacle to the propagation of acoustic waves. We were looking for an abundant species that was easy to find, producing a song that was simple and short enough that its acoustic structure could be studied without too much difficulty. The song also had to produce a frank (i.e., aggressive) territorial reaction when the bird heard an opponent. We wanted to observe the response of birds to songs that were recorded and then broadcast from a loudspeaker.

"*Basileuterus leucoblepharus*," said Jacques in his slow, modulated voice; the white-browed warbler, as it's commonly known in English. This was the model we needed. "A bird absolutely typical of the Atlantic Forest." The die was cast; our bird would be the warbler. And the work began. We spent long hours in the rainforest, first observing the warblers to learn about their territories and their behavior, then recording their songs and testing them with signals emitted from our loudspeaker. This is how Jacques, Thierry, and I were going to unlock the secrets of their sound world.

The Morro Grande Reserve is only an hour and a half from the megalopolis of São Paulo. But what a contrast! Here, the Atlantic Forest, of which there is almost nothing left, is still splendid. When we first arrived, it was late afternoon. Night falls early in the tropics, and the forest was already rustling with the sounds of the night. Once we got out of the car, Jacques solemnly said to us: "Gentlemen, welcome to the *Mata Atlântica*." I was very impressed: it was my first experience in the rainforest, and my first steps were very hesitant. I imagined a hostile environment, where you could step on a dangerous animal at any moment, where any number of unknown threats were waiting for you. The following morning, in the midst of the dawn chorus, Jacques taught me that the characteristic feature of the tropical forest is its biodiversity and not its density of animals. Reassured, I listened, and within a matter of minutes, we spotted an incredible number of different bird species, each represented by a small number of individuals. The same is true of insects, snakes, and any other species living in the tropical forest: an extreme diversity of species but a small population of each.

We got to work. First, we observed and recorded male warblers. The bird is discreet in appearance, but Jacques was right: the white-browed

White-browed warbler

warbler is one of the most common species at Morro Grande. Its song is simple and short: a few very high, tenuous notes, almost difficult to make out, then a succession of small notes becoming lower and lower, with a very regular descent—"twiiii—twiii—twi—twuu—twu—tuu—tu"—all for about 10 seconds.

Why do birds sing? For their pleasure, one might think; and it is quite possible, insofar as the act of singing seems to fulfill an irrepressible need. However, scientific observations and experiments over the last 50 years have shown that singing has two essential functions: attracting sexual partners and repelling competitors.[11] Sometimes the males are the only ones who sing. Sometimes it is the females that sing. Or females and males vocalize together, in duets so synchronized that we don't know which one produces which note. Songs are indeed communication signals; that is, they contain information for other birds of the same species. Like our words, the songs represent a means of calling out to fellow birds. They allow the American robin or Eurasian wren to identify itself as an American robin or a Eurasian wren, and even to say, "My name is So-and-So." This information is encoded in the acoustic properties of the song. In principle, a song consists of a succession of notes, alternating with silences, according to a particular rhythm and sequence. The notes are characterized by their frequency or pitch (high or low), their amplitude or intensity (high or low), and their variation or modulation during emission (i.e., the frequency or intensity of a note can move). To decipher this coding, the analysis of the acoustic structure of the vocalization is an essential step. However, only experiments consisting in emitting the vocalization from a loudspeaker allow us to really understand the role of the acoustic signal, the information it carries, and how it is coded. It's a bit like asking animals questions.

In the white-browed warbler, males establish territories about 100 meters in diameter, where they live with a female, and which they defend against other males. In the forest, where the birds cannot be seen beyond a few meters, it is through song that the male signals his presence, and therefore his territory, and repels possible competitors. However, in the tropical forest, where the plants intertwine in a profusion of leaves, branches, and creepers of all sizes, it is not easy to make his song heard. These obstacles considerably hinder the propagation of sounds by absorbing their energy and reflecting sound waves in multiple directions, creating echoes and distortions. In such an environment, a song suffers significant changes as it is diffused far away, and the degradation of its structure worsens with the distance of propagation. This is nothing extraordinary: we know that it is difficult for us to understand what someone in a distant room or on another floor is saying. In order to evaluate these signal changes imposed by propagation through the forest, we placed a loudspeaker and a microphone at different distances from each other, and at different heights in the vegetation, at the respective positions of a transmitter bird and a receiver bird. Then we played warbler songs from the loudspeaker, while recording them with our microphone after they had propagated through the forest environment. These experiments gave a clear result: at a distance greater than 100 meters in the forest, the first half of the warbler song (the very high notes) became totally inaudible because the high-pitched sounds were not transmitted well. Only the second half of the song could be recorded. And that's not all: the silences were filled by echoes due to the reverberations of the sound waves on the vegetation. How can a male warbler still identify another male of his species at this distance?

Animal sound signals carry two broad categories of information. The first category is related to the stable characteristics of the singer: his belonging to a species or to a group, his morphology and anatomy, his sex, his weight, his age, his past experiences, etc. This information is described as *static* because, even if it sometimes varies with the individual's age, it forms a kind of identity card for the issuing individual. I sing who I am. The second category is related to the state of the moment; for instance, the degree of motivation to defend one's territory or to seduce a partner, the need to warn of the presence of danger, etc. I sing my state of mind in the here and now. This is *dynamic* information, the value of which is likely to vary rapidly in the signal. Our studies of the warbler focused on static information: we wanted to understand how male warblers could signal both their species ("I am a white-browed warbler") and their individual identity ("I am Marco") despite the extreme constraints on sound transmission in a forest with particularly dense vegetation.

The first step in our experiments was to understand how a warbler recognizes over a long distance that a song is being produced by another male of its species.[12] In other words, how can the acoustic structure that has largely been degraded by propagation through the rainforest still convey enough information to be recognized as a warbler song? Displayed on a computer screen, the spectrogram of a birdsong resembles a musical staff, which is very useful for visualizing its structure. The notes of a warbler song seem to hang on a taut wire from a very high note with a frequency of about 8000 Hz to a much lower one around 2000 Hz (if you haven't read chapter 2 on the structure of a sound, just remember that the frequency of a sound corresponds to its pitch—high, medium, or low). This imaginary thread follows what a mathematician or a mountain dweller would call a slope; here it's quite regular—a *glissando*, as the music lovers say. Since the notes descend from high to low, the song is said to follow a descending frequency modulation slope. By comparing the songs of several males that we had recorded, we easily noticed that they share the same construction rule: the songs of all the warblers follow the same slope. We then hypothesized that this slope carried the information "I am a white-browed warbler." To test this idea, we made computer-generated artificial songs that mimicked the warbler's song, but whose frequency slope was either less steep than in the original song or, on the contrary, more pronounced. The acoustic playback experiments could then begin.

We positioned the speaker in the middle of a male warbler's territory. Thierry operated the tape recorder where the prepared sounds were stored, and I held my binoculars fixed on the area around the loudspeaker. Then we played various songs (the original and the modified ones) and observed the bird's reactions. They exceeded all our expectations. Indeed, in response to a natural song, recorded from a warbler living a few kilometers away, the bird that owned the territory would fly directly to our loudspeaker, singing and screeching loudly. Conversely, as soon as the loudspeaker played a modified song, with even a slightly different slope—no reaction. The bird continued to go about its ordinary business, scornful of our calls. On the other hand, its attention increased sharply again when it heard an artificial whistle in a frequency that followed the slope of the natural song, even without alternating notes and silences. In other words, a male warbler does not need to hear the detail of the notes of the song to understand that a rival is present and is trying to take away his territory. Everything in our results indicated that the slope of the frequency modulation of the song is the criterion, the only criterion, used by white-browed warblers to acoustically recognize their fellow creatures. We had understood the rule followed by this bird to identify members of its species. Pleased as punch, we had found the species signature code of a rainforest bird!

Species identity, however, is only one piece of information among many encoded in the bird's song. We then decided to investigate how warblers differentiate between their neighbors and strangers. Let me explain: when we play back to one warbler the song of an individual recorded at a distance, a bird it has never had a chance to hear before, the tested individual responds very aggressively, whereas if it is a song from a closer territory (for example, on the edge of its own territory), its reaction is greatly attenuated. It has recognized its neighbor, the one who occupies the territory next door and has no designs on its own domain. This recognition between neighbors is frequently found in

Black redstart

birds and other animals. Tudor Draganoiu, a researcher at the University of Paris Nanterre, has studied it in a bird species common in France: the black redstart *Phoenicurus ochruros*. These little birds love old stones and establish their territories in small villages in the Massif Central and the Alps. They are easy to observe because the males do not shy away from singing from the rooftops. Their song consists of several successive musical phrases: the first are rapid successions of notes, while the last one sounds like rustling newspaper. Tudor has shown that redstarts from the same hamlet have similar melodies—they share common phrases—and they tolerate each other. They can spot a stranger from his song and then forcibly chase him away. This is known as the dear-enemy effect: it allows the birds not to worry unnecessarily, and to reserve the energy allocated to the defense of the territory for really perilous situations; in other words, for when an unknown intruder arrives and seeks to establish himself. A neighbor whose territory is stable does not represent a great danger.[13]

In the case of our warblers, we first had to establish whether their song has a signature that could differentiate one individual from another. The analysis of the acoustic details of the recorded songs was very instructive: not only did they all follow the species-specific rule of descending pitch, but the descending slope of the notes was also punctuated by small breaks here and there. In other words, the gradual descent of the frequency of the notes—from the highest to the lowest—is in fact not perfectly regular; sometimes one note that follows another is a little lower than might be expected if the regularity were perfect. Remarkably, the songs of the same individual all have the same irregularities, although they differ from one male to another. They could therefore seriously be suspected of bearing the individual signature of a male and allowing recognition between neighbors—an interesting hypothesis that still needed to be tested.

The second stage of our playback experiments would confirm this. When our loudspeaker, placed at the edge of the tested bird's territory, emitted the song of its neighbor, our warbler continued to indulge in his warbler routine. However, if we changed the irregularities of slope in this song, the reaction was immediate: no longer recognizing its usual neighbor, the bird rushed to threaten the intruder. I'm always amazed when a playback experiment provokes an animal's reaction. Isn't it incredible to be able to enter into the intimacy of a nonhuman world in this way?

In this second stage of our exploration, we had discovered something exciting: the irregularities of slope that bear the individual signature are only found in the first part of the male's song, the very part that is transmitted only at short distances—a few dozen meters at the most. A bird will therefore be able to locate each of its close neighbors individually but will be unable to precisely identify males further away, since the information that encodes their individual identities has disappeared, absorbed along with the high notes by the tropical vegetation! The individual identity of the white-browed warbler, therefore, is private information: only the closest neighbors, and probably females in or near the territory of the singing male, have access to it.

Another bioacoustician, Frédéric Sèbe, who has since become a close colleague, was to show some time later that the irregularities of the songs of neighboring warblers sound more similar than those of songs from distant individuals. In other words, the closer the males' territories, the more similar the individual acoustic signatures. This is reminiscent of the shared phrases of the black redstart. In the warbler, we do not know if this resemblance between neighbors is the result of an imitation of a signature, which would allow the structure of the songs of individuals living next to each other every day to converge, or if it is the result of genetic proximity. Perhaps the males of this species tend to establish their territory close to their hatching site and then find themselves close to a brother or a cousin, whose vocal characteristics they share. As for the possible interest of this sharing of vocal signatures, it remains a mystery. One can imagine that it facilitates recognition between neighbors. It seems difficult to test these hypotheses on the warbler of the faraway Atlantic Forest, but we can hope that the French redstarts, less difficult to access, will let us in on their secret.

In our warbler, the coding of the two pieces of information—species identity and individual identity—is thus adapted to the constraints that the tropical forest imposes on the transmission of sound waves: if any listening warbler, even a distant one, is interested in spotting my presence, then he (or she) can, but only my neighbors need to know that I am Marco.[14] We have also shown that a warbler can locate the position of a transmitting individual based on the extent of the changes in its song during transmission. Indeed, the longer the propagation distance, the fewer notes remain in the song. After 20 meters, let's say that, from the original song "twiiii—twiii—twi—twuu—twu—tuu—tu," we only hear "twi—twuu—twu—tuu—tu." At 100 meters from the loudspeaker, only "twuu—twu—tuu—tu" is audible. And Marco knows whether or not to start a fight with the emitter!

Therefore, despite the strong propagation constraints to which it is subjected, the warbler's seemingly simple song contains information enabling these birds to manage their social relationships.[15] This example illustrates how important it is to break through the mechanisms for encoding and decoding the information present in the songs. This is how the mysteries of a sound-based communication system can be deciphered, and how we can understand which environmental constraints have shaped this system over the course of evolution.

Numerous studies have sought to test whether the sound signals emitted by animals, particularly the vocalizations of birds, are precisely adapted to the environment in which they live. The results of these investigations were somewhat contradictory.[16] It was thought that species living in forest environments produce lower notes, degrading less quickly than high notes. Their songs should also be poorly modulated, i.e., made up of notes whose frequency varies slowly, making them less sensitive to the echoes caused by the reverberations of sound waves on trunks and branches. Yet a forest bird like the wren produces a very high-pitched song, made up of a succession of very fast trills. It therefore degrades as it spreads through the forest. To put it another way, the hypothesis of "acoustic adaptation" (which we discuss in chapter 17) was only partially supported. To make sure, Thierry and I asked Jacques to propose a list of birds in the *Mata Atlântica* that represented the different layers of the forest, for which he would have quality recordings in his archives. As Jacques had spent his life recording birds, he had a great sound library and plenty to satisfy us!

After a few days spent preparing our tape recorders, we were ready for new adventures, new propagation experiments. The forest was going to ring with the songs of the brown tinamou, the squamate antbird, the white-browed warbler (again and again), the white-shouldered fire-eye, the swallow-tailed manakin, the giant antshrike, the surucua trogon, and the hooded berryeater.[17] This list is not a random miscellany: it ranges from species that spend most of their time on the ground or at very low altitudes to those that live in, and even way up high on, the canopy. Our hypothesis was that the song of each one propagates better, and at a longer distance, at the height from which the bird usually vocalizes (otherwise,

Hooded berryeater

Surucua trogon

Giant antshrike

White-browed warbler

White-shouldered fire-eye

Swallow-tailed manakin

Squamate antbird

Brown tinamou

why choose to stand there?). As with what we had found in the literature, the results were not obvious to interpret. Take the brown tinamou: it walks on the ground in search of its food and does not perch to sing. Yet the range of its song would be greater if it sang at a height of several meters. Quite the reverse is true for the hooded berryeater, which lives at the very top of the canopy, and it is indeed here that the greatest range of its song is obtained (we hired a tree climber to place the loudspeaker there). In short, some birds sing at positions where their melody spreads efficiently over long distances; others do not. This is probably because a bird's position in the forest strata does not depend solely on acoustics. Many other factors come into play: Where is its favorite food? Where are the predators? Where is it sunny more often? Where is it windy? And who knows what else? Sometimes, however, it seems easy to explain why a species emits a signal that is difficult to hear from a distance. For example, swallow-tailed manakin males gather in assemblies called leks, in small forest clearings. There, females join them—they know the place! Their courtship ritual combines singing and dancing, with the males hopping on branches. No need to be heard from afar. On the contrary, by emitting songs of relatively low intensity, the manakins probably avoid announcing their whereabouts to predators. Clever, those manakins.

Two birds that are very common in Europe were going to teach me that there are other ways to improve communication in the forest other than emitting songs with an optimal acoustic structure to disperse them. Let's take a look at the cute Eurasian wren *Troglodytes troglodytes* and the Eurasian blackcap *Sylvia atricapilla*. The former belongs to a rich French research tradition. Jean-Claude Brémond, the man who introduced me to the world of bioacoustics, as I stated in chapter 2, had cleared up the coding of the identity of the species, showing that male wrens are very sensitive to the rhythm of the notes, i.e., the speed at which they are produced, and that this is what enables them to recognize other male wrens.[18] Michel Kreutzer, from the University of Paris Nanterre, had also shown that each wren population has a particular dialect, with acoustic properties different from the songs produced by other populations, similar to regional patois and accents in humans.[19] As for me, I decided to study the behavior of this bird in the presence

Eurasian wren

of songs propagated over a great distance; in other words, songs whose acoustic structure is very degraded.

The wren is a tiny forest bird. Imagine a small ball of feathers, with a fine beak and a small, erect tail, sneaking through the bushes with elegance and precision. The male sings surprisingly loudly for its size and actively defends a territory about 50 meters in diameter. I made a very simple experiment: I placed a loudspeaker on the ground, in the middle of a wren's territory, and then broadcast the vocalizations of another wren that was unknown to him, having been recorded far from his "home." Two versions of this unfamiliar song were played in succession: one was what in science is called a *control stimulus*, i.e., an unmodified version supposed to provoke a strong response; the other was an *experimental stimulus*, i.e., the same song but recorded after propagation over a long distance in the forest. Each time (because the experiments must, of course, be repeated), the tested bird responded to both versions by vocalizing in response. However, the exploration behavior that followed was very different. In response to the undisturbed control song, the birds came very close to the loudspeaker, some landing squarely on it, frantically searching for their invisible opponent. On the other hand, when they heard the experimental stimulus—the melody coming from far away—they preferred to perch several meters above the ground to sing. Why such a difference? I made two hypotheses. First, like the white-browed warblers, the wrens must have perceived information about the distance separating them from the individual transmitter: the control song (not degraded by the propagation) was to be interpreted as that of an intruder who had entered their territory, while the song propagated over a long distance was attributed to a more distant bird. Second, perching to respond to a distant individual might improve the quality of the signal and the information to be transmitted into the distance.[20] Both hypotheses had yet to be confirmed.

The wren's song is powerful and high-pitched. The hypothesis that it is seriously degraded during long-distance propagation could be tested by experiments of signal propagation in the forest. The results proved quite interesting: the degradation of the acoustic quality of the song depended strongly on the height of the loudspeaker placed in the tree, and especially on the height of the microphone. When the loudspeaker was located high up in the tree, the song was certainly received with greater intensity and underwent fewer transformations (fewer echoes and better preservation of high notes), but these improvements were much better when the *microphone* was perched several meters high. Here one must keep in mind what a forest environment typically inhabited by wrens looks like. Starting from the ground and moving to the treetops, there are several strata: first, an area of fairly dense bushes, with a few shrubs and trees in early stages of growth; then a relatively clear space between the top of the bushes and the underside of the canopy; and, finally, a more or less dense area where the tree branches intertwine. The wren spends most of its time in the bushes, not very far from the ground. Perching when it sings improves the range of its vocalizations: the song will be heard from further away. But the improvement is mainly in the quality of its listening: perched high up, it can more easily perceive another wren chirping in the distance.[21]

Eurasian blackcap

A few years later, together with my colleague Torben Dabelsteen, a professor at the University of Copenhagen, we were to confirm these results by working on another species of European forest bird, the blackcap. Torben is a specialist in sound propagation issues. By making very precise measurements, we showed that when a blackcap moves from a perch at a height of 4 meters to a perch at 9 meters, the horizontal range of its song increases by about 25 meters. In other words, perching a few meters higher is equivalent to getting the blackcap's song 25 meters closer to a possible receiving bird. Note that this 25-meter gain from perching at a greater height applies to all directions. Since the diameter of a blackcap's territory is between 50 and 100 meters, this increase is enormous and has a considerable advantage.[22] Let's do a little math to get the ideas straight. Let's say that the song of a blackcap perched 4 meters high can be heard up to 50 meters from the bird. The area over which the song will be heard around the warbler will be equal to 3.14 (the number pi) multiplied by the square of 50 (remember your college lessons: the area of a disc is equal to the product of the number pi multiplied by the square of the radius of the disc), i.e., $3.14 \times 50^2 = 7850$ square meters. When the blackcap perches at 9 meters, the range of the song (thus the radius of the disc) increases by 25 meters, as we've said. The area over which the song is heard then becomes $3.14 \times (50 + 25)^2 = 17,662$ square meters. That is more than double the initial surface of 7850 m². This means that a blackcap that climbs 5 meters into a tree more than doubles the area in which its song is heard. Playback experiments further showed that the blackcap is able to identify a song of its species even if it has been particularly degraded during its propagation in the forest.[23]

Acoustic communication thus makes it possible to exchange information even if we cannot see each other, which is particularly interesting in an environment with dense vegetation. Clearly, while the echoes created by the reverberation and absorption of the energy of sound waves by vegetation can greatly modify the acoustic structure of signals, birds have developed strategies during their evolution that overcome these drawbacks: perch high up, emit a song with high intensity or propagation-resistant characteristics, use information coding that counteracts or even uses signal degradation.

However, forest environments present yet another difficulty for acoustic communications: competition for sound space, when multiple species—not only birds but also insects and amphibians—use sound to communicate. A real mess! Don't they get in each other's way? How do they navigate this huge jumble of sound? This question has been the subject of many studies and, it must be said, of a certain amount of speculation.[24] The acoustic avoidance hypothesis has thus had a good run and is still being actively discussed (along with the acoustic niche hypothesis[25]). It's the "Let's not get in each other's way" hypothesis. Animals might talk to each other on different frequencies, with different rhythms, even at different times of the day or night, to avoid interference from equally voluble neighbors. This hypothesis has been verified in some animals, particularly crickets, whose noisy stridulations are more or less acute depending on the species.[26] But if you've ever listened to the dawn chorus of multiple species of birds singing together just before sunrise, you'll certainly agree that acoustic avoidance is not a priority for them. In fact, there are several strategies for transmitting and receiving sound signals that allow animals to identify the signal they are seeking in the midst of the ambient cacophony. I'll take you far away from the rainforest, to much colder climates, to see how seabirds use acoustics to communicate in the terribly noisy environment of their breeding colonies.

4

Cocktails between birds

NOISE AND COMMUNICATION THEORY

Hornøya Island, Sea of Barents, Norway. From the lighthouse where I'm staying, you can see the furious sea. To reach the gulls' cliff, we have to cross a windswept plain, where angry gulls dive on me as if they wanted to open my skull; probably convinced that I'm getting too close to their young. Once I reach the cliff, the white birds' antics become dizzying. Nests overflow from every crevice. The cliff is bristling with them. Everywhere, seagulls are nesting. The season will be short, and woe to the adults who are late in breeding as their young have no chance of survival. The incessant screeching drowns out the sound of the waves. Here one individual—female or male?—vociferates as it flies toward the rock. It comes to relieve its partner, who responds just as loudly. How do these gulls find their way through the multitude of nests in this cliff building? No doubt the birds have a precise knowledge of the topography. But chances are that their partners' calls help them find their way, despite the surrounding noise. This is what I came here to study, at the farthest reaches, with Thierry, my usual partner in crime.

In this chapter, we discuss a concept that is essential for all acoustic communication—the *noise*—and see how animals manage to exchange information through sound signals in spite of it. All communication involves three elements: an emitter, a signal (sound waves), and a receiver. The emitter, or *sender*, produces the signal in which information is encoded. The signal propagates in the environment while being transformed to a

Gull

greater or lesser degree. As we have already seen, the sound waves prop-agating through a forest reverberate on the vegetation, creating echoes; the energy of high-pitched sounds is more easily absorbed than that of lower-pitched sounds. The signal is then received by the receiving individual. The receiver hears the signal and decodes its information. Once this information has been integrated into its brain, it will potentially be able to react, for example, by attacking the intruder or responding to its partner. This succession of steps from the sender to the receiver conditions all communication and constitutes the chain of information transmission. It was an American, Claude Shannon, who in 1948 formalized these concepts by establishing the mathematical theory of communication (typically known as information theory).[27] In scientific language, contrary to what is sometimes thought, a theory is not a hypothesis or something that has not yet been proved and could have competing explanations. Instead, a scientific theory is a system that explains a part of the world around us. It is, in a way, the result of a large body of scientific research whose observations and results are consistent. A scientific theory can certainly evolve as a result of new discoveries, but if it is accepted, it is because no knowledge or experience has ever come to oppose it at its foundation. In the life sciences, two theories are particularly well founded: the cell theory, which states that all living beings are made up of cells and that every cell comes from a mother cell (the corollary of which is that spontaneous generation does not exist and that all animals, including humans, share a common ancestor), and the theory of evolution, which explains that living beings change over time, under the combined effect of chance (e.g., genetic mutations) and selective constraints favoring one characteristic or another (natural and sexual selection).

The mathematical theory of communication was not originally intended for biologists. It had been established to provide a framework for the telecommunications research that was taking off at that time. The aim was to characterize both qualitatively and quantitatively the deformations undergone by any signal during its transmission in a propagation channel. More specifically, the aim was to understand how a signal can carry information from a transmitter to a receiver in a noisy channel, and thus to implement strategies for encoding information in telecommunication signals in order to transfer messages as efficiently and quickly as possible. The information can be seen as a cube of a certain volume that must pass from the transmitter to the receiver. At the time it is produced by the sender, the signal corresponds to a certain volume of information. During its propagation in the environment, the signal is deformed. When it reaches the transmitter, the volume of information it encodes has decreased: the volume of the cube has become smaller. In a way, the information cube is stripped down during propagation. This reduction in the amount of information is the result of noise. To understand this, let's take a concrete example. Imagine that you have a breakdown on the side of the road and you call your mechanic and say, "My car has broken down at the side of road 2 at mile marker 103." If the telephone transmission is unsatisfactory, your mechanic may hear "grrr . . . M . . . car . . . broke down . . . grrr . . . road . . . 103." Your mechanic will understand that your car has broken down and you are calling for help, but the original information has not been transmitted in full. The amount of information you sent has been reduced by two things: on the one hand your words have been distorted, and on the other hand some parasitic sounds ("grrr") have been added. These two types of modifications, which in nature correspond to signal modifications due, for example, to reverberations on vegetation and the vocalizations of other animals, contribute to the noise of the signal. In fact, they *are* the noise.

Shannon's goal was to understand noise in order to imagine ways to encode information in telecommunication signals to limit its impact. Let's take the example of the roadside breakdown. How can you make your mechanic hear you despite the noise from the telephone transmission? Shannon's theory shows that three strategies are possible. The first one is obvious: you can repeat the message until you are sure that the mechanic understands it. Emitting the same signal several times in order to improve the transmission of information in a noisy environment is to practice what is called information redundancy. Since the noise of the telephone line is irregular, each time you repeat it, a different part of the signal will reach your mechanic—e.g., first that your car has broken down and then what your location is. A second strategy is to talk louder, or even to shout into your phone. Increasing the strength of the signal makes it stand out against the noise. Finally, you can change the tone of your voice, by making it higher-pitched, for example, especially if the crackling on the line is in the low register. Avoiding sound frequencies occupied by noise is a strategy we often use when the propagation channel is congested, for example, when talking to someone at a noisy party. By using these three tools, you try to make your signal stand out from the noise, which in acoustics jargon means increasing the *signal-to-noise ratio*. These three strategies—redundancy, increasing the intensity, and modulating the tone of the voice—play on the three components of the sound signal: time (increasing the duration of the signal), amplitude (increasing the strength of the signal), and frequency (changing it to avoid the frequency band occupied by the noise). Shannon put all these processes into mathematical equations, and it became possible to make predictions about the coding strategies to be employed based on the characteristics of the noise and the information to be conveyed. I won't take you that far, don't worry. But we will see that nonhuman animals grapple just as much with the problem of noise from acoustic signals during their communications, and that they, like us, develop strategies in accordance with the mathematical theory of communication. This theory actually applies to the entire living world. It is particularly useful for the study of animal and human acoustic communications.[28]

Most seabirds—seagulls, gulls, albatrosses, shearwaters, auks, penguins, and others—breed in colonies of hundreds, thousands, even millions of individuals. Finding a place to nest is not easy when you spend

Black-headed gull

most of your time in the middle of the ocean—or even in the ocean for penguins, which lead a decidedly aquatic life. Seabirds therefore find themselves nesting on the same coasts, sharing the available space. Nesting together limits the risks. While a predator would not hesitate to pounce on an isolated bird, it will hesitate to venture into a colony where individuals are densely packed. Another characteristic of most seabird species is that individuals form breeding pairs. They are said to be monogamous and in principle particularly faithful, at least during the same breeding season, if not for life—so much so that raising young in the marine environment requires unfailing parental cooperation. Being able to recognize your partner among the other birds in the colony is therefore essential. Sometimes, identifying one's young can also prove very useful, as in some penguins where the young gather in a kind of nursery and the risk of confusion is high. Scientific research carried out in the field, as close as possible to the animals, has shown that acoustic communication plays a primordial role during this recognition between partners and between parents and their young.

It was with Isabelle Charrier, my first PhD student and now director of research at the *Centre national de la recherche scientifique* (the famous CNRS), that I began my investigations into parent-offspring recognition in a gull species, the black-headed gull *Larus ridibundus*. Like other gull species, the black-headed gull forms breeding colonies in which each pair establishes its nest sometimes only a few meters from the nest of the neighboring pair. The black-headed gull is quite eclectic in its choice of habitat and is easily found inland, where it establishes breeding colonies on ponds. It was in France, on a pond in the Forez plain, that Isabelle and I studied it. The doors of the pond had been opened to us by Jean-Dominique Lebreton. Director of research at the CNRS, member of the Académie des Sciences, erudite, and passionate about nature and animal behavior, Jean-Dominique had been studying the demography of the black-headed gull for many years and was delighted to see ethologists interested in his favorite bird. We wanted to record the calls of the parents of young gulls and then test with acoustic playback experiments whether the young were able to recognize their parents from their calls. This meant getting as close as possible to the animals. How could this be done? Jean-Dominique had a perfectly successful technique: "You're going to have to build a floating blind. And squeeze into a wetsuit. You will go unnoticed and be able to approach the nests." Imagine our construction: a wooden and polystyrene frame, covered with branches and leaves, with small openings allowing us to observe outside and to pass a pole on which our microphone was fixed. I ordered two diving suits from the administration of my university, who questioned me about the purpose of the operation, suspecting me of organizing a holiday by the sea. We built a blind large enough to house two people. Walking carefully on the bottom of the pond, we could finally approach the birds. "Careful, not too close, though—you mustn't disturb the broods," warned Jean-Dominique, always concerned about preserving nature.

The pond of the Ronze is a magical place. It was a rare privilege for me to enter it incognito, hidden beneath my blind. As for the flora, the water lilies, pondweed, and reeds were reflected in the water and enchanted the eye. A multitude of different species of birds gathered there: coots, teals, mallards, and redheads paddled across the surface. Some grebes too, including the little black-necked variety, whose red tufts and vermilion eye make it look like an operetta marquis. And gulls by the hundreds. Making an unholy racket. To wait near a nest for one of the parents to come back to feed the young, record its calls, then go near another little gull family, such was our first step. The next day, we came back with a camouflaged loudspeaker fixed on the blind, to play parental and nonparental calls a few meters away from the young in "our" nest from the day before, hoping to provoke different reactions that would prove they could distinguish their parents' voices from those of other adults. This merry-go-round had to be repeated day after day—true as it is that, in science, each experiment must be repeated and repeated again for the result to be validated. Busy with my university classes, I could only go to the field once or twice a week, and it was Isabelle who had to provide the bulk of the effort. Within a few weeks, her patience was rewarded with success: we obtained experimental proof that young gulls recognize their parents by voice. Above all, we showed that, in the hustle and bustle of the colony, redundant information is necessary for optimal voice recognition. Sometimes a particularly sensitive chick would be able to recognize its parent if we made it hear a single call. Half of the chicks tested needed three calls. However, when the parental call was repeated at least four times, 100% of the chicks tested responded to the voice of their parent while ignoring the voice of another adult in the colony. The behavior of our gulls was consistent with the mathematical theory of communication.[29]

If the noise of a gull colony is already a major constraint to communication, this constraint can be much more extreme. This is the case with king penguins. To my regret, I have never worked on king penguins. That would have required missions of several months in the Southern Hemisphere, which my position at the university made difficult. I also did not want to stay so long away from my family. So I experienced the penguin adventure vicariously. But Thierry told me so much about it, and I've read so many publications and seen so many films that I almost feel like I've been there! Come with me, let's go to Crozet Island, in the middle of the Indian Ocean, to discover how the king penguins use acoustics to recognize each other. This is certainly the most beautiful example of communication in a noisy environment that nature gives us to observe.

Penguins form a unique and very special family of birds, the Spheniscidae. Their peculiarities? They live only in the Southern Hemisphere, are unable to fly, and spend most of their time swimming underwater—sometimes at very great depths—to fish. They only go ashore to reproduce, which entails nuptial parades, mating, brooding, feeding, and protection of the young against the cold for the species closest to the South Pole. There are about 20 species of penguins. Even if you have never been particularly interested in these birds, you certainly know the large penguins: the king and the emperor (*Aptenodytes patagonicus* and *A. forsteri*, respectively). In these two species, the adults and their young form huge colonies, which can number thousands of birds—up to a million for the largest colony of king penguins. But the penguin parents are faithful and exclusive: they look after only one young at a time. The question of recognizing family members is therefore crucial. First, during the incubation period and the first days posthatching, when one of the two parents is brooding or staying with the newly hatched young, the other is at sea, restoring its health. When it returns, it must find its partner in the colony. This is a difficult task because large penguins do not make nests: the parent can walk around with the egg or chick on its legs, warm in a fold of skin. Later, when the young have gained a little independence and can join other penguins in a kind of mobile crèche, each parent has to be able to find the chick and feed it. If recognition is not forthcoming, the outcome is certain: the chick will die of hunger. The penguin pair can't rely on the other parents. Long ago, barbaric experiments in which penguins had their beaks taped shut to prevent them from vocalizing showed that adult penguins who had been rendered dumb were unable to recognize their partners even if they passed each other. In adult penguins, recognition between partners in a couple or between parents and chicks is only vocal. The problem with this acoustic communication is obvious: at any given moment, there are dozens of birds in the colony looking for their partner or young, shouting at the top of their lungs. The background noise is frightening. It can exceed 95 decibels, which is considerable (between a lawn mower and a jackhammer). Moreover, since this noise is produced by the penguins themselves, it is obviously not possible to avoid it by pitching the frequencies of their voices higher or lower. Nor is it possible to squawk more loudly: everyone is already giving their all! For the emperor penguin, finding the right chick can take more than two hours—a real challenge. And they succeed![30]

Let us return for a moment to the chain of information transmission that characterizes all communication: an emitter, a signal, and a receiver. To communicate in noisy environments, each element of the chain is decisive. With our previous example of the breakdown on the highway, we know that the sender can adopt signal production strategies that will facilitate the transmission of information. Are such strategies found in penguins?[31] One of the scientific studies led by Thierry on this subject was published in a prestigious British journal under the title "How Do King Penguins Apply the Mathematical Theory of Information to Communicate in Windy Conditions?" It has met with great success in the world of bioacoustics and beyond. When the wind is strong—and it is true that it's windy under the latitudes where penguins live—the difficulty of communicating in the colony increases considerably. This wind contributes to the noise of the signal. To counteract it, the king penguins practice redundancy of information. Their song is a succession of syllables, all of which carry the individual identity of the emitter. Researchers have measured wind speed and counted the number of syllables in the penguins' songs. And, hold on tight, the number of syllables in the song of the king penguin is proportional to the wind speed! It's hard to believe. More precisely, as long as the wind speed remains below 8 meters per second, the penguins produce songs of between 4 and 6 successive syllables. Above 8 meters per second, the songs get longer, increasing to 10 syllables when the wind blows at 9 meters per second and easily reaching 14 syllables when the wind speed exceeds 11 meters per second. The relationship between the number of syllables and the wind speed is linear, meaning that the number of syllables can be predicted just by knowing the wind speed.[32]

When vocalizing, penguins use many other strategies to counter the noise of the colony and the resulting confusion. For example, when one individual starts to squawk, all its neighbors within 7 meters fall silent. This courtesy rule limits interference and reduces signal noise. In addition, penguins vocalize by adopting body postures that facilitate the transmission of their signals. For example, the king penguin points his beak toward the sky and extends his neck as far as it will go: the sound waves then propagate above the bodies of the other penguins and avoid being absorbed by the masses of feathers along the way. The emperor penguin uses his chest as a reflective mirror: he sings with his beak lowered against his torso, which reflects the beam of sound waves in the direction of his partner.

The adaptations of penguins to communication in noisy environments cannot be limited to the behavior of the emitter. Penguin song is a signal with interesting characteristics in the face of this constraint. The analysis of the temporal and frequency structure of the song, together with playback experiments with vocalizations that have been previously modified, show a very precise individual vocal signature. This signature is coded differently depending on the species of penguin. In the king penguin, it is the frequency modulations that count. "Wrooinnn— wrooin wrooin wrooin wrooin—wrooin wrooin"—the song of the king penguin sounds like a trumpet blast. On the spectrogram, the frequencies rise and fall several times, rapidly, with a rhythm and acceleration specific to each individual. If you artificially make the song lower- or higher-pitched, it does not alter the recognition. On the other hand, the slightest change in the frequency modulation, and the signal of the partner or parent becomes an unknown voice.[33] If the receiving bird is close, half a single syllable is sufficient for recognition: "wroo. . . ." As with the black-headed gull, redundancy of information is important: the penguin trumpets and trumpets again and again, repeating its simple name: "wroo . . . wroo. . . ." The emperor penguin is a bit of an exception: in his case, three successive syllables are necessary. In these two penguins, the individual signature is coded by the temporal dynamics of the song: frequency modulation in the king's case,[34] amplitude modulation in the emperor's case.[35] Another remarkable fact: both species produce two-voice signals.[36] I'll try to explain the phenomenon in a simplified fashion.

To produce a vocalization, an instrument is required. As in the world of music, there are two main categories: wind instruments, such as the trumpet, where a breath of air makes a flexible structure vibrate; and friction instruments, such as the violin, where solid structures rub or collide, making them vibrate as well. In both cases, the vibrations are transmitted to the air, producing sound waves. Birds and mammals have wind instruments: the syrinx and the larynx, respectively. Other modalities of sound production may exist in these two groups of animals; for example, some birds have feathers that vibrate during nuptial flights and produce sounds.[37] But the larynx and syrinx are the two main vocal instruments of mammals and birds. Larynges and syringes are very complex organs: cartilage, bone, muscles, and the nervous system work together to vocalize. In simple terms, they are first of all membranes placed in the path of the air flow coming out of the lungs—membranes that will be stretched by muscles to a larger or smaller extent, allowing the emitted frequencies to vary. We see this in more detail in the next chapter. For the moment, you should know that the larynx of mammals and the syrinx of birds differ in one essential point: they are not in the same place. While the larynx is located high up on the respiratory tract (at the Adam's apple), the syrinx is placed very deep, just at the exit of both lungs, at the junction of the two primary bronchi—the tubes that carry air into the lungs.[38] In some birds (e.g., the songbirds), there is a functional originality of the bird syrinx: it is a double organ. With two systems of vibrating membranes, controlled by two nerve commands, the syrinx is a musical instrument that can emit two different sounds at the same time. This is in theory.[39] In practice, in most songbirds, either the two voices of the syrinx are perfectly tuned because its two parts are coupled, or only one of the two semisyringes emits a sound while the other remains silent; sometimes the two semisyringes alternate. One could say that there is only one voice.

In large penguins, the two semisyringes produce two sounds of different frequencies at the same time. This is called the two-voice phenomenon. These two frequencies are not very far apart, of course, but they are far enough apart to be noticed on a spectrogram and to play a role in encoding the individual identity of the transmitter. It is in the emperor penguin that the two-voice phenomenon is most remarkable. The two frequencies are close enough to generate a bizarre phenomenon called beats, which results in a regular modulation of the signal amplitude: "WO-wo-WO-wo. . . ." I will not dwell here on the physical explanation of the beats. Just know that they are well known to guitarists: they get them by simultaneously plucking two strings emitting very close notes. In the emperor's case, the rhythm of the beats is characteristic of each individual. It is the double syrinx that controls the vocal signature. This coding of individual identity information is extraordinarily reliable: while intensity modulations managed by variations in the flow of air leaving the lungs would probably be fluctuating, the beats depend directly on the anatomy of the syrinx of the transmitter bird. Since each individual has a unique anatomy, the beats are unique. In addition, an individual's beat remains more or less intact during propagation, because the two frequencies that create it reach the receiving individual with little distortion and thus restore the entire initial beat to its original level. Beats carry the signature of the emitter far and wide with great reliability.[40]

Large penguins therefore have emission strategies and a sound signal that are well adapted to the constraints placed on individual vocal recognition by the noise of the colony. On the signal reception side, they are also very efficient. Chicks have shown an extraordinary ability to extract information from a noisy signal. In playback experiments, the call of one of the parents was mixed (as a DJ might) with the calls of other adults, mimicking ambient noise. This controlled-signal noise allowed researchers to set a very precise signal-to-noise ratio; that is, the emergence (or disappearance) of the parent's voice against the cries of other adults could be precisely controlled. The results of the experiments are breathtaking. King penguin chicks were able to recognize their parent's voice even if its intensity was 6 decibels lower than that of the ambient noise.[41] In other words, even if the noise level is twice as loud as the parent's call, the parent is still heard and identified by its chick. By way of comparison, we humans would have great difficulty identifying a call as soon as its intensity falls to the level of the noise. This astonishing ability to extract the signal from the noise—the cocktail-party effect—is based on the ability to identify the frequency modulations of the parental call lost in the middle of the heterogeneous noise. It ensures the survival of the king penguin chick.

The two large penguins are the ones facing the most important constraints for recognition between partners and between parents and young: terrible ambient noise, birds constantly moving around the colony, and no visual cues. Other species of penguins, such as the Adélie penguin *Pygoscelis adeliae*, form less dense breeding colonies, where each pair incubates its egg in its nest, which is often a simple, small depression in the ground, but a nest nonetheless—a clearly identifiable home—where the young stays and is fed by its parents until it becomes independent. In these species, adults returning from a feeding trip to the ocean will not have much difficulty finding their partners or young. They first return to their nest, probably based on their visual knowledge of the area. It is only on arriving at the nest that they start to vocalize. The colony is therefore much quieter, and the ambient noise is much less pronounced than that of the colonies of large penguins. The same experiment that I have just told you about on the king penguin—the one with the parent's song mixed with vocalizations from other adults—was carried out on the Adélie penguin. Under these constraints, the Adélie penguins are much less successful than their royal counterparts. As soon as the parent's call no longer emerges from the noise (i.e., a signal-to-noise ratio of 0 decibels), the chick becomes unable to recognize its parent's voice. The natural constraint is lighter, as is the sensory performance.[42]

There is much more to be said about penguins, such as the fact, for example, that a king is particularly gifted at finding the exact location of his or her singing partner or parent.[43] Are these differences in the reliability of vocal recognition between species with different levels of risk of confusion found in other animals? Early studies have compared species of American swallows nesting in large, dense colonies with solitary swallows where each pair breeds in isolation. Individual vocal signatures are much more pronounced in the calls of colonial swallows.[44]

The results of Isabelle Charrier's work on acoustic communications in seals are in the same vein: the more densely a species forms its breeding colonies, with a high risk of confusion between individuals, the more reliable the vocal signatures of mother and young are and the more effective the individual recognition.[45] The transition is now in place: in the next chapter, we will observe the communication between mother and young in pinnipeds. We are about to go deeper into the complex world of acoustic communication between parents and their offspring.

5

Family dinner

Igloolik, Foxe Basin, Arctic Province of Nunavut. After pushing the snow clouds southward, the icy wind from the pole has died down and the low Arctic sun is timidly reborn. We will be able to set sail. Aboard a small motorboat, a derisory walnut hull in the immensity of the ocean, I am with Isabelle in search of walruses, these enormous animals with long tusks resting on the blocks of the fractured pack ice. Isabelle studies marine mammals, and she is an outstanding field researcher. "The objective of the mission is very simple," she told me. "To find females with their young and test if the young calves are able to recognize their mothers by voice." But nothing is easy in this inhospitable region, starting with the journey to the camp. It required several flights from France, with the size of the plane becoming progressively smaller at each stopover. In Iqaluit, the gateway to Nunavut, we definitely left the heat of the Canadian summer behind us. As we flew over Foxe Basin, the first ice floes appeared. Finally, we reach Igloolik and its ice pack—an immensity of white. Only the airstrip is clear. Another hour of snowmobiling to the cabins, then . . . 13 long days of Arctic summer follow, confined in our cabins, completely isolated from the outside world by the storm. The Arctic is untamable. It's the most hostile environment I've ever been in.

Walruses, seals, elephant seals, fur seals, and sea lions form a group of marine mammals called pinnipeds.[46] Marine, yes, but not through and through. Unlike whales and dolphins, pinnipeds must go ashore to reproduce. On land . . . or sometimes on the ice, like the walrus when in the Arctic spring the pack ice is just beginning to break up and the mainland shores are still inaccessible. Some pinnipeds, especially eared seals, form large colonies of tens, hundreds, and even thousands of individuals, similar to the penguins we discussed in chapter 4. The young face the same problem: finding a parent who will feed them. In pinnipeds, being mammals, it is the mother who suckles her young, and therefore who feeds, whereas in penguins, both female and male feed the chick. As in the case of penguins, the difficulty depends on the size and density of the colony. It is therefore highly likely that the reliability of recognition between mother and young will be higher in pinniped species, where the risk of being separated from their young is greater. This is the hypothesis that Isabelle is testing as she travels the world to study and compare recognition systems between mothers and their young in these animals.[47] Let's take the case of the walrus first. With these pieces of floating ice drifting on the ocean as the only places to give birth, the risk of the mother being separated from her calf in just moments is significant. All the more so since she is forced to make an occasional deep-sea dive to feed. Although her young are able to swim from birth and try to follow her everywhere, they regularly lose sight of one another and have to find each other again because the young walruses are in great need of their mothers. They continue to suckle for two to three years after birth. We were in Igloolik to test the hypothesis that a walrus mother must recognize her cub's voice accurately.

Igloolik. *The place where there are igloos.* Originally a place where Inuit families met temporarily during their nomadic life. Imagine a single gravel airstrip, a few dozen houses, a kind of grocery store–warehouse, a small church, a curious building resembling a flying saucer pompously named the Igloolik Research Center, and, a little higher up, a few large fuel tanks, probably replenished just once a year. Igloolik, with its 1500 inhabitants and its dirt streets, is not a megalopolis. The surroundings don't have a very pronounced topography. The city is plain, without any particular charm, in a rather desolate landscape. The nearest village, Hall Beach (Sanirajak), is 70 kilometers away as the crow flies. This is definitely the Arctic: sled dogs tied up at the entrances of houses or wandering in the streets and some impressive polar bear skins on drying racks, waiting to be tanned, testify to this. And the omnipresence of white. Not the familiar, jagged white of a mountain with its bare rocks and forest patches, but the boundless, flat, infinite white of the frozen sea blending with the horizon.

Brad Parker is waiting for us on the tarmac. His stocky build, large hands, and trapper's demeanor exude a roughness mixed with a surprising cynicism. He does, however, inspire confidence. In any case . . . we have no choice. The Arctic is another planet. The rainforest may be impressive when you discover it, but life is easy there. In the far north, life is a daily challenge. After ordering us to throw our bags in the back of his pickup, Brad invites us to have a solid breakfast at his place. "There're still a lot of ice floes this year. Walruses won't be easy to find. Plus, the weather here is unpredictable. Bad most of the time. And sometimes a day when you're rewarded beyond your wildest dreams. The advantage is that the day never goes to bed." He adds, "We'll stay in my huts. We won't come back to town. So eat up!" A little later we glide over the ice pack in a sled pulled by a snowmobile, our legs protected by a reindeer hide. We head for the huts. "The greatest danger is the polar bear. He doesn't fear man. When you go out, look everywhere." The year before, a hungry bear had knocked down the wall of the cabin where Brad cooks. "Pakak . . . ," Brad said, pointing to the man who would be our guide, "Pakak was once saved by another hunter. A polar bear had pinned him to the ground and opened his back with its claws." And then to underscore his point he added, "just as easily as that—crash! Like peeling an orange." A polar bear's claws are like daggers, each up to ten centimeters long. Pakak had some impressive scars.

We learn from the dictionary that the Latin name of the walrus, *Odobenus rosmarus*, comes from the Greek *odo*, which means "tooth," and *benus* ("I walk"). This name was built on a legend: it was long believed that this animal moved across the land thanks to its long tusks. The tusks are mainly used as weapons during fights between males. The walruses also use them as a tool in the search for food and to climb onto a block of ice when coming out of the water, which is very handy for getting a

Walrus

good grip. A walrus eats mainly mollusks from the seabed around 80 meters deep, and more rarely at a depth of between 200 and 500 meters. An adult female walrus weighs up to 1300 kg, males 1800 kg, and new-borns 85 kg, which in itself makes for a beautiful baby.[48]

Taking advantage of a day of good weather where the sun undulates endlessly around the horizon, we go in search of walruses. Pakak is at the controls of the motorboat. Minimalist equipment: sandwiches, a rifle with a few cartridges, no navigation system. We suggested Pakak use our GPS. After taking a quick look at it, he got rid of it, disdainfully. We will never understand how Pakak finds his bearings in this universe without landmarks. The point is, he knows where he's going. So we trust him absolutely and we focus on finding the walruses, eyes glued to our binoculars. What a sight! The Arctic spring is a magical season: here, a flight of snow geese, a magnificent bird adapted to the coldest regions; over there, the double breath of a bowhead whale, then the head of a bearded seal emerging from the waves. And on this strip of pack ice, a female polar bear with her two cubs. The walruses are playing hard to get. After long hours of searching, we finally find a group of males and females with their pups, some barely a few days old. Having spotted a mother a little away from the group, we approach and stretch out the pole carrying the microphone. First, we have to record the young. As in many mammals, young walruses regularly ask their mothers for access to their milk-filled teats. Once the first one's calls are in the box, we move on to another baby.

We had to build up a bank of recordings of several individuals before we could test the females with playback experiments. It took many trips out to sea, playing with the vagaries of the weather, to find different females with their young. Finally, the big day arrived when there were enough recordings to start the playback experiments. The principle was simple: have females listen to the calls of their own calf on the one hand and the calls of a non–family member on the other, and compare the mothers' responses to these two signals. For each of the females tested, we had at our disposal the calls of babies recorded from the previous days: easy enough to have them listen to the voice of a stranger. But each time, we had to start by recording the female's baby, before moving the canoe back a good distance so that Isabelle on her laptop could prepare the signals we were going to emit from our loudspeaker. Since the walruses were resting on drifting blocks of ice, it was indeed impossible to find the same females twice in a row. Another constraint: It was unthinkable to kidnap the little one and replace it with our loudspeaker. Walruses are dangerous animals, and they would have plunged into the water if we had gotten too close. We were going to play back the sounds from the canoe. By dint of tenacity, and almost surprisingly, it was a success. Of the 13 females that we managed to test in two successive missions, all of them responded to the calls of their young from our loudspeaker. With variations, of course: Some females just turned their heads in our direction; others barked loudly. On the other hand, they remained motionless as they listened to the calls of an unfamiliar pup. Only one female reacted to it by taking a quick glance. The evidence was there: walrus mothers recognize their young by voice without fail.[49] Two field trips were necessary to obtain the experimental demonstration. Isabelle stayed a good month each time. Thierry and I took turns. I can hear you from here: all that for this? It's true that we would have liked to explore the recognition mechanisms in walruses in greater detail. But the severe conditions in the Arctic decided otherwise. During the second stay, Thierry and Isabelle had even come close to dying there. Their guide didn't have Pakak's know-how and had been caught in the trap of monstrous ice packs! This memorable episode is told in the film *Bonjour les morses* that I invite you to watch.[50]

Our walrus experiments supported Isabelle's hypothesis: when there is a high risk of separation between the mother and her young, speech recognition is reliable. Is this reliability lower when the risk is less? With their varied social structures, pinnipeds are a good model for answering this question. Some species are solitary: the female gives birth to her young alone on a beach or on a floating ice floe. This is the case in many earless seals, such as the bearded seal *Erignathus barbatus*, for example, where the female stays permanently with her pup for several weeks. On the other hand, most eared seals form colonies, the density of which varies from species to species. There, females suckle for several months. During these long lactation periods, they alternate between stays on land, where they look after their young, and trips to sea, where they feed and replenish their reserves. What are the results obtained by Isabelle by comparing the mother-juvenile recognition systems in these different situations? First of all, how reliable is the individual vocal signature? In solitary seals, such as the Weddell seal *Leptonychotes weddellii*, the harbor seal *Phoca vitulina*, or the Hawaiian monk seal *Monachus schauinslandi*, the voice of a pup is rather variable and is therefore not easy to recognize vocally. On the other hand, the voice of a small walrus, a small elephant seal,[51] a small Australian sea lion *Neophoca cinerea*, or a small subantarctic fur seal *Arctocephalus tropicalis* (both species breeding in colonies) is well differentiated and facilitates its identification. It should also be noted that, in species where it has been tested, recognition is often mutual: mother and young are able to identify each other vocally. Another interesting criterion is the speed at which this recognition is established. Investigations into the implementation of mother-juvenile recognition are rare. Only four species of eared seals have been studied, with very different results. It takes between 10 and 30 days for a Galápagos baby sea

Fur seal

lion *Zalophus wollebaeki* to learn how to recognize its mother's voice, compared to 10 days for a Galápagos baby fur seal *Arctocephalus galapagoensis* and between 2 and 5 days for the subantarctic fur seal. The greatest temporal amplitude was observed in the Australian sea lion: if mothers can recognize their cubs 10 days after birth, the cubs take several weeks![52] This is curious, and still unexplained.[53]

The subantarctic fur seal or Amsterdam fur seal *Arctocephalus tropicalis* lives in the Indian Ocean and breeds on small islands in the area (including the island of Amsterdam, which is one of the French Southern and Antarctic Lands). Isabelle spent nine months studying this species as part of her doctoral thesis. After the birth, a mother fur seal from Amsterdam Island stays for a few days with her newborn but has only one thought in mind—to go back to the ocean alone to feed. She has been fasting for several weeks, and there is not much to eat in the immediate vicinity of the island. She will leave her calf for two to three weeks. Unbelievable but true—two to three weeks when the baby is on its own in the colony with no food to eat. After that time, the mother returns. The reunion is an amazing sight to behold, with mother and baby calling each other repeatedly. You have to see this little black ball clawing its way across the rocks that separate it from its mother to understand the sheer strength of its motivation. Once the little one is finally there, the mother sniffs it, probably as a final olfactory check. All this happens very quickly; within 10 minutes after she touches the shore, the mother is in contact with her cub. These reunions are crucial: a lack of recognition means no feeding, and the death of the young. Recognizing the mother's voice is a vital imperative.

Playback tests carried out by Isabelle show that a newborn baby reacts to the voice of any female within hours of birth and that it must learn to recognize its mother. The most surprising result was when, on Isabelle's return to France, we compared the number of days before the mother's recognition with the time of her departure to the ocean. The two were correlated: the faster the baby learned, the sooner the mother abandoned it and vice versa.[54] In other words, a fur seal mother seems to pay extreme attention to her baby's ability to recognize her: "Do you recognize me? Well, I can leave you alone now."[55]

What happens as the length of the mother's absence increases, you might ask? To find out, Isabelle played the barks of female fur seals to pups when their mothers had abandoned them for a short or a long time. The results of her experiments show that the motivation of the pup to shout in response to the barking of females increases with the time of separation.[56] After being left alone for a day, one in five pups react, while if two weeks have passed, three out of four call out in hunger! What if they hear the voice of an unknown female? No reaction at the beginning, of course, but up to 30% of the pups cry out in response after two weeks. It's certainly not because the cubs have become unable to recognize their mother, as borne out by the dramatic reunion when she finally returns. But it is because they are willing to do anything to get milk, including begging for milk from any nipple. In short, they are too hungry to wait, and they let it be known.[57]

The call of the baby fur seal is a *signal*. It carries information not only of the emitter's identity but also about its state of satiety from the sender to a receiver, the mother fur seal. I'm not hungry—I'm not crying. I'm a little hungry—I yell moderately. I'm as hungry as a wolf—I'm screaming all the time. We know that with our own babies, don't we? Listening to this signal, the mother will change her behavior: calling in turn, getting into position to suckle her young, for example. These observations lead us to a slightly more theoretical reflection. In order for a communication system between a transmitter and a receiver to be preserved by natural selection, the information transmitted must generally be of interest to the receiver. Otherwise it would stop responding. It is not necessary for this information to be systematically interesting or truthful. But it does have to be interesting enough for evolution to have selected a receiver that responds to it. In fact, as a rule of thumb, if the transmitter-receiver pair do not find a mutual interest in making the communication work, it will disappear. And let's be clear: the signal has to be honest *enough* for communication to continue.[58]

This hypothesis of *honest communication* is not easy to test. Indeed, it requires being interested in several elements that are not necessarily easy to grasp. Let's stay with our juvenile begging calls, which are signals that are widely used not only in our walruses and seals but in many mammals and birds. First of all, do these food-begging signals accurately reflect the state of the transmitter? In our baby fur seals, this seems to be the case. Experimental tests have been carried out in birds, where it is easier to artificially manipulate the amount of food ingested by the offspring. Rebecca Kilner, a professor at Cambridge University, conducted famous experiments more than 20 years ago showing that the begging calls of the chicks of the Eurasian reed warbler reflect their need for food. Since then, other studies have confirmed this honesty of information in other bird species.[59]

Kilner was interested in the roles of the begging calls in parent-offspring interactions in birds, and how the cuckoo *Cuculus canorus* might exploit them—you know, that bird that lays its eggs in other birds' nests. We'll talk more about that later. So Kilner's model bird was the Eurasian reed warbler *Acrocephalus scirpaceus*. The warbler chicks solicit their parents by opening their beaks wide, displaying the red background of their throats, and cheeping insistently. In response, the parents shove lots of small flying insects and delicious caterpillars down their throats. In a first experiment, Kilner and her collaborators temporarily removed chicks from their nest and placed them in a small, experimental box.

Kilner then fed them until they stopped begging, then kept them warm but without feeding them further. For the next 2 hours, every 10 minutes she mimicked the arrival of a parent by tapping the box, which inevitably provoked the chicks' begging behavior. The results were clear: the longer the chicks had gone without food, the more loudly they would chirp. You want numbers? Here are some. After 20 minutes of waiting, a 3-day-old chick cheeped at the rate of 2 calls every 6 seconds. After 100 minutes, 8 calls. The phenomenon was even more sensitive with older chicks (6 days old). They gave less than 15 calls after 20 minutes of waiting compared to more than 35 after 100 minutes, more than twice as many. In both the youngest and oldest chicks, the number of calls was proportional to the time spent waiting. But Kilner didn't stop there. She had to check that it was hunger that caused the chicks to call more, not the stress of absent parents or any other reason. She measured the amount of food needed to stop the calls of isolated chicks and found a strong correlation between the amount of food and the number of calls. It is probably safe to assume that the begging calls of reed warbler chicks honestly reflect their nutritional needs. However, other studies have shown that this correspondence between nutritional requirements and the intensity of begging behavior is not as rigid as it seems. It is thus possible to train chicks to chirp more or less according to the promptness of the parental response.[60] If they are fed on their first call, they will soon learn that there is no point in forcing their voice to get rewarded. On the other hand, if we wait for them to chirp insistently before feeding them, they will quickly get into the habit of expressing their hunger vehemently. Honest, but not stupid.

Why not beg as loudly as possible right away? After all, having a tantrum as soon as you feel the slightest pang of hunger would allow you to get more from your parents, and faster. You might as well lie a little or a lot about your condition: you'll eat more and get fatter faster than your brothers and sisters. In other words, why stay honest? Scientists who have investigated this question by mathematical modeling have shown that begging behavior can only remain honest if it has a certain cost. Although I am not a proponent of the use of these anthropomorphic terms, I must confess that they are powerful metaphors. It is well understood that the temptation to dishonesty will be greater if it costs nothing and can be profitable! Where could the cost of begging come from? It is in fact of two kinds: a direct energy cost (screeching requires energy) and an indirect cost due to the predators that may be attracted by the cries.

How can we estimate the direct energy cost? A first approach is to put a chick in a respirometer, a device that measures its oxygen consumption. The results of various studies show that this consumption is multiplied by only 1.3 when the chick emits calls compared to resting, which does not seem considerable.[61] Since time is money, time spent begging can also be considered a cost. Video recordings of house wren nests have shown that chicks beg only 4 to 10 times per hour and that each round of cries lasts 4 to 7 seconds. This makes 16 to 70 seconds spent begging per hour, or 0.4% to 2% of the time.[62] The results of all studies show that the energy cost of the begging calls is rather low. Becoming dishonest by chirping more, therefore, would not cost much.

But that measurement does not take into account the other side of the cost: increased predation. A nest, even well hidden, if full of young, bawling chicks, is like a lighthouse in the middle of the night, and some predators apparently understand this. Be aware that, in birds, predation by other birds, mammals, or snakes is responsible for 80% of brood mortality.[63] Do begging calls have anything to do with that? To answer this question, experiments have been carried out that involve placing small loudspeakers in artificial nests to mimic the presence of young chicks begging. Most of these studies have concluded that the cries of the young attract predators.[64] It is even likely that their impact is underestimated. Predators don't just listen; they watch. The more the chicks solicit their parents, the more the parents tend to feed them and thus increase their comings and goings, probably making it easier to locate the nest.[65]

As noted above, chicks can change their begging behavior depending on external conditions. It has been shown that they can lower the sound level of their calls and make them more high-pitched when the number of predators increases, making them somewhat less noticeable.[66] Sometimes they even decide to remain silent if they hear a predator approaching.[67] And if they don't, their parents can order them to shut up or at least turn it down.[68] But we are not at the end of our surprises. It has been suggested that chicks may instead exaggerate their calls despite the risk of predation, forcing their parents to respond to their requests as quickly as possible.[69] A bit like an angry child throwing a tantrum to demand a piece of candy at the supermarket, parents will give in faster so as not to attract attention.

As you can see, the begging calls are honest . . . to a certain extent. The young are able to modulate them to get more or to limit the danger of being eaten by a predator. Another important element to take into account is that very often a nest houses several siblings who may compete for food provided by the parents. Who will receive the most pudding? Studies indicate that competition within nests can be accompanied by an increase in the intensity and number of calls, with each one exaggerating its real needs in an attempt to tilt parental choices in its favor. The correlation between the chick's state of satiety and its motivation to call becomes less reliable.[70] An experiment with the yellow-headed blackbird *Xanthocephalus xanthocephalus*, a North American bird, has shown that a chick begs longer if it is placed in a nest with another hungry chick rather than with a full chick.[71]

As parents of large families know, children can also join forces to get what they want. In our study of black-headed gulls, Isabelle and I highlighted such a strategy.[72] From our floating blind, we observed the begging behavior of gull chicks in nests of one, two, or three chicks. The first interesting result was that the number of calls per hour decreased with increasing clutch size, from more than 7 calls per hour when the chick was alone to less than 3 calls per hour in three-chick nests. When only one chick was present, the parent regurgitated all the more easily as the chick screamed loudly. But only 17% of the chick's calls were followed by parental regurgitation. In nests with several chicks, the number of regurgitations was proportional to the number of chicks crying at the same time: 23% of the cries were followed by regurgitation when only one chick was crying compared to 87% when both were crying. The group effect was even more pronounced in nests with three chicks. If only one of the three siblings was begging alone, it could motivate the parent in only 13.5% of cases. In pairs, 55% of the begging sequences induced regurgitation. And when the whole tribe cheeped together . . . 100% of their collective begging was satisfied. United we stand. However, it is not certain that this behavior is generalizable to many species. In black-headed gulls, the parents do not feed the young directly; they regurgitate on the nest floor, and the young must then retrieve what has fallen out of the parental beak. Competition between chicks is therefore more likely to occur after parental regurgitation than before. In birds where the parent feeds directly in the bill, there may be less incentive to cooperate.

Interactions between chicks are not always limited to the crucial moment of the parent's arrival. Surprisingly, chicks of many species call out when the parents are absent, even though they are easy prey for predators. Sometimes this can be a misinterpretation, as the chicks hear a noise or feel a vibration that makes them think a parent is arriving. The temptation to make themselves known as early as possible is sometimes great. For example, in the tree swallow *Tachycineta bicolor*, the first chick that calls out is the one that will be best fed.[73] It is also possible that the chicks are talking to each other. When competition between chicks is intense, it may be more interesting to agree on sharing resources rather than spending a lifetime fighting. Alexandre Roulin, professor at the University of Lausanne in Switzerland, his colleague Amélie Dreiss, and their collaborators have thus supported the *sibling negotiation hypothesis*.[74] This hypothesis suggests that, in the absence of the parents, chicks inquire about their mutual motivation to obtain food by calling out. By listening to the calls of the other chicks, each one then decides its chances of actually having access to food when the next parent returns to the nest. Alexandre's favorite study model is the barn owl *Tyto alba*. This large owl, with its pale plumage and silent flight, frequently nests in buildings.

A barn owl nest has an average of four chicks, with a maximum of nine. What a party! With their hooked beaks and sharp claws, barn owl chicks are very well equipped to engage in some serious battles. Yet they're quite friendly to each other, smoothing their plumage, warming each other, and sometimes even exchanging food. There's no physical struggle over who gets the next meal. Instead, they negotiate. How's that? In the absence of their parents, a particularly hungry chick produces many long-lasting cries. If its siblings are less food starved, they remain silent when the parent arrives. It is a so-called iterative process: at first, everyone begs; then as the negotiation progresses, some chicks withdraw while others exaggerate their cries. Sort of a poker table for owls.

The chicks pay attention to the details of the calls. If a hitherto dominant individual gradually gives shorter and fewer cries, those that were previously silent try to get the upper hand.[75] Moreover, the one who ultimately dominates the negotiation is the one who forces the others to progressively issue shorter and less frequent calls. To achieve this, it starts by imitating the duration of the calls of the others, who respond by reducing the duration of their own calls. If this is not enough, the hungry chick then increases the rhythm of its calls; and, in principle, if the others are less determined to eat, they give up. When the owl parent enters the nest, it brings only one prey—a small rodent, for example. This prey cannot be shared among the chicks and will therefore be swallowed by only one of them, the one that won the negotiation. Imagine if the little owls fought beak and claw for the vole. It is easy to understand the value of sitting down at the negotiation table to decide beforehand who will eat. The chicks that were not fed on this occasion will be fed next time.[76] The little barn owls know how to wait.

With this example, we can see that the life of the nest is not limited merely to communication between a chick and its parent. In reality, parents and offspring form a real communication network—a concept that we discuss in detail later as it applies to most animal communication systems. The chicks have an intimate knowledge of their siblings. They know that Peter is hungry, while Paula is full, etc. A few studies have shown that the chicks' calls have vocal signatures, and playback experiments have revealed that the chicks recognize each other.[77] In some species, parents also use these signatures, for example, to share the responsibility for feeding their children. Do you remember Tudor Draganoiu, the man who showed that black redstart males have similar songs when they are neighbors? Tudor also studied the relationship between parents and their young. He found this sharing of the brood in his favorite bird.[78] Two or three days after leaving the nest, the chicks are still unable to feed on their own. A young redstart is then fed preferentially by either its mother or father. Tudor has shown through playback experiments that parents recognize which young they are dealing with on the basis of its calls.

Communication between parents and young starts very early on. It has long been known that bird embryos can inform their parents of their temperature by calling through the eggshell:[79] "I'm too hot!"—the parent stops brooding; "I'm too cold!"—it gets back on the eggs. When I discuss crocodiles in chapter 8, we will see that the embryos signal acoustically to their mother and brood mates when they are ready to hatch. What about the parents? Recently, Mylene Mariette, a researcher at Deakin University in Australia, showed that adult zebra finches *Taeniopygia guttata* talk to their chicks when they have not yet hatched.[80] And what do they tell them? They tell them the outside temperature. When it is very hot and dry in the Australian desert, an adult zebra finch is stressed. It then emits a series of little calls, the *heat call* as Mylene described it, especially when it is in the nest incubating its eggs. Mylene and her collaborators had the excellent idea of playing these calls from a loudspeaker to brooding zebra finch embryos. She noticed that their growth was slowed down, resulting in smaller adults than usual; in other words, birds better adapted to withstand harsh weather conditions and poor food resources. When they were able to reproduce, these birds produced more offspring than control individuals who had not heard the heat calls when they were hatched. Unbelievable, isn't it? We don't know, of course, if the zebra finch parents are deliberately calling "It's too hot!" to their eggs—it's even possible that the heat calls are an unintended consequence of altered respiratory activity due to heat stress. Nevertheless, these sounds alter the growth of the embryos, allowing them to prepare for more difficult living conditions than expected.

In the nest's communication network, children talk to parents, they talk to each other, parents talk to them. And do the parents talk to each other while they're both busy feeding their brood? When Mylene was a postdoctoral researcher in our laboratory, she took part in an unprecedented experiment. In the zebra finch's nest, female and male alternate their presence, each one brooding in turn, so the test involved delaying the male's return to the nest (we had a system to capture the male when he came to the area of the aviary where the food was). This delay altered the vocal exchanges that characterize reunions between partners in this species.[81] More precisely, the time that the female would then stay out of the nest was predicted by the rhythm of the duo. The more the female sped up the pace of the exchanges, the less she would stay in the nest waiting for the male the next time. It is likely that these exchanges of information between parents, these kinds of negotiations, are also commonplace. Observations in natural conditions with the great tit have shown that the female in the nest signals her hunger to the male outside the nest with her calls.[82] Presumably, this is taken into account by the male when it has to decide that it is time to take over from his partner.

While communication signals generally need to carry reliable (honest) information in order for the communication to continue (if her baby cries wolf too often, the parent will eventually stop responding), they can also be a means of manipulating the receiving individual rather than simply informing him or her. This concept of a manipulative signal was put forward as early as the 1970s by Richard Dawkins and John Krebs, both professors at Oxford University.[83, 84] The main idea is this. When communicating, the "purpose" of the sender is often to get something from the receiver (food, copulation, etc.). To do this, it distorts its signal by exaggerating the information it carries. Dawkins and Krebs suggest that, in response to this escalation, the receiver will become increasingly resistant to the received signal over time. This arms race, as it is called, between the emitter and the receiver is reflected on the evolutionary scale of a species by the progressive development of increasingly extravagant signals and a growing resistance to react to them.[85] You can see that this conception of animal communication as an endless spiral, an unstable process, is opposed to the stable balance proposed by the signal seen as honest that we were talking about earlier. We return to these important notions later when we look at the mechanisms of the evolution of acoustic communication. For the moment, let us say that the two points of view can be reconciled by assuming that a communication is likely to be fairly honest and stable when the transmitter and receiver have a common interest (which is the case for our chicks and their parents), whereas it is likely to be a manipulation accompanied by an arms race when the degree of conflict between transmitter and receiver is very high. This second situation is typically encountered when one species of bird has its young raised by another species. This is a very instructive case study for understanding the concept outlined by Dawkins and Krebs.

Parasitic birds represent about 1% of all bird species. Some are specialized: the female always lays her eggs in the nest of the same species. Others are eclectic: the brown-headed cowbird, for example, can entrust its eggs to more than 200 different species of birds.[86] This method is accompanied by multiple strategies to facilitate the integration of the eggs, and then the young parasites, into the host nest.[87] First of all, one thing is obvious: when a parent bird feeds a young bird that is not of its species, the interests are divergent. In short, the parent wastes time and energy—even more if its young have been killed by the parasite or die of starvation—while the young freeloader gains its food, survival, and growth. The young parasite is therefore the archetype of the manipulator, and the adoptive parent will have every interest in arming itself against this manipulation. How to manipulate? One method is to imitate the begging signals of legitimate children;[88] another is to reproduce the signals by exaggerating them.[89] Although few studies have compared the acoustic structure of the calls of parasites and parasitized, it is estimated that the calls of parasitic chicks mimic the calls of chicks of the host species in one out of two parasite species.[90] Let's take the example of the Horsfield's cuckoo *Chrysococcyx basalis*, an Australian bird that is in the habit of having its young raised by the superb fairy wren *Malurus cyaneus*. The cuckoo chicks' calls sound similar to those of the fairy wren chicks. However, some chicks are better at imitation than others. Unskilled imposters will soon be spotted by their hosts, and the adoptive parents of a Horsfield cuckoo abandon their nest in 40% of cases, leaving the cuckoo to perish. Better still, if another parasite, the shining bronze cuckoo *Chrysococcyx lucidus*, tries its luck in a fairy wren nest, it never succeeds in forcing its adoptive parents to feed it and is abandoned every time. The reason is simple; this unfortunate cuckoo emits a begging call that is very different from that of the fairy wren chicks.[91] Not content to imitate, parasitic chicks often exaggerate the intensity of their cries of begging. In a way, they lie (a little) about their condition; the adoptive parents get trapped, believe they are dealing with a particularly hungry chick, and redouble their energy to feed it.

In this arms race between parasites and their hosts, host species develop strategies to limit the impact of the parasite.[92] These strategies are not limited to paying attention to the acoustic quality of the calls of the chicks present in the nest. They also include, for example, harassment of parasitic adults to prevent them from laying eggs in the nest and the recognition and elimination of parasitic eggs once laid. Some strategies can be really subtle. A few years ago, Mark Hauber, my colleague when I was a visiting professor at Hunter College in New York, now professor at the University of Illinois, together with Diane Colombelli-Négrel, Sonia Kleindorfer, and their collaborators, made a very unexpected discovery: the parents of some birds teach their children a password while they are still in the nest.[93] Here's the story.

Once again, the host-parasite pair formed by the superb fairy wren and the Horsfield cuckoo served as a study model. After recording the sound activity of parasitized and nonparasitized fairy wren nests throughout the nesting period, they first discovered that the fairy wren female produced an *incubation call* during the days before hatching (so it's not just zebra finches that talk to their eggs!). The scientists then compared the acoustic structure of these calls with the acoustic structure of the begging calls that the chicks emit once they have hatched. They found strange similarities. The calls of the chicks in one nest were more like those of their own mother than those of another fairy wren female—in other words, there are family accents in the fairy wrens. To test whether this resemblance had a genetic component, scientists exchanged fairy wren eggs between different nests at the very beginning of the brooding period. When they hatched, the fairy wren chicks called out like their adoptive mother. So they *learn* this call when they hear it while in the egg. The next step was to see if the parents paid attention to this family signature when they had to feed their young. Playback experiments showed that this was indeed the case. Fairy wren parents respond much better to the calls of their own offspring than they do to those of foreign chicks.

Let's describe the course of events for clarity. In an unparasitized nest, the chicks hatch after a two-week incubation period. Five days before hatching, the female begins to make her incubation calls. In the days following hatching, the nest becomes silent again. The parents feed the young, who beg by opening their beaks without making any audible sounds. Three days after hatching, the chicks begin to vocalize. Their calls resemble their mother's incubation call. They have learned the password. Both the mother and the father, who has also learned to recognize the acoustic key, continue to feed their young until they fly away. In a nest where a cuckoo has laid its eggs, things are different. If the female begins to make her incubation call about ten days after laying, the cuckoo chick hatches on the twelfth day of incubation, three days earlier than fairy wren chicks. It takes advantage of this situation by removing their eggs from the nest. Faced with a nest with only one chick left, the female fairy wren stops making hatching calls. A few days later, when the cuckoo chick starts to call, it will have some difficulty imitating the call of its adoptive mother. It must be said that the lesson will have been short, heard for only two days behind its shell. Perhaps it is less genetically prepared for imitation than legitimate children. If the fairy wren parents are not convinced by the young cuckoo's vocalizations, they will abandon the nest, leaving the cuckoo chick to its sad fate. I told you that fairy wrens abandon their parasitized nest in 40% of cases. Errors in passwords are probably to blame for this. We can make the hypothesis that the next step in the arms race between the superb fairy wren and the Horsfield cuckoo will be for the fairy wren to further refine the acoustic correspondence between the maternal incubation call and the legitimate chick's call, and for the cuckoo chick to hone its study skills.

Teaching your own children a password—this is amazing, to say the least! And it's not only birds that talk to their young before they're born.[94] A study suggests that the female bottlenose dolphin *Tursiops truncatus* vocalizes while she is pregnant and that her young imitate her calls once born.[95] Might she teach them the vocal signature that will enable them to recognize each other among other dolphins?[96] This provides us with a nice transition. Let's give up the birds for a while and go back to the ocean. The whales and dolphins are waiting for us there . . . vocalizing, of course.

6

Submarine ears

UNDERWATER BIOACOUSTICS

Somewhere in the Foxe Basin, between Igloolik and Rowley Island. We're still with Pakak. He shuts off the engine near a block of ice. He lets us drift a few yards and throws the grappling hook on the ice. The hook slips and finally snags. Here we are, moored. The sea is calm. Leaning on the edge of the canoe, Isabelle lets the hydrophone cable slip through her fingers. The little black capsule dives into the icy water. We quickly lose sight of it. Five meters, ten meters . . . "Take the headphones," Isabelle says to me, "listen . . . they're there." I obey. Immediately, lancinating songs reach my ears, like the wind blowing in the trees: "Twwu-huhuhoohou hoohoooooooooooooooooooooooo. . . ." Songs, more songs, endless songs—almost disturbing. Long melodies, at first high-pitched, then descending to the low register, both regular and modulated in trills, like someone gradually losing his breath while blowing into a musical pipe. When one melodic line ends, another has already begun. Or several. Some close and strong, others like an echo in the distance. "These are bearded seals," Isabelle whispers. "What you can hear is the song of courting males." A memory surfaces of an old picture book from one of my grandmothers, with a drawing of a seal singing, head down, its body bowed. Sometimes, they say, the seal songs are so loud that you can hear them even when you are out of the water.

The bearded seal—*Erignathus barbatus* by its Latin name—is a sedentary, solitary inhabitant of the great north. When we see it, it is usually

Bearded seal

lying on a block of ice, waiting for time to pass. It has a large body disproportionate to its small head—and incredible whiskers. In the air, they dry out and bend back. But in the water, this imposing comb spreads out harmoniously to probe the bottom in search of crabs and mollusks. During the breeding season, each male defends an underwater territory. Isabelle had hypothesized that the bearded seal's song plays the same role as that of the birds: to ward off intruders. How would we test this idea? With underwater playback experiments, of course. The method was obviously appealing, but not seeing what the seal does underwater made the operation uncertain, to say the least. We could only record a possible vocal response of the seals to our signals and observe their behavior on the surface. It wasn't much, but we decided to try it anyway! From the canoe, we sent our underwater loudspeaker down sixteen times, a good 3 kilometers apart each time to make sure we would encounter different individuals' territories. Almost every time, the males in the vicinity began to sing less often as they heard the song from the speaker. Also, in half of the experiments, we saw a seal head appear on the surface of the water near the boat, as if the local occupant was trying to spot the intruder on the surface that he hadn't been able to see underwater.[97] It is certainly difficult to say from these experiments alone that singing allows the bearded seal to defend its territory. The fact remains that our playbacks changed their behavior, apparently encouraging them to patrol their territorial waters.

These experiments underscore the difficulty encountered in bio-acoustic research in underwater environments. There, underwater, it is difficult to directly observe the behavior of the animals.[98] Studying the acoustic communications of seals, whales, dolphins, fish, and various shrimp[99] in the sea requires special methods, such as using underwater sensors (hydrophones, recorders, GPS), sometimes placed on the bottom, or even directly attached to the animals if they are big enough. Until recently, due to a lack of suitable technology, the sea was thought to be mostly silent. In the 1950s, didn't Jacques Cousteau make a documentary on the seas and oceans called *The Silent World*? Believe me, the opposite is true. The oceans are teeming with sound-producing beasts. In fact, if you had the courage to read chapter 2 of this book, the one in which we made circles in water, you already know that the aquatic environment is particularly conducive to the propagation of sound waves. You surely still have in mind that they move very fast—much faster than in the air (something like a kilometer and a half per second)—and that they travel very far. The songs of some whales can be heard from dozens, even hundreds, of kilometers away. This property of the marine environment has interesting biological consequences. For example, the coral reefs, inhabited by myriads of shrimp and fish, all chattering, rasping, squeaking, and growling, produce a background sound that can be heard several kilometers away. Attracted by this very particular hubbub, fish or larvae of various animals looking for a place to set up home find the reef of their dreams much more easily than if they had looked for it randomly in the blue immensity. It has even been shown that each type of reef has its own unique sound signature, which varies with the time of day and seasonal cycles. But, you may ask, how has it been proved that fish use the sound of the reefs to get there? Here's the story.

First of all, be aware that most of the myriads of fish that live as adults on coral reefs spent their childhood in the open sea. During this larval phase, the fish are more or less left to drift until they reach an age when nomadic life begins to weigh them down. Then they decide to actively swim toward the reef. In the first of several studies, New Zealand scientists, with the knowledge that light attracts fish, deployed two light traps at sea 500 meters apart. Next to one of the traps they placed an underwater loudspeaker that emitted the sound of a coral reef. By repeating the experiment several nights in a row, the scientists found that the loudspeaker trap attracted on average more small sardines than the silent trap. The sound reinforces the light-attracting effect.[100] Another study went even further. In this instance, the scientists built 24 small patches of artificial reefs by placing pieces of dead reefs on the sand at a depth of 3 to 6 meters. On 12 of these fake reefs, they placed loudspeakers that broadcast recordings of reef noise, previously recorded on living reefs. These recordings consisted mainly of sounds produced by fish as well as the snapping of shrimps as they close their claws at high speed, which usually represents the dominant noise of the coral reefs.[101] What super-practical instruments for stunning or killing prey![102]

Most of the fish arrivals were at night. They all arrived in greater numbers on the sonorized reefs. Scientists renewed the experiment by sonorizing some of their reefs either with only shrimp snapping (high-pitched sounds) or only fish sounds (lower-pitched sounds). They found that some species of fish preferred to settle on "shrimp" reefs while other species chose "fish" reefs.[103] The arriving fish seemed to choose their destination according to the sound environment. These discoveries have interesting applications for reef conservation. It has been shown that fish can be encouraged to come to damaged reefs by using loudspeakers that emit the background sound of a healthy reef.[104] We come back to fish later. For the moment, I propose to focus on the giants of the seas: the whales and their cousins.[105]

Whales, dolphins, and porpoises make up the large group of mammals known as cetaceans. They comprise two very distinct categories: toothed whales (odontocetes: dolphins, porpoises, and sperm whales) and baleen whales (mysticetes: the "real" whales).[106] You don't need to be gifted to understand that toothed whales and baleen whales can be easily differentiated by looking into their mouths. If, instead of finding nice, tidy teeth, you see long, stiff bristles, you are looking at a baleen whale, whose considerable baleens (those huge brushes, if you will) allow the whale to get its meager sustenance (fish, shrimp and other crustaceans, etc.) by filtering it out of huge quantities of seawater.

As I mentioned before, cetaceans are mammals. Like you and me. Their ancestor was terrestrial and must have lived a bit like the hippopotamus, about 50 million years ago. The transition from land to water was accompanied by several important anatomical and physiological changes, such as the acquisition of a hydrodynamic body shape, hair loss to limit water resistance during swimming, and a thicker skin with a strong layer of fat to increase insulation.[107] Yet, cetaceans kept their air breathing, which means they must regularly come to the surface to breathe in and out. The famous whale exhalations are blasts of air accompanied by water vapor expelled from their lungs through slightly special nostrils—the blowholes—which are located on the top of their heads. What is the point of knowing this to understand acoustic communication, you ask me? Well, cetaceans, having kept their mammalian respiratory system, have also kept the vocal apparatus: the larynx and its vocal cords. In terrestrial mammals in the human species, for example, it is the air coming out of the lungs that makes the membranes of the vocal cords vibrate. These vibrations produce sound. So far, nothing extraordinary. But here's the rub: the production of sound is usually accompanied by the exhalation of air from the mouth. Try to talk in front of a mirror and you'll see mist forming on the mirror, a sure sign that you are exhaling air. But when a whale sings, you won't see a single tiny bubble of air coming out of its mouth or blowholes.

Olivier Adam, professor at the Sorbonne University in Paris, is a specialist in whale acoustic communications. Let him explain the theory he and his colleagues are defending: "A whale, like the humpback whale, which is known for its long vocalizations, has a special anatomical feature, a kind of bag—the laryngeal sac—whose opening connects just above the vocal cords."[108] Let's visualize it: first, the lungs, which are extended by a pipe, the trachea; at the top of the trachea, the larynx with its vocal cords;[109] after the larynx, the main pipe continues to the blowholes; last, a secondary pipe escapes just after the larynx and opens into the laryngeal sac. "This is the mechanism we propose. The air comes out of the lungs, vibrates the vocal cords, and then goes into the laryngeal sac instead of coming out through the blowholes. The air can then return from the laryngeal sac to the lungs, and the cycle starts all over again. Such a mechanism makes it possible to produce very long songs without blowing air out of the body or breathing in air." Like ingenious closed-circuit bagpipes.

In dolphins, the mechanisms of vocal production are better known, the more reasonable size of these animals making them easier to study in captivity.[110] These mechanisms are really strange. Just look! Like all toothed whales, dolphins have two pairs of special anatomical structures placed in the animal's nose: the monkey lips. Why are they called monkey lips? Because these structures actually look like primate lips, with their oblong shape and folds. The two pairs of lips are connected to the nasal passages and vibrate as air passes through them, producing sounds. In some species, only one of the pairs (right or left) produces sounds, while the other is silent. In other species, one of the pairs produces very high-pitched clicks used for echolocation (the identification of prey and obstacles through their echoes; we'll talk more about this later), while the other emits whistles for communication between individuals.[111] These sound vibrations are then transmitted to a kind of ball of fat, the melon, which focuses the acoustic waves in a directional beam. It is the melon that gives the domed shape to the front of the dolphins' heads. Note that dolphins also have a larynx and vocal cords, like other mammals, but, amazingly, it is still not established whether they use them for vocalization. In short, baleen whales sing with their larynx, a bit like we do, while dolphins and other toothed whales whistle with their noses thanks to specific anatomical devices. What do whales say to each other while singing and dolphins while whistling? This question interests hundreds of scientists and the general public as well. Yet our knowledge on the subject is very limited.

The vocal repertoire of baleen whales varies greatly from one species to another.[112] Some, like the blue whale *Balaenoptera musculus* or the fin whale *B. physalus*, are quite discreet and their vocalizations are rather stereotyped. Others, like the humpback *Megaptera novaeangliae* and the bowhead *Balaena mysticetus*, are very talkative.[113] The songs of the humpback, made famous by Roger Payne and Scott McVay, can last for hours and have infinite variations.[114] Let's take a closer look at this mythical animal.

The humpback whale is a baleen whale, present in almost all seas and oceans, which can reach a respectable size of 15 meters long and a no less respectable weight of 25 to 30 tons. During the summer, it feeds on small fish or krill, the shrimp-like crustaceans that abound in the icy waters of the Arctic and Antarctic. As winter approaches, humpback whales migrate thousands of kilometers to tropical waters where they breed. It's here, in these warm waters, that the males' songs reach their full expression. Endless melodies that can stretch for hours. Each song is a succession of short notes and long moans, "Oohooohooohooohooohooo," lasting up to 8 seconds, whose pitch varies between 30 and 4000 Hz, making them perfectly audible to our ears. These notes and moans are not emitted randomly, in a disorderly manner. On the contrary, the humpback whale's song is perfectly structured, like a kind of musical ritual. It is organized in cycles of about half an hour. Each cycle contains roughly eight themes that follow one another in a precise order. When the eighth or final cycle is over, the whale stops singing, pauses for a moment, and then starts a new cycle from the beginning. Each theme is itself structured in sentences, which are successions of notes. The number of phrases per theme varies from two to about twenty.

Males from the same corner of the planet ocean sing the same songs at the same time. However, the songs vary from place to place, much like the regional dialects of birds.[115] Sometimes the regional dialect changes abruptly. In 1996, Michael Noad of the University of Brisbane and his colleagues noticed that, in the humpback whale population they were studying on the east coast of Australia, two males were singing a different song from the others, with other phrases and themes.[116] And get this: a year later, most of the 112 males they recorded at the same place were singing this new tune, which must have been catchier than the old one. Noad realized that this new song was similar to the one that had been produced for years by the whales on the west coast of Australia. One can assume that some individual from the West, lost in his humpback whale thoughts, had taken the wrong route back from the

Antarctic Seas, and had brought a new and fashionable hit to the East. A cultural revolution of sorts. Do you realize that this event means that humpback whales imitate each other? So they are capable of vocal learning, just like us humans. This is a topic we'll talk about later.

Although it is unpleasant for a researcher to admit it, we do not know precisely why whales sing.[117] Of course, since males sing mainly in the breeding grounds, it can be assumed that their vocalizations serve to charm females and repel competitors, as in birds. But that remains to be seen. To test this hypothesis, as we did for the bearded seal, scientists have tried to provoke behavioral responses from the whales by having them listen to songs emitted by underwater loudspeakers. Success has been mixed. Faced with a playback of a song of their species, some humpback whales flee, while others approach the loudspeaker.[118] It's hard to interpret this. A currently popular hypothesis suggests that song plays a complex role in the interactions between males. Maybe they are in vocal competition? Another hypothesis: Several males singing together could stimulate the receptivity of females. By joining his voice to that of others, a male would increase his chances of mating. The males' chorus melting the females' hearts! Nothing is certain, though. Some even suggest that songs are not communication signals but rather sonar signals, used by whales to probe their environment.[119] The songs are indeed intense enough to reverberate off the bottom, even at great depths, or off other obstacles, such as rocky shores. It is estimated that a whale could hear echoes of its song from more than 5 kilometers away. Perhaps they are building an acoustic image of their surroundings by singing. In terms of acoustic reverberations, the ocean is quite a cathedral.

It is worth noting that whale vocalizations could be very useful for probing the ocean floor! A recent study reports that part of the energy of the sound waves produced by these animals is transmitted to the ground and makes it vibrate. These waves then become seismic waves and propagate in the ocean floor (remember your geology classes in high school: this floor is the oceanic crust); they can be recorded by seismographs, the ultrasensitive devices humans have placed all over the world to detect the smallest of earthquakes. These seismic waves provide information about the organization, structure, and movements of the earth's crust. Using whale songs to understand the geology of our planet. Who would have thought it?[120]

The incredible songs of male humpback whales should not make us forget other vocalizations, however discreet they may be. Isabelle Charrier, Olivier Adam, and their student Anjara Saloma have been working for several years on the calls exchanged between mother and calf.[121] Recording tags were placed on the backs of the animals using a long pole. Not so easy. But very interesting, because the tag records not only the sounds but also a lot of other information, such as the depth or the speed of the animal's dive. Analyses of these recordings show that mother and calf vocalize especially after a dive, when they come to the surface to breathe and rest. Their calls are of very short duration and low amplitude. Nothing is known about the information that is transmitted—the secrets that the mother and her young exchange remain to be deciphered.

With dolphins, things are (a little) clearer. The vocalizations of these sociable animals play a major role. They allow individuals of the same band to recognize each other and facilitate collective decision making, such as whether to go this way or that.[122] Dolphins are known for their whistles, aren't they? Well, you should know that each dolphin has its own unique whistle, which is recognized by the other dolphins in its close circle. When a dolphin whistles, it's a bit like announcing its first name.[123] Sometimes dolphins imitate the whistles of their fellow dolphins, perhaps as a way of addressing one another.[124] Dolphins have a complex social structure where whistling as a means of address is very useful.[125] For example, males, who form long-term alliances by cooperating to seduce females, recognize and call each other when they whistle.[126] Similarly, dolphin mothers whistle when they wish to call back their calf.[127]

There is one toothed whale whose vocalizations send shivers down the spine if you live under the sea. It's the killer whale. The *killer* whale! *Orcinus orca*—the top of the top predators. One of the most socially, culturally, and cognitively complex creatures on the planet, the killer whale is pure intelligence. Killer whales often live and hunt in packs, sometimes with very elaborate techniques, such as cooperating to rock the block of ice on which an appetizing seal is found until lunch falls into the water. Orcas are divided into several ecotypes, i.e., populations with different lifestyles, that feed on different prey. In the Pacific Northwest, resident killer whales eat salmon, transient killer whales eat marine mammals (such as sea lions), and offshore killer whales eat sharks![128] In the Atlantic, one ecotype has a predilection for other cetaceans, another is more eclectic. In the Antarctic, five ecotypes have been identified: one has a preference for the minke whale *Balaenoptera acutorostrata*, another for the seal, yet another revels in the penguin, and the last two are especially fond of fish. Killer whales are real foodies and very picky about their diet.[129]

Orca ecotypes form well-separated populations that do not reproduce with each other. And, as you might expect, each has its own vocal dialect.[130] We also know that killer whales can imitate others: the transmission of dialects is most certainly cultural.[131] The vocal repertoire of killer whales has three main types of vocalizations: clicks, whistles, and pulsating calls. The clicks are extremely short and are used for echolocation. The whistles are highly modulated, varying between high and low pitches. Pulsed cries are very fast trills, which the human ear hears as a short call of about a second. Let's admit it outright, once again: we don't understand what the killer whales are saying to each other.[132] The only certainty is that acoustics play a major role in the cohesion of social groups and when killer whales cooperate to hunt. They also know how to keep quiet when necessary. Thus, killer whales that hunt marine mammals are less talkative than others. It must be said that their prey has good ears.

Charlotte Curé, one of my PhD students, now a researcher at Cerema,[133] and her colleagues recently conducted playback experiments demonstrating that humpback whales can distinguish between the vocalizations of different killer whale ecotypes. The humpback whales approach an underwater speaker that emits herring-eating orca calls as if it were a bell announcing a meal. On the other hand, the humpback whales avoid a speaker emitting whale-eating orca vocalizations, which is a very good idea.[134] Similarly, Charlotte and her team have shown that

pilot whales *Globicephala melas* in Norway are able to distinguish between killer whales from different populations just by listening to their
voices.[135] Being able to distinguish one from the other helps pilot
whales determine whether they can relax or should prepare their defenses before the arrival of the great predator. Even the sperm whale, as
gigantic as it is, can fall victim to killer whales. In fact, the sperm whale
is very careful: if it hears the calls of killer whales from a loudspeaker, it
stops feeding.[136] Instead of diving as deep as might be expected, the
sperm whale flees at full speed or regroups with other kin members,
perhaps hoping to scare off the terrible predator.[137] United we stand!

Let's talk about the sperm whale: *Physeter macrocephalus*, the big-
headed blower. It's a truly unique animal—the largest toothed whale on
the planet; the *biggest brain* on the planet; and an incredible nose,
which, if it isn't shaped like Cyrano's, is much bigger than Pinocchio's;
up to 30% of its body volume! You read correctly: a third of the sperm
whale's body is just a nose. Inside this nose, the space is essentially occupied by a large pouch filled with a rather special oil, the spermaceti,
which changes consistency with temperature. It was for this oil, widely
used in cosmetics, for example, that sperm whales were once hunted.
For a long time, it was believed that spermaceti allowed the sperm whale
to adapt its buoyancy to the depth of its dive.[138] But in the 2000s, another hypothesis was validated. This pocket with its spermaceti is . . . a
musical instrument!

This enormous nasal device is capable of generating sounds—simple
and short clicks—which are extremely powerful.[139] Two hundred and
thirty-six decibels.[140] Enough to stun any squid. As with dolphins, it is
the famous monkey lips that produce the sound. Caricaturing the situation, here's a brief description of how it works. The valves of the monkey lips, connected to the nasal duct, vibrate as air passes through. The
vibrations are transmitted back to the spermaceti pouch and spread
through this oily substance to the skull bone. You should see the shape
of this bone: it's a huge parabola that reflects the sound waves . . .
forward. The spermaceti then acts a bit like glass lenses do with light; it
concentrates the waves into a beam directed in front of the animal. The
sound waves are thus sent forward into the water—"click-click-click."

Sperm whales vocalize to locate and perhaps stun their prey, to navigate the depths of the ocean, and to maintain social ties with their fellow creatures.[141] For echolocation, they use simple directional clicks emitted in regular series. When they find their prey, the series ends with an acceleration of the clicks, causing quite a stir. Acoustic communications between individuals rely on more elaborate sound productions, such as codas, which are short series of clicks: "click-click-click . . . click-click-click."[142]

If we zoom in on one of the clicks constituting a coda, we can see that clicks are not simple sounds. A click has a very short duration, around 100 to 200 milliseconds. Our ear hears it as a single sound ("click!"), yet that click is actually a burst of pulses.[143] (I hope you are following me: a coda is a series of clicks; a click is a series of pulses.) These pulses are echoes of the click produced by the monkey lips. Imagine the enormous spermaceti as a kind of cathedral, in which every click reverberates, creating pulses. Since the sound waves move very fast in the spermaceti (around 1800 meters per second), there is no chance for our ear to discern these echoes; we only hear a single "click"!

The time interval between the pulses of a click depends on the volume of the spermaceti, just as the return time of the echoes in a cathedral depends on the size of the building. By measuring this interval, we can therefore estimate the body size of the animal.[144] As female and male are of very different sizes, it is possible to know the sex of the animal by looking at the pulses that make up the clicks.

A coda has between four and a dozen clicks, all occurring over a period of a few seconds at most. Scientists usually annotate the clicks in a coda by looking at the time between two successive clicks. For example, the annotation "$2 + 1 + 1 + 1 + 2$" means that the sperm whale made two clicks close together, took its time to produce the next three, and ended with a short burst of two clicks: "click-click . . . click . . . click . . . click . . . click-click." In the early 2000s, Luke Rendell of the University of St. Andrews in Scotland noticed that sperm whales belonging to the same group all produce the same kind of codas.[145] They seem to share the same Morse code, if you will. Rendell suggested that there are acoustic clans among sperm whales. A few years later, it was hypothesized that these codas might instead be used by females and juveniles, perhaps to signal their identity.[146] Again, we have to admit: we still don't understand what the sperm whales are saying to each other.[147]

Perhaps you feel a little frustrated that you haven't learned more about the languages of dolphins and whales. I have, of course, considerably summarized what we know. Nevertheless, our knowledge is still limited: it is extraordinarily difficult to study these animals in their natural environment. However, new technologies, combined with modern methods of analyzing long sound sequences, give us reason to be optimistic. Perhaps we are on the verge of unlocking the secrets of the humpback, the killer whale, and the sperm whale. For now, let's get our heads out of the water. The elephant seals are waiting for us on the beach. There, in our aerial element, it's easier to explore what animals are saying to each other.

7

The tango
of the elephant seals

VOCAL SIGNALS AND CONFLICT RITUALIZATION

Año Nuevo Reserve, near Santa Cruz, California. Jaws open, the two giants clash. Each one wants to push the other back, to bite it, to finish it off—in a word, to kill it. Their flaccid trunks are torn on all sides. Blood flows. The fight is incredibly long: they have been fighting for half an hour already! Their chests stuck together, exhausted, they stop for a few seconds. Time for a breather. Then one of them swings back and picks up momentum, determined to strike again. The enormous daggers of his canines cut into the opponent's thick skin, who steps back for the first time. He will yield. A few more assaults and off he goes, turning tail and running away. A two-ton mass crawls across the beach. The winner is not finished; it's still chasing him. Another bite, a violent bite, on the rump. Then, triumphant, he stops, stands up, and—head high, his trunk thrown backward—launches a hoarse cry, regular as a metronome: "Clork . . . clork . . . clork. . . ." The loser will never fight back again: he will run away.

The first time I visited the elephant seal colony in Año Nuevo, it was raining hard. It was February. Imagine a low sky, a cloud-covered horizon, the Pacific Ocean marbled with large rollers, and hundreds of huge seals on the beach. In *Mirounga angustirostris*, the males are as big as pickup trucks, up to 4 meters long. The females are smaller, 2½ meters at most. Sexual dimorphism (the fact that females and males are not the same size and do not look alike) is obvious at first glance. And here and there, big black tootsie rolls of more than a hundred kilograms each: the babies of the year. Northern elephant seals come to spend the winter in Año Nuevo to breed, give birth, and mate. The adults migrate from their fishing grounds in southern Alaska, arriving here in November–December and leaving again in mid-February. They fast throughout this time. A highly effective weight loss diet: A male or female elephant seal can lose up to 40% of its weight in a single breeding season.[148]

The competition for what is trivially called "access to females" is quite tough between males. Less than 1% of them are able to reproduce in their lifetime. On the beach, relations of dominance are first established through physical confrontations, often of fairly short duration. They are maintained during the breeding season mainly through ritualized behaviors, including characteristic postures (the seal raises its body), vocalizations, and ground tremors (when an individual hits its chest against the sand). Battles such as the one I just described are rare: they only take place between two dominant (alpha) males, both equally motivated to win the competition. The outcome can be fatal.

Why am I interested in elephant seals? As usual, due to a mixture of chance, luck, and curiosity. That year, I was a visiting professor at the University of California, Berkeley, hosted by my colleague Frédéric Theunissen from the Department of Psychology. Together we were conducting a research project on hyena communication, which I will tell you more about later. My visiting professorship was funded by the Miller Institute, a wonderful foundation that invites scientists from all over the world to spend a year at Berkeley. The competition is strong here too, but the reward is great: a year devoted to research, with no constraints other than to attend the Tuesday lunch with the other visiting professors. I decided to take advantage of this opportunity to visit other research laboratories in the region. That's how one day I took the bus to Santa Cruz, a pretty town south of San Francisco well known by surfers. On arrival, Ron Schusterman was waiting for me. Ron had founded the Pinniped Cognition & Sensory Systems Laboratory to explore the cognitive abilities—as scientists call *intelligence*—in pinnipeds. Ron was a pretty amazing guy. "Nic, I have to get my break-fast first. I hope you don't mind." Instead of visiting a lab, I found myself in front of an amazing 80-year-old who quietly started telling me science stories over coffee and pancakes. I liked this unconventional situation, of course.

"I'd like to introduce you to Colleen Reichmuth—my favorite stu-dent, who took over my lab. She's great. She's my spiritual daughter." The old man's eyes twinkled. He was happy to have entrusted his life's work to someone he had confidence in. Ron is gone today, and Colleen and I often remember him.[149] Science is also about transmitting wisdom.

The Pinniped Laboratory is a unique structure where seals, sea lions, and sea otters are trained to respond to sound signals. This method is called operant conditioning. Here's the principle of it. You may be fa-miliar with nonoperant conditioning, or Pavlovian conditioning. It was first discovered by the Russian researcher Ivan Pavlov in the late 1800s in his famous experiment with dogs. When meat is presented to a hun-gry dog, it starts to salivate. In Pavlov's study, conditioning consisted of associating the presentation of a piece of meat with the ringing of a bell. After a few trials where the bell was rung at the same time the meat was presented, the dog salivated as soon as the bell rang, even if the meat did not follow. The dog was conditioned to the bell stimulus. The re-sponse to this stimulus is a reflex over which the dog has no voluntary control. It cannot do anything about it; it salivates. In operant condi-tioning, the aim is to provoke in the animal a voluntary response to the stimulus: the animal must complete an action, such as pressing a button or moving in a certain direction. For example, Colleen's animals learn to hit a target with their muzzle when they hear certain sounds. To bring them to this behavioral response, the seal or sea lion is taught that, if it reacts "correctly" to the sound stimulus, it will be rewarded with a nice fish. Once the seal is trained, it becomes possible to ask questions, for example, to test its hearing. We can see that, if the sound level of the stimulus is lowered too much, the seal stops responding and there-fore stops hitting the target. By systematically exploring different loud-ness levels at different frequencies, Colleen creates audiograms for the animals being tested. If you've ever been to an otolaryngologist, you probably had to listen to sounds of varying intensity to measure your hearing. At the end of the test, your results are plotted onto a curve that shows your hearing ability, ranging from low- to high-pitched sounds. This curve is your audiogram. Using this method, Colleen's goal was to explore the sound world of pinnipeds. When I told her about the elephant seals I had seen at Año Nuevo, she was enthusiastic: "Let's do some field experiments! Fantastic, especially since we already have a connection with the Año Nuevo Reserve; and one of my students, Caroline Casey, will certainly be willing to help us." That same evening I contacted Isabelle Charrier, my former PhD student, now senior researcher at the CNRS and an expert in marine mammals. We had always said that one day we would like to work together again, and her great knowledge of field work with pinnipeds was essential to the success of the project. So a new, small team of passionate scientists—Colleen, Caroline, Isabelle, and me—was formed. We were going to work for several years to explore the sound world of elephant seals (and we still do!). Once again, chance encounters were an essential ingredient in my life as a researcher.

Elephant seals are so large that they have no fear of humans. As long as we didn't get too close, our presence on the beach was mostly tolerated. This made it easier to make recordings and playback experiments—to a certain extent, of course. We had to be very careful because females can become aggressive and try to bite if you get too close to them. Some of them chased us over long distances. And the bite of an elephant seal is more than annoying: their jaws are powerful and their saliva contains bacteria that our immune systems are not equipped to handle. As for the males, they're so big that they can kill a human with a single butt of their head.

The vocalizations produced by males are made up of a succession of clicks, several per second. It has long been known that these signals play a role in the interactions between males: it is common to observe that, when an alpha male begins to vocalize, the males around him beat a hasty retreat. Our first hypothesis was that these calls carry information about the strength and power of the individual transmitter: it is easy to imagine that the strongest males must have the lowest voices or produce

Elephant seal

their calls at a faster rate. An alternative hypothesis was that males produce vocalizations with marked individual characteristics and that they learn to recognize each other by their particular voice. Between the two hypotheses a decision had to be made—or we had to reconcile them.[150]

It was a considerable amount of fieldwork, which required several consecutive missions. Each year, about 40 males were given a mark—a number painted on their lower back—which allowed them to be identified throughout the breeding season. But since the painted marks disappear at sea, Caroline would attach a numbered tag to some of them on one of the rear flippers, hoping to find them the following year. I'll let you imagine Caroline, crouching, approaching one of these large animals from behind. Painting the back of an elephant seal or attaching a tag to its flipper is no small business—not a sport I'd recommend! Each bull was carefully photographed and measured and its voice recorded several times during the season. Tagged males, when found, were recorded over several years.

Isabelle and I had the idea that the most impressive males would be those with the lowest voice pitch, as size and pitch are two characteristics often linked in the animal kingdom. This is indeed what our measurements showed. In addition, the bigger the individual, the faster the rhythm of its clicks. The hypothesis that males can show their strength and thus their ability to win a fight by their voice alone became plausible. We still had to check that the males at the top of the hierarchy were also the biggest.

Estimating the hierarchical rank of the males is not obvious. It takes many hours of observation to describe the interactions between individuals and to record the outcome. Are you familiar with the Elo rating system, named after the chess player who developed it? It is used to rank a group of individuals from the strongest to the weakest, after competitions involving only two players at a time. We used this method to evaluate the dominance rank of male elephant seals. At the beginning of the breeding season, each bull received an Elo score of 1000. Then the Elo scores were modified, up or down, at the end of each male-male interaction. After a few weeks, the scores stabilized: some individuals—the alpha males—were at the top of the rankings because they had won almost every fight they had ever had. Further down in the rankings were the beta males, then the peripheral males who had systematically fled from any opponent.

To test the hypothesis that the males' vocalizations carry information about strength and ability to win a fight, we looked for possible correlations between the acoustic structure of the clicks (their rhythm, pitch) and the males' Elo score. I was very surprised to find nothing conclusive. A very dominant alpha male could have a slow or, on the contrary, a very fast rhythm, a deep or rather high-pitched voice. It was impossible to predict the hierarchical level of an elephant seal just by listening to its voice.

Experiments with acoustic signal playback made it possible to find out for sure. We had males listen to voices of unknown males—voices that we had recorded a hundred kilometers south of Año Nuevo, in the region of Piedras Blancas. We had modified the characteristics of these voices, making them lower or accelerating their rhythm in order to imitate the voice of ultradominant male alphas. The results of the experiments were disappointing: the males mostly ignored our signals or seemed to respond randomly. We were back to where we had started. Now we had to test our second hypothesis: elephant seals probably had to learn to recognize the voices of their fellow elephant seals.

For Caroline, the one of us who spent the most time in the field, there was no doubt that each male had a different voice from the others. Just as you can recognize a familiar person when you hear them on the phone, Caroline could identify each of the males on the beach by hearing them vocalize. Our acoustic and statistical analyses confirmed it: the rhythm of the clicks and the pitch of the voice, higher or lower, were enough to distinguish one individual from another. Even better, the vocal characteristics of each individual were preserved from one year to the next. A male keeps his voice . . . for life. All that remained was to ask the animals the question, Can you recognize each other by voice alone, the way Caroline can?

Asking such a question to an elephant seal is not simple. Caroline was utterly sure of herself: she recognized the males, she differentiated between them very well just by their voices. So we started playback experiments. We played the vocalizations of familiar individuals, who were neighbors on the same beach and with whom our test bull had interacted over the previous weeks, to beta males. For each male tested, two trials were conducted. In one, we made the male listen to the voice of a dominant, i.e., an individual in front of whom the male would back down during an encounter. In the other, we played the voice of a submissive. Imagine the scene: a male elephant seal lying on the beach, the loudspeaker barely 10 meters in front of it. The voice of a dominant reaches its ears, and suddenly the huge mass rises abruptly, turns around, and runs like mad in the opposite direction from the loudspeaker. When we played the voice of a submissive, the animal took its time, reacted only after a while, and then approached the loudspeaker seen, or rather heard, as a weak male. Sometimes, there was even a little breakage and a few welds needing repair in the evening when we returned from the colony. In any case, the results were clear: an elephant seal can recognize Peter, Paul, or James by its voice alone.[151]

Like my teammates, I was now curious to understand the acoustic basis of this recognition. In particular, do elephant seals rely on the rhythm of the successive clicks that make up their calls to identify their fellow bulls? Using the same protocol, we tested 10 males with vocalizations of dominant individuals whose clicking rhythms had previously been modified by making them faster or slower. As long as the change in rhythm remained very small, the males continued to run away from the stimuli: they always recognized their dominant. But as soon as the rhythm was accelerated or slowed down by 10% of its initial value, their behavioral response became more hesitant, or even disappeared altogether. They would lie motionless, seemingly ignoring vocalizations whose sender they no longer recognized. This experimental result is conclusive; it shows that elephant seals memorize and use the rhythm of their fellow elephant seals' calls to identify them. Each individual must remember the voices of several dozen others. What a memory—an elephant's memory, of course![152]

One year, on the beach of Año Nuevo, it was claimed anecdotally that an alpha male was particularly powerful and aggressive. The following November, this individual did not return, and the other males were able to spread their harems over a larger area. A few days later, a new male arrived, poked his head out of the water, and started to vocalize "rlurk . . . rlurk . . . rlurk . . . ," and all the others bolted. It was the alpha male from the previous year who was cleaning up the beach with his calls . . . remembered by his fellow males. If the memory of elephants in Africa and Asia is legendary, the memory of Northern elephant seals is just as brilliant.

Remember: one of Nikolaas Tinbergen's four questions concerns behavioral *development*. How does behavior develop over the course of an individual's life? It is understood that baby bull elephant seals are unable to make the characteristic clicking sounds of the adults. The babies do call out a lot, but they sound like sheep bleating. Also, the male babies are not looking for a fight. In fact, once weaned, they even have a tendency to huddle together. How does the hierarchy of dominance play out? When do they start using their voices to recognize each other? With all these questions in mind, I wasn't finished with elephant seals.

Hierarchies of dominance—the fact that individuals living in groups do not all have the same access to resources—are a characteristic shared by most animals, including humans. What is a resource here? Anything that can be valued by individuals, like food resources. For example, we observe a hierarchy of dominance being set up in groups of hens for access to seeds distributed by a farmer. Sexual partners can also be resources. I grant you that using the term *resource* to designate individuals with whom it would be possible to mate is not the most elegant, but this metaphor from the world of economics has the merit of clarity: if the same resource is desired by many, it can be a source of competition and lead to the establishment of hierarchies of dominance. In elephant seals, the resource prized by the male is the female. This does not mean that female elephant seals do not also desire resources and just sit around idly. For example, they might be competing with each other for a place on the beach that is sheltered from storms, where their young would not be threatened by the ocean. Some are very aggressive toward their peers when they feel they are too close to their young. Moreover, let's not generalize; in animals, including humans, it is not always a case of males competing for females. The animal world is much more complex than that. We will see further on that the situation can be totally reversed, as in spotted hyenas, for example. The female elephant seals themselves may choose to let this or that male approach them. We lack sufficient data here to say for sure.

To understand the establishment of the hierarchy of dominance of male elephant seals and the role of acoustic communication in this development, our small French-Californian research team had its work cut out. We decided to study in parallel the age-dependent changes in space use, social interactions, and vocalizations. The males were classified into five categories: individuals aged four, five, six, seven, and eight years and older. Why start with four-year-old males? Because it is only at this age that young males return to the breeding colony. Before that, they are somewhere in the ocean. Long hours of observation began. The study would last four years.

This fieldwork allowed us to get a clear vision of how social networks work in elephant seals. Our data, obtained through the observation of 1352 interactions involving 207 males, shows that social relationships evolve with age. Younger males come into contact with a smaller number of fellow bulls than older males. In addition, they favor contact with individuals of their own age class. In other words, they avoid the old bulls, probably because they fear them. Moreover, interactions between young males often involve two individuals who have never met before, while older males regularly interact with the same rivals. In short, whereas younger males form fluctuating networks of social interactions, with many unknown individuals, mature males navigate a stable social network made of familiar individuals.

What role does voice play in these interactions? From our earlier work, we knew that adult males have voices with a strong individual signature. And I've told you about our experiences showing that individuals recognize each other vocally. What about younger individuals? Were their voices also different? Were they also able to recognize each other? Our analyses showed that the vocal signature appears gradually, stabilizing when the animal reaches seven years of age. This is, in a way, the age of reason for the elephant seal. Younger males have a voice that is not very personalized, so it is not possible for them to recognize each other with certainty. It is only with age that the individual vocal signature becomes clearer.[153]

In our elephant seals, acoustic communication therefore plays a key role, allowing each seal to navigate within its social network, by adjusting its behavior according to the individuals it encounters and dealing with conflicts mostly by voice rather than in bloody fights. Here's Pierre talking! He's stronger than I am and scares me stiff. I have to get away fast! There, I hear Paul. He's a loser. He doesn't scare me. As soon as he hears me, he'll run away! The life of an elephant seal is no bed of roses: only 5% of male elephant seals survive to adulthood. But without this vocal ritualization, the loss of lives in adulthood would probably threaten the survival of the species.

8

The caiman's tears

ACOUSTIC COMMUNICATION IN CROCODILES

Nhumirim Ranch, Pantanal, Brazil. Standing at the back of the Toyota, Zilca lit the marsh with a powerful torch. Undisturbed by the bumps in the muddy and rutted road, she never took her eyes away from the beam of light. *"Aqui! Uma mae jacaré com seus pequenos!"* Shining like stars, the eyes of the caimans betray their presence. I came to South America, in the Pantanal, near the Bolivian border, to study the acoustic communication between mothers and their young in the jacaré caiman—a species of crocodilian that is abundant in this region, where a wooded savanna, swamps, and magnificent sunsets are accompanied by an incredible diversity of fauna. Nhumirim is a *fazenda*—a farm, with cattle and cowboys—as well as a research station. I have known Zilca for some time, but I had never met her in person before this fieldwork. A few years ago, our first online conversation was an unforgettable moment. To communicate with her, I used a language translator found on the internet, as I don't speak Portuguese and Zilca is not comfortable with English. Since our exchanges were only in writing, I didn't realize until much later that she was a woman. The stereotype of Crocodile Dundee made me think of a big, tough guy when talking about crocodiles. My first memory after landing at the Corumbá airport is of the wrinkled face of a small woman, with her gentle smile and sparkling eyes: Zilca Campos! Sharing the life of this warm and fascinating person during a field mission was a rare human and scientific opportunity.

Reaching the Nhumirim Ranch means a seven-hour trip, mostly on sandy roads. Zilca had organized everything down to the last detail. Once we left the hills of Corumbá, the landscape became flat. The Pantanal savanna is the floodplain of the Paraguay River. During the dry season, from October to April, water is only present in the main rivers and isolated lakes of the Pantanal. But the rainy season floods most of its 100,000 hectares. In Nhumirim, there can be more than a hundred lakes in wet years. Conversely, "some years, it almost doesn't rain and everything is dried up," explains Zilca. This year, luckily, the lakes are full, ensuring good conditions for the caimans.

"This will be your house," she says. "It hasn't been occupied for a while. I lived there for so many years!" With my colleague Nicolas Grimault from the CNRS and our doctoral student Leo Papet, I entered the modest house, delighted by its breathtaking view of a large marsh. Its name, Fauna, was not undeserved. A few days after our arrival, we saw two long snake molts stuck in the grooves of the kitchen door. The animals had probably come out of the cupboard during the night, disturbed by our presence. We finally found one of them coiled up against a wall. The other had evaporated. Every morning, until we decided to tape up all the doors and windows, a new snake came into the house.

The jacaré caiman (*Caiman yacare*) is emblematic of the Pantanal. You have likely already seen them in photographs, either with a colorful butterfly on their snout or throwing a catfish in the air to gobble it up head first, thus preventing it from getting stuck when it passes through the esophagus. It is a sister species to the spectacled caiman (*Caiman crocodilus*), the most common caiman found from Mexico to Argentina. "The number of nests found in Nhumirim varies considerably from year to year. Sometimes we've found more than 60. And sometimes none. The last few years have been the worst," Zilca tells me. "We don't know why these huge variations exist. The decrease in rainfall over the last 10 years may be part of the reason." Zilca is a researcher at Embrapa, a Brazilian research institute, and has been working on the ecology of the jacaré for several decades. She is passionate about this animal. She continues: "The female jacaré lays between 20 and 30 eggs in the middle of the rainy season, from December to February. As with all caimans and alligators, the nest is a mound of vegetation. The eggs take about three months to mature—three months during which the female remains in place, providing protection from predators. The mother defends the nest and takes care of the young." Parental care is a common behavioral characteristic of archosaurs, a group that includes crocodiles, birds, and extinct dinosaurs and pterosaurs. By doing research on crocodiles, I feel like I'm opening a window to the past, a window to these extraordinary animals.[154]

Acoustic communication is a remarkable feature of the mother-young relationship in crocodilians.[155] Females and their young produce vocalizations. In a few species, fathers also participate in watching over the young. In the gharial (*Gavialis gangeticus*), a crocodilian with a very long snout (of which there are still a few rare populations in India), the males group together to watch over newly hatched young while the females go about their business.[156]

When they are ready to emerge, mature crocodilian embryos emit vocalizations that act as signals to their siblings and mother.[157] It was with Amélie Vergne, one of my most passionate students, that we demonstrated this experimentally. The idea came to me during a family visit to a crocodile zoo. On the edge of one of the ponds where dozens of huge Nile crocodiles (*Crocodylus niloticus*) were basking, a sign indicated that crocodiles use sounds to communicate . . . which I didn't know. A little upset by my lack of knowledge, but above all terribly intrigued, I began to gather all the scientific publications on the subject—only to find that no one had really studied the question. Fortunately, Luc Fougeirol, the founder of the zoo *La Ferme aux Crocodiles* in Pierrelatte, France, opened his doors to me to begin the research.

The principle of our first experiment was simple: go into the crocodile pond, move a mother away from her nest (Nile crocodiles lay their eggs in the sand), bury a loudspeaker in place of the eggs, leave the pond, wait for the mother to come back, and play hatching calls from the loudspeaker (or noises of the same duration but without the acoustic structure characteristic of baby crocodile vocalizations). Imagine the scene: two zookeepers with shovels hitting the water to keep away a mother trying to defend her nest, while the zoo veterinarian is digging the hole to bury the loudspeaker. It was a bit chaotic.

Most of the females we tested quickly returned to their nests. In response to the playbacks of the hatching calls, the females began to dig in the sand above the loudspeaker, whereas they remained motionless when we played other noises. It was the calls of the hatching young crocodilians that led to the first maternal care behaviors. In another experiment, where we played the same sound stimuli to eggs about to hatch, we saw that the calls of babies led to the hatching of their siblings.[158] In the wild, everyone hatching together is probably an important advantage: the mother is there to help and to protect the last ones that come out.[159]

In crocodilians, the mother's help doesn't stop at the nest. Have you ever seen animal films that show a crocodile with youngsters in its mouth? The mother (or father in some species) is carrying her young to the river or body of water. After hatching, she stays with them for several weeks. The young crocodiles, which measure about 20 centimeters, are extremely vulnerable to predators such as large birds or carnivorous fish. Cannibalism is also widespread, so one crocodile parent will not tolerate any other adults near its young.

In order to study the acoustic communications between the crocodile parent and its young in the days following hatching, it was necessary to go into the wild with free-living animals. In a zoo, this is not possible because the density of animals is too high, and the young, if left in the basins, would not survive for long. I asked Luc for advice on the best places to approach crocodiles. "Go to Peter Taylor's house," he said. "He is an American who has a field station in Guyana, on the banks of the Rupununi River. He is an expert on the black caiman. You'll see, that caiman is a very big one—a real man-eater." Then he added, "Going to Taylor's isn't easy. So, for a start, you could go to the Kaw marshes. There are black caimans there too." The Kaw marshes: the mythical, virtually inaccessible, and miraculously preserved stretches of swamp in French Guiana. I contacted Eric Vidal, then at the University of Marseille, who had a research program on the Agami herons of French Guiana, precisely in Kaw. With his usual kindness, Eric invited me to join one of his missions, asking me in exchange only to make some recordings of herons. I called Thierry Aubin, who, not losing any opportunity to get close to a wild animal in its natural environment, was immediately willing to accompany me.

It was my first helicopter experience. We took off from the parking lot of the caiman camp hostel, on the road from Roura in the direction of the Kaw marshes. A few minutes later, after diving over the marsh below, the helicopter approached a very small aluminum platform floating on the water. A few cable lengths away, another platform of about 10 square meters protected by a tarpaulin would be our camp. Lost in the middle of the Kaw swamps, we were ready to begin our experiments on the black caiman, the amazing *Melanosuchus niger*.

It was, of course, by boat that we approached our first female. I didn't have a recording of baby caimans, so I decided to play baby Nile crocodile calls; we would see the result. During the first broadcast, the female's response was impressive: she leaped forward, roared, and charged at our boat. Needless to say, we immediately stopped our calls. We were full of hope when we returned to the platform to prepare our signals for the next day's experiments. It was quite something: even coming from a different species of crocodilian, the screams of babies provoked an intense maternal reaction. It opened up new perspectives. But the next day, we were disillusioned. Same individual, same test: no more answers. Yesterday's female ignored us.

This phenomenon is well known to ethologists. It's called habituation: when an animal classifies a sensory stimulus as one to ignore due to lack of reinforcement. Baby crocodile calls in the absence of real babies were indeed of no interest to the female. What was surprising was how quickly the mother caiman became accustomed to our signals. But what was even more surprising was that the other adults in the lagoon that we subsequently tested immediately showed a very weak reaction to the calls of the babies. Let me be very clear: all the black caimans who had witnessed our first experiment and had probably heard our signals had integrated in a very short time that calls coming from a loudspeaker placed on a boat were not worthy of interest. Clearly, we had not taken enough precautions. We had underestimated the animal, its sensory integration capabilities, and its memory. Duly noted. This is a great risk of ethology: not realizing that the animal you are studying has a keen knowledge of its environment—much better than yours—and that it is capable of understanding very quickly that your presence results in negligible stimuli, such as calls played from a loudspeaker. Primatologists pay attention to this in principle because it is easy for us to imagine that monkeys or apes, close to our own species, develop cognitive faculties close to ours. But with our caimans, we should have thought about it: these superpredators, who spend their lives probing what surrounds them with their senses, are quick to realize that experimental stimuli are irrelevant. Plus, they have the memory of an elephant. So we had to go to work in a context where we would have access, for each experiment, to naive individuals. We couldn't work on a lake where all the animals were constantly observing us. We had to look for a river, where the caimans would be spaced a long distance apart. The following year we went to Peter Taylor's house. This time Amélie was on the trip.

Luc was right. Getting to Caiman House, the field station of Peter Taylor and his wife, is part of the expedition: a transatlantic flight from France to Guyana's capital, Georgetown; then several hours in a small bush plane to go deep into the heart of the country, landing on a dirt runway; finally, half a day in a pickup truck before arriving in an indigenous village on the banks of the Rupununi River, where the caimans were waiting for us.

Peter had spent the last few weeks spotting female caimans and their young. He had found a dozen of these small families on the river, about a kilometer apart. It seemed like the perfect situation: we would avoid habituation since the females were far enough apart from each other. "But we'll have to work at night," Peter said. "Caimans have been hunted a lot around here, and they are fearful of humans." The first few nights we were busy recording baby caimans. We decided to do it in two very different contexts. First, a quiet context. The young caimans gather near the shore probably to avoid being swept away by the current. A microphone and a recorder left near them made it possible to record calls of fairly low intensity, which the youngsters frequently emit. Then we moved to a stressful context. Peter grabbed one of the babies by the tail and shook it slightly to mimic a predator attack, and Amélie recorded the calls of the baby. Acoustic analyses showed that the sounds emitted in the quiet context differed from those emitted in the stressful context. The former were short, muffled, and slightly frequency modulated: "Djong! djong!" The stress calls were a little longer, louder, and much more modulated: "Zuh! zuh! zuh!" Even though they sound different, these two calls are built on the same pattern, the same acoustic family in a way: they are frequency-modulated harmonic series, and you can go from the quiet context call to the stressed context call simply by modulating the frequency more—putting more energy into the treble and increasing the volume. Are these calls signals, you may ask? Do they allow individual senders to communicate information to their mothers? To other juveniles? Only acoustic playback experiments could answer these questions. That's what we did.

The loudspeaker is placed on the bank, about 10 meters from the group of young black caimans. The mother is in the middle of the river, motionless. We are in our boat, moored to a tree on the opposite bank, attentive and silent. Judging the moment appropriate, Amélie presses the button on the tape recorder and triggers the emission of calls recorded in a quiet context. The mother does not move an inch. The young ones, however, move toward the speaker, continuing to emit low-intensity calls, and gather around it. Amélie stops the stimulus, and we begin again to wait patiently in the dark. A few minutes later, we play distress calls from the loudspeaker. The mother caiman immediately turns around and rushes in the direction the sounds are coming from. (In follow-up experiments with other individuals, some mothers roar, with a guttural and threatening sound: one of her babies is in danger and she is coming to save it.) The young stop motionless, but carry on calling. The acoustic analysis of their calls will show that they are making distress calls. I'm happy: the Guyana field trip has demonstrated that baby crocodiles have a vocal repertoire that allows them to encode information and target the receiver: contact calls that allow the babies to stay together and distress calls that enrage the mother and alert siblings.[160] I was also delighted for Amélie: her first field trip was a success, despite some problems caused by intolerance to an antimalarial drug. This shows how important it is to observe the animal you are studying in its natural living conditions. This is essential if we want to be able to ask meaningful questions.

Further research showed that all crocodilians share the same acoustic distress code. A few years ago, during fieldwork in the Venezuelan Llanos, Thierry and I tested female spectacled caimans with calls from black caiman, Orinoco crocodile (*Crocodylus intermedius*), and Nile crocodile juveniles. The females responded to all these stimuli.[161] Other experiments using artificial signals that we had created on a computer showed that a simple whistle mimicking the descending frequency of a distress call is enough to provoke a maternal response.[162] Based on these observations, it is likely that crocodile mothers are not able to recognize their own offspring but respond to any newborn call.[163] Moreover, the calls of baby crocodiles do not just inform us about their state of stress. I tell you in chapter 10 how they also code for the size of the emitter. For now, let's go back to the Pantanal, where Nicolas Grimault and Leo Papet, my fellow researchers on this journey, and I have decided to test the ability of an adult crocodilian to locate its young when they vocalize.

"I marked every nest I could find with one of these red flags," says Zairo, smiling broadly. Zairo is Zilca's field assistant. He was born here and is an expert on the wilderness of the Pantanal. To find nesting females, he had to cross thorny bushes surrounding the lakes, well protected by leather clothes. Denis is the team's driver. He is the only one who knows how to start the Toyota engine with a piece of metal when the old truck breaks down. My goal is to test whether caiman mothers can accurately locate a newborn baby based on its vocalizations. Since noise from human activity has become a critical issue for many animals and its impact on crocodilians has never been studied, I wanted to determine if noise could impair their ability to locate a sound source. Nicolas, Leo, and I had hypothesized that, if the noise source weren't placed in the same location as the source of the signal of interest, female caiman crocodilians would locate the signal more easily—a process called spatial unmasking. This is a strategy known in humans and in only a few other animal species, like parakeets, ferrets, frogs, and crickets: the hearing system uses spatial cues to isolate the signal from the noise. Signal detection should be better when noise and signal are emitted at a distance from each other rather than from the same location. When we began our research, knowledge about sound localization in crocodilians was limited. Field data was almost nonexistent, and we had no idea how accurately crocodilians are able to locate a sound source or what acoustic characteristics they use to do so. The first investigations were carried out in our laboratory in France.[164] Leo had trained young Nile crocodiles to go to a loudspeaker emitting an artificial sound to get a food reward. The results showed that these animals locate a sound source very precisely. They quickly turn their heads in the direction of the sound, making almost no directional errors.[165]

To carry out playback experiments in Nhumirim, we first had to record calls from baby jacaré, the local species of caimans. Zairo had spotted a female staying in a large pond where cattle came to drink and where it would be easy to catch young ones. Although the female jumped out of the pond to attack us while we were handling a newborn, we made it out in one piece and got good recordings. Back at the Fauna, our home base during the trip, Nicolas, Leo, and I prepared the stimuli for the next step. The mother's first reaction had already shown us that these calls are emotionally charged for a mother caiman. We were just as moved by her reaction!

In the early morning, Denis, Zairo, and Zilca came to pick us up at the Fauna. "*Nós vamos para a lagoa 42!*" Our companions had decided to make the first attempt on lake "42," a beautiful pond where Zairo had spotted a female and her young a few days before. After a short drive, we parked the truck at a distance and walked to the lake to assess the situation. The caiman family is there, the female immobile and her little ones moving slowly around her. I enter the water without too much apprehension, carrying two speakers. After all, the jacaré caiman is a relatively small species compared to the Nile crocodile or the black caiman, and does not usually attack humans. Still, I'm a bit worried as I walk along, imagining myself walking on a dangerous animal hidden underwater. When I stop, about 20 meters from the shore, the water has already filled my boots. I put the two speakers on their tripods so as to emit the sounds at just above the water level, and we all hide behind the vegetation to observe the female's behavior. Leo turns on the first speaker, which starts to make a continuous background noise. The female

Jacaré caiman

doesn't move. The noise doesn't bother her at all. We then begin to emit a series of distress calls from the same loudspeaker, with an intensity lower than the noise level. Still no response. But when we transfer the distress calls to the second loudspeaker, the mother caiman immediately turns her head, attentively; after a few calls, she starts swimming quickly in its direction. We had just proved that spatial unmasking works in crocodilians. Female caimans are able to detect distress calls when the angle between the loudspeaker emitting the noise and the loudspeaker emitting the calls is just over 4°.[166]

Back at the station, the topics discussed with Zilca began with the behavior of the female caiman we had just tested, and then moved on to the future of the Pantanal. I was mostly just listening. "Poaching is a real problem here," Zilca explained. "Once I met poachers. They thought I was a ranger. I had to convince them I didn't have a gun." I imagined Zilca, with her frail silhouette and hesitant pace, addressing a group of determined poachers. How can we understand each other with such incredibly divergent interests?

One day, we wanted to go to Campo Dora, a large private ranch next to Nhumirim, where Zilca used to work on the jacaré populations. We never made it to the ranch because a branch of the river, too deep to cross, blocked our passage. However, we didn't waste the day, as we saw some nice jaguar footprints on the muddy road. The Pantanal has one of the highest concentrations of wildlife on the South American continent. Every evening, as we shared a glass of caipirinha sitting outside the Fauna, hundreds of ibis and storks returned to their colonies. On one occasion, a big tapir jumped up in the air, surprised from its sleep, right in front of our eyes.

According to Zilca, the development of intensive agriculture is the main threat to the Pantanal. Brazil needs food, water, and money. The Pantanal is a fragile area. Most of its water comes from the Cerrado on its eastern side. And the Cerrado, this vast savanna region, is under great pressure, rapidly turning into an agricultural region with a string of dams and dikes. But "without water, there is no Pantanal," Zilca said.

At Corumbá airport, we were told that the flight to São Paulo had left the day before. So we had to travel a good part of the night by bus to Campo Grande, then wait a few hours before taking a plane to Belo Horizonte, and finally spend the next night waiting for a flight to São Paulo. The Pantanal is far away from everything, a wonderfully remote place in the modern world. It is so threatened, however, that spotting jaguar footprints and surprising a sleeping tapir could soon become impossible.

9

Hear, at all costs

MECHANISMS OF AUDITION

It seems quite commonplace to use sounds to communicate. Most animals that are close to us bark, meow, neigh, sing, whistle, or moo. In short, most of them chatter routinely. And so do we! Spoken language is our preferred means of exchanging information. Yet, in the animal kingdom, acoustic communication is a peculiarity rather than a universally shared characteristic. It is used only by vertebrates (although snakes are not talkative[167]) and arthropods (some crustaceans, some spiders perhaps, and many insects[168]). How is it possible that none of the 100,000 species of mollusks are capable of emitting the slightest sound?[169] Couldn't an animal like the octopus, known for its resourcefulness, make its jaws squeak? Why do dragonflies remain so desperately silent? What about most butterflies? And how about sea anemones? Acoustic signals are very useful for attracting a mate, trying to scare off a predator, or signaling one's presence to others—all things that would be useful to countless species that, despite everything, remain as silent as carp (which we know are not deaf!). The reasons for this silence are certainly varied and difficult to identify. There is no escape, however, from the two following conditions: it is necessary to be able to *hear*, and to *produce sounds*.[170] In this chapter we address the first of these two conditions because, during the evolution of species, hearing has often preceded sound production. In chapter 10, we address the second.

What is hearing? A succession of several stages, from the perception of sound by the ear to the brain's processing of the information contained in the sound signal. The first stage corresponds to what in physics is called a *transduction*—the transformation of a signal of a certain nature into a signal of another nature. When you speak into a voice-activated device, such as your telephone or computer, this device must transform the sound waves into electric waves, which are then processed and lead to the execution of your command. This step of converting sound into an electrical signal is what our ears do when they pick up sound and transform it into a nerve signal that is sent to the brain. However, this is not simple to achieve. The ears have a lot of work to do, and it's a complex task.

All animals' ears (and we will see that there are many different kinds) were formed during evolution from already existing sensory structures whose role was to perceive movements of the animal's body and to inform the nervous system. These structures are called mechanoreceptors. Like all animals, we have many of them inside our body and on the surface of our skin. They allow the brain to know at each instant the position of our limbs and the pressures on our feet, hands, and elsewhere. In short, they are essential sensors for the management of our posture and movements.[171] A mechanoreceptor always works according to the same principle: a vibration, a pressure (that of a finger placed on the skin, for example) causes a part of the body to move in relation to other parts, generating a differential motion that is perceived by the mechanoreceptor. The production of a nervous signal is the task of the sensory cells, which are neurons specialized in the transformation of mechanical vibrations into electrical waves (the nervous impulse). These cells are in fact sensitive to deformation. In a vertebrate ear, these cells have extensions—in the form of cilia, or ciliary tufts if you like. When these cilia move, curve, or bend, the auditory cells feel it and produce the electrical signal in response. Depending on how these cells and their cilia are organized and how the sound vibrations reach them, we have two very different types of ears: short-distance ears and long-distance ears.

I don't want to bother you too much with physics, but there's an important notion to understand before going any further, and it's simple. It's the notion of near and far sound fields. If you stand very close to a speaker and the music is loud and rhythmic, you can feel the vibrations with your body, and even a sort of blast every time there is a "boom" of drumming, for example. In fact, you really feel the air molecules that come and go with the transmission of sound waves. Their vibrations are so strong that if you dangle a piece of string between your fingers, you will see it swing. You are in the *near sound field*. Now move a few meters away. You can still hear the sound. But you no longer feel that physical sensation of vibration. The piece of string hangs motionless between your fingers. Only your ears can hear it. The air molecules still come and go, of course, transmitting the sound pressure waves, but with much less amplitude. You have moved into the *far sound field*. What does this mean for our animal ears? That to perceive sounds in a near sound field, a hair or a filament suffices.[172] It is enough for this hair or filament to be directly connected to the sensory cell to allow the perception of sound to take place. However, in a distant sound field, it is essential to stretch a membrane (an eardrum) to receive pressure variations and to connect this eardrum, more or less directly, to the deformable cilia of the sensory cells.[173] One last little notion of acoustical physics before continuing: since sounds propagate better in water than in air, the near field extends further away from the sound source in this medium. Of course, the extent of the near field also depends on the power at which the sound is emitted and on its wavelength. Here are some numbers to illustrate this: the near field of a cricket extends up to 4 centimeters from the animal; that of a blackbird, up to 10 centimeters; and that of a codfish, up to 5 meters.[174] With these three examples, you can guess that short-distance ears are mainly observed in animals living in the water (crustaceans such as lobster or shrimp,[175] cephalopods,[176] and most fish[177]) and in insects in the air that communicate very closely, while long-distance ears are the prerogative of the majority of animals in the aerial environment (most insects and vertebrates) or aquatic species communicating at long distances (certain fish and marine mammals).[178] Before we talk about how vertebrates hear, let's take a look at insect ears as an example, since both short-distance and long-distance ears are found there.[179]

Insects! What a world. At least a million species have been described, and the total number of species is probably close to six million. Insects are one of the most diverse groups of animals—from species where individuals live in perfect solitude to the densest animal societies on the planet. Insects conquered the aerial environment some 400 million years ago. As their ancestors were already equipped with mechano-receptors, the first insects could probably perceive the vibrations of the substrate, such as a stem or a leaf. The acquisition of eardrums, the membranes needed to hear from afar, took place in a disorderly fashion. Note that hearing is rather uncommon: most insects cannot hear any-thing![180] Nonetheless, over time, the ability to hear sounds developed independently in different groups of insects. As a result, the location of long-distance eardrums on the bodies of insects differs between groups—they can be in as many as 15 different places.[181] In locusts, the eardrums are on each side of the first segment of the abdomen, i.e., on the sides of the animal behind the wings;[182] in grasshoppers, they're on their front legs; and some moths of the Sphingidae family have their ears on the appendages surrounding their mouths.[183] From an anatomi-cal point of view, the tympanum of insects corresponds in principle to a slimming of the cuticle (the external envelope that serves as a skeleton for insects) leaning against a small cavity filled with air derived from the respiratory system[184] and connected to sensory neurons (scolopidia[185]). The sensory neurons end with cilia that the vibrations of the eardrum distort, causing the neurons to become excited and generate the nerve message.

One of the essential evolutionary drivers behind the development of insect eardrums is the bat. Throughout the history of insects, the ear-drum has appeared independently more than 20 times. And in 14 of those cases, the eardrum was (and still is) adapted to perceive ultra-sounds, the high-pitched frequencies produced by bats when they search for their prey. Fossils, molecular analyses, and biogeographical data indicate that these ultrasonic ears were in place very late in the his-tory of insects, only 65 million years ago. The first bat fossils date back to that same moment in time.[186] As my colleague Michael Greenfield says, night flying became far too dangerous for insects at that time, unless they were properly equipped to detect and avoid these terrible new predators.[187] Thus, there was probably considerable selection pressure favoring individuals with anatomical and physiological changes that would allow them to hear a bat coming. Nocturnal or deaf individuals were in danger of rushing headlong into the jaws of death. They had to hear at all costs to survive in the dark. Moths provide the best evidence that there is a parallel evolution (coevolution) between insect hearing and bat predation. In insects, the ultrasonic ear has appeared independently in a dozen different lineages. Butterflies without ultrasonic ears live in areas where bats are absent, or fly at times of the day or year when bats are not active. It should be noted that insects with ultrasonic ears react to hearing bats by turning or diving toward the ground, or even by emitting ultrasonic sounds themselves that interfere with the bat's signal. Clever little buggers.

Bats have not been alone in promoting the evolution of long-distance ears in insects. Sexual selection has also played its part. Insects are small in size, and often low in density. Imagine a female cricket in a meadow, surrounded by tall grass. The probability that she will find a male by moving randomly through the meadow is not very high. In any case, it would take time, a lot of determination, and a lot of energy. But if she hears a male chirping, she has a much greater chance of detecting and locating him. Acoustic signals, therefore, are an efficient means of finding a partner: they can be used day and night, cutting through vegetation; their infinite modulations can easily encode information, such as the identity of the species; the emitting individual has total control over the beginning and end of its emission; and the origin of the message is generally easy to locate (unlike chemical signals, which can stagnate in the air and whose source is more difficult to identify).[188] You are likely familiar with the praying mantis, this magnificent predatory insect with large prehensile forelegs. There are many species of them all over the world. Some have a single long-distance, ultrasound-sensitive ear on the ventral side of their thorax. Well, this ear was already present in mantises 120 million years ago, long before the arrival of bats. Perhaps it was already used for communication, or to detect prey, or to spot predators we don't know about.[189]

Let's move on to the short-distance ears, and stay with the insects. There are many different kinds of these ears in insects, but perhaps the most impressive are those of flies and mosquitoes.[190] These animals have short-distance ears on their antennae.[191] I guess you've never observed a mosquito antenna; here's a project to pencil in your diary. It's kind of like a pretty, fluffy feather duster on the front of the animal's head, filled with sensors that allow the bug to navigate its environment, find its food, locate partners, etc. One of those sensors, known as the Johnston's organ, serves as the mosquito's short-distance ear. Located in the second segment of the antenna, the Johnston's organ has no less than 15,000 sensory neurons, a number comparable to the neurons found in the human ear. This mosquito ear is incredibly sensitive.[192] In the *Drosophila* fly, there are only 500 neurons in the Johnston's organ, but a second system—the arista, positioned on the third segment of the antenna—completes the array. The *Drosophila* fly is therefore equipped with two types of short-distance ears. While the tympanic ears can be sensitive to a wide range of sounds, up to very high-pitched signals whose frequency can exceed 100 kHz, the antennal ears perceive mainly low-pitched sounds (frequencies below 1 kHz). And, as you can see, these ears no longer hear anything beyond a few centimeters, i.e., as soon as you leave the near sound field. So what are these short-distance ears for? Essentially for love conversations.[193] Talking about love requires intimacy, we know that! In flies, these ears allow the female to hear the male's song during the courtship parade. The two individuals are then very close to each other, resting on a leaf or fruit, and the male vibrates his wings to produce his songs. Singing is not the prerogative of male flies; researchers have recently reported that females sing during copulation. Singing with pleasure? We don't know, but their singing starts with the arrival of seminal fluid, modulates the amount of sperm transfer from the male, and lasts for 15 to 20 minutes in an apparent feedback loop coordinating female singing and male sperm transfer. Does the male respond due to the female's song?[194]

Fanny Rybak, an associate professor at the University of Orsay in Paris, and her colleagues have shown that these acoustic signals act in concert with chemical signals during the male courtship.[195] These very short-range communications are also common in bees. During the dance on the hive combs, the worker vibrates her wings at a frequency of 200 to 300 Hz. These sound waves are perceived by her fellow bees through their antennae.

In mosquitoes, the ears hear the sound produced by the beating of the wings of other mosquitoes. We also hear it, by the way, and we know this song that announces the bite only too well, don't we? Male mosquitoes use it to detect, locate, and pursue females. The sensitivity of their Johnston's organ (its *frequency selectivity*, as it is called) is even tuned to the frequency of the beating of their sweetheart's wings.[196] An experiment has shown that male and female mosquitoes can tune the frequency produced by the beating of their wings so that they emit the same sound.[197] The next time you hear the particularly annoying song of a mosquito in flight, think of it as a love song . . . maybe it will help you endure it!

All vertebrates (fish, amphibians, reptiles, birds, and mammals) hear thanks to a particular anatomical structure: the *inner ear*.[198] The inner ear has two distinct parts. The first, the vestibular system, is mainly dedicated to the detection of the movements and positions of the body.[199] It is fundamental for the control of body balance but does not intervene in hearing, so we do not discuss it here. The second part is the place of sound detection—the ciliated sensory cells (or hair cells), which are sensitive to vibrations. As in insect ears, these cells transform sound vibrations into nerve signals sent to the brain. The inner ears of all vertebrates are variations on this theme. The main difference lies in the way in which pressure variations (from the far sound field) or particle oscillations (from the near sound field) are transmitted to the inner ear. Let's clarify. We are no longer dealing with insects here, but with vertebrates, and listening a few centimeters away with hairs attached to antennae is no longer relevant. For the vast majority of vertebrates, their ears are therefore long-distance ears, and it is an eardrum that receives the variations in sound pressure. These variations are then transmitted to the inner ear by a middle ear . . . except in fish! Remember: in the air, the near sound field is not very large; and vertebrates, which are rather large animals compared to insects, are generally too far away from their fellow creatures to perceive their sound emissions.[200] In the water, however, this field can extend for a few meters.

But these poor fish have another problem: they live in the water. (I hope you knew that!) And the density of the water is the same as their bodies. This is annoying for the sounds, because it means that a sound wave that hits a fish will make its whole body oscillate evenly, just like the water around it. It is therefore impossible for the sensory cells in the ear to detect the slightest movement of their cilia if they also oscillate in concert. This is a difficult problem to solve. How can it work? In practice, the cilia of the inner ear cells are stuck in an otolith, a kind of small crystal of calcium carbonate that is much denser than water. When the body of the fish, including the sensory cells, vibrates under the effect of the oscillations of the near sound field, the otolith reacts with a delay. This deforms the cilia, and voilà! Nerve cells are excited by this deformation and inform the fish's brain.[201] This ear is not very efficient, limited to low frequencies of a few hundred hertz. But we haven't reached the end of our surprises: in about a third of fish, the ear can also detect pressure variations and thus perceive sounds in a far sound field. How does the ear do this? By associating a small air pocket with the inner ear. Yes, you read correctly: an air pocket in the body of a fish. This pocket of air changes in volume due to pressure variations, and these movements are transmitted to the nerve cells. More precisely, the wall of this pocket vibrates, acting like the membrane of a long-distance ear. So these fish do have a kind of internal eardrum. Great, isn't it? For some species (more than 8000), the air pocket is simply the swim bladder, the structure that allows the fish to regulate its water depth, which finds here a new function.[202] Definitely a multitask organ. A succession of small ossicles (the *Weberian apparatus*) transmits the oscillations of the bladder to the inner ear. Such structures considerably increase the capacity of the fish's ear: it becomes more sensitive and able to hear frequencies of several thousand hertz. There are even species related to the herring that hear ultrasound up to 180 kHz, which is considerable . . . and perhaps useful in trying to escape the killer whale that is looking to devour them.[203]

What about other vertebrates—amphibians (frogs, newts, salamanders), reptiles (snakes, lizards, crocodiles, tortoises), birds, and mammals?

The great majority have a tympanic membrane (one for each ear) in contact with the external environment, which collects pressure variations due to sound waves. In most frogs, this eardrum is clearly visible, placed just behind the eye. Behind it a small bone transmits the vibrations from the eardrum to the inner ear. Frogs' eardrums work in water as well as in air. An anatomical feature that frogs share with birds, crocodiles, and lizards is that the eardrums of the right and left ears are connected by an air-filled canal.[204] As a result, each eardrum not only vibrates in response to the arrival of sound waves but also transmits its vibrations to the other eardrum via this channel that connects them. An eardrum thus receives vibrations on each of its sides, external and internal. The connecting channel therefore makes it much easier to locate a sound source.[205] Known as *pressure-differential ears*, they are especially beneficial for small animals because the clues usually used for sound localization (the difference in sound amplitude and the difference in sound arrival time between the two ears)[206] are not sufficient. This is because the sound arrives at practically the same time and with practically the same intensity in both ears when the ears are very close together, whether it comes from the left or from the right. It is therefore difficult to use these indices of time and intensity to locate the source of the sound. The pressure-differential ear makes it possible to perceive them more finely. In fact, it is found in many insects.[207]

A quirk of frogs and other amphibians is the relationship between their ears . . . and their lungs. When the frog's lungs are inflated with air, they are set in vibration by sound waves. These lung vibrations are transmitted to the inner surface of the eardrum and oppose the tympanic vibrations due to the sound waves hitting the outer surface of the eardrum. The lung vibrations thus attenuate the intensity of the sounds perceived by the frog. Where the matter becomes subtle is that only certain sound frequencies can vibrate the lungs and are therefore attenuated. Which ones, you may ask? Well, those caused by the vocalizations of other frog species. The lungs act as an attenuator of ambient noise, and by listening less to the croaking of other species of frogs, individuals can better hear the croaking of their own species. This is useful when vocalizing in ponds where several species of frogs live together.[208]

But all frogs don't hear that way. There are frogs that do not even have an eardrum—only the inner ear is present. Renaud Boistel, a French researcher specializing in the bioacoustics of frogs, Thierry Aubin, and their colleagues have studied the *Sechellophryne gardineri*, a little frog which, as its name indicates, lives in the Seychelles.[209] Even if its inner ear, well hidden inside the head, is not connected to any eardrum, the Seychelles frog hears and communicates through vocalizations, as shown by Renaud and Thierry through playback experiments. They used sophisticated methods to measure the densities of various parts of the frog's body and found that the conduction of sound waves to the inner ear involved the bones of the head . . . and the mouth! By simulating the structure of the mouth through mathematical calculations and computer modeling, they demonstrated that the mouth was a *resonator*—a sounding board, like a drum—for the characteristic frequencies of the frog's call (around 5710 Hz). In other words, when sound waves from a call reach the frog, they enter the mouth, which amplifies them through resonance. The wall separating the mouth from the inner ear is extremely thin (about 80 micrometers), making it easier for the amplified sound waves to be transmitted to the sensory cells. Complicated but effective.[210]

In birds, the eardrum is indeed there, at the bottom of a small pipe opening through a hole at the back of the eye, where a system of ossicles connects it to the inner ear.[211] Sometimes the feathers around the auditory hole form a kind of auricle—a sort of outer ear—concentrating the sound waves. This device is clearly visible in owls. Generally speaking, birds have quite good hearing: their range of audible frequencies easily extends from a few hundred hertz to more than 6000. Aquatic birds—especially those that dive deep to pursue prey—have developed specific adaptations to cope with the acoustic properties of the water and the high pressures during deep dives.[212]

In mammals, we find the same configuration as in birds, except that the right and left eardrums are not connected. The mammalian eardrum is a simple pressure receiver. Most species of mammals living in the air have a true external ear, with a clearly visible, sometimes mobile pinna, which allows more efficient collection of sound wave energy. Look at your cat or your dog listening: it can direct the pinna of its ear toward the sound source. In marine mammals (whales and dolphins), there is no outer ear.[213] This would not be very practical for swimming. In addition, they have the same problem as fish: their body vibrates in phase with the water in which it is immersed. The ear canal is narrow, is full of cellular debris, and has lost its role. It has become a useless relic. So how are the auditory sensory nerve cells excited? In toothed whales (odontocetes, dolphins, sperm whales), sound vibrations are transmitted to the ear through a blubber-filled canal in the lower jaw. Well sheltered in the skull, the ear consists of two small bones suspended by ligaments in an air-filled cavity, a bit like in the fish we were talking about. One of these bones, the tympanic bone, vibrates in response to incoming sound vibrations. It is connected by a short chain of ossicles to the inner ear, where the sensory cells are located. It is at this level that the vibrations generate nervous action potentials, which are transmitted to the auditory areas of the brain. This system (which is quite complex[214]) is very efficient, and dolphins are known to hear a wide range of sounds, from medium frequencies (a few kHz) to very high frequencies (200 kHz).[215] In baleen whales (mysticetes), although the ear appears to have roughly the same characteristics as in dolphins, things are much more mysterious. It must be said that it is not very easy to study these massive animals. We don't really understand how sound waves are transmitted to their eardrums—probably through the bones. In any case, although it has never really been measured, their hearing range seems much less extended than that of dolphins—from a few hundred hertz to a few kilohertz.[216] But perhaps we underestimate them.

Perceiving sound waves and producing a nerve signal in response—or signal transduction, discussed earlier—is only the first step toward hearing sounds. The nerve signal must then be processed by the brain. We would have to enter the world of neuroscience to explain these neurophysiological mechanisms, which is not the purpose of this book. However, let's look at a process that allows a receiver to extract useful information from the sound signals it receives: *categorical perception*. The sensory world of animals, including humans, is indeed overloaded with nonessential information. Critical information—an incoming predator, the presence of a fellow animal—can be drowned out. Processing all this information and extracting useful information is a real challenge. Categorical perception is an important mechanism that allows the individual receiver to sort the stimuli into distinct categories. We humans practice categorical perception every day. The most common example is the perception of the two syllables /ba/ and /pa/. For /ba/, the delay between the beginning of the lip movement that produces /b/ and the production of the vowel /a/ by the vocal cords is almost zero. On the other hand, for /pa/, the lips move before the vocal cords vibrate. We can construct by computer synthesis intermediate syllables between /ba/ and /pa/ by which we progressively vary the duration of this delay. When listening to the progression of these variants, human adults identify each variant as either /ba/ or /pa/—they do not perceive a continuum between the two. For them, the signals they hear do not gradually shift from /ba/ to /pa/; the boundary is perfectly clear. They are two distinct categories.

In animals, surprisingly, categorical perception has essentially been studied by looking at their ability to distinguish . . . syllables from human language. Just think where anthropocentrism leads! A few rare studies have considered natural situations, such as the swamp sparrow *Melospiza georgiana*, which perceives the variation between the acoustic elements that make up its song in a categorical manner;[217] the Japanese macaque *Macaca fuscata*, which does the same for two calls in its vocal repertoire;[218] or the túngara frog *Engystomops pustulosus*, when it has to distinguish the parade call of its fellow frogs from that of another frog species.[219] My colleague Nicolas, who traveled with me to the Pantanal, and I decided to investigate this issue in crocodiles. These elite predators spend their lives picking up information from their environment, through sight, smell, and hearing. And they are constantly making important decisions. Am I or am I not going to that calling baby? Could the noise I hear over there be potential prey? We had to find an experimental context to test the crocodiles' ability to categorize sounds. The idea came to Nicolas during one of our stays at *Crocoparc*, the zoological park of Luc Fougeirol (who was also the founder of *La Ferme aux Crocodiles*, discussed in the previous chapter) and his family, in Morocco.

In the ponds of the park, the North African green frog (*Pelophylax saharica*) is very common. It provides the park's soundscape with its croaks. A frog croak has an acoustic structure that is not far removed from a small crocodile's contact call: a short sound, with a fundamental frequency and small harmonic series. On the other hand, it is weakly modulated, sounding like "yhah! Yhah!" instead of the " djong! Djong!" of the crocodile. How does a small crocodile distinguish the croaking of a frog that forms the background of its environment from the call of one of its fellow crocodiles? With Julie Thévenet, our PhD student, we conducted an experiment to find out.

Imagine a little crocodile, a few months old, a good 20 centimeters long, in a basin several meters in diameter. On the edges of the basin, four loudspeakers re-create the background sound of the frogs: "yhah! Yhah!" The crocodile is accustomed to them, ignoring them and going about its business. Suddenly, one of the loudspeakers emits a crocodile contact call: "djong! Djong!" The little one reacts by moving toward what it believes to be a fellow crocodile. It can tell the difference between the two signals, the frog and the crocodile. To evaluate its ability to distinguish between the two categories, we constructed chimeric acoustic signals, such as 10% frog/90% crocodile, 20% frog/80% crocodile, and so on up to 90% frog/10% crocodile: "yhaong," if you like. We presented them to our little crocodile in his pond. The results were clear. As soon as the chimeric signal included more than 20% of a crocodile call, the little one turned its head toward the loudspeaker, sometimes moving a bit. It had noticed that this signal was no longer "pure frog." We observed these same reactions—visible but nevertheless quite weak—each time for all the signals until the call was 80% that of a crocodile. There was no progressive increase in the intensity of the reaction. However, when the chimeric signal was 80% crocodile, suddenly our young crocodile showed a more sensitive interest in it, moving decisively toward the loudspeaker. With this 20% frog/80% crocodile signal, we reached a plateau. With more than 20% frog in the call, the little one classified the signal as "environmental noise that is not too worthy of interest." With more than 80% crocodile, the response became, "It's a fellow crocodile call; I'm going!" Being able to categorize is a great decision aid.

What will the next step in our research be? Identifying the acoustic parameters that allow the crocodile to categorize the two types of signals: Is it the pitch of the calls? The difference in modulation? Or some other characteristic? Julie has already begun to explore this question by constructing chimeric signals in which only the frequency or modulation varies. I must confess that I'm looking forward to the results. Exploring the sensory world of crocodiles in this way is fascinating.

Let's stop there for now. Of course, other physiological and psychological mechanisms deserve to be detailed, such as the *precedence effect*, which characterizes the fact that insects or frogs caught in a chorus of acoustic signals pay attention to the first signal received, ignoring the next.[220] It would also be interesting to explore recent developments combining behavioral and computational approaches used to reveal the acoustic features that receivers rely on for signal recognition and discrimination.[221] But you already know a lot about how animals hear. Now let's see how they make sounds. The Okavango expedition is waiting for us.

10

Tell me what you look like

PRODUCTION OF SOUND SIGNALS

Panhandle, Okavango Delta, Botswana. The motorboat suddenly swerves and we zigzag quickly on the narrow river. "Hippos!" Sven shouts excitedly. At the helm, however, Vince frowns, a shadow of anxiety passing over his face. Hippos are the most dangerous animals in Africa, and they could easily overturn our boat. That was a close call![222] The afternoon is over, the horizon is on fire, and we have been sailing up one of the arms of the Okavango for almost an hour. Now the GPS indicates the place we are looking for. Vince throttles the engine. Sven ties us to a bunch of papyrus and we stop. Silence . . . binoculars pointed . . . "She's here," says Vince with his finger pointed. "Her nest is on the bank, right here; I remember." Of the female Nile crocodile we see only a ridge line, nostrils, eyes, and ears. The rest of her body is submerged. Vince starts the engine again, and we are going to attach our loudspeaker in the vegetation, near the entrance of the nest. Back at our waiting position, we suffer in silence. The mosquitoes arrive in squadrons, tightly packed and buzzing. Thierry is smacking his cheeks, grumbling. "*Le terrain* forever," he had written to me one day. . . . I wouldn't give up my place for anything in the world. We came to the Okavango to test whether Nile females react differently to the distress calls of newborns and older hatchlings. Through our acoustic analyses, we had already learned that the vocalizations of baby crocodiles change as they grow. But do mothers pay attention to this information? The Okavango expedition had to answer that question.

Years before, with Amélie, we had tested the hypothesis that young crocodiles have a vocal signature that allows their mothers to recognize them. Unfortunately, that hypothesis had fallen through. However, analysis of the acoustic characteristics of the calls in the days following hatching showed that the calls change with age.[223] Each day calls become more low-pitched, accompanying the growth of the small crocodile. At the time, we did not test whether this correlation between body size and voice provided information that might be of interest to the mother during the weeks when she stays with her young to protect them.

However, over the years, I had been accumulating recordings of young crocodilians of various species and sizes. Ruth Elsey had welcomed me warmly to the Rockefeller Wildlife Refuge in Louisiana. She had studied the reproductive biology of the American alligator *Alligator mississippiensis* for years, and thanks to her, I was able to record many newborns of various sizes. Thierry and I had also recorded many small spectacled caimans and Orinoco crocodiles during our stays in Hato Masaguaral, Venezuela. Black caiman, American alligator, Nile crocodile, Morelet's crocodile *Crocodylus moreletii*, Orinoco crocodile—for all these species we had found that the acoustic structure of the call is correlated to the size of the individual. We needed to know if this apparently solid information had a biological role. For that, we had to go into the field. Going back to Peter Taylor's house in Guyana was an interesting possibility, but I didn't have a large enough sample of black caiman calls. The most numerous recordings—those with the greatest diversity in hatchling sizes—were Nile crocodile calls. Thanks to Luc Fougeirol, I had been able to access many broods of this species. But in order to experiment on the mothers, I had to go and meet them, obviously in the wild. Where the Niles live. In Africa.

Thierry and I had discovered Vince Shacks while watching a documentary on crocodiles. He was immediately interested in our project. "I know the area very well. We have to go to the Panhandle. That's where we can most easily find females with their young." With his colleague Sven Bourquin, Vince had just spent several years studying the biology of the

Nile crocodile in the Delta. The two partners were a great pair: Vince, thoughtful, verging on anxious, but perfectly organized; Sven, more whimsical and playful, confident—nothing was ever a problem for that Crocodile Dundee. Both men were familiar with the Nile crocodile and its habits. Both professional and friendly, they soon inspired confidence in us. When we arrived, they had organized our expedition down to the last detail. The two all-terrain cars, well equipped, were ready to go. From Maun in the south of the Delta where we had landed, it took us a few hours on a fairly good road to reach the trail leading to the camp. That's when things got serious. Botswana's sandy trails live up to their reputation. The next few weeks would be spent cut off from the world.

The camp is surprisingly luxurious. Thierry and I are staying in a cabin on stilts. It has a small terrace with papyrus as far as the eye can see; bathroom in the open air; comfortable beds. "Not much to fear in the area. The big cats aren't here this time of year," Vince reassures us. I ask nervously, "And what about the snakes?" Vince smiles: "Gentlemen, you are in the region of Botswana with the highest density of black mambas. We found one on the kitchen table one morning a year or two ago!" The mamba is a very nice snake, one whose sudden acquaintance you'd rather not make: a good 3 meters long, agile, of the efficient kind that kills you unceremoniously in about 30 minutes. "And if you get bitten?" "No problem, we'll give you a piece of paper and a pencil to write your last words on," Sven laughed. "But don't worry too much; he's discreet and you're unlikely to meet him . . . and if that happens, back off quietly. No need to run; he'll be much faster than you." The last crocodile hunter in the area was found on the trail dead and still in the driver's seat of his Land Rover. Bitten by a mamba, he had attempted to reach the village, but the powerful venom had quickly paralyzed him. This was an exciting start, but we are full of courage, make no mistake!

The aim of the field trip was to compare the behavioral response of Nile crocodile mothers to calls from very young, and therefore small (30–40 cm, including tail), crocodiles with that of slightly older, and therefore larger (60–90 cm), crocodiles. A few weeks before our arrival, Sven and Vince had made a first visit to locate the nests. "Do you see these GPS points?" Vince asked, pointing to his map. "Each one is a nest. We have about 15 of them. Taking into account the losses due to predators, we should still be able to count on about 10 females with their young." In a period of two weeks, we managed to test 9 females. The females responded to the playback by either moving toward or away from the loudspeaker. Sometimes the approach was fast and furious, sometimes just a swim of a few meters toward the loudspeaker. It was always in response to a call from a very small individual being emitted. But if the call came from a larger individual, in principle the female would not move, or would move backward. Only one female approached the loudspeaker when she heard the vocalizations of a bigger youngster. These individual variations did not mask the general rule: Nile crocodile mothers are attracted to smaller juveniles and ignore larger ones. They therefore pay attention to the size information encoded in their calls.[224]

Do you remember the hypothesis of honest communication, which I first introduced in chapter 5 about the cries of baby fur seals and warbler chicks? I mentioned to you that one of the reasons a communication signal could carry reliable information about the transmitter was the cost it could represent; the energy expenditure or the risk of predation, for example. With this correlation between the body size of the young crocodile and the characteristics of its call, we discover another reason for honesty: a signal can be reliable if the transmitter is unable to lie! This is indeed the case with our crocodile. As its body grows, the vibrating membranes at the back of the crocodile's throat, which produce the sounds, grow with it. Its call becomes lower. It cannot lie; it has no choice; it could only do so if it did not grow. To understand this relationship between anatomy and the coding of information in acoustic signals, let's leave the crocodiles behind and go to the world of mammals.[225] The mechanisms of sound production have been very well studied there.

In all mammals, vocalizations are the result of two independent processes: the production of a source signal by the vocal organ—the larynx—and its modification by the cavities of the vocal tract. This process has been formalized in the *source-filter theory*.[226] The larynx is a very complex organ whose presence is noticeable in men, thanks to their Adam's apple.[227] At the level of the larynx are the vocal cords, which are cords in name only. They are actually a kind of membrane that can be made to vibrate when air comes out of the lungs.[228] These vibrations produce the source signal—sort of a primary sound wave, if you will. Since the vibration of the vocal cords is complex, this primary wave is not a simple hissing sound. It is a complex sound, consisting of a frequency called the fundamental frequency, as well as harmonic frequencies, which are multiples ($\times 2$, $\times 3$, etc.) of the fundamental.

The value of the fundamental frequency is primarily determined by the length of the vocal cords of the individual transmitter. The longer the cords, the slower they vibrate, producing a lower fundamental frequency. This fundamental frequency is therefore the main factor responsible for the pitch of the voice. When it is high, the voice is perceived as high-pitched. When it is low, a low voice is heard. As a first approximation, larger mammals produce the lowest fundamental frequencies.[229] In the 6-ton African elephant *Loxodonta africana*, the fundamental frequency measures about 20 hertz, compared to tens of thousands of hertz in a bat weighing only a few grams. But be careful! The bodily dimensions of a species and the length of the vocal cords are not always so related. In monkeys, for example, some species of macaques will have higher-pitched voices than others of the same size.[230] Within the same animal species, more subtle fundamental frequency differences can be observed depending on the relative size of the individuals, and also sometimes their age and sex in the case of sexual dimorphism (if female and male are of different sizes). However, some research has shown that the fundamental frequency is an unreliable indication of an individual's size. Indeed, this frequency results from a complex interaction between the density of the tissue constituting the vibrating membranes of the larynx, their degree of stretching, and the length of the actual vibrating part of the membrane. An individual sender can to some extent modulate the fundamental frequency by controlling the power of the air flow coming out of its lungs and the tension of its vocal cords. When the sender varies the fundamental frequency, modulations can be heard in the intonation of its vocalizations.

What about harmonic frequencies, you might ask? Harmonic frequencies also contribute to the characteristics of vocalization. But they are modified as the vocalization passes through the vocal tract—the conduit that goes from the larynx to the exit of the mouth or nostrils. The first sound wave produced by the larynx will indeed propagate, and during its journey it is modified. Certain frequencies are reinforced; others will be filtered. Let's draw a picture to visualize things. You probably know that the light that comes to us from the sun is made up of several colors. You can easily see this when raindrops break the light down into a rainbow. Let's say that sunlight is the first wave produced by the larynx and that the different colors are the frequencies that make up the first wave. Place a sheet of paper in the sunlight. It is white because sunlight, with all its colors mixed together, is white. Now place a blue tinted glass over the sheet. The sheet is no longer white but blue. The blue glass plays the role of the vocal tract between the larynx and the mouth. It acts as a filter, allowing only part of the light to pass through it. The light that comes out of the glass is different from sunlight because some of its components (red, green, etc.) have been caught in the glass and are no longer there. When we vocalize, the first wave produced by the larynx undergoes fairly similar treatment as it passes through the vocal tract. Certain sound frequencies, called formants, are reinforced. What determines which particular frequencies are reinforced (in other words, what determines the formants) are the resonance properties of the vocal tract. This tract is in fact made up of a succession of cavities: the pharynx, the oral cavity, and the nose. Each of these cavities can resonate with the sound waves that pass through it, and reinforce one or the other formant. The way in which the formants are distributed allows the characteristics of vocalization to change (for example, in humans, to pronounce the vowel /o/ rather than /a/) and affects the timbre of the voice, so that a voice will be more or less soft, nasal, or metallic.[231] Let's take a specific example with the human voice. Our fundamental frequency can vary between 60 Hz (very low voice) and 300 Hz (high voice). On average, it is around 210 Hz for women and 120 Hz for men, who have slightly longer vocal cords. If the fundamental frequency is 100 Hz, then the associated harmonics are 200, 300, 400, 500 Hz, etc., since they are multiples of the fundamental. This series of frequencies (100 and its multiples) constitutes the first wave, which is then filtered by the vocal tract. In the human species, the vocal tract has four main formants, distributed between 500 and 3500 Hz. Let us consider the first two formants, which are centered, respectively, on 500 and 1500 Hz. The 500 Hz formant reinforces this frequency in the voice, while the lower frequencies become less audible. The 1500 Hz formant reinforces this frequency, which means that frequencies between 500 and 1500 (say, around 1000 Hz) are attenuated. When we change the position of our tongue or move our jaw, we can change these formants and give them more or less importance, which allows us to modulate the timbre of our voice and even to pronounce this or that vowel. Don't worry if you've lost your way a bit. Just remember that the sound produced by the vocal cords is modified as it passes through the throat, mouth, and nose, and that we have the ability to modulate these changes by moving our tongue and lips and changing the opening of our mouth.

This ability to substantially and precisely modulate the physical structure of our vocal tract, and thus the properties of the formants, makes the human species a little special. Most nonhuman mammals do not possess this ability: their formants are somewhat fixed and in principle even quite easy to predict simply by measuring the length of the vocal tract. The vocal tract can be considered to resemble a tube from the larynx to the lips. The longer the tract, the lower the frequency of the formants; therefore (and logically), the bigger an animal is and the longer its vocal tract, the lower its formants will be. This type of information directly related to the physical characteristics of an individual is called a *quality index*. This reliable relationship can be observed not only between species but also between individuals of the same species.[232] Such is the case in our crocodilians, where the membranes at the back of the throat play the same role as the larynx of mammals and the oral cavity plays the same role as the vocal tract, and where small individuals have higher-pitched voices than large ones.[233]

The source-filter theory predicts that the biomechanical characteristics of the vocal cords and of the vocal tract translate into quality indexes in vocalizations. Body size, and sometimes sex, physical condition, or age, is static information imposed by the process of producing vocalizations.

This information can be very useful in the context of mate choice (intersexual selection) or competition between individuals of the same sex (intrasexual selection). Moreover, these selection mechanisms can lead to additional effects. It has been shown, for example, that male terrestrial mammals produce vocal signals whose formants are even lower when the size dimorphism between females and males is pronounced.[234] In these animal species, sexual selection favors males whose voices are lower and therefore sound bigger and scarier.

My colleague David Reby has studied acoustic communications in the Cervidae—the deer family—for many years, and has become a specialist in identifying the mechanisms by which these animals produce their vocalizations and the information they encode.[235] Maybe you've had a chance to hear the deer *Cervus elaphus* roar. It's absolutely breathtaking. It happens in the evening or early in the morning, in the woods; it's almost dark, and the bellow begins, something like a long, powerful, low call, "RooooaarrrhhhHHHHHH!," which the deer repeats over and over again. Let's take a close look at the animal as it bellows. It extends its head and stretches it upward, tilting its antlers toward its back. When it does this, we can clearly see a bump moving on its throat—its Adam's apple, or larynx. The larynx descends toward the sternum as the neck stretches, away from the animal's mouth. At the same time, the bellow becomes deeper. By lengthening its tract, the deer lowers its formants. In fact . . . it is cheating. It exaggerates its vocal size. Like Jean de La Fontaine's frog, it tries to pass itself off as bigger than it is. Of course, the maximum length of the tract, obtained when the larynx is at its lowest point, depends ultimately on the size of the individual. Since all deer lower their larynges when they bellow, eventually their signals reliably reflect the differences in size between individuals. The signal therefore remains an index of quality.[236] We can suppose that this behavior of lowering the larynx, now shared by all males, is the final stage of a lie that has gradually become widespread over the course of evolution. When the ancestors of today's deer began to lower their larynges, males probably had a serious advantage in terms of reproductive success. Like today's deer, those with the deepest voices were preferred by the hinds;[237] moreover, it was easier for them to frighten off an opponent

with their voice. By promoting the success of these males, sexual selection has done its job in order to achieve what we now call a *stable
evolutionary equilibrium*. It's difficult for the male deer to lower his larynx any further as it goes down to touch the sternum. But who knows?
The deer of the future may have developed other tricks to cheat vocally
about their size. However, females don't rely solely on voice quality to
judge a male. They also pay attention to the number of bellows per minute.[238] Contrary to what we might believe, bellowing is a very tiring
exercise, and only males in good shape are able to do it continuously
and win the prize. This situation was explained in the *handicap theory*,
formalized by Amotz Zahavi a long time ago. This theory states that the
cost of sending the signal (in all its aspects, which we have already discussed) represents a handicap that only certain transmitters are able to
bear. In the case of the bull deer, the "bellow" signal becomes reliable
because of its high energy cost. The deer's voice says so much that David
and his colleagues have demonstrated with playback experiments that
a female can tell the difference between a male that has already won a
harem and one that hasn't . . . just by listening to it.[239]

Relying on vocalizations to choose one's partner is very common. In
the animal kingdom, many acoustic signals are produced in the hope of
attracting a soul mate. When sexual selection is intense, which is the
case in polygynous mammalian species (where males have harems and
sexual dimorphism is usually pronounced), anatomical adaptations giving males low-pitched voices have often developed. Being as large as
possible is often an advantage when competing with rivals; but size is
not always part of the equation. A very original example is provided by
the koala *Phascolarctos cinereus*. You probably know this Australian
teddy bear that feeds on eucalyptus leaves.[240] It's not very large, about
15 kilograms at the most. Yet its territorial call, which is used to repel
male intruders and attract females, is incredibly low-pitched, a succession
of frightening breaths in and out: "Dro-he dro-he dro-he"—a Kawasaki
engine noise. The call's fundamental frequency averages 27 Hz, which
is 20 times less than one would expect given the size of the animal. In
fact, this frequency is very close to the one emitted by . . . an elephant.
Ben Charlton, David Reby, and their colleagues have shown that the

larynx of the male koala has no particular ability to produce a low fundamental frequency. But, to their surprise, they discovered a second pair of vibrating membranes higher up in the vocal tract, at the junction between the nasal cavity and the mouth. Koalas have a second vocal organ! And these membranes are long enough to produce frequencies as low as 10 Hz.[241]

Other mammals have adaptations that allow them to produce fundamental frequencies that are abnormally low for their body size. For example, the male hammer-headed fruit bat *Hypsignathus monstrosus*, from the African tropical forest, has a disproportionate larynx that fills the entire thoracic cavity, i.e., more than half of its total body volume (the flying larynx, as some people call it).[242] Its loud "Honk! Honk!" is emitted when males gather and compete for females.[243] In other mammalian species, laryngeal hypertrophy is a little more modest but real.[244] It is sometimes accompanied by vocal pouches (in howler monkeys, for example) or nose expansions (as in the saiga antelope *Saiga tatarica*), which increase resonance and lower formants. Sometimes it is the structure of the vocal cords themselves that is peculiar, with an increase in their mass that lowers their frequency of vibration. This type of device is found in large cats, such as the lion *Panthera leo*, the tiger *P. tigris*, the leopard *P. pardus*, and the jaguar *P. onca*. If you are lucky enough to have heard a lion roar at night, you will agree that it bears little resemblance to the mewing of its cousin, the domestic cat. And in the human species, you might wonder? While our species does not present a very pronounced sexual dimorphism on many traits,[245] the octave difference between men and women is one of the largest among mammals, with only howler monkeys being more extreme.[246] Besides, having a low-pitched voice is not always favored in animals. In bonobos (*Pan paniscus*), the apes closest to the human species, both females and males have very high-pitched voices. Their vocal cords are about twice as short as those of their chimpanzee cousins (*Pan troglodytes*), which are of comparable body size. But unlike chimpanzees, the bonobo society is known for its codominance between females and males. Equality between the sexes? Maybe, maybe not; but in any case there is no selection pressure on them in favor of low-pitched vocalizing in males.[247]

The fact that communication signals can carry reliable information about the sender is of major importance to the receiving individual. Without a minimum of reliability in the transmission of information, the communication would lose all its value and would probably not have been maintained over the evolutionary history of the species.[248] Imagine if every time someone spoke, what they said had nothing to do with reality, was totally unpredictable, and contained no reliable information.[249] After a while, that individual would probably stop being listened to, especially if listening and paying attention represented a cost to their listeners or a disadvantage compared to not knowing what they were saying. This is a simplified picture of what happens in the evolution of species and how they communicate. As we have seen, there are many attempts at cheating. In the green frog, for example, when a small male meets a large male, it starts to sing at a lower pitch.[250] The fact that senders have the opportunity to cheat suggests that communication signals should not be considered completely honest or totally deceptive. As usual in biology, a black-and-white perspective would be wrong, and remember that the balance between honesty and deception in communications is a subtle one. In principle, it results from what the senders can produce and what the receivers are willing to accept. Do you remember the concept of the arms race I mentioned in chapter 5 when I was talking about communication between parents and their young? The principle is simple: the sender exaggerates its signal in order to get something from the receiver, while in return the receiver becomes increasingly reluctant to respond to the received signal. We have seen two processes that ensure the relative reliability of the information encoded in acoustic signals: 1) the cost of the signals, which means that not everyone can afford to do as they please ("a signal has to be expensive to be honest");[251] and 2) the fact that the quality indexes are inseparable from who the sender really is, making it difficult for them to alter the information sent.[252]

My colleagues Kasia Pisanski and David Reby at our ENES Bioacoustics Research Laboratory in Saint-Etienne, France, had the idea to explore these processes in humans. Their starting point was that, while vocal exaggeration is widespread in the animal kingdom, it is difficult to establish its actual capacity to confuse listeners. Kasia and David used the fact that the depth of the human voice is correlated with the size of the individual: taller people have on average a deeper voice than smaller people. However, everyone is able to modify their voice somewhat by forcing it. Kasia and David asked women and men to pronounce the vowels A-E-I-O-U in three different ways: first, in an "honest" way, speaking naturally (first condition), then as if they wanted to sound taller (second condition), and, finally, as if they wished to sound smaller (third condition). Kasia and David then played these recordings to 200 adults, asking them to estimate the real size of the speaker and to guess whether he or she had tried to deceive them. Acoustic analysis of the vocalizations showed that, while it is possible to modify one's voice to appear larger or smaller, the level of cheating is still limited. The body size of an individual imposes an inescapable anatomical constraint on the vocalizations that he or she is able to produce. The playback experiments also showed two interesting things. First, listeners were able to guess quite accurately when the individuals had voluntarily modified their voices: it is not so easy to cheat. However, strangely enough, listeners were still fooled by the altered voices. People's heights were indeed overestimated by about 3 centimeters when they tried to sound taller, and underestimated by 4 centimeters when they tried to sound smaller. How to interpret these results? Well, we have here an example of the arms race that leads to a balance between honesty and cheating. In the human species, the depth of the voice plays a role in the choice of a partner. Thus, for example, women are, on average, more attracted to tall, deep-voiced men (which does not mean, of course, that all women are attracted to tall, deep-voiced men—individual tastes differ, and these are certainly not the only parameters of choice). Therefore, a man may have an interest in forcing his voice lower to be perceived as taller, but this strategy will quickly reach its limits since it will be detected by the listener. These results have been published in the journal *Nature Communications*, and if you are curious about the details, I invite you to read the scientific publication in its entirety. Isn't it exciting to see once again that the human species does not escape the evolutionary processes that run through the living world?[253]

Body size is certainly not the only static information useful to a receiving individual when choosing a partner, or when evaluating an opponent on the basis of its vocalizations. Let's consider birdsong. It provides information on many aspects of the sender. In monogamous bird species where both parents are involved in brooding and rearing the young, it is important for the success of the brood that the male invest time and energy. In short, and without anthropomorphism, it is in the females' interest to pay special attention to the quality of the male they choose. Let's take the example of the canary (*Serinus canaria*). In this species, females are solicited by a particular song of the male, and they may respond favorably by adopting a very explicit posture to invite mating. This song of invitation is the accelerated repetition of a syllable composed of two notes, both very strongly modulated in frequency: "Zee-oop! zee-oop! zee-oop! zee-oop!" This extremely fast trill was named the A phrase by Eric Vallet and Michel Kreutzer, from the University of Paris Nanterre.[254] Through playback experiments, they showed that females prefer the fastest A phrases, suggesting that these versions demonstrate superior vocal performance in males.[255]

Although physiological experiments have shown that singing probably does not cost birds much energy (since oxygen consumption is only slightly increased),[256] it is likely that being able to perform such trills at full speed can be a marker for a female of some interesting qualities in a male. This can be supported by the *developmental stress hypothesis*, which suggests that the complexity of the song in adulthood is partly determined by the development of the embryo in the egg and then of the young bird—a period that coincides with the establishment of the brain structures involved in sound production.[257] If during this period the chick is under particular stress, such as insufficient or unbalanced nutrition, the development of its singing abilities will be disturbed.[258] For a female canary, mating with a male who masters the production of fast A phrases means having a partner who has not had any particular stress during the crucial phases of its development and is therefore in "good working condition," a well-qualified father-to-be. Singing is therefore an easily assessable indicator of the male's history and life characteristics, such as his motor or cognitive abilities. It should be noted that stresses with long-term consequences on an animal's vocalizations are not limited to the perinatal period. An individual may experience dietary restrictions, heat stress, parasites, and other stresses throughout its life. Any unpredictable change in the environment is likely to be stressful to an organism, and the response to any stressor results in physiological changes (such as an increase in blood levels of certain hormones or a decrease in the performance of the immune system).[259] An example? The repertoire size (i.e., the number of different notes in the song) of the sedge warbler *Acrocephalus schoenobaenus* decreases by about 20% when the bird has parasites.[260] By attesting to the physical condition of the sender, communication signals are valuable informants for the receivers. Be careful, though! I wouldn't want you to think that receivers make reasoned choices, such as, "If Pierre sings like that, it's because he's in great shape. . . . Quick, let me take him as a partner!" Receivers are, of course, unaware of the value and meaning of these signals; their choice is the result of sexual selection that has gradually established itself over time.[261]

Appearing to be the biggest or the most beautiful is not always what succeeds in seducing a partner. The males of the superb lyrebird *Menura novaehollandiae* have another strategy: when they try to mate with a female, they play the illusionist. Their vocalizations give the impression that a group of birds is harassing a nearby predator. Everything is there: imitations of calls of various species accompanied by wing noises for good measure. The bird switches from one signal to another at an astonishing speed, superimposing calls of several species. It's a real tour de force. Playback experiments demonstrate the quality of this imitation: when listening to it, individuals of the imitated species start to nag the loudspeaker as they would in response to natural calls. This false danger signal is produced only during the 45 seconds of copulation (which is a long time for a bird). During this moment, the male is perched on top of the female and flaps his wings in front of her, blocking her view. The female can only remain motionless, trapped in this polyphonic sensory trap.[262]

Let's stop for a moment to review. From the previous examples, an important element stands out: sexual selection is one of the essential evolutionary drivers of the mechanisms that produce communication signals. In chapter 5, we saw that parent-offspring interactions are also a driving force in this evolution. Whether it is potential partners, parents, or opponents who judge the qualities of the sender by its voice, in all cases it is always the receiving individual who ultimately determines whether the vocalization produced is a valid communication signal and is worth keeping or needs to be altered. (Let us specify once again that this decision is the result of selection; it is not a reasoned judgment.) A communication signal is a biological characteristic peculiar to an animal species (in the same way as the shape of its head, the color of its fur, or any other characteristic) whose evolution over the history of the species has made it possible to influence the behavior of the receiving individuals. Senders and receivers can therefore be seen as engaged in an evolutionary game, i.e., a relationship where each moves its pawns according to what the other does.[263] And this game can be more or less cooperative or conflictual.

I can feel a question coming. . . . How on earth, at the very beginning, does a signal appear? How can the ability to produce information-carrying sounds be acquired during the evolution of a species? If you remember chapter 1, you know that we are dealing with the most formidable of the questions of ethology formulated by Nikolaas Tinbergen: the fourth question—the question of origins. Behavior unfortunately leaves little or no fossil record;[264] vocalizations even less. We must therefore turn to tools other than those of archaeological excavations to try to understand the beginning of the history of acoustic communication.

Let us first bear in mind that the evolution of biological systems is a gradual process. An animal species is defined as a set of individuals sharing many common characteristics, in principle capable of reproducing among themselves, but all slightly different. This individual variability, partly due to chance, is the raw material of evolution. From one generation of individuals to the next, more complex features may emerge from less complex features in small steps. This is cumulative selection. Evolution is therefore a contingent process, which means that new features do not appear out of thin air. They are always preexisting traits that are gradually transformed over generations through complicated interactions between chance (so-called neutral drift) and natural selection (the fact that the communication signal increases the probability that the individual producing it will have offspring). The evolution of acoustic communication signals is not immune to these mechanisms.

Two main processes are identified that lead to the initiation of new communication. The first is the *precursor sender model*, in which the new signal results from the transformation of an already existing behavior, which produced sound and was already carrying information but almost inadvertently. The second is the *model of sensory bias*, in which the sender exploits sensory capacities already present in the receiver, but which were used for something other than communication. These two models are not always in opposition; they can be complementary. Of course, if nobody ever emits anything or nobody ever responds, there will never be any sound communication. I repeat: evolution always works from elements that already exist. Let's look at it together.

The precursor sender model usually involves *ritualization*.[265] When an animal moves or makes a movement, associated sounds are often produced. Think of the sound of the wings of a bird flying away, or the "knock-knock" of the spotted woodpecker looking for larvae in a tree trunk. These sounds are not real communication signals, but they can still carry information for other birds. "We're flying away! Danger on the horizon? Be careful!" or "Woody Woodpecker's up in the tree, stuffing his face." If individual receivers react by modifying their behavior and this interaction results in a benefit for everyone, senders and receivers alike, then the signals gradually become stronger by exaggerating primitive noises. Their components will often be repeated. They will become more stereotyped and easier to detect, more efficient. The "knock-knock" in the random rhythm of the woodpecker extracting larvae from the tree trunk becomes a characteristic drumroll, heard at a long distance: "TRTRtrtrtrtrtrtrr." Sometimes the very anatomy of the structure that emitted the original sound changes.[266] When frightened, the crested pigeon *Ochyphaps lophotes* will fly away, producing with its wings a series of two notes very quickly repeated: "D-rrrr-d-rrrrr-d-rrrrrr." In response, the other pigeons fly away too. These are real communication signals informing the group of the presence of danger. They are produced by a feather, the eighth flight feather, considerably thinner than the others.[267] Over time, the shape of this feather has changed, diminishing its role in flight but transforming it into a musical instrument. Evolutionary biologists call this an *exaptation*—when a structure that initially devolved and adapted to a certain function (in this case, flight) is more or less reconverted into another function by undergoing transformations. Nature is full of examples.[268] Many other birds have modified feathers that enable them to produce acoustic signals, and many other animals have sound organs whose primary function was quite different. In fact, we may say without too much risk of overstating that everything *always* starts with exaptation.[269] As biologists often say, nothing is created; everything is transformed.

Let's take the case of the bugs. Have you ever observed a cricket singing? Take a good lock at it. It raises its two elytra and rubs them together, producing the well-known "tchip-tchip-tchip-tchip." This mechanism of sound production is called stridulation. In concrete terms, the underside of each elytron has a row of small teeth, known as the rasp, while its edge, called the scraper, is thickened. Stridulation is produced when the rasp of the elytron rubs against the scraper. The surface of the elytron is thinned in two places, the mirror and the harp, which resonate with the vibrations, and something akin to ribs plays a role in their conduction. If you still have some memories of physics, the wings of the cricket function as *coupled oscillators* that finely control the sound produced.[270] In short, the elytra of crickets, which were originally wings, have become a sophisticated musical instrument, producing communication signals. Generally speaking, exaptations are commonplace in insects: modifications of the wings, legs, parts of the thorax, abdomen,[271] or even elements of the reproductive apparatus—all of which have led to the establishment of structures producing acoustic communication signals. It must be said that the rigid external skeleton of insects is an excellent raw material for the creation of musical instruments![272]

Fish provide other interesting examples of exaptation for acoustic communication.[273] We should not be surprised, since we already know that some species of fish can hear. However, it is not enough to be able to hear; there must also be sound production—and vice versa. Acoustic communication requires both hearing ability and sound emission. In fish, therefore, a significant number of species produce sounds and use them to communicate information to other fish.[274] Here there are no larynx or syrinx, but gnashing or snapping jaws, swim bladder vibrations, chirping, and fin movements. Eric Parmentier, a great specialist of fish sound productions and professor at the University of Liège in Belgium, has shown with his team that the clownfish *Amphiprion clarkii* produces a series of a few clicks (between one and eight) by rapid oscillations of its head and clapping of its jaws. More precisely, the clownfish has a particular ligament (which Eric calls the sonic ligament) responsible for the rapid elevation of the lower jaw, which in turn creates the click by forcing the jaws to snap.[275] Another way to produce sounds is to make the membrane of the swim bladder vibrate through particular muscles, whereby the bladder acts as a sound box. Eric and his team have shown that the piranha fish *Pygocentrus nattereri* uses this mechanism to produce a varied repertoire of sounds that males use when competing against each other.[276] The barking sound, called type 1, is emitted when two males face each other. It is probably some sort of warning accompanying an attempt at intimidation. The type 2 sound is usually associated with a competition for food. The type 3 sound is only emitted when one individual chases another and tries to bite him. A piranha bite! Now you know what you have to do if you ever hear a type 3 sound while swimming in an Amazonian river!

Finally, there's nothing very original about the acoustic communication functions of fish. Attracting a partner and scaring away uninvited guests are the main goals.[277] The information conveyed by the acoustic signals therefore depends first of all on the identity of the species: a piranha does not produce the same sound as a clownfish. However, there is more subtle information, such as quality indexes like body size.[278]

It is not easy to demonstrate that the information encoded in the acoustic signals produced by fish is actually perceived and used by the receiving individuals. A few years ago, with my colleagues Marilyn Beauchaud and Joël Attia, both associate professors in our ENES laboratory, we decided to tackle this question, with the primary objective of transposing to fish the experimental approach classically used in mammals and birds; that is, to perform playback experiments with signals

Metriaclima zebra

whose structure could be perfectly controlled. At the time, very few people had tried to make fish listen to sound signals. It has to be said that it is quite complicated technically. We advertised for a student who wanted to prepare a thesis on this subject. Almost immediately, I received a response from a particularly motivated student, Frédéric Bertucci. His enthusiasm and seriousness made it possible to take up the technical challenge. Frédéric succeeded in setting up playback experiments in aquariums, which is not easy.

The animal chosen was a pretty little fish with blue stripes, from the cichlid family—the *Metriaclima zebra*. Frédéric showed that the sounds produced by this fish (a succession of clicks) have the value of territorial signals ("Beware! This is my home!") and bear an individual signature that depends on the length of the body.[279] Only a fish already established in a territory responded to the playback signals—"I'm here, I'm staying put." Frédéric also showed that the receiving fish remained very tolerant to changes in the artificial signal. The rate of clicks could be considerably slowed or accelerated without changing the response of the test individuals.[280] A few other studies have shown that the information carried by the fish sound signals was indeed taken into account by the receiving individuals. For example, Clara Amorim, Paulo Fonseca, and their colleagues at the University of Lisbon in Portugal found that the female painted goby, a small fish you may have seen in tidepools at low tide on the Atlantic coast, prefers the more talkative males but pays little attention to the size-encoded information in the signals. The amount of time spent talking is correlated with the amount of fat reserves in the animal. Females are therefore sensitive to the physical condition of the male and assess it during courtship. To be honest, however, it should be pointed out that acoustic signals are not the only issue here, as the female will only pay attention to sounds if she *sees* the male.[281]

Let's get back to business, so to speak! Let me remind you of the question that guides us: understanding how acoustic communication takes place during the history of an animal species. Bird feathers that become whistles, a woodpecker that drums instead of pecking on a tree trunk, insect wings that turn into musical instruments, fish jaws that snap, swim bladders that become a bass drum: All these exaptations are associated with the establishment of ritualized behaviors that produce well-structured sound signals—behaviors that have been favored throughout the history of each animal species by their ability to transport information between senders and receivers.

However, as we mentioned before, this model of the sender as a precursor is not the only one that explains the implementation of acoustic communication. Let us now consider the model of sensory bias, where the sender exploits a sensory capacity already present in the receiver. You'll see that it's very easy to understand. Let's take the crickets as an example. In the field cricket *Teleogryllus oceanicus*, females are attracted to sounds with frequencies below 16 kHz, whereas they are repelled if the frequency is above 16 kHz (higher-pitched).[282] In chapter 9, we saw that these categorization phenomena are widespread in animals and that they facilitate rapid decision making. Our female crickets choose to go toward a singing male (sound of about 5 kHz), whereas they flee from bat calls (of very high frequency). Keep this information in mind: females process sounds in two categories—one attracts, the other repels. Now let's look at a group of cousins of the field cricket, the *Lebinthini*. In their home, the males' song is very high-pitched, up to 20 kHz.[283] This is quite surprising already. Moreover, when you play this song to the females . . . they don't move. On the contrary, they stop suddenly, in a trembling

motion, exactly like when they hear a bat call. It's a behavior often observed in insects because it allows them to escape from these predators.
So how do *Lebinthini* females meet their partners? Well, the females
shiver, and the vibrations generated are transmitted to the plant on which
they are standing. The male perceives these vibrations and he comes
right away![284] As a result, the high-frequency songs of the males derive
from sensory exploitation. The males of this group of crickets have developed a sound signal that uses a previously existing behavioral response of the females. Perhaps they have simply reinforced the higher
harmonic frequencies of their original signal? Anyway, the result is there,
and they take advantage of a sensory capacity of the females that had
been selected for reasons other than communication between the sexes.

Note that this example of the *Lebinthini* crickets is not isolated.[285] For
example, the same story can be observed in moths. In several species, the
males produce sound signals during courtship that the females cannot
distinguish from bat calls. Male calls and bat calls again provoke the same
response in females: they freeze. The male will take immediate advantage
of the situation. As the female stands still, it's easier for him to approach
her and convince her to mate.[286] Here again, the sender exploits a sensory disposition of the receiver that was there for other reasons.

Notice the difference between this model and the precursor sender
model that we were talking about earlier. In the precursor sender model,
the sensory capabilities of the receiver play a critical role, since it is the
receiver that either does or does not perceive the signal carrying the information and then "decides" whether or not a communication can start.
Here, in the model of sensory bias, the receiver is ready to hear and react
to sounds that the sender does not yet produce. The receiver's sensory
system has evolved to detect predators, as in *Lebinthini* crickets and
moths, for example. For a variety of reasons, some of which are probably
due to chance, the sender one day produces a new signal that exploits
the sensory predispositions of the receiver, initiating a new communication process.

You now have a good idea of how a new acoustic communication can
occur, i.e., how new signals are set up. Let's tackle the next question: How
can we explain the diversity of acoustic signals observed in nature?[287]

Look at birds, for example. There are tens of thousands of species, and almost as many vocal repertoires, even if we exclude the few birds that use little or no acoustic communication. Within each species, there are populations that often have their own dialects and individuals whose vocalizations often differ. How is this possible? Why this extravagant divergence between animals that are, after all, quite similar? It's true that the common ancestor of today's bird species dates back 65 million years,[288] and a lot has happened since then.[289] We've already evoked some of the mechanisms driving this diversification.

We've just seen that sexual selection can be a powerful driver of signal evolution. The senders producing the most convincing vocalizations (for example, the fastest trill of the song or the most elaborate repertoire) are chosen in a privileged way, directing the evolution of communication for this species or that population in this or that direction.[290] When we look at animal species forming social groups in chapter 14, we will see that communication signals can be subject to social selection by kin groups: one alarm call, more effective, will be favored over another. When we were walking in the tropical forest with the white-browed warbler and its cortege of Brazilian species, I told you that some birds emitted vocalizations whose acoustic structure was adapted to the conditions of propagation of sound waves in the environment (for example, songs traveling in a particularly effective way despite the density of vegetation). Acoustic communication signals can thus be under the influence of ecological selection. Those that are most effective in a given environment will be favored. As bird species inhabit a variety of environments, this ecological selection can lead to a diversification of the signals used from one species to another. Moreover, the propagation conditions of sound waves are not the only element that can influence the evolution of signals. It is the entire set of environmental conditions that must be taken into account. In particular, we sometimes observe a real acoustic competition between different animal species for the sound space. It has been suggested that animals can share this space, for example, by emitting lower-pitched or higher-pitched sounds so as not to interfere with each other.[291] You can appreciate that things are very complex—many different factors can come into play.[292]

Sometimes the same signals can be preserved throughout the history of the species, but the role of the signal changes. About 20,000 years ago, astyanax fish living in rivers in Mexico became trapped in underground rivers and lakes, in caves. Since then, this fish has morphed into two forms—a form living in surface rivers with well-developed eyesight and a subterranean form that has lost the use of its eyes. Sylvie Rétaux, a researcher at the CNRS and a specialist in genetics and development, is trying to understand the mechanisms that allowed *Astyanax mexicanus* to change into its cave form. With my colleague Joël Attia and their postdoc Carole Hyacinthe, Sylvie has recently been interested in the acoustic signals produced by astyanax, because this fish talks, of course, like everyone else. Its repertoire is even quite varied, since it can emit at least six different sounds. The two astyanax forms differ in their use of acoustic signals. It appeared to Sylvie and her team that sharp clicking was produced during aggressive interactions by surface fish, whereas cave fish use the clicking signal when feeding.[293] In caves, food is scarce, and recycling an initially aggressive signal into a way of informing each other is probably a beneficial adaptation.

To understand the history of communication signals, evolutionary biologists construct phylogenetic trees, based on the genetic heritage and acoustic characteristics of current signals. The basic idea is to reconstruct all the historical changes from a common ancestor that have led to the diversity observed today. I recently conducted a study with a team of colleagues on the evolution of drumming behavior in woodpeckers.[294, 295] As I mentioned earlier, drumming is an exaptation from foraging behavior, as most woodpeckers spend their time vigorously beating the bark of logs in search of food. It's also how they dig the cavities where they're going to make their nests. Some woodpeckers drum ("DrDrDrDrDr . . .") very regularly; others speed up ("Drrr-Drrr-Drrr-Drr-Dr-Dr-Dr"); others slow down; and some are content with a short "TOK-TOK" that can be heard far away. Like birdsongs, these signals are used to attract a mate or deter an intruder. We first analyzed the acoustic structure of the drumming of 92 species of woodpeckers from around the world, then tried to reconstruct the history of these drummings over the history of the woodpecker family, and then attempted to explain

Great spotted woodpecker

their diversity. Our results showed that drumming is a very ancient technique, which was already mastered by the ancestor of all present woodpecker species 22.5 million years ago. We have also shown that the diversity of drumming types is essentially explained, quite simply, by chance genetic mutations. The closer two species are genetically related, the more similar their drumming patterns are. The drumming of sister species is difficult to distinguish. Those of first cousin species are a little easier to tell apart. As for those of second cousin species, they are quite different. Through playback experiments, Maxime Garcia, then a postdoc on my team, showed that the great spotted woodpecker *Dendrocopos major* cannot distinguish between the drumming of its species and that of a sister species. On the other hand, it does discriminate well between those of a second cousin species (sister or cousin—these are analogies, of course, to illustrate a close or distant relationship). But then, how do woodpeckers of different species living in the same forest manage to distinguish between each other? Fortunately, sister or first cousin species rarely live in the same places. We studied five woodpecker communities (in Switzerland, Guatemala, the US, Malaysia, and Guyana), and in each case the six to eight woodpecker species likely to live together in the same forest were almost always distant cousins. Their drumming patterns are therefore very different from each other, thus limiting the risk of confusion. When two sister species live in the same place, which does happen from time to time, then a very well-known phenomenon in evolutionary biology occurs: *character displacement*. The drumming of each species changes and becomes less and less similar to that of the neighboring species. To say that the character is displaced means that the drumming pattern changes, differs, and moves away from the drumming of its sister species. It can take thousands, even millions of years.[296]

I won't close this chapter without presenting you with one last mechanism explaining the divergence of communication signals between species: the magic trait! You have certainly heard of Darwin's finches. These small, somewhat dull birds inhabit the Galápagos Islands and provided Charles Darwin with a prime example when he established the theory of evolution. There are several species of Darwin's finches. They all come from the same original species and have gradually specialized in the different food sources available on the islands. Darwin's finches are an example of adaptive radiation, as it is called in the jargon of biologists: *radiation* because the new species went in different directions during their evolution (they radiated around the starting area); *adaptive* because the trajectory of each species corresponds to a food specialization for which the bird's morphology, especially its bill, has adapted to pick up food. In short, big beaks, big seeds; small beaks, small seeds. Studies of the songs of the different species show that there are correlations between the shape of the beak and the acoustic properties of the songs. Large-billed finch species have simpler songs (less difficult to produce) than small-beaked finch species. These correlations can be explained by mechanical stresses related to the type of bill and its associated musculature. A large, powerful bill cannot be opened and closed extremely quickly, making it more difficult to produce songs with rapid modulations. Changes in bill morphology during the diversification of finch species are partly responsible for the song diversification.[297] The diversity of the songs, which seems to appear by magic, is in fact nothing magical. It is the consequence of another phenomenon that has nothing to do with acoustic communication.[298]

Do you find all these mechanisms complicated? You're right, because they are! I warned you; it's difficult to reconstruct the history of communication behavior. Now let's get some air in our brains. Let me take you to the Amazon rainforest.

11

Networking addiction

ACOUSTIC COMMUNICATION NETWORKS

Amazonian rainforest, near Belèm, Brazil. Invisible in the high branches, a screaming piha blares out its song, one of the most powerful in the Amazon. First, some very soft trills that make the singer seem far away: "Wuuu—wuuu—wuuu." Then a brutal burst: "Weee-weee-YUUU!!" Right behind me, another one: "Wuuu—wuuu—wuuu … weee-weee-YUUU!!!" Then a third on the right, and several others join the sound demonstration. The forest fills up for several minutes with the calls of about 10 individuals. The chorus of *Lipaugus vociferans* submerges all the other sounds. Suddenly, for no apparent reason, all fall silent. The cicada concert can be heard again. Half an hour later, the *Lipaugus* resume possession of the sound space. Why do these birds congregate like this to sing? Does their apparently chaotic chorus obey a certain logic?

It was Thierry who suggested that we should take an interest in the *Lipaugus*. After our disappointing experiment with black caimans in the Kaw marshes (Do you remember? The female caimans had stopped responding to our signal broadcasts on the second day of the experiments …), we decided to leave our floating platform as soon as the opportunity arose. After a week, a refueling helicopter brought us back to the Camp Caïman hostel, and we got our rental car back, determined to explore French Guiana as tourists during the remaining week before our flight back to France. French Guiana is surprising: you drive on a perfectly maintained road, passing a post office or police cars similar in every way to those encountered in metropolitan France, then you stop at a small parking lot, put on your walking shoes, and after a few meters on a trail, you find yourself transported into the middle of a primary rainforest. It's a brutal, magical jump—and it's magnificent. Gigantic trees. Here, a tree trunk lying across the path being devoured by hundreds of huge beetle larvae—an impressive sound of jaws chomping. There, up in the branches, quiet, a lazy sloth swaying. Everywhere, termite nests, strange insects, big metallic-blue butterflies, colossal caterpillars, plants with huge leaves, furtive birds, and Julida—an order of enormous, astonishing millipedes with shiny black bodies. I was in heaven, chatting away with Thierry in the wilderness. Then, suddenly, "Wuuu—wuuu—wuuu . . . weee-weee-YUUU!!" We were to spend a long time looking for it in the branches, our singer, perched as it was about 10 meters high. However, it is a rather large bird. About 30 centimeters long, its grayish color camouflages it perfectly. "A bird that calls terribly loudly and that we can't see . . . a perfect model for a bioacoustic study!" Thierry has always been skilled at identifying research subjects. A little look in our ornithological guides would increase our curiosity. The *Lipaugus* form exploded leks. In other words, the males gather regularly to sing in particular places in the forest, always the same males but remaining a good distance from each other, between 40 and 60 meters. They stay there for a while, then go back to their other occupations and return a little later. No one understands why they do this. Our microphones, computers, and loudspeakers were not going to stay in our suitcases. That same evening, while we were drinking a little rum at the restaurant and watching the leatherback turtles laying eggs on the beach below, we decided to start the *Lipaugus* project.

Our first day was dedicated to the observation and the first recordings of the vocalizations. As good ethologists, we had to know a little more about the behavior of this animal before trying to ask it any questions. It was quite easy to spot several leks: the Guianese forest is easily accessible, at least at its edge. The pihas' chorus always started in the same way: first a few powerful, isolated calls ("TSSIOO!! . . . TSSIOO!! . . . TSIIOO!!") coming from several individuals; then a series of "wuuu—wuuu—wuuu," muffled sounds whose origin could not

Screaming piha

be located; and, finally, the complete songs, followed by a culminating explosion exceeding 110 decibels, a sound level between a rock concert and a jackhammer. Imagine the bird opening its beak wide, retracting its head, its whole body behaving like a high-fidelity loudspeaker projecting the sound waves as far as possible: "Weee-weee-YUUU!" These calls were first emitted by single individuals and then taken up by the whole chorus: Did the birds alternate among themselves? Was there a kind of organization in time, in space? A chorus conductor, perhaps? And why the different calls? These were the questions we had in mind.

After a few days spent recording these vocalizations, we decided to do some initial playback tests. Our first goal was to test if the order of emission that we had observed—first the "TSSIOO," then the "wuuu—wuuu," and finally the "weee-weee-YUUU!"—had a functional meaning. In other words, was it imperative that the birds first heard "TSSIOO" to start the lek chorus? Our second goal was to determine what happens if birds hear vocalizations from another lek. Do the individuals singing in a lek know each other? Do they react differently when they hear strangers? At that time, we were not sure if the birds in a given lek were always the same, but it seemed likely. So the principle of the experiment was as follows. Arriving at the place where we had spotted a lek the day before, we tied our loudspeaker to a tree; then we moved about 30 meters away, our tape recorder connected to the loudspeaker by a long cable. We waited until the pihas had been silent for at least half an hour before playing a first signal—one of the three types of calls—and then observed the birds' response. After a few days on this regime, we gained our first insight: the "TSSIOO" as well as the "weee-weee-YUUU!" quickly triggers strong behavioral responses from our singers. To our "TSSIOO," individuals responded with their own "TSSIOO." When we played the "weee-weee-YUUU!," the chorus would start. In both cases, it was not uncommon to see a bird silently flying close to our loudspeaker in response to our sound stimuli, then flying away again immediately, before it made its first calls. On the other hand, the "wuuu—wuuu" emission alone did not seem to provoke any particular excitement: the birds generally remained silent. The problem with the pihas was that we could not see them, so we could not tell who was singing. To understand the structure of this communication network, we had to be able to identify who was who in the lek and who was where. We were not equipped for that. In addition, our return flight to metropolitan France was scheduled, so we left French Guiana with no feather in our caps. The calls of the pihas resounded in our heads, and we had only one desire: to get back into the field.

A few months later, we met up with our friend Jacques Vielliard at the International Bioacoustics Conference in Italy. Do you remember Jacques, the professor at the University of Campinas in Brazil, with whom we had studied the white-browed warbler? Malu, Jacques' wife, had just obtained a professorship at the University of Belèm in the Amazon, and Jacques had decided to spend most of his time there. "Folks, I'm building a field station on the edge of a beautiful primary forest . . . you must come and see us," he told us, a glass of beer in hand. We immediately asked him, "Do you have any *Lipaugus* there?" In his ever slow and modulated voice, he answered, "Guys . . . think a little! . . . I'm talking about the Amazonian forest! . . . Of course, we have some. The *cricrió*, as they say in Portuguese . . . *A voz de Amazônia*." The voice of the Amazon! Indeed, its song is frequently used in films to reproduce the ambience of the rainforest. The screaming piha is widespread throughout the Amazon basin. Some even call it the captain of the forest. That's how, a few months later, we joined Jacques in the Amazon.

Belèm is a large city, bordered by a wide arm of the Amazon River. Although the modern part looks like any other, with its large, aesthetically challenged buildings, the old town is charming, a real postcard— with its small port where urubu vultures nonchalantly walk in search of some fishy leftovers, its colorful *Ver-o-Peso* market, and its old colonial houses, whose walls, roofs, and balconies are overrun with epiphytes plants. In the tropics, nature creeps in everywhere. An hour's drive away is the field station built by Jacques. He doesn't do things halfway: a large, beautiful house, with wide terraces where hammocks can be hung, immediately facing the forest. In the evening, we can see colorful flights of parrots and toucans. However, you should not get too idyllic an idea of the situation: mosquitoes, ticks, poisonous spiders, and scorpions are all part of daily life. The scorpion, a sneaky wee beastie, loves to slip into your shoes at night. One morning, Thierry shook his shoe a little by chance and discovered an intruder who hastily scuttled off. He had already put on the other shoe without taking this precaution, and had been lucky, and he never forgot to empty both shoes again before putting them on. Not to mention the antimalaria medication that had to be taken every day, never to be forgotten, and the compulsory vaccination against yellow fever, that very nice disease that makes you die from vomiting streams of black blood. The tropics are teeming with invisible life. Every small wound can quickly take on worrying proportions, and Jacques recommended that we carefully disinfect the slightest scratch.

As our friend had promised, the screaming pihas were on the scene. We had access to several leks on foot from the house. The objective of this first mission was simple: to see if we could identify and locate the individual singers of a lek, to follow the temporal dynamics of their choral group, and to write the score of the chorus. How to locate the position of invisible birds? It's not so difficult when using the principle of *sound triangulation*. Imagine six microphones positioned at different places and at different heights in the forest. When a piha sings, the microphone closest to the bird is the first to receive the sound waves; the furthest away is the last. Since the sound propagates at 340 meters per second, the delay in reception between two microphones corresponds to the difference in the distance to the bird. By using six microphones, whose exact position is known, we can then calculate the position of the sender bird quite accurately. The microphones were wireless, and we could centralize all the recordings on a multitrack tape recorder, which received the signals from each microphone via radio waves. Chloe Huetz, the engineer in Thierry's research team, was in charge of developing the computer program to process all this data. At the same time, it was necessary to establish the vocal signature of each individual participating in the lek. This meant recording each bird, then analyzing the acoustic structure of its song to identify its individual characteristics. It soon became clear that each individual was recognizable by its voice. A comparison with the vocalizations of birds from other leks in the forest further showed that, like the existence of a regional language, birds participating in the same orchestra share a lek signature. And by comparing the Brazilian recordings with those we had obtained in French Guiana, it appeared that these two populations, the Brazilian and the Guianese, did not have quite the same way of singing, even if the difference remained minimal (small variations in the modulation of the "weee-weee-YUUU!"). Despite the presence of these local and individual signatures, the singing of the piha was remarkably similar everywhere. As we will see in chapter 12, some birds learn their songs by copying other individuals, and this cultural transmission is likely to result in significant variations in the acoustic structure of vocalizations between geographically distant populations. On the other hand, species that produce their sound signals without the need for learning by imitation have songs that normally show little variation, which is directly related to the genetic distance between populations. The screaming piha probably belongs to this second category of birds.

Once the multimicrophone recording and triangulation system had been developed, and the individual signatures well characterized, we conducted playback experiments inspired by our first tests in French Guiana. The results confirmed our first impressions: the different calls produced different responses. The "wuuu—wuuu" was rarely followed by a vocal response. On the other hand, the birds reacted strongly to "TSSIOO" and "weee-weee-YUUU!" and even to an isolated "YUUU!" More precisely, when we broadcast this type of call, one individual would fly over the loudspeaker and then call in response. Only the "weee-weeee-YUUU!" seemed to systematically trigger the chorus. I should make one essential point: we were only able to spot the singing males, so, obviously, we didn't know—and still don't know—what the females were doing. Yet, of course, they had to be there too. Otherwise, why else would the guys gather together? Not for a rugby match, that's for sure. One hypothesis is that the "wuuu—wuuu" is addressed to the females, like a drumroll signaling that the males are gathering in the lek and are available. The "weee-weee-YUUU!," higher-pitched and thus allowing a precise location of each individual, would be addressed to both females and males. No discrimination. As for the "TSSIOO," they could be alarm calls, signaling the arrival of a possible danger. To find out more, it would be necessary to spend some time on the spot—a lot of time. Neither Thierry nor I could abandon our duties and our families for that long, so we hired Frédéric Sèbe on a postdoctoral contract. Frédéric, now a researcher on my team, was a former PhD student of Thierry's, a field man par excellence—afraid of nothing, and certainly not of living in the Amazonian forest for long periods. It was the perfect opportunity: Frédéric left to spend a year studying the pihas.

It was a big challenge, and despite extensive work, Frédéric would not, alas, unravel all their secrets. But he did come back with data that made it all a little clearer. Frédéric had been able to follow the activity of several leks throughout the year. By combining the system of locating individual singers by triangulation and their individual identification through vocal signatures, he was able to describe their dynamics precisely. A few surprises awaited us. We thought that the pihas gathered to sing during special seasons, but the recordings showed that the leks were active year-round and that they always had more or less the same number of individuals. Going every day to sing in the lek seemed to be part of the daily routine of the *Lipaugus*. On closer inspection, however, it became clear that the individuals singing together in a given lek were not always the same. At intervals of a few months, a good one-third of the males had been replaced by other individuals we had never recorded before. Let's be precise: this turnover varied between 0% (the same individuals were all still present in the lek) and 75% (three-quarters of the males had been replaced by others). In a way, the leks are like theaters where males enter the performance in turn. Some of these theaters have the same troupe performing year-round, while others see most of their actors change over the months. Frédéric's data also made it possible to describe the performance of these actors. It begins a little more than two hours after dawn, with a fanfare opening where all the singers vocalize at the top of their voices. Then, regularly, new explosions of calls occur, albeit more modest. Finally, two hours before twilight, comes a closing fanfare. Remarkably, each singer has its place on the stage and rarely moves from it. Here again, let's be precise since the data allows it: 90% of the birds found from one month to the next in the same lek remained within a radius of 30 meters around their initial position. An individual can have up to ten singing posts—i.e., trees on which it perches to sing—all very close to each other, but it spends most of its time on two or three of them. Just imagine: the same male will be singing for months from nearly the same perch! Now let's get into the figures. While the surface area occupied by all the individuals of a lek can reach 50,000 square meters (which corresponds to a circle 250 meters in diameter), each individual is confined to a maximum of 700 square meters. Moreover, males are not all equal when it comes to the distribution of roles. The most active singers are always in the center of the lek and can call their "weee-weee-YUUU!" up to six or seven times a minute. Those who are on the edges, less vocal, will be satisfied with making the call once over the same length of time. You can bet that there is no such thing as equality among the pihas.

What happens when a new male, unknown to the others, appears one day in a lek? To find out, Frédéric took up the playback experiments that we had initiated in French Guiana. This time, he was able to follow the behavior of each of the males in the lek. Frédéric mimed the arrival of a new individual by playing the calls of an individual recorded in another lek during the dawn chorus. The loudspeaker was placed in the singing place of an already present individual. The disturbed individual would first react sharply by nervously flying over the loudspeaker, accompanied by a clear increase in the number of calls he made per minute. Then he would move a few dozen meters to perch on an unoccupied tree and resume his vocal activity—no more disturbed than that. The other males of the lek obviously couldn't care less, continuing their vocal routine from their singing places. The next day, if Frédéric stopped the playback, the disturbed individual would return to its initial position. The *Lipaugus* were therefore ready to accept newcomers. When the number of males increases, they simply enlarge the surface area of the lek. In short, what have we learned from the *Lipaugus*? That this dull-colored bird spends a lot of its energy singing a very repetitive repertoire; and that the lek, this aggregate of males, is organized both in time and space. We will probably never know if and how the females make their choice among all these suitors, but it is reasonable to think that not all males are the same for them. Moreover, perhaps their preferences are variable: if we can assume that the stars— the males in the center of the lek—are more attractive to the majority of them, we can also imagine some females preferring the more discreet ones. Loudmouths versus wallflowers.

The *Lipaugus* project had whetted my appetite for studying *communication networks*. It was Torben Dabelsteen, my colleague from the University of Copenhagen, and Peter McGregor, then also a professor at the same university, who had formalized this concept a few years earlier.[299] Starting from the observation that most bioacoustic studies to date have considered any communication process as a simple sender-receiver duo, they had written several articles in scientific journals about the idea that, in nature, individuals are usually in the middle of a network, in which each individual can be alternately a sender and a receiver (not forgetting some who will in fact only be receivers, simply listening to what others say to each other). Communicating in a network, with the possibility that the intended listener is not the only one who hears the communication, can have consequences. We know this—we who modify our speech and way of speaking according to who might be listening in.

I decided to launch a new project to explore this issue, this time in my laboratory, with a species that is easy to breed in captivity: the zebra finch. With a size comparable to that of a sparrow (about 10 centimeters from beak to tail and weighing about 15 grams), this bird, native to the

Zebra finches

Australian semidesert zones, flies in large groups that can number in the hundreds of individuals in search of food or water.[300] The pairs nest in the same bushes, forming breeding colonies. But beware! This species has a very strict moral code: male and female are faithful for life. Let us note, nevertheless, that the life of a zebra finch is short compared to ours—a few years at the most. Females and males are easy to tell apart: the female appears to be in half-mourning, with a grayish plumage and a clearly visible black tear falling from her eye, while the male has an orange cheek, a zebra-striped chest, and a red flank peppered with white dots. Their vocalizations are much more complex than in the piha. A song emitted only by the males, lasting a few seconds, is a succession of squeaky notes, which are not very pleasant to hear: "Tzeek tzeeek tzeek tzeeek didiguezic dziduck." Many short calls, each consisting of only one note, are emitted by both sexes. Julie Elie, one of our students who was passionate about observing animals and who has since become a renowned specialist on the vocal repertoire of the zebra finch, has analyzed the acoustic structure of thousands of vocalizations of adult individuals and has categorized the calls into eight types: the whine, the nest, the tet, the distance call, the wsst, the distress call, the thuk, and the tuck.[301] As you might guess, each of these calls is made in a more or less precise circumstance. For example, the whine and the nest calls are produced when the male and female exchange nest duty. The wsst accompanies aggressive behavior. The thuk and the tuck are alarm calls, the first being intended for the young and the partner, the second for the whole troop. The one that caught my interest was the distance call, which allows the couple to find their partner when they lose sight of each other. This is the easiest to record because, to encourage the birds to produce it, all you have to do is separate the pair by isolating the male in one room and the female in another.

Initially, inspired by Thierry's work on penguins that I told you about in chapter 4, I wanted to test the ability of males and females to recognize each other through the distance call in the chaotic context of a breeding colony. I also imagined that I could identify the neurophysiological bases of this recognition: Are there regions of the brain dedicated to this task? A few observations suggested that this call allowed the couple to identify each other after separation, but experimental evidence was lacking. I could not carry out this project alone and hoped to convince a student to share this adventure. At the end of a course I was giving at the *École normale supérieure de Lyon* to prepare students for the prestigious French exam, the *Agrégation*, a young woman stood in front of me—smiling, determined, and confident: "I want to work with you," she said. I asked her about her motivations. Why would a student of the *École normale supérieure*, who must have had all the research laboratories of this famous school making eyes at her, want to join my team? "Because you do things that nobody else does. I didn't even know you could get paid for it! I'm studying biology because I'm interested in living things, but I find myself in front of pipettes and molecular formulas. Getting to listen to birdsongs— that is what I'm passionate about." Clementine Vignal really was a godsend. We didn't know that we were going to spend more than 10 years exploring the world of the zebra finch together.

We started with the resources we had on hand: a small aviary built by the technical team at the University of Saint-Etienne; a few cages; and some birds bought at the nearest pet shop. In a short time, we had our colony of zebra finches, with our first mating pairs and their nests.

First experiment, first disappointment. When we isolated a male from its female and made it listen to distant calls from females, including those of its partner, the poor male usually remained motionless, seemingly bewildered. We started by checking that the females' calls had an individual signature. The result of acoustic analysis and statistical comparisons between individuals was clear: each female had her own voice. Even we were able to distinguish between them. Our males had no excuse not to recognize their sweethearts. We had to look elsewhere. We knew that the zebra finch is a very social bird and not used to being alone for a long time. So we came up with the idea of placing a few individuals in another cage, next to the cage containing our male and the subject of our experiment—some companions to relax him; some friends from the bar, so to speak. That's when we made what, for us, was the discovery of the century!

"Look," said Clementine. "The results of the experiments are really quite odd. I get the feeling that the male's response to his female's calls varies depending on his companion birds." And so we found ourselves comparing the male's behavior in response to his partner's calls with that of other familiar females—all in different social contexts—all the while changing his companions. Next to some males, we placed a cage with a couple, a female and her male. For others, the two companions were an unmated female and male put in two separate cages. Finally, others were placed in the company of two other males. The first impression was confirmed: the male's response to his female's calls varied according to his audience. If the male was in the presence of a single female and a single male, or two single males, he would respond to any female call played by the loudspeaker, regardless of whether it was his own beloved or another, with no preferential reaction to those of his own female. But when he was accompanied by a well-established female-male couple, he reacted like a madman to his partner's calls (the number of calls emitted was multiplied by five!) and barely reacted to the calls of a familiar female who was not his own. Male zebra finches were therefore perfectly capable of recognizing their female partner by her voice,[302] but they were sensitive to the social context—to their surroundings. There was an effect of the social audience on the behavior of the males.[303] More importantly for us, it was the first time that an experiment had demonstrated a bird's ability to grasp the nature of social bonds between other individuals. In order to explain the results of our experiment, we had to admit that the male, who was made to listen to female calls, was able to distinguish between a couple of single birds and two birds in a couple. This social intelligence had so far only been shown in monkeys.[304] Excited by this discovery, we wrote an article for the famous magazine *Nature*, which accepted it. Champagne!

Champagne, yes. For two reasons. First, because we were extremely happy that we had made a scientific discovery. Second, because we hoped it would enable us to obtain funds to continue our work. In order to explore the world of the zebra finch, we needed money, and the article in *Nature* would be a great asset in finding it. We wanted to build larger aviaries, monitor the behavior of each bird in the colony from a distance, record each individual in its nest . . . and Clementine dreamed of going to work in the field, in Australia. Money is something I haven't told you about yet. Being a researcher is a bit like having the only job where, once you've been hired for life, you're told, "Now it's up to you to find the money for doing the work; no one else will do it for you!" What you don't realize right away is that you're going to be looking for money and then looking for more money for the rest of your life— which means writing more and more research proposals to persuade those who read them that the research work you propose is exciting and new, that it will revolutionize our knowledge, and that it has the flavor of excellence; basically, that you will save the world. For Clementine and me, the publication in *Nature* was to act as a real magic wand: it suddenly transformed a research-team-ignored-by-everyone into a lab-of-potential-international-notoriety. Not being in the habit of shooting myself in the foot, I won't say that it's easy to publish in this kind of journal, but luck has a lot to do with it. As Peter Marler, the high priest of bioacoustics, once told me, "It's almost impossible to predict what will interest them." At this stage, it didn't matter to us. Our little male zebra finch refusing to respond on command to the voice of his female had helped us reach the grail of scientific journals. And the National Research Agency, the main backer of French public research, agreed to finance the continuation of our work.

We were not at the end of our surprises with these birds. We had recruited Julie Elie, who would spend hundreds of hours observing them. She would notice that the monogamous couple—a female and a male "married and faithful"—is the foundation of their social organization. Female and male spend most of their time side by side, touching beaks, grooming each other's plumage, and conversing softly. Nice life as a couple, isn't it? Moreover, it is easy to distinguish a group of single birds from one made up of paired couples: the single birds are considerably noisier, with sudden explosions of sound where everyone is vocalizing at the same time; while the group of couples is calmer, each pair of birds whispering sweet nothings.[305] Can we go so far as to talk of wedded bliss? Moreover, in the case of the zebra finch, forming a couple for life does not seem to be an option but rather an obligation: whatever the circumstances, you have to pair up. When we had isolated single females in one cage and single males in another, we discovered a surprising thing indeed: in the absence of an individual of the opposite sex, the zebra finches soon developed same-sex pair bonds.[306] The males paired up with their brothers, and the females did the same with their sisters. Let's be clear: true single-sex couples are formed where both partners go so far as to mimic copulations and build their nests. And they're faithful to boot. When these birds were put back into mixed aviaries, each one stayed with its partner. With it I am, and with it I stay. As Clementine says, "The zebra finch pair is a true social partnership." To associate with a fellow bird to form a couple is a vital need for the zebra finch. How to explain it? It's possible that the intense constraints to which this bird is subjected in the wild—little water and food in a semidesert environment—favor a rapid pairing, with very strong bonds between partners: a solid and ready-for-anything couple will be better able to reproduce and raise young.

A recent study on a small African bird, the blue-capped cordon-bleu *Uraeginthus cyanocephalus*, has shown that what we observed in the zebra finch is not anecdotal. In the cordon-bleu, the couple performs very special courtship rituals in which both male and female tap-dance with their legs.[307] You read that right: tap dancing! The two birds jump very quickly on their perch by clapping their legs. For a cordon-bleu, that sound is super sexy, as long as the rhythm is perfectly controlled. To complete the scene, male and female sing. In short, the cordon-bleu's courtship is akin to Fred Astaire and Ginger Rogers crossed with Sonny and Cher.

Manfred Gahr, researcher at the Max Planck Institute for Ornithology in Germany; Masayo Soma, professor at the University of Hokkaido in Japan; and their student Nao Ota tested the influence of a social audience on the cordon-bleu's dance. The Max Planck Research Institute is located next to the small village of Seewiesen in Bavaria. This is where Konrad Lorenz studied goose imprinting. Do you remember that? We talked about it in chapter 1. The geese that had hatched in Lorenz's presence followed him everywhere, even when he swam in the lake next to the research station. Today's lake is not Lorenz's lake—it was filled in and dug anew—but the spirit of the man who was one of the founding fathers of ethology still permeates the place.

Manfred, Masayo, and Nao showed that male and female cordon-bleus adjust the modalities of their courtship if they are in the presence of other birds.[308] In particular, they increase the number of dances, combining singing and tap dancing. What is the reason for this? Perhaps to make it clear to the audience that they are courting each other and thus to keep away other possible suitors. Furthermore, during their dance, the partners point their tails toward their lover: while you never know for sure who a song is intended for, in this case there is no room for doubt. Another hypothesis is that the dancers do not want to put all their eggs in one basket and may be trying to charm members of the audience rather than their partner. We don't know. Perhaps marital morality is more elastic in cordon-bleus than in zebra finches.

The *audience effect* in acoustic communication networks hasn't only been studied in these two species of birds.[309] Let's give Caesar back what belongs to him: the first time an audience effect was suggested, it was for . . . chickens. The rooster makes calls to indicate the presence of food only when hens are there. Otherwise, it's silent. The same behavior is true of his alarm calls when a predator approaches.[310] It is now well established that many animals modulate their communication behavior according to the individuals around them. In primates, examples abound. For instance, when a chimpanzee calls out in response to an attack from a fellow chimpanzee, it exaggerates the intensity and duration of its calls if there is another individual in the vicinity that is known to be stronger than the attacker and therefore likely to drive it away.[311] As another example, rhesus macaque mothers (*Macaca mulatta*) respond more quickly to their babies' requests when they are close to adult individuals known to be aggressive toward the young.[312] When threatened by a predator, Thomas's langurs (*Presbytis thomasi*) emit alarm calls until each member of the group has responded with an alarm call, suggesting that they are able to identify who has responded and who has not.[313] In short, it is clear that we are far from animals vocalizing by simple reflex. This is very similar to what we observe in our own species.

Moreover, *social intelligence*, that remarkable ability to analyze the relationships between individuals in the group, as identified in the zebra finch, has since been found in many animals. These discoveries are not very astonishing if we think about it; obtaining an acute awareness of the social relations between the individuals that we are close to makes it possible to limit errors in judgment that can have serious consequences— for example, when you have to form alliances with other individuals. Take the chacma baboons (*Papio ursinus*), large African monkeys living in organized packs—males with canines that never end, and an ironclad hierarchy between individuals. When two males fight, the dominant one grunts while the dominated one squeals at the top of its lungs. With playback experiments, the famous primatologist Dorothy Cheney and her colleagues have shown that the other members of the group are attentive to these vocal exchanges and learn from them. See for yourself: when the researcher used a loudspeaker to emit sequences mimicking a reversal of hierarchy between two individuals (the dominant squealed while the dominated grunted), she observed that the other monkeys were clearly astonished: they immediately turned their heads toward the loudspeaker and stared intensely at it.[314] This behavior shows that chacmas track the evolution of the hierarchical situation within their social group by listening to the vocal exchanges of their fellow primates. They seemed very disturbed by what they were hearing. You might object,

"Yes, but social intelligence—this ability to analyze the relationships between fellow creatures—is probably reserved for animals with big brains." Well, it's not. Evidence of social intelligence can be found in . . . fish.[315] The first studies on this subject were conducted more than 20 years ago by Claire Doutrelant when she was still a student (she is now a researcher at the CNRS). Claire and fellow researcher Peter McGregor had placed a female Siamese fighting fish *Betta splendens* in an aquarium near another aquarium containing two males. In this species, the males fight each other with impressive demonstrations. No acoustics here, but visual signals: The males spread their fins and gill protectors (opercula) apart and twisted their bodies, all to appear as big as possible and impress the opponent. When the female was put in the position of choosing which male to approach, she systematically chose the winning male.[316] She had therefore observed the interaction between the males and had been able to deduce, and memorize, which one was the strongest—the best candidate. As another example, a male fighting fish exaggerates the visual signals he deploys all the more in front of an opponent that he has previously seen surrender to another competitor.[317]

In birds, too, females may listen to discussions between males. Eavesdropping on conversations between males influences the decisions females make when choosing who to breed with. A study of the black-capped chickadee *Poecile atricapilla* has demonstrated this experimentally. Paternity testing has shown that an average of one-third of the chickadee's chicks in a nest have not been sired by the male who looks after them, but by one or more of the males in neighboring territories. Morals are flexible in these birds. The females do not hesitate to look elsewhere—but why? Scientists have tested whether they base their reproductive decisions on information gained by spying on male vocal competitions. The first step in the study was to observe pairs of neighbors to find out which of them showed more ability to dominate the other. This was easy to see: a high-ranking bird did not wait until its neighbor had finished singing before it began singing. It did not think twice about butting in. Then, from their loudspeaker, the scientists produced various situations by playing chickadee songs for six minutes. In the control situation, they mimicked situations where the natural order was preserved: the neighbor entered the territory singing submissively, without barging in, while the owner of the territory sang aggressively. In a first experimental situation, the researchers reversed the established order, this time simulating an intruder singing aggressively in front of a landowner previously identified as being of high rank. In a second situation, they simulated an intruder singing submissively, i.e., without ever interrupting the owner of the territory, even though the latter had been identified as submissive himself (submissive perhaps, but never interrupted).

Some time after the hatchings, scientists took blood samples from the chicks and compared their DNA with that of the local males. The high-ranking males that had been confronted with the control situation (nonaggressive playback of the neighbor) could claim paternity of 90% of the chicks present in their nests. On the other hand, in the nests of the high-ranking males shaken by an aggressive playback of a usually submissive neighbor, 50% of the chicks were fathered by the neighbor! However, the individuals identified as submissive during the observation phase did not gain anything from having, once in their lifetime and for six minutes, passed for dominant near their female.[318] As easy as it is for a male to lose his rank, it's no mean feat to win the favor of female black-capped chickadees.

This ability to listen to signals that are not intended for the "receiver" (a sort of eavesdropping) sometimes extends beyond the subject's own species. It has been extensively studied in the heterospecific groups (i.e., groups with several different species) that birds often form. The red-breasted nuthatch *Sitta canadensis*, an elegant little Canadian bird, has a habit of joining groups of black-capped chickadees and takes advantage of their alarm calls when a predator approaches. But the nuthatch is cautious when it comes to interpreting these calls. Let's take a closer look. The structure of the black-capped chickadee's alarm call varies with the size of the predator. If a large owl arrives, the chickadees that spotted it make a short call ("chick-a-dee-dee-dee"); if it's a small owl, the call is long ("chick-a-dee-dee-dee-dee-dee"). In short, the call codes for the size of the predator.[319] It has even been shown that there is a proportional relationship between the size and the number of "dee's" in the call. Subtle, isn't it? Let's talk a bit about the nuthatches' reaction. If one of them sees the predator, it also modulates its calls: short, high-pitched calls faced with a small owl, longer and lower-pitched calls faced with a big owl. But if it does not see the predator and hears only the chickadee calls, it starts to make alarm calls of intermediate duration, regardless of the duration of the chickadee calls,[320] as if it didn't completely trust that information. You never know: What if the chickadees are wrong? Information gleaned by listening to others has never been worth strict adherence. As one of my grandmothers used to say, it's better to be safe than sorry. The nuthatch won't risk sending the wrong information to its fellow nuthatches![321]

To sum up, the next time you hear a singing bird, you have to imagine it as a sender inserted in a communication network, listened to by many receivers and listening to other senders.[322] There are many other things to say about communication networks.[323] For instance, an emitter individual can vary the size of the network likely to listen to it by modulating its vocalizations. Private conversations can exist, but then you have to be very discreet and produce signals that are inaudible from a distance. We talk about this later, in the story of how some animals living in groups can use acoustic signals to cooperate and form alliances. For now, let's return to another issue mentioned above—that of learning. You now know that some species of birds have to learn to produce their songs by copying others. How does this work? More generally, how does communication behavior develop over the course of an individual's life? This is, you remember, the third question asked by Nikolaas Tinbergen. It's also the focus of the next chapter.

12

Learning to talk

VOCAL LEARNING IN BIRDS AND MAMMALS

Marin County, California. In spite of the intense sunshine, the air is crisp and the wind strong. The ocean can be heard roaring in the distance. Perched on the top of a bush, a small bird with the appearance of a finch is singing its "yeeeee . . . peee peee peee pew pew pew pew pew pew" ritornello. It has a pretty head with black and white stripes, a yellow beak, and elegant marbled wings, brown and beige. "Have you seen this bird? What is it?" I said, turning to my colleague Frédéric Theunissen. "Come on, Nic! It's the white-crowned sparrow. Peter Marler's bird." Peter Marler! The high priest of bioacoustics, the one who unraveled many mysteries concerning the learning of song by imitation in song-birds. In his famous study, published in the prestigious journal *Science* in 1964, he recorded white-crowned sparrows in various locations around the San Francisco Bay Area: in Berkeley to the east, the home of the famous university where Marler once worked and where Frédéric is now; in the Sunset Beach area to the south; and in Marin County to the north.[324] By comparing their songs, Marler showed that the spar-rows had dialects, and it was easy to distinguish the songs of the Berke-ley residents from those of Sunset Beach and Marin County—a bit like how one can tell the French apart from their accents, be they Marseil-lais, Parisians, or inhabitants of Saint-Etienne. Later, Marler raised young sparrows in his laboratory and showed that these birds learned their song by imitating the one broadcast from a loudspeaker; it was possible to teach the Berkeley dialect to a sparrow born in Marin County, provided the bird was very young and had not yet begun to sing. So at this moment I am amazed, thinking that the individual we are watching may be a great-grandson of the sparrows recorded by Marler!

Peter Marler was not the first to work on how birds learn to sing. His studies were inspired by some illustrious predecessors, some of whom go back a long way. If 2000 years ago Pliny the Elder had already noted in his *Natural History* that the parrot was a good imitator, it is to the Austrian Ferdinand Pernau that we owe the first documented observations on the learning of song by birds. That was in 1720! Some 50 years later, in 1773, a man named Daines Barrington raised goldfinches (*Carduelis carduelis*) in the presence of adults of other species and observed that the little goldfinches imitated their guardians. Barrington also noted that, when given a choice, goldfinches preferred to imitate adult goldfinches rather than the adults of another species.[325]

However, it was not until the twentieth century that our knowledge in this area began to skyrocket. This was mainly due to technical progress with recorders (tape recorders) and the sonograph, an instrument that makes it possible to represent sounds graphically, much like musical notes. Two scientists got the ball rolling: the British William Thorpe and the Danish Holger Poulsen.[326] Then, by combining recordings of dialects in the field with experiments in his laboratory, Marler provided compelling evidence that an animal could learn to produce its vocalizations by imitating adults. It must be realized that, at the time, many still firmly believed that this ability was reserved for the human species and that animal songs and calls were innate reflexes, engraved in genetics. Proving through scientific experiment that an animal learned to produce its vocalizations by imitating adults was a real revolution.[327]

From there, songbirds became the preferred model of study for understanding human language learning. Since then, thousands of scientific papers have been published on the subject, and our knowledge of how birds learn to sing has become considerable. Frédéric Theunissen, with whom I was observing Marler's sparrow, is one of the scientists seeking to understand the mechanisms of song learning; that is to say, understanding how birds hear, how they memorize, and how they produce sounds. Let's set the stage. At the risk of repeating myself, we are right in the middle of the third question formulated by Nikolaas Tinbergen: the question of the *ontogeny* (i.e., the establishment) of a behavior. With Marler and Tinbergen by our side, we are making solid progress.

It's important to remember that there is no equality when it comes to vocal learning in birds.[328] Only three groups of birds learn to sing by imitation: a category of passerines (oscines, or singing passerines, such as the sparrow, American robin, wren, or zebra finch), parakeets and parrots, and hummingbirds.[329] To give you an idea of numbers, these learning birds represent more than half of the bird species currently living.[330] There are almost 10,000 species of birds on the planet, including just over 4700 species of oscine passerines, about 350 species of parakeets and parrots, and more than 330 species of hummingbirds.[331] You can see that the number of bird species practicing vocal learning is far from anecdotal. Other birds seem to develop their vocalizations innately.[332] In other words, nonoscine passerines (such as the screaming piha, whose leks I told you about in chapter 11) and all other birds do not learn to sing.[333] Their voice comes to them as they grow up, automatically as it were. A nonoscine passerine, the eastern phoebe *Sayornis phoebe*, if artificially deafened at a very young age (we are sometimes cruel, but rarely) produces a completely normal song as an adult.[334] Admittedly, there are a few examples that suggest that these two categories, capable and not capable of learning, are not totally watertight.[335] For example, the three-wattled bellbird *Procnias tricarunculatus*, a large Central American bird of the same family as the piha, emits versions of a nasal song, "aaaiiar!," which differs from region to region. So there are dialects! It has also been said that a young bellbird in captivity learned to imitate the vocalizations of a Brazilian bird, the chopi blackbird *Gnorimopsar chopi*.[336] Thus, vocal learning in nonoscine birds may be more widespread than previously thought.

How does vocal learning in oscines take place? Classically, it is done in two stages: the chick first memorizes the sounds it will have to produce, and then it has to practice and gradually match what it produces to what it has memorized. In the wild, a young chick is traditionally fed in the nest by its parents and therefore hears its father singing, and sometimes its mother, depending on the species—and sometimes adults in the neighboring nest, for species in a colony, for example. At about three weeks of age, the chick becomes able to memorize the songs it hears. At about five weeks, it is independent and goes off to explore other horizons. It is then exposed to the songs of other individuals. Some time later, sometimes even the following year, on its return from winter migration, the young adult bird will establish a territory by singing in turn. This pattern is, of course, variable—ranging from the length of stay in the nest and the age of passage to adulthood to the tendency to be sedentary or migratory—and many factors may differ from one species to another.[337] But the question remains the same: When, where, and from whom did the bird learn its song?

Studies that have explored these issues abound. Much of the laboratory work, inspired by that conducted by Peter Marler, has provided convincing evidence that young songbirds learn to sing by imitating one or more adult song masters, whether real or simulated by a loudspeaker. Our little zebra finch is an ideal species for these studies, a so-called biological model. The development of singing has been particularly well studied in this species, and we have been able to draw from it some fairly complete knowledge. Let's take a look.

The song of the zebra finch is a short sequence of syllables: "tzeek tzeeek tzeek tzeeek didiguezic dziduck." The song of each adult male is a version of this theme. Very standardized, the song does not change throughout the life of the individual. This bird lives, as we discussed in chapter 11, in the semidesert areas of Australia, which do not experience marked seasonal variations, as is the case in temperate zones. Like the climatic conditions, the song of the zebra finch remains immutable. The pairs are faithful for life. The zebra finch thus offers a remarkable model of social stability. In the male zebra finch, the song appears gradually, between 20 and 90 days after hatching. The process begins with a *sensitive period* (or *sensory period*) during which the young bird needs to hear a song from an adult male. This phase occurs roughly between the twentieth and sixtieth day after hatching. It is during this period that the young bird imprints in its memory the song pattern it will try to copy.[338] At the age of one month, the young chick begins to babble; this is the *sensorimotor phase*. At the beginning, its vocal productions are not a real song but a series of sounds that have neither head nor tail, known as a subsong. During the days and weeks that follow, the young bird will sing and sing again, thousands and thousands of times. At the end of the 90 days, the song will be impeccably mastered. Three months to become a professional singer is not too bad! Basically, this learning process is a lot like the way we humans learn to speak. As babies, we're immersed in a bath of words. Day after day, we memorize and memorize again syllables and words. We keep repeating "ba-ba-ba-pa-pa-pa" and other babblings, and then we get better and better at it. The major differences between birds and humans are the length of learning (a few weeks for the zebra finch, many months for humans) and the complexity of the vocalizations learned.[339]

So baby birds learn to sing by imitating adults. Humans are not the only ones with this ability. However, there's a flaw, isn't there? I can hear you from here saying, "That's all well and good, these zebra finch observations are very interesting, but weren't they made in captivity in the laboratory, in very special conditions, far from what birds experience in the wild?" Now you've touched on a real question. The laboratory is certainly a *controlled* environment, as biologists say. It allows us to ask specific questions and to isolate the role of this or that factor. But captive birds are not aware of the wealth of social interactions that their free counterparts experience on a daily basis. Nor are they aware of all their problems, such as searching for food or escaping from predators. So how does vocal learning take place in real conditions? Let's take a tour to Daniel Mennill's home at the University of Windsor in Canada. Mennill and his team were the first to conduct an experiment on song learning in a natural environment, with wild birds, in absolute freedom. Their work, published in 2018 in the journal *Current Biology*, is a landmark study that confirmed that observations made in the lab are consistent with what happens in the birds' real-life environment.[340] Mennill chose the Savannah sparrow *Passerculus sandwichensis* as his subject of investigation. As its name suggests, it is a species quite similar to Marler's white-crowned sparrow. The song of the Savannah sparrow sounds like this: "zit zit zit ZEEEE zaay." Mennill worked in eastern Canada on Kent Island, a strip of land about 3 kilometers long and 800 meters wide that the sparrows are fond of. Researchers installed 40 loudspeakers on the island, which played songs during the nesting season for six consecutive years; in other words—and this was a brilliant idea—six generations of birds bathed in an artificially modified sound environment. Mennill already knew that a young Savannah sparrow learns its song by mixing what it hears from several adult males. He decided that the loudspeakers would emit two types of songs. One had been recorded from adult males living on Kent Island and thus reproduced the local dialect, and the other came from a remote population of sparrows singing a different dialect. A different foreign dialect was used each year. The bet was risky because laboratory studies had shown that young birds generally learn to sing much better when their tutor is physically present, which is why researchers like Sébastien Derégnaucourt of the University of Paris Nanterre use small, birdlike robots to enhance the impact of the songs produced by the loudspeaker. The probability of Mennill's sparrows copying songs simply emitted by loudspeakers was therefore minimal—especially since the young birds were simultaneously in contact with real adults, which they could both hear and see. A real challenge! At the end of six years of observation, however, the results were clear. Not only had the young birds copied songs from the loudspeakers, but dialects previously unknown to this population were passed on to the next generations. Mennill had created a new sound culture on the island of Kent . . . and won his bet![341]

Now that we know vocal learning is not a laboratory artifact, let's examine the role of social influences. A bird does not learn very well if it is isolated in a cage with only a loudspeaker repeating songs. In this situation, it mostly learns in a somewhat automatic way, without any richness of expression, so to speak—a bit like trying to learn to play the piano with a tutorial on the internet, in the absence of a real music teacher. In a remarkable study, David Mets and Michael Brainard of the University of California, San Francisco, experimentally demonstrated how important the presence of a flesh-and-blood tutor is when learning to sing in the Bengalese finch Lonchura striata domestica.[342]

By comparing different strains of finches with different singing rhythms (some families of Bengalese finches in Japan sing slowly and others sing quickly), Mets and Brainard found that computer-based learning resulted in strong heritability: a bird sings with the same rhythm as its father and grandfather, whatever the rhythm of the song produced by the computer. In other words, genetics ruled. Inversely, when the sparrows were trained by adult birds—real teachers, if you like—each one adopted a new rhythm inspired by that of the teacher. This time, it was not genetics.

Another study shows that the presence of a female, and the way she reacts to the young bird's vocalization attempts, also greatly influences the learning process. This was an experiment in which young zebra finches saw an adult female on a video screen. These young zebra finches more accurately copied the song of the tutor if they saw the adult female react to their song, and they made more copying errors if the female's behavior appeared disconnected to them.[343] When you see that the audience is satisfied, you learn better.[344]

This leads me to point out that songbirds show great diversity in the way songs are learned. For example, the length of time a bird may learn varies from a few weeks (as in the zebra finch), to the first year (as in the common chaffinch, for example), to a lifetime (as in the canary, starling, and many others). Species that only learn early in their lives are called "closed-ended learners." Species that are capable of learning beyond their first year of life—those that do not have a critical period for learning to sing—are known as "open-ended learners."[345] It is among the latter that we find the most virtuosos.

The variation does not stop at the temporal sequence of learning; it also involves the size of the vocal repertoire. A male zebra finch stubbornly sings only one version of the species' song. He produces the same song all his life, the one he learned once and for all, without changing it one iota. In many other birds, the individuals sing several songs. In 80% of these species, this repertoire remains modest (for example, a great tit *Parus major* sings less than five different songs) or moderate (about ten songs). But in some, the repertoire of a single individual can be quite large (more than a hundred different songs in the nightingale

Luscinia megarhynchos) or even absolutely extravagant: the brown thrasher *Toxostoma rufum* from North America can sing more than a thousand songs![346] Well, it had to memorize them. Do I dare speak of an elephant's memory?

Species of birds also differ in the degree of fidelity in imitating their song. If the copy is never perfectly like the original, it is still very successfully replicated by the zebra finch. By rendering almost exactly what it hears, it's safe to say this is a bird that doesn't go in for originality. Depending on the species, the quality of the imitation can be more or less . . . mediocre. Some birds even riff on the theme of the song they hear.

Finally, there are real composers who simply invent their songs and seem to take no notice of what they hear. Moreover, while some only copy the songs of their own species, there are some that are open to diversity and will not hesitate to imitate the songs of another species. This is the case, for example, with the song sparrow *Melospiza melodia* and the swamp sparrow *M. georgiana*, two North American cousins that are very similar. While the song sparrow only imitates the songs of other song sparrows, the swamp sparrow does not think twice about incorporating the songs of its cousin into its repertoire. The highest level of openmindedness is reached in certain species. The brown thrasher is a remarkable case, an outstanding imitator with a fabulous repertoire. With more than a thousand songs, it must really like musical diversity. Some can copy almost any sound from their environment, such as the northern mockingbird *Mimus polyglottos* and the famous common hill myna *Gracula religiosa*[347]—a kind of black starling with a yellow adornment on the back of its neck, which I once heard whistling to perfection the tune of the French national anthem "La Marseillaise." However, the prize probably goes to the superb lyrebird *Menura novaehollandiae*, an Australian species. It is a large bird with a splendid tail whose ability to imitate and incorporate environmental sounds into its vocal repertoire apparently knows no limits. It is estimated that 70%–80% of its vocalizations are imitations of other bird species. But it's not just birds that it imitates. I remember an international bioacoustic conference where one of the presenters played a recording of a lyrebird that faithfully reproduced the sound environment . . . of a building construction site! Hammer noises, saw squeaks, orders issued by the workers, nothing was missing. It was so faithfully reproduced that it sounded like the soundtrack of a movie.[348]

How can we explain the diversity of these learning programs? The question is still unresolved.[349] It has been suggested that one of the interests of birds that build their songs by imitating others could fall under the principle of good neighborliness. The sharing of songs between neighbors, i.e., the fact that individuals with territories close to each other sing in the same way, is a principle common to many species. This makes it easy to differentiate a neighbor from a newcomer. Do you remember? We have already talked about the dear-enemy effect in chapter 3 with the black redstart, the favorite model of our friend Tudor Draganoiu. Birds forming leks—like the piha—also apply this principle of sharing songs between neighbors. It could explain the prevalence of species with small repertoires: it is easier to sing the same thing if the number of songs is not too large. On the other hand, the more songs you learn, the easier it is to adapt to a new neighborhood, for example, when you return from migration. In that case, it's a bit like arriving unexpectedly at a party where people are singing: if your song repertoire is large, you'll be integrated more easily. We should therefore see more variety in the songs of species where individuals regularly change neighbors. This is the hypothesis that Don Kroodsma defends.[350]

Kroodsma is a seasoned maestro in the field of birdsong. Now professor emeritus at the University of Massachusetts, he has spent several decades exploring how and why birds sing. As an extremely rigorous scientist, he is known for his extensive field studies and for writing books on birdsong of major importance to both the scientific community and the general public. The sedge wren, one of Kroodsma's subjects of study, provides the impetus for his hypothesis.[351] Let's take a closer look. Sedge wrens (*Cistothorus stellaris*) living in North America are migratory and seminomadic during the breeding season, so they change neighbors often. When one of these birds is trained to imitate a song, it tends to improvise or even invent. In contrast, its cousin, the marsh wren *Cistothorus palustris*, which is particularly sedentary and never changes neighbors, faithfully copies the proposed model. In the wild, marsh wren neighbors do indeed sing similar songs. To complete the picture, the sedge wrens fortunate enough to live in the tropics are sedentary there, so they always have the same neighbors, and this little world sings the same songs. The constant proximity probably leads them to copy each other.

Let's not forget an important element that has hardly been mentioned so far. The females of many bird species sing, especially those in the tropics. The analysis of the evolution of birds shows that in the ancestral species common to all current oscines, both females and males sang. It was only during the evolutionary history of birds that female song seems to have been lost.[352] In this ancestral species, both sexes may have had a vast repertoire of songs that allowed them to defend their territories more effectively as a pair than alone. Sexual selection is also likely an important driver of this evolution, and many argue that the manner in which a mate is chosen plays a decisive role. Here, a partner capable of producing a wide variety of songs will be chosen; elsewhere in other species, more attention will be paid to the stability and quality of its sound production. Moreover, some researchers have ventured to develop the same kind of hypothesis for the evolution of human language: having a complex language could have been favored either by sexual selection (one prefers a partner mastering a vast repertoire, for various reasons), or by so-called kinship selection, with the sharing of information between members of the community (the more one is capable of exchanging complex information within one's family group, the more everyone's survival increases). There are many other hypotheses attempting to explain the genesis of human language. We talk about them later. So let's be careful—especially since, even if one or the other of these explanations is valid in birds, none of them explains all the observations and the diversity of situations, whether it is the importance of the neighborhood, sexual selection, or information sharing. It must be acknowledged that we do not yet know why in one species individuals learn many songs while in another the vocal repertoire remains limited. In short, it is still a matter of artistic blurring, and there is room for new generations of researchers. All are welcome!

Now, let's go back to the very process of vocal learning—in other words, what happens in the brain when the bird learns its song and produces it.[353] The neurophysiological basis of these mechanisms is beginning to be well understood.[354] In the brains of birds, there are groups of neurons (the cells that transmit and analyze nerve information) that are specialized in learning and song production. These groups of neurons, called "song nuclei," are interconnected by nerve fibers. They have names such as HVC, aire X, lMAN, etc. I will not detail their respective roles here, but you should know that they are the subject of many studies.[355] It's not every day that you find a model to study the brain structures responsible for memory and learning, so researchers seized on the birds. The first nuclei were discovered in the canary by Fernando Nottebohm, a former student of Marler and now a professor at Rockefeller University in New York.[356] This "song system" was then identified in all the oscine passerines that were studied. In species where only males sing, it is present only in males. In those species where females also make themselves heard, song nuclei are also found. Among hummingbirds, it's the same story: a study published in *Nature*, in which Jacques Vielliard took part, reports that hummingbirds have regions in their brains that specialize in vocal control, very similar to those of the oscines.[357] We are probably facing a phenomenon of evolutionary convergence: phylogenetic reconstructions (those studies that make it possible to follow the history of a species) strongly suggest that the common ancestor of oscines and hummingbirds did not learn its songs. Vocal learning would therefore have appeared independently several times in birds. Showing similar brain organizations underscores a structural and functional convergence of brains. This is not uncommon in biological systems: the same problem (learning and producing a song); the same solution (specialized brain nuclei for these functions).[358, 359]

What is interesting and unsettling at the same time is that this convergence also exists between the brains of songbirds and the human brain.[360] Yet our brains are very different in their general organization. Nevertheless, the neural circuit of the songbird brain shows important similarities to the areas that are dedicated to language in the human brain. Like our brain, the bird brain has groups of neurons that control the production of vocalizations stored in the brain's memory, which is called motor control of sound production. You may also know that our brain's production of language is lateralized. To oversimplify, the left hemisphere is more specialized in the production of words and sentences, while the right hemisphere mainly manages intonation. In birds, there is also a dominance of the left hemisphere for the control of vocal production. In any case, it has been found in two well-studied species, the canary and the Bengalese finch. But things are complex. Remember that the syrinx, the sound-producing organ of birds, is a double structure in songbirds—a kind of double whistle, if you like. In the northern cardinal *Cardinalis cardinalis*, the low-pitched part of the notes of the song is produced by the left syrinx, while the high-pitched part is produced by the right syrinx.[361] Let's take a note from the cardinal's song—for example, "piyou!"—which starts high-pitched and ends low-pitched, all in less than a second. Well, "pi" is whistled from the right while "you!" is whistled from the left. Can you imagine the necessary motor coordination? And, of course, this vocal lateralization corresponds to a neural lateralization: each hemisphere controls the half-syrinx on the opposite side. In other species, one of the half-syringes and its hemisphere produce most of the notes of the song; the other side is mute. In the zebra finch, the two hemispheres alternate rapidly, controlling each of the syringes and song production in turn. Hardly simple, that's for sure.

On the hearing side, there are also similarities between birds and humans. In both groups, the processing of sound signals by the brain takes place differently between the right and left hemispheres. Let's skip the details; just remember that the right brain hemisphere of birds seems to mainly process the spectral properties of sound (Is it high-pitched or low-pitched?), while the left hemisphere would rather deal with information encoded in the sound signal, such as the identity of the individual whose song is heard.[362] These results, obtained in a small number of species, such as the canary or the starling, are to be taken with a grain of salt . . . nobody knows if they are valid in other species. However, in the case of humans, we observe similar characteristics: the left hemisphere is more concerned with understanding the meaning of words and sentences, while the right hemisphere deals with context. What I am saying here is, of course, extremely oversimplified, and I will probably be pilloried by my colleagues who are specialists in the neurobiology of language and its perception. The main point is that one can draw both anatomical (brain structure) and functional (process) parallels between the bird brain and the human brain when it comes to studying the production and perception of vocalizations.

In terms of neurons, some amazing things are observed in birds. For example, some neurons activate only in response to certain sounds, certain syllables, or certain combinations of syllables in the song of the species. Other neurons only respond if the bird is listening to the entire song. Frédéric Theunissen has shown that this selectivity appears during the learning process.[363] Birds' brains change when they memorize. In addition, many factors play a role in modulating the activity and development of the song system. Hormones are involved, of course, which in many species will make the number of neurons, and therefore the size of the song nuclei, vary according to the season: more neurons in spring, fewer neurons in winter, for example. Another factor that seems to be essential for song learning is sleep. It is often said that, to memorize something well, you have to sleep on it. It seems that birds follow this rule to the letter. It was in the early 2000s that Ofer Tchernichovski, a professor at Hunter College in New York, and Sébastien Derégnaucourt, then a postdoctoral student in Ofer's laboratory, made this strange discovery—a winning ticket that would earn them a publication in the journal *Nature*.[364]

What had tipped them off was the fact that certain neurons involved in the control of song production in adults were particularly active during the chick's sleep. To be more precise, the neurons of the sleeping chick were as active as the same neurons in an adult who was awake and singing. Would we go so far as to say that the baby bird dreams that it is singing like a grown-up bird? Well, why not? In any case, this result was quite puzzling. Sébastien and Ofer decided to examine in detail how vocal learning progressed during sleep-wake cycles in the young zebra finch. They trained 12 zebra finches to imitate a song from a loudspeaker, and they recorded their vocalizations day after day. One might have expected the chick to progress steadily throughout the learning period. But the whole affair turned out to be more subtle. During the day, as the nestling practiced singing at the top of its voice, its singing became more and more structured; that is to say, it had an increasingly regular and standardized organization of the notes. But every morning, after a good night's sleep, the song produced by the chick had lost much of its organization from the previous day. It's not always best to sleep on it. It is only after two to three hours of morning training that it finds the tune again, and improves on it in the afternoon. And so on every day. But here's the remarkable thing: the birds showing the greatest morning loss of structure were also the ones that, in the end, would best imitate the singing pattern broadcast by the loudspeaker. The morning's vocal doodling that followed sleep was correlated with better learning. It is better to sleep on it! Sébastien and Ofer proposed an interpretation of these results: starting the day with a more variable song could give the bird the opportunity to explore its vocal abilities and improve imitation. Using melatonin, the famous sleep hormone, the researchers put birds to sleep for a few hours during the day. After this forced siesta, the birds had a destructured song. The researchers also found that the singing of young birds that had been made deaf after a few days of learning experienced the same degradation as that caused by sleep. In summary, during the day, the chick trains and improves its performance; in the morning, its highly variable singing allows it to adjust things by realizing how well it is able to control its sound production; at the end of the day, it falls asleep satisfied, having improved its vocalizations; at night, it obviously does not sing and is not likely to hear itself, but it dreams about it! When it wakes up, it is not in top singing form—it does not immediately regain its singing abilities. But its memory has worked well during its sleep. At the end of this new day, by dint of training, it will have perfected its vocal imitation a little more. Sleep helps learning—in birds and humans alike.[365]

Oscine birds are experts at vocal learning. A recent study estimated that the types of syllables sung by the swamp sparrow *Melospiza georgiana* could persist for over 500 years, so efficient is the copying accuracy.[366] The existence of dialects, however, suggests copying errors, and imperfect imitation leads to small changes in the typical song of a population—changes that accumulate, are culturally transmitted from one generation to the next, and may lead to divergence between isolated populations.[367, 368]

An interesting study on this issue has focused on the greenish warbler *Phylloscopus trochiloides*. The range of this small forest bird begins in eastern Europe, extends over western Asia, descends to the Himalayas, skirts the Himalayan massif in the south, rises on its eastern side, and ends in China to come into contact with western Asian populations. I hope you can visualize this kind of loop, circling the Tibetan Plateau, which is too high for forests. It is not for nothing that we speak of this bird as being a ringed complex of species and subspecies.

The song of the greenish warbler gradually changes as you move along the ring. The further you move away from eastern Europe, the more different the song becomes from the typical song of that region. Once the ring is crossed, in the contact zone between the eastern European and western Chinese populations, the dialects are so different that the birds do not recognize each other. Why? Because the colonization of the ring by the warbler populations has taken place gradually from Europe to China via the southern Himalayas. This colonization was accompanied by a progressive variation in song. The acoustic difference in the contact zone is such that the warblers on either side no longer consider themselves to be of the same species: females and males do not pair up.[369]

Copy errors from one generation to the next are probably the main reason for this song drift. Does this mean that nonoscine birds—you know, those that don't learn their song—don't experience any drift in their vocalizations? Of course not. It is genetic drift—the fact that distant populations accumulate small mutations—that causes changes in their vocalizations. Thierry and I took part in a study to compare different populations of the woodcock *Scolopax rusticola*, a very pretty little wader that has left the marshes for forests and moors. We found that, among these nonoscine birds (therefore a priori unable to learn to sing), there were dialects nevertheless: woodcocks from the forests of the Paris region do not sing with the same accent as those from the Azores. It is likely that these populations do not mix and that small genetic differences accumulated over the generations explain their small acoustic differences.

Like human language, the songs of oscine birds are transmitted by imitation from generation to generation, forming true *cultures*. Within

Woodcock

the same bird species, the various dialects that differ between isolated populations are *cultural traditions*, a bit like our different human languages. The parallel with humans does not end there. Just like what happens to human languages, vocal cultures observed in a bird species can disappear. A recent study has shown that an Australian bird, the regent honeyeater *Anthochaera phrygia*, is forgetting the songs of its ancestors.[370] The cause is the decline of regent honeyeater populations and the resulting low density of individuals. Until the middle of the twentieth century, hundreds of honeyeaters used to travel hundreds of kilometers during migrations across all of southeastern Australia. The destruction of their habitat means that there are now only between 200 and 400 individuals left, spread over an area of 300,000 square kilometers! In the regent honeyeater, the young do not learn the songs of their fathers because the latter do not sing during the breeding period. In order to learn the typical songs of the species, the newly emancipated young must join other adult males. And with only about 10 conspecifics in every 10 square kilometers, it is not easy to find company. A recent study of the songs of 146 male regent honeyeaters showed that 18 of them were no longer singing the regent honeyeater song. Instead, they sang the melodies of other bird species, such as those of the little waterbird *Anthochaera chrysoptera*, the noisy friarbird *Philemon corniculatus*, and several others. Deprived of listening to their fellow birds, the honeyeaters learn the language of other bird species—a bit like the character Tarzan who, raised in the fictional *Mangani* ape tribe, had learned their language.[371] As for the other recorded honeyeaters, if they sang the honeyeater melody, it was simplified compared to its original version. In the absence of a sufficient number of tutors, the singing lessons had lost their effectiveness.

Birds are a wonderful model for the study of understanding human speech. But don't make me say what I did not say: birdsong is not the equivalent in every way of human language. Song is a much simpler communication signal than our language. In certain species of birds, songs are certainly complex and show a syntactic organization, appearing as combinations of short elements (notes that combine into syllables, which themselves combine to form sentences whose combination forms songs). However, bird vocalizations show little compositional syntax, i.e., they are not combinations of elementary structures that have meaning (words) that themselves generate meaning (sentences). However, as we will see in chapter 14, birds sometimes combine calls to generate meaning, particularly to designate the identity or size of a predator.

Do you remember the birds that parasitize other species' nests by depositing their eggs in them? Some of these birds are oscine passerines, and the young are supposed to learn to sing by imitating adults. How do they do this when they have been raised by adoptive parents of another species? By studying the brown-headed cowbird *Molothrus ater*, Mark Hauber, a professor at the University of Illinois, showed that these birds have an incredible solution to this problem: the young birds rely on a password to choose who to imitate. As soon as they leave the nest, the young birds recognize the adult cowbirds, without any prior learning, when the latter emit a call (the *chatter*).[372] Then they learn by imitating the entire vocal repertoire.[373]

I can hear you from here: "That's enough for the birds! What about the other animals? The bats? Whales? Dogs? Monkeys? Bugs? What about the others?" It must be said that vocal learning seems to be rare in the animal kingdom.[374] Apart from the categories of birds that I mentioned, learning would seem to be strictly reserved for a few mammals: cetaceans (dolphins and whales), pinnipeds, elephants, bats, and humans.[375] It *would* seem so, considering that for many of these groups the data is still fragile. As a recent paper points out, deciding whether an animal species belongs to the "very select club" of vocal learners is still often a matter of debate.[376] Probably because being capable of vocal learning is a matter of degree: if some species or group demonstrates it without any ambiguity, this capacity is more or less developed in others. Let's take a look at this.[377]

The animal for which we have the most certainty is the humpback whale, which is already familiar to you. You already know that this animal, which is part of the baleen whale group, produces absolutely fabulous, very complex songs. If you've never heard them, listen to *Songs of the Humpback Whale* on the internet. These are the first recordings of humpback whales made by Roger Payne in the 1960s.[378] The humpback whale is a cosmopolitan animal, present in all the oceans of the planet. While the males of a given population all share the same song, each humpback whale population sings in a slightly different way. Moreover, the song of a population is not unalterable; it tends to change. It is a cultural transmission.[379] The whales copy each other; they imitate the songs produced by other whale populations.[380] One population will switch to another's dialect in just a few years' time. How is this possible? It's not that complicated. Whales make colossal migrations, over thousands and thousands of kilometers, which gives them the opportunity to meet individuals from different populations and to borrow some vocalizations that are particularly pleasant to their ears. Also, remember that sound travels well in the water. Even without moving, whales can hear one another singing several hundred kilometers away. It's easy, then, to change their tune![381]

Even though it is one of the most popular and best-studied whales, the humpback whale is not the only one to practice vocal learning. The bowhead whale, to name but one, also produces complex songs that vary from year to year. This species, like many other baleen whales, is most likely a vocal learner. On the other hand, if we look at other cetaceans, such as dolphins, things are a little less clear. However, several observations and experiments have shown that dolphins learn to imitate whistles—artificially produced by computer or emitted by other dolphins—that they did not produce before.[382] There is also the beluga, the famous white whale, which more or less succeeds in imitating human sounds.[383] But we still lack studies that offer results that I believe to be wholly convincing.

However, it is quite likely that dolphins and other toothed cetaceans are capable of learning to produce sounds that are new to them.[384, 385]

Now let's take the pinnipeds and the story of a seal known as Hoover.[386, 387] Having lost his mother, Hoover had been raised from a very young age as a pet, without contact with other seals. When Hoover came in contact with humans, he had transformed his seal bark into humanoid expressions, such as "Hello!" or "Come over here!" While Hoover himself showed some serious vocal plasticity by modulating his vocalizations to "talk," from my point of view he didn't really learn how to vocalize. Instead, this was simply a matter of distorting sounds that the seal was already producing instinctively. You will agree with me, I hope, that in this case we are quite far from the songbird that must imperatively hear an adult to become able to sing—and just as far from the human child who learns to produce thousands of words. In fact, experimental evidence that would demonstrate vocal learning in pinnipeds remains limited. To my knowledge, only one study, conducted on gray seals (*Halichoerus grypus*), has shown that these animals can learn to change the intonation of their voices, to copy human vowels, and even to imitate sequences of sounds of different frequencies.[388] But again, while there is no doubt about vocal plasticity in these animals, vocal learning seems to remain very limited.[389]

In elephants, the situation is somewhat comparable.[390] Since these creatures have brains as big as a large watermelon (5 kilograms, compared to the kilogram and a half we carry with us) and a very complex social life, it is reasonable to think that they have some ability to learn their vocalizations by imitation.[391] But, you see, there is poor evidence of this. Only three anecdotes have been reported, all concerning individuals in captivity. The first was a 10-year-old female African elephant mimicking the sound of a truck. The second was a 23-year-old bull of the same species who had spent his life in the company of Asian elephants—the only other surviving species of elephants—and who reproduced their trumpeting, which was much higher-pitched than that of African elephants. Our bull had taken on the Asian accent, so to speak.[392] The third was a male Asian elephant named Koshik, born in captivity in a Korean zoo in 1990. Koshik had spent the first five years of his life in the presence of two adult females. Then he found himself alone, without an elephant nearby. His keepers and the public who visited him were the only living beings around him. He had been trained to obey several vocal commands, such as "Sit!" and "Down!" When Koshik reached the age of 14, his caretakers noticed an astonishing thing: the elephant was talking! To be exact, he was producing five or six "words." When bioacoustics researchers recorded the elephant, they showed that Koshik mimicked human vocalizations very well. The researchers played the recordings to people who were unfamiliar with the elephant and their work, and the recorded words were perfectly understandable to the listeners. What did he say? Oh, of course, nothing very complex. Koshik simply repeated the injunctions of his caretakers: "hello," "sit," "no," "down," "good." To produce the words, he used the resources available to him: he bent his trunk back and pressed the end to his mouth.[393] The story does not say whether the elephant selected the appropriate word for the situation, but it is likely that by imitating the only people around him, he was trying to establish the social contacts without which every elephant gets depressed. Hoover the seal may have been a similar case. Parrots do the same, as they only learn to imitate us if they are deprived of parrot company. As we say in France, "If there are no thrushes, we will eat blackbirds"; we make do with what we have.

Among bats, the picture is not very clear either.[394] Most of these animals are very social and very talkative. In the Egyptian fruit bat *Rousettus aegyptiacus*, a large bat that you may have seen in a zoo, the vocalizations carry a variety of information: the identity of the sender, the context of the call (aggression, distress, etc.), and even who the vocalization is addressed to. Young fruit bats raised alone with their mothers develop their calls later than when they are in a social group with many individuals.[395] If they listen to calls of their species that have been artificially altered in frequency, they modify their own calls by trying to imitate the artificial sounds. And if you raise young fruit bats in the absence of adults, they take a very long time to develop the vocal repertoire typical of their species. Exposure to playback of adult calls makes their task much easier.[396] Besides, we know that the pale spear-nosed bat *Phyllostomus discolor* requires auditory feedback for normal vocal development and that the baby of the bat species *Saccopteryx bilineata* babbles like human babies.[397, 398] Different dialects between populations of bats of the same species have also been identified. It is not known whether these variations are imposed by genetic differences or whether they emerge from a cultural drift, as in songbirds.[399] Perhaps we are on the eve of great discoveries about bats. Indeed, only 2% of studies on vocal learning concern bats. No one has yet tested whether they are capable of mimicking truly unfamiliar sounds. A while ago, I was visiting my former student Julie Elie, an expert on the calls of the zebra finch, in her new lab at the University of California, Berkeley. Julie is now working with bats and is investigating whether they are capable of vocal learning. During my visit, I was impressed by the experimental device she had set up. She records the individual vocalizations of bats living in groups while filming the animals. She knows who is who, who talks to whom, who talks before or after whom, etc. Her expertise in vocal repertoire analysis makes me feel that, if there's anything interesting to find out about bat vocal learning, she'll find it. Other amazing scientists are also working on the subject.[400] When the results of all of their research converge, we will be able to measure the extent of bats' ability to learn to produce a vocal signal . . . or not. Research is also about confirming that a hypothesis is not validated, that nothing can be found.

Lately, it is the naked mole rat *Heterocephalus glaber* that has been in the spotlight. This small hairless rodent, living in a colony underground, was featured in the magazine *Science* because researchers discovered it has vocal dialects that individuals copy from each other. At last, had we found a mammal that ensures a true cultural transmission of vocalizations like the birds? When we look at the details, the reality is a little less exciting. Mole rats are eusocial mammals, which means that their groups are dominated by a queen and a king, the only ones to reproduce, while the other individuals cooperate in the service of their majesties. These animals are very talkative, constantly peeping, chirruping, and grunting. It must be said that, in the darkness of their galleries, communicating by acoustics is rather a good idea. No less than 17 types of vocalizations have been identified in this animal. The most common is referred to as the soft chirp, which is emitted when two individuals meet at the bend of a meander. It is the mole rat's "hello" in a way, letting the animals know who they are dealing with. In the study, scientists recorded 36,190 soft chirps from 166 animals, belonging to seven different colonies. The analyses showed that it is possible to guess the identity of the colony of an individual mole rat from its calls alone, suggesting that mole rat families speak with the same voice. In addition, playback experiments showed that individuals respond with more calls when they hear a soft chirp produced by a member of their own colony than when they hear a call from an individual from another colony; in other words, mole rats are more likely to say hello to family members than to strangers. Scientists then found that young individuals placed in a colony other than the one where they were born adopt the accent of their new family. This final discovery was made by chance. During the course of the study, one of the colonies observed by the researchers lost its queen twice. In the ensuing periods of anarchy, the individuals in the colony lost their common accent—everyone started to call in their own way. In conclusion, the mole rat does not seem to do better than other mammals: it modifies its vocalizations according to the ambience of the colony, by copying its mates—just to please the queen, moreover. This social conformism provides us with a good example of vocal plasticity, but here again, we are far from the vocal learning capacities observed in birds.[401]

So while most animals don't really learn to talk, many show great vocal plasticity.[402] In fact, it is quite possible that the number of species in this category is vastly underestimated, especially in mammals. David Reby, who knows a lot about the mechanisms of vocal production and who is also a very observational person—an essential quality for an ethologist—recently played me some vocalizations of his own dog. He had recorded the dog's lively barking at the television in response to TV series theme songs—proof that dogs can be distracted, like restless children, by putting them in front of a screen. "Listen carefully," he said. "This is funny. My dog barks along with the melodic line of the theme music!" And it was true. It was quite perceptible to the ear. Computer analysis of the acoustic structure of the music and his dog's barking confirmed David's first impression: a dog singing the theme tune of his favorite show. Try it with yours, to see.

And what about monkeys and apes, you may ask? Nonhuman primates are very close to us, aren't they? Our common ancestor with apes is some six million years old, which is not much by geological standards. On top of that, nonhuman primates are physically similar to us—comparatively speaking, of course. Almost all of them vocalize, and often they have complicated social lives that require complex exchanges of information between individuals. In short, they have everything to make their language resemble ours, don't they? Well, as surprising as it may seem, and according to the current state of our knowledge, we have to admit the reality that we are the only primates alive today who learn to speak by copying individuals who have already mastered language. Human supremacy at last. The gibbon shows no willingness to learn its screams by imitation.[403] Young squirrel monkeys raised in isolation produce the same calls as if they were raised in the presence of other monkeys. Young macaques raised by monkeys of another species do not learn the calls of their adoptive parents. Finally, superhuman efforts were made during several decades of the twentieth century to try to teach chimpanzees—and even an orangutan—to talk. Some of the people involved in this work reported that "their" monkey had learned to master words.[404] For example, a chimpanzee named Vicki is said to have uttered "daddy" and "mug." I have listened to one of her recordings, and let's just say . . . her pronunciation leaves something to be desired. In short, the results were very disappointing.

However, we must nuance things. A human being raised by elephants or orangutans might try to imitate them, but chances are that this would also be without much success. Wait, I can hear you from here taking me to task and quoting the famous "wild children" and "wolf children" who have been the subject of much discussion and have made great movie subjects. About 50 cases of children "raised" by wolves or bears have been recorded over the centuries. None of them spoke, some were characterized as barking, others growling. All of them reportedly had enormous difficulty in acquiring human language. A little bit of anything and everything has been said about them, to support theories about the innate or acquired nature of human language. In any case, the information has always been insufficient to draw the slightest scientifically valid conclusion. We do not know, for example, at what age they were abandoned or how long they lived with animals. What can be said, beyond fantasy, is that for a human to acquire human language, contact with human speech is necessary from a very young age and in a constant manner.

I don't want to leave you with the impression that monkeys are less capable than birds or whales when it comes to their sound communications. First of all, the vocal repertoire of many primates is very complex, using combinations of vocalizations whose secrets we are just beginning to unravel (and I'll come back to that soon). Their vocalizations are not just set-in-stone genetics. In recent years, evidence has been accumulating that certain species of monkeys modify their vocalizations through learning. In particular, studies of marmosets in captivity have shown the importance of hearing adult vocalizations in order for young animals to acquire a normal vocal repertoire.[405]

One of the first in-depth experimental studies on this vocal plasticity was conducted by two field researchers: Marie Charpentier, director of research at the CNRS in Montpellier, France, and my colleague Florence Levréro. Energetic and determined, Florence spent many years in various African countries studying the lives of great apes, gorillas, chimpanzees, and bonobos. Equally energetic and determined, Marie is leading exciting research on the biology of the mandrill *Mandrillus sphinx*, a sturdy African monkey, whose males proudly display an impressive red and blue snout complemented by a short, yellowish beard. Marie invited Florence to set up an experiment on mandrill acoustic communication. I didn't take part in it—we can't do everything—but I regret it a bit because it must be something to work with this animal. The two researchers and their team were about to make a major discovery. The voice of the mandrills is certainly partly genetically determined, since closely related monkeys have more similar voices than individuals who are distant kin. But it is also shaped by experience: the mandrills take on the accent of the social group in which they live. It was the first time that such vocal plasticity based on the imitation of social peers was demonstrated in a monkey. Furthermore, playback experiments showed that mandrills can recognize by voice related individuals they have never seen before: even if the accent is not the same, they can still recognize their family's song. It's the genetic background that speaks. These intriguing results earned Florence and Marie the publication of their work in the journal *Nature Communications* (which is no easy task).[406] Other similar examples exist in monkeys and apes.[407] With Florence and our doctoral student Sumir

Bonobo

Keenan, we went on to show a few years later that cultural influence also permeates the voice of bonobos, the closest cousin of the human species. Besides, a study carried out on several natural sites in Borneo and Sumatra shows that orangutan populations develop "vocal personalities" that depend on the number of individuals: vocal repertoires are more stable and more complex in populations with a low density of individuals.[408] Other observations, carried out in other monkeys, suggest that the dual influence of genetics and learning on vocal signals is found in many, if not all, primates. The vocalizations of monkeys and apes should no longer be seen as innate and totally fixed.[409, 410]

13

Inaudible speech

ULTRASOUNDS, INFRASOUNDS, AND VIBRATIONS

Goegap Nature Reserve, Northern Cape Province, South Africa. The sun is vanishing behind the rocky mountain. "Follow me!" Céline says, "we're going to set the traps." We walk quickly across the small, dry plain dotted with bushes, placing small rectangular cages here and there. A few minutes later, a sudden slam: The first trap closes on an unfortunate little mouse, too greedy, attracted by the smell of the bait. Céline then opens the trap, delicately takes the little animal between her skillful fingers, notes the number on the tag attached to its ear, and releases it. "This one is a regular. I've been catching her for the past two years." Farther on, a student holds up a large antenna, trying to receive the radio signal sent by the transmitter attached a few days earlier to the neck of another mouse, to locate where it is now. And here's one of them running through the bushes! The African striped mice are scurrying about, trying to find a bite to eat. Sometimes two of them argue over a few seeds. I can hear some sporadic, very high-pitched calls. Most of their conversation is ultrasonic and out of my hearing range. The silence of the karoo weighs on the whole space. The celestial vault is a festival of stars. It's getting seriously cold. A few meters away, a large oryx antelope with swordlike horns calmly grazes on the steppe grass. I feel strangely at home. This is just the place to come to work, I think to myself.

A few days before, I had received an enthusiastic email: "I am Céline, postdoc at the Succulent Karoo Research Station in Goegap Nature

Reserve. If I remember correctly, you'll be at the reserve next week. If you want to see what we are doing in the field, at the moment, we are conducting a mouse trapping session from 8:10 in the morning to 5:00 in the evening. You can come with us if you wish. Our field is right in front of the research station. Or you can visit us at another time during the day if you prefer. It would be super nice to put a microphone near a mouse's nest; if possible we should do this at sunset (17h 45). Hope to see you there!"[411]

I was on a family holiday, and we were on our way across southern Africa from Cape Town, at the southern tip of the continent, to Kasane, in northeastern Botswana—several thousand kilometers of travel—where we would experience the Kalahari nights with the roaring of lions and being awakened by the trumpeting of an elephant bursting into the quiet darkness of our camp. On the itinerary, I had planned to sleep two nights at the Goegap Nature Reserve. Intrigued by the place, I had previously contacted Carsten Schradin, director of research at the CNRS, who runs a field station there to study the life of small mammals in extreme conditions. It rarely rains in Goegap, and only in the summer months. The winter is terribly dry. Carsten is trying to understand how the striped mouse's physiology and behavior enable it to adapt to this desert environment. His research has shown that a drop in corticosterone hormone levels allows the mice to reduce their energy expenditure during the dry season, when food resources are reduced.[412] Energy savings everywhere, even in mice. I asked him if he had ever been interested in the acoustic communications of his little animals. "Apart from making a few recordings, we've never studied their vocalizations," he replied. "Come and see us! See for yourself if there is anything you can do." As usual, guided by my curiosity, I responded favorably to his invitation.

According to Carsten, the striped mouse *Rhabdomys pumilio* has one of the most complex and interesting social systems described in rodents. In many ways, this behavior resembles that of some monkeys. Consider this: these mice live in groups of up to 30 individuals of both sexes sharing a nest in a bush. A real community. At night, they sleep together. In the early morning they separate, each one going off to find food for itself.

Striped mouse

At the end of the day, they return to the nest and seem to enjoy each other's company, saying good night and sniffing each other. Mice from two different nests do not like each other very much: when they meet by chance during the day, they chase each other away. Males are aggressive toward foreign males, while females are aggressive toward both sexes. If an outside mouse approaches the nest, it is immediately attacked.

At the Goegap Nature Reserve, a mouse family consists of several males and females living together in the same nest. With the use of cameras, Carsten has seen that the males actively participate in the care of the young. They keep them warm and maintain and care for their coats. So do the other members of the family. Raising children here is community based. In other regions of Africa where the same species of striped mouse is present, the social structure can be quite different: a monogamous nuclear family (only one female and one male in a pair), a polygynous harem (only one male, several females) . . . all cases exist.[413] The striped mouse is a very good example of social flexibility. In this respect, it is very much like us—families can be nuclear or extended, with many variations in between. Our social organizations can also change over time. None of our other primate cousins show such flexibility. Although the common ancestor we share with the striped mouse is older than the one we have with nonhuman primates, this small rodent represents an excellent animal model for understanding the factors that modulate social organization.

I didn't record anything on my first visit to Goegap because I didn't bring the necessary equipment. It was only a few months later, in our ENES Bioacoustics Research Lab in Saint-Etienne, that the first vocalizations were recorded, using striped mice that Carsten had entrusted to us. Nicolas Boyer, our technical manager, always ready to take up the challenge, and Aurélie Pradeau, the animal keeper, had set up a five-star hotel for the striped mice. My colleague Florence Levréro immediately had an interest in studying this species. As a specialist in acoustic communication in great apes, she was more aware than anyone of the challenge of this study: she had just returned from a field mission in the Democratic Republic of Congo that she had organized to work with bonobos. The mission had been difficult: bonobos are constantly on the move and hide very easily in the virgin forest. Recording them and especially recognizing which individual produces the vocalizations is a real challenge. So you can understand that studying an animal with a complex social system while freeing yourself from the immense difficulties encountered with the great apes was a fantastic opportunity. We decided to tackle the striped mouse adventure together.

The main problem was technical: mice produce ultrasound, and we had neither the equipment nor the know-how to deal with this constraint. Fortunately, Michael Greenfield had joined ENES a few months earlier as a research associate. He had been a professor in France and in the US and had spent a good part of his career working on species communicating by ultrasound, particularly insects. He was the right man for the job and enthusiastically agreed to participate in the project. I am telling you this to show you that science is almost always a collective adventure. Find people who are passionate, competent, and driven by great curiosity, and you have the ingredients; then put them together. That's the recipe.

What do mouse vocalizations sound like? There are two main categories, similar to those we find in birds: calls and songs. The calls are short, emitted when the animal is startled, for example. The songs are a series of very fast, very high-pitched notes (I remind you that mice mainly produce ultrasounds, whose frequencies exceed those perceived by our hearing system): "Twup-twid-twup-twidup. . . ." Different notes almost always follow one another, in a kind of continuous babbling.[414] This may disappoint you, but I must confess that, for the moment, we still have no clear idea what striped mice are saying to each other. Deciphering their language is one of my goals for the next few years. The next mission is already planned: I'm leaving soon for Goegap with Florence, Michael, and our new PhD student Leo Perrier to put ultrasound recorders on the mice domain. Our first hope is to map their communications throughout the day, with the hypothesis that the nests and their immediate surroundings are privileged spaces for exchanging information about food-rich places. You may find it strange how little we know about the acoustic communications of a rodent that is quite common in the wild. I see two explanations. First, since we humans do not hear ultrasound, we are not very interested in it. Second, they are more complicated to study than audible sounds. It was not until the late 1930s that Donald Griffin and Robert Galambos discovered that bats emit ultrasound.[415]

Why use ultrasound to communicate? Scientists currently agree on several reasons: to allow precise localization of the origin of a sound, to avoid being heard by predators, to avoid ambient noise (as in the case of ultrasonic frogs living near streams, which we talk about later), and because the transmission of high-frequency sounds is more effective for small animals.

To locate the source of a sound (e.g., a baby mouse who has strayed from its nest), whether it's an ultrasound or a sound audible to our ears, the receiving individual (the mother mouse) compares the sound information coming from each of its ears. If the baby is on the mother's left, she will hear its calls louder with her left ear than with her right ear. The mother's head acts as a filter that absorbs some of the sound waves. Conversely, if the baby is on the right, the mother's right ear will receive more sound energy. And, of course, if the baby is in front of the mother, both ears will receive sound waves of the same intensity. This *interaural difference in intensity* between the two ears provides valuable information about the location of the sender. These differences in intensity are particularly noticeable if the wavelength of the sound vibrations is small compared to the size of the animal's head. In other words, for a small animal such as a mouse, very high-pitched sounds (of very short wavelength) will be useful for perceiving interaural differences in intensity. Ultrasound, therefore, is perfect here.

Sound intensity is not the only parameter that differs between the two ears if the sender is to one side. Indeed, if the sound wave comes from the left ear, it will arrive earlier in the left ear than in the right ear. The time difference between the two ears is simple to calculate. Remember what I told you in chapter 2: sound waves travel at 340 meters per second. Since the two ears of a human head are 22 centimeters apart on average (or 0.22 meters), a sound wave coming from the side that hits the first ear will hit the other ear $0.22/340 = 0.0006$ seconds later. This *interaural difference in time* of a little less than one-thousandth of a second is certainly small, but it is enough to induce a shift between sound waves. This shift is a second cue analyzed by the brain to locate the position of a sound source. Remember this: a sound is a pressure wave that travels. Air pressure rises, then falls, then rises again very quickly. The rhythm of these pressure oscillations is the frequency of the sound. The faster they are, the higher the frequency and the higher the sound. If you have understood correctly, you will realize that the shift between the two ears will be easier to perceive if the frequency is lower. In fact, imagine a sound wave coming from the left. It hits the left ear first and then reaches the right. If the oscillations of the wave are fast (high-pitched sound), in other words, if the pressure changes very quickly, the left and right ears will have difficulty feeling pressure differences. On the other hand, if the oscillations are slow (low-pitched sound), the left ear will feel an increase in pressure before the right ear, and when this increase reaches the right ear, the left ear will already feel a decrease in pressure. And so on. At any time, the difference in pressure between the two ears will be more noticeable with a low sound than with a high sound. So, in mice? Well, the explanation is simple: mice have very small heads. Moreover, the smaller the head of the animal, the smaller the time difference will be. In a mouse whose two ears are only 9 millimeters apart, the time difference is $0.009/340 \sim 0.00000002$ seconds for a sound coming from the side. Therefore, interaural differences in time are not a very useful parameter for mice. It is better for them to rely on interaural differences in intensity.

This is the first reason for using ultrasound: to locate easily and to be easily locatable. The mother mouse will thus find her straying baby very quickly. Another property of ultrasounds, as I mentioned earlier, is that their energy is progressively absorbed as they are propagated in the environment. The amount of energy loss depends on the frequency of the sound; high-pitched sounds lose energy very quickly. This makes them much harder to propagate than low-pitched sounds. In other words, they are attenuated much faster and become inaudible more quickly. Ultrasound is by definition a hyper high-frequency sound, so it does not propagate very far into the environment. To give you an idea, calls from baby mice are no longer recordable beyond a few tens of centimeters. The ultrasonic signals emitted by the babies therefore will not be perceived by a possible predator, unless it has already discovered the nest. This is reinforced by the fact that rodent predators, such as birds and snakes, are not able to hear in the ultrasonic range. Even when the predator is able to hear ultrasounds—which is the case with some carnivora, such as foxes—the signal will only be perceived by listeners close to the youngster calling out: the mother and sometimes other individuals taking care of the young, such as the father. Using ultra-sound to send private messages—that's pretty smart. This may explain the use of ultrasound by many rodents.

We do not yet know what the African striped mice say to each other, but we do have some ideas. Our knowledge of rodent ultrasound communication is becoming substantial, although over the past 20 years, research on rodent ultrasound communication has focused mainly on captive species: white mice and laboratory rats.[416] Let's take the latter first. In the rat, there are three main types of vocalizations: calls emitted by the young when they are isolated (around 40 kHz) and to which the females respond by picking up the young; calls emitted by adults in contexts of fear—in the presence of a predator, for example—or aggression (around 22 kHz); and attraction calls in the context of play in the young or interactions between sexual partners in adults (with a very variable frequency, between 30 and 90 kHz, centered around 50 kHz).[417] In the white mouse, things are very similar. You also find calls of isolation in the young, calls of fright, and other calls made in contexts of social interaction.[418] The males emit long sequences of calls, all different. By analogy with birds, we even talk about the "song" of the mouse! Moreover, for a few years, scientists thought they had found in the white mouse a model for language learning that would have advantageously replaced songbirds: mice are mammals, therefore closer to the human species than birds, and they are much easier to breed in the laboratory. However, despite numerous attempts and experiments, it seems that mice vocalize innately and show no particular plasticity in their calls.[419] Mouse song is therefore not a substitute for birdsong in the study of language learning. But that's not the point. The ingenuity of scientists knows no bounds. If mouse vocalizations are genetically determined, then they can be used as a phenotypic marker of genetics, just like any visible morphological or anatomical characteristic, such as eye color, for example. Vocalizations are in fact a marker of embryonic develop-ment, motor skills, nervous control of muscles, and much more. The study of ultrasound vocalizations in mice thus makes it possible to study cerebral processes and their regulation, emotional states, motivation, and many pathological processes.[420] Did you know that some mice showing signs of autistic or schizophrenic syndromes sing differently from the others? With all the genetic tools available for modifying labo-ratory mice, it is now possible to unravel the processes leading to these syndromes. I won't take you any further on this subject, but you should know that the bioacoustics of laboratory mice probably has a bright future. As for our striped mice, I bet they will reveal some exciting secrets. Maybe we'll be able to understand the meaning of their song. Maybe they will become a model for studying complex acoustic communica-tion networks. It's quite exciting to think that no one has yet been able to understand what they're saying to each other.

In the meantime, what else can I tell you about ultrasonic vocaliza-tions? First of all, rodents aren't the only animals that produce them. Many insects, some frogs, some bats, as well as cetaceans (dolphins and other toothed whales) are also known to produce and detect ultra-sound. While the use of ultrasound by some animals no longer surprises us—after all, ultrasound is simply too high-pitched for our ears—it was only in the twentieth century that it was identified. In the early 1930s, George Pierce, a physics professor at Harvard University who had built an "ultrasound detector," was the first to show that some grasshoppers in the Tettigoniidae family produce ultrasounds.[421] Then, in 1938, Donald Griffin, a Harvard student who was interested in the spatial orientation of birds and bats, brought a few bats into Pierce's laboratory. Using the famous detector, he found that bats emitted ultrasounds when they flew. He then showed that these bats orient themselves in the environment and hunt their prey by relying on the echo of their ultrasonic emissions.[422] Since then, research on bat echolocation has proliferated.[423] I want to be clear that not all bats practice echolocation. But the Microchiroptera, which represents 90% of bat species, uses echolocation, along with fruit bats of the genus *Rousettus*. The other fruit bats (Megachiroptera) are not known for this. The frequency of the echolocation calls varies between 8 and 215 kHz according to the bat species. We are in the ultra of ultrasound at the high end here.

How does echolocation work?[424] The principle is not very complicated. The bat emits a series of calls, each of which is of short duration: "Click-click-click-click-click. . . ." The rate at which clicks are emitted varies greatly from one species to another—from 3 to 200 clicks per second.[425] The rhythm also varies according to the moment: when the bat flies quietly in search of prey, it emits between 3 and 15 clicks per second. Echolocation is a mechanism that primarily allows the bat to navigate *by hearing* in the dark. Very useful for avoiding trees and other obstacles.[426]

When the prey is spotted, the bat approaches the insect. In its terminal phase, when it is about to capture its prey, it considerably accelerates the rate of clicks. Each of these clicks—the term used to designate them is *pulse*—spreads in front of it. If a pulse encounters an obstacle (for example, a tree branch or an insect in flight), some of the sound energy is reflected and an echo is sent back in the opposite direction, reaching the bat's ears. Since the sound propagates at about 340 meters per second, the time between the moment the bat calls and the moment it hears the echo is very short. Imagine that the obstacle reflecting the pulse is 1 meter away from the bat; the sound will travel that distance in 1/340 of a second, or about 3 milliseconds. If we add another 3 milliseconds, corresponding to the time it takes for the echo to return to the bat, we obtain a time delay of 6 milliseconds between the transmission of the call and the reception of the echo by the animal's ears. If you have followed this, you will have realized that by listening to the echo of its call, the bat accesses important information: the distance that separates it from the reflecting obstacle. Numerous experiments have shown that bats are able to gauge extremely short delays between the moment of emission of the pulse and the reception of its echo, and therefore to estimate very close distances. Part of the bat brain is even dedicated to analyzing these *pulse-echo* delays.[427] This auditory brain area is spatially organized, from one end where neurons are found reacting to pulse-echo delays of around 18 milliseconds, to the other end where neurons are specialized in analyzing delays of around 1 millisecond. Those cerebral neurons that perceive pulse-echo delays calculate the distance between the bat and the obstacles around it. The activation of the brain area containing these neurons represents a kind of map of the environment. But the distance to the reflecting obstacle is not the only information the bat extracts from the echo it receives. Experiments conducted in the laboratory on bats trained to distinguish between two reflective targets have shown that their resolving power is extremely high—they are able to differentiate between two objects placed side by side with a time spacing of 2 microseconds (0.000002 seconds).[428] That calculates as $0.000002 \times 340/2 = 0.00034$ meters, i.e., they can differentiate if the objects are separated by a distance of less than half a millimeter. This extreme ability allows the bat to analyze the texture of the reflecting object: the scales on the wings of a moth will not reflect sound waves in the same way as the hairs on a fly—useful information if moths are its favorite dessert. The bat extracts much more information from the echoes it receives, including the size of the reflecting object: the bigger the object, the more energy it will reflect back. This allows it to choose larger prey. Another piece of information is the position in space of the reflecting object, which the bat can determine by comparing the echoes coming to each of its ears. It does this by using the two processes described above: the interaural difference in intensity and the interaural difference in time. We are not going to go much further in the study of bats: it is a complex world that goes beyond our present objective. But I can't resist the temptation to explain to you a little marvel of behavioral adaptation, which I regularly present to my students—always to their amazement. It's a phenomenon known as *Doppler effect compensation*.

You're familiar with the Doppler effect, perhaps without knowing the name. In any case, you've already experienced it. When you are walking around and a car is passing by and honking its horn continuously, or when an ambulance passes you with a screaming siren, you may have noticed that the sound of the horn or siren changes as the vehicle approaches or moves away from you. As the vehicle approaches, the sound seems higher-pitched, and it becomes deeper once it passes you. Yet it is the same horn. When you honk the horn continuously in your car, the sound doesn't get higher or lower over time. It's the fact that the car is approaching and moving away that changes the game. This is the Doppler effect. How do we explain it? Imagine the sound waves of the horn as circles gradually moving away from the car in all directions. If the car is stationary, these circles remain evenly spaced: the wavelength and therefore the pitch (low or high) of the sound remains the same in all directions. On the other hand, if the car moves forward, the sound waves at the front of the car are compressed: the spacing between two waves (the wavelength) is reduced, so the frequency increases and the sound becomes higher. At the same time, the spacing at the rear of the car increases, so the frequency decreases and the sound becomes lower. In short, an approaching sound source will sound higher than a sound source moving away. Let's get back to our bats and look at the next lab experiment. A bat is in the dark, resting on a perch. In front of it is a stationary pendulum—a ball hanging from a string. The animal's calls are recorded ("click-click-click . . .") at a rate of about 15 per second. They are emitted at a constant frequency of 61 kHz, characteristic of this bat species. Then the pendulum is made to oscillate, moving closer, then further away, then closer to the animal again. Surprise! The frequency of the calls then starts to vary with the movements of the pendulum: when the pendulum accelerates as it approaches, the bat calls become more and more low-pitched—down to about 59.5 kHz; then its calls gradually return to their basic frequency (61 kHz) until the pendulum is at its closest. None of this happens when the pendulum moves away: the frequency remains unchanged. It only starts to drop again when the pendulum approaches again. Why this unusual behavior? Because of the Doppler effect. The pendulum is an obstacle that reflects the echo of the calls made by the bat. An echo is always of the same frequency as the original sound. So when the pendulum is stationary, the echoes are at 61 kHz. But when the pendulum approaches, it becomes a moving sound source—like the car coming toward us—and the echoes become higher-pitched. The bat hears it immediately and lowers the pitch of its calls. Its goal: to keep the echo at 61 kHz. Neurophysiological studies have shown that these bats have heightened auditory sensitivity at this frequency. By forcing the echo to remain fixed at this value, the bat's hearing is optimized.[429] As the pendulum accelerates and then decelerates, the bat lowers and then raises the pitch of its calls following the movement to neutralize the Doppler effect. Doppler effect compensation therefore requires hearing the echo from obstacles and correcting the calls produced almost in real time. Let's put ourselves in a real bat's life: When it chases a moth, it must constantly adjust the pitch of its calls according to its relative speed—depending on how fast it is approaching the prey it is aiming at to find out where it is. In the experiment, note that it does not do this when the pendulum moves away, and this is quite normal. If the bat does not approach the reflecting obstacle, it is not interested in it! Other studies have shown the existence of another area in the bat's brain dedicated to measuring the Doppler effect. Concretely, this is where neurons analyze the bat's relative speed in relation to its prey and trigger behavioral responses, such as changing the pitch of the calls emitted, thus enabling the bat to fly and hunt.[430] One last thing: When a bat hears the echo produced by a moth, the Doppler effect also exists on a smaller scale. In fact, during flight, the moth's wings alternately rise and fall back. Hold on tight: if the bat comes from the side, a rising moth wing moves away while a wing that folds down comes closer. So there will be a small extra Doppler effect due to the flapping of the wings. It has been shown that bats are sensitive to this and thus gain information about the type of moth they are chasing: large wings that flap slowly will not produce the same Doppler signature as small wings that flap quickly.[431] Unbelievable, but true.

In the 1950s, a fascinating discovery was made about moths, the favorite victims of many bats: these animals are able to detect the ultrasonic calls of bats and flee from them.[432] These insects have developed "ears" during their evolution that can hear them. This regularly saves their lives. The ears of moths are called *tympanic ears*.[433] What is remarkable is that ultrasound-sensitive tympanic ears have appeared several times, independently, in the history of moths: sometimes on the thorax in Noctuidae, sometimes on the abdomen in Pyrales, or even around the mouth in some Sphinx. The sensitivity of the tympanic organ of a moth species often corresponds to the acoustic characteristics of the vocalizations of bats living in the same place. And the moths with the most sensitive ears are also those that live in environments where bat density and diversity is greatest. These ears are very simple structures: a small membrane to which one to four sensory cells are connected. They are sensitive enough to detect ultrasound before the bat has received its echo.[434] In reaction to the ultrasound, the moth changes direction and sometimes performs acrobatic maneuvers to thwart the attack. In addition, some moths are also capable of producing their own ultrasounds, probably to disrupt bat echolocation by adding noise.[435] These sound signals can also startle the bat or even warn it that the moth is inedible—many species of moths contain toxins. This warning value has been shown experimentally: bats avoid moths that emit ultrasound and prefer to hunt silent species. However, if the sound-producing organ of these moth species is suppressed, bats will start to hunt them just like other moths. Let us note in passing that bats also have their own predators . . . and that they can deal with them acoustically! A recent study shows that the greater mouse-eared bat (*Myotis myotis*), when captured, imitates the sound produced by bees or wasps to discourage birds it could be the victim of.[436]

Faced with these adaptations developed by moths in the course of their evolutionary history, bats have not remained without an answer. For example, some bats have lowered or, on the contrary, considerably raised the pitch of their ultrasonic calls so as to move outside the range of frequencies perceived by moths. A remarkable adaptation is that of the western barbastelle *Barbastella barbastellus*: this bat feeds only on moths that are capable of hearing ultrasound and has become capable of transmitting its signals and receiving the echo at an intensity 10 to 100 times lower than other bats. This ability gives it a definite advantage: it will only be detected by the moth when it is very close to it, ready to bite it. Realize that this battle between bats and moths has been going on for more than 60 million years. And it is not unique: other groups of insects—crickets and grasshoppers, mantises and beetles—have also developed ultrasonic ears under bat pressure.[437] While these adaptations allow prey to escape from predators, there are also systems that have evolved to facilitate the work of bats. Some tropical plants have developed umbrella-shaped flower shafts in the form of ultrasonic reflectors, making it easier for nectar-drinking bats responsible for their pollination to spot them. However, if a small ball of cotton wool is placed in the hollow of the flower's pavilion, the ultrasonic mirror is masked, and the plant will not be visited by bats at night and will not be pollinated.[438, 439]

Among the other animals that produce ultrasound, frogs have long kept their ability secret. The first ultrasonic frog was only discovered in the early 2000s by Peter Narins of the University of California, Los Angeles, and his colleagues at the Chinese Academy of Sciences. *Amolops tormotus* is an endemic species of the Huangshan Hot Springs region of China—it is found nowhere else. While most amphibians— including frogs—have limited hearing ability and generally do not hear beyond 12 kHz, *Amolops tormotus* perceives sounds up to 30 kHz, and possibly beyond. This animal produces melodic songs made up of an incredible variety of notes—vocalizations that are unusual for a frog. Its sound productions are closer to the songs of birds than to the usual repetitive croaking of other frogs and toads. Moreover, its tessitura extends far into the ultrasonic range, i.e., beyond 20 kHz (remember that 20 kHz is the audible limit for the human species).[440] Peter and his colleagues wanted to know if these ultrasounds were simply a side effect of a particular production mechanism or if they were a genuine adaptation to avoid the rather low-frequency noise of the torrents near which this frog lives. The scientists conducted playback experiments in the wild with computer-modified songs, and showed that the ultrasonic part of the songs was sufficient to provoke a vocal response from the frog. Electrophysiological recordings made in their brains showed that they are actually able to hear the ultrasound. Ultrasonic communication of *Amolops tormotus* is therefore probably a means of avoiding the noise of torrents.[441] Along with bats, this frog provides a good example of evolution leading to the production of ultrasonic vocalizations.

While most frogs do not produce ultrasound, almost all make extensive use of acoustic communications. Our knowledge of their sonic world is constantly increasing. Recently, Peter, with his vivid imagination, told me about his research on a tiny yellow frog, endemic to Guyana and belonging to the Dendrobatidae family. While we were together at the International Bioacoustics Conference, he took out the digital tablet that never leaves him and showed me pictures of his latest expedition. "Do you see that tiny bug? It's the golden rocket frog. *Anomaloglossus beebei*, as it is called. It croaks in the air, resting on these broad leaves of bromeliads. But its croaking also makes the leaf on which it's resting vibrate." And then he explained that *Anomaloglossus beebei* males are used to emitting powerful vocalizations—a short series of three repeated calls—from the surface of the large leaves. These calls attract females and repel competing males. The surrounding males, resting on other leaves, respond aggressively. Peter and his colleagues hypothesized that, in addition to audible airborne sound waves, this frog's call was translated into leaf vibrations and that these vibrations were signals to other frogs. They brought a portable laser vibrometer into the field (Peter will stop at nothing) and measured the frequency of leaf vibrations when a frog vocalizes. Better still, after real-time processing on a laptop computer, the recorded vibrations were reproduced at the end of a small stick, a sort of minishaker, placed on a leaf about 20 centimeters away from a frog. Peter and his colleagues hoped to make the animal believe that another male was vibrating its leaf. And indeed, all the frogs subjected to this device were fooled: they all moved toward the vibrating end of the shaker and began to produce longer series of calls. They adapted their response to the vibrating stimulus, waiting until it was over before responding. So the little yellow frog uses both airborne sound waves—which we humans perceive—and leaf vibrations that are inaudible to our ears.[442] Probably, airborne sounds allow long-distance communication, such as "Hey, that's my neighbor across the street," while vibrations signal a real intruder ("What the hell is this one doing on my leaf?"). Many other animals, including cicada-like insects, use the vibrations of a substrate—a leaf, a branch, the ground—to transmit sound waves that are imperceptible to us.[443]

For social insects—ants and termites in particular—drumming can be a way to set off alarms in the colony. Let's take a look at termites, those small blind insects that live in colonies that can easily number tens or even hundreds of thousands of individuals. Termites are vegetarian; their life depends on the exploitation of wood resources and plant debris that workers sometimes look for far away from the center of the colony. The surface area used to find food is usually 2000 square meters. Many species of termites grow and feed on a fungus, and the plant debris they search for is then used to fuel the growth of the fungus. A termite colony consists of breeding individuals (queen, king), workers, and soldiers. It is a kind of superorganism based on cooperation between all the individuals that are part of it. All members of this little world communicate, of course, through chemical signals—called pheromones—and mechanical vibrations. Of the 2600 species of termites, it seems that all of them use vibration as a means of communication.[444] Although our knowledge of vibratory communications in termites is still patchy, it is certain that vibratory signals play a major role in alerting the colony to a danger—whether that danger is a predator that eats termites (and there are many) or a simple fungus that threatens to rot the one being grown by the small animals.

It is usually with the head that individual soldiers (and workers in some species) drum repeatedly when disturbed. By banging their heads against the walls of their gallery, they make them vibrate, and this alarm signal attracts other soldiers as it pushes the workers deeper into the nest. Some termites cause their abdomen to vibrate, and the waves are transmitted by leg or body contact with the gallery. The vibrations that propagate along the gallery walls are a kind of micro drum, "Trtrtrtrtrtr-trtrtrtrtr-trtrtrtrtr-trtrtrtrtrtrtrtrtr, . . ." emitting sounds at a rate of 10 to 30 drumbeats per second, depending on the species. In the termite *Macrotermes natalensis*, the soldier raises his head to a height of 1 centimeter and then lowers it extremely quickly, hitting it on the ground at a speed of 1.5 meters per second. It is not known whether these drummings carry accurate information about the level or nature of the hazard. Ultimately, they appear to be fairly constant, except when the danger is major. When the termite mound is attacked by a predator, for example, pheromones are added to the vibrations, and the combination of the chemical molecules and the vibrations increases the aggressiveness of the soldiers by tenfold.

I said that the vibrations produced by termites are transmitted along the walls of the termite mound galleries. Alas, the vibratory waves grow weaker very quickly during propagation: after 40 centimeters, they can no longer be perceived. How can the information manage to travel through the tens or hundreds of meters of galleries in which the colony lives? Well, you see, termites have a relay system. Are you familiar with the Great Wall of China's warning system? When an enemy was in sight, a fire was lit on the top of the watchtower from where it had been sighted. In response to that fire, the guards on the neighboring tower about a kilometer away would light a fire as well, and so on, spreading information along the entire wall and allowing for quick mobilization of troops. Termites follow the same strategy. Soldiers are distributed along the galleries. When one of them drums the alarm, the closest ones perceive the signal and hurry to drum themselves. The information is transmitted rapidly—between 1 and 2 meters per second—in all directions throughout the colony, which responds to the alarm in a few seconds. The attenuation of the signal during its propagation is thus reduced to zero.[445] Isn't it brilliant? This *social amplification of information* has not often been detected in animals.[446] The day I first heard about this strategy, I regretted not having devoted my career to termites. Their worlds are fascinating. One last thing: The termite that perceives an alarm vibration must make an important decision. In which direction should I go? Up or down the gallery? The soldiers will choose to go to the source of the problem, while the workers will run away from it. But how do they all know where the vibrations are coming from? The solution is found in their legs: the termites perceive the time difference between the vibrations arriving at their left and right legs—a process similar to the perception of the interaural time difference we were talking about earlier, except that here the difference is not perceived by the ears but by the legs.[447]

Besides insects such as termites, spiders (which are not insects but arachnids) are also known to make the ground or the leaves they walk on vibrate, especially during courtship. One day, we had the pleasure of welcoming to the laboratory Noori Choi, a student who was preparing his thesis with Eileen Hebets, a professor at the University of Nebraska in the US and an expert in spider communication. To record the vibrations that the spider produces, Noori places the animal on a special type of paper on which the sound waves will propagate and records them with a laser vibrometer, like the one Peter Narins uses with his frogs. The video recordings that Noori showed me were amazing: depending on their species, the spiders had different dances, waving their legs in a regular and rhythmic way and tapping the paper in a seemingly controlled manner.[448]

And what about bigger animals? Can we find this kind of communication in them?[449] In mammals, an example is found in mole rats, whose vocalizations we talked about in the preceding chapter. There are several species of these rodents living in underground galleries. In some, the individuals make the earth vibrate with their heads, while for others it is with their feet. Their ears show adaptations to hearing low frequencies, and these animals respond to the underground vibrations produced by their fellow mole rats by hitting the ground in reply.[450] But rather than detailing what happens in the mole rats, for which our knowledge remains rudimentary, let's see what happens in another fascinating animal—the elephant. And let's move on to the field of very low-pitched sounds, so low that we cannot hear them. These are infrasounds, which are to low-pitched sounds what ultrasounds are to high-pitched sounds.

Elephants, which as everyone knows are big animals, can produce very powerful sounds. Emitting more than 100 decibels (almost the sound level of a jackhammer) is not a problem for them. They can also make very low-pitched sounds: around 20 Hz.[451] Their vocalizations are so powerful that the sound propagates on the surface of the ground independently of air waves and makes the ground vibrate.[452] These seismic waves can be considered—proportionally speaking—as small earthquakes, which can be perceived from a long distance by other elephants.[453] How do they spot them? Two main mechanisms may be involved. On the one hand, the skin on the soles of the elephants' feet has sensory cells—mechanoreceptors—that are sensitive to pressure variations and therefore to vibrations. On the other hand, the vibrations travel along the leg and shoulder bones up to the jaw and inner ear bones, where the sound waves are analyzed by sensory cells connected to the brain. The elephant's foot has a special anatomical feature: a mass of fat and cartilage between the arch of the foot and the toe bones. This mass may act as an *impedance adapter*, facilitating the transmission of vibrations from the ground to the bones.[454] This structure is very reminiscent of the fatty masses found in the heads of whales, which also facilitate the reception of sounds. When an elephant listens to vibrations from the ground, it adopts a particular posture: motionless, it puts more weight on its forelegs. Ground vibrations propagate in principle a little more slowly than airborne sound waves—around 230 meters per second as opposed to 340 meters per second. The further away the sound source is, the longer the time interval between the arrival of these two components will be. This may inform the receiving elephant of the distance separating it from the sender. In addition, seismic waves propagate over a greater distance than air waves, and with less attenuation. Vibrational communication may allow elephants to communicate at distances of more than 2 kilometers without any problem, and maybe up to 10 kilometers. Is there any advantage in being able to communicate over such distances? Elephants are animals with a very complex social organization. Females form family groups in which individuals know each other well. They also know members of other family groups. These recognitions are based on both acoustics and olfaction. Karen McComb, a professor at the University of Sussex in the UK, has shown that an adult female elephant can know the vocal identity of 14 different families, for a total of about 100 individuals.[455] Females of different families stay in contact through their vocalizations, both aerial and ground. Males, on the other hand, may join a herd of females, group together, or remain solitary. They use acoustic signals in the infrasound as well as those that are audible—we all know the trumpeting elephants are capable of—when searching for sexual partners or during competitions between males.

Ovulating females vocalize more, which attracts males. The vocalizations of males ready to breed drive others away. All this can be done at a distance, without the animals seeing each other. Infrasounds, whether propagated by air or by ground vibrations, are excellent vehicles for long-distance information.[456]

Ultrasounds, infrasounds, vibrations of leaves, branches, or the ground —all are sound worlds inaccessible to our senses. We can only imagine the bat perceiving its environment and its prey through the echoes it receives and the elephant communicating with another elephant family a few kilometers away.

14

The laughing hyena

COMMUNICATIONS AND COMPLEX
SOCIAL SYSTEMS

South Gate Campground, Moremi Game Reserve, Botswana. Night. Half asleep in the tent on the roof of the car, I can hear the fuss being made nearby. The trash can in the campground has been knocked over and the garbage is being searched noisily. There's giggling, squeaking, blowing, and whistling. I cautiously open the tent zipper and turn on my torch. In the beam, I see two of them—honey badgers, savagely plundering everything they can. Suddenly, a growl, both muffled and powerful, threatening, long. Freeze frame. The feast has been cut short. The two angry badgers call out in rage and jump back at the sight of a huge spotted hyena, determined to take advantage of the situation. The badgers are formidable beasts that, it seems, do not hesitate to ward off even an elephant; but the jaws of the queen of the night are the strongest, and they know it. The hyena dashes over to a piece of rubbish and then trots away. It'll probably join its family—I can hear them giggling and whooping in the distance. The badgers, relieved, start their ruckus all over again. The sounds of the African night! On another night in a camp further north, I'll be woken up too, but by the mighty trumpeting of an elephant. In the south, at the Polentswa campsite in the Kgalagadi, the lions had roared all night. My camera trap attached to the bumper of the car showed the next day the passage of a lioness followed by a big male—a nightly stroll of the big beasts close by. During these nights, as

Hyena

during the days, far from the hustle and bustle of the university, endless meetings, administrative paperwork, and my computer screen, I always feel in my element. Excited but serene. My second life will be here, I decided one day.

The spotted hyena *Crocuta crocuta* is carnivorous like its two hyena cousins, the brown and the striped.[457] Contrary to what you might think, it is a great hunter and does not deserve its reputation as a scavenger. Spotted hyenas hunt and kill their prey, just like lions, wild dogs (those beautiful striped carnivores), and other predators. I am not saying that a hyena retreats in front of carrion—my nighttime experience shows that it does not hesitate to look for easy food, which is quite understandable. But most of the prey that hyenas devour they've caught themselves. Hans Kruuk, a famous ethologist from the University of Aberdeen in Scotland, studied hyenas in the wild and showed that 80%

of the prey eaten by the Serengeti lions in Tanzania was actually killed by hyenas.[458] So they get robbed more often than they steal. The respective reputations of these two species ought perhaps to be more nuanced ... the king of the savanna has a few unwitting helpers. Hyenas regularly hunt in groups, skillfully cooperating to isolate and kill their victims. The hyenas can then kill large animals, such as zebras or wildebeests—those antelopes that migrate in vast herds that you've probably seen in animal films crossing rivers where crocodiles are waiting for them.

The spotted hyena is not a solitary animal; it loves large families— very large indeed since a clan of hyenas numbers between 6 and 90 individuals.[459] Family meals are obviously very hectic. Not only do you have to eat before a lion comes and ends the feast, but you have to contend with the members of your family in intense competition over the carcass of the unfortunate wildebeest. Each family member devours as much food as possible. The giggles from frustrated individuals who are struggling to find a place at the table are everywhere.

In many ways, the social organization of spotted hyenas resembles that observed in many species of monkeys of the Old World (*Cercopithecinae*—baboons, macaques, mandrills, etc.). The females born in a clan in principle remain there all their lives, while the young males leave to join another clan. This system is coupled with a very strict matrilineal hierarchy; in other words, in clans of spotted hyenas, the females lead. The males, if you have followed me, are all outsiders. This social organization has been particularly studied by Kay Holekamp, a professor at the University of Michigan and a specialist in the spotted hyena. I was also able to see the hyena's social organization during my stay at the University of California, Berkeley. The university's field station, hidden in the hills overlooking the famous campus, was home to about 30 spotted hyenas. This colony had been established several decades earlier by the late Stephen Glickman, then a professor at Berkeley, to study the hormonal basis of social dominance and aggression. Glickman was particularly interested in the mechanisms of masculinization that affect the genitalia of female hyenas.[460] The latter have a pseudo penis, which makes them look like a male hyena, the rest of the body being very similar between the two sexes. However, females are on average larger and more aggressive than males.

I was introduced to Glickman by my colleague Frédéric Theunissen. Glickman was then a respectable gentleman, of an equally respectable age, who had kept his scientific spirit alive. "Frédéric told me about your joint project to study the vocalizations of hyenas. As far as I know, no one has really looked into the matter. Yet, as you will see, acoustic communication is essential for hyenas." Glickman was obviously right. During the first few hours I spent observing the Berkeley colony, I was constantly surprised by new types of calls. In addition to the famous giggle, of course, the "uh-uh-uh-ooh-ooh-ooh-ooh-hi-hi," there's also the whoop, "woooooooooo ... woo-hoop!," which individuals use when they don't see each other, not to mention multiple variations of growls, rumbles, and whispers that allow hyenas to finely tune their social interactions. Experts agree that there are at least a dozen different basic calls. In addition, each of these calls can be modulated (louder—lower- or higher-pitched) depending on the circumstances. These are called *graded vocalizations*. This feature allows the sender to refine the information it encodes in its call. "I'm just a little frustrated, I'm going to giggle softly." Or "my frustration is mounting because I am very hungry and this very aggressive female is preventing me from getting to eat; I'm going to giggle very loudly." All this at the risk of bringing in lions from elsewhere, which is another story.

Holekamp showed that the hierarchical position of a hyena in its clan is roughly established around 18 months after its birth.[461] For females, daughters of mothers high up in the hierarchy acquire the same rank— they're born with a silver spoon in their mouths. This is probably due to a combination of their own level of aggression (they have been bathed before birth in a particularly testosterone-rich uterine juice) and the attention their mothers give them by protecting them from other members of the clan. For lower-ranked females, it's not the same picture, and the spoon is made of plastic. The situation is quite different for males. Here it is the order of the male's arrival in the clan that roughly determines his rank. Of course, things can be more subtle and reversals can happen sometimes, especially for those who are smart or clever enough to foment coups by forming coalitions with fellow hyenas. The social system of spotted hyenas is more than a simple linear hierarchy. As in ape groups, affinities are created. So are enmities. This one will ally itself with that one, these three will form a coalition against that one, and so on. In short, the organization of a hyena clan is more like a mafia than a military regiment. But the most general rule is that these hierarchies, once well established, last over time.

Imagine that you are a female or male hyena in a clan of 80 individuals. Complicated to know who is who, who is allied with whom, who has just arrived, who is in a good mood today, who you'd better not rub the wrong way, how to behave with this individual or with that one, etc. To answer these questions, you fortunately have an important tool at your disposal: the communication signals that others will send you and those that you are capable of producing. And among hyenas, they are legion: I repeat, more than a dozen basic calls, with variations. But you also have to add in the influence of pheromones, those chemical molecules produced by various glands, as well as the use of many postures that serve as visual signals—for example, the attacking posture (head held high, ears and mane erect, mouth closed, and tail upright) and the fleeing posture (ears flattened against the skull, mane smoothed against the body, and tail hidden between the hind legs). No doubt about it. Communication in hyenas is multimodal, that is to say, it uses several channels of communication, from chemical to visual to acoustic.[462]

But Frédéric and I are bioacousticians, as you must be well aware by now. Our only ambition was to decipher the information contained in the hyenas' vocalizations. One question in particular was bothering us: Do hyenas have a voice? An identifiable voice, like the ones we are able to recognize on the telephone with barely three spoken words? In other words, was it possible to identify an individual signature in a hyena's giggle and to find that same signature in another vocalization? This question may seem trivial. Yet, in the animal kingdom, there is no known example of a well-established voice. Even in the human species, it is not as obvious as we think. Would you be able to recognize someone by their laughter, shout, or whisper, if you have previously only heard their spoken voice? Of course, many animals recognize each other individually on the basis of vocalization. In previous chapters, I have detailed several examples: the white-browed warbler that recognizes its neighbors by listening to their song, the elephant seal that is able to identify each of its opponents by their "clork . . . clork," etc. But these recognitions, however precise and effective they may be, most of the time relate to a single type of vocalization. To my knowledge, only rare studies have explored individual recognition through the vocal repertoire of a species.[463] One of the most thorough was carried out on the zebra finch by Julie Elie. Do you remember her? She is the expert on the vocal repertoire of this little Australian bird. Julie showed that the zebra finch is only able to identify a fellow finch, whatever the type of call it emits, if it has first learned the individual vocal signature carried by each of these calls.[464] In other words, the bird is not able to learn to recognize an individual from one type of call and then transpose it to another call. Why not? Because zebra finches don't have a true individualized voice.[465] If it is assumed that the same is true for hyenas, then just because Kadogo recognizes Ursa or Winnie (three of the Berkeley residents) by hearing their giggles, that doesn't mean it will recognize them when they whoop or growl, unless of course it has learned to recognize them for each of their different calls.

So off I went to record our hyenas. Helped by the technical staff at the field station and by Aaron Koralek, a friendly Berkeley student recruited by Frédéric, it was not very complicated. Mary Weldele, the head of the station, had spent so much time with her animals that she knew each one by heart. So she knew the context in which to put this one or that one so as to have the best chance of hearing it giggling in frustration or whooping to call a mate. A few years before my arrival, Mary had suffered a serious injury: while trying to retrieve a piece of meat that had fallen on the ground, she had her hand bitten by a hyena. Hyenas have tremendous jaws, able to crush big bones like you crush an eggshell. Experiments on hyenas at Berkeley have shown that they can easily develop a force of 1422 newtons, which is enormous.[466] The jaws of hyenas are only surpassed by those of lions (3400 newtons), tigers (3000 newtons), and bears like the polar bear (2400 newtons), all much more imposing animals. By the way, the power of our human jaws is 650 newtons, which allows us to easily crush roast chicken! Mary had been wise not to try to force her hand out of the animal's mouth, which would have certainly made the situation worse, and the hyena finally released her. Mary's hand, however, bore impressive stigmata. So caution was called for—above all, being extra careful when transferring animals from one enclosure to another. Entering one of the enclosures was out of the question.

The easiest vocalization to provoke is the giggle, so we decided to start our investigations there. All you have to do is present a hyena with a piece of meat without letting it have it immediately, and it will giggle with frustration. A few seconds of recording, and voilà! The released piece of meat is gulped down. Extensive acoustic analysis, followed by solid statistical processing—something Frédéric is an expert at—revealed to us that the hyenas' giggle contains previously unsuspected information.[467] First of all, each hyena giggles in a very personal way. It's a bit like in humans—there are frank or hesitant giggles, raucous or discreet, singing or monotonous. As a result, it's not so difficult to distinguish between Kadogo's giggle and Ursa's or Winnie's. But that's not all: the giggle also gives information about the age and hierarchical rank of the sender. Older individuals have lower-pitched giggles. A dominant individual produces a slightly variable giggle ("ho-ho-ho-ho") while a subordinate—who will giggle much more, because it is more often frustrated—has a giggle akin to singing ("oh-ooh-ooh-ooh-ooh-hi-hi-hi-hi"). This information about hierarchical rank could be very useful for a hyena trying to access a carcass. Should I stay or should I go? We then saw that the major information conveyed by the whoops ("wooooooo . . . woo-hoop!") represents the sender's individual identity, age, and gender. Not very surprisingly, this vocalization is emitted to establish contact with distant kin or clan members and can be heard from several kilometers away. It's important to know who you're dealing with.

To test whether hyenas could recognize an individual's voice, we decided to adopt an operant conditioning protocol. Remember? That's the approach Colleen Reichmuth from the Pinniped Laboratory in Santa Cruz uses to test the hearing abilities of seals. The principle is to teach the animal to respond to a sound emitted from a loudspeaker (the target stimulus) with a particular behavior. Finding out how, in practice, to adapt this kind of protocol to hyenas was quite a challenge. Luckily, this challenge fascinated Julie, the zebra finch expert, who had also joined us in Berkeley. There were many attempts, some less successful than others as the hyenas brutally destroyed any device within their reach. Julie devised a system in which the test subject first had to trigger a sound by pulling a rope and then had to move to a food dispenser that only worked if the sound was the target stimulus. Yes, okay, it's not that simple, but a system had to be found. Imagine the scene: one of the hyenas was isolated in a paddock. In one corner was the rope; at the other end of the paddock was the food dispenser. We wanted the hyenas to learn how to pull and pull again on the sound-emitting rope until they identified the target stimulus and went to the dispenser in response—triggering a conditioned reflex, so to speak. What were the stimuli used? We started out by emitting whoops from different individuals, only one of which triggered a food reward. So the hyena had to learn that a whoop from Kadogo, for example, was a reward, while it would get nothing from all the others. We hoped that it would pull the rope to trigger successive whoops without going to the food dispenser until it heard Kadogo's whoop. But it was hard to get anything from the hyenas. While four of the five hyenas included in the experiment obviously understood what we expected of them, they soon became bored and lost motivation after a few tries. Maybe they weren't hungry enough to spend energy playing with us. The meager results of these experiments are interesting, however. First of all, since most of the animals passed the test, we can conclude that hyenas are able to distinguish their fellow hyenas by their whoops. Second, by using recordings of giggles instead of whoops, we found that hyenas can also distinguish between two individuals by this vocalization. The individual signatures carried by both whoops and giggles are therefore perceived by hyenas. On the other hand, no hyena succeeded in the last phase of the experiment, i.e., recognizing an individual's giggle when it had simply learned to identify its whoop. A hyena that learned to identify the vocal signature of a fellow hyena contained in one type of vocalization does not seem to be able to recognize it in another. It would seem that our hyenas are in the same lot as the zebra finches. Unable to transpose their knowledge from one vocalization to another, they must learn to recognize the vocal signature for each of an individual's calls. As I have already said, it is not certain that we ourselves have this ability: Does the same person who speaks, laughs, or calls have the same identifiable vocal signature in each case?

Although they don't seem to have more individualized voices than zebra finches, hyenas nonetheless have a complex and effective system of sound communication. We now know that these animals use acoustics to recognize clan members, to identify a stranger, and to obtain all manner of information about the sender, such as age, sex, hierarchical rank, and even its mood at the time. The diversity of their vocal repertoire is probably due to the complexity of their social organization. This situation is found in all animal species living in groups where strong and lasting social interactions (alliances, cooperation, well-structured hierarchies, etc.) develop. Let's explore some of these worlds together.

Let's start with the example of meerkats (*Suricata suricatta*).[468] You probably know these beautiful and dynamic mongooses of the African savanna, standing on their hind legs with their arms at their sides, their little nose pointing toward the camera in wildlife documentaries. They live in groups of up to 50 individuals, hunting scorpions and beetles, raising their young cooperatively. Their vocal repertoire is particularly extensive. Marta Manser, from the University of Zurich, who studies meerkats in the Kalahari Desert, has been able to distinguish more than 30 different calls. Meerkats are therefore great users of acoustic communication, whether it is during episodes of competition for food, during mutual grooming sessions, or when traveling in groups. A meerkat has a very special life; it spends between five and eight hours a day digging in the ground for food. With its head half buried in the sand, it can't see what's going on around it for most of the time. Sound signals are the main means of keeping in contact with the other members of the group. Sentinel individuals, standing on their hind legs, are on deck keeping watch for the arrival of danger so they can alert burrowers with calls. These small animals are very vulnerable and are a food source for a whole cohort of predators—eagles, jackals, and other carnivores. In addition, our small earthworkers are often several dozen meters from the first burrow. Meerkats take risks! And they pay a heavy price; on average, 60% of adults perish in a single year. This is not only because of predators; it must be said that conflicts between groups of meerkats also cause damage. Under this multiform pressure, the meerkats have developed a sophisticated alarm call system allowing all members of the colony to be quickly informed of danger and above all to know what behavior to adopt. Woe to the distracted individual, who's a goner for sure.

In the late 1990s, Marta spent several months observing 18 groups of meerkats along the Nossob River in what is now the Kgalagadi Transfrontier Park, the vast semidesert region straddling South Africa and Botswana, and further south near the Kuruman River. The beds of these two rivers are mostly dry, except during floods—the famous flash floods—which follow the violent thunderstorms that sometimes occur over the Kalahari. In spite of the drought, the relative proximity of the groundwater allows the existence of some vegetation with its associated food chain. Marta reported that the meerkats often use their alarm calls; during observation sessions, she rarely spent more than 45 minutes without hearing one.[469] To test whether the calls contained any information about what caused them, Marta provoked them, walking a dog on a leash, dragging a stuffed jackal across the field, or flying various objects suspended from balloons. Acoustic analysis of the recordings showed that the meerkats' alarm calls vary depending on two factors: the nature of the predator and the level of threat.[470] With regard to the first factor, predators appear to be identified as either *ground* or *air* based on the corresponding calls made by the guards. For example, those calls emitted at the sight of an aerial predator extend over a wider band of frequencies. As to the second factor, the level of threat, the calls may vary in several ways. For instance, the greater and more immediate the danger, the louder the corresponding calls will be: they take on a chaotic character, a bit like when a human baby goes from a soft cry to an angry scream. The vocalization may escalate from "or . . . or . . . or . . . or . . . or" when the sentinel spots a jackal on the horizon to "ourrh-ourrh-ourrh" when the jackal approaches and becomes really threatening. Another vocal clue to the threat level is that the more immediate the danger, the shorter the time between calls. In aerial alarm calls, the more urgency the situation requires, the fewer calls the individual sentinel makes. If a martial eagle is spotted, the sentinel can't stay out in the open for long. Another vocal variation is the *recruiting call*, which, as its name suggests, calls the other meerkats to the rescue. It is emitted when the sentinel sees a snake or identifies something that it finds odd (such as traces of urine, droppings, or hair from foreign meerkats or predators). The final variation, the panic call, is used as a last resort. This death call is emitted once or twice very quickly by the sentinel, either in reaction to alarm calls from birds (Horror! The birds see something, and I don't see anything!) or when a predator, ground or air, is spotted at the last moment, already very close to the group. This call does not give precise information on the identity of the predator; it is an "everyone for themselves" call, provoking the immediate and desperate flight of the whole meerkat family, diving into the burrows to take shelter.

With the alarm calls of the meerkats, we therefore find a real code, combining a lot of useful information for decision making. The playback experiments conducted by Marta and her team have shown that the meerkats know this code like the back of their little furry hands. They rarely make mistakes and react to different calls in a perfectly appropriate way. They either gather around the sentinel who gave the alert and examine the source of the danger in chorus (if it is not immediate) to evaluate the best strategy to adopt (Is it better to form a group to go scold the trespasser? To wait a little? To hide on the spot?) or they run like mad to take cover if that is obviously the right decision.

Returning to the idea that these calls carry information about the identity of a predator, even if they are not the equivalent of words in human language since they are not constructed on the basis of a combination of syllables, they are nevertheless strongly reminiscent of them. This kind of sophisticated system is called *referential communication*.[471] In other words, a call means or represents something in the environment. These are far from simple calls of fear emitted by reflex. In fact, Marta has shown that young meerkats improve their response to alarm calls over time, suggesting some learning in their behavior.[472]

In recent decades, several referential communication systems, most of them related to alarm calls or calls indicating the presence of food, have been identified in mammals—in particular, in several species of monkeys. For the record, Charles Darwin anticipated this. The father of the theory of evolution was very interested in animal communication. He related that one day he presented a stuffed snake to some monkeys at the London Zoological Gardens, and observed the monkeys emitting "calls of danger that were understood by other monkeys."[473] It took a marriage between bioacoustics and primatology to demonstrate the existence of referential communication in monkeys. The first and most famous example is certainly the one observed in the vervet monkeys (*Chlorocebus pygerythrus*). Do come with me; I'll take you to discover this communication system. It has become mythical.

The vervet is an African monkey that produces three alarm calls: a first ("kof") for leopard-like land predators; a second ("uaho") for aerial predators like the martial eagle, and a third ("cla-clack ... cla-clack") for snakes—very useful because large pythons can attack this monkey. Just as meerkats show behavioral responses consistent with the alarm call heard, so do vervets. They climb a tree when they hear the leopard call. When it is an eagle call, they look to the sky and hide under the nearest bush. And in response to the snake call, the adults stand up on their hind legs and scan the ground, grouping together to harry the trespasser. Thomas Struhsaker, a professor at Duke University in the US, published these observations in 1967.[474] But the experimental evidence was to come in 1980, and it made a pair of primatologists famous: Dorothy Cheney and Robert Seyfarth, two students of Peter Marler, the discoverer of the dialects of the white-crowned sparrow. Cheney, Seyfarth, and Marler confirmed through playback experiments conducted in the field in Kenya that the vervets fully understood the meaning of their various calls. Their study was published in the prestigious journal *Science* and received worldwide media coverage.[475] Monkeys were proving that semantic communication was not restricted to human language! Semantics, a big word, perhaps? Let me explain. Whereas until now it was considered that animal vocalizations were merely a record of the emotional state of their sender, the experiments of Cheney and colleagues showed that animal calls could represent elements of the environment, just as the word *table* means to us the object on which the cutlery is placed. Seyfarth and Cheney reported the details of their investigations in a book that I recommend to all ethology enthusiasts: *How Monkeys See the World: Inside the Mind of Another Species.*[476] Since then, a lot of water has flowed under the bridge and many people have struggled to find out if this remarkable characteristic of human language can really be attributed to vervets. We don't know the cognitive mechanisms that drive the production of calls in meerkats or vervets, and biologists are afraid of being anthropocentric! In my opinion, we are bound to see striking similarities between the way in which animals' calls indicate a type of danger and the way in which human words represent an object in the environment. David Premack, a former primatologist at the University of Pennsylvania, has suggested that a communication system based solely on the reflex expression of emotions may well evolve into a referential system.[477] His argument is this: if you know that I squeal with delight when I see strawberries and scream with fear when I see a snake, you will soon learn to associate each of my calls with the right trigger. In other words, my vocalizations give reliable information about the objects around me. Does perceptual semantics go hand in hand with conceptual semantics in meerkats or monkeys? That is, does a vervet monkey climb a tree because it has really understood that a leopard is approaching? Let's dare to go even further. Could the vervet imagine, in its little monkey head, an image of a leopard? Or is the explanation for its behavior more basic? It climbs a tree because it reacts to a particular acoustic signal ("kof!") and it knows it must climb when it hears it? It's not an easy task to answer these questions. However, a study with Diana monkeys *Cercopithecus diana* by Klaus Zuberbühler, a primatologist at the University of Neuchâtel in Switzerland and an associate of Cheney and Seyfarth, provides some interesting insights.

The Diana monkey inhabits the rainforest of West Africa, between the Gambia and Ghana. A bit like the vervet, the adults produce two different alarm calls, depending on whether the predator approaching is an eagle or a leopard. But strangely enough, here both males and females have their own calls. Moreover, the females react to the male alarm calls with their own alarm calls, which suggests that females might understand the meaning of the male calls. Does an eagle image spring to mind when they hear males call "eagle approaching"? To answer this question, Zuberbühler tested whether the response of Diana females was caused solely by the acoustic characteristics of the male calls or by the existence of a predator concept. Let's look at the protocol of the experiment; hang on tight, because it's a bit complicated. Zuberbühler first recorded or obtained from a sound archive four different calls: an eagle call from a real eagle, a roar from a real leopard, an eagle alarm call from a male Diana monkey, and a leopard alarm call from a male Diana monkey. In each experiment, a group of Diana females heard two of these calls, separated by five minutes of silence. In the so-called baseline condition, the two calls were exactly the same, acoustically and therefore semantically identical (e.g., an eagle call followed by the same eagle call). In the test condition, only the semantics were the same (e.g., an eagle alarm call by a Diana male followed by an eagle call). Finally, in the control condition, the stimuli differed both in their acoustics and meaning (e.g., a leopard alarm call by a Diana male followed by an eagle call). Zuberbühler hypothesized that if the Diana females formed a mental representation of the predator based on the alarm call, they would only show surprise if the call following the alarm call did not match. However, if there was a match between the two successive stimuli, or if the two stimuli were identical, the Diana females would show less interest in the second stimulus. He was right. In the baseline experiment, when the Diana females heard the first eagle call, they responded with their own eagle alarm call. But at the second eagle call, their response was less intense. Since no birds were visible nearby, they seemed to lose interest. Same story with two successive leopard calls. In the test condition, the females initially reacted strongly to the eagle alarm calls from the Diana male by emitting their own alarm calls. However, they hardly called out at all in response to the second stimulus, which was a real eagle call. Presumably, by the time the second stimulus was given, they had already integrated that an eagle was there and had the image of it in their heads. Finally, during the control condition, when the first call was a Diana male's leopard alarm call, the females initially responded with their own leopard alarm call. In addition, when the male's call was subsequently followed by an eagle call, the females responded very strongly to the eagle call with their eagle alarm call. Are you still following me? We can interpret these results by listening to Zuberbühler: Diana females get an idea of the predator by listening to the male's call. If the predator does not match what the males announced, they sound the alarm again, but with the "right" call.[478] Nothing says whether they blame the males for making a mistake, or whether they then take retaliatory action.

Referential communication systems have so far been identified in several other primate species, such as the Campbell's monkey *Cercopithecus campbelli*,[479] the chacma baboon *Papio cynocephalus ursinus*,[480] the black-fronted titi monkey *Callicebus nigrifrons*,[481] the chimpanzee,[482] and some lemurs (for example, the particularly remarkable forest-living ringtailed lemur *Lemur catta* with its orange eyes and long black-and-white striped tail[483]).

Sometimes it is not the acoustic structure of the call that carries information about the type of predator but a combination of calls that encodes referential information. The leopard alarm call of the *Colobus guereza*, a beautiful monkey with a long black-and-white coat, is a succession of many repeated "rrooarr-rrooarr-rrooarr" sequences. Their eagle alarm call is a succession of a few sequences, each containing many roars.[484] And there are more complex systems. In the Campbell's monkey, each of the three identified alarm calls are graded (like hyenas' calls, they vary according to the motivation of the sender), and, surprisingly, they can be given suffixes (i.e., a short call that is added after the alarm call and changes its meaning). Human language is teeming with examples: a motor/a motorist; the suffix *ist* changes the meaning of the word. In response to different contexts or events, the Campbell's monkey emits the calls "boom," "krak," "hok," or "wak." A team of primatologists led by Zuberbühler and Alban Lemasson, a professor at the University of Rennes in France, has shown that each of these calls can take the suffix "oo" and become "wak-oo," "krak-oo," etc. Moreover and amazingly, the monkeys combine these calls in sequences. It is now known that calls with suffixes and call sequences do not appear to be random but could correspond to particular contexts and events, such as a tree falling, a neighbor arriving, an increase in urgency, etc. The basic principle is that each individual call carries specific information ("leopard," "eagle"). If a call is repeated alone, the meaning does not change—it is still "leopard" or "eagle." If a suffix is added, the meaning becomes more general ("predator"). If two calls with suffixes are combined, some of the observations suggest that the overall meaning may change drastically: "leopard" with suffix + "eagle" with suffix is emitted when a tree falls, for example.[485] In the absence of playback experience with Campbell's monkeys, it is currently impossible to say whether this syntactic complexity actually carries any information for the receiver individuals. The future may tell us. The only small piece of evidence that we have today of the role of suffixes comes from playback experiments with male Diana monkeys. They respond more strongly to calls without suffixes (indicating a leopard or an eagle) than to calls with suffixes, whose meaning is less precise.[486]

I've talked a lot about the alarm calls. Don't think I'm done with them yet. It is indeed in the context of predator signaling that referential communication has been most often highlighted. But be aware that there are a small number of animals in which referential communication has been found to signal food. Marmosets, chimpanzees, and bonobos emit calls or particular sequences of calls in the presence of food. For example, a chimpanzee produces specific vocalizations when it discovers "high-quality" food, and its colleagues are particularly sensitive to this.[487] In bonobos, it is instead a certain sequence of calls that indicate the quality of the food.[488] Aside from primates, there are few other examples. The calls made in the presence of food, although they often provoke the arrival of other individuals, do not really disclose the nature of the food.[489]

Referential communication systems do not only exist in meerkats and monkeys.[490] Ground squirrels and marmots also have a repertoire of several calls in response to the arrival of different predators. The Gunnison's prairie dog *Cynomys gunnisoni* has four calls, referring respectively to a bird of prey, a human, a coyote, and a dog.[491] In addition, more than a dozen species of birds are known to produce referential signals.[492] The first example identified was in chickens, which emit a particular call to announce the presence of food and two distinct calls to warn of the arrival of either a land predator or an air predator.[493] (We mentioned this example in chapter 11 when we looked at the audience effect.) Funny thing: The aerial predator has to move sufficiently fast to trigger the chicken's call.[494] If the suspicious thing doesn't exceed a certain speed, the cock won't crow!

Let's stay with the birds for a while. In recent years, research on referential communications has flourished in birds, with astonishing results.[495] Three main systems were highlighted. Like the chicken, some birds use alarm calls that point out this or that type of danger. Others vary the number of notes, the speed of repetition of the calls, or even finer acoustic characteristics, such as the duration of the call, its pitch, or its intensity. Finally, there are those who use combinations of notes. Let's take a few examples to illustrate. The first case involves a study with my usual adventure companion, Thierry Aubin, in the field in Australia. We went there to meet a very small bird, the superb fairy wren *Malurus cyaneus*, known to use an acoustic alarm system with two calls.[496] Off we go to the other side of the world!

The superb fairy wren is one of the favorite study models of Robert Magrath, a professor at the University of Canberra in Australia. Robert has been interested in the vocalizations of the fairy wren for years, and when Thierry suggested that we study the coding of information in its alarm calls, he willingly accepted. Fairy wrens are very common in the southeast of the continent. Imagine two little balls of feathers—one simply dressed, with a brown back, lighter on the chest; the other sporting a more complex outfit, consisting of an iridescent light blue metallic head, a line of jet black encompassing its eyes and widening along its neck, deep blue on the throat, back, and tail, and finally a chalk-colored belly. Guess who the female is. On each ball of feathers is a never-ending tail, often pointing upward. The female and male fairy wren live as a pair on a territory that both defend bitterly with songs and by always being ready to ruffle a few feathers of any overly daring intruders. The breeding adults often team up with helpers—youngsters from the previous year who participate in the rearing of the chicks. Cooperative rearing is not uncommon in birds.

The fairy wren loves eucalyptus forests. We studied it in the Canberra Botanical Garden, which is full of them—a huge and very beautiful park but very different from the wild places where we usually traveled. I was a little disappointed. I hadn't imagined Australia like this, as I dreamed

Fairy wren

of vast stretches of wilderness rather than urban parks. Besides, I had hay fever. It was November, and it hadn't occurred to me that the Australian spring would not escape the grassy outbreaks. I hadn't brought any appropriate medication with me, and I had a strong tendency to be gloomy. That said, the place was pleasant, not very crowded, and I appreciated being able to get out of the university routine once again. Here a kangaroo, there the nest of a bowerbird decorated with colorful bottle caps. Much improved by the exotic animals, the vegetation and the climate, and the pleasure of working together again, our stay was nevertheless excellent. Not to mention that our experiments yielded good results. "Here is the map of the botanical garden; you will find many birds banded by us," he said. Robert Magrath is a highly meticulous person. He let us work alone, but we felt that letting us approach his protégés worried him vaguely.

Thierry was eager to tackle the two calls of the fairy wren. Issued in the presence of a danger—a predator most of the time—they provoke diametrically opposed reactions from the others present on the territory. The mobbing call, as its name suggests, attracts other fairy wrens to harass the predator as a group. The flee call incites everyone to take immediate refuge in the bushes. Why choose one of the calls over the other? Because sometimes it is better to hide from a predator that is too skilled than to try to scare it away by harassing it. The two calls have important acoustic similarities: they are both extremely high-pitched trills, lasting about 90 milliseconds. The flee call ("trreee") is around 9000 Hz. The mobbing call differs from it in two ways. First of all, a hyperfast and rapidly rising high-pitched whistle ("ps") is added in front of the trill. On the spectrogram, it draws a kind of hook on the front of the call, a primer if you like: the sound starts around 7500 Hz, goes down immediately to 6500 Hz, then goes up no less immediately to 9000 Hz. Then the trill starts at around 9000 Hz, just like the flee call, but it goes down regularly to 7000 Hz. The mobbing call could be translated as "pstrreeoo," where "ps" represents the hook and "trreeoo" the declining trill. Because they are short and very high-pitched, the two calls are difficult to distinguish for an untrained human ear. Our problem was to understand how the birds choose between going together to face the danger or hiding quickly. Given the structure of the two signals, we had three hypotheses: either the presence of the initial "ps" was the critical cue that allowed the birds to know immediately which call it was; or the cue was given instead by the frequency of the descending slope of the trill; or the combination of the "ps" and the trill slope were necessary for the recognition of the call.

You are now familiar with our experiments, so you already know our methodology: playback experiments, of course. And fortunately for us, fairy wrens are fairly easy to test. With a loudspeaker placed in the middle of their territory and a bit of waiting hidden behind bushes, we observed the reaction of the fairy wrens to four different signals: the flee call ("trreeee") and the mobbing call ("pstrreeoo") as the control measures of their behavioral reaction, a mobbing call without a hook ("treeoo"), and a flee call to which we had added the hook ("pstrreeee"). As always, we expected these artificially modified signals to give us the key to the information coding. And so they did. As expected, the birds' responses to the flee call and the mobbing call were exact opposites: fast flight to the nearest bush and absolute silence for the flee call; approaching the loudspeaker, erratic flights above, and many calls for the mobbing call. When the mobbing call was missing the hook, it did not lose its meaning: the birds continued to approach and noisily express their displeasure. On the other hand, when the hook was present but the trill remained around 9000 Hz (as in the flee call), the birds would dive into the bushes. The hook does not encode the nature of the predator and therefore does not carry information about the decision to be made: to harry the intruder, or to dive into a bush and lie low. It is the trill that fulfills this role. One hypothesis that we had not yet tested was that the hook provides an excellent location cue for the bird emitting the call. This would not be surprising since we know that a sound that varies extremely rapidly in frequency is easily locatable, and this would be particularly relevant for fairy wrens, which need to know quickly where to go to scold an uninvited guest. One can even venture to imagine that the flee call, the "trreeee," first appeared in the ancestor of the fairy wren, saving many lives. Then it was transformed over time to a "trreeoo" that proved to be a rallying cry, allowing an effective collective struggle in the face of danger. When the "ps" was added to give "pstrreeoo," it further facilitated the operation by helping the birds locate each other. This pretty description quickly sums up what probably happened. The evolution of communication systems, like that of any biological system, is rarely so simple and linear.

The system used by the fairy wren is therefore based on the use of two calls, each designating a different type of danger. Many other systems are based on the use of a single call with varying characteristics. One of the best-known examples is that of the black-capped chickadee *Poecile atricapilla*, the most common chickadee in North America. As we discussed in chapter 11, this bird changes the number of repetitions of the note "dee" in its call "chick-a-dee" to signal to its fellow chickadees the size of a bird of prey perched motionless on a branch. Playback experiments have confirmed that other chickadees understand the message: When the recording of the sender chickadee produces numerous repetitions ("chick-a-dee-dee-dee-dee-dee-dee-dee"), signaling a small owl or other small predator, the chickadees come close to the speaker, scolding bitterly and continuously. If the number of notes is limited ("chick-a-dee-dee"), the chickadees know that it is a large owl, and they are less motivated to attack it because they are more agile and quicker than the owl, so it represents much less of a danger.[497] I admit that, with this chickadee as with the fairy wren, we are in somewhat incomplete systems of referential communication. The chickadee doesn't tell us exactly who the predator is. But the information is precise enough to induce appropriate behavior on the part of the chickadees listening.

Finally, the last system encountered in birds is similar to the combinations of calls I was telling you about in some monkeys. Many species of birds combine calls or notes in sequences of greater complexity to indicate the type of predator or how far away it is, and the receiving individuals respond appropriately. Take the Japanese great tit *Parus major minor*, which looks a lot like the European great tit. It produces alarm calls (called ABC) when it detects the threat of a predator, a recruitment call (D) when it wants to attract other tits in a seemingly harmless environment, and a combination of these calls (an ABC-D sequence) when it wants to motivate other tits to harass a stationary predator, such as a bird of prey perched on a branch. Toshitaka Suzuki, an assistant professor at Kyoto University, recently showed through playback experiments that individuals receiving these signals respond to ABC by scanning the surroundings, to D by approaching the speaker directly, and to ABC-D by mixing these two responses—i.e., by approaching while scanning. If the order is reversed, by presenting D-ABC to the birds, they seem to be annoyed because they do not express either response.[498] These results suggest that tits also use the order of the calls like a kind of syntax, to extract information. I'm not implying that the process observed in tits is strictly identical to the syntax of words in human language. A tit's call is not a word—not in its structure and probably not in what it provokes in the most intimate part of the bird's brain. However, a recent study has shown that Japanese great tits who have listened to the alarm call they usually make at the sight of a snake are more likely to spot a snakelike object later on.[499] These results show that animals can have complex composition rules for their acoustic communication units. The language of the human species is a particularly remarkable example of this, but it is not the only one.[500] Indeed, there are still many more surprises in store for you. I'm going to tell you a few more stories that will make you even more aware of the complexity of acoustic communication systems.

A sophisticated example of information encoded in a sequence of calls is given by the southern pied-babbler *Turdoides bicolor*. Big as a blackbird, black and white, with a slightly curved beak, it lives in the savanna of southern Africa in social groups of 3 to 15 individuals. In one group, only the dominant pair reproduces. The other individuals help defend the territory and the nest, incubate the eggs, and feed the young. The birds also feed in groups, searching the ground for various inverte- brates. Like the meerkats, they spend most of their time with their heads more or less buried in the ground, so acoustics are their preferred means of communication. Sabrina Engesser and her colleagues from Simon Townsend's team at the University of Zurich have observed that bab- blers produce a first type of call, a slightly hoarse one, in response to a sudden but not very intense threat (another approaching animal, for example), and another call, a repeated hissing sound, to attract other animals on the move.[501] Remarkably, southern pied-babblers combine the alert call and the recruitment call in sequence to lead their compan- ions to harass a land predator. Encouraged by this initial result, Engesser and Townsend went much further.

Human language is made up of combinations of meaningless units that we can easily discern by ear. These units can be combined to create a potentially unlimited set of information-carrying signals. For example, consider the two words *arc* (a/r/c) and *car* (c/a/r), which have dis- similar meanings. What differs between the two words from an acoustic point of view is the combination of the sound units /a/, /r/, and /c/, each of which has no meaning in isolation, but which in combination comprise the respective meanings of *arc* and *car*. Do we find similar phenomena in nonhuman vocalizations—meaningful signals consist- ing of combinations of meaningless sounds if they are isolated? This is what Engesser and Townsend sought to find out with the chestnut- crowned babbler *Pomatostomus ruficeps*. Their work was published in the prestigious magazine *PNAS* in September 2019.[502] Let's take a closer look. The chestnut-crowned babbler has two very similar calls in its vocal reper- toire: the flight call, which serves to coordinate group movement, and the prompt call, which is used when feeding nestlings. The flight call is composed of two successive sound units, F1 and F2. First, to find out whether the birds were able to distinguish between F1 and F2, Engesser conducted an experiment known as habituation-dishabituation. If a bird is repeatedly made to listen to the F1 unit, it eventually loses interest in the loudspeaker. But if F2 is then emitted, it looks again in the direction of the speaker, showing its ability to distinguish between the two units. So far, it's clear, but pay close attention to what happens next because things get a bit complicated. The next step was to test the bird's ability to distinguish between the two F units and three other units, P1, P2, and P3, which in combination characterize the prompt call. The results showed that the birds perceived F1 to be identical to P2, and F2 to be indistinguishable from P1 and P3. In other words, this babbler builds its calls with a vocabulary of two sound units (on the one hand, F1 and P2 are equivalent; on the other hand, F2, P1, and P3 are also equivalent). If we denote the first sound unit as A and the second as B, the flight call will be written AB, while the prompt call will be written BAB. So these two calls are different combinations of the same two units. In a playback experiment, Engesser checked that each of the calls induced a different behavioral response. However, when the sound units were played alone, there was no noticeable change in the listeners' behavior. Conclusion? The calls make sense while the units do not—just like in *arc* and *car*. Some nonhuman animals can therefore combine sound units to construct vocalizations that make sense in a way that is reminiscent of how we construct our words.[503]

Besides syntax (combining units into meaningful sets), human language is famous for what is called pragmatics. You may never have heard this term, and yet you know what it is because you practice it every day—just as the famous character Monsieur Jourdain from Molière's play was practicing prose.[504] *Pragmatics* encompasses the processes by which the context, or more generally our knowledge, influences our understanding of words or phrases. Some examples of pragmatics? Here are two. If a parent says to a child, "Your room is a mess," the child will understand "You have to clean your room," because he/she knows that the parent expects no less. If a friend had said the same exact thing, the child might not have come to the same conclusion. Here's another example: "Did you know that Anthony has quit smoking?" asks your companion. If you don't smoke, and you know Anthony, you can deduce a whole range of more or less probable information: Anthony must have made a lot of effort, he's probably in a difficult mood these days, be careful not to upset him, or anything else that seems relevant to you. On the other hand, if you smoke, you will probably understand what is not being said: "You should do the same!" And, sheepishly, you'll put out your cigarette, or you'll shrug off the information like water off a duck's back. It will depend on a whole set of parameters. In short, the same sentence, the same acoustic signal, can give you different information depending on the context and can play differently on your feelings and behavior. So in human language information is not just the acoustic signal. In other words, in accordance with Shannon's mathematical theory of communication that we discussed in chapter 4, the sender is not the only one involved; the receiver is an integral part of the information transmission chain. The passage of information from the sender to the receiver requires two steps: coding, which is the job of the sender who produces the sound signal; and decoding, carried out by the receiver by integrating the information given by the signal and the context. What about nonhuman communication systems? Obviously, at first glance, it seems complicated to explore the world of pragmatics in animals. However, we now have evidence to safely assert that it does play a role. And by no means just any role—a role that is all the more important in animal species living in groups with complex social organizations, since information exchange between kith and kin is more developed and indispensable. Let us again follow Dorothy Cheney and Robert Seyfarth, whose research on vervets we spoke of earlier in the chapter.[505] Cheney and Seyfarth studied chacma baboons from the Okavango Delta in Botswana for years, and they came home with an incredibly detailed knowledge of their social life and communications.[506]

Chacma baboons live in groups of 50 to 150 individuals. If a baboon becomes isolated from the group, even the bravest male feels lost. And not without reason, because it is in great danger of falling prey to a leopard. I remember seeing one, near the Khwai River in Botswana, wandering alone on the edge of the woods after its group had been dispersed by rangers because the monkeys had savagely attacked a camp. The poor male barked once or twice a minute in a loud voice. Its calls lasted most of the night, which moved us a lot. Among chacmas, recognition happens at a distance by voice. The call repertoire is relatively limited, even though each call can vary (the repertoire is graded as in hyenas): grunts, threat grunts (emitted by individuals of high rank toward those of low rank), barking of fear, and screams (emitted by individuals of low rank toward those of high rank). Our lone male, regardless of his initial rank, must have been barking out of fear and looking for his group.

Field experiments conducted by Cheney and Seyfarth showed that the communication system of baboons has three characteristics. First, when a baboon hears a call, it evaluates whether the sender's call is addressed to it before responding. Cheney and Seyfarth give the following example. If two animals have just finished fighting, and then one hears a threatening growl from the other, it will respond as if the threat was intended for it. But if both had previously been enjoying some mutual delousing, the listener will behave as if the growl was directed at another monkey. Same call, two different contexts, two different interpretations. The pragmatics are already there. Second, calls facilitate social interaction. If a female approaches a mother with her cub by emitting growls, the mother will eventually be able to entrust her cub to her. But if she approaches silently, the mother will run away with her baby. It is likely that, in this case, the growls signal peaceful intentions. Again, the receiver's response depends on its assessment of the sender's intentions, like the child being told, "Your room is a mess." Third, the receiver monkey combines the information encoded in the acoustic signal (the type of call, the identity of the sender, etc.) with information from its memory of past events ("we just had a nice grooming session, it's not me that Paul is aggressively talking to"; "I remember Paul, what he says is trustworthy"). Also involved is the knowledge acquired about the sender's relationship with the receiver and with the other members of the group ("I have already confronted Paul—he is stronger than me"; "Paul is Peter's ally, who is the ally of Mark, who is my ally. So it is probably not to me that Paul addresses his threats"). We can see that baboons have a very detailed knowledge of the social network in which they are integrated.

If an individual knows that the female Anna dominates the female Julia, and yet it hears Julia make a threatening growl followed by Anna barking with fear (a situation that Cheney and Seyfarth, and others after them, have obviously mimed by playback), the receiving individual marks its surprise. Wow! It's a bark that violates what it knows about the hierarchical relationships of the group. Many similar experiments have been conducted, always with the same conclusion. If we consider the isolated vocalizations of baboons, they are rather generic signals, associated in a rather vague way with affectionate or, on the contrary, aggressive behaviors, or with fear or alarm. But, as Cheney and Seyfarth point out, these vocalizations are not produced in a social vacuum. Quite the contrary. Each one occurs in a context where sender and receiver know each other personally and share a rich, common history. When one vocalizes, the other completes the information given by the vocalization with information taken from the context. A call carrying very general information will then become very informative. Pragmatics, I told you!

Whether in baboons and other monkeys, meerkats, birds, or of course our beloved hyenas, the communication systems described in this chapter are based on sound units, which the sender combines in a more or less complex way to encode information, according to rules specific to the species but with a large degree of freedom. On the receiver side, the decoding of information depends on the acoustic structure of the signal and practical inferences based on social knowledge. Although each animal species, including humans, has its own communication system, there are common rules. Can we talk about languages? In chapter 18, the last chapter of the book, I answer this question. And I adopt a position. But for now, let's continue our sound journey by going to see how these animal vocalizations—like ours—can express emotions.

15

Ancestral fears

THE ACOUSTIC EXPRESSION OF EMOTIONS

Crocoparc, Agadir, Morocco. Late afternoon. The Nile crocodiles are lying on the banks, enjoying the last rays of sunshine. The weather is still good. Jasmine exhales its sweet fragrance. All is calm. Suddenly a human baby cry breaks the silence: "WAAAAaaa-WAAAAaaaa!" A curious wailing. And our crocodiles, especially the females, throw themselves into the water to swim vigorously toward the guilty speaker. There are 5, 10, and soon 20, irresistibly attracted by the distress signal of a kiddo who is not of their world. But the modulations of the little human's cry—its rapid, high-pitched crescendos, its rough-hearted side—evoke for them the calls of their own children, and from the depths of their hearts comes their protective instinct. Quickly! We must go and rescue this baby fast, it is calling for help! . . . I am in Luc Fougeirol's zoological park, with my colleagues Gérard Coureaud and Nicolas Grimault from the CNRS, our students Julie Thévenet and Leo Papet, and Niko Boyer, our technician. A few hours earlier we were in the middle of a discussion: Can we understand the emotions encoded in the vocalizations of species other than our own? Of course, it is not very difficult to guess from your dog's voice if it is afraid or in pain. But what about crocodiles? They are much more different from us. Do you think they perceive emotions encoded in mammalian vocalizations? I had recordings of human babies' cries on my computer, and the rest was obvious:

"Why don't we ask them?" When we saw these Nile crocodiles grouped in a star shape around our loudspeaker, we had our answer.

You may remember that crocodile mothers (and sometimes fathers in some species) stay with their young after hatching, protecting them from predators. When a small crocodile is isolated or under threat from a predator, it gives a distress call, and the mother drops everything to come to its rescue. I have observed this behavior many times during field playback experiments with different species, from the jacaré caiman to the Nile crocodile, the spectacled caiman, the black caiman, and the Orinoco crocodile. During a trip to Venezuela with Thierry Aubin, we observed that the Orinoco crocodile and the spectacled caiman respond indiscriminately to the distress calls of the young of either species.[507] The acoustic differences, both real and audible to our ears, are certainly not huge. That an Orinoco crocodile understands the distress encoded in the calls of a black caiman is not surprising, that's for sure. After all, they are very similar species—their common ancestor does not go back very far. But the experience at the *Crocoparc* in Agadir with these Nile crocodiles reacting to the cries of human babies suggests that this interspecies communication can extend much further, between very distant animal species. Is this an exceptional case, a curiosity? Or do general rules of coding exist for certain information, especially emotional information, that would cross the animal kingdom? This is what we are going to talk about here.

In *The Expression of Emotions in Man and Animals*, Darwin describes many species that use sounds to express their emotions.[508] He points out that animals that are usually not very talkative, such as rabbits, will produce shrill calls when they are in pain. He says that many also vocalize in comfort situations, such as when they find a lost companion. He was also aware of the isolation calls from mammalian mothers and their young as well as the threatening calls that make an opponent back down in the face of obvious rage. Finally, he also emphasizes that vocalizations are not the only acoustic manifestations of emotion. The porcupine threatened by a predator makes its quills vibrate with fear or rage, we don't know for sure. In short, in sketching a picture of animal sound signals, Darwin suggested that the sound codes used to express emotions may be widely shared, at least within mammals and perhaps beyond (he reports that the buzzing of bees changes when they are angry). Over the past two decades, research on this topic has expanded considerably. It has a practical objective—that of understanding the mental health of animals kept in captivity in order to improve animal welfare. It is possible to estimate how an individual feels by recording and analyzing its vocalizations, without manipulating it and in the absence of invasive procedures (such as blood tests that would otherwise have been necessary to measure stress hormone levels, for example). So what are these acoustic markers and how do they code for emotions?

Elodie Briefer, a professor at the University of Copenhagen in Denmark and former doctoral student of Thierry Aubin, is a world-renowned specialist on this question. Her aim is to identify acoustic markers and assess the extent to which they may be valid between species.[509] According to Elodie, "Emotions are defined as intense, short-lived affective reactions to specific events or stimuli of importance for the organism. Their crucial function is to guide behavioral decisions in response to these triggering events or stimuli, through approach or avoidance. . . . An emotion can be described by two main dimensions: its valence (positive or negative) and arousal (its level of alertness)." To describe emotions, Elodie supports the bipolar model, with an axis of negative and unpleasant emotions ranging from depression to fear, sadness, and anxiety, and an axis of positive and pleasant emotions ranging from a feeling of calm and relaxation to joy and excitement. Elodie and other scientists have shown that this classification of emotions is expressed through measurable neurophysiological and behavioral indicators. Some indicators, most of them neurophysiological, are more indicative of the level of arousal or alertness: heart rate and its variability, breathing rate, skin temperature, skin conductivity (an emotion can lead to an increase in sweating, which, even if weak, results in the skin's ability to conduct electrical current), the secretion of certain hormones, etc. In addition, and as you have probably already seen with dogs, behavioral postures can also provide information about the animal's emotional state and are more linked to the emotional valence. Ear and tail position, bristling fur, eye movements—all of these behaviors leave little doubt as to the mental state of the animal, relaxed or aggressive, as it approaches you. If the dog whines gently or if on the contrary it growls, there will be little room for doubt.

Let's be precise. How can vocalizations encode information about the valence and arousal of an emotional state? An experiment conducted by Elodie with horses is very informative here.[510] Elodie placed horses in four situations, inducing different levels of arousal with negative or positive valences. As everyone knows, horses are very social animals that, when left in the wild, live in harems (a stallion, mares, and their foals) or in packs of young single males. Elodie and her colleagues have worked with groups of horses. Within each group, the individuals knew each other well and had formed social bonds over a long period of time. The horses emitted a varied repertoire of calls, including neighing, snorting, and squealing. Scientists have focused on neighing, which is known to be produced by horses both when they are separated from their companions (negative valence) and when they are reunited with them (positive valence). We have all heard it before. A typical neigh begins with a rather high-pitched introduction ("huuuh . . ."), then continues with a long, strongly frequency-modulated "climax" ("huh huh huh huh huh . . ."), and ends with a low-frequency, low-intensity "finale" ("BRRrrrrr . . ."). Sounds familiar, doesn't it? In a first experimental situation, neighing was recorded when an animal was totally isolated from its companions—a particularly negative situation for an animal that loves company. Other times, the horse was recorded while it was separated from its best companion, while most of the group remained with it. Still a negative situation, but a little less stressful. Finally, horses were recorded during reunions, either with the entire group for those who had been completely isolated, or with their best companion. These last two situations were therefore of positive valence with, Elodie hoped, a different degree of arousal. To check this, she measured the level of arousal objectively, recording the heart rate of the horses in each case. This measurement indicated three levels of arousal: a low level when a horse was reunited with its favorite companion (heart rate around 44 beats per minute), a medium level when its favorite companion was taken away or when all horses from the group came back after a separation

(50 beats per minute), and a high level when the horse was isolated from all of its fellow horses (56 beats per minute). What about neighing? The analyses carried out showed that the acoustic structure of a neigh is complex. It is a kind of double whistle, known as *biphonation*. It's as if two sound sources are combining their effects. The two frequencies are called F0 and G0.[511] In the horse, complex vibrations of the vocal cords are probably responsible for biphonation. Neighing produced at high levels of arousal has a higher pitch, with a higher F0 than that produced at low levels of arousal. This result is consistent with what has been observed in other animals, such as the pig, the squirrel monkey, and even the zebra finch: when an individual experiences a strong emotion, its voice normally becomes higher-pitched.[512] The novelty of this horse study is that Elodie and her colleagues observed neighing of shorter duration, with a lower G0 frequency in emotionally positive situations in comparison to emotionally negative situations. In addition to expressing the level of arousal, the horse's neigh thus contains information about the valence of the emotion experienced by the sender. This dual coding of information about the sender's emotional state, including *arousal* and *valence*, is likely to exist in vocalizations of other animal species (it is found in goats, pigs, and wild boars, for example).[513] It is probably enough to look for it to find it! More work to be done.

While expressing emotions through vocalizations is probably a fairly common feature of animal sound signals, are there common rules for encoding this information throughout the animal kingdom? Deciphering the universal encoding of emotions is the challenge that scientists like Elodie are tackling. This challenge is not new, since Darwin himself proposed, in another of his books, the idea that the vocal expression of emotions could have its roots in the ancestor of terrestrial vertebrates, and that today's animals share its key characteristics.[514] Based on the premise that all vertebrates living in the aerial environment have lungs and a trachea for inhaling and exhaling the outside air, Darwin suggested that, when these early vertebrates were particularly excited and contracting their respiratory muscles tightly, they would inevitably produce sounds. Darwin hypothesized that these primitive sounds may have been useful in signifying to the others the excited state of the individual sender. The process of natural selection to increasingly sophisticated speech devices would thus have been set in motion. A century later, Eugene Morton, a researcher at the Smithsonian Institution in Washington, DC, stated the principle that the acoustic structure of a sound signal should reflect the motivation of the sender (motivation-structural rules).[515] Morton's principle is simple: birds and mammals use rather low-pitched and "rough" sounds when they have hostile intentions, and rather high-pitched and pure tone sounds (whistles, if you like) when they are frightened or animated by friendly intentions. Morton was not really talking about emotions, but he emphasized the relationship between the physical structures of sounds and the motivation behind their use. Recent studies by Elodie and other scientists have clarified these relationships. The vast majority of these studies show that vocalizations are louder and produced at a faster rate, higher-pitched, and more frequency-modulated when arousal increases. These variations in sound correspond closely to the effects of anatomical and physiological changes linked precisely to variations in intensity. Moreover, the valence of an emotion is also encoded by acoustics, with positive vocalizations shorter and less modulated than negative vocalizations. As we know, Darwin was an excellent observer and a brilliant visionary.

Coding emotions into vocalizations is one thing. Being able to *decode* them when you hear acoustic signals is another. If it is easy to distinguish between the threatening bark of an angry dog and the disarming whimpers of your favorite Rex, can we finely decode the information about emotions in the vocalizations we hear? And on what acoustic basis? Research has mainly focused on the distress calls of babies, human and nonhuman, when they feel in danger, bringing parents or any other individual in charge of their protection back to them. For convenience, let's call them cries. This term should not surprise you if you have ever heard the plaintive wails of a lamb or puppy taken away from its mother. They're definitely crying. One of the research projects I'm doing with my colleague David Reby (Remember?), a specialist in mammalian voice production, is looking at the cries of human babies. We want to identify the information carried by these cries, analyze what acoustic traits they are encoded by, and study how the individual receivers (parents among others) decode them—in other words, how information is transmitted from the baby sender to the adult receivers. Alexis Koutseff, then a postdoctoral student at our laboratory, and my colleague Florence Levréro recorded babies aged two to three months while they were bathing (not all babies cry during their bath, but we managed to record some of them anyway) and the same babies when they were vaccinated at the pediatrician's.[516] In addition, the babies were recorded under two different conditions. Some were given a first injection of a vaccine that was supposed to be painful and then a second injection of another vaccine that would be less painful. Other babies received the same two injections, but in reverse order: the mildly painful first, followed by the more painful one. Acoustic analysis showed that the recorded cries can be placed in a two-dimensional acoustic space, and that they move around in that space according to the pain the baby is experiencing. Imagine two lines (two axes) that intersect at right angles and delineate four quadrants: upper right, lower right, lower left, upper left. The horizontal line represents the *roughness* of the cry, ranging from the most harmonic ("ouuinn") on the left to the most guttural ("iiirraahh!") on the right. The vertical line represents the pitch of the cry, ranging from the lowest-pitched cries at the bottom to the highest-pitched at the top. The cries recorded during the bath are in the quadrant at the bottom left ("ouuuiinn"). Let's follow the trajectories of the cries caused by the vaccines. Those following a first injection of the painful vaccine are resolutely on the right ("iiiiRRRRhh!"). Once the injection is finished, the cries move to the point where the axes cross, becoming less rough ("iiiiAaaarr!"). The pain subsides a little. The second vaccine, which is supposed to be less painful, is then injected. The cries start again on the right, without, however, reaching their first position ("iiiiRhhh!"), then return to the center of the acoustic space ("iiiiAaaarr!"). The acoustic characteristics of the cry therefore reflect the level of pain experienced by the baby. Roughness is obviously the best marker: the more intense the pain, the more the cry grates on the ears. The second, less reliable marker is the pitch of the cry, the fact that it is more or less high-pitched, with a tendency to be higher-pitched when the pain is stronger.

The group of babies who received the vaccines in the reverse order (least painful first) would teach us that the order of injection influences the baby's pain. The first cries caused by the least painful vaccine are slightly shifted to the right from the vertical axis. It hurts, but not that much. Once the injection is finished, they cross the vertical axis to the upper left. That's better! The second injection, of the painful vaccine this time, makes them come back to the center, so much less to the right than when this vaccine is injected first. You get used to everything. . . . The last cries are in the lower left quadrant, similar to the bath cries, harmonious and lower-pitched. The baby is still not happy, but it is no longer in pain. The moral of the story is that the least painful vaccine should be injected first. That's a useful bioacoustic experiment, isn't it?

To see if adult listeners can extract pain information as efficiently as our acoustic analysis, we conducted psychoacoustic experiments. The principle is simple; it is a matter of making people listen to cries and asking them to rate them on a pain scale between 1 (no pain) and 7 (very strong pain). Our results showed that adults are able to evaluate the pain perceived by a baby based on the acoustics of its cries. They could easily distinguish between cries caused by an injection and those emitted during bathing. However, the tested adults were not very good at distinguishing variations in the level of pain; they did not detect the difference between the cries provoked by the two vaccines. Perhaps this is because they heard cries from unfamiliar babies. I'm fairly confident that people who know the baby they are caring for can measure the pain level very accurately by hearing their own baby's cries.

Rules for encoding emotions across the animal kingdom suggest the possibility of communication between species. A recent study has confirmed that we humans are able to distinguish between vocalizations marking different levels of arousal, even if these signals come from animals as varied as a monkey, a pig, a panda, an elephant, a bird, an alligator, or . . . a frog![517] Susan Lingle's work at the University of Winnipeg, Canada, points in the same direction: Mule deer and white-tailed deer mothers are attracted to the baby cries of a variety of mammalian species.[518] Unsurprisingly, all these vocalizations follow Morton's principle. But then again, are we able to accurately judge these vocalizations? That's the

Chimpanzee

question Taylor Kelly, one of my students when I was a visiting professor at Hunter College in New York, and I asked ourselves. Taylor made adult women and men listen to the cries of human babies, as well as the cries of baby chimpanzees and bonobos.[519] The results showed that the people tested assessed the degree of distress expressed by the vocalizations based on their pitch (high/low). They applied the same rule to all cries, those of both baby humans and baby apes. Because bonobo calls are extremely high-pitched, all were consistently rated as expressing severe pain. On the other hand, the cries of baby chimpanzees, which were quite low-pitched, were recorded as indicating low levels of distress. These results show that, in the absence of familiarity (our listeners had probably never heard baby chimp or bonobo cries before), our ability to assess the emotional content of our closest cousins' vocalizations is biased. The interspecific value of the signals coding for emotions, if it is real, should not be overestimated.

What about our crocodiles? They have no tail posture, no eye rolling, no bristling fur. Encouraged by our first observations with the Niles of Agadir responding to the cries of human babies, we repeated the experience with the cries of baby chimpanzees, then baby bonobos. Each time the crocodiles reacted, running toward the loudspeaker. On closer inspection, however, we found that the number of crocodiles approaching the loudspeaker and their readiness to move varied according to the sound stimuli. The cries of human babies recorded during a vaccination session were the most attractive; those recorded during a bath were much less effective. In conclusion, crocodiles also perceive the level of stress encoded in the cries of human babies. By comparing their reactions to human and ape cries, it appears that the *rougher* the cry, or more unpleasant to hear, the more irresistible it is.

These reactions to signals that do not belong to one's own species raise the question of a possible contagion of emotions through vocalizations. We humans know this very well. A call or a cry can give us goose bumps or bring tears to our eyes. A happy laugh will put us in a good mood. We are accustomed to these phenomena of empathy—when we feel what others feel. Now, do our female crocodiles, who rush to the cries of a human baby as well as to the calls of their own babies, really feel the distress expressed in these vocalizations? I think it's likely. There is no objective reason why evolution has not allowed or even promoted empathy in crocodiles (although it is possible, of course, that crocodiles are attracted to crying human babies for less friendly reasons—to eat them, for example). It seems to me that a mother Nile will be more willing to spend energy to come to the rescue of her young if she actually feels the distress. To support this point, there is growing experimental evidence in mammals for emotional contagion through the voice.[520] This contagion even appears indispensable in animals living in groups and having to coordinate or cooperate. It is reasonable to assume that the fright experienced by the vervet monkey or meerkat seeing an eagle and triggering the alarm spreads to other members of the group. Primatologist Julia Fischer, director of the German Primate Center, has observed that chacma baboons pay more attention to alarm barks produced in response to dangerous predators than to other alarm vocalizations.[521] You may remember the chacma who had lost his group and whom I heard barking plaintively for hours. The emotion I felt when I heard him cannot be far removed from what one of his group mates would have felt if they had been able to hear him.[522] The contagion of emotions through the voice is probably a very common phenomenon in mammals. It is certainly highly developed in those with complex social relationships, such as monkeys, apes, meerkats, or hyenas, when understanding how the other individual feels is particularly important.

And in other animals? In birds, for example? When we were studying zebra finches in our laboratory, my colleague Clementine Vignal had the idea of looking into this question. Do you remember from chapter 11 that this bird cannot do without its fellow birds? I have already told you that, when we tested the ability of males to identify their partners by their voice, we had to introduce companions into the experimental room. Alone in the room, the male responded poorly to the playback of its female's calls, and did not differentiate between its partner's calls and those of other females.[523] We then hypothesized that the stress caused by loneliness must be significant and must explain this difference in behavior. To test this hypothesis, we repeated the same playback experiment, but this time we paid attention to the blood level of a stress marker hormone: the corticosterone. In this new experiment, some males that were with companions (control group) heard calls from females while in the presence of mates. Other males (isolation group) heard them when they were alone. Finally, other males ate a few seeds soaked in the stress hormone just before hearing the playback signals in the presence of mates (CORT group). Clementine's doctoral student recorded the test males throughout the playback. She then analyzed the acoustic structure of their calls. As expected, the males in the isolation group called less than those in the control group. More importantly, their calls were different—higher-pitched and longer. The CORT males called just as much as the controls, but their calls resembled those of the birds in the isolation group. Blood tests showed that the isolated birds had more of the stress hormone than the control birds. Let's summarize the situation. In response to calls from females, a male that feels stress (either because it is stressed from being alone or because it has just ingested the stress hormone) produces calls that are higher in pitch and longer than a bird accompanied by other birds and therefore not stressed.[524] So the calls of the zebra finch provide information about its emotional state. But the story doesn't end there. According to other experiments, if the female listens to the calls of her stressed male mate, the amount of stress hormone circulating in her blood increases.[525] You read correctly: the partner's voice is enough to stress the female! She doesn't need to see him to feel the same way he does. When she hears her partner call out to her in an anxious voice, the female zebra finch is also frightened. The contagion of emotions through the voice exists in birds. Quite an interesting discovery. Mammalian vocalizations are not the only ones that express emotions.

The fact that emotions are coded according to rules that transcend the animal kingdom has interesting applications. To end this chapter, let's look at two of them: communication between humans and domestic dogs and the use of distress signals to frighten birds.

At the risk of disappointing you, I have no particular fondness for pets. (Nobody's perfect!) However, when I was a professor in New York, I had to find subjects to study. New York is an extraordinary city, of course, but in terms of wilderness it's not ideal. One day, Tobey Ben-Aderet, one of the students whose master's thesis I was supposed to supervise, suggested that I take an interest in the communication between humans and dogs. "To pay for my studies, I work in a kennel. It would be easy for me to record them. It is a subject that fascinates me." So off to the dogs we went! I sought advice from my colleague David Reby, who is a true dog lover. Then we thought about the scientific question that Tobey could tackle.

When we talk to our babies, we use a particular voice register, characterized by a higher and variable pitch, a slower rhythm, and clearer articulation. Why this habit? Perhaps because of the emotion caused by the toddler's tender face. Maybe because it makes it easier to learn language through hearing. Or maybe both? In any case, what we find is that we talk to our dogs in a slightly similar tone. Is this a consequence of the juvenile appearance of some of our four-legged companions, or an attempt to interact with a living being who is not gifted with language and whom we judge to be of limited intelligence? In the first case, we should restrict "baby talk" to puppies. In the second case, we should continue to talk to adult dogs in this way. Interesting hypotheses. Tobey's scientific objective was well established.

First, we had to record "dog-directed speech." Tobey presented photographs of dogs to people who had to say the following sentences,

Dog

while imagining themselves talking to the dog in the photo: "'Hi! Hello cutie! Who's a good boy? Come here! Good boy! Yes! Come here sweetie pie! What a good boy!" The same people then had to say those sentences again, but in a neutral tone. Acoustic analysis of the recordings confirmed that our way of talking to dogs is very similar to "baby-directed speech." And above all . . . that we use this register regardless of the age of the dog.

You're probably thinking, What do the dogs think? Well, Tobey asked them straight out. With the help of another student, Mario Gallego-Abenza, she played her recordings to puppies and adult dogs. Can you imagine the scene? The dog is in a comfortable room, and a loudspeaker on the floor is playing the phrase pronounced in "dog-directed speech" and then the same sentence pronounced in a neutral tone. The results of the experiments have the merit of clarity. When the tone was neutral, both adult dogs and puppies showed very little interest in the loud-speaker. On the other hand, the puppies became very excited when listening to the "dog-directed speech," barking and approaching the speaker, while the adult dogs ostensibly ignored the recording.

What conclusions can be drawn from this? First, dog-directed speech is effective for engaging an interaction with a puppy, but loses its effectiveness for an older dog. The older dog is probably waiting for other signals from us, such as gestures or facial expressions. Or it may not respond to an unfamiliar voice coming out of a loudspeaker. Yet we talk to adult dogs like babies. The reason may be that, consciously or unconsciously, we may feel that the dog may not understand us. As if it were one of our babies, we hope it can make progress. If we talk to our dog in this way, it is because we would like the dog to one day respond to us . . . by talking.[526]

With this example of dog-directed speech, we can see that there is still progress to be made in terms of communication between species.[527] However, sometimes precise knowledge of the coding of information makes it possible to finely, and I would even say artificially, control an acoustic communication system. This is what is done by using distress calls and their synthetic avatars to scare birds at airports.

Birds have always been a challenge for airplanes—unless it's the other way around, you might say, but that's beside the point. On September 7, 1905, Orville Wright's plane inaugurated what was to be a long series of collisions between birds and airplanes in flight.[528] These accidents occur mostly at low altitude or when the aircraft is taking off. The cost is considerable, several hundred million dollars a year, not to mention the loss of human (and bird) life. How can we limit these collisions? The method considered the most effective and least expensive is to use distress calls to disperse the birds. Distress calls are vocalizations emitted by most birds when they are captured or injured by predators. They differ from the alarm calls that we have discussed extensively, both in the circumstances in which they are emitted (an individual producing an alarm call is not in the claws of a predator) and in their acoustic characteristics. Distress calls appear as complex, highly frequency-modulated sounds (i.e., rapidly changing from high-pitched to low-pitched and vice versa), lasting about half a second ("Zeeoooop!"). When birds hear a distress call, they react very strongly, fleeing on the wing. Sometimes, as with seagulls, they fly over the area where the call comes from before flying off into the distance. But the result is there, very clearly; the distress calls are amazing bird scarecrows.

You may remember that I prepared my doctoral thesis in Jean-Claude Brémond's laboratory, where Thierry Aubin also worked. These two researchers devoted part of their careers to the study of these distress calls. The acoustic details differ according to the species of birds—the distress call of the black-headed gull is not exactly the same as that of the starling, for example. However, Aubin and Brémond observed that the similarities are strong and that there is a true acoustic convergence between the distress calls of different bird species. This convergence is accompanied by an interspecific information value: gulls react to starlings' distress calls as they would to those of other gulls. Distant animal species, such as deer or wild boar, do the same. In fact, different species use the same law for decoding distress information.[529] Based on this common code, Aubin and Brémond developed computer-made signals, mimicking distress calls but exacerbating the acoustic parameters carrying the distress information, thus amplifying their effectiveness and limiting habituation. What a sight it was to see how they were able to test the effectiveness of their signals: stationed in a car some 150 meters from a group of about 40 herring gulls, black-headed gulls, lapwings, ravens, and starlings, with the loudspeaker on the roof suddenly emitting about 60 distress calls per minute. All the birds flew away, then dispersed into the distance—exactly the desired effect on an airport runway before a plane takes off.[530]

From the calls of baby crocodiles to the distress signals of birds, from the cries of baby humans to the neigh of joy of the horse reunited with its companions, you can see that acoustic signals powerfully encode the emotions of the sender. Besides the static information related to the stable characteristics of the individual (its size, age, sex, genetic inheritance, etc.) that define its vocal signature, sound signals also carry dynamic information that reflects the individual's current state. The challenge for the receiver is to decode all this information in order to correctly interpret the received signal.

16

The booby's foot

ACOUSTIC COMMUNICATIONS AND SEX ROLES

Village of San Blas, Nayarit Province, Mexico. The rendezvous on the beach where we are to board was scheduled for six o'clock in the morning. But the beach is empty, and the village next door is still asleep after yesterday's party. We are waiting patiently. The sun rises quickly. On the horizon a flight of pelicans materializes—at least they are awake and going fishing! A few hours go by before two men arrive. Not very talkative. One carries jerricans, approaches a boat, and starts to fill the tank, with a cigarette in his mouth. *"Aqui esta tu barco. El es el que te conduce,"* he says, pointing to his massive companion. Moments later we are on our way, quickly lost in the blue vastness. From the village to the island is 70 kilometers. Four hours in the boat! It all started with a phone call from Fabrice Dentressangle, a former student of whom I had very fond memories. "Nicolas, I'm preparing my doctoral thesis with Roxana Torres in Mexico. On the reproductive behavior of the blue-footed booby. I'm working on an island paradise!" And he continued, "In the blue-footed booby, the female and the male have totally different calls. They take turns at the nest, and I'd like to know if they recognize each other vocally. Wouldn't you like to come and pay me a little visit?" You bet! I immediately took the opportunity. A phone call to my friend Thierry, plane tickets, and a few months later, we were there. "Did you see that?" asks Thierry suddenly. "A shark's fin! These waters must be infested with them." We're happy to be away again, disconnected from the world. As in a dream, the mysterious *Isla Isabel* approaches.

The blue-footed booby *Sula nebouxii* is a surprisingly good-looking seabird. You may already know them because they are often seen in documentaries about the Galápagos Islands, where they are present in large numbers. It is a large bird, with a long dagger-shaped beak, perfect for harpooning fish; a big head and neck speckled with brown; a light eye with a distinct black pupil; a brown back; and a white breast. As its name suggests, its feet are blue. You must see the female and male displaying together. They walk side by side, each raising their feet high, their toes as far apart as possible to spread their azure webbing. And each one calls out. The female makes her "quack-quack," the male his "phiew-phiew." This difference in their calls is almost the only clue to telling them apart. Along with pupil size: a pinhead in the male, rounder in the female—not easy to spot from a distance. Sexual dimorphism comes down to very little in the booby.

Isabel Island is a volcanic islet less than a kilometer wide and a kilometer and a half long, where thousands of seabirds breed. Thousands of frigate birds, brown boobies, red-footed boobies, blue-footed boobies, brown noddies, seagulls, and pelicans flock here among the iguanas, omnipresent on the island. The blue-footed boobies settle on the top of the beach, a nest every hundred feet. They are a monogamous species, where female and male share the incubation of the eggs and the rearing of the young. The parents take turns on the nest. While one fishes at sea, the other stays ashore. The reunion is sonorous—both partners start calling before the returning fisher has even landed.

As soon as we disembark, we find Fabrice. Present on the island for many months, he has become the manager of the camp where about 15 students are each working in their own area of study. This small colony is a real-life *Swiss Family Robinson*: a long table for meals taken together; a stove, a few pots and pans regularly visited by the iguanas, an old fridge serving as a pantry; water reserves; solar panels to recharge the batteries of the equipment; tents to sleep in, between which blue-footed boobies wander happily. The atmosphere is both relaxed and hard-working. Thierry and I have to be efficient because we have only planned a fortnight

Blue-footed
booby
♀   ♂

on the spot. Fortunately, Fabrice and his team have prepared the work well. The nests are already carefully labeled and the youngsters are all banded so that we can easily identify the animals. From the first morning, the recordings begin. Equipped with a long pole, we put the microphone right next to the nest, wait for the return of the female or the male that has left for the sea, and record the vocalizations of the pair during their re-union. About 20 pairs are "in the can" within a few days. All that remains is to conduct the playback experiments. That is another kettle of fish.

Placing the loudspeaker near the nest is not complicated, especially since the birds seem to be completely unaware of it. Unroll the cable (we didn't have wireless speakers), put ourselves at a distance, and wait for the right moment—that is, the moment when the bird left in the nest, whose partner's calls (or those of an unknown individual of the same sex) we are going to play, is calm. When the young are not beg-ging. When there isn't a big iguana coming to get who-knows-what that's scaring all these little guys off. . . . Here we are. First try. The fe-male call comes out of the speaker. Surprised, the male glances briefly up at the sky. Then . . . nothing. "Here's a response that is going to make life difficult," says Thierry, used to the exuberant vocal demonstrations of penguins. Invoking one of his favorite phrases, he declares, "We have to make a decision." On the way back to camp walking on the beach, the decision is made: we will conduct these experiments rigorously but blindly. I will play the signals from the tape recorder, without passing comment. Thierry will evaluate the behavior of the bird being tested, without knowing whether it is the call of the partner or that of a stranger that we are playing—seriously frustrating for Thierry who, as the days go by, will never know if the results of the experiments validate or not the hypothesis of a partner's voice recognition. But it's the only way to be sure not to bias our data.

In the blue-footed booby, female and male have very different calls. The female's "quack-quack" is weakly modulated but extends over a wide frequency range. It sounds like a cartoon duck. The male's "phiew-phiew" is wheezy, more muffled, like a failed attempt by someone practicing a whistle. Our acoustic analysis showed that these two calls can identify individual boobies with a success rate of about 45%. According to our calculations, chance would result in only a 6% success rate, so both the female and male calls have a fairly accurate individual signature. Our playback experiments have confirmed that birds in the same pair use these signatures to recognize each other. Of those individuals we tested, 11 of the 14 females and 11 of the 15 males showed a more pronounced behavioral response to their partner's voice. Parity could not be expressed more clearly.[531]

I left *Isla Isabel* with my head full of thousands of birds, whales jumping out of the water just a stone's throw from the beach, friendly conversations with people who know the value of calm, and images of flamboyant sunsets (which earned me a few laughs from Thierry: "You could have taken the same photos in Brittany!"). A feeling of paradise. One day Fabrice suggested we go down into the crater of the volcano. There, forget the sound of waves. Only the calls of the superb frigates—modern-day pterosaurs, hijackers snatching their fish from other birds—broke the silence. *Isla Isabel* takes you back in time to a world where the human species was absent.

Throughout these pages, we have seen that sexual dimorphism—the fact that female and male have morphological and behavioral differences—is quite widespread in the animal kingdom.[532] Sexual selection has a lot to do with that, with its two components, *intrasexual selection* (competition between females or between males) and *intersexual selection* (specific preference for the characteristics of the other sex). Both encourage the expression of extravagance, especially in males: imposing body size (think of elephant seals) or complex sound signals (some birdsongs), to name but a few. They contribute strongly to the exaggeration of differences between females and males. In birds and insects, it is often the males that sing to attract females and repel intruders. But let's not forget that one of the major characteristics of the living world is its diversity. This differentiation in the appearance of females and males and their behavior is not an absolute rule. Far from it. And we are not always aware of it. Among scientists, as everywhere else, the way we perceive things is tainted by ideological biases. This is even more true when it comes to studying and understanding the behavior of females and males.[533] A 2011 study published in the journal *Animal Behaviour* shows that the vocabulary used by scientists to describe female and male behavior is incredibly stereotypical, generally referring to females as passive while males are said to be the real actors.[534] In the blue-footed booby, however, the obvious vocal dimorphism does not translate into asymmetrical recognition. Female and male are equal actors in acoustic communication. There are many examples of this equality. Let's observe the shearwaters, seabirds in which the balance between the roles of females and males during communication is astonishing.

Do you remember Charlotte Curé? I told you about her research in chapter 6 when we were with the whales and dolphins. Well, Charlotte started her career with a thesis on the acoustic communications of Mediterranean shearwaters. Shearwaters are agile seabirds that only come ashore to nest. They choose cliffs where it will not be necessary to move around on the ground because they cannot walk. The nest is usually a burrow, at the bottom of which female and male take turns brooding and caring for the young. When Charlotte asked us to supervise her thesis, Thierry and I were very interested. How does a shearwater find its burrow? Previous work had shown that olfaction could play a role, as it is highly developed in these birds. But the fact that shearwater colonies are noisy suggests that acoustic communication plays a

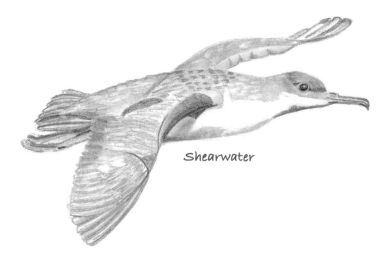

Shearwater

role. Sight, perhaps?, you might think. Wrong! Shearwaters return to their nests . . . at night.

Charlotte had not chosen an easy subject. Crawling over the cliffs in the dark, risking breaking her bones at any moment, putting her microphone or loudspeaker here and there at the entrance to the burrows to record the occupant or test its ability to identify this or that sound signal: that was her daily life! The shearwaters' call is a kind of mewing, high-pitched, strangled "eyoh-eee-eyoh-haaa." Charlotte patiently recorded and analyzed the vocalizations of many individuals and several species of shearwaters. Each time, females and males had different voices.[535] Following the example of what we had observed in boobies, Charlotte showed vocal recognition between the partners of the pair (female and male respond to each other) and detailed the acoustic parameters that supported it.[536] But in her experiments, she observed something unexpected. When she had a bird listen to calls from a stranger, the bird would answer if the stranger was of the same sex, but it would not respond if the stranger was of the opposite sex. In other words, I respond only to my partner and to calls from strangers of my sex. No extramarital relationships among shearwaters. Both female and male only speak to same-sex intruders. Intrasexual competition is mediated by acoustics.[537]

All right, that's understood. Among boobies, shearwaters, and other seabirds, there is a beautiful parity between the roles of females and males in acoustic communication. But still, let's not exaggerate. Many of the songbirds discussed in previous chapters do not fit this pattern, do they? What about the robin and the warbler we see everywhere in our gardens? And the nightingale? And the canary? For all these familiar species, we must recognize that it is indeed the male that sings, bravely exposing itself to predators, and the female that waits, wisely hidden away in the bushes. It is enough to confuse anybody! But the song, however demonstrative it may be, is not the only vocalization in the bird repertoire. Remember our discussion in chapter 11 about the zebra finch and its eight types of calls? Take the distance call, for example— the one that allows partners to reestablish contact when they've lost sight of each other. Here again, female and male produce different versions, and both versions allow equally effective vocal identification of the partner. It was Solveig Mouterde, whose thesis I was supervising with Frédéric Theunissen, who explored this question.[538]

We already knew that distance calls carry an individual signature, but we didn't know how far that signature could be transmitted. To find out, Solveig and I conducted propagation experiments. The principle is simple: a loudspeaker and a microphone are placed at different distances from each other (2, 16, 64, 128, and 256 meters). The biggest difficulty of the experiment was to find the right field—flat and free from road noise. The results showed that, even after more than 200 meters of propagation, there is still enough information in the calls for us to recognize the individual, whether it's a female or a male.[539] Our acoustic and statistical analyses revealed that a call, quite short, and whose intensity is made very weak by a journey of several hundred meters, can still be identified as Jane, Lucy, Matthew, or Mark. But what about the birds? Can they identify it?

Solveig tested female zebra finches. She asked them to distinguish between the calls of two males, according to two different experimental protocols. In the first, the females learned to distinguish between calls from two unknown males recorded at a short distance (2 meters). Once they had successfully learned this, Solveig had them listen to calls from the same males recorded at greater distances, up to 256 meters. In the second protocol, the females were tested several days in a row with different pairs of males and the calls propagated at different distances. This meant that they did not have the opportunity to practice distinguishing between the two males. The results showed that, even in the untrained situation, females were able to distinguish between calls from males up to 128 meters apart. In the first protocol, training increased the performance of the females and they were able to distinguish between the two calls from the males even if they had previously been transmitted over 200 meters. In short, female zebra finches are experts at identifying males by means of a call, even from a great distance.[540] For time reasons (a thesis must be done in three years in France), Solveig could not do the symmetrical experiments, with males. But as the individual signatures coded in the calls remain well pronounced at long distance for both male and female calls, it is quite likely that males would pass the test as well.[541] Moreover, Solveig and Frédéric showed in other experiments that the neurons allowing individual recognition of the partner are equally effective in both female and male brains in zebra finches.[542]

Let's get back to the songs—and ideological bias with regard to male singing. Even Darwin was guilty of it, suggesting that the first role of female birds was to listen and choose "the most melodious and beautiful males."[543] However, the situation is actually highly contested. A study published in *Nature Communications* reports that the species of birds where the female sings are numerous.[544] You want figures? Female song is found in 229 out of 323 species of songbirds, or 70% of the species. That's hardly negligible. By reconstructing the history of songbirds, the authors of the article show a high probability that both females and males could sing in the ancestral species, about 40 million years ago.[545] During the diversification of bird species, most of the lineages have retained singing females and males. Most of the species where females no longer sing live outside the tropics, establish territories for a short period of time, and show pronounced sexual dimorphism both morphologically (e.g., feather color) and behaviorally (females and males have different roles). These characteristics are often found in temperate environments, characterized by a reduced duration of spring and also reduced availability of food resources—constraints that explain the task differentiation between females and males. However, even this does not prevent some birds in these regions from having singing females, such as in the European robin *Erithacus rubecula*.

In many tropical birds, females and males produce equally complex songs.[546] Males and females sing to defend food resources or to attract a mate.[547] Sometimes females and males sing together, forming duets to defend a territory or simply to maintain contact, especially during breeding.[548] Let's go listen to one of these duets in the home of the plain-tailed wren *Pheugopedius euophrys*, which lives in certain regions of the Andes in South America, between two and three thousand meters above sea level. In the dense bamboo groves where these wrens are found, the duets are emitted at a high amplitude (over 80 decibels) and can be heard at over 200 meters. You can't lose your partner! But how do they coordinate when singing together?

During my visit to Eric Fortune's laboratory at the New Jersey Institute of Technology, I was very impressed. Eric is a neuroethologist, and among other topics he details precisely how animal brains can interact with each other during acoustic communication. With the plain-tailed wren, he has an excellent study model. The spectrogram of the song scrolls on his computer screen: "Woopweewoopoopwee . . . woopweewoopoopwee"; the notes flow continuously. "One bird? . . . No, two!" says Eric. "The female and her male! Incredible, right?" Incredible indeed. A remarkable duo of speed and precision, exactly as if a single individual were running through its song. Females and males alternate notes, each for less than 300 milliseconds with intervals of about 20 milliseconds. "Woopwee," begins the male; "woop," answers the female; "oop," takes over the male; "wee," ends the female—"woopweewoopoopwee." And the pair repeats this sound pattern a hundred times for two minutes. Often birds improvise with variations, a change of syllables or their order. You have to know your partner well to succeed in such a tour de force. Newly formed pairs get a bit muddled up, miss notes, and make shorter duets. Like two pianists in front of a four-handed piece, it takes practice. The first attempt is rarely a success—one goes too fast, the other too slow. It is only by listening to the four hands together, many times, that the duet is successful, as if there were only one musician. When wrens sing in a pair, they can hear themselves and their partner. This acoustic feedback allows them to constantly adjust their vocal production.[549, 550] In fact, Eric has tracked the activity of wren brains during these moments, and has shown that the singing area (the HVC) is activated chiefly when the bird hears a full duet, including all the notes. One wren knows not only its own score but also that of the other. Sound recordings that Eric and his team have made in the wild also show that the male makes more mistakes than the female. Every once in a while he forgets a note, especially when the duet lasts a long time. It's as if he can't keep up. The female continues singing anyway, but with longer rests between notes—like the piano teacher slowing down the tempo to give the student time to get back into the melody. (Believe me, I've been through this before.) However, Eric and his team do not say if the female wren gets tired of the males who make mistakes over and over again . . . ! Nevertheless, as you can see, the respective roles of the female and male in acoustic communication in birds are much more varied than is usually assumed.

Now let's jump to the human species. I've been studying the cries of babies for a few years now, and there, too, surprises await us. It was my colleague Philippe Gain, with his thousand ideas per minute, who had strongly encouraged me to take an interest in babies' crying. I quickly noticed in the myriad publications on this subject the same two questions: how to use babies' cries to identify diseases or disabilities and why babies often cry unexplainedly. Very few scientists had looked at the cry as a communication signal—the ethologist's eye had not been focused there. That was enough to arouse my curiosity. Inspired by the research carried out with Isabelle Charrier on baby fur seals, I decided to start by working on the recognition of the baby by its parents. Is it possible to identify one's own baby just by its cries? I talked to David Reby, who was enthusiastic about the idea, and we decided to start a study. We were in luck: Hugues Patural, a former high school classmate, had become the head of the neonatal ward at Saint-Etienne Hospital and was fascinated by the project. Florence

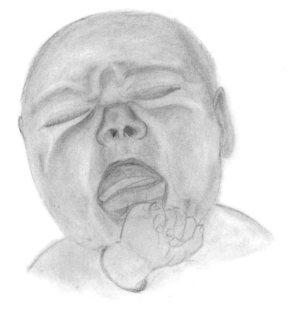

Levréro, who was always interested in the smallest little primate, was also on board. Then we welcomed a postdoc researcher, Erik Gustafsson, who agreed to join us in the laboratory. The new "baby cries" team could get down to work.

Of course, you've heard of maternal instinct—the idea that mothers everywhere have a special gift for caring for babies, a quality that is innate, written in the genes; whereas this gift is absent, or at least very underdeveloped, in fathers. This opinion is widespread, both among the general public and among biologists. Yet nothing particularly obvious supports it. In mammals, many different systems of social organization exist, from species where females and males meet only very briefly, with the female remaining alone to raise her young, to cooperatively bred species, where all members of the group are likely to participate. Take monkeys, for example, where there is great diversity.[551] In yellow baboons, males have very little interaction with their young. But in siamangs, tamarins, marmosets, titi monkeys, and owl monkeys, fathers are very involved, and sometimes other individuals (such as the previous year's young) as well. Among the titis, it is the males that carry the young from the age of three weeks. The baby titis prefer to go with their father rather than with their mother.

What about the human species? Of course, we know that pregnancy, with its considerable hormonal modifications, has important effects on the maternal organism, including the brain, but the extent of these effects and their true nature remain poorly understood.[552] In addition, the mother is usually not the only person to take care of the baby. Among the few remaining hunter-gatherer societies in the world, cooperative breeding is the norm. In the pygmies of central Africa, an average of 14 people take care of the same baby. By the age of four months, the baby will have been carried less by its mother than by other people.[553] But in our "modern" societies, things are different. The basic family is nuclear, and the perception tends to be that it's primarily the mother who takes care of the baby—at least in general, right? Well, it's not as obvious as that. Following the birth, the mother doesn't hesitate to entrust the baby to the good care of the hospital staff; in other words, to leave it in the hands of people she doesn't know. And when she returns home, it is not uncommon for the father, siblings, grandparents, and even neighbors to be able to give the baby bottles, change diapers, or bathe him or her. I'm not even talking about what happens next: day care, school, leisure activities . . . multiple people will take care of the baby and then the child. In short, let's stop beating around the bush—we are a kind of primate with cooperative breeding.

Does this characteristic of the human species have consequences for our ability to recognize a baby by its cries alone? Our first hypothesis was that both mothers and fathers should be able to do this, as equals. However, two previous studies conducted some 40 years ago suggested that mothers were much more gifted than fathers.[554] Not really convinced by these results, we decided to start from scratch, doing what we would have done for any animal species: recording cries (at the time of the baby's bath), looking for the presence of vocal signatures, and testing mothers and fathers through playback experiments. Acoustic analyses of the cries of about 30 babies first showed that the individual signature was indeed there. You probably expected that. The acoustic characteristics of each human baby's cry, as in a fur seal baby, thus make it possible to identify reliably who is crying. To test whether parents can detect and use this vocal signature, we had them listen to the cries of different babies. Of the 30 recordings heard, only 6 were from their offspring. The test results were conclusive: the parents correctly identified more than 5 of the 6 cries on average.[555] The only parent who had some difficulty identifying their baby's cries was spending less than four hours a week with the baby. In other words, they never heard it. (I'll let you guess whether they were a mother or a father.)

In a cooperatively breeding primate species, individuals other than the mothers and fathers are therefore involved with the babies, and this is the case in the human species. But then—if we follow this reasoning—everyone should be able to learn to recognize any baby by its cries. We had to test that hypothesis. Aurélie Cantais, an intern working with Hugues at Saint-Etienne Hospital, and Hélène Bouchet, a postdoc who had just joined our ENES team, were about to do just that. In the first part of the study, Aurélie recorded newborns in the days following their birth and did playback experiments with the mothers. The results showed that the mothers increased from 40% to 80% recognition of their babies between one and three days after birth. I was very amused by this result. Remember: it takes between two and five days for mother-young recognition to be established in the fur seal on the island of Amsterdam. Hélène invited adults—women and men without children—to the laboratory and played them a baby crying. "Pay attention! This is your baby's cry!" she told them. Each person had to listen to a different baby, of course, and a number of cries varying between one and six sequences of about 10 seconds each. People willingly played the game, listening carefully to the cries of a baby they had never seen before. A few hours later, they returned to the lab for a playback test, where they were asked to identify "their" baby every time they heard it. Two main results were obtained. First, those who had cared for a baby in the past performed better, recognizing "their" baby more than 60% of the time compared to 42% for the others. All they had to do was hear two or three sequences of a few seconds each of "their" baby's cry to be able to distinguish it from any other baby. Such rapid learning is remarkable. And there was no difference between women and men.[556]

With these studies, we showed that recognizing a baby's identity through his or her cries can be learned. What about our ability to assess the level of pain encoded in cries? With Siloé Corvin, a PhD student at the lab, we recently hypothesized that, again, experience matters. To check this hypothesis, we tested different categories of adults: people with absolutely no experience with babies; people with a little experience (having babysat or having had younger siblings); parents of children over five years of age; parents of very young children, less than two years of age; and professionals in the field of pediatrics. Each person was first asked to listen to eight cries, recorded during bath time, of a baby who would be considered that person's "familiar baby." The familiar babies were different from one person to another. The next day, each person again heard bath cries of "their" baby (but not the same recording sequences) and also cries of the same baby recorded during a vaccination session at the pediatrician's office. They were also played the bath and vaccination cries of a baby they did not know. For each of these cries, the question asked was simple: Would you say that this cry expresses simple discomfort (bath cry) or real pain (vaccine cry)? Well, what did the results say? That the listeners' ability to categorize cries as "discomfort" or "pain" depends on their past and current experience with babies. And the results couldn't be clearer. People who had no experience with babies decided at random whether the cry they heard was one of discomfort or pain. Those with moderate experience showed some ability to identify the cry, recognizing only the pain cries of the familiar baby. In contrast, adults with strong experience with babies, i.e., those who were parents or pediatric care professionals, were able to identify the familiar baby's discomfort and pain cries. Remarkably, parents of very young children were also able to do the same for the unfamiliar baby! Gaining experience is therefore the key to successfully decoding babies' cries. This is true for both men and women.[557]

We then moved on to explore adults' perceptions of cries by looking at the influence of gender stereotypes. We all know that men have a lower-pitched voice on average than women. Does this knowledge influence how we perceive the cries of baby girls or boys?[558] The first step in our experiments was to see if there are differences between the cries of baby girls and baby boys. In particular, are baby girls' cries higher-pitched than those of baby boys? Acoustic analysis showed that this was not the case. In a second step, we asked adult listeners to give the sex of the babies whose cries they were listening to. Surprise! Instead of answering that they didn't know, people classified the cries as girl or boy based on the pitch of the cry. The higher-pitched the cry, the more likely it was classified as a girl's cry; the lower-pitched it was, the more likely it was classified as a boy's cry. So people did indeed base their sexing of cries on what they knew about adult human voices. It is nevertheless rare that one is led to define the sex of a baby by using its cries; there are other more direct ways! But two experiments followed.

This time, people listened to babies' cries whose sex they knew. More exactly, we told them what the sex of the baby was. They had to assess the baby's degree of masculinity or femininity just by hearing the cry. Again, the pitch of the cry was critical. Whether they were listening to a girl's or a boy's cry, adults rated babies with higher-pitched cries as more feminine or less masculine. What they didn't know was that we played them the same cry, telling them that it was either a girl's or a boy's. You can see that here things are more subtle. Imagine taking care of a baby girl or a baby boy. Depending on whether the cries are higher- or lower-pitched, you may attribute more or less feminine or masculine characteristics to her or him without being aware of it (these are called *gendered traits*). We also asked people to rate the degree of distress coded in these cries. The higher-pitched the cry, the more people thought the baby was in pain. Most importantly, the same cry was rated differently by male listeners depending on whether we told them it was a girl's or a boy's cry. Female listeners did not make a distinction. It is interesting and important to know, therefore, that the perception and interpretation of a sound signal of major importance to our species may be subject to cognitive bias.

Our studies on baby cries are ongoing. We should know more in a few years. With our colleagues Roland Peyron, Isabelle Faillenot, and Camille Fauchon from the University of Saint-Etienne, and François Jouen from the *École pratique des hautes études* (EPHE) we are exploring, for example, how the brain perceives the cries. We are using the techniques of functional magnetic resonance imaging (fMRI) and thermal imaging. Initial results show that specific areas of the brain are activated by the sound of a crying baby—especially the areas of empathy, the ability to make us feel what the other person is feeling. The differences observed at the brain level between men and women are minimal. However, parents' brains appear to differ from those of nonparents, likely reflecting learning from their baby's cries. We are a cooperatively breeding species, you know that. On the other hand, I am willing to bet that the brain of one of those huge male elephant seals doesn't blink much when it hears a baby elephant seal crying.

17

Listening to the living

ECOACOUSTICS AND BIODIVERSITY

Désert de Platé, French Alps. It's freezing. Hindered by our snowshoes, we slowly climb the slope on the hardened snow. Yesterday we arrived in a storm, as is common here in May. On the track, the Land Rover quickly gave up, and we climbed on foot to the hut that would serve as shelter for the night. We left it at three o'clock in the morning. Frédéric walks in front of me, with his long ice axe in his hand: a confident mountain man. He extends his arm in the direction of Mont Blanc. The giant is decorated with a long light trail that scars its flank—city-dwelling mountain climbers lining up to conquer its summit, their headlamps winking as their heads move. We stop on a rocky promontory, drenched in sweat. With his usual foresight, Frédéric pulls on a fresh T-shirt. The wait begins, cold and long. "Rrrrrr-rr-rr"—a first song breaks the silence. A hoarse, rattling sound, right in front of me, soon followed by another one further to the left. I note on the counting sheet the estimated position of the two rock ptarmigans. Back at the hut, all the counters will pool their observations: five or six birds for the whole area. The ptarmigan is disappearing from our mountains. Here, as elsewhere on our planet, mornings are falling silent.

You already know Frédéric Sèbe from chapter 11. He was the one who spent a year observing the communication networks of the screaming pihas in the Amazonian forest. Frédéric now uses bioacoustics as a tool for monitoring biodiversity. One of his research projects concerns

Rock ptarmigan

the rock ptarmigan *Lagopus mutus*, a kind of snow partridge, white in winter and brown in summer. Typical of the Scandinavian tundra, it is still found in France at high altitudes in the Alps and the Pyrenees. Frédéric's aim is to be able to estimate the size of ptarmigan populations, i.e., the number of individuals, by using automatic sound recorders. These are very weather-resistant tape recorders that Frédéric leaves at the test site to record ambient sounds at regular intervals. To validate the method, human observers check the presence and number of ptarmigans visually and especially by ear. Then, in the laboratory, sophisticated techniques for analyzing the recorded acoustic signals must be developed. Since tape recorders remain in place for several weeks in a row and record for several hours a day, I'll let you imagine how many hours of recording have to be processed. Developing analysis techniques was the core of our PhD student Thibaut Marin-Cudraz's work for three long years. He spent many hours in front of his computer, but his tenacity was rewarded. He managed to show that this automatic counting by the bioacoustic method is reliable. He proved that it avoids double counting, as when two observers count two different birds when they both hear the same individual. This technique also compensates for those possible moments of inattention in some observers.[559]

Identifying and counting birds or other animals automatically by re-cording them rather than listening to them is one of the practical ap-plications of bioacoustics.[560] This technique is also used in the ocean to spot whales and even fish.[561] But we can also work on a different scale. Why not take a look at all the species that produce sounds? In other words, why not consider soundscapes as indicators of the animal spe-cies living in the same environment?[562] This approach, which is cur-rently in full expansion, forms the discipline called *ecoacoustics*. The idea is simple: to use bioacoustics to measure biodiversity and assess the condition of an ecosystem.[563] All we had to do was think of it.

Almost at the end of our sound journey, you now have a good idea of the extraordinary profusion of sound signals used in the animal world to exchange information. In aerial environments, birds and insects are at the top of the list, followed by frogs and mammals. Below the water, shrimp, fish, dolphins, and whales compete in their ingenuity to produce sounds. The tropical forest, with its incredible diversity of species, rus-tles with a thousand sounds. Coral reefs, the tropical forests of the sea, are equally rich.[564] It's true that a soundscape only partially represents the diversity of life, but everything comes together in nature. Where the species of insects and birds producing acoustic signals are diverse and numerous, so are other animals, plants, and all other living things. The widespread and ongoing declines in bird populations on Earth cause changes in soundscape quality.[565] The soundscape reflects the diversity of life.

Do you remember Frédéric Bertucci? This is the enthusiastic PhD student I told you about in chapter 10, who developed playback experi-ments in aquariums when we were trying to test how the fish *Metri-aclima zebra* reacted to the sound productions of its fellow fish. Once he had finished his PhD, Frédéric left the semimountainous climate of Saint-Etienne (where our ENES Bioacoustics Research Lab is located) to work in the tropics with Eric Parmentier (the piranha man) and David Lecchini, a specialist in coral reef fish. It is on the island of Moorea, in the Pacific Ocean, that the *Centre de recherches insulaires et observatoire de l'environnement* (CRIOBE) is located, a superb field sta-tion installed on the edge of a lagoon bordered by superb coral reefs.[566]

If coral reefs represent only 2% of the total surface of the oceans, they represent 30% of their biodiversity—and a very interesting sound ambience. In Moorea, the coral reefs are dominated by only a few sounds (between 2 and 6) during the day, while at night it rustles with more diverse signals (up to 19 different sounds). Dusk and dawn are the busiest times. Attributing these sounds to specific fish species is still a difficult task, as our knowledge of the repertoire of each species remains superficial. However, Frédéric and his colleagues were able to identify that the most sonorous fish belong to the families Balistidae, Pomacentridae, Holocentridae, and Serranidae. Some sounds, such as the "whoots," a long and strongly modulated signal, still keep their mystery.[567] Coral reefs are fragile environments because, to survive and develop, they need particular conditions of light, temperature, and acidity level. Any imbalance results in the death of corals, which begins with their bleaching.[568] A bleached coral is a coral that has lost the microscopic algae with which it is in symbiosis and which are essential for its survival. No coral means no shrimp or fish—nor any other marine organism, for that matter. If conditions become favorable again, a bleached coral reef takes about 10 years to recover its colors. To save coral reefs, we must first be able to measure their health. This is what scientists are trying to do using ecoacoustics.[569]

Ecoacoustics is a recent discipline, defined as the ecological study and interpretation of environmental sound. The main idea is to measure and monitor the biodiversity and ecology of a living environment through sound.[570] While the objective of classical bioacoustics, as we have seen, is to study the transfer of information between individuals, the objective of ecoacoustics is to "consider sounds as a component and indicator of ecological processes."[571] Bernie Krause, who has spent more than 50 years recording soundscapes around the world,[572] coined the terms *geophony, biophony,* and *anthropophony* to describe soundscapes.[573] *Geophony* represents all the noise caused by nonliving natural phenomena—the thunderclap, the rustle of the wind in the trees, the murmur of a stream, the roar of a waterfall. Although they may be of short duration, they often provide a remarkably continuous background to the landscape. Sometimes they mark changes in the time of day. In spring in the mountains, the torrent is silent in the morning when every-thing is frozen. Later, when the heat of the sun melts on the peaks, the water starts to cascade again, and it becomes thunderous. *Biophony* en-compasses all the sound productions of living beings—with the excep-tion of those of the human species, which constitute *anthropophony*.[574] Krause separates anthropophony into two constituents: first, what he calls controlled productions—the speech, the songs, the music; and second, the incoherent noises emerging from human activity—air and sea transport, construction activities, mining and industrial activi-ties, wind farms, etc. There is no shortage of sources of anthropogenic noise. Since the Industrial Revolution and the world population boom, anthropophony has become massive. It continues to grow. Both on land and underwater, the places on Earth where it is absent are becoming rare.[575] In Europe, for example, only particularly remote areas are pre-served. A new phenomenon on the geological time scale, anthropo-phony is to be considered as pollution, in the same way as chemical pollution or ocean acidification.[576]

As I write these lines in my house located in a rather quiet neighbor-hood, a common Eurasian blackbird *Turdus merula* sings loudly a few meters away from me ("tooodeee-too-tooo-deeee"). A few sparrows chirp ("tchip-tchip-tchip"). Every two or three minutes, a car drives along the street. Its noise does not completely cover the blackbird's vo-calizations, but it considerably diminishes its apparent power. One of my neighbors in the distance is mowing his lawn. An insect in a hurry is humming as it whizzes by. Ah! Another car. The blackbird is silent, perhaps discouraged? Anthropophony affects biophony. Knowledge accumulated over more than 20 years confirms this with certainty.[577] Arthropods, amphibians, fish, birds, mammals—all are affected,[578, 579] at all ages.[580]

Noise due to human activities affects animals' communication, their distribution in the environment, their feeding behavior, and of course their physiological condition and survival,[581] both in the air and underwa-ter. They even affect animal species that do not use acoustics to communi-cate, such as mollusks.[582] The disturbances caused by noise depend of course on their acoustic characteristics. The continuous noise of highway traffic will reduce the range of birdsong and perhaps cause chronic stress to birds, as it does to many humans.[583] A sudden mining explosion may destroy hearing systems or cause irreversible damage to various organs of the body, including the brain. Underwater, the problem of anthropogenic noise is particularly critical.[584] As you know, sounds travel much faster and farther in the aquatic environment. Where a car will bother singing blackbirds nearby, the sound of a cruise ship can disturb whales or dolphins for miles around. Remember, cetacean sound signals are as low-pitched (up to 20 Hz) as they are high-pitched (over 300 kHz). Any anthropogenic sound, whether the lowest or the highest, will fall within this frequency range and be a serious parasite.[585] Did you know that every year there are numerous underwater explosions along tens of thousands of kilometers to find oil reservoirs in the ocean floor?[586] Whales and dolphins are victims of these sound explosions.[587] But they are not the only ones. Recent studies have shown that anthropogenic noise pollution drastically alters the soundscape of coral reefs. This pollutant, whose consequences are probably underestimated, can mask communication among reef organisms, such as certain fish, and represents a stress factor whose effects have not yet been fully discovered.[588] The increase in anthropogenic noise in the oceans is such that it is estimated that North Atlantic right whales (*Eubalaena glacialis*) may have lost two-thirds of their communication space.[589] Studies show that these underwater sound waves can even kill zooplankton— the small shrimp, larvae, and other animalcules that float in the water and are the main source of food for countless animal species.[590, 591]

Some animals are more or less comfortable with anthropogenic noise.[592] As when we find ourselves in an environment where the noise level is high, one of their strategies is to increase the intensity of their own signals (the so-called Lombard effect, named after the French doctor who described this phenomenon more than a century ago). In the city of Berlin, for example, the nightingale sings more quietly on weekends, and much more loudly during the week when road traffic from the working world increases.[593] Underwater, the humpback whale also increases the power of its song when the background noise due to human activities increases.[594] A recent study shows that the bearded seal is also able to sing louder in the presence of noise. However, scientists who have documented this behavior found that the seal lacks the ability to increase its voice level when the ambient noise becomes too loud.[595] Since anthropophony is characterized by rather low-pitched sounds, frogs, insects, and birds produce higher-pitched sound signals to make their signals stand out from the anthropophonic cacophony.[596] This phenomenon was first observed by comparing the songs of urban great tits with those of individuals living in the countryside. City birds sing higher-pitched songs.[597] In addition, for biomechanical reasons, high-pitched vocalizations are often emitted with greater intensity (to sing high-pitched, one must sing loudly).[598] This strengthens their signal even more against ambient noise. These adaptations can come from individual behavioral plasticity, where the singer modifies the way it vocalizes depending on the background noise, or they can be true evolutionary adaptations. It seems that, in the great tit, it is exposure to noise over several generations that leads the birds to modify the pitch of their song.[599]

A study recently showed that birds start to sing less loudly if anthropogenic noise is reduced. The population containment measures imposed during the COVID-19 crisis in the spring of 2020 resulted in a drastic decrease in human activity, in particular road traffic. Elizabeth Derryberry and her team at the University of Tennessee have studied the consequences of this on the vocal activity of the white-crowned sparrow—Peter Marler's bird, which, we learned in chapter 12, has different dialects depending on its location in the San Francisco Bay Area. Derryberry and her colleagues have shown that, during the COVID-19 crisis, this bird sang less loudly.[600] There was no longer any need to sing loud in an environment free of noise pollution: the range of their song more than doubled. And if the range of each individual's song doubles, four times as many individuals can be heard at the same time, which also explains why people reported hearing more birds than before the crisis. In addition, the birds sang some lower-pitched notes, as did their recorded ancestors in the 1970s, when the San Francisco Bay Area was still relatively quiet. Marler's finches somehow filled the new acoustic space that was available to them. Nature abhors a vacuum, as we all know.

Field cricket

Insects, too, are capable to some extent of adapting to the presence of noise caused by human activity. By combining observations and field playback experiments, Mario Gallego-Abenza, in collaboration with David Wheatcroft and me, has shown that the response of field crickets (*Gryllus bimaculatus*) to anthropogenic highway noise depends on their previous experience with cars. Male crickets living on the edge of a highway reduce the rhythm of their chirping when a car passes by. Mario compared the response of these crickets to car sounds emitted from a loudspeaker with that of other individuals living further away from the highway (up to a mile and a half). The farther the crickets lived from the noise, the more they decreased the rate at which they chirped.[601] So the highway crickets have become a bit used to the noise. That's good for them. However, the study does not say if they are more stressed, if they find partners just as easily, or if they are more or less resistant to predators than their countryside cousins. In female crickets, it was recently discovered that ambient noise disturbs their male preferences. While they are usually attracted by a rather fast song—a sign that the male is vigorous—they no longer show any preference when the sound environment becomes annoying. The noise makes them lose their judgment in a way.[602]

At our laboratory in Saint-Etienne, my colleague Vincent Médoc is studying the effects of boat noise on the behavior of freshwater fish. In one of his experiments, Vincent and his colleagues put European minnows (*Phoxinus phoxinus*), which are small river fish, in aquariums with varying food densities. These delectable food items included small larvae of the *Chaoborus* fly, a prey that minnows love to hunt. In some aquariums, very few prey were served, only 8 *Chaoborus* larvae per tank. In others, the number of prey was higher: 16, 32, 64, 128, and up to 256 larvae—simply huge. Two fish were put in the aquarium at the same time. One was free to move around to hunt prey; the other, the companion fish, was in a transparent plastic tube. It was there only because minnows don't like to be alone—a sort of stooge that could only watch as lunch passed by. Let's summarize: aquariums with prey densities ranging from scarcity to excess, and only one fish with a hunting license. In addition, Vincent varied the sound conditions: in some of the aquariums, it was dead calm; in others, a loudspeaker regularly reproduced the sound of a motorboat passing by. Vincent let the fish feed for an hour; at the end of this time, he counted the number of prey that had been eaten. As one might expect, the results showed that the noise of the boat slowed down the hunting of the minnows. What's remarkable is that the lower the larvae density, the greater the effect. To give you an idea, with 32 larvae in the aquarium, the minnow swallowed about 15 in the absence of boat noise and only about 10 in the presence of noise. The noise seemed to spoil their appetite. On the other hand, if there were 256 larvae in the aquarium, the fish ate as many with or without the noise. However, it must be said that larvae were everywhere and easy to find.[603]

What Vincent's study shows is that the effects of anthropogenic noise on fish behavior depend on other environmental characteristics, in this case, prey density. But how do things work under natural conditions, in a river with minnows, *Chaoborus*, and a whole bunch of other, different prey likely to be eaten by fish? We can hypothesize that the passage of boats may change the hunting strategies of minnows depending on the relative density of the different prey. In calm conditions, minnows may hunt this or that prey instead, because they prefer it, for example. In noisy conditions, stressed minnows may hunt their preferred prey less efficiently, especially if this prey is at low densities. They will then fall back on other, more abundant prey, even if it is less tasty. These variations in hunting strategies will lead to variations in the relative densities of the prey, which in turn will lead to variations in the relative densities of the species the prey feeds on, and so on.

Are you familiar with food webs—the idea that each living being present in an ecosystem is the food of another, and that an ecosystem can be seen as a set of interactions between all the living beings that it comprises? The presence of anthropogenic noise could possibly alter these webs across hitherto unsuspected levels, even to living beings that hear nothing but that undergo modifications from higher levels of the network (e.g., plants or bacteria)—because at the end of the line, the first foods consumed in a food web are not animals. The result, which is a bit frightening, is that it is the whole ecological balance that could be affected by anthropophony. Today, we don't know much about these domino effects.[604]

A change in the sound environment caused by anthropogenic noise is not the only one to impact living organisms. An irruption of silence can have equally deleterious effects. Do you remember that the larvae of coral reef fish lead a planktonic life away from the reef of their birth? When they reach the appropriate age, these larvae are attracted to the reef background and choose to settle here or there, according to their musical taste, so to speak. I explained this in chapter 6, where we had our ears underwater. What happens when the reef is degraded and its soundscape is changed? Well, young fish have a harder time coming to the reef, either because they don't find the soundscape pleasant anymore or because they just don't hear it. Moreover, fish have a delicate ear: it does not take much change for larvae to decide to desert a reef with a depleted sound environment. A study conducted on the Great Barrier Reef in Australia showed that a sound environment that has lost 8% of its attractiveness will result in 40% fewer fish larvae on the reef.[605] Another study conducted by Bertucci and Lecchini, in collaboration with other scientists, has shown experimentally that other larvae—those of corals this time (the small immobile animals of the sea anemone group that build the reefs)—that are usually attracted by the odor emitted by some coralline algae are reluctant to approach these algae in the presence of boat noise.[606]

A key idea of ecoacoustics is that natural soundscapes are *structured*. In other words, the organization of the ecosystem, with its undisturbed trophic chains, is perceptible through the soundscape, which can be recorded and analyzed. The interactions between the organisms that make up an ecosystem would result in the structuring of soundscapes. This is the "great animal orchestra" cherished by Bernie Krause, where each species is like a musical instrument in dialogue with the others. On the other hand, when balances are disturbed by human activities, the soundscape would become chaotic. The structuring of the biophony would be lost. The biophonic orchestra would be stifled, become less complex, or even disappear completely. Anthropophony would take center stage. The soundscape is evidence of these changes. As Krause says: "The soundscape is the most reliable way to assess the state of a natural habitat."[607]

The idea that soundscapes are structured is based on two assumptions, that of the *acoustic niche* and that of *acoustic adaptation*. The *acoustic niche hypothesis* states that each species of animal occupies an acoustic space, different from that of other species living in the same location (known as *syntopic* species). In other words, each of the syntopic species would produce acoustic signals occupying different frequencies from those of the other species, or emitting them at different times. This would avoid acoustic competition. They would not interfere with each other, if you like. This would make the communication mechanisms between individuals within each animal species optimal.[608] This hypothesis is directly inherited from the concept of the *ecological niche*: in an ecosystem, there cannot be two animal species using the resources of the environment (food, space, etc.) in exactly the same way.[609] If this happens, the two species are in competition, and eventually one will exclude the other. Most often, the ecological niche of one or both species will change—the individuals of one species change their diet, for example—and the competition diminishes.

In its original version, as stated by Krause in 1993, the *acoustic niche hypothesis* considers sound frequencies as a limited environmental resource.[610] Animals should therefore "share" the frequencies. For example, some species will have very high-pitched vocalizations, others a little less, others will be quite low-pitched. When you look at the spectrogram of a recording taken in the rainforest, for example, it is easy to see that the different animals actually use different frequency bands. Jérôme Sueur of the *Muséum national d'Histoire naturelle* in Paris, and his collaborators, recorded the soundscape at the biological station of the Nouragues Nature Reserve in French Guiana on December 12, 2010, at 6:30 a.m.[611] On the spectrogram showing the recording, three levels are clearly visible. In the low frequencies, between about 0 and 1 kHz, an almost continuous color band is visible. These are the growls of howler monkeys. Above that, between 1 and 5 kHz, colored spots indicate birdsongs—at least two species, one making trills, the other monotonous whistles. When the trills are there, the whistles are difficult to perceive. Finally, in the upper part of the spectrogram, the one that corresponds to high-pitched sounds, between 5 and 15 kHz, we have insect chirps. Again, there are several species. Some produce chirps between 5 and 7 kHz, which sweep the spectrogram from beginning to end. Other chirps, centered on very high frequencies around 11 kHz, are more rhythmic and mark pause times.

This example from the Nouragues is quite demonstrative. There is indeed a kind of frequency sharing between animal species, but this sharing is imperfect. Many birds, for example, are found in more or less the same frequencies of the spectrogram. Does that mean that they get in the way? That they are competing for acoustic space? Perhaps, to some extent. But not necessarily. The low or high pitch of a vocalization isn't everything—its temporal structure is also very important for encoding information. And when the songs of different species of birds use the same frequencies (i.e., equally high-pitched, equally low-pitched, or equally in the midrange), their different temporal structures—their different rhythms, if you like—allow them to be distinguished from one another. Jean-Claude Brémond showed this phenomenon very well with the European wren *Troglodytes troglodytes*. This tiny little bird with its raised tail is common in European woods and forests and produces a song lasting a few seconds, composed of very fast trills. Brémond tried to artificially mask the song of the wren by mixing it with the song of other species, such as that of the Eurasian blackcap *Sylvia atricapilla*, the willow warbler *Phylloscopus trochilus,* or the dunnock *Prunella modularis,* which has a song close to that of the wren. In playback experiments, the wrens reacted vigorously to these noisy signals, showing that identifying the song of their species mixed with that of another bird was not a major challenge for them. Brémond carried on with his noise experiments by mixing the song of the wren with artificial noise, always in the same frequency range. Once again, the wrens identified the song from their species without difficulty.[612] So a wren is well able to locate the song of another wren even if the frequency band it uses is occupied by other sounds.

The hypothesis of the acoustic niche is therefore nuanced, and it is not surprising that studies about it give conflicting results.[613] As I was describing in the recording of the Nouragues in French Guiana, it is true that there is a sort of layering of the animals' vocalizations from low to high notes. But this layering is rather crude because many animals occupy the same frequency range, and its source is only partially explained by acoustic competition.

Let's remember the drumming of the woodpeckers, which I talked about in chapter 10. This illustrates a rather common situation, in my opinion. Each species of woodpecker drums in its own way. However, the closer two species are to each other, the more similar their drumming is, because this signal is largely determined by genetics. In a forest, the woodpecker species that cohabitate are mostly distant cousins. Two species that are too similar would compete not so much for acoustic space but mainly for food resources: they would look for the same insects on the same trees, for example. Does this mean that acoustic competition does not exist at all in woodpeckers? No, of course not! Because, as I told you, if two sister species with similar drumming patterns live in the same place, then there is a character shift, as evolutionary biologists say. Their drumming evolves and becomes a little bit more different than it was in the beginning. But remember: these changes are rarely rapid. They usually occur on a geological time scale.

So that's the acoustic niche hypothesis. Now let's move on to the acoustic adaptation hypothesis. The *acoustic adaptation hypothesis* states that the sound signals of an animal species are optimized to be transmitted as well as possible in its habitat. Again, you've heard this hypothesis before. I first considered it at the very beginning of our sound journey, when we were with the white-browed warbler and all the other birds of the Brazilian Atlantic Forest. As you may recall, the results of our propagation experiments were not very convincing: while some species of birds did produce songs adapted to high propagation by the height from which they were vocalizing, others did not. But perhaps our study focused on particular cases, giving a distorted view of the general situation. For example, I have told you how great tits singing at a higher pitch are more successful when confronted with the noise of the city. As a lot of work has been done on this issue, it should be possible to get a clear picture.[614] Let's take a look.

First of all, it is not really surprising that animal sound signals are somewhat adapted to the physical properties of the environment. Imagine a situation where an individual has to transmit information over several hundred meters in a forest environment. A low-intensity ultrasonic signal will not do the trick because the ultrasound only propagates a few meters.[615] This type of signal, not adapted, will therefore have no chance of being preserved by natural selection. On the other hand, such a signal will be perfect in a situation of very short-distance communication, transmitting information well and, as another advantage, not likely to attract a predator.

In a more straightforward way, one can make predictions as to which acoustic features should be favored in a given environment. In forests, the ideal long-distance signal would require a relatively low frequency, long duration, and little modulation—a kind of long hissing sound, impervious to the reverberations caused by tree trunks and adept at passing through obstacles. When vegetation is absent, short, highly modulated signals (going from high-pitched to low-pitched or vice versa very quickly), such as trills, would be more appropriate.[616] Since sound waves of very low frequency are strongly attenuated when propagating close to the ground, frequencies above 0.5–1 kHz should be favored. Finally, in a noisy environment, sound signals should be tuned to frequencies as far away as possible from those of the noise, e.g., very high-pitched if the animal lives close to a torrent whose tone is in the lower frequencies.[617]

Slower time rates and lower modulations would also be more conducive to successful signal transmission and reception.

Things can be subtle, of course. For example, in a forest, the acoustic conditions are different depending on whether you are vocalizing from the ground, from midheight, or from the top of the canopy. The microhabitat of the animal must therefore be taken into account if we really want to test the acoustic adaptation hypothesis. That's what we did in the Atlantic Forest, with mixed success, as I mentioned a few moments ago. That's also what Sandra Goutte and her colleagues from the *Muséum national d'Histoire naturelle* did by working with frogs.[618] And their results are in line with ours. They combined data from 79 species of frogs to test the hypothesis that frogs living near streams produce vocalizations adapted to a noisy environment. They looked for correlations between the acoustic structure of the signals and the fine characteristics of the habitat (size of the rivers, slope of the bank, temperature, density of vegetation, ambient noise measured at the exact location where the frog was recorded, etc.). To sum up, although frog vocalizations are mainly determined by genetics, they found that the croaks of frogs living near streams have a higher frequency than the others. On the other hand, their temporal rhythms are not different, and the signals are more frequency modulated, thus varying rapidly between high and low frequencies, which is contrary to predictions.

The study by Sandra Goutte and colleagues, I feel like repeating, only partially validates the acoustic adaptation hypothesis. Similarly, an analysis of the song of 5085 bird species found no relationship between song frequency (whether it was low- or high-pitched) and the density of vegetation in the environments inhabited by the different species.[619] Rather than being an adaptation to travel far, depending on whether one is in a forest or an open area, song pitch is essentially explained by the body size of the bird: the largest species are those that sing the lowest, the smallest the highest. In fact, the adaptation of the characteristics of a sound signal to the acoustic constraints of the propagation environment should not be seen as the only element that counts in the evolution of a communication system. First of all, every system has its own constraints: an animal species is the inheritor of its ancestors—of their genes and of the anatomical, morphological, and physiological constraints that accompany them.[620] A frog cannot modify its croaking at will. Second, the important thing is that the transmission of information between senders and receivers must be sufficiently stable, and not necessarily optimally so. Life is the result of chance and necessity. I bet no engineer has ever thought of that! Nothing is perfect; everything is a precarious balance. Finally, as we have seen with the wren, the capacities of the individual receiver to decode the information of a signal, even if strongly degraded by propagation, can be astonishing, and can compensate for an imperfect acoustic adaptation.

Does the fact that ecoacoustics is based on simple and imperfectly confirmed hypotheses invalidate the overall approach? Certainly not. This does not prevent ecoacoustics from being a powerful tool that is proving more and more indispensable day by day to assess biodiversity and ecosystems—perhaps one of the best tools available. Because two things remain true: each environment has its own sound signature, and it is possible to characterize it. A soundscape integrates a considerable amount of information—much more than a photograph, for example. As Bernie Krause says beautifully, "If a picture is worth a thousand words, a sound is worth a thousand pictures." Ecoacoustics also has a considerable advantage: we can follow the evolution of things over the long term, without the time, effort, expense, and disruption of having observers go out in the field every day to measure biodiversity or assess the presence of this or that animal species. A few discreet recorders, left on site for long periods of time, are a much more advantageous solution.[621]

It was in Piran, Slovenia, at a conference of the International Bioacoustics Society, that Jérôme Sueur and I were outside at a café, chatting.[622] Usually cheerful, Jérôme was pessimistic that day. "I think our work is useless. Studying the songs of cicadas and birds doesn't bring much benefit. Couldn't we do some useful research?" Although I didn't agree with him, I didn't have many counterarguments. Selfishly pursuing my passion was enough for me. But, deep down, I was quite disturbed. "Jérôme is right," I thought to myself, "but what can I do?"

Jérôme found his solution. He is now one of the leading figures in global ecoacoustics. One of his main activities is to participate in the development of *acoustic indices of biodiversity*.[623] Imagine that you want to report on the biodiversity of an environment, such as a forest, through bioacoustics. You're going to place automatic sound recorders in different places and then let them record every day, for weeks, months . . . years! How, from this raw data, can you obtain measurements, figures, notes, representing biodiversity? It would be impossible to identify and count all the animals participating in the soundscapes that succeed one another over time by observing kilometers of spectrograms. You would need to define more global measurements, and this is what we are doing with these famous indices. There are two main families of indices: alpha-type indices, which measure the biodiversity of a given soundscape, and beta-type indices, which compare two soundscapes. I won't go too much further here. You'll find them elsewhere if you want more detail.[624] Just be aware that they allow us to describe the richness, complexity, heterogeneity, regularity, and composition (in terms of geophony, biophony, and anthropophony) of soundscapes, both on land and underwater.[625] No single current index can capture all aspects. Every year, new indices are developed by scientists.[626]

To have a clear idea, let's go back to the coral reefs.[627] I think everyone is up for such a trip! Rather than trying to identify which animal species produces which sound on the reef, scientists can use acoustic indices calculated directly from the soundscape, and this approach allows a global assessment of the state of the reef ecosystem. In their study of the Bora Bora lagoon in French Polynesia, Frédéric Bertucci and his collaborators used two indicators: the signal intensity (sound pressure level, in decibels) and the acoustic complexity index.[628] They found, not surprisingly, that the signal intensity is higher where there are many fish and that the complexity index is higher when the reef is more developed.[629] Acoustic indices thus give varied information about the recorded reefs. Let's be more precise: high sound intensity values or a high complexity index in the low frequencies of the sound environment (<1 kHz) indicates that the reef is healthy, with diverse and abundant populations of fish and other organisms. However, if these indices are high when considering the high-frequency portion of the soundscape, the ecosystem is degraded, with many dead corals and shrimp feeding on detritus.[630] In addition to information on the composition of the biodiversity of an environment, acoustic indices can provide what biologists call *functional information*. To illustrate this, let's take the example of a study carried out by the researcher Simon Elise and his collaborators in the coral reefs of Europa Island, a small piece of land in the southwestern Indian Ocean. These scientists tested whether it was possible to get an idea of ecosystem *functions* using acoustic indices calculated from the reef soundscape. In other words, instead of simply focusing on the diversity and abundance of fish and coral cover, Elise wanted to use acoustics to report on the activity of fish and other animals on the reef. Europa Island being particularly remote, this was quite an expedition and was the subject of a film that I would recommend you watch.[631] To record the underwater sounds, the scientists deployed a large metal tripod onto which they fixed a hydrophone. The tripod was gently placed on the bottom, with the hydrophone suspended at about 1.5 meters. All sounds between 0 and 50 kHz were then recorded for a minimum of two hours. This operation took place at several locations on the reef. At the same time, the scientists carried out meticulous photographic surveys to identify the organisms present, from corals to fish, including shrimps and other shellfish. This was followed by an in-depth acoustic analysis of the recorded sound signals, which required long hours of work in front of the computer. At the end, no less than six different acoustic indices were calculated.[632] The results are remarkable, since Elise and his colleagues have shown that the six acoustic indices can be considered as indicators of six ecosystem functions. They reflect coral cover, with encrusting corals on one side and nonencrusting corals on the other, habitat complexity, grazing fish (which graze or scrape the coral), planktivorous fish (which eat plankton floating in the water), and tertiary consumers (defined in the study as fish-eating species smaller than 50 cm). It is assumed that acoustic cues miss algae, plankton, and invertebrates such as sea slugs, all of which are very quiet. Some acoustic indices are closely correlated with the functions listed above. For example, the acoustic complexity index discussed earlier increases with the amount of grazing fish. The number of tertiary consumers affects the index of temporal variability: the more tertiary consumers that are active, the higher the index. In fact, the values of the six ecosystem functions given by the photographic surveys and the visual counts are closely correlated to at least one of the acoustic indices calculated from the sound recordings. Acoustics can therefore be used to illustrate the functioning of a reef ecosystem.

Of course, these acoustic approaches are not yet fully mastered, and we still need many years of research to refine the way to extract information from acoustic recordings.[633] But things are progressing fast, and it is likely that one day coral reefs will be acoustically monitored everywhere, providing firsthand data for the implementation of effective conservation policies. This monitoring will allow us to evaluate the risks and effects due to various stresses, both environmental (e.g., storms and hurricanes, or arrival of invasive animal species) and human (anthropic noise, chemical pollution, etc.).[634] The recordings made on the coral reefs of Puerto Rico during Hurricane Maria have shown that the shrimps remained silent—and thus inactive—during the storm and only gradually resumed their normal life in the days that followed. On the other hand, the vocal activity of the fish increased, perhaps because the increased turbidity of the water pushed them to use acoustics to communicate or because they became difficult to spot visually and thought they were safe from predators.[635] When listening is done in real time, as is beginning to be the case in some places, the sound ambience of the reef can give an early warning of any change affecting the ecosystem.[636]

The greatest challenge in ecoacoustics is to make sense of very long recordings—to deal with big data, as they say.[637] Researchers like Jérôme spend a lot of time inventing solutions to automate the analysis of recordings. Ecoacoustics depends crucially on new technological developments and is very demanding in terms of mathematical processing capabilities. We are probably in the prehistory of this discipline. In 20 or 30 years, perhaps much less, we can hope to monitor the health of a tropical forest or a coral reef from a distance. The fact that recorders have become reasonably priced, that data transmission systems are more efficient and reliable, that artificial intelligence and its calculations are becoming more accurate by the day, is rapidly increasing our ability to analyze soundscapes and therefore to analyze ecosystems by listening to them.

Soon, only arrays of microphones connected to high-powered computers by radio waves will be needed. Listening to forests and reefs, lakes and mountains, will tell us in real time, and for a modest cost, how they are changing, whether new animal species have joined them, whether others have disappeared, and whether human disturbances are affecting them.[638, 639] Let's only hope that there will still be forests and coral reefs, lakes and mountains left. Bernie Krause has sounded the alarm.[640] According to him, more than 50% of the habitats he has recorded over the last 50 years have already disappeared or have been profoundly modified.

At the sunny table of the café, Thierry has just joined us. "You never know what fundamental research is going to be used for. It's totally unpredictable! Did you know that a sound analysis method my student Ruben Mbu Nyamsi and I had developed to study bird calls was later used by a car manufacturer to identify problems with electric windows on an assembly line? And think of Nobel Prize winner Pierre-Gilles de Gennes, whose fundamental research on the physics of soft matter inspired the development of shampoos and helped maximize the grip of car tires!" [641, 642] Point taken. I was reassured. Jérôme and Thierry are both right.

Some time has passed since this conversation at the Piran café. Jérôme continues to work on the development of acoustic ecology to better understand biodiversity and its dynamics. As for me, I realize that my fundamental research on acoustic communications illustrates the value of the conservation and preservation of biodiversity. I now dream of the moment when artificial intelligence tools will allow us to link soundscapes and information exchange processes between individuals. It will then be possible to describe the structure of soundscapes, to quantify the information flows that cross them, and to extract the network of relationships between the levels of organization of an ecosystem. Bioacoustics has already started this transition toward studying the dynamics and interactions between organisms that characterize ecosystems. The exploration of acoustic biodiversity has only just begun.

18

Words . . . words

DO ANIMALS HAVE A LANGUAGE?

Tanjung Puting National Park, Borneo Island, Indonesia. Yesterday we flew over some gigantic palm oil plantations—primary forests sacrificed to the crazy population explosion and human consumption. Today, under the foliage of the still intact trees, I try to forget the disaster that is taking place over there, and I let myself be engulfed by the concert of insects and birds. Dawn is breaking already. From far away, like long echoes, almost human songs come out of the forest: "Woop woop woop woop wooopooooppooop wooooooop wooooooooop wooooooopp. . . ." The vocalizations follow one another, at first slow, then faster and faster. Simple calls expressing emotion that the sender cannot repress? Or complex vocal construction skillfully mastered? Two individuals seem to answer each other. What mysterious rules of conversation do they follow? What information is exchanged in this way? What a bizarre *language* is that of the Bornean agile gibbon!

When questions come from the audience at the end of my conferences, there's always someone who asks, "Finally, can we say that animals have language?" During the course of the book, you were able to develop your own opinion on this subject. But perhaps you feel that everything is not so clear-cut and that answering this question would leave you uncomfortable. The problem is that there are two sides to this question. The first is, Can animals express anything other than their current emotions through calls that they have no control over? The second is more specific: Do animals have a language of the same type as our spoken language, with rules that allow us to exchange information as we do, through sequences of words and sentences?

Since we cannot enter a nonhuman brain, it is complicated to decide on that first question. However, whether it is the isolated male zebra finch that only responds to its female's calls in the presence of another pair of zebra finches, the bonobo that informs its companions of the presence of food through a series of vocalizations, the superb fairy wrens that cooperate by calling out to scare off a predator, the orcas that adjust their hunting behavior by whistling, or the chacma baboon that responds to the vocalizations of its fellows according to its memory of past events, all suggest that the production of sound signals in animals may be more than an uncontrolled reflex mechanism. Without taking too many risks, we can therefore affirm that at least some animal vocalizations represent something other than an emotional state expressed in a purely spontaneous manner. Especially in species with complex social lives, where acoustic communications allow individuals to manage their interactions astutely, animals can certainly control all or part of their sound production. So why, then, would we forbid ourselves from talking about *language*? It is, after all, a convenient term for an acoustic communication system in which the sender produces sounds voluntarily to send information to the receivers.

However, we must take a precaution and not put all animal species in the same basket. Remember the first chapter of the book. I told you that each animal is in its own world, the human species as well as the others. There is not a single animal language. There are as many animal languages as there are animal species that use sounds to communicate. The spoken language of humans is one of them.

Let's come to our second question: Is there a language in animals comparable to the *spoken language* of humans, that is to say, with an organization based on a combination of units such as words or sentences? For the vast majority of people, scientists included, the question of the originality of spoken language in the living world does not really arise: it would be unique and more complex than any animal acoustic communication system. This led the authors of a famous article in the magazine *Science* some 20 years ago to say that a Martian landing on planet Earth would notice that the spoken language of humans is the only one on Earth to be so complex and to allow the communication of an almost infinite amount of information.[643]

However, as Yosef Prat of Tel Aviv University points out in a recent article, before coming to this conclusion, the Martian would have to identify in the continuous flow of human speech the units that make sense (such as phonemes, syllables, or words) and then understand how these units come together in groups that make sense (expressions, sentences).[644] For us, who have been immersed in spoken language since our earliest childhood, these operations are automatic and easy to perform. But what about the Martian? If it could make sense of the spoken language of humans, maybe it could do it just as well for the song of the humpback whale and the song thrush. Who knows? Maybe it would find in these animal languages degrees of complexity that we don't yet suspect.[645] Or rather would it be disappointed by all its earthly observations—for example, by noting that spoken language takes a long time to say simple things like "Could you pass me the bread, please?" when the transmission of a Martian thought at 10 times the speed of light is sufficient. The Martian would then put us in the same acoustic basket as the other animals, considering all this small terrestrial world as a group of inferior beings quite similar to each other and unable to measure up to Martian standards.

Enough joking. We are among serious people. Here again, we will not take many risks by saying that human spoken language is, on our good old Earth, the one that allows us to communicate the most information. As it is trivially said, bonobos have never built a library and do not seem to give long philosophical speeches. The social and cognitive complexity of the human species has probably evolved along with the complexity of its communication systems, of which spoken language is one of the paragons. The story of the Martian simply points out the difficulty—and the bias—of analyzing animal languages as *external observers*, while we understand human spoken language *from the inside*. Our ability to analyze what other animals mean, what they have in their heads in some way, is limited.

Therefore, we must be wary of drawing conclusions about the sophistication of nonhuman languages. As early as the sixteenth century, Michel de Montaigne had a good sense of this. He wrote about animals: "This defect that prevents communication between them and us, why should it be reserved for us? Whose fault is it that they do not understand us? Because we do not understand them any more than they understand us. Therefore, animals may consider us to be unintelligent, just as we ourselves consider them to be unintelligent."[646]

Another point to consider is that spoken language is only a small part of the acoustic space that humans use to communicate. Through our vocalizations, we do much more than just talk. We call out in fear, joy, and pain, we sing songs or opera arias, we sigh with sadness or impatience . . . these unspoken vocalizations could represent living fossils of our ancestors' vocalizations when spoken language did not yet exist.[647] This diversity is found in nonhuman animal species. Thus, when we compare their languages with human language, we must consider all our vocalizations, not just words. Then the perspective changes. For example, as Prat's article notes, an acoustic analysis would perhaps show that our fear calls contain less precise information than those produced by vervet monkeys, which inform their fellow monkeys of the precise nature of the predator—eagle, leopard, or snake. To compare the relative complexity of human and nonhuman vocalizations, it is therefore essential to use the same methods of analysis.

When employed, this comparative approach highlights interesting similarities. Let's call the Cape penguin (*Spheniscus demersus*, also known as the African penguin) for assistance. The first time I saw African penguins in their natural environment was at their breeding colony on Boulders Beach in Simon's Town near Cape Town, South Africa. About 60 centimeters high and weighing in at 3 kilograms, it's a small penguin. On the beach, beak outstretched toward the sky, neck stretched as far as it can go, and wings spread, the males trumpet their parade call. It's all about pleasing their loved ones and impressing their mates. Nothing very original, is there? This call is a succession of several syllables. First a series of short (A) syllables, increasing in intensity ("honk! honk! honk! hoNK! HONK!"), followed by a long (B) syllable, very strong ("HUURRRR!!!!")—all of which are emitted by expelling air from the lungs. Another lungful of air, and the series cycle of A followed by B starts again—sometimes with short (C) syllables produced when the animal catches its breath between the A syllables or between A and B. No monotony—it is important not to put the females to sleep.

A study led by Livio Favaro, a professor at the University of Turin in Italy, shows that the calls of the African penguin follow two laws, known to characterize human language: *Zipf's law* and the *Menzerath-Altmann law*.[648] Zipf's law, or the *law of brevity*, states that the more frequent an element of a signal is, the shorter it will be. In human spoken language, words like "yes" and "no" that are used very often are indeed very short. Menzerath's law, on the other hand, says that the larger the whole, the smaller its constituents. Thus, in a text, the longest sentences have on average shorter words than short sentences. These two laws are predicted by Shannon's mathematical theory of communication, which we discussed in chapter 4: *information compression* (shortening the duration of the signal) increases the efficiency of coding information and its transmission from the sender to the receiver.

By analyzing 590 songs from 28 penguins, Livio and his collaborators found that the longer a song sequence is, the greater the proportion of short syllables A and C, and the shorter the duration of the C syllables even if they are already very short. In the African penguin's call, the duration of the syllables is therefore inversely correlated with the number of times they are repeated. Zipf's law is respected. Moreover, the number of syllables in a sequence is inversely proportional to the average length of the syllables. Menzerath's law is also present. If you're a little lost, just remember that we find in the calls of the African penguins the principle of information compression: the more we want to say, the shorter the signals are. Other teams of scientists have shown that these same laws can be found in the vocalizations of various animal species, particularly in certain primates, such as the indri lemur *Indri indri*, the gelada *Theropithecus gelada*, the Formosan macaque *Macaca cyclopis*, and two species of gibbons (*Nomascus nasutus* and *N. concolor*).[649] Information compression seems to be a universal principle shared by all languages, human and nonhuman.

African
penguin

Let's come to the rules that characterize vocal exchanges. When two humans enter into a conversation, they follow the unwritten rules of practicing alternation (each one speaks in turn) and observing avoidance (one does not speak when the other speaks).[650] These rules of alternation and avoidance are respected, more or less, as you can see in many examples if you watch televised debates. But when they are broken too often, the conversation stops (except on TV!). During our sound journey, we met animals that vocalized almost alone (the deer that bellows occasionally in solitude) or, on the contrary, all at the same time (the frog choruses). In these two cases, it is not necessarily easy to spot a conversation rule. Besides, are they conversations in the strict sense? However, we have also observed in the plain-tailed wren the duet in which female and male alternate in a perfectly synchronized manner.

Let's not forget the territorial disputes of chickadees, where a bird can wait until its neighbor has finished his song before starting its own. But if it does interrupt, it is precisely to let the other bird know that it intends to dominate the situation. Like all phenomena where the concept of rhythm intervenes, this rule of alternation ("turn-taking") is much studied in animals.[651] It is particularly searched for in nonhuman primates (with the risks of bias that this entails . . . we always end up finding what we are looking for). As a conclusion, we can say that the turn-taking rule can be found just about everywhere, from the starling to the bonobo[652] (but not the chimpanzee[653]), via the elephant and the meerkat[654]— which is not very surprising if we go back to the basics. The mathematical theory of communication predicts that the transmission channel will be noisy if signals interfere with one another. In order for information to be transmitted, it is better to send the signals one after the other and each one in turn.

But the question that interests us is more subtle: Do individuals who practice the turn-taking rule consider it to be a social rule that everyone must respect? In other words, are they surprised when they hear others violating the rule? To find out, my colleague Florence Levréro and a team of researchers conducted playback experiments with gorillas. The idea was to play the apes exchanges of calls between familiar individuals, by creating three different situations. In the first one, the sequence broadcast by the loudspeaker was a succession of two calls between two adults, the two calls being separated by a 500-millisecond silence, which reproduced a normal and familiar situation. In the second situation, a voice exchange was broadcast, still between adults, but where the second call started while the first call was not finished. This is not at all normal in gorillas. Finally, in the third case, the vocal exchange emitted by the loudspeaker was a series of two calls separated by half a second of silence, but the first call was that of a young individual while the second was that of an adult—an event that apparently rarely happens in natural situations in gorillas (when gorilla children talk, they almost never get an answer from the adults).[655] In short, one playback mimicked a completely conventional situation, while the other two were out of the ordinary and supposed to surprise the gorillas. The results of this experiment are a bit confusing but still interesting. First, there was evidence that the gorillas paid more attention to the first sequence (an ordinary succession of two adult calls) than to the second (where the calls overlapped). Perhaps this overlap caused some confusion, and the gorillas preferred to ignore this unexpected vocal exchange? Does this situation make them uncomfortable? We don't know, but it is clear that not following the alternation rule makes a difference to them. Further, scientists did not observe any significant difference between the gorillas' response to the third situation and their response to situations one and two. Apparently, a young gorilla talking before an older gorilla does not shock them that much.

To be clear, let's summarize our answers to the original questions: (1) Can animals express anything other than their current emotions through calls they have no control over? and (2) Do animals have a language of the same type as our spoken language? As to the first question, the answer is yes! It is now established that some animals can express something other than their emotions of the moment, as we do; they therefore control some of their vocalizations according to the context and the memory they have of the events and their interlocutors. The answer to the second question is also yes! But it's not a single language. There are in fact as many languages as there are animals using sounds to exchange information. These nonhuman languages are more or less complex and follow general rules that are very similar to ours. However, the structure and organization of the sound signals used for communication remain specific to each animal species, including humans.[656]

I can feel one last question burning on your lips: That's all well and good, but how did human spoken language come about? My answer is going to disappoint you a little; our certainties in this matter are few, and I'm not going to write a long essay on the subject here. But hang in there; there are a few things worth considering.

First of all, it is clear that our ability to master spoken language differentiates us from other apes.[657] Our ability to express any concept goes far beyond what any other animal species can do.[658] Relieved, right? During the second half of the twentieth century, however, there were a number of attempts to teach apes to talk.[659] None of them were very successful. Here is an outline, although I find these experiments ethically questionable. The animals lived in conditions that were poorly adapted to their species, separated from their fellow apes, sometimes dressed as humans. In short, these experiments were far away from the Tinbergen-style ethological approach.

In the 1950s, Cathy and Keith Hayes were among the pioneers. They raised a female chimpanzee to monitor her intellectual development, and in particular to see if she could learn to speak. Her name was Vicki, and the Hayes's raised her as if she had been their child. Although they present things in a rather positive way in their published paper,[660] it was a failure: after six years, Vicki still couldn't pronounce words correctly (she was able to say four words, with the greatest of difficulty).[661] Then other scientists—Allen and Beatrix Gardner, David Premack, and Duane Rumbaugh, to name a few—conducted the same experiments with other chimpanzees, using different approaches. For example, instead of trying to teach their chimpanzee (a female named Washoe) to talk, the Gardners taught her sign language. They had relative success: Washoe actually learned to produce about 30 words in 22 months of training.[662] Did the chimpanzee go further, combining words to construct expressions or phrases, as the Gardners maintain? On that point, things are not very clear.

To get to the heart of the matter, Herbert Terrace, a professor at Columbia University in New York, also raised a chimpanzee, Nim.[663] After long training sessions, Nim could name over a hundred objects—and seemed to be able to combine words. But alas! By carefully watching the videos of the training sessions, Terrace realized that, in the vast majority of situations, the person training Nim was giving him clues. Nim was content to imitate his coach, hoping to be rewarded. This was a great disappointment for Terrace. He decided to watch the films made by the other scientists who had taught other apes to talk. Terrace says that, again, in most situations, the person training the monkey gave them clues.[664]

Does that mean apes are incapable of humanlike language? Not exactly. All these experiments and others, such as those with the bonobo Kanzi or the chimpanzee Ai,[665] have shown that they are able to learn to associate objects, colors, numbers, and even concepts such as "the same" and "different" with words. They are even able to name them in sign language or using graphic symbols (after extensive training) or to associate these learned signs in original combinations to express their current desires.[666] But, in any case, no acoustic production! The apes don't articulate a sentence, or even a word. They are not well equipped for that.[667]

The same kind of experiment, in which an attempt was made to teach an animal to speak "human," was carried out with a parrot named Alex. Irene Pepperberg, a professor at Brandeis University near Boston, had managed to teach it a hundred words. Alex would point to objects and colors and answer questions such as "How many yellow cubes are there?"[668] However, Alex wasn't having big conversations. Only innovations in the history of the human lineage could therefore lead to the emergence of spoken language. The question is, Which ones? Because when it comes to vocal communication, we have a lot in common with other animals.

First of all, we share our auditory abilities, which aren't very original. The hearing machinery is quite similar in all four-legged vertebrates, from the anatomy of the ear to the information-processing mechanisms in the brain, including the ciliated neurons that convert sound waves into nerve signals.[669] So our hearing apparatus probably predates the development of the ability to speak. It was already in place.

Human originality must therefore lie elsewhere. For a long time, people thought it was due to the low position of our larynx.[670] Remember the larynx? The organ with the vocal cords that allows mammals to make sounds? In reality, other mammals also have a descended larynx, so at first glance it is not technically impossible for them to be able to speak.[671] Moreover, the larynx is not absolutely essential for conversing: human beings can talk . . . by whistling. *Whistling languages*, which imitate the acoustic form of the words of the locally spoken language, are indeed used for long-distance communication in forests or mountainous areas.[672] Another quirk of our species is the absence of laryngeal sacs, the diverticula that allow monkeys to make their vocalizations lower-pitched and more powerful. But no one knows if this played a role in language acquisition.

Of course, researchers have also looked at genetics. Whoever finds the "gene for language" will win the Nobel Prize for sure. And some people thought they had reached the grail when the FOXP2 gene was identified as an important contributor.[673] *Forkhead box P2* is its full name. This gene makes it possible to manufacture a protein made up of 715 amino acids that is involved in the regulation of DNA transcription. In other words, it is a gene that regulates the activity of other genes. Some of its mutations, one of which has been identified in a British family, lead to severe language deficits.[674] Present in all vertebrates, and even in the small *Drosophila* fly, FOXP2 exists in a modified form in humans. But not modified by much—only 2 amino acids differ from that of the chimpanzee and 3 from that of the mouse.[675] Apparently, FOXP2 is especially important for language acquisition. It plays a key role in the synaptic plasticity that accompanies learning, when neural networks form and connect.[676] Experiments with songbirds have shown that a mutation of FOXP2 renders them unable to imitate their tutor—a feat of science.[677] It has also been possible to isolate and analyze the FOXP2 gene from our Neanderthal cousins, who died out about 30,000 years ago; they possessed the same version of FOXP2 as we do, except for a very small part of the gene that allows it to regulate its activity—its switch, if you will. Maybe the switch that changed everything . . . and the beginning of The Word?

But FOXP2 is probably just one of many genes that played a role in language acquisition.[678] Recently, researchers have started to do big things on the genetic side, and we may be on the verge of major discoveries. For example, a team of scientists led by Eric Jarvis of the Howard Hughes Medical Institute has compared the activity of all the genes (yes, all the genes) in the brains of humans and different species of birds. They showed that the brains of songbirds (those that learn their song by imitating a tutor) showed gene activities in some of their structures that are very similar to those of humans. In particular, the genes in the neurons of the region involved in the control of the syrinx (the vocal organ of birds) are activated in a manner quite similar to the genes in the region of the human brain involved in the control of the larynx.[679] Although this is all very promising, we are still far from having put all the pieces of the puzzle together.[680]

While waiting for advances in genetics and new fossil discoveries, it is probably the neural control of speech that we should be looking at. In any case, this is the hypothesis defended by Tecumseh Fitch, a professor at the University of Vienna and a specialist in the evolution of language[681]—a hypothesis in the footsteps of Darwin, who said, "The fact that superior apes do not use their vocal organs to speak depends, no doubt, on the fact that their intelligence has not progressed sufficiently."[682] It is true that two interesting innovations for spoken language distinguish the human brain from that of other primates. First, our species has direct connections between the part of the brain called the motor cortex and the neurons that drive the muscles involved in the act of speaking. Second, our brain is characterized by a considerable development of neural connections between the auditory and motor areas. In other words, there are very many neurons that make connections between what we hear and what we say. That seems to be unique in the animal kingdom. It's good to feel unique, isn't it?

Over the course of human history, our brains have gradually organized themselves to give us the capacity to speak. This did not happen in a regulated way—there was no design office with the plan in hand to oversee its realization—but in a way in which every small change to better manage our vocal communications with others gave our ancestors an advantage over those who could not. In our human lineage, natural selection has favored smooth talkers.[683]

This cerebral reorganization has been accompanied by anatomical modifications of the larynx, the organ of vocal production. For example, with the loss of air sacs present in our ancestors, the human larynx has become simpler, allowing it to produce more stable and more easily controllable sounds.[684]

In conclusion, the acquisition of spoken language by the human line probably required relatively few innovations. Most of the structures and mechanisms involved in human spoken language have very ancient roots.[685] Everything was there before. Spoken language simply required some neurophysiological (and a bit of anatomical) reorganization, which probably went hand in hand with the increasing complexity of our social relationships.

Robin Dunbar at Oxford University thus defends the idea that spoken language (as well as laughter and singing) is our way of engaging in social grooming[686]—you know, the habit monkeys have of picking lice out of each other's fur that serves as a way to build and maintain social bonds. One thing is obvious: the more people to be groomed, the longer the grooming takes. Dunbar defends the idea that it would be impossible to groom all members of a human social group one by one in the monkey fashion because there's not enough time and the group is too big. So language would have been a good substitute.

"Communication is the glue that holds animal groups or societies together, and in general sociality goes hand in hand with sophisticated communication systems."[687] Given the complexity of our human societies, that glue must be strong. Between the emergence of the first *Homo* about two million years ago and modern humans *Homo sapiens*, the size of social groups has increased considerably. In traditional societies, the number of individuals that a human is connected to is around 150. However, according to calculations by Dunbar and his colleagues, direct grooming connects to 50 individuals at best.[688] By telling stories around the fire, we can "groom" many people at the same time, better, and for longer.[689] And what about the possibilities offered today by social networks—a super-grooming!

There are many other hypotheses about the origin of human spoken language. Indeed, disputes between scientists have been raging for a long time on this subject, and it's not over yet. To give you a little anecdote, you should know that the prestigious *Société linguistique de Paris* already wrote in its statutes of 1866 that it would not accept any communication dealing with the origin of language. We can guess that in the mid-1800s it was already necessary to calm down opinionated zealots. Since the 1990s, scientists have been working to unify positions, or at least to look at the bigger picture by avoiding overly simplistic points of view.

As Chris Knight and Jerome Lewis of University College London argue, it may not be possible to base a theory on the origins of human spoken language by isolating it from theories on the origins of morality, law, religion, and so on.[690] For Knight, spoken language is out of the ordinary and it is not enough to study it as a "classic" acoustic communication system. Considering that, alongside the real world governed by the laws of physics, humans build a world based on beliefs (in deities, in the value of money, etc.), he proposes the idea that spoken language is there to navigate this virtual world. Words and grammar allow extraordinarily developed levels of interaction within a human group, as well as the establishment of rituals more or less disconnected from the real world. The question is complicated, and if you want to delve deeper into it, I invite you to read his article, coauthored with Lewis and published in *Current Anthropology*.[691]

Without going into detail, let's retain one of the major ideas of Knight and Lewis: the extraordinary flexibility of our vocal apparatus allows us to produce sounds aimed at deceiving the individual receiver (let's note, by the way, that Darwin had already raised this idea: the high priest of evolution was a visionary[692]). The primary receivers are other animals. When we were vulnerable beings with limited weaponry, as we were until very recently, we kept predators at bay by increasing the range and diversity of our vocalizations. And predators were plentiful in the Pleistocene savanna. More than 12 species of saber-toothed tigers, 9 species of hyenas, not to mention all the others—I will spare you the list. According to Knight and Lewis, this type of antipredator strategy would explain the polyphonic songs sung, especially on moonless nights, in certain hunter-gatherer societies (among the Hadza and Ba Yaka of sub-Saharan Africa, for example).[693] We find singing in the dark reassuring, that's for sure. And if each one of us pitches in with some polyphonic variations, the predator will have the impression that these humans are particularly numerous and will not dare approach. On the other hand, when it comes to hunting, we imitate the animals' calls to attract them. This is a strategy practiced daily in hunter-gatherer societies. And even in the present day, don't we do everything we can with our voice or our whistles to imitate the animal whose attention we wish to attract?

Scaring away predators, hunting prey—there is another situation where humans use sounds to communicate with wild animals. This is when they wish to cooperate with them. The case of the greater honeyguide, or indicator bird, is inspiring in this respect. The Latin name of this small African bird is doubly insistent—*Indicator indicator*—as if to make it clear that it will do anything to show us the way. Did you know that, curiously enough, the honeyguide loves to eat beeswax, the substance that bees produce to build the combs of their hives? The problem is that the wax combs are normally inaccessible, hidden at the bottom of some tree trunk, and boldly defended by bees that are as determined as they are armed. The honeyguide has understood, apparently for a very long time, that humans are also looking for hives and that they can be valuable allies provided the reverse is true. The clever bird is thus in the habit of leading hive hunters to the object of their common lust. How does it proceed? By calling and flying from one tree to another, patiently waiting for the humans to catch up. When all have arrived at their destination, the hunters, protected from the bees by clothing and equipped with tools, open the hive to extract the honey from it, leaving the wax combs in plain sight, which their winged guide delights in as they leave. In Mozambique, honey hunters produce a special cry that attracts the indicator bird, something like "brrr-hm!" Claire Spottiswoode of Cambridge University and her colleagues at the FitzPatrick Institute of African Ornithology at the University of Cape Town have shown through playback experiments that this "brrr-hm!" prompts the indicator bird to guide the honey hunters to a hive.[694] The bird does not respond to other sounds. Hunters say that this sound code was taught to them by their fathers. As for the honeyguides, no doubt they learn to recognize the "brrr-hm!" by observing experienced honeyguides feasting on the wax combs provided by the honey hunters. A wild animal, in its natural environment, responding to a call for cooperation from humans: This scene is a rare example of *mutualism* between a wild animal and our human species mediated by acoustics.

Mimicking the cries of animals, finding sound codes to which they are sensitive—our vocal flexibility seems to have been primitively selected to facilitate our interactions with species other than our own. It would then have been put at the service of social interactions within our own species, with some success! From animal languages to human language, the loop is complete.[695]

Dear reader, we are coming to the end of our journey among the voices of nature. I hope I've entertained you sometimes and interested you often. I could have talked to you about many more things, as the world of bioacoustics is so vast. Recently, it's been discovered that it even extends beyond the animal kingdom! Plants might be sensitive to sound waves or even themselves produce acoustic signals carrying information.[696, 697] The flowers of the beach evening primrose *Oenothera drummondii* produce a sweeter nectar within three minutes of "hearing" the sound of bee wings.[698] There are certainly plenty of surprises still in store for the younger generations of bioacousticians, and the next few years should be rich in discoveries.

But . . . let's not forget the essential things! Last but not least, let's evoke a last impression. A last ambience. Our last common bioacoustics moment. . . .

ACKNOWLEDGMENTS

Vallon de la Fauge, Vercors Massif, French Alps. It's five o'clock in the morning and the concert is already starting again. It had stopped late yesterday, well after sunset. But the musicians have decided that the night rest had lasted long enough, and the great symphony resumes. Still lying in my tent, I try to identify who is who. The robin is the first to warm up its voice. Closely followed by a wren. With its explosive singing, this one is a first-rate alarm clock. I imagine it perched on its bush unless it is already snooping here and there while singing, restless as it is. The chaffinches are getting under way—there's more than one of them. Three maybe? And here comes the blackcap. This one brings back memories. I can still see it screeching at the top of my loudspeaker, perched on a branch. In the rising hubbub, I watch for the first part of its sentence, this slightly disordered babbling that makes it possible to distinguish it from its garden cousin. A willow warbler is reeling off its river of notes. And there, a Eurasian blackbird? No, the song is too complex, too full of flourishes—probably a song thrush. Getting a little lost in this festival, I let myself bask in the pleasure of listening to the symphony without trying to isolate the instrumentalists. Last evening, things were simple. Tits and warblers were leading the dance. While it was still light, an owl had begun to hoot before waiting for everyone to shut up. Alone, it had tried to fill the silence of a night still too cold for insects, barely helped by the rare barking of rutting deer. No nightingale, alas, to keep it company. Are we too high for the spring herald? In the afternoon, on the flank of the *Grande Moucherolle* mountain, we came across a herd of 70 Alpine ibex. One of them whistled as we approached, signaling its irritation. We had also heard the whistles of the marmots, frightened by a black kite. I had hoped for wolf howls during the night, but the packs are probably far away, on the uplands over there. Two days later we'll see their droppings, full of fur and bones. I crawl out of my sleeping bag and get out of the tent, still sore from the previous day's long hike. The meadow is soaked with dew. The thrush is there, on the tip of the nearest tree, singing its melodies. Blocked by the peaks, the sun will take a long time to reach the valley. I think I still have two more chapters to write, and I light the little stove to make tea. Tonight we must be at the foot of the *Grand Veymont* mountain . . . the book will wait!

Writing a book is a way to explain things in more depth than giving lectures, where time is always short. In particular, it helps to emphasize the human aspect of research. When writing this book, I have tried to remember all the people who have marked my scientific adventures. Still, an unpredictable memory combined with a well-dimensioned ego can play tricks. Might I have omitted someone who had an idea or initiated a project for which I claim to be the bearer? I would blame myself . . . less for fear of offending someone than for the feeling of having failed in an essential duty: to show that researchers do not forge themselves alone, that no one makes discoveries in a vacuum. In this book, none of the personal stories rested just on my shoulders.

Joëlle Ayats is the second pen of this book. I would like to thank her warmly! I loved discovering her uncompromising corrections, their relevance, their intelligence. They added dynamism and humor. They were also accurate, removing heaviness and errors. "She doesn't miss a thing!" Eliane Viennot had warned me, our mutual friend who had put me at Joëlle's mercy.

The third pen is a pencil. My father, Bernard Mathevon, took up the challenge of illustrating each of the animals I worked with (only the flamingo, the eagle owl, and the Antarctic skua did not find their place in the text; that's why they are here!). I thank him wholeheartedly for his meticulous work. Believe me, the original plates are even more beautiful than the printed ones. His wren is my favorite. Which one was yours?

I was deeply honored when Bernie Krause agreed to write the foreword for the English edition of my book. I warmly thank him for writing a text that is both personal and inspiring. I share with him this vision of

Flamingo

science: the most important thing is the path one takes. Krause is a gentleman of great stature. His expertise on the world's soundscapes is unparalleled. It was a real pleasure to read his kind words about my book: "Like a fine novel, this notable work is impossible to put down once you begin."

Bethan Wakeling did a fabulous job of correcting my English. I am very grateful for her generosity and dedication. The fact that she described her commitment as an "enriching experience" and that she found the book "great" warmed my heart.

I am very grateful to David, Eilean, and Elizabeth Reby for translating the animal sound onomatopoeia into English (except for the Savannah sparrow song, translated by Dan Mennill). I heard it wasn't easy to agree. You be the judge.

Eliane Viennot played an important role in the design of this book, first by opening my eyes a few years ago to the thought biases that mark

Eagle owl

all scientific research, then by helping me write the first chapters, and then by proposing Joëlle as a reviewer.

From Olivia Recasens, my French editor, I retain the passionate interest in science and the soft firmness of her "I can't wait to read you," which punctuated her emails. Writing this book was an exciting adventure, and I am deeply grateful to her for giving me the opportunity.

It was my friend Mark Hauber who encouraged me to send my manuscript to Alison Kalett, acquisitions editor at Princeton University Press. I was thrilled when Alison wrote me that she and her colleagues would be excited to publish *The Voices of Nature*. I sincerely thank her and the PUP team for this opportunity to bring my book to the English-speaking readership. I hope that many vocations will be born from this reading.

I am indebted to all those who reviewed all or part of the manuscript, saving me from serious errors or omissions: Olivier Adam, Thierry Aubin, Renaud Boistel, Elodie Briefer, Isabelle Charrier, Charlotte Curé, Etienne Danchin, Sébastien Derégnaucourt, Tudor Draganoiu, Michael Greenfield, Isabelle Horwath, Florence Levréro, Anne Mathevon, Etienne Parizet, David Reby, Oscar Roman, Dominique Rouger, Fanny

Antarctic
skua

Rybak, Frédéric Sèbe, Marc-André Sélosse, Jérôme Sueur, Frédéric Theunissen, and Laurent Villermet, as well as the two PUP referees, Daniel Blumstein and Tecumseh Fitch, whose careful reading and constructive criticism and advice have greatly improved the book. A huge thank-you for their advice, corrections, and encouragement!

Throughout the book, you have discovered the people with whom I have been fortunate enough to experience these scientific adventures. I continue my journey with some of them. With others, trajectories have diverged. They are all dear to me and I am grateful for the moments I have shared here and there, reflecting, discussing, trudging along, or simply waiting patiently for the right moment to do the experiment.

I would also like to thank my students, my colleagues at the ENES Bioacoustics Research Lab, and my collaborators from France and elsewhere, as well as all those who welcome our research (I have a special sense of gratitude for the people of *Crocoparc* in Agadir, Morocco).

Here I must solemnly thank the University of Saint-Etienne and the people who make it what it is today. By agreeing to host my small team more than 20 years ago, this university has made everything possible.

It took a lot of energy to initiate the ENES Bioacoustics Research Lab (eneslab.com), but through stubbornness and luck, too, things gradually improved. Many people helped me on the way. I am especially indebted to three of them: Professor André Giret, a field geologist, at one time responsible for research in the French Southern and Antarctic Territories; Professor Hervé Barré, one of my former professors, a demanding scientist who headed an important research laboratory at the CNRS; and Professor Jean-Marc Jallon, director of a famous CNRS research lab on memory, learning, and animal communications at the University of Paris–Sud.

A bioacoustics laboratory such as the ENES must actively participate in the training of new generations of bioacousticians. The Bioacoustics Winter School (https://www.eneslab.com/bioacoustic-winter-school) and the International Master of Bioacoustics (https://www.masterofbioacoustics .com/) have been born thanks to the enthusiasm of ENES researchers, in particular David Reby and Frédéric Sèbe, and of many bioacoustician colleagues and friends from France and abroad. I thank all of them warmly here.

I also owe a lot to the *Institut universitaire de France*. The IUF is an organization allowing academics to devote more time to research.[699] Getting in for 15 years as a junior and then as a senior member was (and still is) an incredible opportunity. In the French research landscape where new administrative behemoths are constantly being built, the IUF is a benevolent UFO. May it always escape the mania of forms and reports that paralyzes researchers by depriving them of their time— their most precious commodity.

I would like to thank the funding organizations that support my research projects. The University of Saint-Etienne and the *Institut universitaire de France*, again, but also the *Agence nationale de la recherche* (ANR), the *Centre national de la recherche scientifique* (CNRS), the *Institut national de la santé et de la recherche médicale* (Inserm), the city of Saint-Etienne, the *Labex CeLyA*, and the Miller Institute for Basic Research in Science, to name but a few. The research with crocodiles was made possible by the National Geographic Society, which funded several field expeditions, and by the unfailing support of the *Crocoparc* zoological park in Morocco.

Science allows me to work with amazing, brilliant, singular, enthusiastic people who have shaped both my thinking and my behavior. All of them different. Here I want to pay a special tribute to Thierry Aubin. Over the years, his sharp intelligence, his qualities as a scientist, his talent as an observer of nature, his vision of the world and of life, his humor, and even his culinary tastes have shaped me in many ways. I wouldn't be who I *am* today if our paths hadn't crossed. You have seen in these pages that, without him, nothing would have been possible. I hope that he finds this book worthy.

Last but not least, I end by warmly thanking my wife and children, who sometimes find bioacoustics a bit invasive, as well as my parents who, according to the time-honored-but-true formula, have always supported me in my projects.

And now it's time to start your journey among the voices of nature for real! The white-browed warbler, the jacaré caiman, the spotted hyena, the elephant seal, the Atlantic walrus, and the procession of babbling, singing, barking, jabbering, and talking animals will be with you.

GLOSSARY

Note: The definitions in this glossary are deliberately brief and therefore could appear incomplete. They are written to reflect a bioacoustics perspective.

Acoustic adaptation A process that promotes sound signals that are optimally transmitted in a given environment.

Acoustic avoidance A process that increases the differences between sound signals from animals living at the same time in the same area.

Acoustic biodiversity The totality of animal sound productions characteristic of an environment.

Acoustic biodiversity indices Quantitative measures derived from the analysis of sound recordings made in an environment and indicative of the diversity of living things in that environment.

Acoustic signals Mechanical disturbances (vibration waves) propagating in water or air. Vibration signals propagating through solids (e.g., soil) can also be considered as sounds.

Adaptive (or evolutionary) radiation Diversification of several species from a common ancestor. Each new species occupies, and is adapted to, a new ecological niche of its own.

Amplitude modulation The variation in the intensity (weak/strong) of a sound signal over time.

Amplitude of a sound The intensity of a sound. The amplitude (high, medium, low) of a sound is measured in decibels.

Arms race The evolutionary escalation between a communication signal and the resistance of the individual receiver to respond to this signal. The sender produces a signal with increasingly exaggerated characteristics, while the receiver becomes increasingly reluctant to respond to the signal. A common metaphor for this phenomenon is the Red Queen's instruction to Alice that they must run to stay in the same place.

Audience effect When an animal reacts to a communication signal or changes its signal production based on the individuals around it.

Behavioral ontogeny The development of a behavior over the course of an individual's life.

Bioacoustics The scientific discipline that studies animal and human acoustic communications.

Biological adaptation The process by which a population of individuals becomes better able to survive and reproduce in a particular environment.

Categorization (categorical perception) The mechanism by which an animal responds to the continuous variation of a physical stimulus by forming categories.

Chain of information transmission The sender-signal-receiver unit, as well as the processes of encoding information (by the sender), transmitting information (propagation of the signal in the environment), and decoding information (by the receiver). All elements of the information transmission chain can be affected by noise.

Character shift The increase in differences that allow a distinction, for receivers, between animal populations living in the same location.

Communication The transfer of information between one or several senders and one or several receivers.

Communication network The set of individuals that can transmit and receive sound signals between them.

Cooperative rearing The involvement of individuals other than the parents in the care of the young.

Cost of a signal The processes that cause the emission of a signal to decrease the probability of survival of the sender or its ability to have offspring.

Dear-enemy effect When a territorial animal reacts less violently to the acoustic signals of a familiar neighbor than to those of an unknown individual.

Decibel A unit of measurement (abbreviated dB) for the intensity of a sound.

Dialect The geographic variation in the vocalizations of an animal species.

Doppler effect A phenomenon characterized by a sound being higher-pitched when the sound source is approaching and lower-pitched when it is moving away.

Dynamic information Information about the characteristics of the sender that may vary quickly (e.g., emotions or aggression).

Eavesdropping When an individual listens to a sound signal that was not intended for them and obtains information from it.

Ecoacoustics A discipline derived from bioacoustics that uses natural sounds to address issues of ecology.

Ecological selection (or ecological adaptation) A process that favors sound signals that are best propagated in the environment where they are used.

Evolutionarily stable equilibrium A communication behavior that characterizes the majority of individuals in a population and is never superseded by another: any variation is counter-selected. Also referred to as an evolutionarily stable strategy.

Evolutionary causes of communication Processes that explain the development of communication throughout the history of a species. The evolutionary causes, in the long term, concern the processes of natural selection in particular (including sexual selection), as well as genetic and cultural drifts.

Exaptation A mechanism occurring during evolution where the function of a preexisting trait is redirected to another function (e.g., the drumming behavior of woodpeckers, originally used to find food, has become a means of emitting an acoustic communication signal for recognition within the species).

Formants Reinforced voice frequencies when sound waves travel through the vocal tract.

Frequency modulation The change in pitch (high/low) of a sound signal over time.

Frequency spectrum The set of frequencies present in a sound. A pure sound has a single frequency; a complex sound has several.

Fundamental frequency The lowest frequency of a complex sound (called F0 or "first harmonic"). It determines the pitch of a sound.

Graded vocalizations Vocalizations that vary along an acoustic continuum, with this variation accompanying changes in the information encoded by the sound signal.

Habituation The gradual loss of a receiver's responsiveness to a signal when it is repeatedly perceived.

Handicap theory The theory that certain features that are heavy to bear are evidence of the quality of those who bear them; for example, signals that impose a cost to survival.

Harmonic frequencies Multiples of the fundamental frequency.

Hertz A unit of measurement (abbreviated Hz) for the frequency of a sound. One hertz corresponds to one cycle of vibration per second.

Honest communication When a signal carries reliable information about certain characteristics of the sender (e.g., body size). This signal cannot manipulate the receiver.

Infrasounds Sound waves too low-pitched to be heard by humans (< 20 Hz approximately).

Kin selection Process favoring altruistic (cooperative) behaviors between genetically close individuals, with a reciprocity all the stronger as they are more closely related.

Language Fundamentally, any communication system based on the production of sound signals. However, this term is often reserved for human-articulated language.

Lek Area where males gather to court females.

Mathematical theory of communication The theory that all communication is based on the sender-signal-receiver triad, and is subject to noise. The coding of information by the signal can be optimized to meet the constraint imposed by noise. Also known as information theory.

Morton's principle (or motivation-structural rules) The correspondence between the structure of an acoustic signal and the context of the sender's emission and motivation. For example, and in general, a dull rumble is more indicative of aggressive intentions, while small, high-pitched, low-intensity calls are typical of a submissive attitude.

Natural selection A process that favors the survival and offspring of individuals best adapted to the environment in which they live. In the context of bioacoustics, natural selection favors communication behaviors and sound signals that increase the survival and/or genetic contribution to the next generation (direct and/or indirect progeny) of the sender and/or receiver.

Noise Any process that results in the loss of information as it is transmitted from the sender to the receiver.

Phylogenetic tree A representation of relationships between species (or between populations, individuals, etc.).

Pitch The property of a sound as determined by the fundamental frequency. Ranges from low to high.

Proximal causes of communication Morphological, anatomical, and physiological bases of communication behavior. Proximal causes, in the short term, concern all the mechanisms of production and perception of sound signals.

Quality index (or index signal) A signal that carries reliable information about the sender because its production is subject to an unavoidable physical constraint that reflects its anatomy, morphology, physiological capabilities, or health (e.g., the lowest fundamental frequency that can be emitted by a mammal is constrained by the size of its vocal cords).

Referential communication The use of signals that designate an environmental feature (such as the presence of a particular predator).

Ritualization The modification of a behavior (aggression, search for food, etc.) that becomes a communication signal by stereotypical simplification.

Sensory bias (or sensory exploitation) A situation in which a receiver's preexisting sensitivity to a signal may be exploited by the sender. This sensitivity has emerged through evolution for reasons other than communication.

Sexual dimorphism Differences (e.g., in size or color) between females and males.

Sexual selection Cases of natural selection imposed by individuals of the same sex (intrasexual selection) or of the opposite sex (intersexual selection).

Social grooming Behavior consisting of searching the coat or plumage of a congener for parasites, which plays an important role in maintaining and strengthening social bonds in primates.

Social intelligence The ability to analyze the relationships between members of one's social group and the social status of others.

Sound frequency The number of cycles of vibration per second. A low-frequency sound is low-pitched; a high-frequency sound is high-pitched.

Sound library A collection of sound recordings.

Sound waves Pressure variations that propagate through air or water.

Source-filter theory The theory that the production of sound signals in mammals (and most four-legged vertebrates) is based on an anatomical structure producing a sound wave (the source: larynx in mammals, syrinx in birds), and a system of pipes and cavities modifying this wave (the filter: vocal tract).

Spatial unmasking The ability to locate the origin of a sound signal despite ambient noise when the source of the signal and the source of the noise are not in the same location.

Spectrogram A graphical representation of a sound in which the time scale is given by the horizontal axis, the frequency scale by the vertical axis, and the intensity scale by a palette of colors.

Static information Information relating to the stable characteristics of the sender; the constituents of a sender's vocal signature.

Timbre The sound of the voice as determined by a combination of many characteristics of sound, including the distribution of energy in the frequency spectrum.

Ultrasounds Sound waves that are too high-pitched to be heard by humans (> 20,000 Hz approximately).

Vocal learning The mechanisms by which an animal learns to produce vocalization by imitation.

Vocal plasticity The ability for an individual to rapidly modulate its sound signals.

Vocal repertoire The set of sound signals that an animal can produce.

Vocal signature The information encoded in an audible signal indicating the identity of the sender.

Wavelength The distance between two amplitude maxima of the sound wave. The wavelength is inversely proportional to the frequency of the sound: when it is short, the sound is high-pitched; when it is long, the sound is low-pitched.

NOTES

Chapter 1

1. Bradbury J. W., and Vehrencamp S. L., *Principles of Animal Communication*, Sinauer, 2011.

2. Lorenz K., "Der kumpan in der umwelt des vogels," *J Ornithol*, vol. 83, 1935, 137–289.

3. Von Frisch K., *The Dance Language and Orientation of Bees*, Harvard University Press, 1967.

4. Tinbergen N., "On aims and methods of ethology," *Zeit Tierpsychol*, vol. 20, 1963, 410–433.

5. Bateson P., and Laland K. N., "Tinbergen's four questions: An appreciation and an update," *Trends Ecol Evol*, vol. 28, 2013, 712–718.

6. On dinosaurs' vocalizations, see Weishampel D. B., "Dinosaurian cacophony: Inferring function in extinct organisms," *BioScience*, vol. 47, 1997, 150–159.

7. Hsieh S., and Plotnick R. E., "The representation of animal behaviour in the fossil record," *Anim Behav*, vol. 169, 2020, 65–80; Senter P., "Voices of the past: A review of Paleozoic and Mesozoic animal sounds," *Hist Biol*, vol. 20, 2008, 255–287.

Chapter 2

8. Maurice S., et al., "In situ recording of Mars soundscape," *Nature*, vol. 605, 2022, 653–658.

9. Some animals are able to produce very loud sounds, up to 125 dB! See Jakobsen L., et al., "How loud can you go? Physical and physiological constraints to producing high sound pressures in animal vocalizations," *Front Ecol Evol*, vol. 9, 2021, 657254.

10. To deepen your knowledge of the physics of sound waves, I advise you to read the following chapters, all written for bioacousticians: Larsen O. N., "To shout or to whisper? Strategies for encoding public and private information in sound signals," in Aubin T., and Mathevon N. (eds.), *Coding Strategies in Vertebrate Acoustic Communication*, Springer, 2020, 11–44; Larsen O. N., and Wahlberg M., "Sound and sound sources," in Brown C. and Riede T. (eds.), *Comparative Bioacoustics: An Overview*, Bentham Science, 2017, 3–62; Wahlberg M., and Larsen O. N., "Propagation of sound," in Brown and Riede (eds.), *Comparative Bioacoustics*, 63–121.

Chapter 3

11. Marler P., and Slabbekoorn H. (eds.), *Nature's Music: The Science of Birdsong*, Elsevier, 2004; Catchpole C. K., and Slater P.J.B., *Bird Song: Biological Themes and Variations*, Cambridge University Press, 2008. For a refreshing and interesting perspective on birdsong, see this review: Rose E. M., et al., "The singing question: Re-conceptualizing birdsong," *Biol Rev*, vol. 97, 2022, 326–342.

12. Species recognition is an important process for mating purposes or territorial defense, and many animal signals support information about species identity. For a nonbird example, look at this paper: Fonseca P. J., and Revez M. A., "Song discrimination by male cicadas *Cicada barbara lusitanica* (Homoptera, Cicadidae)," *J Exp Biol*, vol. 205, 2002, 1285–1292.

13. Draganoiu T. I., et al., "Song stability and neighbour recognition in a migratory songbird, the black redstart," *Behaviour*, vol. 151, 2014, 435–453. There are many other articles dealing with the dear-enemy effect; see, for example, Briefer E., Rybak F., and Aubin T., "When to be a dear enemy: Flexible acoustic relationships of neighbouring skylarks, *Alauda arvensis*," *Anim Behav*, vol. 76, 2008, 1319–1325; Briefer E., et al., "How to identify dear enemies: The group signature in the complex song of the skylark *Alauda arvensis*," *J Exp Biol*, vol. 211, 2008, 317–326; Tumulty J. P., et al., "Ecological and social drivers of neighbor recognition and the dear enemy effect in a poison frog," *Behav Ecol*, vol. 32, 2021, 138–150; and Amorim P. S., et al., "Out of sight, out of mind: Dear enemy effect in the rufous hornero, *Furnarius rufus*," *Anim Behav*, vol. 187, 2022, 167–176. See also our recent paper on hippos: Thévenet J., et al., "Voice-mediated interactions in a megaherbivore," *Cur Biol*, vol. 32, 2022, R55–R71.

14. Aubin T., et al., "How a simple and stereotyped acoustic signal transmits individual information: The song of the white-browed warbler *Basileuterus leucoblepharus*," *An Acad Bras Cienc*, vol. 76, 2004, 335–344.

15. Mathevon N., et al., "Singing in the rain forest: How a tropical bird song transfers information," *PLoS ONE*, vol. 3, 2008, e1580.

16. Slabbekoorn H., "Singing in the wild: The ecology of birdsong," in Marler and Slabbekoorn (eds.), *Nature's Music*, 178–205.

17. Their Latin names, respectively, are as follows: *Crypturellus obsoletus, Myrmeciza squamosa, Basileuterus leucoblepharus, Pyriglena leucoptera, Chiroxiphia caudata, Batara cinerea, Trogon surrucura,* and *Carpornis cucullatus*.

18. Brémond J.-C., "Specific value of syntax in the territorial defense signal of the troglodyte (*Troglodytes troglodytes*)," *Behaviour*, vol. 30, 1968, 66–75; Brémond, "Role of the carrier frequency in the territorial songs of oscines," *Ethology*, vol. 73, 1968, 128–135; Brémond, "Acoustic competition between the song of the wren (*Troglodytes troglodytes*) and the songs of other species," *Behaviour*, vol. 65, 1978, 89–98.

19. Kreutzer M., "Stéréotopie et variations dans les chants de proclamation territoriale chez le Troglodyte (*Troglodytes troglodytes*)," *Rev Comport Anim*, vol. 8, no. 2, 1974, 70–286.

20. Mathevon N., and Aubin T., "Reaction to conspecific degraded song by the wren *Troglodytes troglodytes*: Territorial response and choice of song post," *Behav Proc*, vol. 39, 1997, 77–84.

21. Mathevon N., Aubin T., and Dabelsteen T., "Song degradation during propagation: Importance of song post for the wren *Troglodytes troglodytes*," *Ethology*, vol. 102, 1996, 397–412.

22. Mathevon N., et al., "Are high perches in the blackcap *Sylvia atricapilla* song or listening posts? A sound transmission study," *JASA*, vol. 117, 2005, 442–449.

23. Mathevon N., and Aubin T., "Sound-based species-specific recognition in the blackcap *Sylvia atricapilla* shows high tolerance to signal modifications," *Behaviour*, vol. 138, 2001, 511–524.

24. Ey E., and Fischer J., "The 'acoustic adaptation hypothesis'—a review of the evidence from birds, anurans and mammals," *Bioacoustics*, vol. 19, 2009, 21–48.

25. I discuss this in more detail in chapter 17 on ecoacoustics.

26. Balakrishnan R., "Behavioral ecology of insect acoustic communication," in Pollack G. S., Mason A. C., Popper A., and Fay R. R. (eds.), *Insect Hearing*, Springer, 2016, 49–80.

Chapter 4

27. Shannon C. E., and Weaver W., *The Mathematical Theory of Communication*, University of Illinois Press, 1949. Claude Shannon first developed and published this theory in 1948. His paper was republished in book form in 1949, with an introduction by Warren Weaver (which I recommend reading).

28. Mathevon N., and Aubin T., "Acoustic coding strategies through the lens of the mathematical theory of communication," in Aubin and Mathevon (eds.), *Coding Strategies in Vertebrate Acoustic Communication*, 1–10.

29. Charrier I., et al., "Acoustic communication in a black-headed gull colony: How do chicks identify their parents?," *Ethology*, vol. 107, 2001, 961–974.

30. Aubin T., and Jouventin P., "How to vocally identify kin in a crowd: The penguin model," *Adv Stud Behav*, vol. 31, 2002, 243–277.

31. Aubin T., and Jouventin P., "Cocktail-party effect in king penguin colonies," *Proc R Soc B*, vol. 265, 1998, 1665–1673.

32. Lengagne T., et al., "How do king penguins (*Aptenodytes patagonicus*) apply the mathematical theory of information to communicate in windy conditions?," *Proc R Soc B*, vol. 266, 1999, 1623–1628.

33. Jouventin P., Aubin T., and Lengagne T., "Finding a parent in a king penguin colony: The acoustic system of individual recognition," *Anim Behav*, vol. 57, 1999, 1175–1183.

34. The frequencies—the pitch, low or high, of the song—change over time.

35. The intensity—the fact that the song is more or less loud—changes over time.

36. Aubin T., Jouventin P., and Hildebrand C., "Penguins use the two–voice system to recognize each other," *Proc R Soc B*, vol. 267, 2000, 1081–1087.

37. Gomez-Bahamon V., et al., "Sonations in migratory and non-migratory fork-tailed flycatchers (*Tyrannus savana*)," *Integr Comp Biol*, vol. 60, 2020, 1147–1159.

38. Kingsley E. P., et al., "Identity and novelty in the avian syrinx," *PNAS*, vol. 115, 2018, 10209–10217.

39. Riede T., et al., "The evolution of the syrinx," *PLoS Biol*, vol. 17, 2019, e2006507.

40. Robisson P., Aubin T., and Brémond J.-C., "Individuality in the voice of the emperor penguin *Aptenodytes forsteri*: Adaptation to a noisy environment," *Ethology*, vol. 94, 1993, 279–290.

41. Aubin and Jouventin, "Cocktail-party effect in king penguin colonies."

42. Jouventin P., and Aubin T., "Acoustic systems are adapted to breeding ecologies: Individual recognition in nesting penguins," *Anim Behav*, vol. 64, 2002, 747–757.

43. Aubin T., and Jouventin P., "Localisation of an acoustic signal in a noisy environment: The display call of the king penguin *Aptenodytes patagonicus*," *J Exp Biol*, vol. 205, 2002, 3793–3798. See also this paper reporting that king penguins' calls inform receivers about the age of the sender: Kriesell H. J., et al., "How king penguins advertise their sexual maturity," *Anim Behav*, vol. 177, 2021, 253–267.

44. Loesche P., et al., "Signature versus perceptual adaptations for individual vocal recognition in swallows," *Behaviour*, vol. 118, 1991, 15–25; Medvin M. B., Stoddard P. K., and Beecher M. D., "Signals for parent-offspring recognition: Strong sib-sib call similarity in cliff swallows but not barn swallows," *Ethology*, vol. 90, 1992, 17–28; Medvin M. B., Stoddard P. K., and Beecher M. D., "Signals for parent-offspring recognition: A comparative analysis of the begging calls of cliff swallows and barn swallows," *Anim Behav*, vol. 45, 1993, 841–850.

45. Charrier I., "Mother-offspring vocal recognition and social system in pinnipeds," in Aubin and Mathevon (eds.), *Coding Strategies in Vertebrate Acoustic Communication*, 231–246; Charrier I., "Vocal communication in otariids and odobenids," in Campagna C., and Harcourt R. (eds.), *Ethology and Behavioral Ecology of Otariids and the Odobenid*, Springer, 2021, 265–289.

Chapter 5

46. Miller E. H., and Kochnev A. A., "Ethology and behavioral ecology of the walrus (*Odobenus rosmarus*), with emphasis on communication and social behavior," in Campagna and Harcourt (eds.), *Ethology and Behavioral Ecology of Otariids*, 437–488.

47. Charrier I., "Mother-offspring vocal recognition and social system in pinnipeds"; Charrier I., and Casey C., "Social communication in phocids," in Costa D. P., and McHuron E. (eds.), *Ethology and Behavioral Ecology of Phocids*, Springer, 2022, 69–100.

48. Stewart B., "Family Odobenidae," in Wilson D. E., and Mittermeier R. A. (eds.), *Handbook of the Mammals of the World IV, Sea Mammals*, Lynx Edicions, 2014, 102–119.

49. Charrier I., Aubin T., and Mathevon N., "Mother-calf vocal communication in Atlantic walrus: A first field experimental study," *Anim Cogn*, vol. 13, 2010, 471–482.

50. https://videotheque.cnrs.fr/doc=2019.

51. Linossier J., et al., "Maternal responses to pup calls in a high-cost lactation species," *Biol Let*, vol. 17, 2021, 20210469.

52. For a review of recognition mechanisms in the Australian sea lion, see Charrier I., et al., "Mother-pup recognition mechanisms in Australian sea lion (*Neophoca cinerea*) using uni- and multi-modal approaches," *Anim Cogn*, vol. 25, 2022, 1019–1028.

53. Studies show that vocal recognition between mother and pup in mammals can last a long time: Briefer E. F., et al., "Mother goats do not forget their kids' calls," *Proc R Soc B*, vol. 279, 2012, 3749–3755; Charrier I., et al., "Fur seal mothers memorize subsequent versions of developing pups' calls: Adaptation to long-term recognition or evolutionary by-product?," *Biol J Lin Soc*, vol. 80, 2003, 305–312.

54. Charrier I., Mathevon N., and Jouventin P., "Mother's voice recognition by seal pups," *Nature*, vol. 412, 2001, 873.

55. The dynamics of voice recognition between mother and pup has been studied in other mammal species, including humans. See, for instance, Sèbe F., et al., "Establishment of vocal communication and discrimination between ewes and their lamb in the first two days after parturition," *Dev Psychobiol*, vol. 49, 2007, 375–386; and Bouchet H., et al., "Baby cry recognition is independent of motherhood but improved by experience and exposure," *Proc R Soc B*, vol. 287, 2020, 20192499.

56. Charrier I., et al., "The subantarctic fur seal pup switches its begging behaviour during maternal absence," *Can J Zool*, vol. 80, 2002, 1250–1255.

57. Mother-pup vocal recognition in mammals has been studied in other models. See, for instance, Sèbe F., et al., "Early vocal recognition of mother by lambs: Contribution of low- and high-frequency vocalizations," *Anim Behav*, vol. 79, 2010, 1055–1066.

58. Maynard Smith J., and Harper D., *Animal Signals*, Oxford University Press, 2003; Searcy W. A., and Nowicki S., *The Evolution of Animal Communication*, Princeton University Press, 2005.

59. Anderson M. G., Brunton D. H., and Hauber M. E., "Reliable information content and ontogenetic shift in begging calls of grey warbler nestlings," *Ethology*, vol. 116, 2010, 357–365; Caro S. M., West S. A., and Griffin A. S., "Sibling conflict and dishonest signaling in birds," *PNAS*, vol. 113, 2016, 13803–13808.

60. Kedar H., et al., "Experimental evidence for offspring learning in parent-offspring communication," *Proc R Soc B*, vol. 267, 2000, 1723–1727.

61. McCarty J. P., "The energetic cost of begging in nestling passerines," *The Auk*, vol. 113, 1996, 178–188; Leech S. M., and Leonard M. L., "Is there an energetic cost to begging in nestling tree swallows (*Tachycineta bicolor*)?," *Proc R Soc B*, vol. 263, 1996, 983–987.

62. Bachman G. C., and Chappell M. A., "The energetic cost of begging behaviour in nestling house wrens," *Anim Behav*, vol. 55, 1998, 1607–1618.

63. Husby M., "Nestling begging calls increase predation risk by corvids," *Anim Biol*, vol. 69, 2019, 137–155.

64. Magrath R. D., et al., "Calling in the face of danger: Predation risk and acoustic communication by parent birds and their offspring," *Adv Stud Behav*, vol. 41, 2010, 187–253; Husby, "Nestling begging calls increase predation risk."

65. Magrath et al., "Calling in the face of danger."

66. Briskie J. V., Martin P. R., and Martin T. E., "Nest predation and the evolution of nestling begging calls," *Proc R Soc B*, vol. 266, 1999, 2153–2159.

67. Haff T. M., and Magrath R. D., "Calling at a cost: Elevated nestling calling attracts predators to active nests," *Biol Let*, vol. 7, 2011, 493–495.

68. Anderson M. G., Brunton D. H., and Hauber M. E., "Species specificity of grey warbler begging solicitation and alarm calls revealed by nestling responses to playbacks," *Anim Behav*, vol. 79, 2010, 401–409.

69. Kilner R. M., and Hinde C. A., "Information warfare and parent–offspring conflict," *Adv St Behav*, vol. 38, 2008, 283–336.

70. Caro, West, and Griffin, "Sibling conflict and dishonest signaling in birds"; Bowers E. K., et al., "Condition-dependent begging elicits increased parental investment in a wild bird population," *Am Nat*, vol. 193, 2019, 725–737.

71. Price K., "Begging as competition for food in yellow-headed blackbirds," *The Auk*, vol. 4, 1996, 963–967.

72. Mathevon N., and Charrier I., "Parent-offspring conflict and the coordination of siblings in gulls," *Proc R Soc B*, vol. 271, 2004, S145–S147; Blanc A., et al., "Coordination de la quémande entre les jeunes de mouette rieuse," *CR Biol*, vol. 333, 2010, 688–693.

73. Leonard M., and Horn A., "Provisioning rules in tree swallows," *Behav Ecol Sociobiol*, vol. 38, 1996, 341–347; Leonard M. L., Horn A. G., and Mukhida A., "False alarms and begging in nestling birds," *Anim Behav*, vol. 69, 2005, 701–708.

74. Roulin A., "The sibling negotiation hypothesis," in Wright J., and Leonard M. L. (eds.), *The Evolution of Begging*, Kluwer, 2002, 107–126; Ducouret P., et al., "The art of diplomacy in vocally negotiating barn owl siblings," *Front Ecol Evol*, vol. 7, 2019, 351.

75. Dreiss A. N., et al., "Social rules govern vocal competition in the barn owl," *Anim Behav*, vol. 102, 2015, 95–107.

76. Dreiss A. N., et al., "Vocal communication regulates sibling competition over food stock," *Behav Ecol Sociobiol*, vol. 70, 2016, 927–937.

77. Ligout S., et al., "Not for parents only: Begging calls allow nest-mate discrimination in juvenile zebra finches," *Ethology*, vol. 122, 2016, 193–206.

78. Draganoiu T. I., et al., "Parental care and brood division in a songbird, the black redstart," *Behaviour*, vol. 142, 2005, 1495–1514; Draganoiu T. I., et al., "In a songbird, the black redstart, parents use acoustic cues to discriminate between their different fledglings," *Anim Behav*, vol. 71, 2006, 1039–1046.

79. Bugden S. C., and Evans R. M., "Vocal solicitation of heat as an integral component of the developing thermoregulatory system in young domestic chickens," *Can J Zool*, vol. 75, 1997, 1949–1954.

80. Mariette M. M., and Buchanan K. L., "Prenatal acoustic communication programs off-spring for high posthatching temperatures in a songbird," *Science*, vol. 353, 2016, 812–814; Mariette M. M., "Acoustic cooperation: Acoustic communication regulates conflict and cooperation within the family," *Front Ecol Evol*, vol. 7, 2019, 445; Pessato A., et al., "A prenatal acoustic signal of heat affects thermoregulation capacities at adulthood in an arid-adapted bird," *Sc Rep*, vol. 12, 2022, 5842.

81. Elie J. E., et al., "Vocal communication at the nest between mates in wild zebra finches: A private vocal duet?," *Anim Behav*, vol. 80, 2010, 597–605; Boucaud I.C.A., et al., "Vocal negotiation over parental care? Acoustic communication at the nest predicts partners' incubation share," *Biol J Lin Soc*, vol. 117, 2016, 322–336.

82. Boucaud I.C.A., et al., "Incubating females signal their needs during intrapair vocal communication at the nest: A feeding experiment in great tits," *Anim Behav*, vol. 122, 2016, 77–86.

83. Dawkins wrote, among other books, *The Selfish Gene*, *The Blind Watchmaker*, and *The Extended Phenotype*.

84. Dawkins R., and Krebs J. R., "Animal signals: Mind-reading and manipulation," in Krebs J. R., and Davies N. B. (eds.), *Behavioural Ecology: An Evolutionary Approach*, Blackwell, 1978, 381–402.

85. This "arms race" can be illustrated by the metaphor of the Red Queen in Lewis Carroll's novel *Through the Looking Glass*, in which Alice and the queen have to run constantly to stay in the same place.

86. Davis N. B., *Cuckoos, Cowbirds and Other Cheats*, T & A Poyser, 2000.

87. Soler M. (ed.), *Avian Brood Parasitism*, Springer, 2017.

88. Anderson M. G., et al., "Begging call matching between a specialist brood parasite and its host: A comparative approach to detect coevolution," *Biol J Lin Soc*, vol. 98, 2009, 208–216; Ursino C. A., et al., "Host provisioning behavior favors mimetic begging calls in a brood-parasitic cowbird," *Behav Ecol*, vol. 29, 2018, 328–332.

89. Kilner R. M., et al., "Signals of need in parent-offspring communication and their exploitation by the common cuckoo," *Nature*, vol. 397, 1999, 667–672.

90. Samas P., et al., "Nestlings of the common cuckoo do not mimic begging calls of two closely related *Acrocephalus* hosts," *Anim Behav*, vol. 161, 2020, 89–94; Jamie G. A., and Kilner R. M., "Begging call mimicry by brood parasite nestlings: Adaptation, manipulation and development," in Soler (ed.), *Avian Brood Parasitism*, 517–538.

91. Langmore N. E., Hunt S., and Kilner R. M., "Escalation of a coevolutionary arms race through host rejection of brood parasitic young," *Nature*, vol. 422, 2003, 157–160; Langmore N. E., and Kilner R. M., "The coevolutionary arms race between Horsfield's bronze-cuckoos and superb fairy-wrens," *Emu*, vol. 110, 2010, 32–38.

92. Wright and Leonard (eds.), *Evolution of Begging*; Royle N., et al., *The Evolution of Parental Care*, Oxford University Press, 2012.

93. Colombelli-Négrel D., et al., "Embryonic learning of vocal passwords in superb fairy-wrens reveals intruder cuckoo nestlings," *Cur Biol*, vol. 22, 2012, 2155–2160.

94. For a review on prenatal acoustic communication and its consequences on developmental processes, see Mariette M. M., et al., "Acoustic developmental programming: A mechanistic and evolutionary framework," *Trends Ecol Evol*, vol. 36, 2021, 722–736.

95. Ames A. E., et al., "Pre- and post-partum whistle production of a bottlenose dolphin (*Tursiops truncatus*) social group," *Int J Comp Psychol*, vol. 32, 2019.

96. King S. L., et al., "Maternal signature whistle use aids mother-calf reunions in a bottlenose dolphin, *Tursiops truncatus*," *Behav Proc*, vol. 126, 2016, 64–70.

Chapter 6

97. Charrier I., Mathevon N., and Aubin T., "Bearded seal males perceive geographic variation in their trills," *Behav Ecol Sociobiol*, vol. 67, 2013, 1679–1689.

98. Montgomery J. C., and Radford C. A., "Marine bioacoustics," *Cur Biol*, vol. 27, 2017, R502–R507.

99. And even birds underwater. A recent paper shows that penguins vocalize underwater: Thiebault A., et al., "First evidence of underwater vocalisations in hunting penguins," *PeerJ*, vol. 7, 2019, e8240.

100. Tolimieri N., et al., "Ambient sound as a navigational cue for larval reef fish," *Bioacoustics*, vol. 12, 2002, 214–217.

101. Lillis A., and Mooney T. A., "Sounds of a changing sea: Temperature drives acoustic output by dominant biological sound-producers in shallow water habitats," *Front Mar Sci*, vol. 9, 2022, 960881.

102. Versluis M., et al., "How snapping shrimp snap: Through cavitating bubbles," *Science*, vol. 289, 2000, 2114–2117; Dinh J. P., and Radford C., "Acoustic particle motion detection in the snapping shrimp (*Alpheus richardsoni*)," *J Comp Physiol A*, vol. 207, 2021, 641–655; Kingston A.C.N., et al., "Snapping shrimp have helmets that protect their brains by dampering shock waves," *Cur Biol*, vol. 32, 2022, 3576–3583.

103. Simpson S. D., et al., "Homeward sound," *Science*, vol. 308, 2005, 221.

104. Gordon T.A.C., et al., "Acoustic enrichment can enhance fish community development on degraded coral reef habitat," *Nat Com*, vol. 10, 2019, 5414.

105. Learn more about whales and dolphins: Whitehead H., and Rendell L., *The Cultural Lives of Whales and Dolphins*, University of Chicago Press, 2015.

106. Wilson and Mittermeier (eds.), *Handbook of the Mammals of the World* IV.

107. Huelsmann M., et al., "Genes lost during the transition from land to water in cetaceans highlight genomic changes associated with aquatic adaptations," *Sc Adv*, vol. 5, 2019, eaaw6671.

108. Adam O., et al., "New acoustic model for humpback whale sound production," *App Acoust*, vol. 74, 2013, 1182–1190; Damien J., et al., "Anatomy and functional morphology of the mysticete rorqual whale larynx: Phonation positions of the U-fold," *Anat Rec*, vol. 302, 2019, 703–717.

109. Things are actually a little more complex. The vocal cords of mysticetes are membranes covering the two cartilages that hold the laryngeal sac to the trachea. When the whale makes a sound, it opens these two cartilages (which are parallel to each other), and the flow of air from the lungs to the laryngeal sacs makes the membranes vibrate. By orienting these cartilages differently, the whale can choose which surface of the membranes starts to vibrate, thus controlling the pitch (low or high frequencies) of the sounds produced. See Damien et al., "Anatomy and functional morphology of the mysticete rorqual whale larynx."

110. Reidenberg J. S., and Laitman J. T., "Anatomy of underwater sound production with a focus on ultrasonic vocalization in toothed whales including dolphins and porpoises," in Brudzynski S. (ed.), *Handbook of Ultrasonic Vocalization* XXV, Elsevier, 2018, 509–519.

111. Ames A. E., Beedholm K., and Madsen P. T., "Lateralized sound production in the beluga whale (*Delphinapterus leucas*)," *J Exp Biol*, vol. 223, 2020, jeb226316.

112. Clark C. W., and Garland E. C. (eds.), *Ethology and Behavioural Ecology of Mysticetes*, Springer, 2022.

113. Stafford K. M., et al., "Extreme diversity in the songs of Spitsbergen's bowhead whales," *Biol Let*, vol. 14, 2018, 20180056; Stafford K. M., "Singing behavior in the bowhead whale," in Clark and Garland (eds.), *Ethology and Behavioural Ecology of Mysticetes*, 277–295.

114. Payne R. S., and McVay S., "Songs of humpback whales," *Science*, vol. 173, 1971, 587–597; Rothenberg D., *Thousand Mile Song: Whale Music in a Sea of Sound*, Basic Books, 2010.

115. Payne K., and Payne R. S., "Large-scale changes over 19 years in songs of humpback whales in Bermuda," *Zeit Psychol*, vol. 68, 1985, 89–114.

116. Noad M. J., et al., "Cultural revolution in whale songs," *Nature*, vol. 408, 2000, 537.

117. Herman L. M., "The multiple functions of male song within the humpback whale (*Megaptera novaeangliae*) mating system: Review, evaluation, and synthesis," *Biol Rev*, vol. 92, 2017, 1795–1818.

118. Tyack P., "Differential response of humpback whales, *Megaptera novaeangliae*, to playback of song or social sounds," *Behav Ecol Sociobiol*, vol. 13, 1983, 49–55.

119. Mercado III E., "The sonar model for humpback whale song revised," *Front Psychol*, vol. 9, 2018, 1156; Mercado III E., "Spectral interleaving by singing humpback whales: Signs of sonar," *J Acoust Soc Am*, vol. 149, 2021, 800–806.

120. Kuna V. M., and Nabelek J. L., "Seismic crustal imaging using fin whale songs," *Science*, vol. 371, 2021, 731–735.

121. See the film by Antonio Fischetti (in French): https://www.youtube.com/watch?v=hbPRRl0Q7hw.

122. King S. L., et al., "Evidence that bottlenose dolphins can communicate with vocal signals to solve a cooperative task," *R Soc Open Sc*, vol. 8, 2021, 202073; King S. L., et al., "Cooperation-based concept formation in male bottlenose dolphins," *Nat Com*, vol. 12, 2021, 2373; Barluet de Beauchesne L., et al., "Friend or foe: Risso's dolphins eavesdrop on conspecific sounds to induce or avoid intra-specific interaction," *Anim Cogn*, vol. 25, 2022, 287–296; Ames A. E., et al., "Evidence of stereotyped contact call use in narwhal (*Monodon monoceros*) mother-calf communication," *PLoS ONE*, vol. 16, 2021, e0254393; Chereskin E., et al., "Allied male dolphins use vocal exchanges to 'bond at a distance,'" *Cur Biol*, vol. 32, 2022, 1657–1663.

123. Janik V. M., Sayigh L. S., and Wells R. S., "Signature whistle shape conveys identity information to bottlenose dolphins," *PNAS*, vol. 103, 2006, 8293–8297.

124. King S. L., and Janik V. M., "Bottlenose dolphins can use learned vocal labels to address each other," *PNAS*, vol. 110, 2013, 13216–13221. The belugas, the famous white whales, also imitate their vocal signatures: Morisaka T., Nishimoto S., et al., "Exchange of 'signature' calls in captive belugas (*Delphinapterus leucas*)," *J Ethol*, vol. 31, 2013, 141–149.

125. Janik V. M., and Sayigh L. S., "Communication in bottlenose dolphins: 50 years of signature whistle research," *J Comp Physiol*, vol. 199, 2013, 479–489.

126. King S. L., et al., "Bottlenose dolphins retain individual vocal labels in multi-level alliances," *Cur Biol*, vol. 28, 2018, 1993–1999.

127. King et al., "Maternal signature whistle use aids mother-calf reunions."

128. Rice A., et al., "Spatial and temporal occurrence of killer whale ecotypes off the outer coast of Washington State, USA," *Mar Ecol Prog Series*, vol. 572, 2017, 255–268.

129. Barrett-Lennard L., "Killer whale evolution: Populations, ecotypes, species, Oh my!," *J Amer Cetac Soc*, vol. 40, 2011, 48–53.

130. Ford J.K.B., "Vocal traditions among resident killer whales (*Orcinus orca*) in coastal waters of British Columbia," *Can J Zool*, vol. 69, 1991, 1454–1483.

131. Deecke V. B., Ford J.K.B., and Spong P., "Dialect change in resident killer whales: Implications for vocal learning and cultural transmission," *Anim Behav*, vol. 40, 2000, 629–638.

132. Poupart M., et al., "Intra-group orca call rate modulation estimation using compact four hydrophones array," *Front Mar Sci*, vol. 8, 2021, 681036.

133. *Centre d'études et d'expertise sur les risques, l'environnement, la mobilité et l'Aménagement* (Center of Studies and Expertise on Risks, Environment, Mobility and Development).

134. Benti B., et al., "Indication that the behavioural responses of humpback whales to killer whale sounds are influenced by trophic relationships," *Mar Ecol Prog Series*, vol. 660, 2021, 217–232.

135. Curé C., et al., "Evidence for discrimination between feeding sounds of familiar fish and unfamiliar mammal-eating killer whale ecotypes by long-finned pilot whales," *Anim Cogn*, vol. 22, 2019, 863–882.

136. Isojunno S., et al., "Sperm whales reduce foraging effort during exposure to 1–2 kHz sonar and killer whale sounds," *Ecol Appl*, vol. 26, 2016, 77–93.

137. Curé C., et al., "Responses of male sperm whales (*Physeter macrocephalus*) to killer whale sounds: Implications for anti-predator strategies," *Sc Rep*, vol. 3, 2013, 1579.

138. Clarke M. R., "Function of the spermaceti organ of the sperm whale," *Nature*, vol. 228, 1970, 873–874.

139. See Joy Reidenberg's animated explanation: https://www.youtube.com/watch?v=sW7o5IC2io0.

140. Huggenberg S., et al., "The nose of the sperm whale: Overviews of functional design, structural homologies and evolution," *J Mar Biol Assoc*, vol. 96, 2016, 783–806.

141. Oliveira C. et al., "The function of male sperm whale slow clicks in a high latitude habitat: Communication, echolocation, or prey debilitation?," *JASA*, vol. 133, 2013, 3135–3144; Tonnesen P. et al., "The long-range echo scene of the sperm whale biosonar," *Biol Let*, vol. 16, 2020, 20200134.

142. For a full explanation of clicks and codas and their roles in the lives of sperm whales, see Whitehead H., and Rendell L., *The Cultural Lives of Whales and Dolphins*, University of Chicago Press, 2015, 146–158.

143. Gordon J.C.D., "Evaluation of a method for determining the length of sperm whales (*Physeter catodon*) from their vocalizations," *J Zool*, vol. 224, 1991, 301–314; Møhl B., et al., "Sperm whale clicks: Directionality and source levels revisited," *JASA*, vol. 107, 2000, 638–648; Møhl B., "Sound transmission in the nose of the sperm whale *Physeter catodon*: A post mortem study," *J Comp Physiol A*, vol. 187, 2001, 335–340.

144. Growcott A., et al., "Measuring sperm whales from their clicks: A new relationship between IPIs and photogrammetrically measured lengths," *JASA*, vol. 130, 2011, 568–573.

145. Rendell L. E., and Whitehead H., "Vocal clans in sperm whales (*Physeter macrocephalus*)," *Proc R Soc B*, vol. 270, 2003, 225–231.

146. Gero S., Whitehead H., and Rendell L., "Individual, unit and vocal clan level identity cues in sperm whale codas," *R Soc Op Sc*, vol. 3, 2016, 150372.

147. Sperm whales are also able to produce nonclick sounds. These include squeals, with a possible communicative social function; pips; short trumpets; and trumpets. For details, see Pace D. S., et al., "Trumpet sounds emitted by male sperm whales in the Mediterranean Sea," *Sc Rep*, vol. 11, 2021, 5867.

Chapter 7

148. Le Bœuf B. J., and Laws R. M., *Elephant Seals, Population Ecology, Behavior and Physiology*, University of California Press, 1994.

149. Dr. Colleen Reichmuth, senior researcher at the Long Marine Lab, University of California, Santa Cruz.

150. Elephant seal males are not the only pinnipeds to emit rhythmic sounds during their reproductive displays. Walrus males also produce some: Larsen O. N., and Reichmuth C., "Walruses produce intense impulse sounds by clap-induced cavitation during breeding displays," *R Soc Open Sc*, vol. 8, 2021, 210197.

151. Casey C., et al., "Rival assessment among northern elephant seals: Evidence of associative learning during male-male contests," *R Soc Open Sc*, vol. 2, 2015, 150228.

152. Mathevon N., et al., "Northern elephant seals memorize the rhythm and timbre of their rivals' voices," *Cur Biol*, vol. 27, 2017, 2352–2356.

153. Casey C., et al., "The genesis of giants: Behavioural ontogeny of male northern elephant seals," *Anim Behav*, vol. 166, 2020, 247–259.

Chapter 8

154. To know everything about crocodilians: Grigg G., and Kirshner D., *Biology and Evolution of Crocodylians*, CSIRO, 2015.

155. Vergne A. L., Pritz M. B., and Mathevon N., "Acoustic communication in crocodilians: From behaviour to brain," *Biol Rev*, vol. 84, 2009, 391–411.

156. Reber S. A., "Crocodilia communication," in Vonk J., and Shackelford T. K. (eds.), *Encyclopedia of Animal Cognition and Behavior*, Springer, 2018, 1–10.

157. Marquis O., et al., "Observations on breeding site, bioacoustics and biometry of hatchlings of *Paleosuchus trigonatus* (Schneider, 1801) from French Guiana (Crocodylia: Alligatoridae)," *Herp Notes*, vol. 13, 2020, 513–516.

158. Vergne A. L., and Mathevon N., "Crocodile egg sounds signal hatching time," *Cur Biol*, vol. 18, 2008, R513–R514.

159. Hatching synchronization seems to occur in other animals, even insects: Tanaka S., "Embryo-to-embryo communication facilitates synchronous hatching in grasshoppers," *J Orthop Res*, vol. 30, 2021, 107–115.

160. Vergne A. L., et al., "Acoustic signals of baby black caimans," *Zoology*, vol. 114, 2011, 313–320.

161. Mathevon N., et al., "The code size: Behavioural response of crocodile mothers to offspring calls depends on the emitter's size, not on its species identity," in *Crocodiles, Proceedings of the 24th Working Meeting of the Crocodile Specialist Group IUCN*, 2016, 79–85.

162. Vergne A. L., et al., "Acoustic communication in crocodilians: Information encoding and species specificity of juvenile calls," *Anim Cogn*, vol. 15, 2012, 1095–1109.

163. Vergne A. L., et al., "Parent-offspring communication in the Nile crocodile *Crocodylus niloticus*: Do newborns' calls show an individual signature?," *Naturwissen*, vol. 94, 2007, 49–54.

164. Papet L., et al., "Influence of head morphology and natural postures on sound localization cues in crocodilians," *R Soc Open Sc*, vol. 6, 2019, 190423; Papet L., et al., "Crocodiles use both interaural level differences and interaural time differences to locate a sound source," *JASA*, vol. 148, 2020, EL307.

165. For a comparative perspective on sound localizations, see this paper: Carr C. E., and Christensen-Dalsgaard J., "Sound localization strategies in three predators," *Brain Behav Evol*, vol. 86, 2015, 17–27. I also talk about sound localization in chapter 9, "Hear, at all costs."

166. Thévenet J., et al., "Spatial release from masking in crocodilians," *Com Biol*, vol. 5, 2022, 869.

Chapter 9

167. Some snakes do produce sounds, however (scraping scales, rattling the end of their tails, whistling sounds). The royal cobra *Ophiophagus hannah* produces a whistling sound at a frequency of about 600 Hz thanks to diverticula in its trachea, which act as resonators. See Young B. A., "The comparative morphology of the larynx in snakes," *Acta Zool*, vol. 81, 2000, 177–193; Young B. A., et al., "The morphology of sound production in *Pituophis melanoleucus* (Serpentes, Colubridae) with the first description of a vocal cord in snakes," *J Exp Zool*, vol. 273, 1995, 472–481; Young B. A., "Snake bioacoustics: Toward a richer understanding of the behavioral

ecology of snakes," *Q Rev Biol*, vol. 78, 2003, 303–325; and Gans C., "Sound producing mechanisms in recent reptiles: Review and comment," *Amer Zool*, vol. 13, 1973, 1195–1203.

168. Greenfield M. D., "Evolution of acoustic communication in insects," in Pollack G. S., Mason A. C., Popper A., and Fay R. R. (eds.), *Insect Hearing*, Springer, 2016, 17–47.

169. Vermeij G. J., "Sound reasons for silence: Why do molluscs not communicate acoustically?," *Biol J Lin Soc*, vol. 100, 2010, 485–493. The snail emits sounds, but we do not know if they are used to communicate: Breure A.S.H., "The sound of a snail: Two cases of acoustic defence in gastropods," *J Mollusc Stud*, vol. 81, 2015, 290–293. The mussel perceives sound vibrations: Charifi M., et al., "The sense of hearing in the Pacific oyster, *Magallana gigas*," *PLoS ONE*, vol. 12, 2017, e0185353.

170. Bradbury J. W., and Vehrencamp S. L., *Principles of Animal Communication*, Sinauer, 2011.

171. Some internal mechanoreceptors serve a completely different purpose, like measuring blood pressure or the amount of food currently distending our stomachs.

172. These "short-distance" ears are sensitive to variations in the speed of air molecules.

173. "Long-distance" ears are sensitive to variations in air pressure.

174. Larsen and Wahlberg, "Sound and sound sources," in Brown and Riede (eds.), *Comparative Bioacoustics*, 3–61. This chapter provides very detailed explanations about these notions, which, as you might have guessed, are far more complex than what I've told you here.... Allergic to physics? Best sit this one out.

175. Budelmann B. U., "Hearing in crustacea," in Webster D. B., and Fay R. R. (eds.), *Evolutionary Biology of Hearing*, Springer, 1992, 131–139; Jezequel Y., et al., "Sound detection by the American lobster (*Homarus americanus*)," *J Exp Biol*, vol. 224, 2021, jeb240747.

176. Kaifu K., et al., "Underwater sound detection by cephalopod statocyst," *Fish Sc*, vol. 74, 2008, 781–786.

177. Ladich F., and Winkler H., "Acoustic communication in terrestrial and aquatic vertebrates," *J Exp Biol*, vol. 220, 2017, 2306–2317.

178. Clack J. A., Fay R. R., and Popper A. N., *Evolution of the Vertebrate Ear*, Springer, 2016; Montealegre-Z F., Robert D., et al., "Convergent evolution between insect and mammalian audition," *Science*, vol. 338, 2012, 968–971.

179. Römer H., "Acoustic communication," in Córdoba-Aguilar A., González-Tokman D., and González-Santoyo I. (eds.), *Insect Behavior: From Mechanisms to Ecological and Evolutionary Consequences*, Oxford University Press, 2018, 174–188; Balakrishnan R., "Behavioral ecology of insect acoustic communication," in Pollack, Mason, Popper, and Fay (eds.), *Insect Hearing*, 49–80; Göpfert M. C., and Hennig R. M., "Hearing in insects," *Ann Rev Entomol*, vol. 61, 2016, 257–276.

180. Matthews R. W., and Matthews J. R., *Insect Behavior*, Springer, 2010.

181. Since ears are derived from preexisting mechanoreceptors, and since these receptors are present throughout the bodies of insects, finding ears in different positions is not surprising. This is evidence that the ear has been "invented" several times, independently, throughout the history of insects.

182. The bodies of insects have three successive parts: head, thorax, and abdomen. Each part is a succession of rings, the metameres. The thorax has three metameres, each with a pair of legs. The abdomen has a greater number of metameres.

183. Göpfert M. C., et al., "Tympanal and atympanal 'mouth-ears' in hawkmoths (Sphingidae)," *Proc R Soc B*, vol. 269, 2002, 89–95.

184. The respiratory system of insects is a genuine piping system (made up of the trachea), bringing air everywhere in the animal's body.

185. A scolopidium is a fairly complex structure consisting of one or more sensory neurons, each having an extension (a dendrite) ending in a cilium (connected to the eardrum in the case of the tympanic ear), and two protective cells. Scolopidia are grouped into chordotonal organs.

186. Brown E. E., et al., "Quantifying the completeness of the bat fossil record," *Palaeontology*, vol. 62, 2019, 757–776.

187. Greenfield M. D., "Evolution of acoustic communication in insects," in Pollack, Mason, Popper, and Fay (eds.), *Insect Hearing*, 17–47.

188. Gerhardt H. C., and Huber F., *Acoustic Communication in Insects and Anurans: Common Problems and Diverse Solutions*, University of Chicago Press, 2002.

189. Yager D. D., and Svenson G. J., "Patterns of praying mantis auditory system evolution based on morphological, molecular, neurophysiological, and behavioural data," *Biol J Lin Soc*, vol. 94, 2008, 541–568.

190. The ears of bush crickets (katydids) are also amazing, with specific anatomical features that enhance the ability of these animals to pinpoint a sound source. Have a look at this recent study: Veitch D., et al., "A narrow ear canal reduces sound velocity to create additional acoustic inputs in a microscale insect ear," *PNAS*, vol. 118, 2021, e2017281118.

191. Albert J. T., and Kozlov A. S., "Comparative aspects of hearing in vertebrates and insects with antennal ears," *Cur Biol*, vol. 26, 2016, R1050–R1061.

192. On hearing by mosquitoes, see Feugere L., et al., "Behavioural analysis of swarming mosquitoes reveals high hearing sensitivity in *Anopheles coluzzii*," *J Exp Biol*, vol. 225, 2022, jeb243535; and Feugere L., et al., "The role of hearing in mosquito behaviour," in Hill S., Ignell R., Lazzari C., and Lorenzo M. (eds.), *Sensory Ecology of Disease Vectors*, Wageningen Academic Publishers, forthcoming.

193. Baker C. A., et al., "Neural network organization for courtship-song feature detection in *Drosophila*," *Cur Biol*, vol. 32, 2022, 3317–3333.

194. Kerwin P., et al., "Female copulation song is modulated by seminal fluid," *Nat Com*, vol. 11, 2020, 1430.

195. Rybak F., Sureau G., and Aubin T., "Functional coupling of acoustic and chemical signals in the courtship behaviour of the male *Drosophila melanogaster*," *Proc R Soc B*, vol. 269, 2002, 695–701.

196. Jackson J. C., and Robert D., "Nonlinear auditory mechanism enhances female sounds for male mosquitoes," *PNAS*, vol. 103, 2006, 16734–16739.

197. Gibson G., and Russell I., "Flying in tune: Sexual recognition in mosquitoes," *Cur Biol*, vol. 16, 2006, 1311–1316.

198. Manley G. A., "Cochlear mechanisms from a phylogenetic viewpoint," *PNAS*, vol. 97, 2000, 11736–11743.

199. Moffat A.J.M., and Capranica R. R., "Auditory sensitivity of the saccule in the American toad (*Bufo americanus*)," *J Comp Physiol*, vol. 105, 1976, 1–8.

200. Although the near sound field of some elephant vocalizations can extend to 20 meters. We humans can't hear them with our ears, but we do feel the particle oscillations they produce.

201. Sisneros J. A., *Fish Hearing and Bioacoustics*, Springer, 2016.

202. Parmentier E., and Diogo R., "Evolutionary trends of swimbladder sound mechanisms in some teleost fishes," in Ladich F., Collin P. S., Moller P., and Kapoor B. G. (eds.), *Communication in Fishes*, Science Publishers, 2006, 43–68.

203. Ladich F., and Schulz-Mirbach T., "Diversity in fish auditory systems: One of the riddles of sensory biology," *Front Ecol Evol*, vol. 4, 2016, 28; Chapuis L., and Collin S. P., "The auditory system of cartilaginous fishes," *Rev Fish Biol Fisheries*, vol. 32, 2022, 521–554.

204. van Hemmen J. L., et al., "Animals and ICE: Meaning, origin, and diversity," *Biol Cybern*, vol. 110, 2016, 237–246.

205. Vedurmudi A. P., et al., "How internally coupled ears generate temporal and amplitude cues for sound localization," *Phys Rev Let*, vol. 116, 2016, 028101; Klump G. M., "Sound localization in birds," in Dooling R. J., Fay R. R., and Popper A. N. (eds.), *Comparative Hearing: Birds and Reptiles*, Springer, 2000, 249–307.

206. Schnupp J.W.H., and Carr C. E., "On hearing with more than one ear: Lessons from evolution," *Nat Neurosc*, vol. 12, 2009, 692–697.

207. Robert D., "Directional hearing in insects," in Fay R. R. (ed.), *Sound Source Localization*, Springer, 2005, 6–35; Montealegre-Z F., et al., "Convergent evolution between insect and mammalian audition," *Science*, voi. 338, 2012, 968–971.

208. Lee N., et al., "Lung mediated auditory contrast enhancement improves the signal-to-noise ratio for communication in frogs," *Cur Biol*, vol. 31, 2021, 1488–1498.

209. Boistel R., Aubry J.-F., et al., "How minute sooglossid frogs hear without a middle ear," *PNAS*, vol. 17, 2013, 15360–15364.

210. For a very interesting, lively, and still modern introduction to frog acoustic communication, see Narins P., "Frog communication," *Sc Amer*, vol. 273, 1995, 78–83.

211. On the origin of the avian ear, see Hanson M., et al., "The early origin of a birdlike inner ear and the evolution of dinosaurian movement and vocalization," *Science*, vol. 372, 2021, 601–609.

212. Zeyl J. N., et al., "Aquatic birds have middle ears adapted to amphibious lifestyles," *Sc Rep*, vol. 12, 2022, 5251.

213. Supin A. Y., et al., *The Sensory Physiology of Aquatic Mammals*, Springer, 2001.

214. Hemilä S., Nummela S., and Reuter T., "Anatomy and physics of the exceptional sensitivity of dolphin hearing (Odontoceti: Cetacea)," *J Comp Physiol A*, vol. 196, 2010, 165–179.

215. Churchill M., et al., "The origin of high-frequency hearing in whales," *Cur Biol*, vol. 26, 2016, 2144–2149.

216. Au W.W.L., and Fay R. R., *Hearing by Whales and Dolphins*, Springer, 2000; Cranfor T., and Krysl P., "Fin whale sound reception mechanisms: Skull vibration enables low-frequency hearing," *PLoS ONE*, vol. 10, 2015, e0116222; Park T., et al., "Low-frequency hearing preceded the evolution of giant body size and filter feeding in baleen whales," *Proc R Soc B*, vol. 284, 2017, 20162528.

217. Nelson D. A., and Marler P., "Categorical perception of a natural stimulus continuum: Birdsong," *Science*, vol. 244, 1989, 976–978.

218. May B., et al., "Categorical perception of conspecific communication sounds by Japanese macaques, *Macaca fuscata*," *J Acoust Soc Am*, vol. 85, 1989, 837–847.

219. Baugh A. T., Akre K. L., and Ryan M. J., "Categorical perception of a natural, multivariate signal: Mating call recognition in túngara frogs," *PNAS*, vol. 105, 2008, 8985–8988.

220. Greenfield M. D., "Mechanisms and evolution of communal sexual displays in arthropods and anurans," *Adv Stud Behav*, vol. 35, 2005, 1–62.

221. Paul A., et al., "Behavioral discrimination and time-series phenotyping of birdsong performance," *PLoS Comput Biol*, vol. 17, 2021, e1008820.

Chapter 10

222. Hippos have a vocal repertoire, with several calls. During a recent field trip in the Maputo Special Reserve in Mozambique, we started to study hippo acoustic communication: Thévenet et al., "Voice-mediated interactions in a megaherbivore," *Cur Biol*, vol. 32, 2022, R55–R71.

223. Vergne A. L., et al., "Parent-offspring communication in the Nile crocodile *Crocodylus niloticus*: Do newborns' calls show an individual signature?," *Naturwissen*, vol. 94, 2007, 49–54.

224. Chabert T., et al., "Size does matter: Crocodile mothers react more to the voice of smaller offspring," *Sc Rep*, vol. 5, 2015, 15547.

225. For an extensive review on sound production in nonavian reptiles, see Russell A. P., and Bauer A. M., "Vocalization by extant nonavian reptiles: A synthetic overview of phonation and the vocal apparatus," *Anat Rec*, vol. 304, 2021, 1478–1528.

226. Taylor A. M., et al., "Vocal production by terrestrial mammals: Source, filter, and function," in Suthers R. A., Fitch W. T., Fay R. R., and Popper A. N. (eds.), *Vertebrate Sound Production and Acoustic Communication*, Springer, 2016, 229–259.

227. The Adam's apple is larger, and therefore more visible, in most men. But everyone has one. Put your finger to your throat and swallow; you'll feel your Adam's apple move.

228. Or when the air comes in. But it's harder. Try to talk while you breathe, just to see.

229. Bowling D. L., et al., "Body size and vocalization in primates and carnivores," *Sc Rep*, vol. 7, 2017, 41070.

230. Garcia M., et al., "Acoustic allometry revisited: Morphological determinants of fundamental frequency in primate vocal production," *Sc Rep*, vol. 7, 2017, 10450. On the evolution of the larynx, see also Bowling D. L., et al., "Rapid evolution of the primate larynx?," *PLoS Biol*, vol. 18, 2020, e3000764.

231. For an in-depth discussion of the concept of timbre, see Piazza E. A., et al., "Rapid adaptation to the timbre of natural sounds," *Sc Rep*, vol. 8, 2018, 13826; and Elliott T. M., et al., "Acoustic structure of the five perceptual dimensions of timbre in orchestral instrument tones," *JASA*, vol. 133, 2013, 389–404.

232. Garcia M., et al., "Honest signaling in domestic piglets (*Sus scrofa domesticus*): Vocal allometry and the information content of grunt calls," *J Exp Biol*, vol. 219, 2016, 1913–1921; Garcia-Navas V., and Blumstein D. T., "The effect of body size and habitat on the evolution of alarm vocalizations in rodents," *Biol J Lin Soc*, vol. 118, 2016, 745–751.

233. Reber S. A., et al., "A Chinese alligator in heliox: Formant frequencies in a crocodilian," *J Exp Biol*, vol. 218, 2015, 2442–2447; Reber S. A., et al., "Formants provide honest acoustic cues to body size in American alligators," *Sc Rep*, vol. 7, 2017, 1816.

234. Charlton B. D., and Reby D., "The evolution of acoustic size exaggeration in terrestrial mammals," *Nat Com*, vol. 7, 2016, 12739.

235. Reby D., and McComb K., "Vocal communication and reproduction in deer," *Adv St Behav*, vol. 33, 2003, 231–264; Frey R., et al., "Roars, groans and moans: Anatomical correlates of vocal diversity in polygenous deer," *J Anat*, vol. 239, 2021, 1336–1369.

236. See this paper on birds: Francis C. D., and Wilkins M. R., "Testing the strength and direction of selection on vocal frequency using metabolic scaling theory," *Ecosphere*, vol. 12, 2021, e03733.

237. Charlton B. D., Reby D., and McComb K., "Female red deer prefer the roars of larger males," *Biol Let*, vol. 3, 2007, 382–385.

238. McComb K., "Female choice for high roaring rates in red deer, *Cervus elaphus*," *Anim Behav*, vol. 41, 1991, 79–88.

239. Reby D., et al., "Red deer (*Cervus elaphus*) hinds discriminate between the roars of their current harem-holder stag and those of neighbouring stags," *Ethology*, vol. 107, 2001, 951–959.

240. In reality, the koala is a marsupial mammal, closer to the kangaroo than to the polar bear.

241. Charlton B. D., et al., "Koalas use a novel vocal organ to produce unusually low-pitched mating calls," *Cur Biol*, vol. 23, 2013, R1035–R1036.

242. Fitch W. T., "Vertebrate vocal production: An introductory overview," in Suthers, Fitch, Fay, and Popper (eds.), *Vertebrate Sound Production and Acoustic Communication*, 1–18.

243. Bradbury J. W., "Lek mating behavior in the hammer-headed bat," *Ethology*, vol. 45, 1977, 225–255.

244. Frey R., and Gebler A., "Mechanisms and evolution of roaring-like vocalization in mammals," in Brudzynski S. (ed.), *Handbook of Mammalian Vocalization*, Elsevier, 2010, 439–450.

245. Mathevon N., and Viennot É., "Avant-propos," in Mathevon N., and Viennot É. (eds.), *La différence des sexes*, Belin, 2017, 7–28.

246. Titze I. R., *Fascinations with the Human Voice*, National Center for Voice and Speech, 2010; Dunn J. C., et al., "Evolutionary trade-off between vocal tract and testes dimensions in howler monkeys," *Cur Biol*, vol. 25, 2015, 2839–2844.

247. Grawunder S., et al., "Higher fundamental frequency in bonobos is explained by larynx morphology," *Cur Biol*, vol. 28, 2018, R1171–R1189.

248. Greenfield M., "Honesty and deception in animal signals," in Lucas J., and Simmons L. (eds.), *Essays in Animal Behaviour*, Elsevier, 2006, 279–300.

249. Interspecific signals also follow the *honesty principle*. Look at this amazing paper: Forsthofer M., et al., "Frequency modulation of rattlesnake acoustic display affects acoustic distance perception in humans," *Cur Biol*, vol. 31, 2021, 4367–4372.

250. Bee M. A., et al., "Male green frogs lower the pitch of acoustic signals in defense of territories: A possible dishonest signal of size?," *Behav Ecol*, vol. 11, 2000, 169–177.

251. Zahavi A., "Mate selection—a selection for a handicap," *J Theor Biol*, vol. 53, 1975, 205–214; Bradbury J. W., and Vehrencamp S. L., *Principles of Animal Communication*, Sinauer, 2011.

252. Vocalizations are not the only sound signals that can inform about the size of the sender. In mountain gorillas, for example, a recent study shows that the acoustic characteristics of the chest beat are correlated with body size: larger males have significantly lower peak frequencies than smaller ones: Wright E., et al., "Chest beats as an honest signal of body size in male mountain gorillas (*Gorilla beringei beringei*)," *Sc Rep*, vol. 11, 2021, 6879.

253. Pisanski K., and Reby D., "Efficacy in deceptive vocal exaggeration of human body size," *Nat Com*, vol. 12, 2021, 968. See also Raine J., et al., "Human listeners can accurately judge strength and height relative to self from aggressive roars and speech," *iScience*, vol. 4, 2018, 273–280.

254. Vallet E., and Kreutzer M., "Female canaries are sexually responsive to special song phrases," *Anim Behav*, vol. 49, 1995, 1603–1610.

255. Vallet E., et al., "Two-note syllables in canary songs elicit high levels of sexual display," *Anim Behav*, vol. 55, 1998, 291–297.

256. Oberweger K., and Goller F., "The metabolic cost of birdsong production," *J Exp Biol*, vol. 204, 2001, 3379–3388.

257. Casagrande S., Pinxten R., and Eens M., "Honest signaling and oxidative stress: The special case of avian acoustic communication," *Front Ecol Evol*, vol. 4, 2016, 52; Spencer K. A., et al., "Parasites affect song complexity and neural development in a songbird," *Proc R Soc B*, vol. 272, 2005, 2037–2043.

258. Gil D., and Gahr M., "The honesty of bird song: Multiple constraints for multiple traits," *Trends Ecol Evol*, vol. 17, 2002, 133–141.

259. Buchanan K. L., "Stress and the evolution of condition-dependent signals," *Trends Ecol Evol*, vol. 15, 2000, 156–160.

260. Buchanan K. L., et al., "Song as an indicator of parasitism in the sedge warbler," *Anim Behav*, vol. 57, 1999, 307–314.

261. Whether and how acoustic features drive mate choice is still an active field of research. For instance, see Wang D., et al., "Is female mate choice repeatable across males with nearly identical songs?," *Anim Behav*, vol. 181, 2021, 137–149.

262. Dalziall A. H., et al., "Male lyrebirds create a complex acoustic illusion of a mobbing flock during courtship and copulation," *Cur Biol*, vol. 31, 2021, 1970–1976.

263. Maynard Smith J., and Harper D., *Animal Signals*, Oxford University Press, 2003.

264. See this paper dealing with sound production in fossil insects: Schubnel T., et al., "Sound vs. light: Wing-based communication in Carboniferous insects," *Com Biol*, vol. 4, 2021, 794.

265. Tinbergen N., "'Derived' activities: Their causation, biological significance, origin, and emancipation during evolution," *Q Rev Biol*, vol. 27, 1952, 1–32.

266. Clark C. J., "Locomotion-induced sounds and sonations: Mechanisms, communication function, and relationship with behavior," in Suthers, Fitch, Fay, and Popper (eds.), *Vertebrate Sound Production and Acoustic Communication*, 83–117; Clark C. J., "Ways that animal wings produce sound," *Integr Comp Biol*, vol. 61, 2021, 696–709.

267. Murray T. G., Zeil J., and Magrath R. D., "Sounds of modified flight feathers reliably signal danger in a pigeon," *Cur Biol*, vol. 27, 2017, 3520–3525.

268. Clark C. J., "Signal or cue? Locomotion-induced sounds and the evolution of communication," *Anim Behav*, vol. 143, 2018, 83–91.

269. Gould S. J., *The Structure of Evolutionary Theory*, Harvard University Press, 2002.

270. Montealegre-Z F., et al., "Sound radiation and wing mechanics in stridulating field crickets (Orthoptera: Gryllidae)," *J Exp Biol*, vol. 214, 2011, 2105–2117.

271. Fonseca P. J., "Cicada acoustic communication," in Hedwig B. (ed.), *Insect Hearing and Acoustic Communication*, Springer, 2014, 101–121.

272. Godthi V., et al., "The mechanics of acoustic signal evolution in field crickets," *J Exp Biol*, vol. 225, 2022, jeb243374.

273. Parmentier E., Diogo R., and Fine M. L., "Multiple exaptations leading to fish sound production," *Fish & Fisheries*, vol. 18, 2017, 958–966.

274. Looby A., et al., "A quantitative inventory of global soniferous fish diversity," *Rev Fish Biol Fisheries*, vol. 32, 2022, 581–595; Rice A. N., et al., "Evolutionary patterns in sound production across fishes," *Ichthyology & Herpetology*, vol. 110, 2022, 1–12.

275. Parmentier E., Herrel A., et al., "Sound production in the clownfish *Amphiprion clarkii*," *Science*, vol. 316, 2007, 1006.

276. Millot S., Vandewalle P., and Parmentier E., "Sound production in red-bellied piranhas (*Pygocentrus nattereri*, Kner): An acoustical, behavioural and morphofunctional study," *J Exp Biol*, vol. 214, 2011, 3613–3618.

277. Amorim P., "Diversity of sound production in fish," in Ladich, Collin, Moller, and Kapoor (eds.), *Communication in Fishes*, 71–105.

278. Amorim P., "Fish sounds and mate choice," in Ladich F. (ed.), *Sound Communication in Fishes*, Springer, 2015, 1–33.

279. Bertucci F., et al., "Sounds produced by the cichlid fish *Metriaclima zebra* allow reliable estimation of size and provide information on individual identity," *J Fish Biol*, vol. 80, 2012, 752–766.

280. Bertucci F., et al., "Sounds modulate males' aggressiveness in a cichlid fish," *Ethology*, vol. 116, 2010, 1179–1188; Bertucci F., et al., "The relevance of temporal cues in a fish sound: A first experimental investigation using modified signals in cichlids," *Anim Cogn*, vol. 16, 2013, 45–54.

281. Amorim M.C.P., et al., "Mate preference in the painted goby: The influence of visual and acoustic courtship signals," *J Exp Biol*, vol. 216, 2013, 3996–4004.

282. Wyttenbach R. A., May M. L., and Hoy R. R., "Categorical perception of sound frequency by crickets," *Science*, vol. 273, 1996, 1542–1544.

283. Robillard T., Grandcolas P., and Desutter-Grandcolas L., "A shift toward harmonics for high-frequency calling shown with phylogenetic study of frequency spectra in Eneopterinae crickets (Orthoptera, Grylloidea, Eneopteridae)," *Can J Zool*, vol. 85, 2007, 1264–1274.

284. Hofstede H. M., et al., "Evolution of a communication system by sensory exploitation of startle behavior," *Cur Biol*, vol. 25, 2015, 3245–3252.

285. Benavides-Lopez J. L., Ter Hofstede H., and Robillard T., "Novel system of communication in crickets originated at the same time as bat echolocation and includes male-male multimodal communication," *Sci Nat*, vol. 107, 2020, 9.

286. Römer H., "Insect acoustic communication: The role of transmission channel and the sensory system and brain of receivers," *Func Ecol*, vol. 34, 2020, 310–321.

287. Wilkins M. R., Seddon N., and Safran R. J., "Evolutionary divergence in acoustic signals: Causes and consequences," *Trends Ecol Evol*, vol. 28, 2013, 156–166.

288. To give you an idea, the common ancestor of the human species and the great apes of today goes back 6 or 7 million years. Our species *Homo sapiens* was distinguished from other human species only 300,000 years ago. See Hublin J.-J., et al., "New fossils from Jebel Irhoud, Morocco and the pan-African origin of *Homo sapiens*," *Nature*, vol. 546, 2017, 289–292.

289. Prum R. O., et al., "A comprehensive phylogeny of birds (Aves) using targeted next-generation DNA sequencing," *Nature*, vol. 526, 2015, 569–573.

290. Mason N. A., et al., "Song evolution, speciation, and vocal learning in passerine birds," *Evolution*, vol. 71, 2016, 786–796.

291. Amezquita A., et al., "Acoustic interference and recognition space within a complex assemblage of dendrobatid frogs," *PNAS*, vol. 108, 2011, 17058–17063; Tobias J. A., et al., "Species interactions and the structure of complex communication networks," *PNAS*, vol. 111, 2014, 1020–1025.

292. See, for example, regarding the evolution of mammalian sound signals, Charlton B. D., et al., "Coevolution of vocal signal characteristics and hearing sensitivity in forest mammals," *Nat Com*, vol. 10, 2019, 2778. See also Leighton G. M., and Birmingham T., "Multiple factors affect the evolution of repertoire size across birds," *Behav Ecol*, vol. 32, 2021, 380–385.

293. Hyacinthe C., Attia J., and Rétaux S., "Evolution of acoustic communication in blind cavefish," *Nat Com*, vol. 10, 2019, 4231.

294. My colleagues included Frédéric Theunissen and Frédéric Sèbe, whom I will tell you more about, as well as Maxime Garcia, who was a postdoctoral researcher on my team at the time, and a few other scientists.

295. Garcia M., et al., "Evolution of communication signals and information during species radiation," *Nat Com*, vol. 11, 2020, 4970. See also the post on the *Nature Ecology & Evolution* blog: https://natureecoevocommunity.nature.com/posts/information-tinkering-in-an-animal-communication-system.

296. Freeman B. G., et al., "Faster evolution of a premating reproductive barrier is not associated with faster speciation rates in New World passerine birds," *Proc R Soc B*, vol. 289, 2022, 20211514; Arato J., and Fitch W. T., "Phylogenetic signal in the vocalizations of vocal learning and vocal non-learning birds," *Phil Trans R Soc B*, vol. 376, 2021, 20200241.

297. Podos J., "Correlated evolution of morphology and vocal signal structure in Darwin's finches," *Nature*, vol. 409, 2001, 185–188; Podos J., and Nowicki S., "Beaks, adaptation, and vocal evolution in Darwin's finches," *Bioscience*, vol. 54, 2004, 501–510; Servedio M. R., et al., "Magic traits in speciation: 'Magic' but not rare?," *Trends Ecol Evol*, vol. 26, 2011, 389–397.

298. An interesting paper on the evolution of signals when evolutionary constraints are relaxed is Rayner J. G., et al., "The persistence and evolutionary consequences of vestigial behaviours," *Biol Rev*, vol. 97, 2022, 1389–1407.

Chapter 11

299. McGregor P. K. (ed.), *Animal Communication Networks*, Cambridge University Press, 2005; Reichert M. S., Enriquez M. S., and Carlson N. V., "New dimensions for animal communication networks: Space and time," *Integ Comp Biol*, vol. 61, 2021, 814–824.

300. Zann R. A., *The Zebra Finch*, Oxford University Press, 1996.

301. Elie J. E., and Theunissen F. E., "The vocal repertoire of the domesticated zebra finch: A data-driven approach to decipher the information-bearing acoustic features of communication signals," *Anim Cogn*, vol. 19, 2015, 285–315.

302. Vignal C., et al., "Mate recognition by female zebra finch: Analysis of individuality in male call and first investigations on female decoding process," *Behav Process*, vol. 77, 2008, 191–198.

303. Vignal C., et al., "Audience drives male songbird response to partner's voice," *Nature*, vol. 430, 2004, 448–451.

304. Perry S., et al., "White-faced capuchin monkeys show triadic awareness in their choice of allies," *Anim Behav*, vol. 67, 2004, 165–170; Jolly A., "Lemur social behavior and primate intelligence," *Science*, vol. 153, 1966, 501–506; Tomasello M., and Call J., *Primate Cognition*, Oxford University Press, 1997.

305. Elie J. E., et al., "Dynamics of communal vocalizations in a social songbird, the zebra finch (*Taeniopygia guttata*)," *JASA*, vol. 129, 2011, 4037–4046.

306. Elie J. E., et al., "Same-sex pair-bonds are equivalent to male-female bonds in a life-long socially monogamous songbird," *Behav Ecol Sociobiol*, vol. 65, 2011, 2197–2208.

307. Geberzhan N., and Gahr M., "Undirected (solitary) birdsong in female and male blue-capped cordon-bleus (*Uraeginthus cyanocephalus*) and its endocrine correlates," *PLoS ONE*, vol. 6, 2011, e26485; Ota N., et al., "Tap dancing birds: The multimodal mutual courtship display of males and females in a socially monogamous songbird," *Sc Rep*, vol. 5, 2015, 16614; Ota N., et al., "Songbird tap dancing produces non-vocal sounds," *Bioacoustics*, vol. 26, 2017, 161–168.

308. Ota N., et al., "Couples showing off: Audience promotes both male and female multi-modal courtship display in a songbird," *Sc Adv*, vol. 4, 2018, eaat4779.

309. Zuberbühler K., "Audience effects," *Cur Biol*, vol. 18, 2008, R189–R190.

310. Marler P., et al., "Vocal communication in the domestic chicken: II. Is a sender sensitive to the presence and nature of a receiver?," *Anim Behav*, vol. 34, 1986, 194–198.

311. Slocombe K. E., and Zuberbühler K., "Chimpanzees modify recruitment screams as a function of audience composition," *PNAS*, vol. 104, 2007, 17228–17233.

312. Semple S., et al., "Bystanders affect the outcome of mother-infant interactions in rhesus macaques," *Proc R Soc B*, vol. 276, 2009, 2257–2262.

313. Wich S. A., and de Vries H., "Male monkeys remember which group members have given alarm calls," *Proc R Soc B*, vol. 273, 2006, 735–740.

314. Cheney D. L., and Seyfarth R. M., *How Monkeys See the World: Inside the Mind of Another Species*, Chicago University Press, 1990. Look also at this recent paper, which shows that marmosets socially evaluate vocal exchanges between congeners and are able to distinguish between cooperative and non-cooperative conspecifics: Brugger R. K., et al., "Do marmosets understand others' conversations? A thermography approach," *Sc Adv*, vol. 7, 2021, eabc8790.

315. Grosenick L., et al., "Fish can infer social rank by observation alone," *Nature*, vol. 445, 2007, 429–432.

316. McGregor P., and Doutrelant C., "Eavesdropping and mate choice in female fighting fish," *Behaviour*, vol. 137, 2000, 1655–1668. See also Doutrelant C., et al., "The effect of an audience on intrasexual communication in male siamese fighting fish, *Betta splendens*," *Behav Ecol*, vol. 12, 2001, 283–286.

317. Clotfelter E. D., and Paolino A. D., "Bystanders to contests between conspecifics are primed for increased aggression in male fighting fish," *Anim Behav*, vol. 66, 2003, 343–347.

318. Mennill D. J., et al., "Female eavesdropping on male song contests in songbirds," *Science*, vol. 296, 2002, 873; Otter K., et al., "Do female great tits (*Parus major*) assess males by eavesdropping? A field study using interactive song playback," *Proc R Soc B*, vol. 266, 1999, 1305–1309.

319. Templeton C. N., et al., "Allometry of alarm calls: Black-capped chickadees encode information about predator size," *Science*, vol. 308, 2005, 1934–1937.

320. Templeton C. N., and Greene E., "Nuthatches eavesdrop on variations in heterospecific chickadee mobbing alarm calls," *PNAS*, vol. 104, 2007, 5479–5482; Carlson N. V., et al., "Nuthatches vary their alarm calls based upon the source of the eavesdropped signals," *Nat Com*, vol. 11, 2020, 526.

321. Another example of eavesdropping is when animals listen to the vocalizations of their predators to assess the risk they represent: Hettena A. M., et al., "Prey responses to predator's sounds: A review and empirical study," *Ethology*, vol. 120, 2014, 427–452.

322. Demartsev V., et al., "Signalling in groups: New tools for the integration of animal communication and collective movement," *Met Ecol Evol*, forthcoming.

323. An interesting research question is the brain basis of communication networks. An excellent example of research in this field is given in this paper, coauthored by Julie Elie: Rose M. C., et al., "Cortical representation of group social communication in bats," *Science*, vol. 374, 2021, eaba9584.

Chapter 12

324. Marler P., and Tamura M., "Culturally transmitted patterns of vocal behavior in sparrows," *Science*, vol. 146, 1964, 1483–1486.

325. Baker M. C., "Bird song research: The past 100 years," *Bird Behav*, vol. 14, 2001, 3–50; Barrington D., "Experiments and observations on the singing of birds," *Phil Trans R Soc*, vol. 63, 1773, 249–291.

326. Thorpe W. H., "The learning of song patterns by birds, with especial reference to the song of the chaffinch *Fringilla coelebs*," *Ibis*, vol. 100, 1958, 535–570; Poulsen H., "Inheritance and learning in the song of the chaffinch (*Fringilla coelebs* L.)," *Behaviour*, vol. 3, 1951, 216–228.

327. Actually, it was Darwin who first announced it. In *The Expression of Emotions in Man and Animals*, he wrote that, among birds, it is the father who teaches the young to sing.

328. Searcy W. A., et al., "Variation in vocal production learning across songbirds," *Phil Trans R Soc B*, vol. 376, 2021, 20200257.

329. Marler P., and Slabbekoorn H. (eds.), *Nature's Music: The Science of Birdsong*, Elsevier, 2004; Araya-Sala M., and Wright T., "Open-ended song learning in a hummingbird," *Biol Let*,

vol. 9, 2013, 2013062; Johnson K. E., and Clark C. J., "Ontogeny of vocal learning in a hummingbird," *Anim Behav*, vol. 167, 2020, 139–150.

330. For a recent review on vocal learning in nonoscine birds, see Ten Cate C., "Reevaluating vocal prodution learning in non-oscine birds," *Phil Trans R Soc B*, vol. 376, 2021, 20200249.

331. On hummingbirds, see this review: Duque F. G., and Carruth L. L., "Vocal communication in hummingbirds," *Brain Behav Evol*, vol. 97, 2022, 241–252.

332. It should be noted that the songs of songbirds, although culturally transmitted, still have a strong genetic component: Arato J., and Fitch W. T., "Phylogenetic signal in the vocalizations of vocal learning and vocal non-learning birds," *Phil Trans R Soc B*, vol. 376, 2021, 20200241.

333. This statement needs to be tempered a bit since ducks show some ability to imitate sounds: Ten Cate C., and Fullagar P. J., "Vocal imitations and production learning by Australian musk ducks (*Biziura lobata*)," *Phil Trans R Soc B*, vol. 376, 2021, 20200243.

334. Kroodsma D. E., and Konishi M., "A suboscine bird (eastern phoebe, *Sayornis phoebe*) develops normal song without auditory feedback," *Anim Behav*, vol. 42, 1991, 477–487.

335. Liu W., et al., "Rudimentary substrates for vocal learning in a suboscine," *Nat Com*, vol. 4, 2013, 2082.

336. Kroodsma D., et al., "Behavioral evidence for song learning in the suboscine bellbirds (*Procnias* spp.; Cotingidae)," *Wilson J Orn*, vol. 125, 2013, 1–14.

337. Marler P., "Three models of song learning: Evidence from behavior," *J Neurobiol*, vol. 33, 1997, 501–516; Tchernichovski O., et al., "Dynamics of the vocal imitation process: How a zebra finch learns its song," *Science*, vol. 291, 2001, 2564–2569; Carousio-Peck S., and Goldstein M. H., "Female social feedback reveals non-imitative mechanisms of vocal learning in zebra finches," *Cur Biol*, vol. 29, 2019, 631–636.

338. A recent study suggests that vocal learning may begin earlier, when the embryo is still in its egg, at least in certain bird species: Colombelli-Negrel D., et al., "Prenatal auditory learning in avian vocal learners and non-learners," *Phil Trans R Soc*, vol. 376, 2021, 20200247.

339. Bloomfield T. C., et al., "What birds have to say about language," *Nat Neurosci*, vol. 14, 2011, 947–948; Lipkind D., et al., "Stepwise acquisition of vocal combinatorial capacity in songbirds and human infants," *Nature*, vol. 498, 2013, 104–108; Brainard M. S., and Doupe A. J., "What songbirds teach us about learning," *Nature*, vol. 417, 2002, 351–358; Doupe A. J., and Kuhl P. K., "Birdsong and human speech: Common themes and mechanisms," *Ann Rev Neurosc*, vol. 22, 1999, 567–631.

340. Mennill D. J., et al., "Wild birds learn songs from experimental vocal tutors," *Cur Biol*, vol. 28, 2018, 3273–3278.

341. If you want to know more about the Savannah sparrow's story, check out this paper: Williams H., et al., "Cumulative cultural evolution and mechanisms for cultural selection in wild bird songs," *Nat Com*, vol. 13, 2022, 4001.

342. Mets D. G., and Brainard M. S., "Genetic variation interacts with experience to determine interindividual differences in learned song," *PNAS*, vol. 115, 2018, 421–426.

343. Carouso-Peck S., and Goldstein M. H., "Female social feedback reveals non-imitative mechanisms of vocal learning in zebra finches," *Cur Biol*, vol. 29, 2019, 631–636.

344. For a recent study on the effect of visual cues on song learning, see Varkevisser J. M., et al., "Multimodality during live tutoring is relevant for vocal learning in zebra finches," *Anim Behav*, vol. 187, 2022, 263–280.

345. Note that there is still a critical period during the first year.

346. Kroodsma D., "The diversity and plasticity of birdsong," in Marler and Slabbekoorn (eds.), *Nature's Music*, 108–131.

347. Klatt D. H., and Stefanski R. A., "How does a mynah bird imitate human speech?," *JASA*, vol. 55, 1974, 822–832.

348. On the flexibility of vocal learning, see this paper, which demonstrates that Bengalese finches can learn to rapidly modify the order of syllables in their song: Veit L., et al., "Songbirds can learn flexible contextual control over syllable sequencing," *eLife*, vol. 10, 2021, e61610.

349. Nowicki S., and Searcy W. A., "The evolution of vocal learning," *Cur Opin Neurobiol*, vol. 28, 2014, 48–53; Beecher M. D., and Brenowitz E. A., "Functional aspects of song learning in songbirds," *Trends Ecol Evol*, vol. 20, 2005, 143–149.

350. See also Osiejuk T. S., et al., "Songbird presumed to be age-limited learner may change repertoire size and composition throughout their life," *J Zool*, vol. 309, 2019, 231–240.

351. Kroodsma, "Diversity and plasticity of birdsong."

352. Odom K. J., et al., "Female song is widespread and ancestral in songbirds," *Nat Com*, vol. 5, 2014, 3379; Riebel K., et al., "New insights from female bird song: Towards an integrated approach to studying male and female communication roles," *Biol Let*, vol. 15, 2019, 20190059.

353. Bolhuis J. J., and Gahr M., "Neural mechanisms of birdsong memory," *Nat Rev Neurosci*, vol. 7, 2006, 347–357.

354. Sakata J. T., et al., *The Neuroethology of Birdsong*, Springer Handbook of Auditory Research, vol. 71, Springer, 2020.

355. Mooney R., "The neurobiology of innate and learned vocalizations in rodents and songbirds," *Cur Opin Neurobiol*, vol. 64, 2020, 24–31; Theunissen F. E., and Shaevitz S. S., "Auditory processing of vocal sounds in birds," *Cur Opin Neurobiol*, vol. 16, 2006, 400–407; Van Ruijssevelt L., et al., "fMRI reveals a novel region for evaluating acoustic information for mate choice in a female songbird," *Cur Biol*, vol. 28, 2018, 711–721.

356. Nottebohm F., and Arnold A. P., "Sexual dimorphism in vocal control areas of the songbird brain," *Science*, vol. 194, 1976, 211–213.

357. Jarvis E. D., et al., "Behaviourally driven gene expression reveals song nuclei in hummingbird brain," *Nature*, vol. 406, 2000, 628–632.

358. Theunissen F. E., and Elie J. E., "Neural processing of natural sounds," *Nature Rev Neurosc*, vol. 15, 2014, 355–366.

359. For an interesting perspective on song production in birds, see During D. N., and Elemans C.P.H., "Embodied motor control of avian vocal production," in Suthers, Fitch, Fay, and Popper (eds.), *Vertebrate Sound Production and Acoustic Communication*, 119–157.

360. Jarvis E. D., "Neural systems for vocal learning in birds and humans: A synopsis," *J Ornithol*, 2007, 35–44; Prather J. F., et al., "Brains for birds and babies: Neural parallels between birdsong and speech acquisition," *Neurosc Biobehav Rev*, vol. 81, 2017, 225–237.

361. Suthers R. A., "How birds sing and why it matters," in Marler and Slabbekoorn (eds.), *Nature's Music*, 272–295; Goller F., and Larsen O. N., "A new mechanism of sound generation

in songbirds," *PNAS*, vol. 94, 1997, 14787–14791; Riede T., et al., "The evolution of the syrinx: An acoustic theory," *PLoS Biol*, vol. 17, 2019, 1–22.

362. Moorman S., et al., "Human-like brain hemispheric dominance in birdsong learning," *PNAS*, vol. 109, 2012, 12782–12787.

363. Amin N., et al., "Development of selectivity for natural sounds in the songbird auditory forebrain," *J Neurophysiol*, vol. 97, 2007, 3517–3531.

364. Derégnaucourt S., et al., "How sleep affects the developmental learning of bird song," *Nature*, vol. 433, 2005, 710–716.

365. Stickgold R., "Sleep-dependent memory consolidation," *Nature*, vol. 437, 2005, 1272–1278; Nusbaum H. C., et al., "Consolidating skill learning through sleep," *Cur Opin Behav Sc*, vol. 20, 2018, 174–182; Saletin J. M., "Memory: Necessary for deep sleep?," *Cur Biol*, vol. 30, 2020, R234–R236.

366. Lachlan R. F., et al., "Cultural conformity generates extremely stable traditions in bird song," *Nat Com*, vol. 9, 2018, 2417.

367. Kroodsma D., *The Singing Life of Birds*, Houghton Mifflin, 2005; Price T., *Speciation in Birds*, Roberts & Co., 2008. See also this recent paper, which investigates the dynamics of vocal cultural transmission in a songbird: Tchernichovski O., Eisenberg-Edidin S., and Jarvis E. D., "Balanced imitation sustains song culture in zebra finches," *Nat Com*, vol. 12, 2021, 2562.

368. Wang D., et al., "Machine learning reveals cryptic dialects that explain mate choice in a songbird," *Nat Com*, vol. 13, 2022, 1630.

369. Irwin D. E., et al., "Speciation in a ring," *Nature*, vol. 409, 2001, 333–337.

370. Crates R., et al., "Loss of vocal culture and fitness costs in a critically endangered songbird," *Proc R Soc B*, vol. 288, 2021, 20210225.

371. Burroughs E. R., *Tarzan of the Apes*, A. C. McClurg & Co., 1914.

372. Hauber M. E., et al., "A password for species recognition in a brood-parasitic bird," *Proc R Soc B*, vol. 268, 2001, 1041–1048.

373. Louder M.I.M., et al., "An acoustic password enhances auditory learning in juvenile brood parasitic cowbirds," *Cur Biol*, vol. 29, 2019, 4045–4051.

374. Tyack P. L., "A taxonomy for vocal learning," *Phil Trans R Soc B*, vol. 375, 2019, 20180406; Nieder A., and Mooney R., "The neurobiology of innate, volitional and learned vocalizations in mammals and birds," *Phil Trans R Soc B*, vol. 375, 2019, 20190054.

375. For a recent perspective on vocal learning in mammals, see Janik V. M., and Knörnschild M., "Vocal production learning in mammals," *Phil Trans R Soc B*, vol. 376, 2021, 20200244.

376. This paper presents an interesting perspective on the different facets of vocal learning: Vernes S. C., et al., "The multi-dimensional nature of vocal learning," *Phil Trans R Soc B*, vol. 376, 2021, 20200236.

377. See this special issue on vocal learning: Vernes S. C., et al., "Vocal learning in animals and humans," *Phil Trans R Soc B*, vol. 376, 2021, 20200234.

378. Payne R., *Among Whales*, Simon & Schuster, 1995.

379. Whitehead H., and Rendell L., *The Cultural Lives of Whales and Dolphins*, University of Chicago Press, 2015; Payne K., and Payne R. S., "Large-scale changes over 19 years in songs of humpback whales in Bermuda," *Zeit Psychol*, vol. 68, 1985, 89–114; Noad M. J., et al., "Cultural revolution in whale songs," *Nature*, vol. 408, 2000, 537; Garland E. C., et al., "Dynamic horizontal cultural transmission of humpback whale song at the ocean basin scale," *Cur Biol*, vol. 21, 2011, 687–691.

380. This view has been recently challenged. See the following papers: Mercado III E., and Perazio C. E., "Similarities in composition and transformations of songs by humpback whales (*Megaptera novaeangliae*) over time and space," *J Comp Psychol*, vol. 135, 2021, 28–50; Mercado III E., "Song morphing by humpback whales: Cultural or epiphenomenal?," *Front Psychol*, vol. 11, 2021, 574403; and Mercado III E., "The humpback's new songs: Diverse and convergent evidence against vocal culture via copying in humpback whales," *Anim Behav Cogn*, vol. 9, 2022, 196–206.

381. Zandberg L., et al., "Global cultural evolutionary model of humpback whale song," *Phil Trans R Soc B*, vol. 376, 2021, 20200242.

382. Richards D. G., et al., "Vocal mimicry of computer-generated sounds and vocal labeling of objects by a bottlenosed dolphin, *Tursiops truncatus*," *J Comp Psychol*, vol. 98, 1984, 10–28; Janik V. M., and Sayigh L. S., "Communication in bottlenose dolphins: 50 years of signature whistle research," *J Comp Physiol*, vol. 199, 2013, 479–489; Luis A. R., et al., "Vocal universals and geographic variations in the acoustic repertoire of the common bottlenose dolphin," *Sc Rep*, vol. 11, 2021, 11847.

383. Eaton R. L., "A beluga whale imitates human speech," *Carnivore*, vol. 2, 1979, 22–23; Murayama T., et al., "Vocal imitation of human speech, synthetic sounds and beluga sounds, by a beluga (*Delphinapterus leucas*)," *Int J Comp Psychol*, vol. 27, 2014, 369–384.

384. Janik V. M., "Cetacean vocal learning and communication," *Cur Opin Neurobiol*, vol. 28, 2014, 60–65; Ridgway S., et al., "Spontaneous human speech mimicry by a cetacean," *Cur Biol*, vol. 22, 2012, R860–R861; Abramson J. Z., et al., "Imitation of novel conspecific and human speech sounds in the killer whale (*Orcinus orca*)," *Proc R Soc B*, vol. 285, 2018, 20172171.

385. Bernie Krause made a recording of an orca, *Orcinus orca*, imitating pinniped vocalizations, likely to attract the prey or at least get close enough to attack. Bernie sent me this recording, and the imitation is indeed truly amazing.

386. Reichmuth C., and Casey C., "Vocal learning in seals, sea lions, and walruses," *Cur Opin Neurobiol*, vol. 28, 2014, 66–71.

387. Rawls K., Fiorelli P., and Gish S., "Vocalizations and vocal mimicry in captive harbor seals, *Phoca vitulina*," *Can J Zool*, vol. 63, 1985, 1050–1056.

388. Stansbury A. L., and Janik V. M., "Formant modification through vocal production learning in gray seals," *Cur Biol*, vol. 29, 2019, 2244–2249.

389. Not everyone agrees. See, for example, Schusterman R., "Vocal learning in mammals with special emphasis on pinnipeds," in Oller D. K., and Griebel U. (eds.), *Evolution of Communicative Flexibility: Complexity, Creativity, and Adaptability in Human and Animal Communication*, MIT Press, 2008, 41–70.

390. Stoeger A. S., and Manger P., "Vocal learning in elephants: Neural bases and adaptive context," *Cur Opin Neurobiol*, vol. 28, 2014, 101–107.

391. Stoeger A. S., and Baotic A., "Operant control and call usage learning in African elephants," *Phil Trans R Soc B*, vol. 376, 2021, 20200254.

392. Poole J. H., et al., "Elephants are capable of vocal learning," *Nature*, vol. 434, 2005, 455–456.

393. Stoeger A. S., et al., "An Asian elephant imitates human speech," *Cur Biol*, vol. 22, 2012, 2144–2148. See also this recent paper reporting that elephants can use lip buzzing (like human

brass players) to produce high-pitched sounds: Beeck V. C., et al., "A novel theory of Asian elephant high-frequency squeak production," *BMC Biol*, vol. 19, 2021, 121.

394. Vernes S. C., and Wilkinson G. S., "Behaviour, biology and evolution of vocal learning in bats," *Phil Trans R Soc B*, vol. 375, 2019, 20190061; Knörnschild M., "Vocal production learning in bats," *Cur Opin Neurobiol*, vol. 28, 2014, 80–85.

395. Jones G., and Ransome R. D., "Echolocation calls of bats are influenced by maternal effects and change over a lifetime," *Proc R Soc B*, vol. 252, 1993, 125–128.

396. Prat Y., et al., "Vocal learning in a social mammal: Demonstrated by isolation and playback experiments in bats," *Sc Adv*, vol. 1, 2015, e1500019. See also Prat Y., et al., "Crowd vocal learning induces vocal dialects in bats: Playback of conspecifics shapes fundamental frequency usage by pups," *PLoS Biol*, vol. 15, 2017, e2002556.

397. Lattenkamp E. Z., et al., "The vocal development of the pale spear-nosed bat is dependent on auditory feedback," *Phil Trans R Soc*, vol. 376, 2021, 20200253.

398. Fernandez A. A., et al., "Babbling in a vocal learning bat resembles human infant babbling," *Science*, vol. 373, 2021, 923–926.

399. Knörnschild M., et al., "Complex vocal imitation during ontogeny in a bat," *Biol Let*, vol. 6, 2010, 156–159.

400. For instance, look at the works of Mirjam Knörnschild at the Museum of Natural History in Berlin, and Sonja Catherine Vernes at the University of St. Andrews, Scotland.

401. Barker A. J., et al., "Cultural transmission of vocal dialect in the naked mole-rat," *Science*, vol. 371, 2021, 503–507.

402. Taylor D., et al., "Vocal functional flexibility: What it is and why it matters." *Anim Behav*, vol. 186, 2022, 93–100.

403. Brockelman W. Y., and Schilling D., "Inheritance of stereotyped gibbon calls," *Nature*, vol. 312, 1984, 634–636.

404. Hayes K. J., and Hayes C., "Imitation in a home-raised chimpanzee," *J Comp Physiol Psychol*, vol. 45, 1952, 450–459.

405. Takahashi D. Y., et al., "The developmental dynamics of marmoset monkey vocal production," *Science*, vol. 349, 2015, 734–738; Gultekin Y. B., and Hage S. R., "Limiting parental feedback disrupts vocal development in marmoset monkeys," *Nat Com*, vol. 8, 2017, 14046.

406. Levrero F., et al., "Social shaping of voices does not impair phenotype matching of kinship in mandrills," *Nat Com*, vol. 6, 2015, 7609.

407. Fischer J., et al., "Vocal convergence in a multi-level primate society: Insights into the evolution of vocal learning," *Proc R Soc B*, vol. 287, 2020, 20202531.

408. Lameira A. R., et al., "Sociality predicts orangutan vocal phenotype," *Nat Ecol Evol*, vol. 6, 2022, 644–652.

409. Ruch H., et al., "The function and mechanism of vocal accommodation in humans and other primates," *Biol Rev*, vol. 93, 2018, 996–1013; Watson S. K., et al., "Vocal learning in the functionally referential food grunts of chimpanzees," *Cur Biol*, vol. 25, 2015, 495–499; Fischer J., et al., "Vocal convergence in a multi-level primate society."

410. On the interactions between genetics and learning, see Mets and Brainard, "Genetic variation interacts with experience."

Chapter 13

411. Céline Rochais is conducting studies on the cognitive performance of striped mice in the wild. See, for example, Rochais C., et al., "How does cognitive performance change in relation to seasonal and experimental changes in blood glucose levels?," *Anim Behav*, vol. 158, 2019, 149–159.

412. Schradin C., "Seasonal changes in testosterone and corticosterone levels in four social classes of a desert dwelling sociable rodent," *Horm Behav*, vol. 53, 2008, 573–579.

413. Schradin C., and Pillay N., "Intraspecific variation in the spatial and social organization of the African striped mouse," *J Mammal*, vol. 86, 2005, 99–107.

414. Holy T. E., and Guo Z., "Ultrasonic songs of male mice," *PLoS Biol*, vol. 3, 2005, e386.

415. Griffin D. R., and Galambos R., "The sensory basis of obstacle avoidance by flying bats," *J Exp Zool*, vol. 86, 1941, 481–506; Galambos R., and Griffin D. R., "Obstacle avoidance by flying bats: The cries of bats," *J Exp Zool*, vol. 89, 1942, 475–490; Griffin D. R., "Supersonic cries of bats," *Nature*, vol. 158, 1946, 46–48.

416. For a review of our knowledge of rodents' vocalizations (including production mechanisms), look at Fernandez-Vargas M., et al., "Mechanisms and constraints underlying acoustic variation in rodents," *Animal Behav*, vol. 184, 2021, 135–147.

417. Simola N., and Brudzynski S. M., "Repertoire and biological function of ultrasonic vocalizations in adolescent and adult rats," in Brudzynski S. (ed.), *Handbook of Ultrasonic Vocalization* XXV, Elsevier, 2018, 177–186.

418. Ehret G., "Characteristics of vocalization in adult mice," in Brudzynski (ed.), *Handbook of Ultrasonic Vocalization*, 187–196.

419. Fischer J., and Hammerschmidt K., "Ultrasonic vocalizations in mouse models for speech and socio-cognitive disorders: Insights into the evolution of vocal communication," *Genes Brain Behav*, vol. 10, 2010, 17–27.

420. Portfors C. V., and Perkel D. J., "The role of ultrasonic vocalizations in mouse communication," *Cur Opin Neurobiol*, vol. 28, 2014, 115–120; Miranda R., et al., "Altered social behavior and ultrasonic communication in the dystrophin-deficient mdx mouse model of Duchenne muscular dystrophy," *Mol Autism*, vol. 6, 2015, 60; Faure A., et al., "Dissociated features of social cognition altered in mouse models of schizophrenia: Focus on social dominance and acoustic communication," *Neuropharmacol*, vol. 159, 2019, 107334.

421. Pierce G. W., *The Songs of Insects*, Harvard University Press, 1948.

422. Pierce G. W., and Griffin D. R., "Experimental determination of supersonic notes emitted by bats," *J Mammal*, vol. 19, 1938, 454–455.

423. Fenton M. B., Grinnell A. D., Popper A. N., and Fay R. R. (eds.), *Bat Bioacoustics*, Springer, 2016.

424. Ulanovsky N., and Moss C. F., "What the bat's voice tells the bat's brain," *PNAS*, vol. 105, 2008, 8491–8498.

425. Leiser-Miller L. B., and Santana S. E., "Functional differences in echolocation call design in an adaptive radiation of bats," *Ecol Evol*, vol. 11, 2021, 16153–16164.

426. Bats are not the only animals to echolocate. In the book, we have already seen that dolphins and other sperm whales are capable of it. It is the same for some birds, and even . . . humans! Indeed, it has been shown that trained blind people can locate objects in their environment

by producing "clicks" and listening to their echoes. See Brinklov S., "Echolocation in oilbirds and swiftlets," *Front Physiol*, vol. 4, 2013, 123; and Thaler L., et al., "Human echolocators adjust loudness and number of clicks for detection of reflectors at various azimuth angles," *Proc R Soc B*, vol. 285, 2018, 20172735. Recently, researchers have discovered another lineage of echolocating mammals: He K., et al., "Echolocation in soft-furred tree mice," *Science*, vol. 372, 2021, eaay1513.

427. Carew T. J., *Behavioral Neurobiology*, Sinauer, 2000.

428. Simmons J. A., et al., "Echo-delay resolution in sonar images of the big brown bat, *Eptesicus fuscus*," *PNAS*, vol. 95, 1998, 12647–12652.

429. Schnitzler H. U., "Control of Doppler shift compensation in the greater horseshoe bat, *Rhinolophus ferrumequinum*," *J Comp Physiol*, vol. 82, 1973, 79–92; Metzner W., "A possible neuronal basis for Doppler-shift compensation in echo-locating horseshoe bats," *Nature*, vol. 341, 1989, 529–532; Metzner W., et al., "Doppler-shift compensation behavior in horseshoe bats revisited: Auditory feedback controls both a decrease and an increase in call frequency," *J Exp Biol*, vol. 205, 2002, 1607–1616.

430. Simmons J. A., "Response of the Doppler echolocation system in the bat, *Rhinolophus ferrumequinum*," *JASA*, vol. 56, 1974, 672–682; Zhang Y., et al., "Performance of Doppler shift compensation in bats varies with species rather than with environmental clutter," *Anim Behav*, vol. 158, 2019, 109–120.

431. Schnitzler H.-U., and Denzinger A., "Auditory fovea and Doppler shift compensation: Adaptations for flutter detection in echolocating bats using CF-FM signals," *J Comp Physiol A*, vol. 197, 2011, 541–559.

432. Yack J. E., and Fullard J. H., "Ultrasonic hearing in nocturnal butterflies," *Nature*, vol. 403, 2000, 265–266.

433. Greenfield M. D., "Evolution of acoustic communication in insects," in Pollack G. S., Mason A. C., Popper A., and Fay R. R. (eds.), *Insect Hearing*, Springer, 2016, 17–47.

434. Yack J. E., et al., "Neuroethology of ultrasonic hearing in nocturnal butterflies (*Hedyloidea*)," *J Comp Physiol A*, vol. 193, 2007, 577–590.

435. Kawahara A. Y., and Barber J. R., "Tempo and mode of antibat ultrasound production and sonar jamming in the diverse hawkmoth radiation," *PNAS*, vol. 112, 2015, 6407–6412; Barber J. R., et al., "Anti-bat ultrasound production in moths is globally and phylogenetically widespread," *PNAS*, vol. 119, 2022, e2117485119.

436. Ancillotto L., et al., "Bats mimic hymenopteran insect sounds to deter predators," *Cur Biol*, vol. 32, 2022, R399–R413.

437. Conner W. E., and Corcoran A. J., "Sound strategies: The 65-million-year-old battle between bats and insects," *Ann Rev Entomol*, vol. 57, 2012, 21–39; Ter Hofstede H. M., and Ratcliffe J. M., "Evolutionary escalation: The bat-moth arms race," *J Exp Biol*, vol. 219, 2016, 1589–1602.

438. von Helversen D., and von Helversen O., "Acoustic guide in bat-pollinated flower," *Nature*, vol. 398, 1999, 759–760.

439. The bat ultrasonic world is an active field of research. Look at this paper, reporting that differences among bats in hearing abilities contribute to the diversity in foraging strategies: Geipel I., et al., "Hearing sensitivity: An underlying mechanism for niche differentiation in gleaning bats," *PNAS*, vol. 118, 2021, e2024943118.

440. Feng A. S., and Narins P. M., "Ultrasonic communication in concave-eared torrent frogs (*Amolops tormotus*)," *J Comp Physiol*, vol. 194, 2008, 159–167.

441. Feng A. S., et al., "Ultrasonic communication in frogs," *Nature*, vol. 440, 2006, 333–336; Shen J.-X., et al., "Ultrasonic frogs show hyperacute phonotaxis to female courtship calls," *Nature*, vol. 453, 2008, 914–916; Gridi-Papp M., et al., "Active control of ultrasonic hearing in frogs," *PNAS*, vol. 105, 2008, 11014–11019.

442. Narins P. M., et al., "Plant-borne vibrations modulate calling behaviour in a tropical amphibian," *Cur Biol*, vol. 23, 2018, R1333–R1334; Lewis E. R., and Narins P. M., "Do frogs communicate with seismic signals?," *Science*, vol. 227, 1985, 187–189; Narins P. M., "Seismic communication in anuran amphibians," *Bioscience*, vol. 40, 1990, 268–274.

443. A recent paper explores the richness of the vibrational environment that can be perceived and exploited by insects: Sturm R., et al., "Hay meadow vibroscape and interactions within insect vibrational community," *iScience*, vol. 24, 2021, 103070. Another paper deals with crocodiles: Grap N. J., et al., "Stimulus discrimination and surface wave source localization in crocodilians," *Zoology*, vol. 139, 2020, 125743.

444. Hager F. A., et al., "Vibrational behavior in termites (Isoptera)," in Hill P., Lakes-Harlan R., Mazzoni V., Narins P., Virant-Doberlet M., and Wessel A. (eds.), *Biotremology: Studying Vibrational Behavior*, Springer, 2019, 309–327.

445. Hager F. A., and Kirchner W. H., "Vibrational long-distance communication in the termites *Macrotermes natalensis* and *Odontotermes* sp.," *J Exp Biol*, vol. 216, 2013, 3249–3256.

446. Similar patterns can be observed in bees; see, for example, Grüter C., et al., "Propagation of olfactory information within the honeybee hive," *Behav Ecol Sociobiol*, vol. 60, 2006, 707–715.

447. Hager F. A., and Kirchner W. H., "Directional vibration sensing in the termite *Macrotermes natalensis*," *J Exp Biol*, vol. 217, 2014, 2526–2530.

448. Spiders can develop amazing abilities to detect vibrations and sounds. In the orb-weaving spider *Larinioides sclopetarius*, the web acts as an acoustic antenna that captures the sound-induced air particle movements: Zhou J., et al., "Outsourced hearing in an orb-weaving spider that uses its web as an auditory sensor," *PNAS*, vol. 119, 2022, e2122789119.

449. Narins P. M., et al., "Infrasonic and seismic communication in the vertebrates with special emphasis on the Afrotheria: An update and future directions," in Suthers, Fitch, Fay, and Popper (eds.), *Vertebrate Sound Production and Acoustic Communication*, 191–227.

450. Schleich C., and Francescoli G., "Three decades of subterranean acoustic communication studies," in Dent M. L., Fay R. R., and Popper A. N. (eds.), *Rodent Bioacoustics*, Springer, 2018, 43–69.

451. Elephants also emit high-pitched sounds: Beeck et al., "Novel theory of Asian elephant high-frequency squeak production."

452. O'Connell-Rodwell C. E., "Keeping an 'ear' to the ground: Seismic communication in elephants," *Physiology*, vol. 22, 2007, 287–294; O'Connell-Rodwell C. E., et al., "Wild African elephants (*Loxodonta africana*) discriminate between familiar and unfamiliar conspecific seismic alarm calls," *JASA*, vol. 122, 2007, 823–830.

453. Parihar D. S., et al., "Seismic signal analysis for the characterization of elephant movements in a forest environment," *Ecol Infor*, vol. 64, 2021, 101329.

454. O'Connell C. E., et al., "Vibrational communication in elephants: A case for bone conduction," in Hill, Lakes-Harlan, Mazzoni, Narins, Virant-Doberlet, and Wessel (eds.), *Biotremology: Studying Vibrational Behavior*, 259–276.

455. McComb K., et al., "Unusually extensive networks of vocal recognition in African elephants," *Anim Behav*, vol. 59, 2000, 1103–1109.

456. The importance of infrasound in the animal world may have been underestimated. For instance, although they do not communicate in the infrasound range, some birds are able to hear it: Zeyl J. N., et al., "Infrasonic hearing in birds: A review of audiometry and hypothesized structure-function relationships," *Biol Rev*, vol. 95, 2020, 1036–1054.

Chapter 14

457. Holekamp K. E., and Kolowski J. M., "Family Hyaenidae (hyenas)," in Wilson D. E., and Mittermeier R. A. (eds.), *Handbook of the Mammals of the World* I, *Carnivores*, Lynx Edicions, 2009, 234–251.

458. Kruuk H., *The Spotted Hyena: A Study of Predation and Social Behaviour*, Chicago University Press, 1972.

459. Watts H. E., and Holekamp K. E., "Hyena societies," *Cur Biol*, vol. 17, 2007, R657–R660.

460. Yalcinkaya T. M., et al., "A mechanism for virilization of female spotted hyenas in utero," *Science*, vol. 260, 1993, 1929–1931; Glickman S. E., et al., "Androstenedione may organize or activate sex-reversed traits in female spotted hyenas," *PNAS*, vol. 84, 1987, 3444–3447.

461. Holekamp K. E., and Smale L., "Dominance acquisition during mammalian social development: The 'inheritance' of maternal rank," *Amer Zool*, vol. 31, 1991, 306–317; Engh A. L., et al., "Mechanisms of maternal rank 'inheritance' in the spotted hyaena, *Crocuta crocuta*," *Anim Behav*, vol. 60, 2000, 323–332.

462. Smith J. E., and Holekamp K. E., "Spotted hyenas," in Breed M. D., and Moore J. (eds.), *Encyclopedia of Animal Behavior*, 2nd ed., Academic Press, 2019, 190–208.

463. On the importance of individual recognition in animal social networks, see Gokcekus S., et al., "Recognizing the key role of individual recognition in social networks," *Trends Ecol Evol*, vol. 36, 2021, 1024–1035.

464. Elie J. E., and Theunissen F. E., "Zebra finches identify individuals using vocal signatures unique to each call type," *Nat Com*, vol. 9, 2018, 4026.

465. Frédéric Theunissen's team recently showed that the zebra finches have fast and high-capacity auditory memory for vocalizer identity. They can remember a mean number of 42 different vocalizers based solely on the individual signatures found in their songs and distance calls. The learning is very efficient, taking only a few trials, and is maintained for up to a month. See Yu K., et al., "High-capacity auditory memory for vocal communication in a social songbird," *Sc Adv*, vol. 6, 2020, eabe0440.

466. Christiansen P., and Adolfseen J. S., "Bite forces, canine strength and skull allometry in carnivores (Mammalia, Carnivora)," *J Zool*, vol. 266, 2005, 133–151.

467. Mathevon N., et al., "What the hyena's laugh tells: Sex, age, dominance and individual signature in the giggling call of *Crocuta crocuta*," *BMC Ecol*, vol. 10, 2010, 9.

468. Ross-Gillespie A., and Griffin A. S., "Meerkats," *Cur Biol*, vol. 17, 2007, R442–R443; Gilchrist J. S., et al., "Family Herpestidae (mongooses)," in Wilson and Mittermeier (eds.), *Handbook of the Mammals of the World* I, 262–328.

469. Manser M. B., "The acoustic structure of suricates' alarm calls varies with predator type and the level of response urgency," *Proc R Soc B*, vol. 268, 2001, 2315–2324; Manser M. B., et al., "The information that receivers extract from alarm calls in suricates," *Proc R Soc B*, vol. 268, 2001, 2485–2491.

470. Townsend S. W., et al., "Acoustic cues to identity and predator context in meerkat barks," *Anim Behav*, vol. 94, 2014, 143–149.

471. Manser M. B., "Semantic communication in vervet monkeys and other animals," *Anim Behav*, vol. 86, 2013, 491–496; Rauber R., Kranstauber B., and Manser M. B., "Call order within vocal sequences of meerkats contains temporary contextual and individual information," *BMC Biol*, vol. 18, 2020, 119.

472. Hollén L. I., and Manser M. B., "Ontogeny of alarm call responses in meerkats, *Suricata suricatta*: The roles of age, sex and nearby conspecifics," *Anim Behav*, vol. 72, 2006, 1345–1353.

473. Darwin C., *The Descent of Man and Selection in Relation to Sex*, John Murray, 1871.

474. Struhsaker T. T., "Auditory communication among vervet monkeys (*Cercopithecus aetthiops*)," in Altmann S. A. (ed.), *Social Communication among Primates*, University of Chicago Press, 1967, 281–324.

475. Seyfarth R. M., et al., "Monkey responses to three different alarm calls: Evidence of predator classification and semantic communication," *Science*, vol. 210, 1980, 801–803; Price T., et al., "Vervets revisited: A quantitative analysis of alarm call structure and context specificity," *Sc Rep*, vol. 5, 2015, 13220.

476. Cheney D. L., and Seyfarth R. M., *How Monkeys See the World*, University of Chicago Press, 1990.

477. Premack D., "Concordant preferences as a precondition for affective but not for symbolic communication (or how to do experimental anthropology)," *Cognition*, vol. 1, 1972, 251–264.

478. Zuberbühler K., "Referential labelling in Diana monkeys," *Anim Behav*, vol. 59, 2000, 917–927.

479. Zuberbühler K., "A syntactic rule in forest monkey communication," *Anim Behav*, vol. 63, 2002, 293–299.

480. Rendall D., et al., "The meaning and function of grunt variants in baboons," *Anim Behav*, vol. 57, 1999, 583–592.

481. Caesar C., et al., "The alarm call system of wild black-fronted titi monkeys, *Callicebus nigrifrons*," *Behav Ecol Sociobiol*, vol. 66, 2012, 653–667.

482. Slocombe K. E., and Zuberbühler K., "Functionally referential communication in a chimpanzee," *Cur Biol*, vol. 15, 2005, 1779–1784.

483. Pereira M. E., and Macedonia J. M., "Ringtailed lemur anti-predator calls denote predator class, not response urgency," *Anim Behav*, vol. 26, 1991, 760–777.

484. Schel A. M., and Zuberbühler K., "Predator and non-predator long-distance calls in Guereza colobus monkeys," *Behav Process*, vol. 91, 2012, 41–49.

485. Ouattara K., et al., "Campbell's monkeys concatenate vocalizations into context-specific call sequences," *PNAS*, vol. 106, 2009, 22026–22031.

486. Zuberbühler K., et al., "Diana monkey long-distance calls: Messages for conspecifics and predators," *Anim Behav*, vol. 53, 1997, 589–604.

487. Slocombe K. E., and Zuberbühler K., "Functionally referential communication in a chimpanzee," *Cur Biol*, vol. 15, 2005, 1779–1784; Slocombe K. E., et al., "Chimpanzee vocal communication: What we know from the wild," *Cur Opin Behav Sc*, vol. 46, 2022, 101171.

488. Clay Z., and Zuberbühler K., "Food-associated calling sequences in bonobos," *Anim Behav*, vol. 77, 2009, 1387–1396; Shorland G., et al., "Bonobos assign meaning to food calls based on caller food preferences," *PLoS ONE*, vol. 17, 2022, e0267574.

489. Understanding how the sequences of calls produced by primates are organized and what they mean is an active field of research. See, for example, Girard-Buttoz C., et al., "Chimpanzees produce diverse vocal sequences with ordered and recombinatorial properties," *Com Biol*, vol. 5, 2022, 410.

490. Townsend S. W., and Manser M. B., "Functionally referential communication in mammals: The past, present and the future," *Ethology*, vol. 119, 2013, 1–11.

491. Kiriazis J., and Slobodchikoff C. N., "Perceptual specificity in the alarm calls of Gunnison's prairie dogs," *Behav Process*, vol. 73, 2006, 29–35.

492. Smith C. L., "Referential signalling in birds: The past, present and future," *Anim Behav*, vol. 124, 2017, 315–323.

493. Marler P., et al., "Semantics of an avian alarm call system: The male domestic fowl, *Gallus domesticus*," *Behaviour*, vol. 102, 1987, 15–40; Evans C. S., and Evans L., "Chicken food calls are functionally referential," *Anim Behav*, vol. 58, 1999, 307–319; Evans C. S., and Evans L., "Representational signalling in birds," *Biol Let*, vol. 3, 2007, 8–11.

494. Evans C. S., et al., "Effects of apparent size and speed on the response of chickens, *Gallus gallus*, to computer-generated simulations of aerial predators," *Anim Behav*, vol. 46, 1993, 1–11.

495. Gill S. A., and Bierema A.M.K., "On the meaning of alarm calls: A review of functional reference in avian alarm calling," *Ethology*, vol. 119, 2012, 449–461.

496. Magrath R., et al., "Interspecific communication: Gaining information from heterospecific alarm calls," in Aubin T., and Mathevon N. (eds.), *Coding Strategies in Vertebrate Acoustic Communication*, Springer, 2020, 287–314.

497. Templeton C. N., et al., "Allometry of alarm calls: Black-capped chickadees encode information about predator size," *Science*, vol. 308, 2005, 1934–1937.

498. Suzuki T. N., "Communication about predator type by a bird using discrete, graded and combinatorial variation in alarm calls," *Anim Behav*, vol. 87, 2014, 59–65; Suzuki T. N., "Referential mobbing calls elicit different predator-searching behaviours in Japanese great tits," *Anim Behav*, vol. 84, 2012, 53–57; Suzuki T. N., et al., "Experimental evidence for compositional syntax in bird calls," *Nat Com*, vol. 7, 2016, 10986.

499. Suzuki T. N., "Alarm calls evoke a visual search image of a predator in birds," *PNAS*, vol. 115, 2018, 1541–1545.

500. For a review on bird calls and their interest for the understanding of language evolution, see Suzuki T. N., "Animal linguistics: Exploring referentiality and compositionality in bird calls," *Ecol Res*, vol. 36, 2021, 221–231.

501. Engesser S., et al., "Meaningful call combinations and compositional processing in the southern pied babbler," *PNAS*, vol. 113, 2016, 5976–5981.

502. Engesser S., et al., "Chestnut-crowned babbler calls are composed of meaningless shared building blocks," *PNAS*, vol. 116, 2019, 19579–19584.

503. On songbird syntax, see Searcy W. A., et al., "Long-distance dependencies in birdsong syntax," vol. 289, 2022, 20212473; and Backhouse F., et al., "Higher-order sequences of vocal mimicry performed by male Albert's lyrebirds are socially transmitted and enhance acoustic contrast," *Proc R Soc B*, vol. 289, 2022, 20212498.

504. In Molière's famous comedy *Le bourgeois gentilhomme*, Mr. Jourdain learns from his philosophy teacher that all language is classified according to the way it is spoken, as poetry or prose (the ordinary form of language). Mr. Jourdain is delighted to realize that he has been speaking in prose for a very long time without even knowing it.

505. Seyfarth R. M., and Cheney D. L., "Precursors to language: Social cognition and pragmatic inference in primates," *Psychon Bull Rev*, vol. 24, 2017, 79–84.

506. Cheney D. L., and Seyfarth R. M., *Baboon Metaphysics*, University of Chicago Press, 2007.

Chapter 15

507. Mathevon N., et al., "The code size: Behavioural response of crocodile mothers to offspring calls depends on the emitter's size, not on its species identity," in *Crocodiles, Proceedings of the 24th Working Meeting of the Crocodile Specialist Group IUCN*, 2016, 79–85.

508. Darwin C., *The Expression of Emotions in Man and Animals*, Penguin Classics, 2009.

509. Briefer E. F., "Coding for 'dynamic' information: Vocal expression of emotional arousal and valence in non-human animals," in Aubin and Mathevon (eds.), *Coding Strategies in Vertebrate Acoustic Communication*, 137–162.

510. Briefer E. F., et al., "Segregation of information about emotional arousal and valence in horse whinnies," *Sc Rep*, vol. 5, 2015, 9989.

511. Things are in fact even more complex, because each of these frequencies, F0 and G0, is accompanied by its harmonic series. For details, see Briefer et al., "Segregation of information about emotional arousal."

512. Briefer E. F., "Vocal expression of emotions in mammals: Mechanisms of production and evidence," *J Zool*, vol. 288, 2012, 1–20.

513. Briefer E. F., "Classification of pig calls produced from birth to slaughter according to their emotional valence and context of production," *Sc Rep*, vol. 12, 2022, 3409.

514. Darwin C., *Descent of Man*.

515. Morton E. S., "On the occurrence and significance of motivation-structural rules in some bird and mammal sounds," *Am Nat*, vol. 111, 1977, 855–869.

516. Koutseff A., et al., "The acoustic space of pain: Cries as indicators of distress recovering dynamics in pre-verbal infants," *Bioacoustics*, vol. 27, 2018, 313–325.

517. Filippi P., et al., "Humans recognize emotional arousal in vocalizations across all classes of terrestrial vertebrates: Evidence for acoustic universals," *Proc R Soc B*, vol. 284, 2017, 20170990.

518. Lingle S., and Riede T., "Deer mothers are sensitive to infant distress vocalizations of diverse mammalian species," *Am Nat*, vol. 184, 2014, 510–522.

519. Kelly T., et al., "Adult human perception of distress in the cries of bonobo, chimpanzee, and human infants," *Biol J Lin Soc*, vol. 120, 2017, 919–930.

520. Briefer E. F., "Vocal contagion of emotions in non-human animals," *Proc R Soc B*, vol. 285, 2018, 20172783.

521. If you want to learn a lot about primates' acoustic communication, I advise you to read Fischer J., *Monkeytalk: Inside the Worlds and Minds of Primates*, University of Chicago Press, 2017.

522. The response of chacma baboons to barks of conspecifics is not easy to understand. See these two papers: Fischer J., et al., "Baboon responses to graded bark variants," *Anim Behav*, vol. 61, 2001, 925–931; and Cheney D., et al., "The function and mechanisms underlying baboon 'contact' barks," *Anim Behav*, vol. 52, 1996, 507–518.

523. Vignal C., et al., "Audience drives male songbird response to partner's voice," *Nature*, vol. 430, 2004, 448–451.

524. Perez E. C., et al., "The acoustic expression of stress in a songbird: Does corticosterone drive isolation-induced modifications of zebra finch calls?," *Horm Behav*, vol. 61, 2012, 573–581.

525. Perez E. C., et al., "Physiological resonance between mates through calls as possible evidence of empathic processes in songbirds," *Horm Behav*, vol. 75, 2015, 130–141.

526. Ben-Aderet T., et al., "Dog-directed speech: Why do we use it and do dogs pay attention to it?," *Proc R Soc B*, vol. 284, 2017, 20162429. See also Massenet M., et al., "Nonlinear vocal phenomena affect human perceptions of distress, size and dominance in puppy whines," *Proc R Soc B*, vol. 289, 2022, 20220429.

527. On human-dog vocal communication, see, for instance, Gabor A., et al., "The acoustic basis of human voice identity processing in dogs," *Anim Cogn*, vol. 25, 2022, 905–916; and Massenet et al., "Nonlinear vocal phenomena affect human perceptions of distress."

528. Azhari F., "Bird strike case study at airport level to include take off, landing and taxiways," *Adv J Tech Voc Educ*, vol. 1, 2017, 364–374.

529. Aubin T., and Brémond J.-C., "Parameters used for recognition of distress calls in two species: *Larus argentatus* and *Sturnus vulgaris*," *Bioacoustics*, vol. 2, 1999, 22–33; Aubin T., and Brémond J.-C., "Perception of distress call harmonic structure by the starling (*Sturnus vulgaris*)," *Behaviour*, vol. 120, 1992, 3–4; Brémond J.-C., and Aubin T., "Responses to distress calls by black-headed gulls, *Larus ridibundus*: The role of non-degraded features," *Anim Behav*, vol. 39, 1990, 503–511; Maigrot A. L., et al., "Cross-species discrimination of vocal expression of emotional valence by Equidae and Suidae," *BMC Biol*, vol. 20, 2022, 106.

530. Aubin T., "Why do distress calls evoke interspecific responses? An experimental study applied to some species of birds," *Behav Proc*, vol. 23, 1991, 103–111.

Chapter 16

531. Dentressangle F., et al., "Males use time whereas females prefer harmony: Individual call recognition in the dimorphic blue-footed booby," *Anim Behav*, vol. 84, 2012, 413–420.

532. Janicke T., et al., "Darwinian sex roles confirmed across the animal kingdom," *Sc Adv*, vol. 2, 2016, e1500983.

533. Vignal C., "Biologie du comportement animal," in Mathevon N., and Viennot É. (eds.), *La différence des sexes*, Belin, 2017, 53–80.

534. Green K. K., and Madjidian J. A., "Active males, reactive females: Stereotypic sex roles in sexual conflict research?," *Anim Behav*, vol. 81, 2011, 901–907.

535. Bourgeois K., et al., "Morphological versus acoustic analysis: What is the most efficient method for sexing yelkouan shearwaters *Puffinus yelkouan*?," *J Ornithol*, vol. 148, 2007, 261–269.

536. Curé C., et al., "Sex discrimination and mate recognition by voice in the yelkouan shearwater *Puffinus yelkouan*," *Bioacoustics*, vol. 20, 2011, 235–250; Curé C., et al., "Mate vocal recognition in the Scopoli's shearwater *Calonectris diomedea*: Do females and males share the same acoustic code?," *Behav Process*, vol. 128, 2016, 96–102.

537. Curé C., et al., "Acoustic cues used for species recognition can differ between sexes and sibling species: Evidence in shearwaters," *Anim Behav*, vol. 84, 2012, 239–250; Curé C., et al., "Acoustic convergence and divergence in two sympatric burrowing nocturnal seabirds," *Biol J Lin Soc*, vol. 96, 2009, 115–134; Curé C., et al., "Intra-sex vocal interactions in two hybridizing seabird species (*Puffinus* sp.)," *Behav Ecol Sociobiol*, vol. 64, 2010, 1823–1837.

538. Mouterde S. C., "From vocal to neural encoding: A transversal investigation of information transmission at long distance in birds," in Aubin and Mathevon (eds.), *Coding Strategies in Vertebrate Acoustic Communication*, 203–229.

539. Mouterde S. C., et al., "Acoustic communication and sound degradation: How do the individual signatures of male and female zebra finch calls transmit over distance?," *PLoS ONE*, vol. 9, 2014, e102842.

540. Mouterde S. C., et al., "Learning to cope with degraded sounds: Female zebra finches can improve their expertise in discriminating between male voices at long distances," *J Exp Biol*, vol. 217, 2014, 3169–3177.

541. Interestingly, unlike the call, the zebra finch's song seems to be more of a "short-distance" signal: Loning H., et al., "Zebra finch song is a very short-range signal in the wild: Evidence from an integrated approach," *Behav Ecol*, vol. 33, 2022, 37–46.

542. Mouterde S. C., et al., "Single neurons in the avian auditory cortex encode individual identity and propagation distance in naturally degraded communication calls," *J Neuro*, vol. 37, 2017, 3491–3510.

543. Darwin C., *On the Origin of Species by Means of Natural Selection, or the Preservation of Favoured Races in the Struggle for Life*, John Murray, 1859.

544. Odom K. J., et al., "Female song is widespread and ancestral in songbirds," *Nat Com*, vol. 5, 2014, 3379.

545. Oliveros C. H., et al., "Earth history and the passerine superradiation," *PNAS*, vol. 116, 2019, 7916–7925.

546. Gahr M., "Seasonal hormone fluctuations and song structure of birds," in Aubin and Mathevon (eds.), *Coding Strategies in Vertebrate Acoustic Communication*, 163–201.

547. On this topic, see this recent paper reporting a 17-year study on the vocal behavior of female and male rufous-and-white wrens, *Thryophilus rufalbus*, where both the female and the male sing. For both sexes, the authors found variation in vocal behaviors with time of day and time of year. However, some behavioral differences were noted between sexes, e.g., female wrens change song types more often in areas with more neighbors, whereas the male vocal behavior did not change with the number of neighbors. This result suggests that female wrens may be

using song-type switching in territorial defense against conspecifics more than males: Owen K. C., and Mennill D. J., "Singing in a fragmented landscape: Wrens in a tropical dry forest show sex differences in the effects of neighbours, time of day, and time of year," *J Ornithol*, vol. 162, 2021, 881–893.

548. Odom J. K., et al., "Differentiating the evolution of female song and male-female duets in the New World blackbirds: Can tropical natural history traits explain duet evolution?," *Evolution*, vol. 69, 2015, 839–847; Elie J. E., et al., "Vocal communication at the nest between mates in wild zebra finches: A private vocal duet?," *Anim Behav*, vol. 80, 2010, 597–605; Lemazina A., et al., "The multifaceted vocal duets of white-browed sparrow weavers are based on complex duetting rules," *J Avian Biol*, vol. 52, 2021, e02703.

549. Fortune E. S., et al., "Neural mechanisms for the coordination of duet singing in wrens," *Science*, vol. 334, 2011, 666–670; Coleman M., and Fortune E., "Duet singing in plain-tailed wrens," *Cur Biol*, vol. 28, 2018, R1–R3; Coleman M., et al., "Neurophysiological coordination of duet singing," *PNAS*, vol. 118, 2021, e2018188118.

550. Another study, conducted in the wild, looked at the vocal coordination mechanisms in the white-browed sparrow-weaver, *Plocepasser mahali*: Hoffmann S., et al., "Duets recorded in the wild reveal that interindividually coordinated motor control enables cooperative behavior," *Nat Com*, vol. 10, 2019, 2577.

551. Levréro F., "Éthologie des primates non humains," in Mathevon and Viennot (eds.), *La différence des sexes*, 277–304.

552. Brunton P. J., and Russel J. A., "The expectant brain: Adapting for motherhood," *Nat Rev Neurosci*, vol. 9, 2008, 11–25; Hoekzema E., et al., "Pregnancy leads to long-lasting changes in human brain structure," *Nat Neurosci*, vol. 20, 2017, 287–296; Kim P., et al., "The maternal brain and its plasticity in humans," *Horm Behav*, vol. 77, 2016, 113–123.

553. Hrdy S. B., "The neurobiology of paternal care: Cooperative breeding and the paradox of facultative fathering," in Bridges R. S. (ed.), *Neurobiology of the Parental Brain*, Elsevier, 2008, 407–416.

554. Wiesenfeld A. R., et al., "Differential parental response to familiar and unfamiliar infant distress signals," *Infant Behav Dev*, vol. 4, 1981, 281–295; Green J. A., and Gustafson G. E., "Individual recognition of human infants on the basis of cries alone," *Dev Psychobiol*, vol. 16, 1983, 485–493.

555. Gustafsson E., et al., "Fathers are just as good as mothers at recognizing the cries of their baby," *Nat Com*, vol. 4, 2013, 1698.

556. Bouchet H., et al., "Baby cry recognition is independent of motherhood but improved by experience and exposure," *Proc R Soc B*, vol. 287, 2020, 20192499.

557. Corvin S., et al., "Adults learn to identify pain in babies' cries," *Cur Biol*, vol. 32, 2022, R807–R827.

558. Reby D., et al., "Sex stereotypes influence adults' perception of babies' cries," *BMC Psychol*, vol. 4, 2016, 19.

Chapter 17

559. Marin-Cudraz T., et al., "Acoustic monitoring of rock ptarmigan: A multi-year comparison with point-count protocol," *Ecol Indic*, vol. 101, 2019, 710–719. See also Guibard A., et al., "Influence of meteorological conditions and topography on the active space of mountain birds assessed by a wave-based sound propagation model," *JASA*, vol. 151, 2022, 3703.

560. Ulloa J. S., et al., "Screening large audio datasets to determine the time and space distribution of screaming piha birds in a tropical forest," *Ecol Infor*, vol. 31, 2016, 91–99; Ulloa J. S., et al., "Explosive breeding in tropical anurans: Environmental triggers, community composition and acoustic structure," *BMC Ecol*, vol. 19, 2019, 28; Ducrettet M., et al., "Acoustic monitoring of the white-throated toucan (*Ramphastos tucanus*) in disturbed tropical landscapes," *Biol Conserv*, vol. 245, 2020, 108574; Perez-Granados C., and Schuchmann K. L., "Passive acoustic monitoring of the diel and annual vocal behavior of the black and gold howler monkey," *Am J Primatol*, vol. 83, 2021, e23241; Stowell D., et al., "Automatic acoustic detection of birds through deep learning: The first bird audio detection challenge," *Meth Ecol Evol*, vol. 10, 2018, 368–380; Stowell D., et al., "Automatic acoustic identification of individuals in multiple species: Improving identification across recording conditions," *J R Soc Interface*, vol. 16, 2019, 20180940; Desjonqueres C., "Passive acoustic monitoring as a potential tool to survey animal and ecosystem processes in freshwater environments," *Fresh Biol*, vol. 65, 2020, 7–19; Poupard M., et al., "Passive acoustic monitoring of sperm whales and anthropogenic noise using stereophonic recordings in the Mediterranean Sea, North West Pelagos Sanctuary," *Sc Rep*, vol. 12, 2022, 2007.

561. Parmentier E., et al., "How many fish could be vocal? An estimation from a coral reef (Moorea Island)," *Belg J Zool*, vol. 151, 2021, 1–29.

562. For a discussion on the definition of *soundscape*, see Grinfeder E., et al., "What do we mean by 'soundscape'? A functional description," *Front Ecol Evol*, vol. 10, 2022, 894232.

563. Sethi S. S., et al., "Characterizing soundscapes across diverse ecosystems using a universal acoustic feature set," *PNAS*, vol. 117, 2020, 17049–17055.

564. Lin T. H., et al., "Exploring coral reef biodiversity via underwater soundscapes," *Biol Conserv*, vol. 253, 2021, 108901; Raick X., et al., "From the reef to the ocean: Revealing the acoustic range of the biophony of a coral reef (Moorea Island, French Polynesia)," *J Mar Sc Eng*, vol. 9, 2021, 420.

565. Morrison C. A., et al., "Bird population declines and species turnover are changing the acoustic properties of spring soundscapes," *Nat Com*, vol. 12, 2021, 6217.

566. The CRIOBE is a laboratory bringing together researchers from the CNRS, the University of Perpignan, and the École pratique des hautes études (EPHE). Founded in 1868, the EPHE has counted among its professors the famous Claude Bernard and Claude Lévi-Strauss.

567. Bertucci F., et al., "Local sonic activity reveals potential partitioning in a coral reef fish community," *Oecologia*, vol. 193, 2020, 125–134.

568. Gross M., "Can science rescue coral reefs?," *Cur Biol*, vol. 26, 2016, R481–R492.

569. Lamont T. A., et al., "The sound of recovery: Coral reef restoration success is detectable in the soundscape," *J App Ecol*, vol. 59, 2022, 742–756.

570. Aran Mooney T., et al., "Listening forward: Approaching marine biodiversity assessments using acoustic methods," *R Soc Open Sc*, vol. 7, 2020, 201287.

571. Sueur J., and Farina A., "Ecoacoustics: The ecological investigation and interpretation of environmental sound," *Biosemiotics*, vol. 8, 2015, 493–502.

572. I invite you to watch these few interviews and lectures by Bernie Krause: https://www .ted.com/talks/bernie_krause_the_voice_of_the_natural_world/transcript?language=fr; https://www.youtube.com/watch?v=osgERQKVrhA; https://www.youtube.com/watch?v =RRM5lDPgQXg.

573. Krause B., *The Great Animal Orchestra: Finding the Origins of Music in the World's Wild Places*, Profile Books, 2012.

574. *Anthropos*: Greek word for "human."

575. Duarte C. M., et al., "The soundscape of the Anthropocene ocean," *Science*, vol. 371, 2021, eaba4658; Grinfeder E., et al., "Soundscape dynamics of a cold protected forest: Dominance of aircraft noise," *Lands Ecol*, vol. 37, 2022, 567–582.

576. Sueur J., Krause B., and Farina A., "Climate change is breaking Earth's beat," *Trends Ecol Evol*, vol. 34, 2019, 971–973. In addition, natural soundscapes have been shown to have beneficial effects on human health; on this topic, see Buxton R. T., et al., "A synthesis of health benefits of natural sounds and their distribution in national parks," *PNAS*, vol. 118, 2021, e2013097118. In addition, anthropogenic noise can act in synergy with artificial light: Wilson A. A., et al., "Artificial night light and anthropogenic noise interact to influence bird abundance over a continental scale," *Glob Change Biol*, vol. 27, 2021, 3987–4004.

577. Brumm H., *Animal Communication and Noise*, Springer, 2013.

578. Kunc H. P., and Schmidt R., "The effects of anthropogenic noise on animals: A meta-analysis," *Biol Let*, vol. 15, 2019, 20190649; Raboin M., and Elias D. O., "Anthropogenic noise and the bioacoustics of terrestrial invertebrates," *J Exp Biol*, vol. 222, 2019, jeb178749.

579. The rise of temperature due to climate change may also impact acoustic communication in animals. See Coomes, C. M., and Derryberry, E. P., "High temperatures reduce song production and alter signal salience in songbirds," *Anim Behav*, vol. 180, 2021, 13–22.

580. Faria A., et al., "Boat noise impacts early life stages in the Lusitanian toadfish: A field experiment," *Sc Total Environ*, vol. 811, 2022, 151367.

581. Recent research shows that noise pollution interferes with cognitive functions in birds: Osbrink A., et al., "Traffic noise inhibits cognitive performance in a songbird," *Proc R Soc B*, vol. 288, 2021, 20202851. Another study demonstrates that bats use more sonar pulses when hunting in a noisy environment: Allen L. C., et al., "Noise distracts foraging bats," *Proc R Soc B*, vol. 288, 2021, 20202689. See this study, which demonstrates that birds and bats avoid areas with high sound levels: Gomes D.G.E., et al., "Phantom rivers filter birds and bats by acoustic niche," *Nat Com*, vol. 12, 2021, 3029. Also have a look at this study, which shows that while urbanization affects territorial and vocal behaviors in southern house wrens, *Troglodytes aedon musculus*, noise does not seem to alter birds' vocal behavior: Diniz P., and Duca C., "Anthropogenic noise, song, and territorial aggression in southern house wrens," *J Avian Biol*, 2021, e02846. Finally, this review examines the effect of anthropogenic noise on a range of vertebrates: Gomes L., et al., "Influence of anthropogenic sounds on insect, anuran and bird acoustic signals: A meta-analysis," *Front Ecol Evol*, vol. 10, 2022, 827440.

582. Wale M. A., et al., "From DNA to ecological performance: Effects of anthropogenic noise on a reef-building mussel," *Sc Total Environ*, vol. 689, 2019, 126–132.

583. Buxton et al., "A synthesis of health benefits of natural sounds."

584. Popper A. N., et al., "Taking the animals' perspective regarding anthropogenic underwater sound," *Trends Ecol Evol*, vol. 35, 2020, 787–794; Tougaard, J., "Thresholds for noise induced hearing loss in marine mammals: Background note to revision of guidelines from the Danish Energy Agency," Technical report no. 28, Aarhus University, DCE—Danish Centre for Environment and Energy, 2021.

585. Wensveen P. J., et al., "Northern bottlenose whales in a pristine environment respond strongly to close and distant navy sonar signals," *Proc R Soc B*, vol. 286, 2019, 20182592.

586. For a review on the effect of marine noise pollution, see Di Franco E., et al., "Effects of marine noise pollution on Mediterranean fishes and invertebrates: A review," *Mar Pollut Bull*, vol. 159, 2020, 111450.

587. A recent study shows by using playback experiments that cetacean species that are more responsive to predator presence (orcas) are also those that react the most to anthropogenic noise: Miller P.J.O., et al., "Behavioral responses to predatory sounds predict sensitivity of cetaceans to anthropogenic noise within a soundscape of fear," *PNAS*, vol. 119, 2022, e2114932119.

588. Ferrier-Pages C., et al., "Noise pollution on coral reefs? A yet underestimated threat to coral reef communities," *Mar Pollut Bull*, vol. 165, 2021, 112129; Van der Knaap I., et al., "Effects of a seismic survey on movement of free-ranging Atlantic cod," *Cur Biol*, vol. 31, 2021, 1555–1562; Leduc A.O.H., et al., "Land-based noise pollution impairs reef fish behavior: A case study with a Brazilian carnival," *Biol Conserv*, vol. 253, 2021, 108910; Vieira M., et al., "Boat noise affects meagre (*Argyrosomus regius*) hearing and vocal behaviour," *Mar Pollut Bulletin*, vol. 172, 2021, 112824.

589. Hatch L. T., et al., "Quantifying loss of acoustic communication space for right whales in and around a US national marine sanctuary," *Conserv Biol*, vol. 26, 2012, 983–994. For an example in fish, see Alves D., et al., "Boat noise interferes with Lusitanian toadfish acoustic communication," *J Exp Biol*, vol. 224, 2021, jeb234849.

590. McCauley R. D., et al., "Widely used marine seismic survey air gun operations negatively impact zooplankton," *Nat Ecol Evol*, vol. 1, 2017, 0195.

591. A study suggests that anthropogenic noise may negatively impact the seagrass *Posidonia*: Sole M., et al., "Seagrass *Posidonia* is impaired by human-generated noise," *Com Biol*, vol. 4, 2021, 743.

592. Klingbeil B. T., et al., "Geographical associations with anthropogenic noise pollution for North American breeding birds," *Glob Ecol Biogeog*, vol. 29, 2020, 148–158.

593. Brumm H., "The impact of environmental noise on song amplitude in a territorial bird," *J Anim Ecol*, vol. 73, 2004, 434–440.

594. Guazzo R. A., et al., "The Lombard effect in singing humpback whales: Source levels increase as ambient ocean noise levels increase," *JASA*, vol. 148, 2020, 542.

595. Fournet M.E.H., et al., "Limited vocal compensation for elevated ambient noise in bearded seals: Implications for an industrializing Arctic Ocean," *Proc R Soc B*, vol. 288, 2021, 20202712.

596. Zhao L., et al., "Differential effect of aircraft noise on the spectral-temporal acoustic characteristics of frog species," *Anim Behav*, vol. 182, 2021, 9–18.

597. Slabbekoorn H., and Peet M., "Birds sing at a higher pitch in urban noise," *Nature*, vol. 424, 2003, 267.

598. Nemeth E., et al., "Bird song and anthropogenic noise: Vocal constraints may explain why birds sing higher-frequency songs in cities," *Proc R Soc B*, vol. 280, 2013, 20122798.

599. Zollinger S. A., et al., "Higher songs of city birds may not be an individual response to noise," *Proc R Soc B*, vol. 284, 2017, 20170602.

600. Derryberry E. P., et al., "Singing in a silent spring: Birds respond to a half-century soundscape reversion during the COVID-19 shutdown," *Science*, vol. 370, 2020, 575–579.

601. Gallego-Abenza M., et al., "Experience modulates an insect's response to anthropogenic noise," *Behav Ecol*, vol. 31, 2020, 90–96.

602. Bent A. M., et al., "Anthropogenic noise disrupts mate choice behaviors in female *Gryllus bimaculatus*," *Behav Ecol*, vol. 32, 2021, 201–210.

603. Hanache P., et al., "Noise-induced reduction in the attack rate of a planktivorous freshwater fish revealed by functional response analysis," *Fresh Biol*, vol. 65, 2020, 75–85; Rojas E., et al., "From distraction to habituation: Ecological and behavioural responses of invasive fish to anthropogenic noise," *Fresh Biol*, vol. 66, 2021, 1606–1618.

604. See, however, Francis C. D., et al., "Noise pollution alters ecological services: Enhanced pollination and disrupted seed dispersal," *Proc R Soc B*, vol. 279, 2012, 2727–2735; Francis C. D., et al., "Noise pollution changes avian communities and species interactions," *Cur Biol*, vol. 19, 2009, 1415–1419; and Francis C. D., et al., "Noise pollution filters bird communities based on vocal frequency," *PLoS ONE*, vol. 6, 2011, e27052.

605. Gordon T.A.C., et al., "Habitat degradation negatively affects auditory settlement behavior of coral reef fishes," *PNAS*, vol. 115, 2018, 5193–5198.

606. Lecchini D., et al., "Boat noise prevents soundscape-based habitat selection by coral planulae," *Sc Rep*, vol. 8, 2018, 9283.

607. See, for instance, Morrison C. A., et al., "Bird population declines and species turnover are changing the acoustic properties of spring soundscapes," *Nat Com*, vol. 12, 2022, 6217; and Rappaport D. I., et al., "Animal soundscapes reveal key markers of Amazon forest degradation from fire and logging," *PNAS*, vol. 119, 2022, e2102878119.

608. Farina A., *Soundscape Ecology: Principles, Patterns, Methods and Applications*, Springer, 2014; Mullet T. C., et al., "The acoustic habitat hypothesis: An ecoacoustics perspective on species habitat selection," *Biosemiotics*, vol. 10, 2017, 319–336.

609. Begon M., Townsend C. R., and Harper J. L., *Ecology, from Individuals to Ecosystems*, 4th ed., Blackwell, 2006.

610. Krause B. L., "Niche hypothesis: A virtual symphony of animal sounds, the origins of musical expression and the health of habitats," *Sound Newslet*, vol. 6, 1993, 6–10.

611. Sueur J., et al., "Acoustic indices for biodiversity assessment and landscape investigation," *Acta Acust Acust*, vol. 100, 2014, 772–781.

612. Brémond J.-C., "Acoustic competition between the song of the wren (*Troglodytes troglodytes*) and the songs of other species," *Behaviour*, vol. 65, 1978, 89–98.

613. Amezquita A., et al., "Acoustic interference and recognition space within a complex assemblage of dendrobatid frogs," *PNAS*, vol. 108, 2011, 17058–17063; Tobias J. A., et al., "Species interactions and the structure of complex communication networks," *PNAS*, vol. 111, 2014, 1020–1025; Sueur J., "Cicada acoustic communication: Potential sound partitioning in a multispecies community from Mexico (Hemiptera: Cicadomorpha: Cicadidae)," *Biol J Lin Soc*, vol. 75, 2002, 379–394; Luther D., "The influence of the acoustic community on songs of birds in a neotropical rain forest," *Behav Ecol*, vol. 20, 2009, 864–871; Schmidt A.K.D., and Balakrishnan R., "Ecology of acoustic signalling and the problem of masking interference in insects," *J Comp Physiol A*, vol. 201, 2014, 133–142; Ruppé L., et al., "Environmental constraints drive

the partitioning of the soundscape in fishes," *PNAS*, vol. 12, 2015, 6092–6097; Schmidt A.K.D., et al., "Spectral niche segregation and community organization in a tropical cricket assemblage," *Behav Ecol*, vol. 24, 2012, 470–480; Balakrishnan R., "Behavioral ecology of insect acoustic communication," in Pollack G. S., Mason A. C., Popper A., and Fay R. R. (eds.), *Insect Hearing*, Springer, 2016, 49–80; Chitnis S. S., et al., "Sympatric wren-warblers partition acoustic signal space and song perch height," *Behav Ecol*, vol. 31, 2020, 559–567; Bertucci F., et al., "Local sonic activity reveals potential partitioning in a coral reef fish community," *Oecologia*, vol. 193, 2020, 125–134; Allen-Ankins S., and Schwarzkopf L., "Spectral overlap and temporal avoidance in a tropical savannah frog community," *Anim Behav*, vol. 180, 2021, 1–11; Allen-Ankins S., and Schwarzkopf L., "Using citizen science to test for acoustic niche partitioning in frogs," *Sc Rep*, vol. 12, 2022, 2447.

614. Boncoraglio G., and Saino N., "Habitat structure and the evolution of bird song: A meta-analysis of the evidence for the acoustic adaptation hypothesis," *Funct Ecol*, vol. 21, 2007, 134–142; Ey E., and Fischer J., "The 'acoustic adaptation hypothesis'—a review of the evidence from birds, anurans and mammals," *Bioacoustics*, vol. 19, 2009, 21–48.

615. Or the sender uses a strategy that reinforces the range of its signals, such as certain insects that position themselves judiciously in the environment. See Montealegre-Z F., et al., "Generation of extreme ultrasonics in rainforest katydids," *J Exp Biol*, vol. 209, 2006, 4923–4937.

616. Morton E. S., "Ecological sources of selection on avian sounds," *Am Nat*, vol. 109, 1975, 17–34; Morton E. S., "On the occurrence and significance of motivation-structural rules in some bird and mammal sounds," *Am Nat*, vol. 111, 1977, 855–869; Marten K., and Marler P., "Sound transmission and its significance for animal vocalization," *Behav Ecol Sociobiol*, vol. 2, 1977, 271–290; Wiley R. H., and Richards D. G., "Physical constraints on acoustic communication in the atmosphere: Implications for the evolution of animal vocalizations," *Behav Ecol Sociobiol*, vol. 3, 1978, 69–94; Richards D. G., and Wiley R. H., "Reverberations and amplitude fluctuations in the propagation of sound in a forest: Implications for animal communication," *Am Nat*, vol. 115, 1980, 381–399.

617. Klump G. M., "Bird communication in the noisy world," in Barth F. G., and Schmid A., *Ecology of Sensing—Ecology and Evolution of Acoustic Communication in Birds*, Cornell University Press, 1996, 321–338; Zhao L., et al., "Noise constrains the evolution of call frequency contours in flowing water frogs: A comparative analysis in two clades," *Front Zool*, vol. 18, 2021, 37.

618. Goutte S., et al., "How the environment shapes animal signals: A test of the acoustic adaptation hypothesis in frogs," *J Evol Biol*, vol. 31, 2017, 148–158.

619. Mikula P., et al., "A global analysis of song frequency in passerines provides no support for the acoustic adaptation hypothesis but suggests a role for sexual selection," *Ecol Let*, vol. 24, 2020, 477–486.

620. Riondato I., et al., "Allometric escape and acoustic signal features facilitate high-frequency communication in an endemic Chinese primate," *J Comp Physiol A*, vol. 207, 2021, 327–336.

621. This integrative side of ecoacoustics is reminiscent of ecological monitoring methods, such as biotic indices, where insect larvae present in a watercourse are counted. The quality of the water is deduced from the species present and their density. The advantage of ecoacoustics is that no living organisms are taken from the water.

622. The International Bioacoustics Society (IBAC) organizes an international scientific conference every two years. If you are interested, you can have a look on the society website: https://www.ibac.info/.

623. For a recent review on acoustic indices, see Alcocer I., et al., "Acoustic indices as proxies for biodiversity: A meta-analysis," *Biol Rev*, vol. 97, 2022, 2209–2236.

624. Sueur J., et al., "Acoustic indices for biodiversity assessment and landscape investigation," *Acta Acust Acust*, vol. 100, 2014, 772–781; Sueur J., et al., "Acoustic biodiversity," *Cur Biol*, vol. 31, 2021, R1141–R1224.

625. See, for example, these studies: Depraetere M., et al., "Monitoring animal diversity using acoustic indices: Implementation in a temperate woodland," *Ecol Indic*, vol. 13, 2012, 46–54; Gasc A., et al., "Assessing biodiversity with sound: Do acoustic diversity indices reflect phylogenetic and functional diversities of bird communities?," *Ecol Indic*, vol. 25, 2013, 279–287; Linke S., et al., "Freshwater ecoacoustics as a tool for continuous ecosystem monitoring," *Front Ecol Environ*, vol. 16, 2018, 231–238; Ulloa J. S., et al., "Estimating animal acoustic diversity in tropical environments using unsupervised multiresolution analysis," *Ecol Indic*, vol. 90, 2018, 346–355; Van der Lee G. H., et al., "Freshwater ecoacoustics: Listening to the ecological status of multi-stressed lowland waters," *Ecol Indic*, vol. 113, 2020, 106252; and Flowers C., et al., "Looking for the -scape in the sound: Discriminating soundscapes categories in the Sonoran Desert using indices and clustering," *Ecol Indic*, vol. 127, 2021, 107805.

626. Efforts are currently being made to develop automated monitoring of soundscapes using global ecosystem monitoring. For instance, see Sethi S., et al., "Characterizing soundscapes across diverse ecosystems using a universal acoustic feature set," *PNAS*, vol. 117, 2020, 17049–17055.

627. Ferrier-Pages C., et al., "Noise pollution on coral reefs? A yet underestimated threat to coral reef communities," *Mar Pollut Bull*, vol. 165, 2021, 112129.

628. The acoustic complexity index (ACI) was developed based on the fact that most sounds produced by animals have varying intensities over time, while noises—and in particular anthropogenic noises—are more constant. When biodiversity is poor, the ACI is low. It rises with increasing numbers of sound-producing animal species: Pieretti, N., "A new methodology to infer the singing activity of an avian community: The acoustic complexity index (ACI)," *Ecol Indic*, vol. 11, 2011, 868–873; Bolgan M., et al., "Acoustic complexity of vocal fish communities: A field and controlled validation," *Sc Rep*, vol. 8, 2018, 10559.

629. Bertucci F., et al., "A preliminary acoustic evaluation of three sites in the lagoon of Bora Bora, French Polynesia," *Environ Biol Fish*, vol. 103, 2020, 891–902.

630. Elise S., et al., "An optimised passive acoustic sampling scheme to discriminate among coral reefs' ecological state," *Ecol Indic*, vol. 107, 2019, 105627.

631. https://youtube/WK0sR7e4F0Y.

632. Each index was calculated on five different frequency bands: 0.1–0.5 kHz; 0.5–1 kHz; 1–2 kHz; 2–7 kHz; and the full bandwidth, 0–50 kHz. For details, see Elise S., et al., "Assessing key ecosystem functions through soundscapes: A new perspective from coral reefs," *Ecol Indic*, vol. 107, 2019, 105623.

633. Dimoff S. A., et al., "The utility of different acoustic indicators to describe biological sounds of a coral reef soundscape," *Ecol Indic*, vol. 124, 2021, 107435.

634. The assessment of stress due to anthropogenic noise can lead to noise mitigation measures: Nedelec S. L., et al., "Limiting motorboat noise on coral reefs boosts fish reproductive success," *Nat Com*, vol. 13, 2022, 2822.

635. Gross M., "Listening to the sounds of the biosphere," *Cur Biol*, vol. 28, 2018, R847–R870.

636. Lin T. H., et al., "Exploring coral reef biodiversity via underwater soundscapes," *Biol Conserv*, vol. 253, 2018, 108901.

637. Stowell D., "Computational bioacoustics with deep learning: A review and roadmap," *PeerJ*, vol. 10, 2022, e13152.

638. Such ecoacoustic programs are already in place. See the site http://ear.cnrs.fr/ for examples.

639. The ecoacoustic approach allows us to evaluate functional aspects of ecosystems. See, for example, Folliot A., et al., "Using acoustics and artificial intelligence to monitor pollination by insects and tree use by woodpeckers," *S Tot Envir*, vol. 838, 2022, 155883. Soundscapes may even be useful for assessing soil biodiversity: Maeder M., et al., "Temporal and spatial dynamics in soil acoustics and their relation to soil animal diversity," *PLoS ONE*, vol. 17, 2022, e0263618.

640. Sueur, Krause, and Farina, "Climate change is breaking Earth's beat."

641. Mbu Nyamsi R. G., et al., "On the extraction of some time dependent parameters of an acoustic signal by means of the analytic signal concept: Its application to animal sound study," *Bioacoustics*, vol. 5, 1994, 187–203.

642. True. See De Novion C., et al., "L'impact des concepts de Pierre-Gilles de Gennes sur l'innovation en France dans le domaine des matériaux," *Reflets Phys*, vol. 56, 2018, 10–19.

Chapter 18

643. Hauser M. D., et al., "The faculty of language: What is it, who has it, and how did it evolve?," *Science*, vol. 298, 2002, 1569–1579.

644. Prat Y., "Animals have no language, and humans are animals too," *Pers Psychol Sc*, vol. 14, 2019, 885–893.

645. Recent research is attempting to find large-scale organizational rules in animal vocalizations. See, for example, Markowitz J. E., et al., "Long-range order in canary song," *PLoS Comput Biol*, vol. 9, 2013, e1003052; and Sainburg T., et al., "Parallels in the sequential organization of birdsong and human speech," *Nat Com*, vol. 10, 2019, 3636.

646. I have translated into modern-day English what Montaigne wrote. The original quote is as follows: "*Ce defaut qui empesche la communication d'entre elles et nous, pourquoy n'est il aussi bien à nous qu'à elles? C'est à deviner à qui est la faute de ne nous entendre point: Car nous ne les entendons non plus qu'elles nous. Par ceste mesme raison elles nous peuvent estimer bestes, comme nous les estimons*" (Montaigne 1595, Essais Livre II).

647. Pisanski K., et al., "Voice modulation: A window into the origins of human vocal control?," *Trends Cogn Sci*, vol. 20, 2016, 304–318.

648. Favaro L., et al., "Do penguins' vocal sequences conform to linguistic laws?," *Biol Let*, vol. 16, 2020, 20190589.

649. Huang M., et al., "Male gibbon loud morning calls conform to Zipf's law of brevity and Menzerath's law: Insights into the origin of human language," *Anim Behav*, vol. 160, 2020, 145–155; Valente D., et al., "Linguistic laws of brevity: Conformity in *Indri indri*," *Anim Cogn*, vol. 24, 2021, 897–906.

650. Pougnault L., Levréro F., and Lemasson A., "Conversation among primate species," in Masataka N. (ed.), *The Origins of Language Revisited*, vol. II, Springer, 2020, 73–96.

651. Ravignani A., et al., "Interactive rhythms across species: The evolutionary biology of animal chorusing and turn-taking," *Ann NY Acad Sc*, vol. 1453, 2019, 12–21; Banerjee A., and Vallentin D., "Convergent behavioral strategies and neural computations during vocal turn-taking across diverse species," *Cur Opin Neurobiol*, vol. 73, 2022, 102529.

652. Cornec C., et al., "A pilot study of calling patterns and vocal turn-taking in wild bonobos *Pan paniscus*," *Ethol Ecol Evol*, vol. 34, 2022, 360–377.

653. Pougnault L., et al., "Temporal calling patterns of a captive group of chimpanzees (*Pan troglodytes*)," *Int J Primatol*, vol. 42, 2021, 809–832.

654. Demartsev V., et al., "Vocal turn-taking in meerkat group calling sessions," *Cur Biol*, vol. 28, 2018, 3661–3666.

655. Pougnault L., et al., "Breaking conversational rules matters to captive gorillas: A playback experiment," *Sc Rep*, vol. 10, 2020, 6947.

656. An interesting perspective that reviews linguistic laws and explores the potential relevance of these laws across all biological levels is Semple S., et al., "Linguistic laws in biology," *Trends Ecol Evol*, vol. 37, 2022, 53–66.

657. Schlenker P., et al., "What do monkey calls mean?," *Trends Cogn Sc*, vol. 20, 2016, 894–904; Fischer J., "Primate vocal communication and the evolution of speech," *Cur Dir Psychol Sc*, vol. 30, 2021, 55–60; Pougnault L., et al., "Social pressure drives "conversational rule" in great apes," *Biol Rev*, vol. 97, 2022, 749–765.

658. Fitch W. T., "Animal cognition and the evolution of human language: Why we cannot focus solely on communication," *Phil Trans R Soc B*, vol. 375, 2019, 20190046.

659. Lyn H., et al., "Apes and the evolution of language: Taking stock of 40 years of research," in Vonk J., and Shackelford T. K. (eds.), *Oxford Handbook of Comparative Evolutionary Psychology*, Oxford University Press, 2012, 356–378; Krause M. A., and Beran M. J., "Words matter: Reflections on language projects with chimpanzees and their implications," *Am J Primat*, vol. 82, 2020, e23187.

660. Hayes K. J., and Hayes C., "The intellectual development of a home-raised chimpanzee," *Proc Amer Phil Soc*, vol. 95, 1951, 105–109.

661. Hayes C., *The Ape in Our House*, Harper, 1951.

662. Gardner R. A., and Gardner B. T., "Teaching sign language to a chimpanzee," *Science*, vol. 165, 1969, 664–672.

663. Terrace H. S., *Why Chimpanzees Can't Learn Language and Only Humans Can*, Columbia University Press, 2019.

664. Terrace H. S., "Can an ape create a sentence?," *Science*, vol. 206, 1979, 891–902.

665. There were also experiments with an orangutan and a gorilla. See Shettleworth S. J., *Cognition, Evolution and Behavior*, Oxford University Press, 2010.

666. Matsuzawa T., "Use of numbers by a chimpanzee," *Nature*, vol. 315, 1985, 57–59.

667. On the different projects to teach apes how to talk, see this review: Ristau C. A., and Robbins D., "Language in the great apes: A critical review," *Adv St Behav*, vol. 12, 1982, 141–255. For a recent point of view, see Bergman T. J., et al., "The speech-like properties of nonhuman primate vocalizations," *Anim Behav*, vol. 151, 2019, 229–237.

668. Pepperberg I. M., *The Alex Studies*, Harvard University Press, 2000.

669. Bodin C., et al., "Functionally homologous representation of vocalizations in the auditory cortex of humans and macaques," *Cur Biol*, vol. 31, 2021, 4839–4844.

670. Lieberman P. H., et al., "Vocal tract limitations on the vowel repertoires of rhesus monkey and other nonhuman primates," *Science*, vol. 164, 1969, 1185–1187.

671. Fitch W. T., et al., "Monkey vocal tracts are speech-ready," *Sc Adv*, vol. 2, 2016, e1600723; Grawunder S., et al., "Chimpanzee vowel-like sounds and voice quality suggest formant space expansion through the hominoid lineage," *Phil Trans R Soc*, vol. 377, 2021, 20200455.

672. Meyer J., "Coding human languages for long-range communication in natural ecological environments: Shouting, whistling, and drumming," in Aubin T., and Mathevon N. (eds.), *Coding Strategies in Vertebrate Acoustic Communication*, Springer, 2020, 91–113.

673. Vargha-Khadem F., et al., "FOXP2 and the neuroanatomy of speech and language," *Nat Rev Neurosc*, vol. 6, 2005, 131–138.

674. Lai C. S., et al., "A forkhead-domain gene is mutated in a severe speech and language disorder," *Nature*, vol. 413, 2002, 519–523; Fisher S. E., et al., "Localisation of a gene implicated in a severe speech and language disorder," *Nat Genet*, vol. 18, 1998, 168–170; Liégeois F., et al., "Language fMRI abnormalities associated with FOXP2 gene mutation," *Nat Neurosc*, vol. 6, 2003, 1230–1237.

675. Enard W., et al., "Molecular evolution of FOXP2, a gene involved in speech and language," *Nature*, vol. 418, 2002, 869–872.

676. Fisher S. E., and Scharff C., "FOXP2 as a molecular window into speech and language," *Trends Gen*, vol. 25, 2009, 166–177.

677. Haesler S., et al., "FOXP2 expression in avian vocal learners and non-learners," *J Neurosc*, vol. 24, 2004, 3164–3175.

678. For instance, a study has just shown that FOXP1, the little brother of FOXP2, plays a role in the cultural transmission of vocalizations in the zebra finch: Garcia-Oscos F., et al., "Autism-linked gene FoxP1 selectively regulates the cultural transmission of learned vocalizations," *Sc Adv*, vol. 7, 2021, eabd2827.

679. Pfenning A. R., et al., "Convergent transcriptional specializations in the brains of humans and song-learning birds," *Science*, vol. 346, 2014, 1256846.

680. Fisher S. E., "Human genetics: The evolving story of FOXP2," *Cur Biol*, vol. 29, 2019, R50–R70.

681. If you wish to delve deeper into the evolutionary history of human spoken language, I recommend Fitch's clear, well-documented, and very accessible article, "The biology and evolution of speech: A comparative analysis," *Annu Rev Linguist*, vol. 4, 2018, 255–279; and the book by the same author, *The Evolution of Language*, Cambridge University Press, 2012.

682. Darwin C., *The Descent of Man and Selection in Relation to Sex*, John Murray, 1871.

683. For a discussion about our understanding of human origins and evolution, see Richerson P. J., et al., "Modern theories of human evolution foreshadowed by Darwin's *Descent of Man*," *Science*, vol. 372, 2021, eaba3776.

684. Nishimura T., et al., "Evolutionary loss of complexity in human vocal anatomy as an adaptation for speech," *Science*, vol. 377, 2022, 760–763; Gouzoules H., "When less is more in the evolution of language," *Science*, vol. 377, 2022, 706–707.

685. A recent study suggests that auditory abilities of Neanderthals were comparable to those of modern humans, allowing a similar vocal communication system: Conde-Valverde M., et al., "Neanderthals and *Homo sapiens* had similar auditory and speech capacities," *Nat Ecol Evol*, vol. 5, 2021, 609–615.

686. Dunbar R.I.M., "Group size, vocal grooming and the origins of language," *Psychon Bull Rev*, vol. 24, 2017, 209–212.

687. Mathevon N., and Aubin T., "Acoustic coding strategies through the lens of the mathematical theory of communication," in Aubin and Mathevon (eds.), *Coding Strategies in Vertebrate Acoustic Communication*, 1–10.

688. Lehmann J., et al., "Group size, grooming and social cohesion in primates," *Anim Behav*, vol. 74, 2007, 1617–1629.

689. Dunbar R., *Grooming, Gossip, and Evolution of Language*, Harvard University Press, 1996.

690. Knight C., and Lewis J., "Towards a theory of everything," in Power C., Finnegan M., and Callan H. (eds.), *Human Origins: Contributions from Social Anthropology*, Berghahn, 2017, 84–102.

691. Knight C., and Lewis J. D., "Wild voices: Mimicry, reversal, metaphor, and the emergence of language," *Cur Anthropol*, vol. 58, 2017, 435–453.

692. Darwin, *Descent of Man*.

693. I won't go into the complex topic of the origin of music. For an interesting perspective, see Leongomez J. D., et al., "Musicality in human vocal communication: An evolutionary perspective," *Phil Trans R Soc B*, vol. 377, 2021, 20200391. Also see this paper: De Gregorio C., et al., "Categorical rhythms in a singing primate," *Cur Biol*, vol. 31, 2021, R1379–R1380.

694. Spottiswoode C. N., et al., "Reciprocal signaling in honeyguide-human mutualism," *Science*, vol. 353, 2016, 387–389; van der Wal J. E., et al., "Awer honey-hunting culture with greater honeyguides in coastal Kenya," *Front Conserv Sc*, vol. 2, 2022, 727479. See also, from the BBC: https://www.bbc.co.uk/sounds/play/b07z43f8.

695. The study of the origins of human language is a very active field because many questions remain unanswered. See, for example, Taylor D., et al., "Vocal functional flexibility: What it is and why it matters," *Anim Behav*, vol. 186, 2022, 93–100.

696. Cardinal S., et al., "The evolution of floral sonification, a pollen foraging behavior used by bees (*Anthophila*)," *Evolution*, vol. 72, 2018, 590–600.

697. Khait I., et al., "Sound perception in plants," *Sem Cell Dev Biol*, vol. 92, 2019, 134–138.

698. Veits M., et al., "Flowers respond to pollinator sound within minutes by increasing nectar sugar concentration," *Ecol Let*, vol. 22, 2019, 1483–1492. But see Pyke G. H., et al., "Changes in floral nectar are unlikely adaptive responses to pollinator flight sound," *Ecol Let*, vol. 23, 2020, 1421–1422.

Acknowledgments

699. French academics devote many hours to teaching (more than twice as many as their American, British, or Swiss colleagues) and are in great demand for administrative tasks.

INDEX

Académie des Sciences, 37

acoustic, 7

acoustic adaptation, 270

acoustic adaptation hypothesis, 24, 272–73

acoustic avoidance hypothesis, 30–31

acoustic communication, 2; bioacoustics, 3; deer family, 120; emitter, 32; finch species, 137–38; fish, 129–32; forest environment and, 30–31; information exchange, 30; noise, 32; processes, 128; receiver, 33, 127; sender, 32, 127

acoustic communication networks: audience effect in, 153–54; blue-capped cordon-bleu (*Uraeginthus cyanocephalus*), 152–53; characterization of signatures, 144–45; eavesdropping on conversations, 155–57; *Lipaugus* project and, 147; *Lipaugus vociferans* (piha), 139–42; social intelligence, 154–55; sound triangulation, 143–44; zebra finches, 147–52. See also *Lipaugus vociferans* (piha)

acoustic complexity index (ACI), 350n628

acoustic indices, coral reefs, 276–77

acoustic indices of biodiversity, Sueur and development of, 275–76

acoustic niche, 270; nuance of hypothesis, 272

acoustic niche hypothesis, 30, 270; sound frequencies, 270–71

acoustic signals: amphibians, 106–8; bats, 103; categorical perception, 109–10; courtship, 104–5; crocodiles, 111–12; ecological selection and, 134; evolution of ears, 100; far sound field, 101, 105; frogs, 107–8, 111–12; inner ear, 105–6; insects, 102; long-distance ears, 101, 103; near sound field, 101, 105; physics of, 100–101; precedence effect, 112; pressure-differential ears, 107; producing, 99; resonator, 108; short-distance ears, 101, 104; Sphingidae family, 102; transduction, 100; use of, 99; vertebrates, 105–9

Acrocephalus scirpaceus (Eurasian reed warbler), begging calls of chicks, 54–55

Adam, Olivier: mother-offspring communication, 73; whale acoustic communications, 69

Adam's apple, 42, 116, 120, 323n227

Adélie penguin (*Pygoscelis adeliae*), 44. See also penguins

African elephant (*Loxodonta africana*), 117; vocal learning, 176

African penguin (*Spheniscus demersus*): calls of, 283–84; illustration of, 285

Agami herons, research program on, 91

Agrégation, French exam, 149

air pressure variations, sound waves and, 11–12

Amazonian rainforest, Brazil, 139

American alligator (*Alligator mississippiensis*), recordings of, 114

American robin (*Turdus migratorius*), 4, 6, 17

Amolops tormotos, frog, 196–97

Amorim, Clara, fish sound signals, 131

amphibians: inner ear, 105–6; relationship between ears and lungs, 107–8

Amphiprion clarkii (clownfish), sonic ligament of, 130

amplitude, intensity of sound, 12–13

Amsterdam fur seal (*Arctocephalus tropicalis*), 52. *See also* seals

Animal Behaviour (journal), 248

animal language, external observers of, 282

animals: communication, 2; expressing emotions in calls, 287; language in, 280–81; mimicking cries of, 294; rules for encoding emotions, 236–37

animals' communication, noise of human activities, 264–65

animal sound, categories of information, 19

animal species, definition, 127

Anomaloglossus beebei (golden rocket frog), 197

Año Nuevo Reserve. *See* elephant seals

Anthochaera chrysoptera (waterbird), 173; regent honeyeater singing song of, 173

Anthochaera phrygia (regent honeyeater), songs of, 173

anthropogenic noise: acoustic complexity index (ACI), 350n628; animal comfort with, 265–66; assessment of stress, 351n634; birds and, 266; sound environment change, 269

anthropogenic sound, 265

anthropophonic cacophony, 266

anthropophony, 276; ecological balance, 269; soundscape, 270; term, 263, 264

ants, social insects, 198–99

apes, attempts teaching, to talk, 287–89

A phrases, 125

Aptenodytes patagonicus (king penguins), 38–39, 40, 43, 44

aquariums, noise and fish behavior, 268

Arctocephalus galapagoensis (Galápagos baby fur seal), 52. *See also* seals

Arctocephalus tropicalis (Amsterdam fur seal), 52. *See also* seals

arms race, 314n85; between emitter and receiver, 61; between honesty and cheating, 124; between parasites and their hosts, 63; between parents and young, 123; between superb fairy wren and Horsfield cuckoo, 64; definition, 305

arousal, dimension of emotions, 231, 233

Asian elephants, vocal learning, 176–77

Astaire, Fred, 153

Astyanax mexicanus (astyanax fish), 135

atmospheric pressure, 10, 11

Attia, Joël: acoustic signals by astyanax, 135; acoustic signals by fish, 130

Aubin, Thierry, 91, 114, 231, 244; acoustics course, 8; bioacoustics of frogs, 108; bird recordings, 24; blue-footed booby, 247; calls of fairy wren, 220–22; on distress calls of young of other species, 230; distress information, 242–43; fairy wren, 219; on feeling for sound, 9; interest in *Lipaugus*, 139; Nile crocodile vocalizations, 113; penguins, 38, 40, 149; scorpion in shoe, 143; spectacled caimans, 95; walrus mothers and young, 50–51; warblers' territories, 15, 16, 20

audience effect, 218; acoustic communication networks, 153–54

audition, hearing, 99, 100

Australian sea lion (*Neophoca cinerea*), 51

Balaena mysticetus (bowhead whale), 70. *See also* baleen whales

Balaenoptera acutorostrata (minke whale), 74

Balaenoptera musculus (blue whale), 70. *See also* baleen whales

Balaenoptera physalus (fin whale), 70. *See also* baleen whales

baleen whales, 68–69; blue whale (*Balaenoptera musculus*), 70; bowhead (*Balaena mysticetus*), 49, 70; fin whale (*Balaenoptera physalus*), 70; humpback, 71–72, 73; humpback (*Megaptera novaeangliae*), 70;

vocal learning, 175; vocal repertoire of, 70. *See also* whales

Balistidae family, 263

Barbastella barbastellus (western barbastelle), ultrasound, 195

barn owl (*Tyto alba*), parent-offspring study model, 58–59

barometric pressure, 10

Barrington, Daines, goldfinches, 159

Basileuterus leucoblepharus. See white-browed warbler (*Basileuterus leucoblepharus*)

bats: Doppler effect, 193–94; echolocation, 191–92, 335n426; emitting ultrasound, 187; greater mouse-eared bat (*Myotis myotis*), 195; long-distance ears in insects, 103; moths and, 195–96; night flying, 102–3; ultrasonic world of, 336n439; vocal learning, 177–78; western barbastelle (*Barbastella barbastellus*), 195

beach evening primrose (*Oenothera drummondii*), 295

bearded seal (*Erignathus barbatus*), 65–66, 72; illustration of, 66; mother and young, 51

Beauchaud, Marilyn, acoustic signals by fish, 130

bees: short-range communication in, 104–5; von Frisch on dance of, 3

begging behavior, gull chicks, 57–58

beluga, vocal learning, 175

Ben-Aderet, Tobey, communication with dogs, 240–41

Bengalese finch (*Lonchura striata domestica*): learning to sing, 163–64; vocal control, 169

Bernard, Claude, 345n566

Bertucci, Frédéric: coralline algae and noise, 269; playback experiments in aquariums, 131, 262; signal intensity and acoustic complexity index, 276–77

Betta splendens (Siamese fighting fish), visual signals, 155

bioacoustics, 7, 13; Brémond, 26; discipline of, 3; identifying and counting species, 262; Marler on, 151, 158; Sèbe using, 260–61. *See also* ecoacoustics

biodiversity: acoustic indices of, 276; bioacoustics to monitor, 260–61; conservation and preservation of, 279; coral reefs, 262–63

biological systems, evolution of, 127–28

biophony, 276; anthropophony and, 264; term, 263, 264

biphonation, horses, 233

birds: anthropogenic noise and, 266; collisions with airplanes, 242; convergence between brains of songbirds and human brain, 168–69; distress calls of, 242–43; eardrum, 108; hearing side, 169–70; hormones and learning, 170; learning to sing, 158–60; singing by female of species, 167; singing of, 17; sound communications, 181; vocal repertoires, 134; wind instruments of, 42. *See also* vocal learning

black caiman (*Melanosuchus niger*), 91–94, 139, 230; recordings of, 114. *See also* crocodiles

black-capped chickadee (*Poecile atricapilla*): breeding decisions, 155–56; repetitions in call, 222–23

black-fronted titi monkey (*Callicebus nigrifrons*), referential communication system, 217

black-headed gull (*Larus ridibundus*): begging behavior of chicks, 57–58; choice of habitat, 37; illustration of, 36; parent-offspring recognition, 37–38; redundancy of information, 41

black redstart (*Phoenicurus ochruros*): communication network, 59; dear-enemy effect model, 166; illustration of, 21; territories of, 21–22

blue-capped cordon-bleu (*Uraeginthus cyanocephalus*), courtship rituals, 152–53 blue-footed booby (*Sula nebouxii*): illustration of, 246; Isabel Island, 245; male and female calls, 246–47; role of females and males in acoustic communication, 250; sexual dimorphism, 245, 247–48

blue whale (*Balaenoptera musculus*), 70. *See also* baleen whales

Bonjour les morses (film), 51

bonobos (*Pan paniscus*): cries of human babies, 237; illustration of, 182; vocal cords, 122; voice of, 182

booby. *See* blue-footed booby (*Sula nebouxii*)

Bora Bora lagoon, coral reefs of, 276

Borneo Island, Indonesia, 280

Botswana, 113, 203

bottle dolphin (*Tursiops truncatus*), 64

Bouchet, Hélène, identifying babies by their cry, 256

Bourquin, Sven, 114–15

bowhead whale (*Balaena mysticetus*), 49, 70. *See also* baleen whales

Boyer, Nicolas: emotions in vocalizations, 229; striped mice, 186

Brainard, Michael, learning to sing, 163–64

Brazilian Atlantic Forest, 273

Brémond, Jean-Claude: bioacoustics, 26; CNRS director, 8; distress information, 242–43; European wren (*Troglodytes troglodytes*), 271–72

Briefer, Elodie: dimensions of emotions, 231; horse vocalizations, 231–33

brown-headed cowbird (*Molothrus ater*): eggs of, 62; vocal learning, 174

brown thrasher (*Toxostoma rufum*), learning to sing, 165

brown tinamou: illustration of, 25; propagation experiments, 24, 26

Caiman House, 93

caimans. *See* crocodiles

Calanques National Park, 1

Cambridge University, 54, 294

Campbell's monkey (*Cercopithecus campbelli*), referential communication system, 217; syntactic complexity, 218

Campo Dora, 97

Campos, Zilca, on caiman, 88–89, 95–98

canary (*Serinus canaria*): song of invitation, 125; vocal control, 169

Canberra Botanical Garden, 219

cannibalism, 91

Cantais, Aurélie, identifying babies by their cry, 256

Cape penguin (*Spheniscus demersus*), calls of, 283–84

Cardinalis (northern cardinal), notes of song, 169

Carduelis (goldfinches), learning, 159

Casey, Caroline, sound world of elephant seals, 81, 83, 84

categorical perception, sounds, 109–10

cats, large, vocal cords, 122

cell theory, 33

Centre de recherches insulaires et observatoire de l'environnement (CRIOBE), 262, 345n566

Centre national de la recherche scientifique (CNRS), 8, 37, 89, 135, 181

Cercopithecinae, monkeys of Old World, 205

Cercopithecus campbelli. See Campbell's monkey (*Cercopithecus campbelli*)

Cercopithecus diana (Diana monkeys), alarm calls, 215–17

Cervus elaphus (deer), acoustic communications, 120

cetaceans, 68–69; dolphins, 68, 70, 73, 75, 77; playback experiments, 347n587. *See also* whales

chacma baboons (*Papio ursinus*): communication system, 226–28; emotions through vocalizations, 238–39; referential communication system, 217; social intelligence, 154

Champollion, hieroglyphics study, 3

Chaoborus fly, 268

character displacement, evolutionary biology, 137

Charlton, Ben, larynx of koala, 121–22

Charpentier, Marie, vocal plasticity, 181

Charrier, Isabelle: acoustic communications work, 45; baby fur seal and mother, 53; baby fur seals, 253; bearded seals underwater, 65–66; begging behavior of gull chicks, 57; field work with pinnipeds, 81; mother-offspring communication, 73; parent-offspring recognition, 37, 46, 47; walrus mothers and young, 50–51

chatter, 174

chemical pollution, 264

Cheney, Dorothy: communication systems of baboons, 226–28; primate study, 154; vervet monkeys, 214, 215, 226

chestnut-crowned babbler (*Pomatostomus ruficeps*), vocal repertoire of, 224–25

chimpanzee (*Pan troglodytes*): audience effect, 154; cries of human babies, 237; illustration of, 237; language, 180; research, 288; vocal cords, 122

Chinese Academy of Sciences, 196

Choi, Noori, spider communication, 200

chopi blackbird (*Gnorimopsar chopi*), learning to sing, 160

Cistothorus palustris (marsh wren), song learning, 166

Cistothorus stellaris (sedge wren), song learning, 166–67

clownfish (*Amphiprion clarkii*), sonic ligament of, 130

CNRS. *See* Centre national de la recherche scientifique (CNRS)

Colobus guereza (monkey), leopard alarm call of, 217

Colombelli-Négrel, Diane, parent and offspring password, 63

Columbia University, 288

communication: brain basis of networks, 329n323; elephants and vibration, 201–2; mathematical theory of, 33–34, 284, 286; parent-offspring, 59–60; semantic, 214; signals, 2; zebra finches, 60–61. *See also* acoustic communication networks

communication signals: drumming behavior of woodpeckers, 135–37; sender and receiver, 123

control stimulus, wren experiment, 27

coral reefs: acoustic indices, 276–77; biodiversity of, 262–63; fish on, 67–68; shrimp, 67, 68

Corvin, Siloé, identifying familiar babies, 257

coupled oscillators, crickets, 129

Coureaud, Gérard, emotions in vocalizations, 229

courtship, *Drosophila* fly, 104

Cousteau, Jacques, 67

COVID-19 crisis, 266

crested pigeon (*Ochyphaps lophotes*), communication signals in danger, 128

crickets: coupled oscillators, 129; ears of bush, 321n190

CRIOBE. *See* Centre de recherches insulaires et observatoire de l'environnement (CRIOBE)

crocodile(s): acoustic communication, 90–95; acoustic signals of frogs and, 111–12; black caiman (*Melanosuchus niger*), 91–94, 96; experiment in nests, 90–91; gharial (*Gavialis gangeticus*), 90; habituation, 92–93; jacaré caiman (*Caiman yacare*), 89; Nhumirim Ranch, 88–89; Nile crocodiles (*Crocodylus niloticus*), 90–91, 95, 96; Orinoco crocodile (*Crocodylus intermedius*), 95; quality index, 119; signals and, 111–12; spectacled caiman (*Caiman crocodilus*), 89

Crocoparc, 110, 229, 230.

Crocuta crocuta. See spotted hyena (*Crocuta*)

cuckoo (*Cuculus canorus*), begging calls in parent-offspring interactions, 54

culture, oscine bird song transmission, 172–74

cumulative selection, 127

Curé, Charlotte: acoustic communications of shearwaters, 248–49; humpback and killer whale vocalizations, 74–75

Current Anthropology (journal), 293

Current Biology (journal), 162

Cynomys gunnisoni. See Gunnison's prairie dog (*Cynomys gunnisoni*)

Dabelsteen, Torben: communication network concept, 147; Eurasian blackcap study, 29

dance of bees, von Frisch, 3

Darwin, Charles, 214, 234; adaptive radiation, 137; bill morphology, 138; expressing emotions, 230–31; finches on Galápagos Islands, 137; on primitive sounds, 233; on role of female birds, 251; on sexual selection, 5

Dawkins, Richard, manipulative signal, 61, 62

Deakin University, 60

dear-enemy effect, 22; black redstart as model, 166

decibels, 12

deer (*Cervus elaphus*): acoustic communications, 120; stable evolutionary equilibrium, 121

Dendrobatidae family, 197

Dendrocopos major (great spotted woodpecker), drumming behavior, 136–37

Dentressangle, Fabrice, blue-footed booby (*Sula nebouxii*), 244, 245–47

Derégnaucourt, Sébastien: song system, 170–71; using birdlike robots, 163

Derryberry, Elizabeth, vocal activity of white-crowned sparrow, 266

Désert de Plate, 260

developmental stress hypothesis, 125

Diana monkeys (*Cercopithecus diana*): alarm calls, 215–17; eagle alarm calls, 215, 216; playback experiments, 218

distress calls, research on, 242–43

diversity, 262

dog-directed speech, humans and, 240–42

dogs, 1; communication of humans and, 240; illustration of, 241

dolphins, 68; bottlenose (*Tursiops truncatus*), 64; echolocation, 335n426; language of, 75, 77; monkey lips of, 70, 75, 76; tympanic bone, 109; vocalizations of, 70, 73; vocal learning, 175

Doppler effect, 193–94

Doppler effect compensation, phenomenon, 192

Doutrelant, Claire, Siamese fighting fish (*Betta splendens*), 155

Draganoiu, Tudor: on black redstart (*Phoenicurus ochruros*), 21; black redstart as model, 166; songs of black redstart males, 59

Dreiss, Amélie, sibling negotiation hypothesis, 58

Drosophila fly: FOXP2 gene, 290; Johnston's organ in, 104

Duke University, 214

Dunbar, Robin, social grooming, 292

dunnock (*Prunella modularis*), 272

dynamic, information in sound, 19

ears: evolution of, 100; mechanoreceptors, 320n181. *See also* acoustic sounds

Earth, gravity on, 10–11

eastern phoebe (*Sayornis phoebe*), learning to sing, 160

eavesdropping: acoustic communication networks, 155–57; assessing risk, 329n321

echolocation: bats, 191–92, 335n426; dolphins, 335n426; Doppler effect compensation, 192; pulse, 191; pulse-echo delays, 192

ecoacoustics, 262, 349n621; dealing with big data of recordings, 278; definition of, 263; Krause on, 275; natural soundscapes, 270; technological developments, 278

École normale supérieure de Lyon, 149

École pratique des hautes études (EPHE), 258, 345n566

ecological niche, concept of, 270

ecosystem(s): coral reefs of Puerto Rico during Hurricane Maria, 278; ecoacoustic approach, 351n639; food webs, 269; functions, 277–78

Egyptian fruit bat (*Rousettus aegyptiacus*), vocal learning, 177

Egyptian hieroglyphics, Champollion on, 3

elephants: impedance adapter, 201; near sound field of, 322n200; vibrational communication, 201; vocalizations, 200–202; vocal learning, 176–77

elephant seals: Año Nuevo Reserve, 78–79, 81, 84, 85; behavioral development, 85–86; breeding season, 79, 82–83; Elo rating score, 83; hierarchy of dominance, 86–87; illustration of, 82; *Mirounga angustirostris*, 78; rhythmic sounds, 318n150; sound world of, 81; vocalizations of, 81–85

Elie, Julie: individual signatures of hyenas, 210–11; vocal repertoire of zebra finch, 148, 152, 208; zebra finch calls, 178

Elise, Simon, coral reefs of Europa Island, 277

Elo rating system, hierarchical rank, 83

Elsey, Ruth, reproductive biology of American alligator, 114

emitter, signal, 32

emotions: animals expressing, 287; arousal and valence, 231, 233; coding into vocalizations, 234–35; dimensions of, 231, 233; distress calls research, 242–43; horses vocalizations and, 232–33; human babies vocalizations, 234–36; humans and dog communication, 240–41; stress hormones and, 239–40; vocalizations and, 229–30, 233–36

emperor penguins (*Aptenodytes forsteri*), 39, 41, 43. *See also* penguins

ENES Bioacoustics Research Laboratory, Saint-Etienne, France, 123, 130, 185

Engesser, Sabrina, calls of babblers, 224–25

Engystomops pustulosus (túngara frog), 110

EPHE. *See* École pratique des hautes études (EPHE)

Erignathus barbatus: bearded seal, 51. *See also* bearded seal (*Erignathus barbatus*); seals

Erithacus rubecula (European robin), singing, 252

Eubalaena glacialis. See North Atlantic right whales (*Eubalaena glacialis*)

Eurasian blackbird (*Turdus merula*), 264

Eurasian blackcap (*Sylvia atricapilla*), 271; acoustics, 26; illustration of, 29; sound propagation of, 29–30

Eurasian reed warbler (*Acrocephalus scirpaceus*): begging calls of chicks, 54–55; intensity of begging behavior, 55–56

Eurasian wren (*Troglodytes troglodytes*), 17; bioacoustics, 26; Brémond and song of, 271–72; illustration of, 27

Europa Islands, coral reefs of, 277

European minnows (*Phoxinus phoxinus*), 267–68; larvae of *Chaoborus* fly and, 268–69

European robin (*Erithacus rubecula*), singing, 252

evolution: animal communication, 61–62; biological systems, 127–28; Darwin, 214; exaptation, 129; human language, 167; theory, 33

evolutionary biology, character displacement, 137

evolutionary convergence, phenomenon, 168

evolutionary history, communication, 6–7

exaptation, change of feather shape, 129; fish, 129–30

experimental stimulus, wren experiment, 27–28

Expression of Emotions in Man and Animals, The (Darwin), 230, 329n327

Faillenot, Isabelle, brain perception of cries, 258–59

fairy wren (*Malurus cyaneus*): acoustic alarm system, 219; Aubin studying calls of, 220–22; Horsfield cuckoo (*Chrysococcyx basalis*) and, 62–63; illustration of, 220; incubation calls, 64; shining bronze cuckoo (*Chyrsococcyx lucidus*), 62; study in Canberra Botanical Garden, 219–20; study model of Magrath, 219

far sound field, pressure variations, 105

Fauchon, Camille, brain perception of cries, 258–59

Favaro, Livio, calls of African penguin, 284

fazenda, Nhumirim, 88, 89

field cricket (*Gryllus bimaculatus*): illustration of, 267; noise by human activity and, 267

field cricket (*Teleogryllus oceanicus*), acoustic communication, 132

finches: Darwin's, 137–38; goldfinches (*Carduelis carduelis*), 159. *See also* zebra finches

fin whale (*Balaenoptera physalus*), 70. *See* baleen whales

Fischer, Julia, chacma baboons' response to alarms, 238–39

fish: acoustic communication, 129–32; inner ear, 106

Fitch, Tecumseh, evolution of language, 291

FitzPatrick Institute of African Ornithology, 294

Fonseca, Paulo, fish sound signals, 131

food webs, ecosystem, 269

forest experiment: control stimulus, 27; experimental stimulus, 27–28; wrens, 27–28

formant(s): frequency of, 119; sound frequencies, 118; voice frequency, 119

Formosan macaque (*Macaca cyclopis*), 284

Fortune, Eric, plain-tailed wren as study model, 252–53

Fougeirol, Luc, 114; Crocoparc, 229; zoo founder, 90, 110

Foxe Basin, 65

FOXP1, little brother of FOXP2, 353n678

FOXP2 gene (*Forkhead box P2*), 290

French Alps, 260

French Guiana, 139, 142, 144

frequency, sound, 14

frequency selectivity, mosquitoes, 105

freshwater fish, noise and behavior of, 267–69

frogs: acoustic adaptation hypothesis, 274–75; *Amolops tormotus*, 196–97; bioacoustics of, 108; eardrums, 107; golden rocket frog (*Anomaloglossus beebei*), 197; inner ear, 105; North African green frog (*Pelophylax saharica*), 111; relationship between ears and lungs, 107–8; signal recreation for crocodiles, 111–12; ultrasound by, 196–97

fruit bats, genus *Rousettus*, 191

functional information, 277

functional magnetic resonance imaging (fMRI), 259

functions, ecosystems, 277–78

fundamental frequencies, value of, 117

Gahr, Manfred, cordon-bleu's dance, 153

Gain, Philippe, interest in babies' crying, 253

Galambos, Robert, bats emitting ultrasound, 187

Galápagos baby sea lion (*Zalophus wollebaeki*), 51–52

Galápagos Islands: blue-footed booby (*Sula nebouxii*), 245; Darwin's finches, 137–38

Gallego-Abenza, Mario: dog-directed speech, 241–42; field crickets and noise, 267

Garcia, Maxime, 327n294; great spotted woodpecker, 136

Gardner, Allen, chimpanzee work, 288

Gardner, Beatrix, chimpanzee work, 288

Gavialis gangeticus (gharial), 90. *See also* crocodiles

gelada (*Theropithecus gelada*), 284

gendered traits, identifying babies, 258

genetics, language and, 290

Gennes, Pierre-Gilles de, physics of soft matter, 279

geophony, 276; term, 263

gharial (*Gavialis gangeticus*), 90. *See also* crocodiles

giant antshrike: illustration of, 25; propagation experiments, 24

gibbons, *Nomascus nasutus* and *N. concolor*, 284

Glickman, Stephen, hyenas' social organization, 205–6

glissando, 19

Globicephala melas (pilot whales), 75

glossary, 305–308

Gnorimopsar chopi (chopi blackbird), learning to sing, 160

Goegap Nature Reserve, 183–85, 187

golden rocket frog (*Anomaloglossus beebei*), 197

goldfinches (*Carduelis carduelis*), learning, 159

gorillas: playback experiments with, 286–87; vocalizations of mountain, 325n252

Goutte, Sandra: acoustic adaptation hypothesis, 274–75; working with frogs, 274

Gracula religiosa (hill myna), song imitation, 165

graded vocalizations, spotted hyenas, 206

grasshoppers, 1

gravity, Earth, 10–11

gray seals (*Halichoerus grypus*), vocal learning, 176

great animal orchestra, Krause on, 270

Great Barrier Reef, 269

greater honeyguide (*Indicator indicator*), 293–94

greater mouse-eared bat (*Myotis myotis*), ultrasound, 195

great spotted woodpecker (*Dendrocopos major*): drumming behavior, 136–37; illustration of, 136

great tit (*Parus major*): learning to sing, 164

Great Wall of China, 199

Greenfield, Michael: communication by ultrasound, 186–87; night flying for insects, 102–3

green frog, communication signals, 123

greenish warbler (*Phylloscopus trochiloides*), song of, 172

Griffin, Donald: bats emitting ultrasound, 187; spatial orientation of birds and bats, 190–91

Grimault, Nicolas: crocodilian study, 89, 95, 96; emotions in vocalizations, 229

Gryllus bimaculatus. See field crickets (*Gryllus bimaculatus*)

gulls, 32; begging behavior of chicks, 57–58; black-headed gull, 36; illustration, 33. *See also* black-headed gull (*Larus ridibundus*)

Gunnison's prairie dog (*Cynomys gunnisoni*), referential communication system, 218

Gustafsson, Erik, new "baby cries" team, 254

habituation, phenomenon, 92

Halichoerus grypus (gray seals), vocal learning, 176

hammer-headed fruit bat (*Hypsignathus monstrosus*), fundamental frequencies, 122

handicap theory, 121

harbor seal (*Phoca vitulina*), 51. *See also* seals

harmonic frequencies, vocalization, 117–18

Harvard University, 190

Hauber, Mark: imitation of vocal repertoire, 174; parent and offspring password, 63

Hawaiian monk seal (*Monachus schauinslandi*), 51. *See also* seals

Hayes, Cathy, chimpanzee project, 288

Hayes, Keith, chimpanzee project, 288

hearing: audition, 99, 100; birds and humans, 169–70. *See also* acoustic signals

hearing machinery, 289

heat call, zebra finches, 60

Hebets, Eileen, spider communication, 200

Heterocephalus glaber (naked mole rat), vocalizations, 178

hierarchy of dominance, elephant seals, 86–87

hieroglyphics, Champollion on, 3

hill myna (*Gracula religiosa*), song imitation, 165

hippos, 113

Holekamp, Kay, hyenas' social organization, 205, 206–7

Holocentridae family, 263

Homo sapiens, 327n288; social groups, 292

honest communication, hypothesis of, 54, 116

honesty principle, 324n249

honey badgers, 203

hooded berryeater: illustration of, 25; propagation experiments, 24, 26

hormones, learning process of birds, 170

Hornøya Island, 32

horses: biphonation, 233; vocalizations and emotions, 232–33

Horsfield cuckoo (*Chrysococcyx basalis*): fairy wren (*Malurus cyaneus*) and, 62–63; incubation calls, 64

Howard Hughes Medical Institute, 290

How Monkeys See the World (Seyfarth and Cheney), 215

Huangshan Hot Springs, 196

Huetz, Chloe, computer program for piha data, 144

human(s): acquisition of spoken language, 291–93; birds as model for understanding voice of, 174; communication with dogs, 240–41; convergence between brains of songbirds and, 168–69; dog-directed speech, 240–42; echolocation, 335n426; emotions and vocalizations of babies, 234–36; evolution of language, 167; fundamental frequency of voice, 118–19; hearing side, 169–70; identifying babies by their cry, 255–56; language, 180; laws characterizing language, 286; mothers caring for babies, 255; perception of cries

by gender stereotype, 257–58; pragmatics of language, 225–26; vocal exaggeration, 123–24

hummingbirds, 160; vocal control, 168

humpback whale: anthropogenic noise and, 265–66. *See also* baleen whales

Hunter College, 63, 170

Hurricane Maria, coral reefs of Puerto Rico, 278

Hyacinthe, Carole, acoustic signals by astyanax, 135

hyena, 203–4; illustration of, 204. *See also* spotted hyena (*Crocuta crocuta*)

Hypsignathus monstrosus (hammer-headed fruit bat), 122

IBAC. *See* International Bioacoustics Society (IBAC)

Igloolik, 46, 47, 65; Igloolik Research Center, 47

impedance adapter, 201

imprinting experiments, Lorenz, 3

incubation call, superb fairy wren and Horsfield cuckoo, 63, 63–64

indicator bird (*Indicator indicator*), 293–94

indri lemur (*Indri indri*), 284

Industrial Revolution, 264

information, species identity in song, 20–22

information redundancy, 35

information theory, mathematical theory of communication, 33, 284

information transmission, penguins, 40–41

infrasound, 14; importance of, 338n456

inner ear, anatomical structure, 105–6

insects: acoustic sounds, 99, 102; anthropogenic noise and, 267; bodies of, 320n182; courtship and acoustic signals, 104–5; eardrum development, 102; Johnston's organ, 104; respiratory system of, 321n184; short-distance ears, 104; Sphingidae family, 102

intelligence, cognitive abilities, 79–80

intensity, sound, 12–13

International Bioacoustics Conference, 142, 197

International Bioacoustics Society (IBAC), 275, 350n622

intrasexual selection, 5, 120; blue-footed booby, 248

Iqaluit, 46

Isabel Island, blue-footed booby, 245

jacaré caiman (*Caiman yacare*), 89, 96, 230; illustration of, 97. *See also* crocodiles

jaguar (*Panthera onca*), vocal cords, 122

Japanese great tit (*Parus major minor*), alarm calls of, 223–24

Japanese macaque (*Macaca fuscata*), 110

Jarvis, Eric, gene activity, 290

Johnston's organ: frequency selectivity, 105; sensor in insects, 104

Jouen, François, brain perception of cries, 258–59

Jurassic Park, 7

Kgalagadi Transfrontier Park, 212

killer whale (*Orcinus orca*): vocalizations, 73–75. *See also* toothed whales

Kilner, Rebecca, begging calls of Eurasian reed warbler chicks, 54–55

king penguins (*Aptenodytes patagonicus*), 38–39, 40, 43, 44, 311n43. *See also* penguins

kinship selection, 167

Kleindorfer, Sonia, parent and offspring password, 63

Knight, Chris, origins of spoken language, 292–93

koala (*Phascolarctos cinereus*), 324n240; call of, 121–22

Koralek, Aaron, recording hyenas, 208

Koutseff, Alexis, recording babies, 235

Krause, Bernie: acoustic niche hypothesis, 270–71; changing habitats, 279; ecoacoustics, 275; great animal orchestra, 270; recording of orca, 333n385; recording

soundscapes, 263; separating anthropophony, 264

Krebs, John, manipulative signal, 61, 62

Kreutzer, Michel: acoustics of wren population, 26; A phrase by, 125

Kroodsma, Don, on birdsong, 166

Kruuk, Hans, spotted hyenas, 204–5

Kyoto University, 223

La Ferme aux Crocodiles, zoo, 90, 110

La Fontaine, Jean de, frog, 120

Lagopus mutus. See rock ptarmigan (*Lagopus mutus*)

language: acquisition of, 290; animals and, 280–81; external observers of animals, 282; genetics and, 290; neurobiology of, 169–70; nonhuman primates, 179–80; songbirds as model for understanding human learning, 159–60; sophistication of nonhuman, 283; whistling, 289. *See also* spoken language

Larus ridibundus. See black-headed gull (*Larus ridibundus*)

laryngeal hypertrophy, 122

laryngeal sac, whales, 69–70

larynx, 42, 120, 289; hammer-headed fruit bat, 122; male koala, 121–22

law of brevity, 284

learning. *See* vocal learning

Lebinthini (cousins of crickets), females processing sounds of males, 132–33

Le bourgeois gentilhomme (Molière), 341n504

Lebreton, Jean-Dominique, floating blind for black-headed gull study, 37

Lecchini, David: coralline algae and noise, 269; coral reef fish, 262

Lemasson, Alban, suffix in monkey calls, 217–18

leopard (*Panthera pardus*), vocal cords, 122

Leptonychotes weddellii (Weddell seal), 51. *See also* seals

Lévi-Strauss, Claude, 345n566

Levréro, Florence: interest in baby crying, 253–54; playback experiments with gorillas, 286–87; recording babies, 235; striped mice, 186, 187; vocal plasticity, 181

Lewis, Jerome, origins of spoken language, 292–93

light, speed of, 10

lightning, thunder and, 10

Lingle, Susan, deer mothers and mammalian baby cries, 236–37

lion (*Panthera leo*), vocal cords, 122

Lipaugus vociferans (piha): analysis of, 145–47; characterization of signatures, 144–45; communication networks, 147; illustration of screaming piha, 141; project, 139–42; recording vocalizations of, 140–42

locusts, 1, 102

Lombard effect, 265

Lonchura striata domestica (Bengalese finch), learning to sing, 163–64

London Zoological Gardens, 214

long-distance ears, 101, 103, 320n173

Lorenz, Konrad: goose imprinting, 153; imprinting experiments, 3

Luscinia megarhynchos (nightingale), learning to sing, 164–65

Macaca fuscata (Japanese macaque), 110

Macaca mulatta (rhesus macaque), audience effect, 154

Macrotermes natalensis (termites), 198

Magrath, Robert: botanical garden detail, 220; fairy wren study model, 219

mamba, 115

mammals: mothers' gift of caring for babies, 254–55; vocalizations, 116–17; wind instruments of, 42

mandrill (*Mandrillus sphinx*): biology of, 181; voice of, 181

manipulative signal, concept of, 61

Manser, Marta, meerkats, 211–12

Mariette, Mylene, adult zebra finches, 60

Marin County, California, 158

Marin-Cudraz, Thibaut, automatic counting by bioacoustic method, 261

Marler, Peter, 159, 160; bioacoustics, 151; songbirds learning to sing, 161; vervet monkeys, 214; white-crowned sparrow, 158, 266

marmosets, 254; food and, 218; vocal exchanges, 181, 328n314

Mars (planet): Martian identifying flow of human speech, 282; recording sound on, 10

marsh wren (*Cistothorus palustris*), song learning, 166

Mata Atlântica, 14, 15, 16, 24. *See also* white-browed warbler (*Basileuterus leucoblepharus*)

Max Planck Institute of Ornithology, Germany, 153

Max Planck Research Institute, 153

McComb, Karen, female elephant vocalizations, 201

McGregor, Peter: communication network concept, 147; Siamese fighting fish (*Betta splendens*), 155

McVay, Scott, humpback whale, 71

mechanoreceptor, 100

Médoc, Vincent, boat noise and freshwater fish behavior, 267–69

meerkats (*Suricata suricata*): alarm calls, 212–13; recruiting call, 213; referential communication, 213; vocal repertoire of, 211–13

Megachiroptera, fruit bats, 191

Megaptera novaeangliae: humpback whale, 70. *See also* baleen whales

Melanosuchus niger: black caiman, 91–94. *See also* crocodiles

melatonin, sleep hormone, 171

Melospiza georgiana (swamp sparrow), 110; learning to sing, 165; vocal learning, 171

Melospiza melodia (song sparrow), learning to sing, 165

memory, learning and, 168

Mennill, Daniel, song learning environment, 162, 163

Menura novaehollandiae (superb lyrebird): song imitation, 165; vocalizations, 126

Menzerath-Altmann law, 284

Metriaclima zebra (cichlid family), 131; illustration of, 131; sound productions, 262

Mets, David, learning to sing, 163–64

mice. *See* striped mouse (*Rhabdomys pumilio*)

Microchiroptera, bats, 191

microphone, signal propagation in forest, 28

Miller Institute, 79

Mimus polyglottos (northern mockingbird), song imitation, 165

minke whale (*Balaenoptera acutorostrata*), killer whale preference, 74

minnows. *See* European minnows (*Phoxinus phoxinus*)

model of sensory bias, 132; communication, 128; Lebithini crickets, 133

mole rats, vocalization, 178–79

Molothrus ater (brown-headed cowbird), vocal learning, 174

Monachus schauinslandi (Hawaiian monk seal), 51. *See also* seals

monkeys, semantic communication, 214. *See also* bonobos; chimpanzees; primates

Morelet's crocodile (*Crocodylus moreletii*), recordings of, 114

Moremi Game Reserve, 203

Morro Grande Reserve, 15, 16, 17

Morton, Eugene, acoustic structure of signal, 234

Morton's principle (motivation-structural rules), 234, 236, 307

mosquitoes, frequency selectivity, 105

moths: bats and, 194–96; Doppler effect, 194; tympanic ears, 195

mouse, ultrasound communication, 187–89

Mouterde, Solveig: propagation experiments, 250; testing female zebra finches, 250–51

Muséum national d'Histoire naturelle, Paris, 271, 271

mutualism, 294

Myotis (greater mouse-eared bat), ultrasound, 195

mysticetes, vocal cords of, 316n109

naked mole rat (*Heterocephalus glaber*), vocalizations, 178

Narins, Peter, 200; ultrasonic frog, 196–97

National Research Agency, 151

Natural History (Pliny the Elder), 159

natural selection, 127

Nature (magazine), 151, 168, 170

Nature Communications (journal), 124, 181, 251

near sound field, particle oscillations, 105

Neophoca cinerea (Australian sea lion), 51

networks. *See* acoustic communication networks

neurobiology, language, 169–70

neutral drift, 127

New Jersey Institute of Technology, 252

Nhumirim Ranch, 88–89, 97

nightingale (*Luscinia megarhynchos*), learning to sing, 164–65

Nile crocodiles (*Crocodylus niloticus*), 90–91, 92, 95; behavioral response of mother to offspring, 115–16; cries of human babies, 237–38; emotions in vocalizations, 229–30; recordings of, 114; vocalizations of baby, 113. *See also* crocodiles

Noad, Michael, humpback whale song, 71

noise: animal communications and human activities, 264–65; behavior of freshwater fish and, 267–69; gull colony, 38–39; penguin colony, 39–40; penguins, 43–44; signal, 32; signal-to-noise ratio, 35. *See also* penguins noisy friarbird (*Philemon corniculatus*), regent honeyeater singing song of, 173

Nomascus concolor (gibbon), 284

Nomascus nasutus (gibbon), 284

nonhuman primates, language and speaking, 179–80

North African green frog (*Pelophylax saharica*), 111

North Atlantic right whales (*Eubalaena glacialis*), 265

northern cardinal (*Cardinalis cardinalis*), notes of song, 169

northern mockingbird (*Mimus polyglottos*), song imitation, 165

Norway, 32

Nottebohm, Fernando, song nuclei, 168

Nouragues Nature Reserve, French Guiana, 271, 272

Nunavut, 46

Nyamsi, Ruben Mbu, sound analysis method, 279

ocean acidification, 264

Ochyphaps lophotes (crested pigeon), communication signals, 128

Odobenus rosmarus (walrus), 48

Oenothera drummondii (beach evening primrose), 295

Okavango Delta, 113

Old World monkeys, *Cercopithecinae*, 205

operant conditioning, Pavlov on, 80

Ophiophagus hannnah (royal cobra), whistling sound, 319n167

orangutan, vocal personalities, 182

orb-weaving spider (*Larinioides sclopetarius*), 337n448

orcas, ecotypes of, 74

Orcinus orca: killer whales, 73–75; Krause recording of, 333n385

Orinoco crocodile (*Crocodylus intermedius*), 95, 230; recordings of, 114. *See also* crocodiles

oscine birds: song transmission forming cultures, 172–74; vocal learning, 160–61, 171

Ota, Nao, cordon-bleu's dance, 153

Oxford University, 61, 292

Pan paniscus (bonobos), vocal cords, 122

Panthera leo (lion), vocal cords, 122

Panthera onca (jaguar), vocal cords, 122

Panthera pardus (leopard), vocal cords, 122

Panthera tigris (tiger), vocal cords, 122

Pan troglodytes (chimpanzee), vocal cords, 122

Papet, Leo: crocodilian study, 89, 95, 96; emotions in vocalizations, 229

Papio ursinus. *See* chacma baboons (*Papio ursinus*)

parakeets, 160

parental cooperation, seabirds, 36

parent-offspring communication: barn owls, 58–59; birds, 54–57; black-headed gull, 37–38; black redstarts, 59; bottlenose dolphin, 64; cowbird, 62; cuckoos, 62–63, 64; evolution of communication, 61–62; fur seals, 53–54; gulls, 57–58; parasites and hosts, 63; seals, 51–53; walruses, 48–51; work of Charrier, 37, 46, 47; wrens, 62–63, 64; zebra finches, 60–61

Parker, Brad, 47

Parmentier, Eric: fish sound productions, 130; piranha man, 262

parrots, 160

Parus major (great tit), learning to sing, 164

Parus major minor. *See* Japanese great tit (*Parus major minor*)

Passerculus sandwichensis (Savannah sparrow), learning to sing, 162–63

passerines, learning to sing, 160

Patural, Hugues, identifying babies by their cry, 253, 256

Pavlov, Ivan, operant conditioning, 80

Payne, Roger, humpback whale, 71, 175

Pelophylax saharica (North African green frog), 111

pendulum, Doppler effect, 193–94

penguins: acoustics for recognition, 38–40; adaptations to noisy environments, 41–42; Adélie (*Pygoscelis adeliae*), 44; background noise and communication, 39–40; emission strategies, 43–44; emperor (*Aptenodytes forsteri*), 39, 41, 43; king (*A. patagonicus*), 38–39, 40, 43, 44; parents and offspring, 47; two-voice phenomenon, 42–43; vocalization by, 42–43

Pepperberg, Irene, parrot language training, 289

Pernau, Ferdinand, song learning, 159

Perrier, Leo, ultrasound and mice, 187

Perseverance rover, microphones of, 10

Peyron, Roland, brain perception of cries, 258–59

Phascolarctos cinereus (koala), call of, 121–22

pheromones, termites, 198

Pheugopedius euophrys (plain-tailed wren), male and female singing, 252

Philemon corniculatus (noisy friarbird), regent honeyeater singing song of, 173

Phoca vitulina (harbor seal), 51. See also seals

Phoenicurus ochruros. See black redstart (*Phoenicurus ochruros*)

Phylloscopus trochiloides (greenish warbler), song of, 172

Phylloscopus trochilus. See willow warbler (*Phylloscopus trochilus*)

Phyllostomus discolor (spear-nosed bat), vocal learning, 177

phylogenetic reconstructions, 168

Pierce, George, ultrasound detector, 190–91

piha: sharing songs with neighbors, 166. See also *Lipaugus vociferans* (piha)

pilot whales (*Globicephala melas*), 75

Pinniped Cognition & Sensory Systems Laboratory, 79, 80

Pinniped Laboratory in Santa Cruz, 209

pinnipeds, 46–47; reproduction, 47; vocal learning, 176

piranha fish (*Pygocentrus nattereri*), repertoire of sounds, 130

Pisanski, Kasia, vocal exaggeration, 123–24

plain-tailed wren (*Pheugopedius euophrys*), male and female singing, 252

Pliny the Elder, 159

PNAS (magazine), 224

Poecile atricapilla (black-capped chickadee): breeding decisions, 155–56. See also black-capped chickadee (*Poecile atricapilla*)

pollution, 264

Pomacentridae family, 263

Pomatostomus ruficeps. See chestnut-crowned babbler (*Pomatostomus ruficeps*)

porpoises, 68

Poulsen, Holger, learning vocalizations, 159

pragmatics, human language, 225–26

Prat, Yosef: acoustic analysis, 283; on spoken language, 282

precedence effect, acoustic signals, 112

precursor sender model: communication, 128; model of sensory bias and, 133

predation, chick begging behavior and, 56

predators, acoustic signals, 103

Premack, David: chimpanzee work, 288; emotions in communications, 215

Presbytis thomasi (Thomas's langurs), audience effect, 154

pressure-differential ears, frogs, 107

primates: identifying babies by their cry, 256; sequence of calls, 340n489; social intelligence, 154–55; sound communications, 181

Procnias tricarunculatus (three-wattled bellbird), learning to sing, 160

propagation media, sound waves, 13

Prunella modularis (dunnock), 272

ptarmigan. *See* rock ptarmigan (*Lagopus mutus*)

Puerto Rico, Hurricane Maria and, 278

pulse, echolocation, 191

pulse-echo delays, echolocation, 192

Pygocentrus nattereri (piranha fish), repertoire of sounds, 130

Pygoscelis adeliae. See Adélie penguin (*Pygoscelis adeliae*)

quality index, 119

Reby, David: acoustic communications of deer family, 120; communication with dogs, 240; interest in baby crying, 253; larynx of koala, 121–22; mammalian voice production, 234; vocal exaggeration, 123–24; vocal production, 179

receiver, signal, 33

recruiting calls, meerkats, 213

red-breasted nuthatch (*Sitta canadensis*), eavesdropping, 156–57

redstarts. *See* black redstart (*Phoenicurus ochruros*)

reef ecosystem, acoustics of, 277–78

referential communication: ground squirrels and marmots, 218–19; marmosets, chimpanzees, and bonobos, 218; primate species, 217; sophisticated system, 213

regent honeyeater (*Anthochaera phrygia*), songs of, 174

Reichmuth, Colleen, 80; hyenas and operant conditioning protocol, 209–10; sound world of pinnipeds, 81; training dogs, 80

Rendell, Luke, sperm whale acoustics, 76–77

reproduction, pinnipeds, 47

resonator, mouth as, 108

resource, term, 86

Rétaux, Sylvie, genetics and development, 135

Rhabdomys pumilio. See striped mouse (*Rhabdomys pumilio*)

rhesus macaque (*Macaca mulatta*), audience effect, 154

ringtailed lemur (*Lemur catta*), referential communication system, 217

ripples, throwing pebble into pond, 9–10

ritualization, precursor sender model, 128

robins. *See* American robin (*Turdus migratorius*)

Rochais, Céline, striped mice, 335n411

Rockefeller University, 168

Rockefeller Wildlife Refuge, Louisiana, 114

rock ptarmigan (*Lagopus mutus*), 261; estimating population size, 260, 261; illustration of, 261

Rogers, Ginger, 153

Rosetta Stone, 3

Roulin, Alexandre, sibling negotiation hypothesis, 58

Rousettus aegyptiacus (Egyptian fruit bat), vocal learning, 177

Rousettus genus, fruit bats, 191

Rowley Island, 65

royal cobra (*Ophiophagus hannah*), whistling sound, 319n167

rufous-and-white wrens (*Thyrophilus rufalbus*), vocal behavior of, 343–344n547

Rumbaugh, Duane, chimpanzee work, 288

Rybak, Fanny, short-range communication in flies, 104–5

Saccopterys bilineata (bat species), vocal learning, 177

saiga antelope (*Saiga tatarica*), 122

Saiga tatarica (saiga antelope), 122

Saloma, Anjara, mother-offspring communication, 73

San Francisco Bay Area, 158

Santa Cruz, 78, 79, 209

Savannah sparrow (*Passerculus sandwichensis*), learning to sing, 162–63

Sayornis phoebe (eastern phoebe), learning to sing, 160

Schradin, Carsten: research at CNRS, 184; striped mouse study, 185–86

Schusterman, Ron, exploring cognitive abilities, 79–80

Science (magazine), 158, 178, 214, 282

Scolopax rusticola (woodcock): illustration of, 173; song vocalization, 172

scolopidium, 321n185

screaming piha, 143, 160; illustration of, 141; Sèbe studying, 145–47, 260; sound triangulation for, 143–44. See also Lipaugus vociferans (piha)

seabirds: breeding of, in colonies, 35–36; breeding pairs, 36

seals, 45; bearded seal (Erignathus barbatus), 51, 65–66; Galápagos baby fur seal (Arctocephalus galapagoensis), 52; illustration of fur seal, 52; parent-offspring communication, 51–53; Weddell seal (Leptonychotes weddellii), 51. See also elephant seals

Sea of Barents, 32

Sèbe, Frédéric, 327n294: songs of neighboring warblers, 23; studying screaming pihas, 145–47, 260

Sechellophryne gardineri (Seychelles frog), 108

sedge warbler (Acrocephalus schoenobaenus), repertoire size, 126

sedge wren (Cistothorus stellaris), song learning, 166–67

semantics, language, 214

semisyringes, penguins and, 42–43

sender, signal, 32

Serranidae family, 263

sexual dimorphism, 117; blue-footed booby, 245, 247–48

sexual selection: blue-footed booby, 247–48; communication signals, 126–27; Darwin on, 5; driver of evolution, 167; signal evolution, 134

Seychelles frog (Sechellophryne gardineri), 108

Seyfarth, Robert: communication systems of baboons, 226–28; vervet monkeys, 214, 215; vervets, 226

Shacks, Vince, 113, 114–15

Shannon, Claude: information theory, 33, 226, 284; strategies for encoding information, 34–35; theory development and publishing, 311n27

shearwaters: acoustic communications of, 248–49; illustration of, 249; role of females and males in acoustic communication, 250

shining bronze cuckoo (Chrysococcyx lucidus), fairy wren (Malurus cyaneus) and, 62

short-distance ears, 101, 104, 320n172

shrimp, coral reefs, 67, 68

Siamese fighting fish (Betta splendens), visual signals, 155

sibling negotiation hypothesis, 58

signal: call of baby fur seal, 53–54; communication, 2; modifications, 34; propagation, 34

signal-to-noise ratio, 35; penguins and, 43–44

Silent World, The (documentary), 67

Sitta canadensis (red-breasted nuthatch), eavesdropping, 156–57

Smithsonian Institution, 234

snail, sound of, 320n169

snakes, 16, 56, 99, 106, 115; alarm calls, 214; black mambas, 115; ultrasonic signals, 189; whistling sound of royal cobra, 319n167

social amplification of information, 199

social flexibility, striped mouse, 185

social grooming, 292

social insects, ants and termites, 198–99

social intelligence, acoustic communication networks, 154–55

social systems, rodents, 184–85

Société linguistique de Paris, 292

Soma, Masayo, cordon-bleu's dance, 153

songbirds, model for human language learning, 159–60

song nuclei, 168

Songs of the Humpback Whale (recording), 175

song sparrow (*Melospiza melodia*), learning to sing, 165

song system, 168

Sonny and Cher, 153

Sorbonne University in Paris, 69

sound(s): frequency of, 14; intensity of, 12–13; physical nature of, 9; physics of, 100–101; producing, 99. *See also* acoustic sounds

sound frequencies, formants, 118

sound triangulation, principle of, 143–44

sound waves, 9–10; air pressure variations, 11–12; analogy of circles in water, 10; propagation media, 13; speed of, 10; variation in pressure, 13–14

source-filter theory, 116; vocalizations, 119–20

South Africa, Goegap Nature Reserve, 183–85, 187

southern pied-babbler (*Turdoides bicolor*), information encoded in call, 224

South Gate Campground, 203

sparrow, 147, 160; Savannah sparrow (*Passerculus sandwichensis*), 162–63; song sparrow (*Melospiza melodia*), 165

spear-nosed bat (*Phyllostomus discolor*), vocal learning, 177

species recognition, 310n12

spectacled caiman (*Caiman crocodilus*), 89, 95, 230; recordings of, 114. *See also* crocodiles

speed of light, 10

speed of sound, 10

sperm whale (*Physeter macrocephalus*), 75; codas of, 76–77; echolocation, 335n426; nonclick sounds, 318n147; vocalization of, 76

Spheniscidae family, penguins, 39

Spheniscus demersus (Cape penguin/African penguin), calls of, 283–84

Sphingidae family, 102

spiders: communication, 200; vibrations and sounds, 337n448

spoken language, 281; acoustic space of humans, 283; acquisition of, 291–93; attempts to teach apes, 287–89. *See also* language

spotted hyena (*Crocuta crocuta*), 204; graded vocalizations of, 206; hunting and killing prey, 204–5; influence of pheromones, 207; operant conditioning protocol for voice recognition, 209–10; recording vocalizations of, 210–11; social organization of, 205–6; social system of, 207; vocalizations of, 207–8, 211

Spottiswoode, Claire, indicator bird, 294

squamate antbird: illustration of, 25; propagation experiment, 24

stable evolutionary equilibrium, deer, 121

static, information in sound, 19

stochastic processes, 6

striped mouse (*Rhabdomys pumilio*): illustration of, 185; social flexibility of, 185; social systems, 184–85; ultrasound communication, 189–90; vocalizations of, 185–87

Struhsaker, Thomas, vervet monkeys, 214

Succulent Karoo Research Station, Goegap Nature Reserve, 183–84

Sueur, Jérôme: acoustic indices of biodiversity, 275–76; automating recordings, 278; development of acoustic ecology, 279; recording soundscape at biological station, 271; studying songs of cicadas and birds, 275

Sula nebouxii. *See* blue-footed booby (*Sula nebouxii*)

superb lyrebird (*Menura novaehollandiae*): song imitation, 165; vocalizations, 126

surucua trogon: illustration of, 25; propagation experiments, 24

Suzuki, Toshitaka, Japanese great tit signals, 223

swallow-tailed manakin: illustration of, 25; propagation experiments, 24, 26

swamp sparrow (*Melospiza georgiana*), 110; learning to sing, 165; vocal learning, 171

Sylvia atricapilla. *See* Eurasian blackcap (*Sylvia atricapilla*)

syntopic species, 270

syrinx, 42; voices of the, 42

Tachycineta bicolor (tree swallow), 58

Taeniopygia guttata (zebra finches), 60

Tanjung Puting National Park, Indonesia, 280

Taylor, Peter, 114; black caiman, 91–94; Caiman House, 93

Tchernichovski, Ofer, song system, 170–71

Tel Aviv University, 282

termites: drumming carrying information, 198–99; *Macrotermes natalensis*, 198; social insects, 198–99; vibrations by, 199

Terrace, Herbert, chimpanzee training, 288

Tettigoniidae family, 190

Theunissen, Frédéric, 158, 208, 327n294; auditory memory of zebra finches, 338n465; hyena communication, 79; hyenas' giggle, 209; learning process of birds, 170; song learning, 159; testing female zebra finches, 250–51; vocalizations of hyenas, 206, 207

Thévenet, Julie, emotions in vocalizations, 229

Thomas's langurs (*Presbytis thomasi*), audience effect, 154

Thorpe, William, learning vocalizations, 159

three-wattled bellbird (*Procnias tricarunculatus*), learning to sing, 160

thunder, lightning and, 10

Thyrophilus rufalbus (rufous-and-white wrens), vocal behavior of, 343–344n547

tiger (*Panthera tigris*), vocal cords, 122

timbre, 118, 119, 308, 323n231

Tinbergen, Nikolaas, 3; animal behavior, 3–4; behavioral development question, 85, 157; development over life, 4, 6; evolutionary causes, 4, 5–6; evolutionary history, 4, 6–7; mechanisms, 4–5; ontogeny of behavior, 160; question of origins, 127; questions for sound communication, 4

toads, 1, 196

toothed whales, 68; dolphins, 68, 70, 73, 75, 77; killer whale (*Orcinus orca*), 73–75; sound vibrations, 109; sperm whale (*Physeter macrocephalus*), 75. *See also* whales

Torres, Roxana, blue-footed booby, 244

Townsend, Simon, calls of babblers, 224–25

Toxostoma rufum (brown thrasher), learning to sing, 165

transduction, hearing, 100

tree swallow (*Tachycineta bicolor*), chick competition, 58

Troglodytes. *See* Eurasian wren (*Troglodytes troglodytes*)

Troglodytes aedon musculus (southern house wrens), noise and, 346n581

túngara frog (*Engystomops pustulosus*), 110

Turdoides bicolor. *See* southern pied-babbler (*Turdoides bicolor*)

Turdus migratorius. *See* American robin (*Turdus migratorius*)

turn-taking, vocal exchange, 286

Tursiops truncatus (bottlenose dolphin), 64

two-voice phenomenon, penguins and, 42–43

tympanic ears, moths, 195

tympanic membrane, frogs, 107

tyrannosaurus, 6, 7

Tyto alba (barn owl), 58–59

ultrasound, 14; communication, 187; definition, 189; detector by Pierce, 190–91; locating source of sound, 187–89; mouse communication, 187–89; mouse producing, 186; sound intensity, 187–88 underwater bioacoustics: bearded seal, 66–67; cetaceans, 68–69; dolphins, 68, 70, 73, 75, 77; fish on coral reefs, 67–68; killer whale (*Orcinus orca*), 73–75; sperm whales (*Physeter macrocephalus*), 75, 76; whale vocalizations, 70–73. *See also* baleen whales; whales

University College London, 292

University of Aberdeen in Scotland, 204

University of Belèm in Amazon, 142

University of Brisbane, 71

University of California: Berkeley, 79, 178, 205; Los Angeles, 196; San Francisco, 163

University of Campinas in Brazil, 15, 142

University of Canberra in Australia, 219

University of Cape Town, 294

University of Copenhagen, 29, 147

University of Hokkaido in Japan, 153

University of Illinois, 63, 174

University of Lausanne, 58

University of Liège in Belgium, 130

University of Lisbon in Portugal, 131

University of Marseille, 91

University of Massachusetts, 166

University of Michigan, 205

University of Nebraska, 200

University of Neuchâtel in Switzerland, 215

University of Orsay, 104

University of Paris Nanterre, 21, 26, 125, 163

University of Pennsylvania, 215

University of Saint-Etienne, 149, 258

University of St. Andrews in Scotland, 76

University of Sussex, 201

University of Tennessee, 266

University of Turin, Italy, 284

University of Vienna, 291

University of Windsor, Canada, 162

University of Winnipeg, Canada, 236

University of Zurich, 211, 224

Uraeginthus cyanocephalus (blue-capped cordon-bleu), courtship rituals, 152–53

valence, dimension of emotions, 231, 233

Vallet, Eric, A phrase by, 125

Vergne, Amélie, crocodiles, 90, 93–94

vertebrates: amphibians, 106–8; birds, 108; categorical perception, 109–10; frogs, 107–8; inner ear structure of, 105–6; mammalian eardrum, 108–9; marine mammals, 109

vervet monkeys (*Chlorocebus pygerythrus*), alarm calls, 214

Vidal, Eric, Agami heron program, 91

Vielliard, Jacques: birds of South America, 15; hummingbirds, 168; on *Lipaugus* recordings, 142–43; recording birds, 24; on white-browed warbler, 16

Vielliard, Malu, 142

Vignal, Clementine: zebra finches, 149–52; zebra finches and stress hormone, 239–40

vocal exchanges: rules characterizing, 285–86; turn-taking, 286

vocalizations, 2; body size and, 124–25; decoding emotions of, 234; elephants, 200–202; emotions in, 229–30; fundamental frequency, 117; harmonic frequencies, 117–18; mouse communication, 186–88; side of sender, 325n252; spotted hyenas, 206; ultrasound and striped mice, 189–90

vocal learning: closed-ended learners, 164; embryo of certain birds, 330n338; hormones, 170, 171; imitation of other species, 165–66; memory and, 168; open-ended learners, 164; oscine birds, 171; oscines, 160–61; process of, 167–69; Savannah sparrow (*Passerculus sandwichensis*), 162–63; sensitive (sensory) period, 161; sensorimotor phase, 162; songbird diversity, 164; wrens, 166–67; zebra finches, 161–62

vocal plasticity: animals, 179; study of, 181–82

voice recognition: dynamics of, 312n55; mother-pup, 312n55, 313n57

von Frisch, Karl, 3; dance of bees, 3

walrus (*Odobenus rosmarus*), 58; illustration of, 49; parent-offspring communication, 48–51; reproduction, 47

warbler, 1. *See also* white-browed warbler (*Basileuterus leucoblepharus*)

water: analogy of circles in, 10; throwing pebble into pond, 9–10

waterbird (*Anthochaera chrysoptera*), regent honeyeater singing song of, 173

Weberian apparatus, ossicles transmitting oscillations, 106

Weddell seal (*Leptonychotes weddellii*), 51. *See also* seals

Weldele, Mary: injury of, 208; recording hyenas, 208–9

western barbastelle (*Barbastella barbastellus*), ultrasound, 195

whales: acoustic communications, 69–70; baleen whales, 68–69; blowholes, 69; humpback recordings, 175; laryngeal sac, 69–70; songs of humpback whales, 71–73; sound vibrations, 109; toothed whales, 68; vocal cords of, 316n109; vocalizations, 72–73. *See also* baleen whales; toothed whales

Wheatcroft, David, field crickets and noise, 267

whistling language, 289

white-browed warbler (*Basileuterus leucoblepharus*), 14, 16, 134, 273; illustration of, 17, 25; information for managing social relationships, 24; males establishing territories, 18, 20; most common at Morro Grande, 16–17; neighbor recognition, 22–23; propagation of sound in forest, 18; recognition of song over distance, 19–20

white-shouldered fire-eye: illustration of, 25; propagation experiment, 24

willow warbler (*Phylloscopus trochilus*), 272

wind instruments, birds and mammals, 42

woodcock (*Scolopax rusticola*): illustration of, 173; song vocalization, 172

woodpeckers: drumming behavior in, 135–37; drumming of, 272; illustration of great spotted woodpecker, 136

wrens: forest experiment for, 27–28; noise and *Troglodytes aedon musculus* (southern house wrens), 346n581; signal propagation in forest, 28; video recordings of house wren, 56; vocal behavior of rufous-and-white wrens, 343–344n547; vocal learning, 166–67. *See also* Eurasian wren (*Troglodytes troglodytes*)

Wright, Orville, 242

Xanthocephalus (yellow-headed blackbird), 57

yellow-headed blackbird (*Xanthocephalus xanthocephalus*), 57

Zahavi, Amotz, handicap theory by, 121

Zalophus wollebaeki (Galápagos baby sea lion), 51–52

zebra finches (*Taeniopygia guttata*), 160; analysis of, 149–52; communication network in nest, 60–61; communication network of, 147–48; females distinguishing between male calls, 250–51; heat call of, 60; illustration of, 148; imitating song from loudspeaker, 170–71; learning to sing, 161–62, 164, 165; response to calls, 281; stress hormones and, 239–40; Theunissen's team, 338n465; vocal repertoire of, 148–49

Zipf's law, 284

Zuberbühler, Klaus, alarm calls of Diana monkey, 215–17

TO RAISE UP A NATION

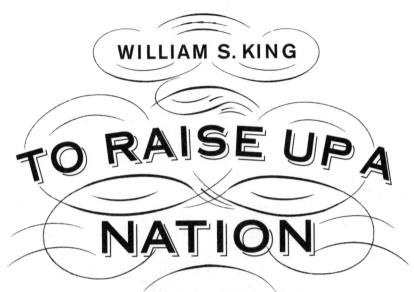

WILLIAM S. KING

TO RAISE UP A NATION

JOHN BROWN,
FREDERICK DOUGLASS
AND THE MAKING OF
A FREE COUNTRY

WESTHOLME
Yardley

Westholme Publishing, LLC
904 Edgewood Road
Yardley, Pennsylvania 19067
Visit our Web site at www.westholmepublishing.com

First Printing October 2013
10 9 8 7 6 5 4 3 2 1
ISBN: 978-1-59416-191-9
Also available as an eBook.

Printed in the United States of America.

I consider the War of Attempted Secession . . . not as a struggle of two distinct and separate peoples, but a conflict (often happening and very fierce) between passions and paradoxes of one and the same identity—perhaps the only terms on which that identity could become fused, homogeneous and lasting.

—Walt Whitman

CONTENTS

List of Maps   ix

Prologue   xi

  1. Life and Times   1

  2. Bleeding Kansas   32

  3. Raising an Army of One Hundred Volunteers   64

  4. Carrying the War into Africa   97

  5. An American Spartacus   129

  6. A Setting Possessed of Imposing Grandeur   163

  7. The Battle at Harper's Ferry   196

  8. Year of Meteors   239

  9. 1860   274

10. Terrible Swift Sword   315

11. His Truth Is Marching On   354

12. Buffeted by Sleet and by Storm   390

13. Lincoln's Emancipation   426

14. Men of Color, To Arms!   454

15. Battles for Liberty, Battles for Union   500

16. War for the Total Expiation of Slavery   535

17. One More River to Cross   575

Epilogue   614

Notes   622

Bibliography   642

Illustration Credits   654

Acknowledgments   655

Index   657

List of Maps

1. Distribution of Slaves in 1859   18–19

2. The Midwest, Northeast, and Canada   92–93

3. Harper's Ferry and Its Vicinity   214–215

4. Battles of the Civil War   410–411

PROLOGUE

SEVERAL EVENTS AT THE BEGINNING OF THE NINETEENTH CENTURY gave the impetus for the expansion of slavery in the southern portion of the new American Republic. The first was the so-called "cotton gin"—a mechanical invention allowing the easy separation of the cotton seed from the fiber, increasing by a hundredfold the yield of one day's labor cleaning cotton bolls. Another, at about the same time, was the black revolution in St. Domingo, which was to deprive Napoleon of his essential base in the New World. These proceedings were followed in 1803 by the United States purchase of France's vast holdings on the northern continent—the Louisiana Territory. All these developments were consequent to the beginning of large-scale cultivation of cotton as a cash crop.

Extensive planting in cotton began in South Carolina and Georgia; by 1820 these states were growing half the cotton in the United States. By the 1830s cotton production was ranging into the Gulf states, which in only five years would exceed the output in the Atlantic states; an astonishing development that brought as its twain redoubled demand for black slaves. By 1830 the number of slaves utilized in the burgeoning Southern economy surpassed two million, twenty years later there were three million, and on the eve of the contest for their freedom, there were nearly four million blacks laboring in bondage in the American South.[1]

Production of staple crops for export—whether cotton, tobacco, rice, sugar, or hemp—could only be remunerative using slave labor if carried out by large gangs on expanses of naturally fertile soil. But as this system flourished, the sullen presence of a black population formed a shadow over the minds of the slave owners and other whites in the South. This was not just a manifestation of guilt, if such existed, but sprang from recognition of the fact that as blacks continued to increase in proportion to whites, no amount of coercion could make them tamely submit.

A fateful decision had been made in the early decades of the nineteenth century. As the slave economy flourished, both to maintain its extraordinary profitability and in a contest for political hegemony in the national government, the planters needed an outlet in the opening of new territory. This became the basis for the famous compromise measures adopted between 1820 and 1850, providing a new slave state to offset the addition of any new free state, which, along with the South's increasing sway over the northern Democratic Party, would ensure the region's dominance in the Senate, and control of the presidency. But an equally portentous factor was strikingly formulated by Senator Robert Toombs of Georgia when he declared at the Montgomery secessionist Congress in 1860 that "in 1790 we had less than eight hundred thousand slaves. Under our mild and humane administration of the system they have increased above four millions. The country has expanded to meet this growing want, and Florida, Alabama, Mississippi, Louisiana, Texas, Arkansas, Kentucky, Tennessee, and Missouri, have received this increasing tide of African labor; before the end of this century, at precisely the same rate of increase, the Africans among us in a subordinate condition will amount to eleven millions of persons. What shall be done with them? We must expand or perish."[2]

As the decades passed there developed across the South broad areas of adjacent counties where blacks predominated. Throughout these areas there was, without doubt, an extraordinary and growing communication among black communities—the holler, voices, signals (conditions and understandings)—passing through fields, across plantations, among families and into neighborhoods, and into the towns and villages along all the roads of the South. A "mysterious spiritual telegraph," as one observer called it, linking slave quarter to slave quarter, spoken without the hearing of the whites—as it were, out the back door of the big house.[3]

As the whites fled into their obdurate denial of black humanity, everywhere they turned they faced this quandary, and they were not prepared to accept it. But even as he sought to repose behind the façade of supposedly more advanced civilization, the planter was ever mindful of the warning in the slave song:

> You might be rich as cream
> And drive you coach and four-horse team,
> But you can't keep de world from movin' round
> Nor Nat Turner from gainin' ground.

In W. E. B. DuBois's biography of John Brown—an alternative to a study of Nat Turner his publisher had rejected—one reads an intriguing tale of a clandestine rendezvous in an abandoned stone quarry outside Chambersburg, Pennsylvania, between Frederick Douglass and John Brown. Also in attendance were the young adjunct and confidant of Brown named John Henry Kagi, a talented journalist, and a fugitive slave from South Carolina and young protégé of Douglass who went by the name Shields Green. This quartet had settled down among the stones and detritus of an old quarry fifteen miles above the Mason-Dixon Line to discuss Brown's pending campaign into the Blue Ridge Mountains of Virginia. He proposed to fortify and hold an area of the southern states with an army of freed slaves and their abolitionist allies, black and white, from the northern states and Canada.

The only firsthand account of this "council of war," as Douglass later called it, is to be found in his autobiography, *Life and Times of Frederick Douglass*. There one reads it was only at this meeting, after extensive planning and cooperation with Douglass, that Brown designated the seizure of the United States Arsenal at Harper's Ferry, Virginia, as the inaugural stroke for a war for the liberty of the American slave. The blow was intended to resound as a tocsin across a divided nation—alarming southern slaveholders into carrying out their threats of secession, while rousing the northern people to the threat posed by slavery. It was also designed to be heard by the slaves across the South and serve to rally them to his standard.

Not only did Brown want Douglass's opinion about the foray; if favorable, he wanted him to join in it as co-leader. This, indeed, was stuff portending a strong relationship between the men. The denouement came after what was surely a comprehensive and far-ranging debate over many hours of two days—August 20 and 21, a Saturday and Sunday—eight weeks before the Harper's Ferry raid. Finally convinced he will not be able to dissuade Brown from his course, Douglass declares he cannot countenance such an action. In announcing his decision he rises, saying he will be returning to his home in Rochester. Turning to Green, whom he had brought as a recruit to Brown's original plan, Douglass tells him because that plan has changed in favor of the one discussed, Green should feel free to return with him. It was evident to Douglass, and he said as much, that Brown and any who joined him would only be hastening to their graves in Virginia.

Then Douglass recounts Brown's words—words revealing a deeply intimate bond between them. As Douglass wrote, putting his arm around him "in a manner more than friendly," Brown said in his inimitable way: "Come with me Douglass; I will defend you with my life. I want you for a special purpose. When I strike, the bees will begin to swarm and I shall want you to help hive them." But Douglass was unmoved by his friend's entreaty. Turning to Green, whom he expected to leave with him, Douglass was surprised as Green coolly said, "I b'leve I'll go wid de ole man."

One

Life and Times

The romance of American history will of course be found by posterity in the lives of fugitive slaves.
—Thomas Wentworth Higginson, *The Liberty Bell,* 1858

ALTHOUGH HE IS FAR FROM NEGLECTED IN AMERICAN HISTORY and letters, even the best treatments of John Brown have a curious hollowness. Often strewn with the incisive pronouncements and deeds of a man radiating a strange magnetism—like molten ore thrown up through fissures opened across geological time—this unreality comes not from their subject, but from the tone deafness of nearly all biographers and historians to the whole man. In the most original chapter of his *John Brown,* titled "The Black Phalanx," W. E. B. DuBois wrote: "To most Americans the inner striving of the Negro was a veiled and unknown tale. They had heard of Douglass, they knew of fugitive slaves, but of the living, organized, struggling group that made both these phenomena possible they had no conception." It is with this narrative that a fuller understanding of John Brown begins, for as the scholar noted, of all white Americans he came "nearest to touching the real souls of black folk."

Born in 1800, Brown was in religious training and belief from a Calvinists lineage, interpreting events in the light of predestination, and viewing earthly life as but a probation for a higher life. So as he began to formulate and struggled with his plan of emancipation at last he came to feel he was chosen by God to deliver those in thralldom in America, but what distinguished him, above all, was his conviction that it was a moral and political imperative to liberate the veritable Africa enslaved in the midst of the United States, and of establishing blacks' full political rights as citizens on an equal footing with their oppressors.

In broad outline the narrative begins in the second decade of the nineteenth century as slavery withered in the northern states, from whose growing free black population recognized leaders began to emerge. The first black paper, *Freedom's Journal*, was established in 1827. Two years later, as slavery became the linchpin of the new industrialism, a Bostonian black named David Walker startled the South with the publication of his *Appeal to the Coloured Citizens of the World*. A sustained truncheon-like assault on slavery and racism, Walker's polemic predicted a revolt whose fulfillment came in another two years, when a slave with a monosyllabic first name led a revolt in Southampton, Virginia, still perplexing to many today.

The 1830s opened with the first national Colored Convention in Philadelphia; the decade was also one of severe trial for blacks and persecution for northern abolitionists. In several towns riots were organized against the growing black presence, and in 1835, four years after the inaugural issue of the *Liberator* was brought out by Benjamin Lundy and editor William Lloyd Garrison, the latter was pulled by a mob through the streets of Boston with a halter around his neck. Shortly after this too in St. Louis the editor of the *Observer*, Rev. Elijah Lovejoy, was driven from the city for protesting the lynching of a black man. Establishing himself across the river in Alton, Illinois, he saw his printing presses destroyed three times. Persisting a fourth time, he was shot and killed while defending them.[1] At the decade's end, Africans led by Joseph Cinqué destined for enslavement in Cuba mutinied aboard the *Amistad*; conveyed to New London, Connecticut, they stood trial for murder and were acquitted.

In the 1840s a new era opened as the Underground Railroad was up and running, while in many towns and communities in the North blacks were making giant strides in self-organization—forming societies and associations for self-help and protection, and founding churches, schools, and businesses. By then Frederick Douglass too was free and began taking a prominent part in the antislavery campaigns of William Lloyd Garrison. All of this and more was known to John Brown; in the decade and a half before the consummation of "his work," as he would say, he was ever anxiously looking for blacks in whom to confide. When they first met in Springfield, Massachusetts, in 1847, as Douglass tells it, Brown empathically remarked to him, that "now he was encouraged, for he saw heads of such rising up in all directions."

Foremost among these daring individuals "rising up" to make the new self-articulating black leadership was Harriet Tubman, recently self-emancipated from the eastern shore of Maryland, beginning the first of more than a dozen forays as the "Moses of her people." In Troy, New York—after escaping together with his entire family from New Market, Maryland—was Henry Highland Garnet, the radical cleric beginning his first ministry as a Presbyterian in 1843. In Pittsburgh there was the champion of African manhood and nationality, Martin Delany, whose education would include a year at Harvard training as a surgeon. In Syracuse, New York, there was a fugitive slave from Tennessee, now of the African Methodist Episcopal Church, the Rev. Jermain W. Loguen, who would dedicate himself after 1850 to assisting fugitive slaves full-time.

There was as well the magnificent William Still, born in New Jersey, who became secretary of the Vigilance Committee in Philadelphia, and through whose hands would pass more than three thousand bondsmen. Men like Lewis Hayden in Boston's Beacon Hill, William Lambert in Detroit, and William Parker of the Christiana Fugitive Slave Rescue, later in Buxton, Ontario, and Mary Shadd Cary in Chatham, Ontario, all of whom, and many like them, became confidants of John Brown.

All of these, and the many like them but for their prominence, became contributors to what DuBois described as the African American's "rich inner development." "Unsensed and despised though it be," the scholar wrote, America's greatest destiny "is to give back to the first of continents the gifts which Africa of old gave to America's fathers' fathers."[2] That bequest has been our deepest striving for freedom: for had there been no Nat Turner there would have been no William Lloyd Garrison; as had there been no Elijah Lovejoy there would have been no John Brown. For it was Turner who gave Garrison the impetus to build organizational support for the idea of immediate emancipation; as it was at a memorial meeting for Lovejoy, seeing that the crisis he awaited had come, that Brown first arose publicly to avow his hatred for slavery and dedicate his life to its destruction.

Foremost among the "heads" Brown saw "rising up" was a slave born in 1818 in the vicinity of Easton, Maryland, who toiling in Baltimore at age twenty-one emancipated himself from his master by running away. Traveling with borrowed documents and in the uniform of a free black seaman, he made his way aboard the railroad cars for destinations north of the Mason-Dixon Line. His betrothed, Anna, also a fugitive from Baltimore who had provided the money for both of them to escape, soon joined him in New York City, where they wed. Passing on to Newport in Rhode Island, the newly emancipated couple were met by abolitionists from New Bedford, Massachusetts, a magnet for escaping slaves, who conveyed them to that town. The man's name was Frederick Bailey. It was in that seaport town and center for whaling in the northeast, and for shipbuilding and copper foundries with ready work for all takers, that Bailey soon found employment as a laborer. It was a free black in New Bedford, Nathan Johnson, who suggested to him the name Douglass from a character in Sir Walter Scott's *Lady of the Lake*, the "black Douglass."

After several months in freedom, offered a subscription to the *Liberator*, Douglass was thus "brought into contact with the mind of Mr. Garrison, and his paper." Soon, too, he had an opportunity to hear its editor in person, and began mastering for himself the ideas of the emergent antislavery movement. It was not long before Frederick Douglass began speaking before his new community in "a little schoolhouse on 2nd Street." In 1841, attending an antislavery convention on Nantucket Island, the novitiate was invited to make some remarks by a member of the convention who had heard him speak. His halting, stammering, but heartfelt words had a great effect, immediately followed by Garrison, who made the fugitive slave his text.

Thus was formed a compact, and the younger man induced to join the Massachusetts Anti-Slavery Society, and enlisted to take on duties on the speaking circuit. "Here opened to me a new life," Douglass wrote. His first assignment: to travel the western counties of Massachusetts, together with a mentor, securing new subscriptions for the *Anti-Slavery Standard* and the *Liberator*. Billed as a fugitive slave recounting his own story in his own words, he was a rarity at the time—the first one out. In the beginning his speeches were narratives. "Give us the facts," his colleagues would urge him, "we will take care of the philosophy." But as he mastered the principles of the antislavery debate—and knowing his own story—it did not entirely satisfy him to narrate wrongs. He felt like denouncing them.

Two difficulties arose as Douglass came to public prominence. By speaking about his life as a slave and advertising himself as a fugitive, he might come to the attention of his former master who could come to claim his "property" under United States law. Second, as his abilities grew, his delivery becoming more confident and polished, those who heard him wouldn't believe he

Frederick Douglass, circa 1847.

had ever been an illiterate slave. Hence his mentors urged him to keep his speeches plain, even encouraging him to "throw in a little plantation speech."

What they feared finally happened; audiences began to doubt a man of such bearing and eloquence could have been a slave. This induced Garrison and his colleagues to encourage Douglass to write his superb *Narrative of the Life of Frederick Douglass*, one of the first of its kind in American letters, and the first of three autobiographies he was to write. It was then decided given the danger to him he should go to England, where his autobiography had been immediately popular. The year was 1845.

Abroad Douglass would travel throughout the British Isles, appearing at numerous antislavery engagements in England, Scotland, Wales, and Ireland, where he listened to Temperance lecturers, taking part in their campaigns, and heard the anti–Corn Law agitation. During these months too he was to develop friendships in reform and antislavery circles, meeting some of the old-line abolitionists like Joseph Sturges, who had founded the British and Foreign Anti-Slavery Society, and George Thompson of Edinburgh. Fascinated by those with the gift of tongue, Douglass listened to the best orators of the day, including Ireland's Daniel O'Connell. The greatest impression on him, however, as he remarked in *Life and Times of Frederick Douglass*, was the unquestioned authenticity of his manhood. Unlike in America, on the omnibus he was not required to ride on the exterior, nor in the cars in that designated for smoking, nor in steerage on the steamship.

After twenty months abroad, in a controversial gesture, when
about to leave, his new friends raised money to buy his freedom, con-
tacting his former owner who agreed to settle for £150 for his property, or $700. Viewed as a tacit acknowledgment of ownership of a
person as a commodity, this arrangement aroused controversy among
abolitionists; but with the sum paid, Douglass was handed his manumission papers and the threat of his apprehension as a fugitive slave
in the United States was removed. At a testimonial given on his departure, these same friends presented him with a gift of $2,200 to buy a
printing press to start a paper for the benefit of the struggle for emancipation in America. It was also an acknowledgment of his God given
intellect and of his developing effectiveness.

In spring 1847 Douglass returned to the "many trials and perplex-
ities" of American life. At the end of a fifty-year career he wrote in
his *Life and Times*: "My friends in Boston had been informed of what
I was intending, and I expected to find them favorably disposed. . . .
In this I was mistaken. They had many reasons against it. First, no
such paper was needed; secondly, it would interfere with my useful-
ness as a lecturer; thirdly, I was better fitted to speak than write."

His colleagues looked at him with astonishment: "a wood-
sawyer" offering himself as an editor. But more than that, his proposal was seen as a sign of an unseemly ambition, a disregard for their
well-meant advice. For a few months Douglass relented. But convinced of its necessity, and more determined than ever to bring out a
paper, though not in competition with the *Liberator* or the *Anti-slavery Standard*, he soon moved himself and his young family—a wife,
a daughter, and two sons—to Rochester, New York. After the completion of the Erie Canal the city was a boomtown not unlike Chicago
would become in the following decade; it was here he would begin his
North Star. But Douglass also had a practical reason for coming to
reside in Rochester; the Underground Railroad to which he was
attached had its stations in Baltimore, Wilmington, Philadelphia,
New York, Albany, Syracuse, Rochester, and St. Catharines, Ontario.
The Douglass home would be a way station for over a decade to hundreds of fugitive slaves.

In Massachusetts Douglass had worked with eminent women,
including Abby Kelly, Maria Weston Chapman, and Lydia Maria
Child, the editor of the *Standard*. In Rochester, where Elizabeth Cady
Stanton resided along with many persons sympathetic to reformers,
he became as full a participant in the women's rights campaigns as

any man in the nineteenth century, beginning a long relationship with many principals in the movement. Stanton invited Douglass to the seminal convention on women's rights in Seneca Falls in May 1848, and he served there, as in subsequent conventions as vice president, and took the floor. The first edition of his paper began with the phrase, "Right is of no sex." Douglass was soon joined, too, in his new endeavor by a young English woman who would be indispensable as his editorial and intellectual coadjutant, Julia Griffiths, a collaboration that has yet to receive the attention it merits.

Shortly after establishing himself in Rochester and starting his paper, Douglass was scheduled to lecture in Springfield, Massachusetts. Arriving in that city he received an invitation to meet a man whose name he had heard mentioned "by several prominent colored men." In speaking of him, Douglass wrote in *Life and Times,* "their voices would drop to a whisper . . . and what they said made me very eager to see and to know him. Fortunately, I was invited to see him in his own house."

The man to whom he referred was John Brown, who had conveyed an invitation to Douglass to meet him at his business, an apparently thriving wool dealership on a prominent and busy street. When Douglass arrived, as the hours of business were ending, he was invited to accompany Brown to his home, where they could talk. Coming to the residence Douglass was surprised to discover it was not well appointed like the business, nor in an elegant neighborhood; instead he found that his host and his family lived in a plain wooden house on a back street populated by laborers and mechanics. Douglass wrote:

> It is said, a house in some measure reflects the character of its occupants; this one certainly did. In it there were no disguises, no illusions, no make-believe. Everything implied stern truth, solid purpose, and rigid economy. I was not long in the company with the master of this house before I discovered that he was indeed the master of it, and was likely to become mine too if I stayed long enough with him. He fulfilled St. Paul's idea of the head of the family. His wife believed in him, and his children observed him with reverence. . . . Certainly I never felt myself in the presence of a stronger religious influence than while in this man's house.

After partaking of a meal of beef, potato, and cabbage "passed under the misnomer of tea," the two went into the family parlor to talk privately. He had, Brown told his guest, been following his course both at home and abroad, and had been eager to meet him. He had long contemplated a way to compel the speedy emancipation of the slaves and had been looking for trustworthy coadjutors among his race in which he could confide, and he had invited him into his home to lay that plan before him. Douglass wrote: "He denounced slavery in look and language fierce and bitter, thought that slaveholders forfeited their right to live, that the slaves had the right to gain their liberty in any way they could, [and] did not believe that moral suasion would ever . . . abolish the system. . . . When I suggested that we might convert the slaveholders, he became much excited, and said that could never be, 'he knew their proud hearts and that they would never be induced to give up their slaves, until they felt a big stick about their heads.'"

A forcible separation of master and slave was necessary; American slavery, he held, was a systematic war carried out by one portion of its citizens against another, and could only be destroyed by war. Peaceful abolition would only lead to renewed subjugation under another form. As no people could have self-respect nor be respected who did not fight for their freedom, the practice of carrying arms would be a good one for blacks to adopt; it would give them a sense of their manhood. Debate on these issues, moreover, was something that should be encouraged, but it had long been his opinion, the sooner this contest was begun the better. Sooner or later, too, the question would have to be contested on the very ground where the slaves were actually held.

Brown has been called a man of one idea, but an idea of the magnitude he held could only develop in a struggle and within a context not of the author's making. He must test himself; thus the idea grows, maturing through an activity that will later make it manifest. There is no doubt that after Lovejoy's murder he began to contemplate how to affect the liberation of the slaves of the South, and began examining the issue from a military perspective. But for any plan conceived he would need the cooperation of trustworthy collaborators, and he would require financial backing to procure arms and equipment. So in the early years while he was carrying out this work his activity can only be seen in his contacts with a carefully chosen few, in confidences taking place behind closed doors, and if his plans were passed along at all, it was only in hushed tones. Beyond this there is very little to discern. But the parameters of Brown's thinking took shape within the bold defense of the liberty of fugitive slaves, proposals for a broad educational and social philanthropy, and the strategic extension of the Underground Railroad.

There is evidence that by the early 1840s Brown began studying the historical precedents for carrying out guerrilla war, including the slave revolts, particularly in the United States, and in Haiti and Jamaica. No better example of the efficacy of armed resistance at that time could be found than that of the maroons on the island of Jamaica.

John Brown in 1856.

While oratorical philanthropy had played a part, the emancipation in 1833 was the result of decades-long struggle, as Brown and others had seen, and that it had been concluded the maroons were militarily unconquerable. Brown's eye, then, was drawn to the Allegheny range, as he keenly assessed its strategic importance. His cousin, the Rev. Edward Brown, near whom he resided in Hudson, Ohio, from 1835 to 1840, related that he "often remarked that, with a good leader, the slaves escaping to those fastnesses and fortifying themselves, could compel emancipation." Another early confidant was George B. Delamater, who knew Brown from 1830 to 1845, and related sometime after Brown's death that "an outline of a plan given me by John Brown in 1842" had resemblance to that which ultimately transpired.

Douglass gives one of the most extended accounts of the plan in his *Life and Times*, writing that as they discussed the issue in the parlor of his home Brown drew his attention to a map of the United States. Pointing to the far-flung Allegheny range, he said: "These mountains are the basis of my plan. God has given the strength of these hills to freedom; they were placed here for the emancipation of the Negro race; they are full of natural forts, where one man for defense will be equal to a hundred for attack; they are full also of good hiding-places, where large numbers of brave men could be concealed, and baffle and elude pursuit for a long time."

In spare Brown proposed to take a body of selected men into the mountains between the slave and the free states to establish a secure base. The most persuasive and judicious of these were to go down to the fields, as opportunity offered, and induce slaves to join them, seeking and selecting the most restless and daring. The true object to be sought, Brown maintained, was to undermine the value of slave property by making it insecure.

"But," objected Douglass, "suppose you succeed in running off a few slaves, and thus impress the . . . slaveholders with a sense of the insecurity in their slaves, the effect will be only to make them sell their slaves further south." "That," Brown answered, "will be what I want first to do; then I would follow them up. If we could drive slavery out of one county, it would be a great gain—it would weaken the system throughout the state."

"But they would employ bloodhounds to hunt you out of the mountains."

"That they might attempt, but chances are, we should whip them, and when we should have whipped one squad, they would be careful how they pursued."

"But you might be surrounded and cut off from your provisions or means of subsistence." That, Brown thought, could not be done so they could not cut their way out. But even if the worst came, he had no better use for his life than to lay it down in the cause of the slave.

To the discerning reader of Douglass's account of this meeting in 1847 it is apparent—and many years later, first Franklin Sanborn and then both Richard Hinton and John Brown Jr., would draw attention to this discrepancy—that Douglass unconsciously mixed elements from this discussion with subsequent discussions with Brown a decade later. The knowledgeable reader may notice that some of these amplifications, that only the experience of a decade would provide, have been dampened here. But thus began a relationship that would deepen into something more than friendship, ending in their final meeting outside Chambersburg, Pennsylvania.

Through the intervening years Douglass and Brown were to remain in regular contact and correspondence, neither passing the other's town of residence without paying a call; and Brown was never without a subscription to Douglass's paper, ensuring that members of his large family also had subscriptions. The importance of the relationship is evident from Douglass's own words, when he wrote: "From this night spent with John Brown . . . while I continued to write and speak against slavery, I became all the same less hopeful of its peaceful abolition. My utterances became more and more tinged by the color of this man's strong impressions."

One of those bringing Brown to Douglass's attention was Henry Highland Garnet, who had been a student at the Oneida Theological

Institute, a Presbyterian-affiliated school
marked as innovative in its day. Ordained a
pastor at the Liberty Street Presbyterian
Church in Troy, New York, Garnet was one
of a number of black leaders who began
enunciating an opposition to American slavery beyond the ethical and moral grounds of
most abolitionists, mounting a critique of the
institutional basis of slavery and racism. At
the antislavery convention in Buffalo, New
York, in 1843, which brought abolitionists
of the eastern and western states together for
the first time, Garnet had delivered a significant discourse titled "Address to the Slaves

Henry Highland Garnet
in 1865.

of the United States of America." Pronouncing for the right of the
enslaved to forcibly resist their oppressors, the reaction of the delegates, including the already prominent Frederick Douglass, had been
a strong one. Some considered it inflammatory, and Garnet was
denounced as a demagogue, while the *Liberator* published an editorial suggesting Garnet had fallen under "bad counsel."

Garnet's speech rightly was considered revolutionary, and it
became an important milestone in the antislavery debate. American
slaves, he held, had the right to resist, and all others had a duty to
assist them. In April 1848 in a publication that he noted had been
paid for by a person with the initials "J. B.," Garnet was to republish
David Walker's 1830 *Appeal* together with his own address. Then in
the second week of May, an article appeared from the hand of Garnet
in Douglass's *North Star* that read: "This is a revolutionizing age; the
time has been when we did not expect to see revolutions; but now we
expect them, and they are daily passing before our eyes; and change
after change, and revolution after revolution will undoubtedly take
place, until all men are placed upon equality."

While it may appear that the movements culminating in Europe in
1848–49 had no counterpart on the western shore of the Atlantic,
there is no doubt that there was reciprocity between the continents in
the minds of contemporaries. Particularly among abolitionists, they
saw their movement akin to these revolutions. Following the first
news of renewed strife in France in 1851, which proved to be
Napoleon III's coup d'état, Brown wrote his wife, in a letter dated
November 22: "There is an unusual amount of very interesting things

happening in this and other countries at present, and no one can fore-
see what is yet to follow. The great excitement produced by the coming of Kossuth, and the last news of a new revolution in France, with
the prospect that all Europe will soon again be in blaze seem to have
taken all by surprise. I have only to say in regard to these things that
I rejoice in them from the full belief that God is carrying out his eternal purpose in them all."[3]

But regardless of the sympathy expressed for democratic revolution by Garnet, Brown, and others, in the wake of their defeat the
connection between the continents was most forcefully felt as the
world's model republic became the refuge for many of Europe's revo-
lutionaries, especially Germans.

The coming of the Hungarian revolutionist Louis Kossuth particularly occasioned excitement and anticipation throughout the United
States, especially among abolitionists who hailed him as a natural
ally, expecting that he would champion the cause of the American
slave as fully as that of the Hungarian people. Five days after his
arrival, Kossuth met with a reception committee on behalf of "the
Negro people" and was addressed by George T. Downing. Calling the
struggle of blacks in the United States the kindred struggle of those
taking place in Europe, Downing declared, "We would express the
deep sympathy we feel in you, because of the relation you sustain to
Liberty." The appeal ended with this: "Respected sir, your mission is
too high to be allied with party or sect, it is the common cause of
crushed, outraged humanity. May you, when you leave our shores . . .
carry with you the sympathy of all, the active countenance of all."[4]

As "press, politicians, and society," howled down upon the
"uncouth" abolitionists for annoying a minister on a state visit, saying he wished to escape "entanglements," Kossuth disdained speaking on the subject of slavery. Embarking on a whirlwind tour through
both North and South, with a meeting with President Millard
Fillmore and an address before the Congress, Kossuth did not violate
his vow of silence. Undaunted abolitionists continued their appeal,
producing a remarkable pamphlet titled "A Letter to Louis Kossuth."

Frederick Douglass wrote anyone wishing to escape "entanglements" must blind himself to "the lineaments of our national face."
He went on: "The letter to Kossuth is a most searching production,
and if he be not insensible to the claims of that justice he so eloquently advocates, he must be convinced by it, that he has bestowed eulogies on this nation not deserved; that he has been playing into the

hands of tyrants worse than Austria ever knew, and that he has inflicted a wound on the cause of freedom which he cannot too speedily do his utmost to heal."

With its business office in Springfield, Massachusetts, Perkins and Brown, through its manager John Brown, sought to initiate a system for grading wool in an attempt to bring higher prices for American growers. Hundreds of growers from New York, Pennsylvania, and northern Ohio had consigned their wool to the firm to be sold at discretion. Brown wrote of these affairs with an acknowledgment toward other matters in a letter dated March 24, 1846, to his eldest son, John Jr.: "I am out among the wool-growers, with a view to next summer's operations; our plan seems to meet with general favor. Our unexampled success in minor affairs might be a lesson to us of what unity and perseverance might do in things of some importance."

One of Brown's early biographers who became one of his intimates, Franklin Sanborn, remarked, "But New England manufactures combined against him," forcing him to seek markets in England. Embarking in August 1849, Brown would be disappointed when American wool was coolly received by British markets. As grading of the wool for sale was postponed, Brown took the opportunity for an excursion to the continent. August 29 and 30 found him in Paris, then journeying swiftly through Germany and Austria, where another of Brown's intimates to become a biographer, Richard Hinton, wrote that he "inquired into moral, social and economic conditions and results," noting with a critical eye European agricultural practices. Visiting some fortifications along his route as well as some battlefields of Napoleon, Brown made his own comments on European military systems and equipment, thinking Napoleon "wrong in several points of strategy." The stronger ground, Brown maintained, should be along a ravine rather than a hilltop; an insight he would adapt for his mountain campaign in the United States, together with modifications of a type of earthen redoubt he had inspected in Europe.

In business outlook, his early biographers suggest, Brown was philanthropic. In the wool dealership in which he was co-proprietor he "proposed nothing Quixotic or impractical," DuBois surmises, "but he did propose a more equitable distribution of the returns of the whole wool business between the producers of the raw material and the manufacturers." With regard to social and economic forms, few have delved into Brown's thinking. In an interview some years later with the Kansas correspondent to the *New York Tribune*, William Addison Phillips, Brown said "He thought society ought to be organized on a less selfish basis; for while material interests gained something by the deification of pure selfishness, men and women lost much by it." All great reforms, "like the Christian religion," he held, "were based on broad, generous, self-sacrificing principles," and "he condemned the sale of land as a chattel, and thought that there was an infinite number of wrongs to right before society would be what it should be."[5]

Brown's venture into the British wool market met with disaster. When no buyers could be found at the prices sought, he was forced to sell at a loss. After his return to the States he would spend several years defending the firm against lawsuits from growers. But he had also returned having garnered valuable lessons. Although revolution had everywhere been defeated by the time of his visit, the mass insurrectionary upheavals and partisan fighting in the previous year, together with the observations he had made, stimulated Brown's theorizing on the American conditions for such warfare.

In commemoration of the twelfth anniversary of emancipation in Britain's West India colonies, Gerrit Smith, a wealthy landowner, philanthropist, and leading political abolitionist, donated 120,000 acres in 30- and 40-acre lots in Franklin and Essex counties in the Adirondack region of New York State to black families. He did this not only to give them a chance to become independent, self-sustaining farmers, but so that each proprietor could qualify to vote under New York's $250 property requirement for black males. But the land was in an inhospitable climate and most of the settlers were city dwellers accustomed for the most part to working in service trades. Worse, some of the settlers were cheated by a land surveyor and by hostile neighbors, and the settlement—called Timbuktu by the blacks—soon began to fail. Hearing of this and desiring to put it right, and undoubtedly to develop a base for his projected operations, Brown had written to Smith at his estate in Peterboro in April 1848: "I am something of a pioneer; I grew up among the woods and wild Indians of Ohio and am used to the climate and the way of life that your colony find so trying. I will take one of your farms myself, clear it up and plant it, and show my colored neighbors how such work should be done; will give them work as I have occasion, look after them in all needful ways and be a kind of father to them."[6]

Obtaining agreement with Smith, Brown went up to the Adirondack colony and surveyed the lands, establishing proper boundaries on the farms that remained, and also a deed for himself and his posterity for 244 acres. On one of his trips he also arrived with a fugitive slave, a boy named Cyrus. His greater purpose was to undermine slavery by drawing off the slave wealth of the South, and so he may have sought a secure base for operations four hundred miles north of the Mason-Dixon Line. Brown's object too, as Thomas Wentworth Higginson pointed out, was finding among the men who would come to live in the colony "coadjutors" for his plan. Higginson added: "He was not wholly wrong, and yet he afterwards learned something more. Such men as he needed are not to be found ordinarily; they must be reared."[7] Thus the nucleus he came to rely on would be drawn first from his own family; four of them sons from his marriage to his first wife who had died in childbirth—John Jr., Owen, Frederick, and Jason; and three sons from a second marriage—Salmon, Watson, and Oliver. To these would later be added by the marriage of two of his daughters, Ruth and Anne, their husbands William and Henry Thompson, and their brother Adolphus, whose family farm was near the Brown farm at North Elba in Essex County, New York. DuBois wrote that Brown "probably looked toward the use of Negro allies almost exclusively outside his own family. This was eminently fitting," as he surmises Douglass must have urged, "but impractical." For "white men could move where they would . . . , but to introduce an armed band . . . of Negroes from the North into the South was difficult, if not impossible. Nevertheless, some Negroes of the right type were needed and to John Brown's mind the Underground Railroad was bringing North the very material he required."[8]

When he returned from England in the fall of 1849, Brown moved his family from Springfield to a leased farm in the Adirondacks. Involved for the next several years both in the northern wilderness and in contesting various lawsuits brought in different cities as a result of the collapsed wool venture, he wrote: "I can think of no place that I think I would sooner go; all things considered than to live with these poor despised Africans to try & encourage them & show them a little as far as I am capable how to manage." The Brown family was considerably dispersed in the early 1850s.

Brown, engaged in several legal claims, appeared at various times in Springfield, Boston, Troy, and Akron. Three of his older sons, and a daughter and son-in-law, lived in Ohio, near Hudson and Akron, respectively. Remaining on the Adirondack farm under the care of Owen, were Brown's wife and three younger daughters, and two boys nearing manhood.

In all this seemingly diverse activity Brown never lost sight of his main interest, and was particularly keen about remaining in touch with Douglass. He wrote in a letter to his wife, dated December 22, 1851, "I wrote to Fred'k Douglass a few days since to send on his paper, which I suppose he has done before this, as he very promptly acknowledged the receipt of my letter." In the following year from Troy, he wrote his daughter and son-in-law in Akron on the same subject: "My journeys back & forth this winter have been very tedious. If you find it difficult for you to pay for Douglas [sic] paper, I wish you would let me know as I know I took some liberty in ordering it continued."[9]

The Brown progeny were abolitionists after their father, but none was as strong as he in matters of religious faith. Of all his children, John Jr. and perhaps Owen were his closest confidants. John Jr. tells in a reminiscence how the family came to join the Congregational Church in 1837 in the town of Franklin, which is now Kent, Ohio. Free blacks and fugitive slaves came to the services, sitting in segregated pews. Seeing this, Brown invited them to take his family's pew, which they did, and his family took their seats in the back of the church. Soon afterward the family moved to Hudson, near Cleveland, and found themselves cut off from the church. When a letter arrived informing them of this, John Jr. wrote, his father became white with anger: "This was my first taste of proslavery diabolism that had intrenched itself in the church, and I shed a few uncalled for tears over the matter, for instead I should have rejoiced in my emancipation. From that date my theological shackles were a good deal broken, and I have not worn them since (to speak of)—not even for ornament."[10]

The escape of slaves to the northern states and Canada by the 1850s represented an annual loss to the southern economy in excess of $200,000. From a slaveholder's point of view, these runaways were also affecting a growing antipathy for slavery in the North. Making matters worse, many fugitives were becoming active participants in the antislavery crusade. To staunch this hemorrhaging of wealth and prestige the South demanded a more effective Fugitive Slave Law than that constitutionally implemented by the act of 1793. The resulting legislation, known as the Compromise of 1850, was intended to strike a new balance between sectional demands by providing for the admission of California—wrested by war from Mexico—as a free state, but stipulating that henceforth the question of slavery in new territories would be decided by popular vote. A second fulcrum was sought in the abolition of the slave trade in the District of Columbia as slaveholders were given a more reliable means of recovering of their absconding property.

The Fugitive Slave Law called for the appointment of federal commissioners to facilitate the reclamation of that peculiar specie of property. These commissioners were to appoint marshals who could "call to their aid" anyone for assistance who happened to be at the scene where an alleged fugitive was arrested, thereby charging the entire population with complicity for defending the slave system. A slave owner could reclaim his property without warrant, and the commissioner would judge the case without jury, and without the testimony of the alleged fugitive. Upon ascertaining a fugitive's identity a certificate of authorization would be issued to an owner allowing the employment of all "reasonable force" to return the fugitive to his or her place before escaping, including that necessary to prevent attempts at "rescue." A marshal who failed to execute the law could be fined $1,000 and held liable for the full value of the property; a fine of equal amount and imprisonment of six months was to be meted out to those assisting in a fugitive's escape, with liability for the value of the property not to exceed $1,000. An officer making an arrest was to be paid a fee of $5, with a fee of $10 paid to the commissioner for each slave returned to its owner.

Scholarship generally acknowledges that, however onerous these provisions were, they were ineffective in stopping the escape of fugitives to the North. In their real purpose, however, they succeeded, as Frederick Douglass pointed out—by spreading "alarm and terror" among those placed at the mercy of its provisions. In his *Life and Times*, Douglass wrote:

> Bishop Daniel A. Payne of the African Methodist Episcopal Church came to me about this time to consult me as to whether it was best to stand our ground or flee to Canada. When I told him I could not desert my post until I could not hold it, adding that I

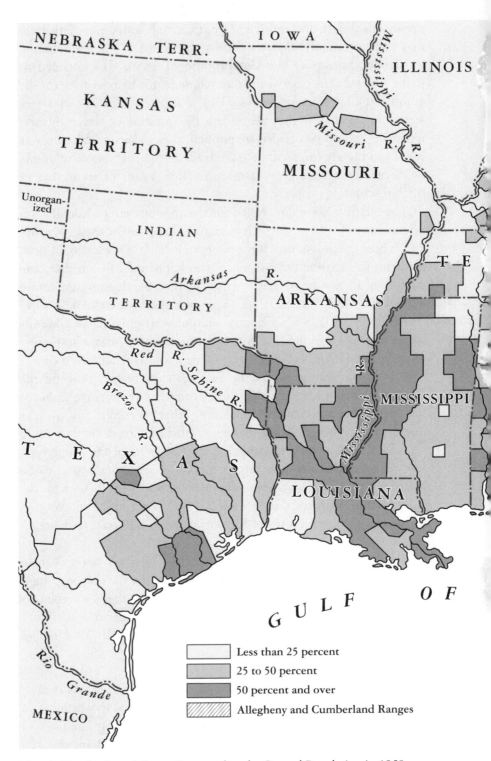

Map 1. Distribution of Slaves Compared to the General Population in 1859.

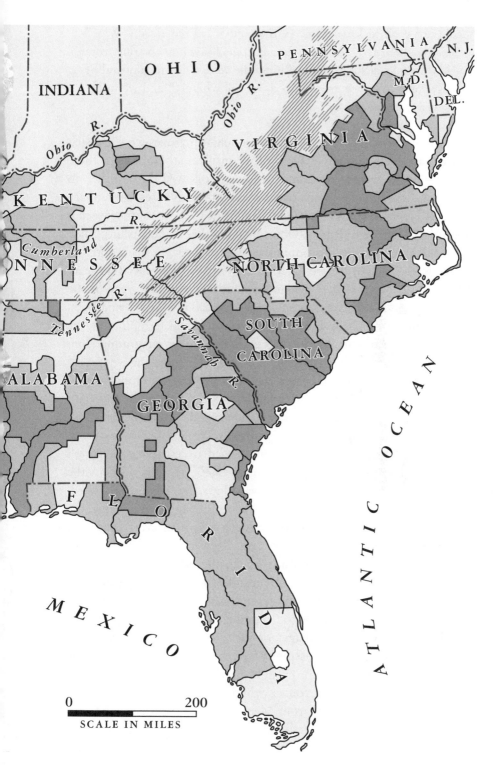

did not wish to leave while Garnet and Ward remained. "Why,"
said he, "Ward? Ward, he is already gone. I saw him crossing from
Detroit to Windsor." "I asked him if he were going to stay and he
answered. "Yes, we are whipped, we are whipped, and we might
as well retreat in order."

Two months after the law's enactment eight fugitives were arrested
and remanded to slavery in Bedford, Pennsylvania. Days later, three
fugitives were seized in Quincy, Illinois, followed by another in
Shawneetown, Illinois. In one of the most infamous cases, on January
8, 1851, Henry Long was taken in New York City. In another week a
fugitive was seized in Ripley, Ohio; then another in Harrisburg; then
in Philadelphia; then in Boston. In the years of the law's operation 333
blacks were brought before federal commissioners; twenty-two were
released, the balance were remanded south. In the first year of the
law's enforcement three thousand blacks crossed over into Canada, as
the enactment made the Lion's Dominion the final terminus for the
Underground Railroad. By decade's end the number emigrating from
the northern states to Canada would exceed fifty thousand.

Concurrently there was to be a huge increase in the number of persons associated with the activities of the Underground Railroad,
while in the cities and towns where blacks resided in sufficient numbers vigilance committees were organized to protect fugitives and to
assist them on to Canada. The routes used by escaping slaves fell
along the chief geographical figurations of the eastern third of the
continental United States. The first was along the Atlantic seaboard,
with slaves escaping by sea or by way of swamps and plying the
rivers, through Virginia and the eastern shore of Maryland and
Delaware up to Philadelphia, New York, New Haven, New Bedford,
and Boston. A second route ran through the hills, skirting western
Maryland into Pennsylvania and upstate New York. A third ran
through the Cumberland region of Tennessee and Kentucky, crossing
the Ohio River, continuing on to Columbus, Cleveland, and Sandusky
and into the towns and villages of the Western Reserve. A fourth
operated in the upper Mississippi River Valley, whose terminus was
in Iowa, Illinois, and Wisconsin and the towns and cities along the
way. The last of the major routes would be opened with the contest
for Kansas, running slaves from Missouri up to Nebraska and into
Iowa. In the southwest, in Texas, many slaves used the coast to escape
into northern Mexico.

In his *Plea for Captain John Brown* Henry David Thoreau would declare, "The only free road, the Underground Railroad, is owned and managed by the Vigilant Committee. They have tunneled under the whole breadth of the land." Henretta Buckmaster was to write: "No slave sharpened to freedom failed to sense the unenforceability of the Fugitive Slave law meant friends waited for him in the North."[11] Speaking to an overflow audience set to vote in favor of a resolution to make Syracuse "an open city," Reverend Loguen declared he would resist any attempt to take him, and he would not flee to Canada. Thundering his defiance, Loguen called out: "Your decision to-night in favor of resistance will give vent to the spirit of liberty, and it will break the bands of party, and shout for joy all over the North. Your example only is needed to be the type of action in Auburn, and Rochester, and Utica, and Buffalo, and all the west, and eventually in the Atlantic cities. Heaven knows that this act of noble daring will break out somewhere—and may God grant that Syracuse be the honored spot, whence it shall send an earthquake voice through the land."[12] When Brown read Loguen's speech of October 4, 1850, he promised the next time he saw his friend he would shake his hand so hard he might wring it off.

The resistance exemplified by the "Jerry" rescue in Syracuse, followed by the rescue of "Shadrack" in Boston, and then the armed resistance led by William Parker on September 11, 1851, at Christiana, Pennsylvania, was to allow Douglass to declare the Fugitive Slave Law would "soon be a dead letter in much of the North." In a letter to his wife dated November 28, 1850, Brown wrote the following: "It now seems that the fugitive slave law was to be the means of making more abolitionists than all the lectures we have had for years. . . . I of course keep encouraging my colored friends to 'trust in God and keep their powder dry.' I did so today at thanksgiving meeting publicly."

Then on January 17, 1851, again to his wife, we read: "Since the sending off to slavery of Long from New York, I have improved my leisure hours quite busily with colored people here, in advising them how to act, and in giving them all the encouragement in my power. They very much need encouragement and advice; and some of them are so alarmed that they tell me they cannot sleep on account of either themselves or their wives and children."[13]

The meetings alluded to, both in public and behind closed doors, led to the formation of the United States League of Gileadites, a name

Brown acquired from the biblical injunction, "whoever is fearful or afraid, let him return and depart early from Mount Gilead." Forty-four persons, all black except for Brown, among a community in Springfield, Massachusetts, numbering 300 persons, many of them fugitive slaves, met on January 15 to affix their names to the agreement the Gileadites would uphold. The purpose of the league was to ensure an organized resistance to any attempt at the enforcement of the Fugitive Slave Law. If there was an arrest, members were to assemble quickly, assuring they would outnumber the slave-catchers and federal marshals, and to thoroughly protect those threatened with seizure. All members were to be armed, but arms were to be concealed, and all were sworn not to betray any plan laid by the league. In the event of trial, members were advised to create distractions, to give the "hoist" to their enemies, and free the defendant. If there was trouble, members were advised to seek refuge in the homes of prominent whites, thereby compelling them to take sides. In his words of "Advice," Brown wrote: "Nothing so charms the American people as personal bravery. Witness the case of Cinques, of everlasting memory, on board the *Amistad*. The trial for life of one bold and to some extent successful man, for defending his rights in good earnest, would arouse more sympathy throughout the nation than the accumulated wrongs and sufferings of more than three millions of our submissive colored population. We need not mention the Greeks struggling against the oppressive Turks, the Poles against Russia, nor the Hungarians against Austria and Russia combined, to prove this."

On January 9, 1854, Brown wrote an extended letter to Frederick Douglass with scathing remarks intended as a suggestion to that editor, who could "take it up and clothe it in the suitable language to be noticed and felt." Who are the persons in the press, church and state, Brown asked, "who use their influence to bring law and order and good government, and courts of justice into disrespect and contempt?" In ending his letter he wrote, "I want to have the enquiry everywhere raised—Who are the men that are undermining our truly republican and democratic institutions at their very foundations?"

The answer to Brown's interrogative came from several directions, as on May 24, 1854, one day before the U.S. Congress enacted the Kansas-Nebraska Bill, a fugitive slave named Anthony Burns was arrested in Boston. The Anthony Burns case was to be the most publicized and notorious of the fugitive slave cases, and taken together with the so-called squatter sovereignty of Illinois senator Stephen Douglas, marks the turning point in the political realignment taking place in the North.

Burns had been seized as a result of collusion between his master, the marshal, and city officers, who agreed to charge him in a robbery in order to make an arrest. Burns himself was to relate the details in the following year when he was again free: "When I was going home one night I heard some one running behind me; presently a hand was put on my shoulder, and somebody said: 'Stop, stop; you are the fellow who broke into a silversmith's shop the other night.' I assured the man that it was a mistake, but almost before I could speak, I was lifted from off my feet by six or seven others, and it was no use to resist. In the Court House I waited some time, and as the silversmith did not come, I told them I wanted to go home to supper. A man then came to the door; he didn't open it like an honest man would, but kind of slowly opened it, and looked in. He said 'How do you do, Mr. Burns?' and I called him, as we do in Virginia, 'master!'"

The morning after Burns's arrest a hearing was held to determine the identity of the fugitive; something that had already been tacitly admitted. Happening upon the scene were Richard Dana, Jr. and Theodore Parker. Soon news of the affair began circulating and a public meeting was called in Faneuil Hall, and as word reached Worcester and other points, a general congregation was assured for that Friday evening's protest. In Troy, in his lawyer's office, on receiving the news, John Brown began pacing the floor. Turning to his counsel, he said, "I'm going to Boston." "Going to Boston? Why do you want to go to Boston?" asked the astonished lawyer. Continuing his pacing, Brown replied, "Anthony Burns must be released, or I will die in the attempt." It took considerable effort for his counsel to dissuade Brown.

That Friday morning at the meeting of Boston's Vigilance Committee, the plight of Burns was extensively discussed; Lewis Hayden and Thomas Wentworth Higginson advocated that the gathering that night at Faneuil Hall should serve as a cover to converge on the courthouse to forcibly remove the fugitive. This, however, was voted down three to one, and without settling on an agenda the meeting adjourned, while Higginson and Hayden and about twenty others remained behind to organize a plan. That night in an atmosphere of tense excitement, Samuel G. Howe, Theodore Parker, and Wendell

Phillips, among others, addressed the hall. When Phillips began, his
• intent was to set the assembly in motion: "Mr. Chairman and Fellow
Citizens—You have called me to this platform—for what? Do you
wish to know what I want? I want that man set free in the streets of
Boston. . . . When law ceases, the sovereignty of the people begins. I
am against the squatter sovereignty in Nebraska, and I am against kid-
napper sovereignty in the streets of Boston. . . . The question tomorrow is fellow citizens whether Virginia conquers Massachusetts. If
that man leaves the city of Boston, Massachusetts is a conquered
State. . . . Tomorrow the question is, which way will you stick?"14

As the meeting continued, from the back of the hall a motion of
adjournment was made. The summons, however, had not been properly understood and so there was to be only a desultory exodus in the
direction of the courthouse, where Higginson, Hayden, and the others had tried to force the doors on the west side of the building.
Finding this too difficult they came around to the opposite side and
began ramming the doors. Stones were thrown and windows broken;
finally two men with axes rendered an opening big enough for a man
to push through. In the melee a guard was killed and Higginson
received a saber gash across his face. But the prisoner had not been
reached and the assault was repulsed.

The mayor now ordered out two companies of artillery, which took
up positions at midnight around the city hall and the courthouse. At
two that morning troops from nearby Fort Independence and a company of marines from the Navy Yard reinforced them. Later the marshal telegraphed President Pierce, informing him of the situation and
of the action that had been taken. The president telegraphed his reply:
"Your conduct is approved; the law must be executed."

When Anthony Burns's hearing resumed as many as three thousand persons crowded the square, as the mayor urged all must "preserve the public peace and sustain the law." When the facts had been
established before the court and the owner issued a certificate of his
right to claim his property, the rendition of Anthony Burns became a
spectacle unto the age. On the day of his delivery a company of
United States infantry and a detachment of artillery guarded the
entrances to the courthouse, as several thousand thronged the square
and the streets leading to it. All of the stores fronting the square were
closed, and several were draped with black crepe and a coffin suspended above the corner of Washington and State streets. As Burns
was brought out, troops cleared the square, two hundred escorting

him to Central Wharf, where thousands watched as he boarded a steamer for Norfolk. The banner of the *Liberator* shouted, "The Triumph of the Slave Power, THE KIDNAPING LAW ENFORCED AT THE POINT OF THE BAYONET—Massachusetts in Disgraceful Vassalage."

The South, it was said, should have congratulated itself that Frederick Douglass had become a fugitive from its domain, for if he remained it would not have been safe on its plantations. In the introduction to his second autobiographical work in 1855, *My Bondage and My Freedom,* J. McCune Smith eloquently sketched his subject's growing prominence.

> If a stranger in the United States would seek its most distinguished men . . . he will find their names mentioned, and their movements chronicled, under the head of "By Magnetic Telegraph," in the daily papers. . . . During the past winter—1854–55—very frequent mention of Frederick Douglass was made under this head . . . , his name glided as often—this week from Chicago, next week from Boston—over the lightning wires, as the name of any other man. . . . And the secret of his power, what is it? He is a representative American man—a type of his countrymen . . . to the fullest extent, has Frederick Douglass passed through every gradation of rank comprised in our national make-up, and bears upon his person and upon his soul every thing that is American.

After a brief tutelage Douglass began to show himself as one of the better orators of his day, in the class of Wendell Phillips, who had a patrician's education. With a logical mind and a sharp memory, Douglass on the speaker's platform was a perfect mimic. Ever combative, he had marshaled all the facets of the antislavery argument, and had a comprehensive philosophy of reform. For sixteen years, too, he would produce a top-quality journal, the *North Star* to be published after 1851 as *Frederick Douglass' Paper,* and in its last years as *Douglass' Monthly.* His early friends in Massachusetts had been quick to recognize his abilities, but by the beginning of the second decade of his freedom he had substantially loosened the strings that bound him to them.

The first of the tenets let go was a commitment to moral suasion as an effective instrument for winning slaveholders to emancipation.

Speaking at a convention in Salem, Ohio, he expressed "apprehension that slavery could only be destroyed by bloodshed," when he was interrupted by Sojourner Truth, an adherent to the school of nonresistance. "Frederick," she asked sharply, "is God dead?" "No," he replied, and "because God is not dead slavery can only end in blood."[15]

During his first years in Rochester Douglass remained in close contact with his mentors, and relations had been cordial. But he had not been forgiven for starting his paper against their advice; now he was coming under other influences, notably Gerrit Smith and John Brown, and soon began to repudiate one after another of his former positions. Garrison always characterized the United States Constitution as "a covenant with death and an agreement with hell" for its proslavery provisions and intent, but now Douglass began to question this interpretation. Arguing that the Constitution was by its *preamble* antislavery, and that Congress had power to legislate against slavery to abolish it; Douglass reasoned, with other political abolitionists, that slavery never was and never could be anything but a system of lawless violence, and it was the first duty of every citizen, whose conscience would permit it, to use his political as well as his moral power for its overthrow.

Douglass was not one either to shun "the terse rhetoric of the ballot box," and would show evidence of a shrewd tactician. While still an adherent of the Garrisonian school Douglass participated in the significant Free Soil convention held in Buffalo, where despite the low antislavery standard of the party, he was called upon to speak, as was Garnet. Douglass was later to observe that by causing the demise of the Whig Party and bringing about a split within the Democratic Party, the Free Soil convention was the beginning of the chain of events whereby slavery was abolished.

Upon Douglass's forthright disavowal of the proslavery character of the Constitution, Garrison exclaimed, "There is roguery somewhere!" A bitter personal dispute was now manifest between the two that they would never reconcile. However in Douglass's view their differences had more to do with whether his race had the ability to deal with the complexities of the antislavery cause without the tutelage of whites. Though Garrison was an ardent advocate of the African's elevation in the United States, he nevertheless wrote: "The anti-slavery cause, both religiously and politically has transcended the ability of the sufferers from American slavery and prejudice, as a class, to keep pace with it, or to perceive what are its demands, or to understand the philosophy of its operations."[16] Taking Garrison to task, Douglass caustically wrote in his paper on April 13, 1855:

> We look upon the past as precedent for the future. Our oppressed people are wholly ignored, in one sense, in the generalship of the movement to effect our Redemption. Nothing is done—no, nothing as our friend Ward asserts, to inspire us with the Idea of our Equality with the whites. We are a poor, pitiful, dependent and servile class of Negroes, "unable to keep pace" with the movement, to which we have adverted—not even capable of "perceiving what are its demands, or understanding the philosophy of its operations!" Of course, if we are "unable to keep pace" with our white brethren, in their vivid perception of the demands of our cause, those who assume the leadership of the Anti-slavery Movement; if it is regarded as having "transcended our ability," we cannot consistently expect to receive from those who indulge in this opinion, a practical recognition of our equality. This is what we must receive to inspire us with confidence in the self-appointed generals of the Anti-Slavery host, the Euclids who are theoretically working out the almost insoluble problem of our future destiny.

Breaking onto the scene of this controversial decade came a marvelous work of rare power Douglass would call "the master book of the nineteenth century"—*Uncle Tom's Cabin* by Harriet Beecher Stowe. "Mrs. Stowe at once became an object of interest and admiration," he wrote in his *Life and Times*: "She had made fortune and fame at home, and had awakened a deep interest abroad. Eminent persons in England, roused to antislavery enthusiasm by her *Uncle Tom's Cabin*, invited her to visit that country, and promised to give her a testimonial. Before sailing for England, however, she invited me from Rochester, N.Y., to spend a day at her house in Andover, Mass."

Received with genuine cordiality, Douglass wrote his hostess made "a nice little speech in announcing her object in sending for [him]." She expected to have a considerable sum placed in her hands by English friends and wanted to use it to raise some useful institution after *Uncle Tom's Cabin*, which would "show that it produced more than a transient influence." She wanted to know what could "be done for the free colored people of the country," particularly for fugitive slaves, thought favorably of establishing an industrial school herself, but wanted to know Douglass's opinion.

Douglass put his views for her in a lengthy letter before her departure, strongly endorsing the idea of a school where blacks could learn industrial trades and become mechanics. Immediately upon her departure he published an account of their meeting and discussion, followed by publication of the entire letter outlining his position. This notice attracted attention, and controversy soon followed. A letter from Martin Delany said: "Now I simply wish to say, that we have always fallen into great errors in efforts of this kind, going to others than the intelligent and experienced among ourselves; and in all due respect and deference to Mrs. Stowe, I beg leave to say, that she knows nothing about us, 'the Free Colored people of the United States,' neither does any other white person—and consequently can contrive no successful scheme for our elevation; it must be done by ourselves."[17]

Two weeks prior to this letter's publication, Douglass also published a letter from Henry O. Wagoner of Chicago in response to the recently enacted Black Law in Illinois prohibiting blacks from entering the state and stipulating the extradition of fugitive slaves. Wagoner proposed a convention of "Free Colored people of the northern states." Delany promptly endorsed Wagoner's suggestion, saying "something must be done and that speedily." Delany's letter concluded: "The so called free states, by their acts, are now virtually saying to the South, 'You shall not emancipate; your blacks must be slaves; and should they come north, there is no refuge for them.' I shall not be surprised to see, at no distant day, a solemn convention called by the whites of the North, to deliberate on the propriety of changing the whole policy to that of slave states. This will be the remedy to prevent dissolution; and it will come mark that!"

In his reply Douglass recalled he once put forth a proposition for a "National League," to effect cooperation and consultation for "free persons of color," and his proposal had not been taken up. Now in rebuttal to Delany's objection to the proposal of Harriet Beecher Stowe, Douglass pointed out that she had supported his idea of an all-northern "colored convention" at that time. Douglass continued: "The fact is, we are a disunited and scattered people, and very much of the responsibility of this disunion must fall upon such colored men as yourself and of the writer of this. We want more confidence in each other, as a race—more self-forgetfulness." Why didn't Delany issue a call for a convention? Douglass asked. In reiterating his idea, he challenged "friend Delany" to draw up a proposal, which he would then publish.

A call was duly issued for a convention to assemble July 6, 1853, at Rochester, signed by Douglass and others, absent Delany. When it convened, the 114 delegates attending took up the questions of establishing a "national council," and disposition of the funds promised by Mrs. Stowe. J. W. C. Pennington was chosen president, and vice presidents were Douglass, William H. Day, and John Jones. In his address, "The Claims of Our Common Cause," Douglass made a strong case that blacks were Americans and should possess all the rights of citizens, "Notwithstanding the impositions and deprivations which have fettered us—not-with-standing the cunning, cruel and scandalous efforts to blot out that right, we declare that we are, and of right ought to be American citizens. We claim this right, and we claim all the rights and privileges and duties which, properly, attach to it."

The delegates went on to discuss the plan for a national council. This body, to consist of two members from each of the ten states represented at the convention, would provide for the election for state councils for each state and set up four working committees. One committee was to have as its purview the manual training school, to procure funds and to locate and establish it. Another on business relations was to establish an employment office. A committee on publication was to compile records that were of statistical, historical, and social value, and to make them available through a public library. Finally, a committee was to establish a cooperative union "for the purchase and sale of articles of domestic consumption." Then the plan for an industrial school was considered. Here Douglass read the delegates his letter outlining the proposal put to Harriet Stowe. Douglass wrote in *Life and Times*: "This convention warmly approved the plan of a manual labor school . . . and expressed high appreciation of the wisdom and benevolence of Mrs. Stowe." He continued: "After her return to this country I called again on Mrs. Stowe, and was much disappointed to learn from her that she had reconsidered her plan for the industrial school. I have never been able to see any force in the reasons for this change. It is enough, however, to say that they were sufficient for her, and that she no doubt acted conscientiously, though her change of purpose was a great disappointment, and placed me in an awkward position before the colored people of this country, as well as to friends abroad, to whom I had given assurances that the money would be appropriated in the manner I have described."

The failure of the central platform, relying as it did on Douglass's pledges from the famed author, damaged the convention's goals. Although many other circumstances of the time militated against its success, the meeting was in itself an impressive achievement and accordingly received a positive appraisal from many quarters. The next meeting, scheduled in 1854, did not produce a quorum, however, and the movement collapsed.

Following the close of the Rochester convention a new call was issued for a National Emigration Convention to convene in Cleveland in August 1854, signed by a dozen prominent blacks including Martin Delany of Pittsburgh, William Lambert of Detroit, and H. Ford Douglass, formerly of Louisiana, now of Illinois. The issuers purposefully sought to differentiate themselves from the proceedings just concluded, which had come out strongly against emigration as a solution to the African's proscribed condition within the United States. Douglass hastened to denounce the new call in his paper on July 25, 1853: "Our enemies will see in this movement, a cause for rejoicing, such as they could hardly have anticipated so soon, after the manly position assumed by the Colored National Convention in this city. They will discover in this movement a division of opinion among us upon a vital point."

"Emigrationists" contended forcefully that slavery was not really a "foreign element" in the government of the United States; that slavery was sanctioned by the Constitution, and not at all "antagonistic to the feeling of the American people." Although Douglass was not at all adverse to this strongly felt denunciation of black proscription within the American polity, he became one of the chief critics of its proponents. "Here, then really is the point at issue," he wrote in a letter to J. M. Whitfield, on December 2, 1853, "not exactly, as some of our friends would declare, whether or not our condition 'can be made worse by emigration' but whether or not we have implicit confidence in the ultimate success of the anti-slavery movement, whether Might or Right shall triumph in the present conflict." In Douglass's view nothing should be abandoned to slaveholders in the fight against slavery, especially not the presence of free blacks in the northern states.

When the convention assembled the next year, otherwise little noticed amidst the jeers and catcalls of the white public outside, delegates heard the keynote speaker, H. Ford Douglass, declare emigration to be a fact of history. Emigration was open to every people and to every nation. He went on: "To stand still is to stagnate and die . . . shall we . . . refuse to follow the light which history teaches, and be doomed . . . ? No! In spite of the vapid anathemas of 'Eastern Stars,' who have become so completely dazzled by their own elevation, that they can scarcely see any of the dark realities below; or the stale commonplace of 'Western Satellites,' the expediency of a COLORED NATIONALITY is becoming self-evident to Colored men more and more every day."[18]

This divergence of opinion, subsequently seen as disclosing integrationist versus nationalist tendencies, has been ever present with regard to the African American's situation in the United States. Indeed on the eve of civil war Frederick Douglass was extended an invitation to visit Haiti to study the feasibility of emigration to that island black republic, and had made preparation to go there for that purpose, but was to stop short of departure as the batteries of South Carolina opened on Fort Sumter in Charleston Harbor in April 1861. These rival postures, too, would have direct bearing on the roads John Brown traveled to Harper's Ferry.

Two

Bleeding Kansas

We cross the prairie as of old
The pilgrims crossed the sea,
To make the West, as they the East
The homestead of the free!
—John Greenleaf Whittier,
"Song of the Kansas Emigrants" (1854)

I N THE FIRST MONTHS OF 1854 A BILL WAS REPORTED IN THE
United States Congress that divided the Nebraska Territory in
two, leaving the settlers of the territories free to "form and regulate
their domestic institutions." It was the express understanding of the
framers of the Kansas-Nebraska Bill that emigrants from Missouri
would quickly settle the Kansas Territory, securing it to slavery. As
one of the leading proponents of a southern conquest of Kansas,
David R. Atchison, put it, "If Abolitionism is established in Kansas
there will be constant strife and bloodshed between Kansas and
Missouri. Negro stealing will be a principle and a vocation. It will be
the policy of philanthropic knaves until they force the slaveholder to
abandon Missouri." Once Missouri was lost, so the reasoning went,
west Texas would be the next to go; then the South's institutions
would be confined to ever narrower limits until their overthrow was
only a matter of time.

Southern politicians and their supporters knew they risked a lot in
courting increased antipathy in the North to slavery, but in any event

they could not accept its restriction from the new territories and ensure its survival. What they had not counted on, however, was that armed resistance would arise to oppose them—a collision whose consequences extended to the last days before the slaveholder-engineered secession in 1860–1861.

The "Squatter Sovereignty Bill," as the act became known, was seen in the North as a naked bid to secure the continued predominance of slaveholding interests in the national government. Challenging this scheme, in what was first assumed to be a contest of votes, was an organized movement of settlers from the northern states into the newly opened territory, initially under the auspices of the New England Emigrant Aid Company.

More than ninety years later the muralist John Steuart Curry undertook a commemorative painting at the Kansas state capitol in Topeka titled *Tragic Prelude*. In the center of his depiction is the figure of a man larger than life, with flowing beard and wild hair. His arms extended, he exerts himself above the noise and agony of battle; in one hand he holds an open Bible, in the other a Sharps rifle. Towering over the fallen bodies of a Union and a Confederate soldier, he straddles two warring groups bearing the flags of the opposing armies of 1861–1865. Several fugitive slaves are seen enmeshed among the struggling figures. In the background a steady stream of settlers in covered wagons is forging ahead against an ominous prairie fire, while on the opposite horizon is a gigantic tornado. But in portraying his subject analogous to a natural force—as irrational in effect as that whirlwind—the muralist has given us, not John Brown in his historical determinations, but a portrait of bathos. The civil war on the Kansas and Missouri border, and its importance as an overture to the larger conflict, is nowhere better told than in the biographies of John Brown. Accordingly, it was on the Kansas plains, as if a summons from God, that Brown saw a presage of the greater war to come.

As the territory opened to immigration on July 1, 1854, many turned their attention toward Kansas, and not least among them were Brown's sons and son-in-law living in Ohio. After hearing from John Jr., the elder Brown wrote him on August 21 of that year: "If you or any of my family are disposed to go to Kansas or Nebraska, with a view to help defeat Satan and his legions in that direction, I have not a word to say; but I feel committed to operate in another part of the field. If I were not so committed, I would be on my way this fall." By

October, five of the Brown sons were making arrangements for the Kansas emigration; with Owen, Frederick, and Salmon leaving first, followed in the spring by John Jr. and Jason, together with their wives and children. Of his journey out, John Jr. was to write: "A box of fruit trees and grape-vines which my brother Jason had brought from Ohio, our plough, and the few agricultural implements we had on the deck of that steamer looked lonesome; for these were all we could see which were adapted to the occupation of peace. Then for the first time arose in our minds the query: Must the fertile prairies of Kansas, through a struggle of arms, be first secured to freedom before freemen can sow and reap? If so, how poorly we were prepared for such work will be seen when I say that for arms five of us brothers had only two small squirrel rifles and one revolver."[1]

With the prospect that free-state settlement of Kansas would be swept aside, John Brown began to deliberate as to his best course of action. Should he return to the Adirondack colony after closing out his remaining business in Ohio, or should he go to Kansas? Seeking advice simultaneously from his family in Akron and from several of the black families in North Elba, he also wrote to Frederick Douglass, Dr. McCune Smith, and Gerrit Smith. Just as the first of his family was setting out for Kansas he wrote from Akron that he felt "pretty much determined to go back to North Elba" and added, "Gerrit Smith wishes me to go (back to North Elba); from Douglass and Dr. McCune Smith I have not yet heard."[2]

As the year closed it was becoming apparent that, by advantage of proximity, Missouri would gain the upper hand. In November's territorial election Missourians crossed the border, seizing ballot boxes and intimidating free-state voters, guaranteeing a proslavery delegate would be elected to Congress. Then in March 1855, another invasion of nearly five thousand Missourians easily succeeded in electing a proslavery territorial legislature. This body would convene in July at a location near the border, where they copied verbatim from a thick volume of Missouri laws, taking care to enact a code specifically for Kansas. These made it a felony even to deny the legality of slavery, stipulating that only men with proslavery views, or as the saying went, "men sound on the goose," could hold office or serve as jurors, and prescribing death for anyone inciting slaves to rebel, vying with the strictest codes of the South.

Regrouping after this onslaught, the free-state settlers held counsel and soon began forming military companies and sending out appeals to the North for arms. The northern settlers, however, were predominately Democrats, with some Free-Soilers who may have wanted to exclude slavery, but only to reserve the land for white labor. Slightly in advance of these was a coalescing of opinion opposed to the extension of slavery which would soon take the name Republican, but who may also have cared little for the plight of the slave. There were also a handful of abolitionists, prominent among whom were Brown's sons and their families who had settled near the town of Osawatomie. The more thoroughgoing among these would meet in June 1855, when they passed a resolution declaring they would not abide by the "bogus enactments" of a fraudulently elected legislature, and where John Brown Jr. served as a vice president. Putting aside differences, men of all persuasions agreed to meet in Big Springs, Kansas, in September, where they would organize a state constitution, set elections, and appeal on behalf of the free-state cause to the Congress and the nation.

Giving a vivid portrait of the situation in Kansas, after his arrival John Jr. sent a letter to his father, dated May 20, 1855: "I tell you the truth, when I say that while the interest of despotism has secured to its cause hundreds and thousands of the meanest and most desperate of men, armed to the teeth with Revolvers, Bowie Knives, Rifles & Cannon—while they are not only thoroughly organized, but under the pay from Slaveholders—the friends of freedom are not one-fourth of them half-armed. . . . The result of this (is) that the people here exhibit the most abject and cowardly spirit, whenever their dearest rights are invaded and trampled down by the lawless bands of Miscreants which Missouri has ready at a moment's call to pour in upon them."[3]

Could not his father, he asked, obtain and send arms and ammunition, as war of "some magnitude" seemed inevitable? "We need them more than we do bread," he wrote. Sensing that Kansas now presented the best opportunity to deal a blow to slavery, John Brown determined to lay all else aside; after securing his wife and young daughters in their needs, he would join his sons in Kansas.

Just at that juncture, after the founding of the Republican Party that spring at a schoolhouse meeting in Ripon, Wisconsin, the adherents of the Liberty Party, including Douglass, Gerrit Smith, and James McCune Smith, were convening in Syracuse under a new banner—the Radical Abolition Party. Most voting abolitionists had promptly united in the ranks of the emerging "anti-slavery extensionists," the

Republicans, while the Liberty Party's three hundred members still thought it best to stay out. Under their new banner the party could keep the transcendent principle of immediate universal emancipation before the nation; coming events, they believed, would drive all peripherally antislavery voters into acceptance of the Radical Abolition platform. John Brown now hastened to join them.

The convention had assembled at the behest of Dr. McCune Smith to formulate its response to the Republican Party, which would come in the form of a pamphlet produced by the convention titled *Exposition of the Constitutional Duty of the Federal Government to Abolish Slavery*. Arriving on June 26, the first day of a three-day meeting, Brown was invited to speak on the last day. He rose to say he had four sons in Kansas, and three others who intended to join them, and that he wanted to arm them. His poverty, however, prevented him from obtaining these, and so he must appeal for aid; as for himself, he would not go unless he could go armed "to aid in the battles of freedom." Gerrit Smith read two of John Jr.'s letters, introduced to the gathering with "such effect . . . as to draw tears from numerous eyes."[4] Funds were contributed on the spot, amounting to sixty dollars in all.

That men were willing publicly to approve of his intentions was a great revelation, and Brown wrote his wife that the convention was "one of the most interesting meetings I have ever attended in my life." Next he was in Springfield, Massachusetts, where he used the money raised in Syracuse to buy guns and powder flasks. While there he met with his long-time confidant, Thomas Thomas, a fugitive slave from Maryland's Eastern Shore whom he had employed in his business there as a porter. Brown attempted but failed to persuade Thomas to go to Kansas with him. In August he was in Akron where again he spoke publicly, successfully obtaining donations of ammunition, clothing, a box of muskets, and ten double-edged swords. In Cleveland he collected his son-in-law Henry Thompson, and the two traveled to Detroit to pick up sixteen-year-old Oliver Brown, while his older brother Watson remained at North Elba. Finally the trio was in Chicago, which they soon left with a wagon laden with arms and supplies, drawn by "a nice young horse" for which they paid $120. Walking much of the time to relieve the burden on the animal, Brown and his companions arrived in Kansas on October 7, 1855.

In a speech before the convention of Colored Citizens of the State of New York, in Troy on September 4, 1855, Frederick Douglass

ating a pretext to sack the town. But before the free-state strong-
ld was reached, eight men with Sharps rifles met the sheriff and his
sse, who freed the man. The governor now called out the proslav-
territorial militia, appealing directly to Missouri for aid. As a
ce of sixteen hundred was marching on Lawrence, free-state men
m surrounding districts hastened to defend the town. With
keshift barricades and redoubts being thrown up, John Brown and
ir of his sons arrived in a large lumber wagon bristling with
apons. With each corner of the wagon fitted with a pole with a
yonet, while from one fluttered the Stars and Stripes, they drew up
front of the Free State Hotel. Each man had two revolvers tucked
o his belt and a broad sword buckled on, with another small pis-
resting in a pocket, while each also carried a rifle. On the bed of
wagon lay more arms ready for use; in aggregate Brown's party
rried arms to fire a hundred rounds without reloading. The *Herald*
Freedom, one of two newspapers published in Lawrence, reported
December 7 that Brown, having more arms "than he could use for
vantage, a portion of them were placed in the hands of the more
stitute. A company was organized and the command given to Mr.
own for the zeal he had exhibited in the cause of freedom both
fore and since he had arrived in the Territory."

Disconcerted by this demonstration, the territorial governor has-
ied to Lawrence to hold talks with its newly formed Committee of
blic Safety. The principals of the committee, Charles Robinson,
ent of the Massachusetts Emigrant Aid Society, and James Lane, a
rmer Democratic Party politician from Indiana, convinced a drunk-
and now solicitous governor to withdraw authority from the
oslavery horde and disperse them. For their part the free-state lead-
s agreed to disassociate themselves from the rescue of the falsely
rested man, pledging to aid in the execution of the territorial laws
vhen called upon by proper authority."[5] With this deal struck, the
rties emerged from the Free State Hotel.

On the following afternoon at an open-air meeting the three prin-
pals in the negotiations spoke. In his remarks the governor declared
ere was perfect unanimity between himself and the leaders of the
mmittee. Just as Robinson was finishing remarks that downplayed
e importance of the concession, John Brown mounted the stand to
nounce the "treaty" as a betrayal to the proslavery administration;
spit on it and would never obey. Calling for ten men with Sharps
fles, he offered to lead them in a night attack on the "border ruffi-

lamented that the nation was in a state of anarchy, "th
ment of the United States has resigned its function to tl
lawless border ruffians of Missouri." In the struggle tha
in Kansas, the free-state cause had, Douglass observed,
ballot-boxes and its liberties with an ease which puts
fighting before Sebastopol. The reason is obvious. '
Sebastopol are of granite. The walls of Kansas are of dot
tiousness upon the attempts being made to drive the free
out the country." Among the resolutions being adapted
state politicians was the so-called Black Law, stating "th
bond or free, shall be permitted to come to the Territory.

It was shortly after the meetings at Big Springs that Brc
James Redpath, a Scotsman and one of a number of cor
from eastern papers in Kansas, who would be instrumen
larizing Brown's Kansas exploits in a biography publish
titled *The Public Life of Captain John Brown*, describe
encounter with him this way:

> The first time that I heard of Old Brown was in connectio
> caucus at the town of Osawatomie. . . . The politician
> neighborhood were carefully pruning resolutions so as
> every variety of anti-slavery extensionists; and more especi;
> class of persons whose opposition to slavery was founded c
> diency—the selfishness of race, and caste, and interest; m
> were desirous that Kansas should be consecrated to fre
> labor only, not to freedom for all and above all. The res
> that aroused the old man's anger declared that Kansas shou
> free white state, excluding Negroes whether slave or free. H
> to speak and soon alarmed and disgusted the politicians by
> ing the manhood of the Negro race, and expressing his e
> anti-slavery convictions with a force and vehemence little lil
> suit the hybrids then known as Free State Democrats.

In reaction to all the activity on the part of the free-state (
proslavery leaders, with the cooperation of the territorial g
called a convention of their own, forming the Law and Ord
With tensions rising, a free-state settler was brutally murde
instead of arresting the man who committed the crime, the
arrested the only witness to the murder. The sheriff's intentior
convey his prisoner through Lawrence, hoping to provoke a

ans" bivouacked around their campfires. Over a flurry of voices disparaged the agreement, Robinson and Lane sought to prevail, assuring everyone the negotiations were a "triumph of diplomacy."

At first Brown feared the "peace treaty" would eventuate in the victory of the proslavery party, but as the days passed he took a more optimistic view. In a letter to the editor of the *Summit Beacon* in Akron, dated December 20, 1855, but published in the following year, he wrote:

> How much truthfulness or sincerity characterized these negotiation I will not say. The invaders soon left, covered over with glory, after suffering great expenses, hardships and privations; not having fulfilled any of their threatenings, or fought any battles, or accomplished anything for which they set out, save commission of certain robberies, and a few acts of violence on the defenseless inhabitants in the neighborhood of their camps, and killing of one unarmed man, leaving the territory entirely in the power of Free State men, with an organized militia, armed, equipped, and in full force for its protection; but the result is not yet fully known. What remains for the Free State men of Kansas, and their friends in the States, and the world to do, is to hold the ground they now possess, and Kansas is free.

Severe weather brought a comparative lull in hostilities as winter came on, but events continued the slide toward confrontation. In January the Free State elections were held, making Charles Robinson governor and John Brown Jr. a member of the legislature. The proslavery settlers boycotted the elections but intimidated voters, savagely hacking a man to death in Leavenworth and dumping his body on his doorstep for his widow. In February the president of the United States, Franklin Pierce, arrayed the power of the federal government behind the proslavery territorial officials, issuing a proclamation that the Free State legislature was illegal and the Free State movement treasonous. With spring two regiments made up of men from Alabama, Georgia, and South Carolina arrived "to see Kansas through." There came too the men to lead them, named Titus, Jackson, Wilkes, and Pate, and a planter from Eufaula, Alabama, Jefferson Buford, who departed Montgomery with a Bible inscribed "Providence, may change our relations to the inferior race, but the principle is eternal—the supremacy of the white race."[6]

Soon after arriving in Kansas, Brown began organizing a force that could strike the "border ruffians"—a phrase coined by James Redpath writing in the *New York Tribune*—in the field. Brown had five sons and a son-in-law in Kansas, but beyond these there were youths, some with exceptional ability—like Redpath, Richard Hinton, John Henry Kagi, John Cook, and Richard Realf—all practicing journalists and abolitionists that would be attracted to Brown's flame. And there came, too, veterans of Europe's shattered revolutions—a German, an Austrian Jew, a Hungarian follower of Kossuth—who would also fight with John Brown. To "Friend Louisa," John Brown Jr. wrote:

> To those who contemplate coming, I would say to them by all means come thoroughly armed with the most efficient weapons they can obtain and bring plenty of ammunition. The question here is—shall we be free men or Slaves? The South is arming and sending in her men. The North is doing the same thing. It is now decreed and certain that the Slave Power must desist from its aggressions in Kansas or if they do not, the war-cry heard upon the plains will reverberate not only through the hemp and tobacco fields of Missouri but through the Rice swamps, the Cotton and Sugar plantations of the Sunny South. From the present appearance the first act of the Drama of insane Despotism is to be performed here.[7]

Brown already had organized men under arms; he knew the use of weaponry and had been studying military arts for more than a decade. Already, too, he had a definite view on how best to attack the "slave power," with a settled military policy. In Brown's theory a small number of men, well armed and organized, were sufficient to initiate the antislavery war he proposed to introduce on the American scene. His practice would be to bring the antagonists to a clash of arms as quickly as possible, and once the fighting began, press his adversaries to close quarters. The most "ready and effectual way," he maintained, "to retrieve Kansas would be to meddle directly with the peculiar institution." It would force the proslavery men "to look to their own neighbor-hoods," dividing and scattering their forces, and because their property had legs and could run away, their basis for power was unstable.[8]

Now a grand jury convened under the chief justice of the Territorial Court. In his charge to the jury he insisted the derided "bogus enactments" were of United States authority, and all who resisted them were "guilty of high treason." The jury promptly indicted most of the prominent free-state leaders and ordered an "abatement" for the two newspapers in Lawrence, commanding the destruction of the Free State Hotel because it "could only have been designed as a stronghold for resistance to law." Meanwhile, near the Brown settlement, at Dutch Henry's Crossing, another court convened under Judge Sterling Cato preparing secret indictments for the arrest of all the Browns.

John Brown Jr., circa 1845.

To carry out the orders of the grand juries, the United States marshal issued a proclamation assembling the proslavery territorial militia, while eight miles from Osawatomie Major Buford and his company, swearing to "clear out that damned Brown crowd," made their encampment. On the morning of May 21, the people of Lawrence (formerly New Boston) awoke to the sight of eight hundred armed men on Mount Oread overlooking the town. Robinson and other free-state leaders, under arrest for "treason" and held twelve miles away in the proslavery territorial capital, Lecompton, had decreed at the urging of eastern politicians that there be no resistance. Accordingly in Lawrence there were no preparations to fight.

First the marshal descended with his posse and made a few arrests. After breakfasting with his men in the Free State Hotel, they returned to the hill. A short time later the sheriff came down with some men and demanded that all arms be surrendered. With one cannon and a few arms, they returned to the hill. Now the proslavery militia was dismissed by the marshal and reconstituted under the sheriff. Taking positions at the head of their column were David R. Atchison, co-author of the Kansas-Nebraska Bill and past president pro tem of the United States Senate; George W. Clarke, Territorial Delegate for Kansas and Indian Agent; Dr. J. H. Stringfellow, editor of the *Squatter Sovereign* and pillar of Weston, Missouri; Henry Clay Pate, editor of the *Border Star* in Westport, Missouri; Jefferson Buford, proslavery filibusterer; and Harry Titus, ostentatious and grandiose slave baron. Atchison addressed the throng: "Boys, this day. . . . We have entered the damned town and taught the damned Abolitionists a Southern lesson they will remember until the day they die. And now, boys, we will go in again with our highly honorable Jones and test the strength of the damned Free State Hotel and teach the Emigrant Aid Company that Kansas will be ours. If man or woman dare stand before you, blow them to hell with a chunk of cold lead."⁹

Hoisting a flag with the inscription "Southern Rights," and a banner that read "South Carolina," and "Supremacy of the White race," on one side and "Kansas, the Out Post," on the other, the proslavery flock descended into the town. Entering the two buildings housing the *Kansas Free State* and the *Herald of Freedom*, they smashed presses, throwing type in the Wakarusa River and books and other articles into the streets. The next target was the Free State Hotel, where an artillery piece was trained at a distance from the structure. Atchison ignited the first shot. Sailing cleanly over the building, more shots were fired, having little effect. Finally it was decided the hotel should be blown up, and four kegs of gunpowder were brought up and placed inside. When these were blown, although the building remained standing, the resulting fire destroyed it. Robinson's home was burned, citizens were robbed, and the town plundered. The charge of the grand jury was complete as the banner "Southern Rights" floated from the corner of the building formerly housing the *Herald of Freedom*. This was the famous sacking of Lawrence.

As word of the impending attack on Lawrence reached Osawatomie, a company commanded by the younger Brown, which included his father, had set out. But before they went far, news of the town's fate was racing through the countryside. The company halted to hold counsel, where the elder Brown was the principal speaker, urging that a blow must now be struck in such a way as to produce a restraining fear. They must show that free-state men could resist, and that men could die on both sides. Brown asked, "How many men will go with me and obey my orders?" Salmon Brown would relate in later years that as the company encamped a rider came with news that Massachusetts senator Charles Sumner had been brutally beaten and maimed on the Senate floor for his speech "The Crime Against Kansas." "That was the finishing, decisive touch," the Brown son told an interviewer in 1913.¹⁰

Owen, Oliver, and Frederick Brown, together with Henry Thompson, a man named Townsely, and an Austrian immigrant named Weiner, volunteered to serve under the elder Brown. Till the end of daylight the men were occupied with sharpening the broad swords Brown had gotten in Akron. At dusk, saying he intended to "steal a march on the slave hounds," Brown and his company moved off toward Pottawatomie Creek, as the remaining free-state men gave a rousing cheer.

The retaliation for the wanton murder of free-state men and for the sacking of Lawrence is still contentious among those who contemplate these scenes, but at the time there was near unanimity about it. The next morning the bodies of five men were found short distances from their cabins, all bearing the lacerations caused by sharp blades. Visiting Kansas several months after what became known as the Pottawatomie Massacre, Thomas Wentworth Higginson reported: "There appeared to be but one way of thinking among the Kansas Free State men. . . . I heard of no one who did not approve of the act, and its beneficial effects were universally asserted." Franklin Sanborn was to sum the matter up twenty years later: "Upon the swift and secret vengeance of John Brown in that midnight raid hinged the future of Kansas, as we can now see it, and on that future again hinged the destinies of the whole country."[11] The men killed were not proslavery leaders—they were merely abettors and enablers who facilitated the terror against the free-state settlers. But with this blow the proslavery settlement of Dutch Henry's Crossing was swept out of existence and Judge Cato's court knocked out, and henceforth, too, the free-state settlers would be compelled to fight.

Several days afterward the first of the standup fights of the Kansas "war" occurred. Over in Westport, Missouri, at the head of the Shannon Sharpshooters, Henry Clay Pate (who distinguished himself days before in Lawrence by riding about on a fine horse decorated with ribbons) set out vowing to capture John Brown. His only fear, he boasted, was not being able to locate the "old man." Not finding the elder Brown, Pate seized two of his sons, John Jr. and Jason, whom he charged with murder and placed in irons. Continuing on his mission, Pate turned his prisoners over to the custody of the man Missouri had elected Territorial Delegate of Kansas, committed some robberies, and took more prisoners. Later drawing his wagons up on the prairie a dozen yards from a deep ravine at the bottom of which was a thick timber called Black Jack, the Virginian had made his night encampment a strong one. Two free-state companies arrived on the scene, the direction of the fight devolving on Brown. Positing one company to the left, in the lower part of the ravine, Brown went with seven men to the right, gaining the ravine's head, thus having his opponents in crossfire. A spirited fight quickly ensued and after a few minutes, with some seven or eight wounded, the Missourians were compelled run for cover in the ravine. Meanwhile the free-state company in the lower part of the gorge, its ammunition spent, abandoned the ground, as Brown continued pressing the fight. Sustaining more casualties, Pate finally sent out his lieutenant behind a free-state prisoner under a flag of truce.

As they approached Brown demanded whether he was captain of the company. "No," came the reply. "Then you stay with me, and let your companion go and bring him out. I will talk with him." As Pate came up he began saying he was a United States marshal, and was continuing when Brown cut him short. "Captain, I understand exactly what you are; and do not want to hear more about it. Have you a proposition to make to me?" As the Virginian floundered, Brown interrupted him—"Very well, Captain, I have one to make to you; your unconditional surrender."

Brown's victory at Black Jack had an electrifying effect on the free-state resisters, whose ranks would be augmented in the days following by youths, who "with no recognized leaders, or temporary leaders," engaged in skirmishes with remnants of the invading "Southern army." William Addison Phillips, in his *The Conquest of Kansas by Missouri and Her Allies*, published in 1857, wrote: "It was at this time that the Free State guerrilla companies sprang up. Finding that armed bands of pro-slavery men were prowling about the Territory, a handful of persons, chiefly youths, took the field. One company, under a young printer named [Charles] Lenhart, was particularly active and bold. Capt. John Brown, senior, who lived near Osawatomie, immediately on the sacking of Lawrence concluded that the war was begun, and that it ought not to terminate."

Phillips asserted these guerrillas could have secured freedom for Kansas had they not been dispersed in the first weeks of June by United States troops. The direct involvement of government troops in backing the proslavery territorial officials came with the developing interest on the part of the Pierce administration, whose secretary of war was Jefferson Davis. The commander of the Department of the West, General Smith, wrote: "Patriotism and humanity alike require that rebellion be promptly crushed."12

Shortly after the battle Brown made his encampment in the dense thickets on an island in Middle Creek. He was there with his prisoners in early June when Colonel Edwin Sumner, the commander of the

First United States Cavalry, and his adjunct, Lieutenant J. E. B. Stuart, advanced with fifty mounted troops. By proclamation of President Pierce, Sumner was authorized to command "all illegally formed companies" to disperse, and he brought with him a U.S. deputy marshal with warrants for Brown's arrest for murder. Brown met the officers outside his camp. Attempting to parley, he said he wanted to arrange an exchange of prisoners, and in particular had in mind gaining liberty for his two sons. Sumner replied he was not there to make terms but to command their dispersal and secure the release of Brown's own prisoners. Having too few men on hand, Brown acquiesced, leading the two officers into his camp and allowing the release of Pate, his men, and their horses. The deputy marshal was now told he should make his arrest, but, evidently losing his nerve, he couldn't find the warrants. "You are a damn liar and a coward! I saw the warrants in your hand last night," fumed the officer.

Pate began a harangue to the effect that Brown was getting away with murder, when Sumner cut him off. "I don't want to hear a word from you, sir, not a word."[13] Brown raised the objection that Sumner had commanded his dispersal, but not that of the Missourians encamped at Osawatomie, when the officer replied that he had also read the president's proclamation there. In spite of that, after his release Pate reunited with these "border ruffians," and together they burned half a dozen buildings in the town, including one housing a valuable library belonging to John Brown Jr. Colonel Sumner would later justify not holding Brown by saying that it would have taken a thousand men to storm Brown's ingeniously fortified island stronghold.

Pate and Brown subsequently dueled again in the pages of the *New York Tribune*, Pate appearing on June 13, and Brown with his rejoinder on July 11. The Virginian alleged his capture had been the result of treachery, as his flag of truce was violated, but had praise for his subordinate who carried it, albeit behind the free-state prisoner. Brown dismissed this complaint as one having "personal cognizance of what then occurred," and wrote of Pate's lieutenant, "I think him as brave as Capt. Pate represents. Of his disposition and character in other respects I say nothing now. The country and the world may probably know more here-after."

It is not always the apparent leaders who do the world's work, remarked DuBois in *John Brown*. "More often those who sit in high places, whom men see and hear do but represent or mask public opinion or social conscience, while down in the blood and dust of battle stoop those who deliver the master stroke." Brown's leadership, the scholar contended, lay not in office, riches, or power "but in the white flame of his devotion to an ideal." The contest to which he had committed his life—the annihilation of slavery in America—he believed had been "predetermined by God in eternity," and *He* had selected him and his family to accomplish this goal. There never has been any doubt, consistent with his Calvinist beliefs, that Brown was in earnest, but this alone did not constitute his totality. Just as clearly he was committed to republican liberty in America, and taken together with his keen military and organizational ability, this accounts for the formidable way in which he has been portrayed. He "was a presence in Kansas," wrote R. C. Elliot of Lawrence, "and an active presence all through '56. Yet it was his presence more than his activities, that made him a power—the idea of his being. . . . No man in Kansas was more respected."14

Many have drawn attention to Brown's remarkable gamesmanship, which enabled him to improvise tactically while keeping his opponents off balance. An expert surveyor, he had keen sense for ground topography, allowing him to screen his movements, and to come up on his wary opponents unexpectedly. It was fruitless, people said, to search for him, so cleverly could he conceal himself to elude pursuers. After Colonel Sumner freed Pate and his men, Brown learned his two sons were to be moved to Lecompton under U.S. Army escort. Chained together at both ankles, with padlocks on each end, the two were driven twenty-five miles a day in the sweltering sun, a total of sixty-five miles. "What a humiliating, disgusting sight in a free government!" wired a correspondent to the *New York Times*.15

As he moved on a parallel course, hoping to effect a rescue, Brown ordered a raid on a store owned by a man with proslavery sympathies, obtaining clothes, blankets, and provisions. Their father would show himself from time to time, Jason Brown later wrote, so that a company of dragoons would be ordered after him, invariably returning exhausted and fuming after a rough ride through river bottoms and tangled thickets, their horses done in. At age fifty-six, Salmon Brown was to write, in his father's life now "paternalism of necessity gave way to comradeship."

With the free-state legislature scheduled to convene in Topeka on the Fourth of July, proslavery politicians were again threatening to lead an army from Missouri, but after the skirmishing of the past month the seventy-five-mile trek looked much more daunting. The program was then changed as President Pierce issued a proclamation declaring the assembly a treasonable body and commanding its dispersal. Now John Brown determined to be nearby when the legislature met; stopping in Lawrence, he invited the journalist William A. Phillips to accompany him and his men on their journey to Topeka. Years later in the *Atlantic Monthly* of December 1879, Phillips recounted their bivouac under the night sky.

The next day as the Free State legislature began to assemble at Topeka's Constitution Hall, five companies of United States troops with two pieces of artillery entered the town, surrounding the building. Colonel Sumner entered and walked down the aisle. He said: "Gentlemen, this is the most disagreeable duty of my whole life. My orders are to disperse the legislature, and I am here to tell you that it cannot meet. God knows I have no partisan feelings in the matter. I have just returned from the border where I have been driving out bands of Missourians. You must disperse. This body cannot be permitted to meet. Disperse." As the members complied with the order as the federal officer rode off, the crowd gathering in the street gave three cheers for "Frémont" and three groans for "Pierce."[16]

In October 1861, the Viennese newspaper *Die Presse* published an article titled "American Civil War" by Karl Marx, in which he asked: "Whence came, on the one hand, the preponderance of the Republican Party in the North? Whence came, on the other hand, the disunion within the Democratic Party, whose members North and South, had operated in conjunction for more than half a century?"

The answer to both interrogatives was *Kansas*. All through the summer in 1856 a spontaneous uprising flared through the major centers in the North—Boston, New Haven, New York, Buffalo, Philadelphia, Cleveland, Milwaukee, and scores of towns and rural communities alike—as "Kansas Societies" had been organized to assist the Free State cause with men, arms, money, and supplies. On July 1 a National Kansas Committee was formed in Buffalo to coordinate this work, raising $200,000 throughout the North, with Massachusetts alone contributing $80,000. Eli Thayer, founder of the New England Emigrant Aid Society, now selected general agent for the organization, wrote: "From this time no further effort was required to raise colonies. They raised themselves."

Thousands of northerners began making preparations for the trek to Kansas. In New Haven that summer a company departed for Kansas armed with "Beecher's Bibles," so called because Brooklyn clergyman Henry Ward Beecher supplied them with the Sharps rifle—a new breach-loading weapon with an effective range of seventeen hundred yards. Massachusetts emigrants bound for Kansas joined hundreds from western New York congregating in Buffalo. A Chicago saloonkeeper, James Harvey, sold his business to lead a train of Illinois emigrants, as that bustling metropolis became the hub for embarkation, where covered wagons were loaded on flatcars to be unloaded at railheads in Iowa for the final journey.

Into this maelstrom now strode a man, originally from Indiana, who'd been an elector for Franklin Pierce and a supporter of the Kansas-Nebraska Bill, who would gain a place of prominence among the Kansas free-state settlers equal to that of Charles Robinson. He was James Lane, who had entered the territory in 1854 espousing proslavery views, but soon switched his allegiance. Opportunistic and temperamental, he was withal a gifted orator, and he hastened out of Kansas in the early summer to raise an army of sorts to flank the blockaded Missouri River by an overland route through Iowa. Thomas Wentworth Higginson—in Chicago on a fact-finding tour for the Massachusetts Kansas Committee—heard Lane exhorting a crowd, after he had spoken in a number of northern cities. Higginson wrote, "Never did I hear such a speech, every sentence like a pistol bullet, such delicacy and lightness of touch, such natural art . . . he had every nerve in his audience at the end of his muscles." The eastern newspapers reported on his progress, as the emigrants he gathered became a marching demonstration for "Free Soil, Free Men and Fremont." By the end of July, Lane's "Northern Emigrant Army"—their trail through Iowa to Nebraska City marked by cairns, or "Lane's chimneys," to guide later parties—was approaching the Kansas border. John Brown rode up to meet them, bringing with him his badly wounded son-in-law, Henry Thompson, and two sons, Salmon, who had been severely kicked by a mule, and Owen, suffering from fever that had considerably reduced him. Among Lane's company was a young Scotsman named Richard J. Hinton, and another was Brown's son-in-law William Thompson. In his book published in 1894, *John Brown and His Men*, with some account of the roads they traveled to reach Harper's Ferry, Hinton recalled the words addressed to him by an elderly gentleman as he approached:

"Have you a man in your camp named William Thompson? You are from Massachusetts, young man, I believe, and Mr. Thompson joined you at Buffalo." Hinton continues:

> These words were addressed to me by an elderly man, riding a worn-looking, gaunt gray horse. It was on a late July day, and in its hottest hours. I had been idly watching a wagon and one horse, toiling slowly northward across the prairie, along the emigrant trail that had been marked out by free-state men under the command of Sam Walker and Aaron D. Stevens, who was then known as "Colonel Whipple." John Brown, whose name the young and ardent had begun to conjure with and swear by, had been described to me. So, as I heard the question, I looked up and met the full, strong gaze of luminous, questioning eyes. Somehow I instinctively knew this was John Brown, and with that name I replied, saying that Thompson was in our company. It was a long, rugged featured face I saw. A tall, sinewy figure, too (he had dismounted), five feet eleven, I estimated, with square shoulders, narrow flank, sinewy and deep-chested. A frame of nervous power, but not impressing one especially with muscular vigor. The impression left by the pose and the figure was that of reserve, endurance, and quiet strength. The questioning voice-tones were mellow, magnetic, and grave. On the weather-worn face was a stubby short, gray beard, evidently of recent growth. . . . This figure—unarmed, poorly clad, with coarse linen trousers tucked into high, heavy cowhide boots, with heavy spurs on their heels, a cotton shirt opened at the throat, a long torn linen duster, and a bewrayed chip straw hat he held in his hand as he waited for Thompson to reach us, made up the outward garb and appearance of John Brown when I first met him. In ten minutes his mounted figure disappeared over the north horizon.[17]

When he arrived at Nebraska City, Brown had opportunity to exchange his tattered clothes for a new suit, appearing now in a white cotton duster with long tails and a broad straw hat. Some youths without knowing him declared he must be a "distinguished man in disguise." It was at this time that Brown met Aaron D. Stevens, a former United States sergeant, court-martialed for striking an officer abusive to his troops, who escaped from prison and was leading a free-state military company under the *nom de guerre* "Colonel

Charles Whipple." Stevens would be second in command of the company that Brown led to Harper's Ferry.

When he arrived with his caravan, Lane, who had taken and issued military commissions, met with representatives of the National Kansas Committee in Nebraska City, with whom he did not have an official standing. After an interview they decided not to recognize him and sent a letter with their emissary, Samuel Walker, to that effect. After reading the letter, Lane said: "Walker, if you say the people of Kansas don't want me, it's all right, and I'll blow my brains out. I can never go back to the states, and look the people in the face, and tell them that as soon as I got the Kansas friends of mine fairly into danger I had to abandon them. I can't do it. No matter what I say in my own defense, no one will believe it. I'll blow my brains out and end the thing right here."[18]

Walker, a staunch antislavery man from the Midwest who commanded his own free-state company, said, "General, the people of Kansas would rather have you than all the party at Nebraska City. I have got fifteen good boys that are my own. If you will put yourself under my orders I'll take you through all right." Samuel Walker, James Lane, and John Brown now left Nebraska City with thirty armed and mounted men, determined to reignite resistance in Kansas. In thirty hours of hard riding they covered the hundred and fifty miles of prairie.

At that time the "border ruffians," and their abettors in Kansas, beleaguered the free-state settlers by maintaining a ring of forts around Lawrence. A vigorous campaign was now launched to bring relief that carried one after the other of these forts. Brown's company was mounted for the campaign and Walker's became the foot soldiers, while Lane, with no troops of his own until the attack on Fort Titus a few days later, assisted after his manner, when Harvey's Illinois contingent came up. But regardless of leadership the Missouri press saw "John Brown" at the head of every engagement.

The first attack came on the night of August 12 against a blockhouse in Franklin. Two days later, Fort Saunders was besieged and carried. Then on the road to Lecompton, on August 16, Fort Titus came under attack. This was another heavily fortified blockhouse, near which Harry Titus had his pillared mansion, complete with slave quarters. Both the blockhouse and the dwelling were fired, and Titus, the most brutal of the proslavery commanders, captured. Titus's arrival in Lawrence as a prisoner became the object of intense excitement. Walker was to relate, in the *Transactions* of the Kansas State Historical Society:

> The citizens swarmed around us, clamoring for the blood of our prisoner. The committee of safety had a meeting and decided that Titus should be hanged, John Brown and other distinguished men urging the measure strongly. At four o'clock in the evening I went before the committee and said that Titus had surrendered to me, that I had promised him his life and that I would defend it with my own. . . . Captain Brown and Doctor Avery were outside haranguing the mob to hang Titus despite my objections. They said I had resisted the committee of safety and was, myself, therefore a public enemy. The crowd was terribly excited, but the sight of my three hundred solid bayonets held them in check. (Vol. 6)

During a lull in the fighting, in his last act before being fired by President Pierce, the territorial governor rode into Lawrence with a United States military escort, where an exchange of prisoners was arranged. John Brown addressed the parolees on the proslavery side, a speech recorded in Jonathan Winkley's *John Brown the Hero*. It should, however, be stressed Brown never expected or sought position as paramount leader of the free-state resistance. Rather he always maintained his independence, consenting to act in cooperation or conjunction with other leaders and their companies. Moving largely in secret, he sought ways to aid in the general result. Like many warriors, too, he quickly adduced and resorted to the necessity of exciting terror in his enemies, holding it to be one of the effective arms of warfare. One of Brown's more interesting biographers, a nephew of William Lloyd Garrison, Oswald Garrison Villard, alleges—in his *John Brown, A Biography Fifty Years After*—that Brown displayed a veritable "fetish" for organization, as if the authority of a state were behind him, despite the "questionable" nature, in the author's view, of some of his activities. It should rather be seen that Brown's organizational forms were crafted with the view to providing military discipline, with observance of the rules of war, for orderly camp life and for the general deportment of enlistees. The by-laws he drafted specified procedures for electing officers, the care and trial of prisoners, and the disposal of captured property—as it was his practice to strip the enemy of its movable wealth, particularly of transport and agricultural animals.

August Bondi, a Viennese Jew and a member of Brown's company, recounted some of his impressions of Brown as commander in *Transactions* of the Kansas State Historical Society:

> His words have ever remained firmly engraved in my mind. Many and various were the instructions he gave during the days of compulsory leisure in this camp. He expressed himself to us that we should never allow ourselves to be tempted by any consideration to acknowledge laws and institutions to exist as of right, if our conscience and reason condemned them. He admonished us not to care whether a majority, no matter now large, opposed our principles and opinions. The largest majorities were sometimes only organized mobs, whose howling never changed black into white, or night into day. A minority conscious of its rights, based on moral principles, would, under a republican government, sooner or later become the majority. Regarding the curse and crimes of the institution of slavery, he declared that the outrages committed in Kansas to further its extension had directed the attention of all intelligent citizens of the United States and of the world to the necessity of its abolishment, as a stumbling-block in the path of nineteenth century civilization. (Vol. 8)

By the 1880s there was to be a vitriolic and concerted effort to downgrade the importance, as seen by his contemporaries, of Brown's role in the Kansas War. He was nothing but "a cold blooded murderer" and "a horse-thief," some maintained, who had used the higher motives of abolitionism and the Free State cause as a cover. Charles Robinson, who in a letter dated September 15, 1856, commended Brown for his role, saying, "History will give your name a proud place in her pages, and posterity will pay homage to your heroism in the cause of God and humanity," became a leader in the chorus of detractors. After he declared his decades-old letter a forgery, John Brown Jr. wrote to Franklin Sanborn, in October 1884, just completing his valuable biography of his father, *The Life and Letters of John Brown, Liberator of Kansas, and Martyr of Virginia*—"I can have no more doubt of its genuineness than I have that Robinson has either broken down mentally or is a contemptible scoundrel." Frederick Douglass, undoubtedly responding to the campaign of vilification against his friend, wrote in his autobiography at the same time:

It would be a grateful task to tell of his exploits in the border struggle—how he met persecution with persecution, war with war, strategy with strategy, assassination and house-burning with signal and terrible retaliation, till even the blood thirsty propagandists of slavery were compelled to cry for quarter. . . . The question was not merely which class should prevail in Kansas, but whether free-state men should live there at all. . . . John Brown was therefore the logical result of slaveholding persecutions. Until the lives of tyrants and murderers shall become more precious in the sight of men than justice and liberty, John Brown will need no defender. In dealing with the ferocious enemies of the free-state cause in Kansas, he not only showed boundless courage but eminent military skill.

Less than a week after the brutal beating of Charles Sumner and a week after the sacking of Lawrence, the Radical Abolition Party met to nominate its candidates for president and vice president, Gerrit Smith and Samuel McFarland, respectively. The Republican Party held its first meeting in Pittsburgh that February 1856, and at its nominating convention on June 17 drew up *An Address to the People of the United States*, where Republican leaders limited the parties appeal from an antislavery perspective, maintaining that the Congress had no authority to legislate against slavery and refused to challenge the validity of the Fugitive Slave Law, limiting their platform statement exclusively to preventing slavery's extension into Kansas. As Douglass announced for the Radical Abolition candidates, placing their names at the head of the editorial column in his paper, he said the platform of this new party was too narrow to admit true antislavery voters. Criticized for maintaining an abstract position and standing in the way, by way of his influence, of the newly formed coalition, another of the antislavery papers called upon Douglass to join in the Republican ranks. Should he and other abolitionists do that, Douglass responded, "we should feel that we were retrograding instead of advancing." To the charge that abolitionists by shunning the Republican candidates would aid in the election of a Democratic president and assure the loss of Kansas, Douglass replied: "We deliberately prefer the loss of Kansas to the loss of our anti-slavery integrity." But, he added, that was arguing at a disadvantage: "It is by no means certain that Kansas can be saved even with the votes of abolitionists."[19]

In his speech before the Radical Abolition Party convention, Douglass warned: "The anti-slavery movement . . . is every hour liable to be entirely superseded by a movement to uphold the political strength of the North—to promote the freedom of white men, without in any way promoting the freedom of black men. And this is the danger of the Republican movement. Its design is—what? To put down the slave oligarchy in Kansas; to limit slavery to the states in which it is and confine it there. When this is said, all is said. It does not even propose to emancipate the slaves in the arsenals and forts that are under the control of the Federal government. It aims simply to limit slavery, and drive it from one point; and that is Kansas and Nebraska."[20]

Five days prior to the party convention in Syracuse its importance to Douglass could be measured by consideration of a letter he wrote to Smith: "Now I want your counsel. Your unceasing interest in me and in my paper, and in the cause to which it is devoted, makes it right that I should seek your counsel." Given the continuing strain in balancing the demands of the lecture circuit, and editing a weekly with dwindling subscriptions against increasing debt, Douglass argued he was "almost convinced" his paper could not be sustained. His credit was still good, but he felt he had done all he could, deep into the paper's ninth year, "toward putting it on a permanent footing."

"Shall the paper go down and be a total wreck—or shall it be saved—by being merged into the Radical Abolitionist? . . . I am sick at the thought of the failure of my paper and humbled by the thought that no Negro has yet succeeded in establishing a press in the United States. . . . I ask nothing for myself in this business, I do not even ask for a place in the paper but simply ask that you will help me to save my paper from positive failure. . . . Yet it might be an element of strength to the concern to have me in some way connected with it."[21]

Yet by August Douglass abruptly removed the names of the candidates of the Radical Abolition Party from the head of his paper and endorsed the Republican candidates, John C. Frémont and William L. Dayton, opening a breach in the normally cordial relations between himself and Smith. In defending this changed outlook, he announced "that upon radical Abolition grounds, the final battle against slavery must be fought out." The divergence in his paper "this week and last week, is a difference of Policy, not of Principle." As Douglass now saw, the struggle for Kansas had brought a fundamental realignment in the American political landscape that would only admit two parties. Since this was the case, even despite the limited vision of Republican politicians, it was his duty "to be with the natural division for Freedom, in form, as well as in fact." In doing this Douglass felt he could "uphold the Radical Abolition platform in the very ranks of the Republican Party."[22]

Despite the increasing burdens, he would continue to publish his paper as the independent editor and sole proprietor, and by September he and Smith were again acting as colleagues. Douglass wrote: "I am just home from Ohio, where I have been lecturing, and find your kind letters for which please accept my thanks. What I think of your letter to our friend William Goodell, will be seen in my paper of yesterday. . . . Yes! I get it all around. Mr. Garrison tries his hand upon my case this week, the most skilful of them all. The Liberator and Standard seem more shocked at my apostasy from the Radical Abolition Society, than at Mr. May's apostasy from the American Society."[23]

Then on December 16 Douglass would conclude a letter to Smith with this—"Please accept my thanks for your generous donation of twenty dollars. I am happy to know by this expressive sign that you still desire to see my paper afloat. . . . No my dear sir, I am not a member of the Republican party. I am still a Radical Abolitionist, and shall as ever, work with those whose Anti-Slavery principles are similar to your own."[24] The stark divisions sundering old political structures, recoiling in the party conventions in the summer of 1856, was but the recurrence of partisan clashes on the Kansas plain; and what was still to happen would only make them deeper.

On August 20 Brown drew into Osawatomie, wrote Bondi, well supplied with "a spic-an-span four-mule team, the wagon loaded with provisions. . . . Contributed by Northern friends of Free State Kansas, men like Thaddeus Hyatt." Joining forces with two other companies under the leadership of James B. Cline and James H. Holmes—the latter a youth whom Brown especially held in high esteem, calling him "my little hornet"—on August 25 Cline's company swept into Linn County to the south, taking prisoners and capturing military equipment, while on the 26th Brown led a raid to the east, returning two nights later with a hundred and fifty head of cattle. Recoiling under the ferocity of the onslaught, the proslavery side began divulging gangs onto the roads and burning dwellings, including the cabin of Samuel Walker. Political leaders in Missouri called for a meeting in Kansas City "of all friends of law and order,"[25] as elsewhere in Missouri meetings were held to gather men and arms.

"Civil War has begun," announced the Weston *Argus*. The acting territorial governor issued a proclamation calling out the militia of Kansas, and Missourians to act as "Kansas Volunteers," to quell "a state of open insurrection and rebellion." "Let the watch word be 'extermination total and complete'"— the proclamation trumpeted.[26]

On August 25 two thousand men mustered in Westport, Missouri, under the tripartite command of Atchison, Reid, and Clarke. As the Missourians crossed the border, one detachment under George Washington Clarke rode into Linn County to overrun the free-state settlements there. The larger wing under Atchison set out intending on destroying Lawrence, the hated "Boston abolitionist town"; while John W. Reid, veteran of the Mexican War, led two hundred and fifty mounted men to "wipe out" John Brown.

As he paused to take his noon rest, Clarke was overtaken by a freestate company commanded by James Montgomery, a preacher and member of a sect known as the Disciples of Christ founded by Alexander Campbell, who had already made a reputation as a fighter. Thoroughly surprised, the enemy fled, leaving most of its horses behind. Then on the morning of August 31 Atchison's columns parried with the "army" commanded by James Lane under the alias "General Joe Cook." The deciding skirmish of this campaign had occurred a day earlier as Reid's force swept into Osawatomie, overwhelming and all but destroying it. But this barren victory was achieved not before recoiling under the scourging violence directed by John Brown. John J. Ingalls, Charles Robinson's counterpart in the United States Senate for Kansas in the 1880s, was to write even as Robinson was disparaging Brown's contribution: "The battle of Osawatomie was the most brilliant and important episode of the Kansas war. It was the high divide of the contest. It was our Thermopylae. John Brown was our Leonidas with his Spartan band."[27]

On its approach Reid's column had wheeled around Osawatomie, halting on a ridge northwest of the town. Further west, Brown was in his encampment when he received word from a messenger, a young boy, of the Missourians' arrival. The boy also brought shocking news for a father that their advance guard had shot and killed his son

Frederick, and his cousin. With fifteen men, Brown hurried into the town, determined, as he later wrote his wife, to "save the women and children first, and then ourselves if we can." Directing men to a blockhouse to divert attention while their families were removed from danger, Brown hurried with his company and a few others into the timber in line with the enemy's advance, "to annoy them," as he wrote. With Reid's metal glistening in the morning sun, Brown instructed his men to keep out of sight to conceal their number, ordering them to spread out forty feet, as he cautioned, "keep cool, take good aim, shoot low." As the enemy approached, Brown directed a withering fire, throwing their entire line into disorder, killing and wounding thirty to forty. Regrouping, the stunned Missourians finally discharged their cannon and charged as Brown withdrew his men across the Osage River. The town was then fired—some sixty structures in all—and prisoners seized to be summarily shot. One of those captured was a German emigrant, Charles Kaiser, who was hurried to a ditch to face his murderer's gunshot.

The story told later in the Missouri press by John Reid was much different. Said he, the abolitionists had been shot down, thirty-one in all, as easily as "shooting quail." It was reported the "notorious John Brown was also killed . . . in attempting to cross the river," and "the pro-slavery party have five wounded."[28]

Brown indeed had crossed the river. Overlooking the smoke and flames on the opposite bank, his breast heaving, tears streaming from his eyes, in a pose no doubt befitting the Curry mural—his son Jason at his side (he had been released earlier from the Lecompton prison)—Brown vowed (in an oath recorded in Villard), "God sees it. I have only a short time to live—only one death to die, and I will die fighting for this cause. There will be no more peace in this land until slavery is done for. I will give them something else to do than to extend slave territory. I will carry the war into Africa."

Days later back in Lawrence, Lane, Walker, and Harvey decided it was an auspicious moment to free the rest of the "treason prisoners" held in Lecompton, and at the same stroke end the proslavery government there. Their plan called for Harvey to march the twelve miles separating the two garrisoned towns, cross the Kaw River, and invest Lecompton on the north, as Lane, with Walker second in command, led a column to attack from the south. On September 4 the Chicagoan marched up the north bank of the river, positioning his men undercover at the ferry and cutting off retreat from the town.

Waiting through the night and into the morning, with no attack coming, Harvey withdrew his force to Lawrence. Only then did Lane,
twelve hours late, come up with his force on the south.

At the edge of town, the commander in the United States Army
camp nearby ordered his bugler to sound "boots and saddles," riding
out to investigate. As the federal officer approached, Lane dismounted, grabbed a private's rifle, and disappeared into the ranks, leaving
his second in command to parlay alone.

"What in hell are you doing here?" Lt. Col. Phillips St. George
Cooke demanded.

Walker replied, "We are after our prisoners and our rights."

"How many men have you?"

"About four hundred foot and two hundred horseback," said
Walker

"Well, I have six hundred men and six cannon, and you can't fight
here—except with me."

"I don't care a damn how many men you have. We are going to
have our prisoners, or a big fight!"

"Don't make a fool of yourself, Walker. You can't fight here. Show
me to General Lane."

Lane was not in command, Walker replied, but if the lieutenant
colonel wanted to confer with the leaders, he would call a "council of
war." Turning to deliberate with the other free-state leaders, they
soon returned as a group, drawing their horses around the officer.

"You have made a most unfortunate move for yourselves," Cooke
told them. "The Missourians, you know, have gone and the militia
have nearly gone, having commenced crossing yesterday to my
knowledge. As to the prisoners, whilst I will make no terms with you,
I can inform you that they were promised to be released yesterday
morning."[29]

Two days after Walker stood down the U.S. Army, John Brown
entered Lawrence to a tumultuous welcome. With Brown were his
son Jason and a wounded member of his company. It was as if
Frémont himself had appeared, someone said. But none could discern
whether the warrior paid the slightest attention as he walked through
the cheering throng. It was about this time, on the last leg of his journey of fact-finding, riding atop a stagecoach from Nebraska City, that
Higginson observed: "Never before in my life had I been distinctive-
ly and unequivocally, outside the world of human law—it had always
been ready to protect me even when I disobeyed it. Here it had ceased

to exist." Also coming in was the new territorial governor, John W. Geary. He wrote to the secretary of state: "Desolation and ruin reigned on every hand; homes and firesides were deserted; the smoke of burning dwellings darkened the atmosphere; women and children, driven from the habitations wandered over the prairies; the highways were infested with numerous predatory bands and the towns were fortified and garrisoned by armies of conflicting partisans, each excited almost to frenzy, and determined upon mutual extermination."[30]

On the night Brown arrived in Lawrence a council was held that would appoint Harvey to lead an expedition against Leavenworth, and "Whipple" Stevens to lead one against Ozawkie. Harvey's force staged some raids, carrying off property, but without reaching Leavenworth. "Whipple's" expedition resulted in the appropriation of eighty "pro-slavery" horses, in the partisan way people on both sides designated even dumb animals. On September 10 the "treason prisoners" were released, and there was a joyous reunion in Lawrence. The *New York Times* called the meeting held in front of Lane's headquarters "the most enthusiastic and heart cheering of any that has ever been held in Kansas." Six speakers were invited to the platform to address the assembly, including "General" Lane, "Captain" Brown, and the "Free State Governor," Charles Robinson.

On September 12 when Geary arrived in Lecompton, after establishing his headquarters he immediately asked for the removal of three hated federal officials. Refusing the proffered services of Henry Clay Pate, he issued a proclamation for the militia and all armed bands to disperse. Secretary of War Jefferson Davis had urgently ordered more U.S. Army troops to Kansas, and President Pierce had requisitioned militia companies from Illinois and Kentucky. On the evening of the new territorial governor's appearance runners brought word that Atchison and other proslavery leaders were approaching Lawrence with twenty-five hundred men, infantry and cavalry combined. With most of the free-state companies scattered in the surrounding country, John Brown was asked if he would command the fighting men remaining in Lawrence. Deferring, he consented to "advise" them and mounted a dry-goods box in the center of town.

> Gentlemen—it is said that there are twenty-five hundred Missourians down at Franklin, and that they will be here in two hours. You can see for yourselves the smoke they are making by setting fire to the houses in that town. Now is probably the last

opportunity you will have of seeing a fight, so that you had better do your best. If they should come up and attack us, don't yell and make a great noise, but remain perfectly silent and still. Wait until they get within twenty-five yards of you; get a good object; be sure you see the hind sight of your gun—then fire. A great deal of powder and lead and very precious time is wasted by shooting too high. You had better aim at their legs than at their heads. In either case, be sure of the hind sights of your guns. It is for this reason that I myself have so many times escaped; for if all the bullets which have ever been aimed at me had hit me, I would have been as full of holes as a riddle.[31]

Circulating among the barricades and redoubts, Brown conferred with the men, offering his encouragement. At five o'clock four hundred enemy horsemen presented themselves in line of battle at two miles' distance. Redpath wrote: "Brown's movement now was a little on the offensive order; for he ordered out all the Sharp's riflemen from every part of the town . . . marched them a half mile into the prairie, and arranged them three paces apart, in a line parallel with that of the enemy; and then they lay down upon their faces in the grass, awaiting the order to fire."

As a lone scout dashed forward, when within range, a shot rang out, raising the sod in front of his horse. Turning away at a gallop, the rifles blazed up and down the line. Redpath continued, "In a few moments the firing became general and in the darkness and otherwise stillness of the night, the continued flash, flash, flash, of these engines of death along the line of living fire presented a scene the appearance of which was at once not only terrible but sublimely beautiful."[32] Their effect, Sanborn was to write in his account, was seen in riderless horses seen crossing in the night. Brown later wrote:

I know of no possible reason why they did not attack and burn the place except that about one hundred free state men volunteered to go out on the open plain before the town and there give them the offer of a fight, which they declined after getting some few scattering shots from our men, and then retreated back toward Franklin. I saw the whole thing. The government troops at this time were with Governor Geary at Lecompton, a distance of twelve miles only from Lawrence, and, not-withstanding several runners had been to advise him in good time of the approach or of the setting out of the enemy, who had to march some forty miles to reach Lawrence, he did not on that memorable occasion get a single soldier on the ground until after the enemy had retreated back to Franklin, and had been gone for about five hours.[33]

At two in the morning Geary left Lecompton with four hundred dragoons and four cannon in tow. At sunrise, when he reached Lawrence he walked through its streets to observe the hastily thrown up barricades, the crowds of loitering armed men, and the wrecked remains of the Free State Hotel. Robinson, who had known Geary half a dozen years earlier in California, summoned him to an outdoor meeting where he excused the display as "necessary to ward off the invasion." Geary acknowledged that "Americans of spirit will protect their property."

As they spoke, word reached Lawrence that Atchison's army was returning to Missouri. Calling for a mass meeting, Geary read his proclamation ordering the warring parties to lay down their arms and disperse. He would make Missouri obey it too, he said, and he hoped all would return to their fields and benches "in this fair and blooming land of opportunity." Then he got into his carriage and rode back toward Lecompton.

When a runner reporting that the exit of the Missourians had been erroneous, that they were again marching on Lawrence, reached Geary, he wheeled about with his military escort and headed straight for Atchison's line. There he found twenty-seven hundred men, mounted and on foot, led by an entourage including Atchison, Judge Cato, General Reid, Colonel Titus, and other leading territorial and governmental officials. Convening a meeting in Franklin in a clapboard house, Geary began: "Though held in a board house, the present is the most important council since the days of the Revolution, as its issues involve the fate of the Union now formed." As a representative of the president, he said, he knew the administration's wishes, and as a good Democrat he wanted to remind other good Democrats that another attack upon Lawrence might cost them the election. Atchison explained that their intentions were peaceful; they had only come to enforce the laws and to apprehend an "organized band of murderers and robbers" under the command of Jim Lane. Geary said he knew Lane was not in Lawrence for he had just come from there. Go home, he urged, and he would take care of the abolitionists.

Seeing that the new governor was in line with the president's policy, Atchison ordered his men to return to their homes, telling them, the governor "promised us all we wanted." In fact, Geary would try to restore the peace by separating the warring parties and suppressing armed bands on both sides. But he had recognized that the free-state settlers had a right to self-defense; and he also understood that the proslavery party had gained by fraud. He would try to appear even-handed while upholding the proslavery government by preventing further persecution of free-state settlers. By September wagon trains were bringing three to four hundred new settlers into Kansas daily. Although the U.S. troops were patrolling the border, arms were still getting in along with plows and other crated supplies. Richard Hinton arrived with a party of five hundred men, armed with money donated by Boston's Theodore Parker; James Redpath was leading a party of a hundred and fifty fighters—all "to ignite the flames of civil war." Riding up to meet them, Lane gave instructions on the trail ahead and described the best places to camp, reminding everyone that "a vote for Fremont was a vote for a free Kansas."[34]

Before he could announce "peace reigns," Geary had to induce John Brown to leave the territory. A warrant had been issued for his arrest, with the promise of immunity should he depart. Again in Osawatomie at the end of September, Brown said good-bye to his company, placing them under the command of James Holmes. Traveling to Topeka, then to Nebraska City, before traveling on to Tabor, Iowa—Brown and his sons rode in a four-mule wagon, with another one-horse covered wagon, heavily laden with arms and ammunition. "Faithful to their cause," DuBois remarked, a fugitive slave traveled with them. Hunted to the Nebraska border by U.S. troops, Lt. Col. Philip St. George Cooke wrote: "I arrived here yesterday at noon. I just missed the arrest of the notorious Osawatomie outlaw, John Brown. The night before, having ascertained that after dark he had stopped for the night at a house six miles from the camp, I sent a party who found at 12 o'clock that he had gone."[35] The women and children of the Brown family had begun the journey back east up the Missouri River, since the blockade was now lifted, on board a steamer.

There were others, too, who were leaving. In Leavenworth at the Planters House, and again in Kansas City at the Gillis House, Harry Titus had been fêted. President Pierce had recently recognized William Walker as dictator of Nicaragua, and Titus was going to raise a "filibuster" company to join him. With the election over and a new administration coming in, there was no doubt that the decision had been made that the Atchison machine was to control Kansas. Geary had been pointedly criticized for statements he had made about the injustice of the incarceration of the "treason prisoners," and for disparaging remarks about the Lecompton government. He had been warned by Samuel Walker when he first came to Kansas: "Mark my word, you'll take the underground railroad out of Kansas in six months." On the day James Buchanan took office, Geary's resignation took effect.

Meanwhile back east John Brown began raising the inquiry: "It has cost the United States more than a half million, for a year past, to harass poor free state settlers in Kansas, and to violate all law, and all right, moral and constitutional, for the sole and only purpose of forcing slavery upon that territory. I challenge this whole nation to prove before God or mankind the contrary. Who paid this money to enslave the settlers of Kansas and worry them out? I say nothing in this estimate of the money wasted by Congress in the management of this horrible, tyrannical, and damnable affair."[36]

Three

Raising an Army of One Hundred Volunteers

Old Captain Brown, the Ethan Allen, the Israel Putnam of today, who has prayers every morning and then sallies forth, with seven stalwart sons.

—Thomas Wentworth Higginson, *The Liberator*,
January 16, 1857

T HE RESISTANCE IN KANSAS TO THE USURPATIONS OF THE "SLAVE power" had had a profound impact, and not only in the North. In a debate at a meeting of the Massachusetts Anti-Slavery Society in January 1857, in view of recent insurrectionary movements across the South, Henry C. Wright asked: "Henry Ward Beecher and his coadjutors consider Sharp's rifles the most efficient and *only* gospel salvation for Kansas. Why not preach Sharp's rifles as the only gospel of salvation to Virginia and Alabama? They say the only efficient gospel to the Free State men and Border Ruffians is a Sharp's rifle, and they raise funds to furnish them with this religion. Why do they not raise money to send the same gospel to the slaves of Kentucky and Maryland, and teach them how to read and practice that gospel? If these are the best means of grace for Kansas much more so for Louisiana."[1]

Dimly perceived and little recollected in history are the rumors and reports that began to trickle up from the South that draw a picture of insurrectionary plots, with widespread incidents of "insubordination" among the slaves. Happening during the presidential contest and continuing into the winter of 1856–1857, it is still possible to cull suggestions of these extraordinary incidents that flickered across the southern landscape.

Starting in September 1856, disturbances of a large scale were reported in Texas, in Colorado County, followed in October by disturbances in Ouchita, Arkansas, and Union counties. Then in November, "an extensive scheme" was discovered in southwest Texas taking in slaves in three counties, La Vaca, DeWitt, and Victoria.[2] In the following week, in Louisiana a plot was uncovered in St. Mary Parish, followed by reports of strife in Fayette County in Tennessee. Next, came report of unrest to the northwest in Montgomery County, then there was trouble in Obin, Tennessee, and in Fulton, Kentucky, followed by the like in Missouri in New Madrid and Scott counties. The following week there was serious consternation among whites regarding the disposition of slaves in New Orleans. A correspondent of the *New York Weekly Tribune* wrote on December 20, "The insurrectionary movement in Tennessee obtained more headway than is known to the public—important facts being suppressed in order to check the spread of the contagion and prevent the true condition of affairs from being understood elsewhere."

Rebellion flared during the Christmas holiday in Kentucky, Tennessee, Arkansas, Mississippi, Louisiana, and Texas. Arms and ammunition were found hidden in some places; here and there telegraph poles were cut down, in others all the dogs of the neighborhood were killed as if in preparation for an uprising. At one place charges were discovered for blowing a bridge, and in a few locales, assemblages of blacks were broken up. Many slaves were arrested and whipped, and many were executed across the South; while free blacks were driven out of more than a few southern cities and towns.

The story was told how in late August John Brown was observed going through the spoils brought in after James Cline's sweep into Linn County, Kansas. Looking over the items piece by piece, Brown was seen intently examining a rifle stock. After he set it down and walked on, in curiosity the onlooker picked it up and saw the inscription "Made in Harper's Ferry, Virginia, United States Arsenal." While Brown was cognizant of the U.S. government armory, it is certain he had not yet determined on seizing it. For the moment his goal was to arrange for the financial and organizational support for his Kansas company—and the creation of the Kansas Committees across the north was encouraging him.

Also buoying him was the presidential race that year that had shown the nation's political platforms increasingly aligning into two opposing camps. The Democratic Party, by shunting aside a discredited president in favor of an older and more experienced politician, had just been able to hold on to power. Lamenting the widening chasm over slavery in national politics, so heartening to Brown, President-elect James Buchanan hoped his administration would be able to reunite the Democratic Party and check the alarming growth of antislavery sentiment. Although the trend for Kansas seemed to favor a free state, Missouri had just elected two new proslavery senators and the United States Supreme Court was deliberating a case that would have important bearing on the entire conundrum—these, Buchanan, who had been a senator from Pennsylvania, a secretary of state, and ambassador to both Russia and Great Britain, could reasonably hope, might yet solve the question in favor of maintaining the interests that rested on slavery.

When he came out of Kansas, therefore, John Brown had the purpose of securing arms and financial backing to enable him to equip, maintain, and expand his company from among the young men he had become acquainted with that summer. To provide for anything approaching reasonable means he calculated needing a minimum of twenty to thirty thousand dollars. And he carefully enumerated all the necessities—"horses, holsters, spurs, wagons, tents, harness, saddles, bridles, belts, camp equipage . . . blankets, knapsacks, entrenching tools, axes, shovels, spades, mattocks, crow-bars . . . money sufficient to pay freight and traveling expenses." Ill with dysentery and with fever and chills associated with malaria, called *ague* by his contemporaries, Brown lay on the bed of his wagon until reaching Tabor, Iowa. There he recuperated between October 10 and 18, then traveled by stagecoach to Chicago. Arriving at the headquarters of the National Kansas Committee, he brought with him letters of introduction and a proposal. Not being able to act on Brown's request for backing for "volunteers" in Kansas, the committee referred it to a full meeting of the members to be held in New York City in January 1857.

But the committee had a request of Brown. A wagon train was headed for Kansas under commission of Dr. J. P. Root; would he agree to provide an escort, as a previous train had been hijacked by border ruffians with the loss of the equipment? Brown agreed, perhaps because he wanted to intercept two of his sons, Salmon and Watson, headed for the Kansas Territory, but also because on the train was a shipment of two hundred Sharps rifles. He was interested in procuring these for himself. Once in Tabor the arms would be secured in the cellar of the home of Rev. John Todd, and were destined for use at Harper's Ferry.

Before leaving for Tabor, Brown received the following note: "Reverend Theodore Parker of Boston is at the Briggs House and wishes very much to see you." When he returned, Brown found another letter waiting for him from George L. Stearns, Parker's colleague and coworker and the chairman of the Massachusetts Kansas Committee. Stearns wrote that if Brown could come to Boston his expenses would be paid; the letter stated that "friends of freedom" there were eager to meet with him.

Accepting the invitation, Brown now traveled east. On December 20 he was in Columbus, Ohio, where Congressman Joshua Giddings introduced him to Governor Salmon Chase. Giddings had known Brown in the mid-1840s when he sought his help in instituting a prize to encourage American woolgrowers. Receiving the governor's endorsement and a contribution of twenty-five dollars, he also obtained a note of introduction. Next Brown was in Ashtabula, Ohio, on the shore of Lake Erie, where he had a reunion with all of his sons except Oliver, who was back in North Elba. In a day or two Brown was in Rochester, where, perhaps during a layover between trains, he visited with Frederick Douglass. Douglass had encountered Brown's two sons on a train through New York, whom he helped by giving them money on their journey to Kansas. While there may be no record of a meeting, the continued correspondence between Brown and Douglass that can be documented during these years was undoubtedly important in renewing and furthering their cooperation and friendship. Douglass was to write in later decades: "I met him often (during the Kansas struggle), and saw deeper into his soul than when I met him in Springfield seven or eight years before, and all I saw of him gave me a more favorable impression of the man, and inspired me with a higher respect for his character. In his repeated visits to the East to obtain necessary arms and supplies, he often did me the honor of spending hours and days with me at Rochester. On more than one occasion I got up meetings and solicited aid to be used by him for the cause, and I may say without boasting that my efforts in this respect were not entirely fruitless."

On Christmas Day Brown was at the estate of Gerrit Smith in Peterboro, New York. Tendered letters of referral, Smith responded with a note of his own. Next, Brown was in Springfield renewing contacts with acquaintances there among the members of the Gilead League he had formed in 1851. He also solicited another letter of introduction from George Walker, an abolitionist and head of the local Kansas committee. Preceded by the moniker "Osawatomie Brown," which had been "a household name in many homes that summer," and with four letters of introduction, he arrived in Boston on January 4, 1857.

The son of William Lloyd Garrison, Wendell Phillips Garrison, was to ask, "What would John Brown have been without Boston?" For it was in Boston that men with standing and influence gave him the money and arms to allow him to continue his work "in the cause"; as it was from Boston, in the voice and by the pen of these same men, that his apotheosis would be pronounced. Along with Stearns, Howe, Sanborn, Higginson, and Theodore Parker, he was to meet with that uniquely American coterie of intellectuals based in Concord known as the Transcendentalists—Emerson, Thoreau, and A. Bronson Alcott. But those who suppose, as Garrison suggests, that Brown rose into history's pages beyond his own utility merely as a tool in their hands, do not understand the movement emerging through the rhythm of present to past into future that discerns its authentic sequel in thought.

Several years later, speaking on the Emancipation Proclamation, Emerson said, "Every step in the history of political liberty is a sally of the human mind into the untried Future. . . . Liberty is a slow fruit. It comes, like religion, for short periods, and in rare conditions, as if awaiting a culture of the race which shall make it organic and permanent. . . . At such times it appears as if a new public were created to greet the new event." When John Brown came to Boston and Concord, therefore, he attracted many of the leading figures of nineteenth-century American intellectual life as a magnet catches iron. He met Wendell Phillips, William Lloyd Garrison, Judge Thomas Russell, Dr. Samuel Gridley Howe, Theodore Parker, John Andrew, Amos Lawrence, Eli Thayer, and many others. It was as if "some Cromwellian hero suddenly dropped down before them," remarked the wife of Judge Russell. Emerson thought him "the most ideal of men, for he wanted to put all his ideas into action." Wearing a brown military cape and a fur cap with visor, his suit beneath of brown broadcloth, the man stepping before them was described as having "an erect military bearing," "a fine courtesy of demeanor and grave earnestness." One observer wrote: "His figure was tall, slender and commanding. . . . and he gave a singular blending of the old soldier and the deacon." Another wrote: "His mien was serious and patient rather than cheerful; it betokened the 'sad, wise valor' which Herbert praises."[3]

He was received into the homes of George Stearns, Wendell Phillips, and Theodore Parker. It was at Parker's that Brown met William Lloyd Garrison and the two had a long talk, each trading quotes of Scripture against the other, Brown urging attack by force and Garrison moral suasion and nonresistance. One scripture Brown liked to cite was "without the shedding of blood there is no remission of sin." He did not believe in moral suasion, he told Garrison, he believed "in putting the thing through." On January 7 he visited the home of Amos Lawrence, who wrote in his diary—"Captain Brown, the old partisan hero of Kansas warfare, came to see me. I had a long talk with him. He is a calm, temperate and pious man, but when roused is a dreadful foe." Lawrence handed the "old Covenanter" seventy dollars.[4]

When he arrived in Boston the first person Brown called upon was Franklin Sanborn, for whom he had a letter of introduction from George Walker. Just twenty-five years old and a graduate of Harvard, Sanborn was the secretary of the Massachusetts Kansas Committee and became a staunch supporter and one of Brown's biographers, publishing in 1885 a substantial portion of his extensive correspondence. A protégé of Emerson, Sanborn had taken a job at the philosopher's behest at a coeducational school for the children of Concord's wealthier families. Slender and six-foot-three, he had delicate feminine features, modeled, someone would remark, "like an early portrait of Raphael." In his presentation before the Boston committee, Brown argued that only fighting could keep Kansas free from the curse of slavery. Sanborn wrote, "His mission now was to levy war on it, and for that to raise and equip a company of a hundred well-armed men who should resist aggression in Kansas, or occasionally carry the war into Missouri."[5] Such a development would be fitting, said Brown: since Missouri had tried to make a slave state of Kansas and failed, Kansas should make a free state of Missouri. All the committee members embraced Brown's tract, some becoming strong enthusiasts.

George Stearns, the chairman, had been a leading figure in the Kansas aid movement, contributing many thousands of dollars of his own money. A prosperous merchant, producing lead pipe and ship chandlery, he would head the recruitment of black regiments in Massachusetts and then for the Union army during the war, and would be a founder of the Freedman's Bureau; he was also the man most deeply committed in terms of material and monetary resources to John Brown's Harper's Ferry expedition. Dr. Samuel Gridley Howe, another committee member, was a noted humanitarian with an international reputation. In his youth, influenced by Byron, he participated in the national struggle of Greece, both as a fighter and as a surgeon. While in Europe he helped organize a committee to aid Polish refugees, and was imprisoned briefly by the Prussian government.

Returning to Boston, he hung the plumed azure helmet of Byron on his hat rack, founded the New England Asylum for the Blind, and during the war became a member of the United States Sanitary Commission. His wife, Julia Ward Howe, would publish, in February 1862, her "Battle Hymn of the Republic," written at the suggestion of a friend to the tune of the "John Brown Song," then a favorite in camp life and in field marching.

The two remaining members of the committee taking a stock in the enterprise were Theodore Parker and Thomas Wentworth Higginson. Suffice it to say Higginson was the most inclined to the military side of the affair, seeing as well as anyone the tendency of the time. Tall, athletic, and muscular, he began preparing himself for conflict by taking up fencing. A graduate of Harvard at seventeen, he taught school, then entered Harvard Divinity School. As a pastor in Newburyport, Massachusetts, he voiced strongly antislavery views and denounced degrading factory conditions that commanded labor at long hours at low pay. Exchanging pulpits with Theodore Parker, and inviting a fugitive slave to preach to his congregation, led to his resignation as the pews of his church began to empty. Throwing himself henceforth into the antislavery movement, he became minister of the Free Church in Worcester. Higginson is best known today for a liaison that began when he received a peculiar missive dated April 16, 1862. It began: "Mr. Higginson—Are you too deeply occupied to say if my verse is alive?" It was from a young woman in Amherst, Emily Dickinson.

Theodore Parker, stout and balding, was cast in the Socratic mold. With knowledge in sixteen languages, he was as well known in

The "Secret Six" members of the Massachusetts Kansas Committee. From the left, top row: George Stearns, Gerrit Smith, Theodore Parker; bottom row: Franklin Sanborn, Samuel Gridley Howe, Thomas Wentworth Higginson.

England and Germany as in his own country. As much a reformer as a preacher, Parker may be the most outstanding of the Transcendentalists. After graduating from Harvard Divinity School he began his career at the Roxbury Unitarian Church; in May 1842, when he discoursed on "The Transient and Permanent in Christianity," denying the divinity of the Bible and the supernatural in Christ, his association with the church was ended. As his attack on slavery intensified, he became "as much an outcast from society as if . . . a convicted pirate."[6] Starting his own pulpit, he drew twenty-three hundred to hear him each week, until he moved to a larger hall housing five thousand. With the passage of the Fugitive Slave Law he became one of the most active members of Boston's Vigilance Committee.

Without the financial reserves of these men of the Massachusetts Kansas Committee, Brown would not have been able to proceed. But just as importantly he needed the Sharps rifles he had conveyed to

Tabor. These had been purchased with Stearns as donor; the committee now voted to entrust them to Brown's possession, with the stipulation he hold them subject to their orders. They also voted him four
thousand ball cartridges, thirty-one thousand percussion caps, four
iron ladles, and an appropriation of five hundred dollars. Howe was
to give Brown two pistols and a rifle, Higginson handed him thirty
dollars, and Stearns another hundred of his own money.

Since the arms voted to Brown's care were now formally in the
custody of the National Committee, the action of the Boston committee had to be followed up by a similar vote by the larger group.
Sanborn was delegated to go to New York where the committee was
meeting at the Astor house, and he met Brown, who had gone there
in advance. H. B. Hurd of Chicago closely questioned whether Brown
intended to invade Missouri or any other slave state. Refusing to
pledge himself to use the arms solely in Kansas, Brown said, "I am no
adventurer. You are acquainted with my history. You know what I
have done in Kansas. I do not expose my plans. No one knows them
but myself, except perhaps one. I will not be interrogated. If you wish
to give me anything, I want you to give it freely." The committee then
voted to restore custody of the rifles to the Massachusetts committee,
knowing they would be turned over to Brown. They also voted him
an appropriation of five thousand dollars, for "any defensive measure that may be necessary," stipulating it could only be drawn in five-
hundred-dollar allotments.

From February 10 to 16 Brown was in North Elba for a reunion
with his wife and daughters, whom he hadn't seen in a year and a
half. On the 18th he appeared with Sanborn before the Joint
Committee on Federal Relations in the Massachusetts legislature.
Sanborn had prepared a bill of relief for Kansas's settlers and introduced Brown to the committee. "He has been in Kansas what Miles
Standish was to the Plymouth Colony," said Sanborn. Brown read a
paper detailing the economic losses to the free-state settlers; but their
effort was unsuccessful, and the committee declared Kansas was not
the responsibility of the state of Massachusetts, shelving the bill.

Setting out to raise funds, Brown spoke in succession in New
Haven, Hartford, Canton, and Collinsville, in Connecticut, and in
Worcester, Concord, and Springfield, in Massachusetts. In Hartford
he said: "I was told that the newspapers in a certain city were dressed
in mourning on hearing that I was killed and scalped in Kansas, but
I did not know of it until I reached the place. Much good it did me.

In the same place I met a more cool reception than in any other place where I have stopped. If my friends will hold up my hands while I live, I will freely absolve them from any expense over me when I am dead. I do not ask for pay, but shall be most grateful for all the assistance I can get."

Brown published an appeal for contributions on the editorial page of the *New York Tribune* addressed "To the Friends of Freedom" two days before the U.S. Supreme Court rendered its Dred Scott decision.

Dred Scott was the slave of an army surgeon who had been taken in the 1830s from Missouri to Illinois, and then to Fort Snelling in the Northwest Territory, for the duration of two years, respectively. Returning to Missouri, Scott was urged to sue for his freedom on grounds he had been a resident in a free state and in a territory declared free by the Missouri Compromise. He did so in 1846 and lost, but won on retrial in 1850. The decision was appealed to the Missouri Supreme Court, which overturned it on grounds that Missouri law took precedence, and remanded Scott to slavery. In an appeal in 1854, the case moved to District Federal Court, which affirmed Scott's status as a citizen but upheld Missouri's denial of his suit for freedom. The suit then headed for the United States Supreme Court, which first heard arguments in 1856, but held the decision over for its next session to avoid the dissension it was sure to raise during the presidential election.

Two days after Buchanan's inauguration the court rendered its decision. Under political pressure, the court, led by Chief Justice Roger Taney and with a southern and Doughface majority, decided not to rely on a previous ruling to hear the petitions of slaves taken from Kentucky into Ohio—instead issuing a comprehensive fifty-five-page ruling addressing three broad issues. Was Scott a citizen with the right to sue; did residence in a free territory confer on him his freedom; was the federal post actually free territory—did Congress have the right in 1820 to ban slavery in territories beyond 36 degrees 30 minutes north latitude? The decision set out a sweeping counterattack upon the antislavery movement, asserting that persons of African descent were utterly separate under all the provisions of the Constitution and the rights it bestowed. Slaves within the framework of the United States, so the opinion of Chief Justice Taney maintained, were property, and the right of the slaveholder was secure wherever the Constitution extended, and the Congress had no right to prohibit it. Slaves, therefore, could safely be taken anywhere under the Stars and Stripes, and they neither were nor could become citizens. In a stinging phrase, Taney declared "that persons of African descent have no rights a white man is bound to respect."[7]

Denouncing the decision in a speech later that spring before the American Anti-Slavery Society in Rochester, May 11, 1857, Frederick Douglass called it the "judicial incarnation of wolfishness." He cast his eye to the distant future, to "the precise speck of time . . . at which . . . the long entombed millions rise from the foul grave of slavery and death." Douglass conceded he could know nothing. On that point all was uncertain. The only thing certain was that slaveholders would "give up only when they must do that or do worse."

He continued, observing that slaveholders had the advantage of complete organization over all opposition. The state governments and the church organizations where the system of slavery existed were completely at the service of slavery, while the federal government "is pledged to support, defend, and propagate the crying curse of human bondage . . . from the chief magistracy in Washington, to the Supreme Court, and thence to the chief marshalship at New York." Nonetheless, "The pen, the purse, and the sword, are united against the simple truth, preached by humble men in obscure places." But, said Douglass, there was another view than this. The fact is, he pointed out, the more the slavery question had been thought settled the more it needed settling, and the space between these settlements was decreasing. The first settlement came when Missouri was admitted as a slave state, and slavery was prohibited north of its border. Fifteen years later came a new settlement when the right of petition and free discussion in Congress were gagged. In another ten years came renewed settlement with the annexation of Texas and war with Mexico. Then came the final settlement with the Compromise of 1850 by which "slavery was virtually declared the equal of Liberty." Four years after this the Kansas-Nebraska Bill was enacted, "a settlement which unsettled all former settlements." The first stood fifteen years—the second, ten—the third, five—the forth, four—the fifth, two years.

And already a new gleam of hope had appeared following the presidential election, said Douglass. First was an unaccountable sickness attributed to the staff of the National Hotel in Washington, then there was an extensive plan of insurrection laid by slaves in three counties in Texas, uncovered the previous November.

"Twenty or thirty of the suspected were put to death. Some were shot, some hanged, some burned, and some died under the lash. One brave man owned himself well acquainted with the conspiracy, but said he would rather die than disclose the facts. He received seven hundred and fifty lashes, and his noble spirit went away to the God who gave it. The name of this hero has been by the meanness of tyrants suppressed. Such a man redeems his race. He is worthy to be mentioned with the Hoffers and Tells, the noblest heroes of history. These insurrectionary movements have been put down, but they may break out at any time, under the guidance of higher intelligence, and with more invincible spirit."

Finally Douglass drew attention to a recent demonstration against the Taney decision in St. Louis. "The wedge has entered. Dred Scott, of Missouri, goes into slavery, but St. Louis declares for freedom."

"Who does not know that Cotton is King?" crowed a propagandist of the South. The meaning of this axiom in its most fundamental sense was estimated by the worth of a prime field hand. Not how much work the man could perform on a plantation in a day's labor, not how much wood he would cord or how many acres of grain he could cradle—but whether he would fetch two thousand dollars from a trader to be taken south for sale to a cotton planter. The market value of a slave's labor was determined by what he was worth for producing cotton.

The systematic cultivation of large plots of naturally fertile land worked by large gangs of slaves, resembled a "Roman lottery" in which "a few very large prizes were promised" and "many very small ones." With a limited number of tickets, speculators could buy tickets at exorbitant prices "when they know perfectly well that the average chance is not worth a tithe of what they must pay for it," wrote Frederick Law Olmsted in *The Cotton Kingdom.*

> I have been on plantations on the Mississippi, the Red River, and the Brazos bottoms, whereon I was assured that ten bales of cotton to each average prime field-hand had been raised. . . . The soil was a perfect garden mould, well drained and guarded by levees against the floods; it was admirably tilled: I have seen but few Northern farms so well tilled: the labourers were, to a large degree, tall, slender, sinewy, young men, who worked from dawn to dusk,

not with spirit, but with steadiness and constancy. They had good tools, their rations of bacon and corn were brought to them in the field, and eaten with efficient dispatch between the cotton plants. They had the best sorts of gins and presses, so situated that from them cotton bales could be rolled in five minutes to steam-boats, bound direct to the ports on the gulf. They were super-intended by skilful and vigilant overseers. These plantations were all large, so large as to yet contain fresh land, ready to be worked as soon as the cultivated fields gave out in fertility. If it was true that ten bales of cotton to the hand had been raised on them, then their net profit for the year had been, not less than two hundred and fifty dollars for each hand employed. Even at seven bales to the hand the profits for each hand employed are enormous.

Planters that produced cotton at a rate of seven to ten bales per hand could well afford, observed Olmsted, to buy fresh hands at fourteen hundred dollars a head. They could even afford to wait for their expected profits a year or two while new land was cleared, ditches dug, levees and fences built, and the necessary stocks of corn and bacon, tools and cattle procured, and afford to pay 15 percent per annum interest on the outlay of capital required. With practically limitless land available on which cotton could be grown cheaply, one lucky crop would repay all the outlay for land and improvement, if not for the "hands"; for the supply of slaves was limited and could not increase at the same rate as demand for cotton. Olmsted wrote, "If cotton should double in price next year, or become worth its weight in gold, the number of Negroes in the United States would not increase four per cent unless the African slave-trade were re-established."

Now Olmsted offers this vignette: Step into a dealer's "jail" in Memphis, Montgomery, Vicksburg, or New Orleans, and you will hear the Mezzano of the cotton lottery crying his tickets in this way: "There's a cotton nigger for you! Genuine! Look at his toes! Look at his fingers! There's a pair of legs for you! If you have got the right sile (*sic*) and the right sort of overseer, buy him, and put your trust in Providence! He's just as good for ten bales as I am for a julep at eleven o'clock."

A cotton bale weighed from three hundred seventy-five to four hundred pounds. Half a million of these bales were produced by 1820. Production reached a million bales in 1831, and doubled by the early part of the next decade. Thereafter census reports reveal three million bales were gathered in 1852, three and a half million in 1856, and an extraordinary yield of five million bales in 1860. Substantially the whole of the economic activity of the South was thrown into this "cotton lottery" in pursuit of bigger and more luxuriant profits, consequent upon the frightful consumption of human beings and of the soil.

Frederick Douglass averred to the reality of the American slave trade and to the sights and scenes of his boyhood in Baltimore where he had seen its horrors, in his great oration of July 5, 1852, delivered in Corinthian Hall, Rochester, "The Meaning of the 4th of July for the Negro": "We are told by the papers, [it] is especially prosperous just now. Ex-Senator Benton tells us that the price of men was never higher. . . . He mentions the fact to show that slavery is in no danger. This trade is one of the peculiarities of American institutions. It is carried on in all the large towns and cities in one-half of this confederacy; and millions are pocketed every year by dealers in this horrid traffic. In several states this trade is a chief source of wealth. It is called (in contradistinction to the foreign slave-trade) 'the internal slave-trade.'"

And while this trade between the border slave states and areas of intensive cultivation of cotton in the Deep South proved inadequate, the call was heard for a reopening of the African slave trade to meet the demand. If it was right to buy slaves in Virginia and carry them to New Orleans, why wasn't it right to buy them in Cuba, Brazil, or Africa, and carry them there?

Now as the large-scale machine industry, concurrent with the rise of capitalism, began to gain a hold in the northeastern region of the United States in the cotton and wool-spinning industries, and in England and elsewhere, attacks on it began to be heard in the South because of the fierce antagonisms occasioned. Slavery was the superior system because it reconciled capital and labor. George Fitzhugh notably articulated this view in *Sociology of the South* in 1854, and again in 1857 with his *Cannibals All!* The emerging capitalist economy with its liberal democracies was merely a "little experiment," possibly only a temporary aberration in world history, that could not survive because of the periodic paroxysms of economic and social crisis attendant to them. Experience had shown, southern propagandists held, that when the black and white races were thrown together, the relation of master and slave resulted in a society "capable of great development, morally, socially, and politically." Such relation was best for both races, and if it was destroyed, both would decline and industry decay. There is not "a respectable system of civilization known to history, or at the present period, whose foundations were not laid in this institution of domestic slavery," Robert Hunter declared in the United States Senate on January 31, 1860. Abolitionists were only maturing what had already been advanced by political economy, warned Fitzhugh.

In a speech preceding the repeal of the Missouri Compromise before the American and Foreign Anti-Slavery Society in May 1853 in New York, Frederick Douglass argued it was becoming increasingly evident, as reflected in this growing ideology, that there was in the United States "a purely slavery party." While it went by no particular name and had assumed no definite shape, the party was "not intangible" in important respects. Its objects and its designs and the center around which it was coalescing had been "forcibly presented . . . in the stern logic of passing events." The country was dividing on its issues; old party ties were being broken as like was finding like on both sides, and the battle was at hand. Said Douglass, the Compromise of 1850 had specified all the objects of slave-holder policy. That policy comprehended "first, the complete suppression of all anti-slavery discussion; second, the expulsion of the entire free colored people of the United States; third, the nationalization of slavery; fourth, guarantees for the endless perpetuation of slavery and its extension over Mexico and Central America." But, remarked Douglass, as bad as the case appeared, "I do not despair for my people," as "neither the principle nor the subordinate objects, here declared, can be at all gained by the slave power" for they involved "the proposition to padlock the lips of whites, in order to secure the fetters on the limbs of blacks." Free speech could not be suppressed, finally, "because God had interposed an insuperable obstacle to any such result," pillowed as the slaveholder was "on the heaving bosoms of ruined souls."

After the portentous election of 1856, as the Republican Party began to see much of its recently garnered support wane, its leading politicians took up opposition to the Dred Scott decision as a new *cause célèbre*. In Illinois in the following year's senatorial contest, matching Stephen Douglas against Abraham Lincoln, the only issue was to be slavery's future. While Douglas mined for electoral capital trying to exploit the issue into an anti-Negro concern, Lincoln would make his reputation in the ensuing debates, and rise to national prominence. Abolitionists, however, continued to raise their banner outside of the Republican Party's tent, holding a Disunion Convention, January 15, 1857, with Higginson presiding and Garrison as vice president. Higginson would there declare that the duty and destiny of proper abolitionism was to be found in disunion—"It is our only hope." Appealing for a National Disunion Convention to be held in Cleveland the following October, over six thousand persons signed the call. Charles Lenox Remond, one of the most forceful of the "disunionists," not long after this meeting spoke at Israel Church in Philadelphia, hotly contending that it would be "craven . . . for colored people . . . to persist in claiming citizenship under the United States Constitution. . . . We owe no allegiance to a country which grinds us under its iron hoof and treats us like dogs." Mary Shadd Cary, calling to northern blacks from Chatham, Ontario, said: "Your national ship is rotten and sinking, why not leave it?"[8]

Foremost among opponents of both of these tendencies—emigration and disunion—Frederick Douglass was fiercely criticized, even ostracized in some quarters, for his views. But Douglass always gave as good as he got, admitting at the Rochester meeting of the American Anti-Slavery Society, May 17, 1857, that the demonstration in St. Louis protesting the Dred Scott decision could not be taken as an evidence of sympathy with the slave, that it was purely a white man's protestation. "Yet," said Douglass, "I am willing to accept a judgment against slavery, whether supported by white or black reasons."

There was nothing in the present situation to drive the antislavery movement to advocating dissolution of the Union, he maintained. It was not "in changing the dead form of Union, that slavery" was to be abolished, but in the living present. Those who sought to fight slavery by disunion were "fighting a dead form instead of a living and powerful reality."

After the Taney decision was handed down, a public meeting with Frederick Douglass as speaker was scheduled in Worcester while Brown was there as a guest at the home of Eli Thayer. Coming to Thayer's house to ask if he had agreed to sit on the platform, an organizer of the meeting wrote: "Here I found a stranger, a man of tall, gaunt form, with a face smooth-shaven, destitute of full beard, that later became a part of history. The children were climbing over his knees; he said, 'the children always come to me.' I was then introduced to John Brown of Osawatomie. How little one imagined then that in less than three years the name of this plain homespun man would fill America and Europe!"[9]

Invited to occupy a seat on the platform, Brown consented, where, at the insistence of the audience, he gave a brief recital after Douglass spoke.

On March 2 Brown delivered one of his lectures with an appeal for contributions at Canton, Connecticut. Charles Blair, a blacksmith and forge master was among his audience. The following morning Brown met him in the village drug store and displayed a two-edged dirk with an eight-inch blade, explaining he'd taken it from Pate at the Black Jack skirmish. Blair later remembered Brown saying that, attached to a six-foot pole, "they would be a capital weapon of defense for the settlers of Kansas to keep in their log cabins to defend themselves against any sudden attack."[10] He wanted to know what Blair would charge for making them? Offhand the toolmaker guessed he could work them for $1.25 a piece for five hundred and at $1 a piece for a thousand. The next day Brown visited Blair's shop in nearby Collinsville, arranging for the manufacture of one thousand weapons.

On the morning of March 11 Brown arrived in Concord, going directly to the school of Franklin Sanborn. At noon the two dined where Sanborn ate daily, at the table in the Thoreau family house. Engaging Sanborn's companion in long conversation, Henry David Thoreau considered Brown "a man of rare commonsense and directness of speech, as of action; a transcendentalist above all, a man of ideas and principles. . . . Not yielding to a whim or transient impulse, but carrying out the purpose of a life. I noticed that he did not overstate anything, but spoke within bounds." When lunch was served, Thoreau also observed his guest was scrupulous about his diet, saying he must "eat sparingly and fare hard."[11] At two o'clock Sanborn left Brown and Thoreau discussing Kansas affairs.

A little later Ralph Waldo Emerson called on his protégé, as he often did. Emerson would later say of the new man: "For himself, he is so transparent that all men see through him. . . . He believed in two articles—two instruments shall I say?—the Golden rule and the Declaration of Independence; and he used this expression in conversation here concerning them, 'Better that a whole generation of men, women and children should pass away by a violent death than that one word of either should be violated in this country.' There is a Unionist—there is a strict constructionist for you. He believes in the Union of the States, and he conceives that the only obstruction to the Union is slavery, and for that reason, as a patriot, he works for its abolition." Sanborn wrote, "The three men, so celebrated, each in his own way, thus first met under the same roof and found they held the same opinion of what was uppermost in the mind of Brown."[12]

That evening Brown spoke in the Concord Town Hall, where, as he had on other occasions, he displayed the chains used to bind John Jr. in Kansas. "The cruelties he there endured, added to the anxieties and sufferings, incident to his position rendered him a maniac—yes, a maniac," said Brown, testifying to the sufferings of his son. Sanborn recorded—"His words rose to a thrilling eloquence and made a wonderful impression on his audience."[13] On the second day of his visit Brown was a guest at the Emerson home.

After this stop in Concord, Brown, Sanborn, and the Reverend Conway of Lawrence, Kansas, traveled together to Pennsylvania, where they met with Kansas's ex-territorial governor, Andrew Reeder, who had failed in that tenuous position before Geary had failed. Brown urged him to return to Kansas where he had stood for election as the free-state delegate to Congress in the spring of 1855. But Reeder would not reenter the fray.

April 1 Brown was in Springfield, DuBois surmises, "among his Negro friends." Several days later, back in Boston, Brown told Theodore Parker that he had learned federal marshals were searching for him. Parker promptly arranged that Brown go into hiding at the West Newton home of Judge Russell and his wife. Sequestered in his room much of the time, or sitting in the family parlor, he was to remain for over a week, April 6 to April 15. While at West Newton he wrote Eli Thayer: "I am advised that one of Uncle Sam's hounds is on my track, and I have kept myself hid for a few days to let my track get cold. I have no idea of being taken, and intend (if God will) to go back with irons in rather than upon, my hands." Even if he was to soil his hostesses' linen and fine rugs, Brown felt no compunction about using his arms. He was, however, deeply disappointed in his fund-raising, securing far less than he needed by half. It was while at the Russell home Brown composed his plaintive leave-taking, "Old Brown's Farewell: to the Plymouth Rocks; Bunker Hill Monuments, Charter Oaks and Uncle Toms Cabins." When he had completed its composition, he came down from his room asking his hostess to summon Mrs. Stearns from her home in Medford. When she arrived he read the composition to both women, with his unusual but effective punctuation, emphasis, and syntax. Finishing the recital, Brown said: "Oh, if I could have the money that is smoked away in Boston during a single day, I could strike a blow that would make slavery totter from its foundations."[14] Should he send it to Theodore Parker as the subject for a Sunday sermon, he asked? Mrs. Stearns thought not, she would show it to her husband.

After reading Brown's "Farewell," the next morning Stearns rode out to West Newton and handed Brown a letter of credit for seven thousand dollars. Leaving the Russells' home Brown sent a letter off to John Jr.: "My collection, I may safely put down at $13,000. I think I have got matters so much in train that it will soon reach $30,000." In this letter the elder Brown also adverted to a previous letter from his son: "Your remarks about the value & importance of discipline I fully appreciated; & I have been making arrangements to secure the assistance & instruction of a distinguished Scotch officer & author quite popular in this country. I am quite sanguine of my success in this matter."[15]

The man was Hugh Forbes, whom Brown had met while in New York. A trained soldier who had once been a silk merchant in Italy and also an advisor and lieutenant to Garibaldi, sharing the perils of the retreat from Rome to the Adriatic in 1849, Forbes published a book on partisan warfare, and had come to New York City after Garibaldi's defeat. Eking out a living editing an Italian-language newspaper and giving fencing lessons, he also wrote an occasional article for the *Tribune*. Brown hired him to instruct and drill his company for one hundred dollars per month, and authorized that Forbes draw six hundred dollars in advance to revise, translate, and print copies of his *Manual for the Patriotic Volunteer*.

In the first week of May Brown arranged to buy two hundred revolvers, toward which he was granted a 50 percent discount by the Massachusetts Arms Company in Chicopee for "aiding in your project of protecting the free state settlers of Kansas," spending $1,300 in the transaction. He wrote to Stearns: "Now if Rev. T. Parker & other good people of Boston would make up that amount I might at least be well armed."[16]

Few people, Brown had remarked to Thoreau, "had any conception of the cost, even the pecuniary cost, of firing a single bullet in war."[17] Although the funds he had received amounted to less than necessary to equip and mount "one hundred volunteers," he now had ample arms worth over seven thousand dollars, and also had received donations of sundry supplies: percussion caps, cartridges, clothing sufficient for sixty men, blankets, grinding mills, and so on. His efforts finally amounted to enough to assure a tolerable beginning.

Adding to the stress Brown had been feeling that spring was a letter he had received from his wife. Answering from Springfield on March 31, he wrote: "Your letter of the 21st inst. is just received. I have only to say as regards the resolution of the boys to 'learn, & practice war no more;' that it was not at my solicitation that they engaged in it at the first: & that while I may perhaps feel no more love of the business than they do; still I think there may be possibly in their day that which is more to be dreaded: if such things do not now exist. I have all along intended to make the best provision I may be capable of; for all my family. What I may hereafter be able to effect; I cannot now say."[18]

Brown was back in North Elba for what he thought would be the last time, by the second week of May. On May 12 he set out, traveling across the northern tier of New York State. On May 14 he was in Conastota, near Syracuse, where he wrote the Connecticut toolmaker, instructing him not to assemble the first pikes until the wood for the poles was properly seasoned to prevent the blades from loosening from the shafts. The next day he was in Peterboro on Smith's estate where the two carried through with arrangements for a "subscription" to be paid toward the support of his wife and daughters, with money coming from Amos Lawrence, Stearns, and Smith. This was followed very likely by a visit with Douglass in Rochester. There is a note of uncertain date sent by Douglass sometime in 1857: "My Dear Captain Brown: I am very busy at home. Will you please come up with my son Fred and take a mouthful with me?" Next, he was in Cleveland, and would stay with his sons in Hudson from May 27 through June 12, much of the time in ill health from his familiar periodic ailment. He would be gratified to learn his third son, Owen, had decided to return with him to Kansas. June 16 he was in Milwaukee, then on the 22nd in Chicago, and backtracking, in Tallmadge, Ohio, on June 24. He had been obliged, as he was to write to Sanborn, to stop at these different points along the way, "and to go to others off the route to solicit help."[19] There were unavoidable costs, such as freights and for provisions and supplies, and expenses had nearly exceeded his income.

When Brown reached Tabor with his son on August 7, he found funds waiting for him sent by Edmund Whitman of Kansas, agent of the National Kansas Committee. The money to the amount of one hundred and ten dollars and a letter had arrived a day before by a courier named Richard Realf, a young Englishman and a journalist and poet, who after deducting for his expenses from the committee's funds returned to Kansas. Whitman's letter stated: "Your friends are desirous of seeing you. The danger that threatened the Territory and individuals have been removed in the shape of quashed indictments. Your furniture can be brought in and safely stored while you are looking for a location."[20]

Brown immediately began sending out correspondence, writing to William Addison Phillips, to James Holmes, and to Augustus Wattles, a long-time abolitionist and friend living outside Lawrence. To Wattles, Brown wrote: "There are some half a dozen men I want a visit from at Tabor, Ia., to come off in a quiet way; I have some very important matters to confer with some of you about."[21]

Two days after Brown's arrival in Iowa, elections for the Lecompton legislature were held, with Kansas free-state voters participating, electing a majority despite ballot box stuffing by Missouri. Hugh Forbes arrived in Tabor on the same day, bringing with him copies of his pamphlet—with a new chapter suggested by Brown titled "The Duty of the Soldier," appealing to United States troops to desert. Before coming out Forbes had gone to see Horace Greeley, who, hearing he was going to Kansas, donated, together with other sympathizers, seven hundred and twenty dollars. Appearing at Gerrit Smith's door, Forbes introduced himself as Brown's associate, soliciting one hundred and fifty dollars from Smith. By the time Forbes was on hand, Brown and his son were taking possession of the Sharps rifles stored in Reverend Todd's cellar, moving them to a barn. Brown wrote to Sanborn he'd found them "in middling good order—some a little rusted. Have overhauled and cleaned up the worst of them, and am now waiting to know what is best to do next, or for a little escort from Kansas, should I and the supplies be needed. I am now at last within a kind of hailing distance of our Free-State friends in Kansas."[22]

A letter from Wattles dated August 21 reached Brown, stating he did not see a pressing need for the resort to military means. "Those who had entertained the idea of resistance, have entirely abandoned the idea," he wrote, adding, "Come as quickly as possible, or not come at present, as you choose." William A. Phillips also replied that he saw no need for military measures and would not come to Tabor to confer with Brown, and knew no one who would. Holmes, however, wrote, "Several times we have needed you very much." Referring "to business for which I believe you have a stock of material with you," he added, "If you wish other employments, I presume you will find just as profitable ones."23

After being impressed by the caliber of Brown's backers, Forbes must have been surprised to find only Brown and his son Owen in Tabor, albeit with a large cache of arms. Brown himself had wearied somewhat of Forbes, hearing he was incessantly bent on begging money, but he wrote Sanborn, "we have, we think, a capable teacher." It was in his discussions with Forbes that Brown revealed for the first time that his true intention was to make an attack in the South. As a result of his experiences in Kansas and the reserves he saw could be mobilized in the North, Brown's theorizing had undergone a decisive transformation. No longer thinking of the strategic extension of the Underground Railroad, he would now "carry the war into Africa," as he vowed at Osawatomie. His intention was to make an incursion along the border of the slave states, ideally he thought, at the head of the Blue Ridge Mountains of Virginia. Now that he had sufficient arms, equipment, and financial backing, he was presented with two problems militarily: How to generate an effective force from among freed slaves, and how to give it sufficient impetus to make it self-sustaining. Both issues would largely rely upon the organization he would introduce at the outset. At no point, said Brown, was the South more vulnerable than in its fear of servile insurrection. Sweeping down upon the plantations and farms abutting the mountains, the freed slaves would be subdivided into three, four, or five distinct parties under the men he would bring with him. The best of these would be mounted, he proposed, when they would make a dash on the arms manufactory at Harper's Ferry, on account of the arms there, destroying what they could not carry off. Then his intention was to retreat into the South, reaching into those counties with an overwhelming predominance of slaves. The arms manufactory was a point to be seized, not held, but if retreat became advisable, it furnished the egress to the North.

In their strategizing sessions Forbes countered that Brown would be disappointed in his expectations; since no notice could be given, "the invitation to rise, unless, they were already in a state of agitation, would meet with no response, or a feeble one." What is more, said Forbes, "a slave insurrection being from the very nature of things deficient in men of education and experience would under such a system (as Brown proposed) . . . be either a flash in the pan or would leap beyond his control, or any control, when it would become a scene of mere anarchy and would assuredly be suppressed." The European tactician argued further to go anywhere near Harper's Ferry would be folly and would surely bring United States troops down upon them. Convincing Brown, or so he thought, to drop his grab at Harper's Ferry, Forbes suggested they should sally out of the mountains, as Brown had proposed, but only to get up "slave stampedes," carrying freed slaves off to safety in the North.[24]

By October relations between the two had completely soured, and Brown fired Forbes. Despising Brown's ability and feeling defrauded, Forbes demanded additional pay beyond that already advanced. Angry and nearly destitute; he would denounce Brown to his eastern backers, and he intended to have reparations for his trouble.

When Buchanan took office he appointed Robert Walker the new territorial governor for Kansas. Walker had served as senator from Tennessee and as secretary of the treasury under Polk; having won and lost a fortune in Mississippi, he was a trusted entity in the South. Yet he alarmed the administration by refusing to accept appointment until the president and the Senate majority leader, Stephen Douglas, assured him the majority of the settlers in Kansas would be permitted to determine the state's institutions. In his inaugural address he recognized the proslavery government, but urged free-state voters to seek restitution by running candidates for election to its legislature, warning that the United States Army would put down any extra-legal resistance. The territory was booming, he said, and he invited all to partake in the prosperity.

On taking office Walker promptly found that the election for delegates to the state's constitutional convention scheduled to meet in September had been rigged to assure only members of the Atchison machine would have seats. The Lecompton legislature had also specified that the new constitution would go into effect without a voter referendum, while Walker took the position there should be a referendum. To everyone's surprise word soon came from Washington that Buchanan would back him on this.

As free-state voters gained a majority in the Lecompton legislature in the August voting, the proslavery party resorted to a last line of defense. Just completing its handiwork, the constitutional convention emerged with a document declaring, "the right of property is before and higher than any constitutional sanction, and the right of the owner of a slave and its increase is the same and as inviolable as the right of the owner of any property whatever."[25] The convention decreed the constitution must be sent directly to the United States Congress for approval, together with a petition for statehood. With the free-state press and public infuriated at this brazen attempt to impose slavery, Governor Walker now called a special session of the newly elected legislature, suggesting it pass an act for a referendum on the constitution.

Brown wrote to Stearns, "I am now waiting further advice from Free State friends in Kansas with whom I have speedy private communications." Then he followed with this: "I am in immediate want of from Five-Hundred to One Thousand Dollars for secret service and no questions asked." "Rather interesting times" were expected in Kansas he wrote, "But no great excitement is reported. Our next advices may entirely change the aspect of things, I hope the friends of Freedom will respond to my call & 'prove me here-with.'"[26]

Earlier Jim Lane had written Brown that he should be in Kansas by October 5: "We want you with the material you have. I see no objection to your coming to Kansas publicly. I can furnish you just such a force as you may deem necessary for your protection here & after you arrive." Brown replied, "As to the job of work you enquire about I suppose that three good teams with well covered wagons & ten really ingenious, industrious men (not gassy) with about $110 in cash, could bring it about in the course of eight or ten days." In a message on September 29 Lane disclosed he was sending "ten true men," and enclosed fifty dollars, saying, "it is all important to Kansas that your things should be in at the earliest possible moment & that you should be much nearer at hand than you are."[27] But Brown demurred in his reply on the next day, returning the money Lane sent. He had no intention of letting the weapons slip from his grasp, nor to engage in an action where he could lose them, and in any event he still held them subject to Stearns's orders.

Brown replied, "it will be next to impossible in my poor state of health to go through on such very short notice, four days only remaining to get ready, load up & go through. I think, considering all the uncertainties of the case, want of teams etc. that I should do wrong to set out. I am disappointed in the extreme."[28]

Several weeks passed before Brown sent a young man from Maine named Charles Plummer Tidd to see Edmund Whitman about sending more money. On October 24 Whitman wrote that Tidd would be returning to Tabor with one hundred and fifty dollars, and "General Lane will send teams from Fall City so that you may get your guns in." Clearly anxious to obtain the arms, Lane wrote to Brown from Nebraska, "I trust this money will be used to get guns to Kansas or as near as possible. If you will get them to this point we will try to get them on in some way. The probability is that Kansas will never need the guns. One thing is certain, if they are to do any good, it will be in the next few days."[29]

As controversy swirled over the imposition of the Lecompton Constitution, a remedy was proposed. The convention made a modification in its position, in line with a suggestion from Stephen Douglas. It would not allow a referendum on the entire constitution, but would allow a vote for it "with Slavery" or "with no Slavery," specifying that if voters choose the latter "the right of property in slaves now in this Territory shall in no manner be interfered with." This subterfuge signaled that the proslavery party had no intention of dealing fairly, and Walker denounced it as "a vile fraud, a bare counterfeit."[30] Buchanan had no choice but to back the convention, recalling Walker from office.

On November 5 Brown arrived at the Lawrence home of Edmund Whitman, who forthwith supplied him with money, tents, and bedding. In two days Brown was gone. Whitman reported to Stearns: "[He] . . . left, declining to tell me or anyone where he was going or where he could be found, pledging himself, however, that if difficulties should occur he would be on hand and pledging his life to redeem Kansas from slavery. Since then nothing has been heard of him and I know of no one, not even his intimate friends, who know where he is."[31]

On November 7 George Stearns wrote Brown, "In my opinion the Free State party should wait for the Border-ruffian moves and checkmate them as they are developed. Don't attack them, but if they attack you, 'Give them Jessie' and Fremont besides." Brown disposed himself to Stearns as follows—, "I find matters quite unsettled; but am decidedly of the opinion that there will be no use for the Arms or ammunition here before another Spring. I have them all safe &

together unbroken & mean to keep them so until I am satisfied they are really needed."[32]

While at Whitman's Brown had met again a young man named John Cook, who served briefly under him at Black Jack. Cook would later relate: "I was told that he intended to organize a company for the purpose of putting a stop to the aggressions of the proslavery men. I agreed to join him and was asked if I knew of any other young men who were perfectly reliable whom I thought would join also. I recommended Richard Realf, L. F. Parsons, and R. J. Hinton."[33] While at breakfast the following Sunday, Cook received a note telling him to come to a meeting together with the men he had named. Realf and Hinton were not in town, so he and Parsons went together and had a long talk with Brown.

A few days later Cook received the following note: "Dear Sir—you will please get everything ready to join me at Topeka, and bring your arms, ammunition, clothing and other articles you may require. Bring Parsons with you if he can get ready in time. Please keep quiet about this matter."

John Cook was twenty-seven years old, five feet five inches in height, with long blond hair that curled around his neck. From Haddam, Connecticut, he had attended Yale and studied law in New York City. Salmon Brown said of him: "Cook was more than high-strung—he was highly erratic and not overly stocked with morality. He was the best pistol-shot I ever saw. . . . He was just as much of an expert in getting into the good graces of the girls wherever we stopped. He would have a girl in a corner telling them stories or repeating to them poetry in such a high-faluting manner that they would laugh to kill thunder."[34]

Richard Realf, another of those joining Brown's company at this time, was twenty-four years old. Born in Sussex County, England, he had, at sixteen, been a promising poet under the guidance of Lady Byron, publishing *Guesses at the Beautiful*. He had made his way to New York City and thence to Kansas, where he wrapped himself in the free-state movement.

Cook, Realf, and Parsons now traveled with Brown to Topeka, arriving November 14. Brown was to interview and recruit his most valued lieutenants, Aaron Dwight Stevens, still known as "Col. Charles Whipple," and John Henry Kagi, at Topeka. Stevens, from New London, Connecticut, was twenty-six and an experienced soldier, expert with a saber. He would become the company's drillmaster and second in command to Brown in military matters. Hinton described him as "handsome and active as a young, Greek gladiator," six feet two inches in height and powerfully built, he had raven black hair and wore a full beard. He loved to sing and had a rich baritone. Kagi, by all accounts was a youth of exceptional ability, who at twenty-two had been a correspondent for several eastern and midwestern newspapers, among them the *Tribune*, Brown calling him "our Horace Greeley." He had been a member of Whipple's company before being imprisoned for four months at Lecompton with the "treason prisoners." When he met Brown in Topeka, he was recovering from a head wound inflicted by a proslavery judge, Rush Elmore. Kagi had written an article critical of the judge, and to retaliate Elmore assaulted him outside his courthouse with his gold-plated cane. As Kagi retreated between two pillars he ended the brutal assault, and the house of Elmore, as one writer remarked, with a pistol shot into the judge's groin. George Gill, who would be recruited to the company later in Springdale, Iowa, described Kagi as "a logician of more than ordinary ability. . . . He was an agnostic of the most pronounced type, so grounded in his convictions that he gave but little thought to what he considered useless problems. His disposition was a model one. No strain or stress could shake his unruffled serenity. His fertility of resources made him a tower of strength to John Brown." All understood Brown's purpose was to make reprisals on slavery "whenever the opportunity offered," but any details of his subsequent movements "were matters of after confidence."[35]

Richard Hinton tells of meeting with Realf just before the company left Topeka in his *John Brown and His Men*, and the two talked without reserve. Hinton wrote: "He assur(ed) me that the purpose was just to prepare a fighting nucleus for resisting the enforcement of the Lecompton Constitution, which it was then expected Congress might try to impose on us. Through this, advantage was to be taken of the agitation to prepare for a movement against slavery in Missouri, Arkansas, the Indian Territory and possibly Louisiana."

Brown had recruited six young men, and together they now traveled for Nebraska City. It was as they camped on the prairie northeast of Topeka that they learned they were to leave Kansas to attend a school for military instruction in Ashtabula County, Ohio. When they reached Tabor they were joined with four other young men, including Brown's son Owen, and Richard Richardson, a fugitive slave from Lexington, Missouri, who had been with Brown in the

Six of those who joined John Brown's fight against slavery. From the left, top row: John Cook, Richard Realf, Aaron Dwight "Col. Charles Whipple" Stevens; bottom row: John Henry Kagi, Owen Brown, Barclay Coppoc.

previous months. At Tabor they learned for the first time that their "ultimate destination was the state of Virginia."[36]

In early December as the company started for Ohio, John Brown said "Good-by," to a Quaker friend he had known there: "You will hear from me. We've had enough talk about 'bleeding Kansas.' I will make a bloody spot at another point to be talked about." The men walked across the plains into the cold alongside two heavily laden ox-drawn wagons. On the snow-swept night of December 8, four days after they set out, Owen Brown recorded in his diary: "Warm argument upon the effects of the abolition of slavery upon the Southern States, Northern States, commerce and manufacture, also upon the British provinces and civilized world; whence came our civilization and origin?"[37]

On December 21, when the referendum came on the Lecompton Constitution, the free-state voters abstained, thereby seeing it approved "with slavery" by a vote of 6,226 for and 569 against. This

Map 2. The Midwest, Northeast, and Canada.

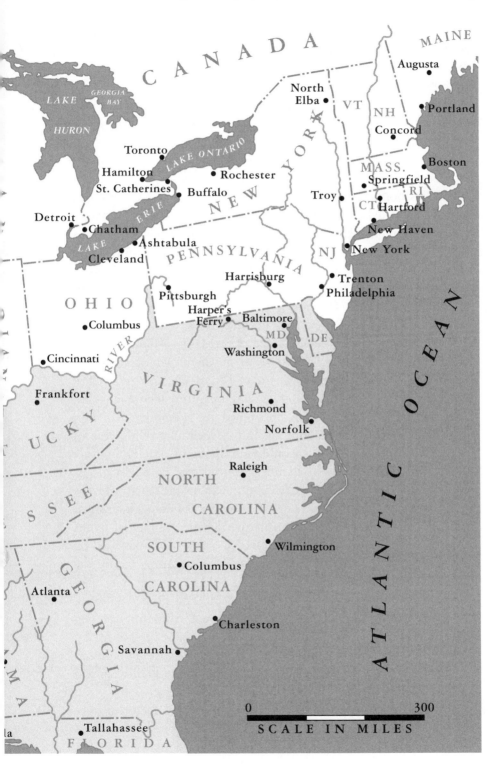

vote was followed on January 4, 1858, by a referendum approved by the free-state majority in the Lecompton legislature, with proslavery voters abstaining, whereby a vote was obtained of 138 for the constitution "with slavery," and 24 "with no slavery," and 10,226 against the constitution altogether. Now Congress had two votes to consider, but the president and his cabinet were not going to let majority rule stand in the way. On February 2 the Lecompton Constitution was submitted to the Congress with the recommendation that Kansas be admitted to the Union as the sixteenth (slave) state.

With his political future hanging in the balance and Congress in the throes of a debate more intense even than the Kansas-Nebraska controversy, Stephen Douglas decided he could no longer support his president, declaring he could not vote to "force this constitution down the throats of the people of Kansas, in opposition to their wishes and in violation to our pledges." For this defection on an issue of paramount importance for the South, Alexander Stephens of Georgia pronounced Douglas "a dead cock in the pit."[38]

The Senate approved the admission of Kansas as a slave state in a vote on March 23, but in the House, as northern Democrats joined Republicans, it was rejected by a vote of 120 to 112 on April 1. The Democratic Party had split, each faction representing sectional issues.

As the end of the decade approached the economies of both North and South were entering an unprecedented twelfth year of expansion. California gold was adding millions of dollars monthly to the national wealth. Railroad mileage had tripled in the decade, as mills, factories, and foundries were running near capacity. In the West, American grain exports were booming, bringing intensified speculation in land. During this same period the number of banks doubled, as did deposits, and the money in circulation increased by 50 percent. But the Crimean War in 1854–56 began to reduce the European capital being invested across the Atlantic. In the immediate aftermath of the war interest rates in England and France doubled, and then tripled, and capital was withdrawn from low-yielding American securities. Then in August 1857 a panic was set off by reports of embezzlement at the New York branch of an Ohio investment firm, and as a consequence of the telegraph, volatility spread with lightning speed. In September a shipment of gold from California was lost in a storm, and by mid-October factories were shutting as bankruptcies and business failures multiplied. Thousands of workers in several cities marched demanding "work or bread"; a threatening crowd on Wall Street was dispersed by soldiers and marines to prevent forced entry into the U.S. Customs House, whose vaults contained twenty million dollars.

Against this background John Brown had taken leave of Tabor, intending to convey his arms and men to Ashtabula, Ohio. On Christmas Day this troop passed quietly through Marengo, then Iowa City, finally arriving fifteen miles beyond at the small Quaker settlement of West Branch, near Springdale. The Quakers were friendly to the man known as Osawatomie Brown, but said to him, "Thou art welcome to tarry among us, but we have no use for thy guns."[39] The two-hundred-and-fifty-mile trek across Iowa had been a hard one as winter was closing in, and so the men took quarters indoors for the first time in weeks—in an inn in West Branch, and others a few miles away in a house near Springdale owned by John Painter. Brown's intention was to sell his teams for rail fare through to Ohio, but the financial panic that had begun in August was in full swing then, and the money could not be raised. It was decided the men would winter in Springdale, while Brown traveled east alone to make arrangements, when the company would join him in spring.

On December 30, Brown wrote his wife in North Elba, "The persons I have with me are mostly well-tried men. Some of them have acted with me before; & all of them are pledged to stand by the work." Four youths from Springdale and near-by West Branch would be recruited and join the others in the cause—these were the son of the doctor and founder of Springdale, George Gill, who had been educated in Canada, and his Canadian friend who had stopped by on his way to Kansas, Stewart Taylor; and two Quaker brothers, twenty-three-year-old Edwin, and nineteen-year-old Barclay Coppoc. With Stevens superintending, the company would study and drill: all would rise at five, breakfast, and from six to ten take classroom instructions with discussion. Along with tactics and ordinance, the other topics engaging the company included abolitionism, equal rights for women, temperance, natural philosophy, theology, and history. In the afternoon there was physical culture, with gymnastics, fencing, and drill. Sham battles were fought in nearby hills, maneuvers practiced, with instruction in the mechanics of entrenching and fortification. Realf also gave lessons to Richardson in his ABC's, teaching him to read. Twice a week in the evenings a mock legislature was held with townspeople participating, with debates featuring the talents of Kagi, Realf, and Cook. Hinton wrote: "There was no attempt to make a secret of their drilling . . . the neighborhood folks all understood that this band of earnest young men were preparing for something far out of the ordinary. Of course Kansas was presumed to be the objective point. But generally the impression prevailed that when the party moved again, it would be somewhere in the direction of the slave states. The atmosphere of those days was charged with disturbance."[40]

Before leaving, Brown sat up several nights with his landlord and two others discussing his plans. One of them later related in Richman's *John Brown Among the Quakers*:

> [Brown] had not then decided to attack the armory at Harper's Ferry, but intended to take some fifty to one hundred men into the hills near the Ferry and remain there until he could get together quite a number of slaves, and then take what conveyances were needed to transport the Negroes and their families to Canada. And in a short time after the excitement had abated, to make a strike in some other Southern state; and to continue on making raids, as opportunity offered, until slavery ceased to exist. I did my best to convince him that the probabilities were that all would be killed. He said that, as for himself, he was willing to give his life for the slaves. He told me repeatedly, while talking, that he believed he was an instrument in the hands of God through which slavery would be abolished. I said to him: "You and your handful of men cannot cope with the whole South." His reply was: "I tell you, Doctor, it will be the beginning of the end of slavery."

Four

Carrying the War into Africa

Courage, courage, courage!—the great work of my life (the unseen Hand that "guided me, and who has indeed holden my right hand, may hold it still," though I have not known him at all as I ought) I may yet see accomplished (God helping), and be permitted to return and "rest at evening."

—John Brown to his wife, Mary, and daughter, Ruth,
January 30, 1858

I N THE EARLY MONTHS OF 1858 JOHN BROWN TURNED HIS attention to a plan he believed would shake American slavery to its foundations. To begin he would need to call on the backers he had gained in New England the previous year, explain his true orientation, and obtain even greater commitments from them. But "more especially," wrote DuBois, his object was to "definitely organize the Negroes for his work. ... He particularly had in mind the Negroes of New York and Philadelphia, and those in Canada."

On the list of those black leaders whom Brown knew, many had risen through the Underground Railroad into abolitionist circles, and increasingly he would begin to feel he would not be able to adequately martial a "Black Phalanx," as DuBois termed it, without Frederick Douglass's considerable commitment, as well as his personal and organizational largesse. Although firsthand evidence of the relation between the two—eyewitnesses, autobiography, letters—is slight, Douglass offers his readers three settings with Brown as principal.

First is their meeting in Springfield in 1847; second, in his home in 1858; and last, the meeting in the stone quarry outside Chambersburg in 1859. Written two decades after Brown's death, Douglass's accounts, while evocative, throw but guarded illumination upon crucial questions regarding their twelve-year alliance, overwhelmed by that closing scene before the culmination in the tragedy at Harper's Ferry. Two of Brown's early biographers, Franklin Sanborn and Richard Hinton, as well as his son John Jr., have drawn attention to an intriguing discrepancy noted apropos Douglass's recollection in the first instance. One must ask—is there something there that will allow further clarification about his commitment, which he was not entirely forthcoming about in later years? The fact that it was from his own home that Brown conducted the initial planning and preparation for his "Virginia Campaign," with Douglass's full knowledge and assistance, already gives some measure to the degree and manner of his assurance and of any pledge he may have given as to continuing this support. In his autobiography, seemingly concluding his remarks regarding his 1847 meeting with Brown, Douglass wrote: "Hating slavery as I did, and making its abolition the object of my life, I was ready to welcome any new mode of attack upon the slave system which gave any promise of success. I readily saw that this plan could be made effective in rendering slave property in Maryland and Virginia valueless by rendering it insecure. . . . In the worse case, too, if the plan should fail, and John Brown should be driven from the mountains, a new fact would be developed by which the nation would be kept awake to the existence of slavery. Hence, I assented to this, John Brown's scheme or plan for running off slaves."[1] Sounding somewhat like an apologia and misplaced by a decade, Douglass, with these words, intimates he was prepared to support, and did support Brown's plan as proposed to him a decade later at the end of January and the early part of February 1858.

Few have ventured beyond Douglass's remarks, and the only witness/participant outside of the immediate Brown family to comment was to be John Cook in his so-called "Confession" after his capture in November 1859. Although there is a wealth of material from which to draw, conclusions remain largely conjectural and circumstantial; nonetheless, the two men were clearly of one mind on the efficacy of violence in forcing an end to slavery. Six months before this January meeting, at a gathering celebrating the twenty-third anniversary of West India's emancipation, Douglass stood before an audience at Canandaigua, New York, where he avowed: "The general sentiment of mankind is, that a man who will not fight for himself when he has the means of doing so, is not worth being fought for by others. . . . Virginia was never nearer emancipation than when General Turner kindled the fires of insurrection at Southampton."

On his journey east Brown's first stop had been at West Andover, Ohio, to consult with John Jr., who would henceforth become his indispensable coadjutor and agent. Preceding Brown eastward by more than a month was Hugh Forbes, who, destitute and nursing resentments, was threatening to expose what he knew of the plan. Insisting he was owed six hundred dollars, he had written Brown at his son's address demanding payment. The elder Brown instructed that his son reply he was returning the letter unseen by his father— John Jr. was also instructed to say he was willing to send Forbes forty dollars to ease the plight of his family, but first would like to hear further of his intentions. In this way Brown sought to "chase Forbes off," and avoid him until he had "a better account of him." From Douglass too Brown was to learn that Forbes had preceded him east; Douglass reporting he received him coolly, putting him up in a hotel for several days, and then paying his rail fare through to New York City. But he had also given Forbes a letter of introduction to his German friend in Hoboken, Miss Ottilia Assing—a woman deeply interested in John Brown. She would introduce Forbes to the Germans gathered around the *Tribune* staff, and into the circle of James McCune Smith, a physician noted for the emphasis he put on black racial pride.

"After the close of his Kansas work, Captain Brown came to my house in Rochester, and said he desired to stop with me several weeks, but he added, 'I will not stay unless you will allow me to pay board.' Knowing that he was no trifler and meant all he said, and desirous of retaining him under my roof, I charged three dollars a week."

So wrote Douglass in *Life and Times*; the date was January 28. During these three weeks Brown was to spend much of his time in correspondence, and when not writing letters he was engaged many hours, according to his host's account, "writing and revising a constitution which he meant to put in operation by means of the men who should go with him into the mountains." Correspondents were to address letters in reply to "N. Hawkins" care of Frederick Douglass, or put a sealed letter inside another addressed to his host. Remittances were to be made out to Douglass, while his youngest son

Charles was engaged in posting outgoing and picking up incoming mail.

Brown's correspondents included the Boston committee members George Stearns, Theodore Parker, and Franklin Sanborn, and Thomas Wentworth Higginson in Worcester as well as Gerrit Smith in New York, in view of sounding them out and inducing even greater financial backing from each than in the previous year. But more especially Brown wrote to the Reverends J. W. Loguen in Syracuse, James Gloucester in Brooklyn, and Henry Highland Garnet, now in New York City after returning from a four-year sojourn in Jamaica. From these, and from Douglass, Brown sought cooperation in convening a series of meetings to include both free and fugitive blacks in several states of the North and in Canada, together with the young men he had recruited in Kansas. The purpose of the meetings was to deliberate on his proposed campaign, and, if they approved of it, institute a legitimating process to culminate in a convention without whose directive it is inconceivable Brown would have been able to proceed.

In his discussions with Douglass at the time, Brown's proposal, like that of a decade before, was to take a body of men into the rim of mountains known in Virginia as the Blue Ridge and in their fastness secure a series of easily defended retreats. Posting small bands on a line of twenty-five miles, they would avoid battle and violence unless compelled to fight in self-defense, subsisting on the country round-about. Thus established, they would begin by drawing slaves off the plantations in such a way that the first movement would have no other "appearance to the masters than a slave stampede, or local outbreak at most."[2] The planters would pursue their chattels and be defeated; the militia would then be called out, and would also be defeated. Richard Hinton related his understanding of Brown's plan in 1858 to be the following—"It was not intended that the movement should appear to be of large dimensions, but that, gradually increasing in magnitude, it should as it opened, strike terror into the heart of the slave States by the amount of organization it would exhibit, and the strength it gathered." The constitution he had devised, Hinton wrote, "was intended as the framework of organization among the emancipationists, to enable the leaders to effect a more complete control of their forces. Ignorant men, in fact all men, were more easily managed by the forms of law and organization than without them. This was one of the purposes to be subserved by the Provisional Government. Another was to alarm the oligarchy by discipline and the show of organization. In their terror they would imagine the whole North was upon them pell-mell, as well as all their slaves." In this initial phase Brown still projected broadening and extending the operations of the Underground Railroad; adding to his force any freed slaves who would be willing to remain and endure the hardships and dangers of mountain life, and sending others north by means of stations from the Pennsylvania to the Canadian border. These refugees would be supplied with food, clothing, and shelter, and forwarded to the next station till they reached safety. Those that did remain, Brown thought. "if properly selected could, on account of their knowledge of the surrounding country, be made valuable auxiliaries."[3]

Soon after becoming a guest in his home, Douglass wrote, Brown "asked me to get for him two smoothly planed boards, upon which he could illustrate with a pair of dividers, by a drawing, the plan of fortification which he meant to adopt in the mountains. These forts were to be so arranged as to connect one with the other, by secret passages, so that if one was carried another could be easily fallen back upon, and be the means of dealing death to the enemy at the very moment when he might think himself victorious." He also thought it would be difficult for their enemies to cut them off from their provisions or means of subsistence, and that it could not be done "so they could not cut their way out."[4]

Indicating, no doubt sincerely, that he was only tepidly attached to these military matters, Douglass remarked his children were more interested in these drawings than he. Brown however would write his son in West Andover about the tenor of their conversations—"He has promised me $50 and what I value vastly more he seems to appreciate my theories and my labours." Lastly Douglass was to write, "Once in a while, he would say he could, with a few resolute men capture Harper's Ferry, and supply himself with arms belonging to the government at the place, but he never announced his intention to do so. It was, however, very evidently passing in his mind as a thing he might do."[5]

On February 2 Brown addressed the following to Theodore Parker:

I am again out of Kansas, and am at this time concealing my whereabouts. . . . I have nearly perfected arrangements for carrying out an important measure in which the world has a deep inter-

est, as well as Kansas. . . . I have written to some of our mutual
friends in regard to it, but they none of them understand my views
as well as you do. . . . I have heard that Parker Pillsbury and some
others in your quarter hold out ideas similar to those on which I
act; but I have no personal acquaintance with them, and know
nothing of their influence or means. Cannot you either by direct or
indirect action do something to further me? Do you not know of
some parties whom you could induce to give their abolition theories a thoroughly practical shape? . . . Do you think any of my
Garrisonian friends, either at Boston, Worcester, or any other
place, can be induced to supply a little "straw," if I will absolutely make "bricks"?[6]

On February 12 Brown made this reply to a letter from Higginson:
"I have just read your kind letter of the 8th instant, & will now say
that Rail Road business on a somewhat extended scale, is the identical object for which I am trying to get means. I have been connected
with that business as commonly conducted from my boy-hood &
never let an opportunity slip. I have been operating to some purpose
the past season: but I now have a measure on foot that I feel sure
would awaken in you something more than a common interest; if you
could understand it."[7]

Gerrit Smith shared his letter from Brown with Edwin Morton, a
classmate of Sanborn's who was then on an extended stay on the philanthropist's estate. Morton wrote to his friend in Concord: "This is
news—he expects to overthrow slavery in a large part of the country." Sanborn then communicated what he had learned with
Higginson: "I should not wonder if his plan contemplated an uprising of slaves—though he has not said as much to me—The Union is
evidently on its last legs and Buchanan is laboring to tear it into
pieces. Treason will not be treason much longer but patriotism."[8]
Brown did not, in point of fact, think an "insurrection" of slaves
advisable, nor is that what he sought, as is usually supposed.
Insurrection would only defeat the object, he said; however, he did
anticipate, as he gained a hold in the Virginia mountains and expanded operations, that he could inspire confidence in slaves and induce
them to rally.

With the business at hand completed, and satisfied with what he
had accomplished, Brown left Rochester on February 17. Through
the auspices of Henry Highland Garnet, meetings were scheduled for

a week in March to be convened at the Philadelphia home of Stephan Smith—among whose conferees were to include Smith, William Still, J. W. Loguen, James Gloucester, Frederick Douglass, as well as other selected men.

The entry in Gerrit Smith's diary for February 18 reads: "Our old and noble friend, Captain John Brown of Kansas, arrives this evening." In his correspondence with Smith, Brown had requested that George Stearns and Theodore Parker meet him at his estate. Brown wrote: "It would be almost impossible for me to pass through Albany, Springfield or any of these points on my way to Boston; & not have it known: & my reasons for keeping it quiet were such that when I left Kansas: I kept it from every friend there."[9]

Neither Stearns nor Parker could make the trip, so they sent Sanborn, who would report back. In private conversation in the meantime Brown detailed his plan to Smith, eliciting such enthusiasm that he wrote his son on February 20 that Smith was "ready to go in for a share of the whole trade." Sanborn arrived on the afternoon of the 22nd. Also a guest at Smith's at the time was Charles Stuart, a soldier from Wellington's army that defeated Napoleon at Waterloo. When Sanborn entered, Brown and the old lieutenant were discussing points of theology. Later, at Brown's request, Edwin Morton played the piano. When he came to his recital of Schubert's Serenade, Sanborn reports Brown sat unashamedly "weeping at the air."

In the late hours of the evening, Sanborn, Smith, Morton, and Brown withdrew to talk at length. Sketching out the plan, Brown indicated the thrust of his movements and exhibited the constitution that he had drawn for the government of the men who would come with him and those joining them; the middle of May was designated as the time operations would commence. Sanborn wrote: "Being questioned and opposed by his friends, he laid before them in detail his methods of organization and fortification; of settlement in the South, if that were possible, and of retreat through the North, if necessary; and his theory of the way in which such an invasion would be received in the country at large." This "little council" sat past midnight "proposing objections and raising difficulties," but nothing could shake Brown. "Every difficulty had been foreseen and provided against in some manner," wrote Sanborn, "the grand difficulty of all—the manifest hopelessness of undertaking anything so vast with such slender means—was met with the text of Scripture: 'If God be for us, who can be against us?'"[10]

Astonished by what they heard, they were also deeply moved. Here was "a poor, obscure old man, uncertain at best of another ten years lease of life and yet calmly proposing an enterprise which, if successful, might require a whole generation to accomplish." As the discussion renewed the next day and as Brown continued to prevail over all objections, the others realized "they must either stand by him, or leave him to dash himself alone against the fortress he was determined to assault." As the sun was setting over the snow-covered hills at the end of the day, Smith and Sanborn walked together among the woods and fields of the estate for an hour. Sanborn wrote: "Mr. Smith, restated in his eloquent way the daring propositions of Brown, whose import he understood fully; and then said in substance: 'You see how it is; our dear old friend has made up his mind to this course, and cannot be turned from it. We cannot give him up to die alone; we must support him.'"[11] Smith said he would make his contribution and Sanborn must lay the case before his Massachusetts friends and perhaps they would now do the same.

After Sanborn's departure, Brown sat down to compose a letter to him, dated February 24:

> Mr. Morton has taken the liberty of saying to me that you felt inclined to make a common cause with me. I greatly rejoice at this, for I believe when you come to look at the ample field I labor in, and the rich harvest which not only this entire country but the whole world during the present and future generations may reap from its successful cultivation, you will feel that you are out of your element until you find you are in it, an entire unit. What an inconceivable amount of good you might so effect by your counsel, your example, your encouragement, your natural and acquired ability for active service! And then, how very little we can possibly lose! Certainly the cause is enough to live for, if not to — for.

Many years later Sanborn wrote he was never able to read this letter without being overcome by emotion.

Extended an invitation with a donation of twenty-five dollars enclosed, Brown traveled to Brooklyn where he would stay in the home of the Rev. James Gloucester and his wife. In his letter Gloucester vowed not to falter in his support for Brown's work, but cautioned, "You speak in your letter of the people. I fear there is little to be done in the masses. The masses suffer for the want of intelligence and it is difficult to reach them in a matter like you propose as far as it is necessary to secure their cooperation. The colored people are impulsive, but they need sagacity, sagacity to distinguish the proper course. They are like a bark at sea without a commander or rudder." In regard to the tenor of his conversations and correspondence thus far in a letter to his wife from Brooklyn dated March 2, Brown wrote: "I am having a constant series of both great encouragements and discouragements, but am yet able to say, in view of all 'hitherto the Lord hath helped me.' . . . I find a much more earnest feeling among the colored people than ever before; but that is by no means unusual. On the whole, the language of Providence to me would certainly seem to say, 'Try on.' . . . I had a good visit with Mr. Sanborn at Gerrit Smith's a few days ago. It would be no very strange thing if he should join me."[12]

As Sanborn arrived in Boston, he hastened to communicate with Parker and Higginson. At Parker's invitation Brown now consented to travel to Boston, having grown the lengthy beard for which he is now remembered. Under the alias "Nelson Hawkins" he took a room at the America House, where he remained for the most part of his four-day stay, March 4 to 8. After two days of consultations he wrote his eldest son: "My call here has met with a hearty response, so that I feel assured of at least tolerable success. I ought to be thankful for this. All has been effected by quiet meeting of a few choice friends, it being scarcely known that I have been in the city." Parker was "deeply interested in the project," "believing that it must do good even if it failed." Samuel Howe, who had seen a few hundred Greek partisans repulsing an army of Turks in the mountains of Macedonia, felt the plan was absolutely feasible on a military basis. Higginson felt an event of this kind, even with only temporary success, "would do more than anything else to explode our present political platforms." Each had pledged to contribute one hundred dollars individually, and they would raise altogether a thousand. Sanborn officially recorded their affirmative decision, and then wrote, "Hawkins goes to prepare agencies for his business." He pledged "to bring the thing off within the next sixty days."[13]

On March 7, while still in his room at America House, Brown addressed a lengthy letter to Theodore Parker asking that he undertake a literary assignment. He wanted a section of Forbes's manual rewritten from an address he himself had proposed in the previous year urging United States Army soldiers to desert. Brown wrote: "The address should look to the actual change of service from that of Satan to the service of God. It should be, in short, a most earnest and powerful appeal to men's sense of right and to their feelings of humanity. Soldiers are men, and no man can certainly calculate the value and importance of getting a single 'nail into old Captain Kidd's chest.' It should be provided before hand, and be ready in advance to distribute to all persons, male and female, who may be disposed to favor the right." He also contemplated a second address "to be sent out broadcast over the entire nation. . . . I know that men will listen, under such circumstances. Persons will hear your antislavery lectures and abolition lectures when they have become virtually slaves themselves."[14]

During his stay in Boston, Brown took the opportunity to call on Charles Sumner, still recuperating from wounds he had received from a caning by Congressman Preston S. Brooks of South Carolina on the Senate floor in 1856. Going to his residence accompanied by James Redpath and the Rev. James Freeman Clarke, the trio stood at Sumner's bedside. Brown finally said to him, "Do you have the coat you were wearing when you were attacked by Brooks?" "Yes," Sumner replied, "it is in the closet. Do you want to see it?" "I recall the scene vividly," Redpath wrote, "Sumner standing slightly bent, supporting himself by keeping his hand on the bed, Brown erect as a pillar, holding up the blood-besmirched coat in his right hand and examining it intently. The old man said nothing, I believe, but I remember that his lips compressed, and his eyes shone like polished steel."[15]

In early February Brown had written his eldest son before leaving Douglass's home: "I have been thinking that I would like to have you make a trip to Bedford, Chambersburg, Gettysburg, and Uniontown in Pennsylvania, traveling slowly along, and inquiring of every man on the way, or every family of the right stripe, and getting acquainted with them as much as you could. When you look at the location of those places, you will readily perceive the advantage of getting some acquaintance in those parts."[16]

On March 4 he again instructed his son to "hunt out friends . . . even at Harper's Ferry." Whatever the status of this trip, whether accomplished at this time or later, the line of travel would have taken John Jr. from western Ohio to Pittsburgh, and thence by rail across the southern tier of the state before reaching Philadelphia, where his father's appointment with notables in that city was scheduled for March 10, and which he too would attend.

Although Brown's meetings in Philadelphia have drawn at most cursory comment, DuBois correctly indicates their importance: "Brown seems to have stayed nearly a week in that city, and probably had long conferences with all the chief Philadelphia Negro leaders." Convened on Lombard Street at the well-appointed residence of the lumber merchant, abolitionist, and philanthropist Stephen Smith—"headquarters" for Brown and his son during their stay— also in attendance were Garnet, William Still, and Douglass, together with other selected men; the Reverend Gloucester was expected but was "detained" in Brooklyn, and Loguen reported "ill." The meetings attest to the significance Brown gave that city, which contained the largest black population of any northern city; as it also attests to the standing and regard of those under whose auspices the gatherings proceeded, and to the reputation of John Brown.

"Slavery! How much misery is comprehended in that single word," Garnet once intoned. Certainly the men at the meetings had counted all its miseries, and every man knew Brown felt its injustice as deeply as any; as Garnet would remark, "He is the only white man who really understands slavery."[17] The thing to do, Brown proposed, was to break its power. The practice of carrying arms would guarantee blacks their freedom, give them a sense of their manhood, and prepare them for equality and self-government. Reverberations of this kind must have shocked the meeting, as did the echoes of solemn words spoken elsewhere.

Although Garnet and Douglass had engaged in rancorous debate over the years, however sustained the argument, it was Douglass who had moved toward Garnet's positions. First came the debate regarding the efficacy of political action, which saw Douglass becoming a Liberty Party member after Garnet. Then on the issue of moral suasion versus the sword, Douglass opposed Garnet's stand, before himself despairing of slavery's overthrow by peaceful means. However Douglass would continue as Garnet's adversary on emigration, although both believed that to meet the ever-growing critical situation in the United States the struggle must be internationalized. Garnet too had been among Brown's first black confidants. In 1849, after receiving an invitation and fees to lecture, he went to England, speaking and traveling in the following years throughout the British Isles; also visiting the Continent, where he became fluent in French and ʌan. As an affiliate of the Scottish Presbyterian Church, in 1853 ʌt to Jamaica where he began to study West Indian society with a view to the effect of emancipation on the former slave population. Returning to the States in 1856 with the intention of lecturing as "a missionary of liberty," he became, instead, pastor of the Shiloh Presbyterian Church in New York City, made vacant by the death of the Rev. Theodore Wright.

William Still, another of those taking part in the meetings on Lombard Street, was born in New Jersey in 1821, of a free father and a fugitive slave mother. Arriving in Philadelphia in 1844, he became a clerk for the Pennsylvania Anti-Slavery Society, and rose into leadership of the Underground Railroad, becoming the corresponding secretary of its important Philadelphia branch. During his tenure he was instrumental in the escape of over a thousand slaves, keeping detailed records of his experiences, published in 1872 as a book titled *Underground Railroad*.

It is often remarked that Brown never revealed more of his plan than was necessary to adhere men to it. While it was not something to be freely broadcast, the outline of his proposal was undoubtedly presented to those in Philadelphia, and although it has never been brought into the historical record, Philadelphia had a significant bearing on what ultimately transpired. In the successive accounts, from Douglass to Smith to the various individuals of the Boston coterie to Higginson, a picture emerges that Brown intended to exploit circumstances to pursue an ascending intensity of operations; in his view the mere fact of the ongoing movement, even if moderately successful, would have a withering effect on slavery. On the other hand, a successful demonstration of this kind, displaying humanitarian restraint to avoid loss of life and destruction of property, would have a thoroughly salutary effect in the North. Not one effective voice, he held, could be raised for the reclamation of the specie in human property to the South once it had been freed.

Brown sought above all from his auditors their sanction and cooperation in bringing selected men from the northern states and Canada, together with his own company, to deliberate on and to ratify his constitution and then provide the framework and organizational support his campaign would entail. It was evidentially the consensus of the meetings that such a convention could not safely be convened in a northern state. It would enrage the South and its northern supporters, if it were known (and even if not), making it impossible to draw anyone who would not desire unwanted attention on their activities. But even as some in the meetings may not have been overly optimistic that it could succeed militarily, others felt it was something that should be tried. To bring it off, Brown was unequivocally told that he must go to Canada and that the man he must seek out was Dr. Martin R. Delany. It was certain he would be able to carry out his plans then.

While Brown received this judgment with disappointment, characteristically viewing it as a lack of resolve, it had been his intention to visit the fugitive slave communities in Canada from the beginning. Now that trip would be doubly necessary. Stopping in New York City and in New Haven to hunt up other possible donors, Brown sent notice to Loguen in Syracuse, who would accompany him to Ontario—"I expect to be on the way by the 28th or 30th inst."

Brown now took the time between his Philadelphia conferences and his call in Canada, to take leave of his family in North Elba. Travel to the Adirondacks and the remote location of the Brown farm in those days was made most expeditiously by taking a train to Vergennes, Vermont, crossing Lake Champlain by ferry to Westport, New York, and hiring a horse or traveling on foot. Brown chose the latter to save on expenses. This visit, so far as he knew, was to be his last, and perhaps he used these moments of his long walk to gather his thoughts and his strength. Among his own sons, the husband of his daughter, and his brothers, Brown sought the recruits he needed for the realization of his plan. He had addressed a letter to his oldest daughter before this trip about her husband, Henry Thompson: "O my daughter Ruth! Could any plan be devised whereby you could let Henry go 'to school' (as you expressed it in your letter to him while in Kansas), I would rather now have him 'for another term' than to have a hundred scholars. I have a particular and very important, but not dangerous, place for him to fill in the 'school,' and I know of no man living so well adapted to fill it. I am quite confident some way can be devised so that you and your children could be with him, and be quite happy even, and safe; but God forbid me to flatter you into trouble!" In the same letter he also wrote: "I would make a similar inquiry to my own dear wife; but I have kept her tumbling here and there over a stormy and tempestuous sea for so many years that I cannot ask her such a question. The natural ingenuity of Salmon in connection with some experiences he and Oliver have both had, would point him out as the next best man I could now select; but I am dumb in his case, as also in the case of Watson and all my other sons."[18]

Black settlement in Canada received its first great impetus when two thousand Cincinnati blacks emigrated in the spring of 1829 in reaction to riots organized to drive them out of that city. After bloodletting and the torching of homes, a deputation was sent to Canada where the governor extended a welcome: "Tell the republicans on your side of the line, that we royalists do not know men of their color. Should you come to us you will be entitled to all the privileges of the rest of His Majesty's subjects." By 1840 the black population of "Canada West," as it was called, stood at around 10,000, with another 10,000 joining these before the end of the decade. But most of the refugees in Ontario Province emigrated after the United States Congress enacted the Fugitive Slave Law, when over a thousand refugees were added to its population thereafter each year. Most of these were newly escaped slaves, or fugitives that had lived in the north for a number of years and no longer felt safe. Of a population of nearly 60,000, by 1860 fully two thirds were fugitives; making a sizable minority in communities like Buxton and Mt. Elgin, Ingersoll, Hamilton, and St. Catharines, where, DuBois writes, "farms had been bought, schools established and an intricate social organization begun." The center of these settlements was Chatham, among whose residents were mechanics and merchants, professionals and farmers, together with hundreds of fugitive slaves; making it an especially cohesive community was an organization known as the True Band with four hundred members.

Brown was in Peterboro on April 2 where Smith gave him a draft of twenty-five dollars intended for Harriet Tubman, whom he would see in Canada. The next day he was at Douglass's, where he was joined by John Jr. On April 4 the Rev. J. W. Loguen arrived, and Brown and the minister crossed over into Ontario, calling first on William Howard Day, the printer residing in Hamilton. Day was to provide Brown with letters of introduction to other notables in Ontario and arranged for an immediate interview with Tubman, who was working as a domestic and living in nearby St. Catharines. Certainly Brown had heard much about her and had come to Canada eager to see her. So impressed was he when they met that the singular was trifurcated, as Brown said: "I see General Tubman once, I see General Tubman twice, I see General Tubman three times."[19] Finding complete unanimity between them, he was to stay two days a guest in her home, where they held long interviews and would meet twice more during his Canadian sojourn. Brown gave Harriet fifteen dollars out of his own purse at the conclusion of their first meeting, and on April 14 would give her the twenty-five dollars sent in a draft from Gerrit Smith, changed for a gold piece.

In 1849 Harriet Tubman had escaped from slavery, fleeing Maryland's Dorchester County at the farthest extremity of the Eastern Shore near Bucktown. Finding herself free but alone, she returned to Baltimore in the following year to rescue her sister and her two children, who had, by prearrangement, sailed across the Chesapeake Bay from Cambridge. Thus began a career that would see Tubman return to the Eastern Shore nineteen times in a decade, bringing as many as three hundred fugitives away with her, including many of her own relatives. But the "heroine of the age," as she was billed, was to all appearance an ordinary woman of the South. No more than five feet tall, she had a rough muscularity gained by doing the hardest work since her early teens. With closely cropped hair; her chin was round and receding, with ponderous eyelids, large protruding lips, with her upper front teeth missing. Commonly "attired in coarse, but neat apparel," as William Wells Brown described her, "with an old-fashioned reticule or bag suspended by her side," she always wore a bandana headdress. In the next few years, as her fame grew, this illiterate woman would have the most educated people in the nation gathering around her, listening spellbound as she told her eventful stories in the plainest manner seasoned with good common sense. Adept at disguise, concealment, and maneuver, she conducted her charges through "middle passage"—as William Still called escape from bondage to freedom—as if it were a military campaign, enforcing strict discipline, always armed with a revolver. Often traveling in the Maryland-Delaware districts for months at a time, she spread the gospel of freedom and imparted to her people the knowledge that "friends" in the North were eager to help escaping slaves. Loguen remarked to Higginson, "Among the slaves she is better known than the Bible, for she circulated more freely." When she had instilled confidence in a particular group she would spirit them north, accompanying them as far as Philadelphia or New York and sometimes all the way to Canada.[20]

Although her reputation was increasing, and she had begun addressing antislavery conventions, Tubman had not yet gained the prominence she would have by 1860. To be sure she was well known to Douglass, Loguen, and Still, as well as to many other abolitionists operating in the eastern branch of the Underground Railroad; but she was unknown in the West and had only furtively appeared in New England. Moreover she had been obliged to drop out of sight for periods of time, working and saving money for her rescue missions, and to support her aging parents, who lived in Auburn, New York, in a home provided on generous terms by William H. Seward.

After accompanying Brown to Hamilton and St. Catharines, and making two important introductions, Loguen evidently returned to the States. After leaving St. Catharines, Brown made the rounds, stopping first in Ingersoll, then Chatham and Buxton, then Toronto, then again in Chatham and back to St. Catharines, and again back to Chatham. On April 8 Brown wrote his son, who was now in Ohio, "I came on here direct with J. W. Loguen the day after you left Rochester. I am succeeding to all appearance beyond expectations. Harriet Tubman hooked on his whole team at once. He is the most of a man, naturally, that I ever met with. There is the most abundant material, and of the right quality, in this quarter, beyond all doubt."[21]

Among those whom Brown sought and with whom he stayed while in Chatham were an influential group at the *Provincial Freeman*, an abolitionist and pro-emigration newspaper whose editor at that time was I. D. Shadd. These included Shadd's sister, Mary Shadd Cary, and her husband, Thomas Cary, Thomas Stringer, and the young printer's devil Osborne Perry Anderson. Another inhabitant of Ontario's black settlements, who would be among those coming out to meet Brown, wrote: "Mr. Brown did not overestimate the state of education of the colored people. He knew that they would need leaders, and require training. His great hope was that the struggle would be supported by volunteers from Canada, educated and accustomed to self-government. He looked on our fugitives as picked men of sufficient intelligence, which combined with a hatred for the South, would make them willing abettors of any enterprise destined to free their race."[22]

When Brown called upon the man he had especially come to see, Dr. Martin Delany, he found the doctor was away and not expected for several days more. Brown ventured on; first he was in Buxton where meetings were held at the homes of Abraham Shadd and Ezikial Cooper, and where Brown, it can be surmised, made the significant introduction to William Parker, then living in the "bush" near there. Traveling on to Toronto, Brown found another eager audience; meetings were held for him at Temperance Hall and at the home

of Dr. A. M. Ross, an abolitionist and naturalist with whom he stayed. After a few days
Brown returned to Chatham. Calling again
on Delany but finding him absent, Brown
promised his wife he would "be back in two
weeks time."[23]

Satisfied that preparations for a convention were in order, Brown had to return to
Iowa to bring on his company who would
serve as the nucleus of the "fighting emancipationists" the gathering was to sanction. On
April 25 Brown was in Chicago at the home
of his friend John Jones, a well-off tailor, who
had been receiving his mail during his two weeks in Canada. Hurrying
on, he arrived in Springdale two days later.

Harriet Tubman in 1880.

Soon after Brown had left for the East the previous winter, the
company had moved into a large concrete house owned by William
Maxton, who charged each man a dollar and a half per week for
board, "not including washing and extra lights." One witness wrote:
"The leave-taking, between them and the people of Springdale was
one of tears. Ties which had been knitting through many weeks were
sundered, and not only so, but the natural sorrow at parting was
intensified by the consciousness of all that the future was full of hazard for Brown and his followers."[24] Before leaving each man wrote
his name in pencil on the wall of the parlor.

Traveling through Chicago and Detroit, the company arrived in
Chatham and took up lodging at the Villa Mansion, a hotel under
black proprietorship. Soon Brown called at the home of Dr. Delany,
who was now within.

Martin Delany had been born in 1812 in Charlestown, Virginia, to
a free mother and a slave father. In 1822 the family moved across the
Mason-Dixon Line to Chambersburg, Pennsylvania. At nineteen
Delany crossed the mountains to Pittsburgh, where he found work
and education, rising into abolitionism and newspaper editing—the
Mystery from 1843 to 1847, and co-editing the *North Star* with
Douglass from 1847 to 1849. An important theorist of the African
diaspora in the United States, in 1852 Delany published his polemi-
cal tour de force, *The Condition, Elevation, Emigration and Destiny
of the Colored People of the United States, Politically Considered.*
Attacked as an apostate for vehemently denouncing the patronizing

and hypocritical attitudes of white abolitionists and for recommending emigration to free blacks, preferably to Africa, Delany envisioned developing a region that would gain a rising economic and political potency to undermine American slavery. In the winter of 1856 he made his word deed by moving with his family to Chatham, Ontario.

For a decade Delany and Douglass had sparred, each becoming the foil to the other and neither letting the other off without a rejoinder. While Delany was dedicated to the premise that blacks must wholly rely upon their own resources to overcome the adversity of their condition, Douglass retorted, "I thank God for making me a man simply, Delany always thanks Him for making him a black man."[25]

Delany recollected his initial meeting with John Brown to his biographer—Frances Rollin Whipper, using the pseudonym Frank A. Rollin, for her *Life and Public Service of Martin R. Delany*, published in 1868. Brown began: "I came to Chatham expressly to see you, this being my third visit on the errand. I must see you at once, sir, and that, too, in private, as I have much to do and but little time before me. If I am to do nothing here, I want to know it at once." Delany continues the explication:

> Going directly to the private parlor of a hotel near by he at once revealed to me that he desired to carry out a great project in his scheme of Kansas emigration, which to be successful, must be aided and countenanced by the influence of a general convention or council. That he was unable to effect in the United States, but had been advised by distinguished friends of his and mine, that, if he could but see me, his object could be attained at once. On my expressing astonishment at the conclusions to which my friends and himself had arrived, with a nervous impatience, he exclaimed, "Why should you be surprised? Sir, the people of the Northern states are cowards; slavery has made cowards of them all. The whites are afraid of each other, and the blacks are afraid of the whites. You can effect nothing among such people." On assuring him if a council was all that was desired, he would readily obtain it, he replied, "That is all, but that is a great deal to me. It is men I want, and not money; money can come without being seen but men are afraid of identification with me, though they favor my measures. They are cowards, Sir! Cowards," he reiterated. He then fully revealed his designs. With these I found no fault, but fully favored and aided in getting up the convention.

In the edition of the *Weekly Anglo African*, dated October 26, 1861, Mary Shadd Cary wrote, "Some of us who knew dear old John Brown . . . well enough to know his plans, and who were thought 'sound' enough to be entrusted with them by him . . . have the greatest opinion of fighting anti-slavery—give us 'plucky' abolitionism." Several years later as she was preparing a talk on Brown's relation to Chatham, her brief outline contains these phrases: "He taught something—He acted. Lessons of endurance, of Charity—of humanity, of zeal in a good cause. He wanted pure politics. Pure religion." Making evident the notion that even after emancipation this work was not finished, Mary Shadd Cary noted as a theme for her talk, "What we are doing in this and other places to carry out his plans." It was clear to her Brown aimed a blow not only against the strategic vulnerabilities of American slavery, but one that recoiled equally upon an entire nation, creating a breach that could become the basis for revolutionizing its politics. This is the explanation of how a man who excoriated other blacks for looking to whites for guidance, was the one man, who, if not quite going all the way with John Brown, nevertheless threw himself wholeheartedly into his Chatham meetings. One participant wrote that Martin Delany was "one of the prominent disputants or debaters."[26]

Some forty to sixty persons met with Brown in Canada, and many, particularly among the fugitive slaves received him eagerly. Above all Brown desired to attract other important black leaders; and Loguen, Douglass, and Charles Lenox Remond would receive invitations. While Loguen and Douglass had assisted in a crucial way in setting up the proceedings, both declined to attend; Douglass reasoning his presence would arouse suspicions, attracting unwanted attention, and however sympathetic, neither may have been eager to participate in meetings where Delany would have a prominent role. Remond, a Garrisonian abolitionist but of independent mind and outspoken, received an invitation signed by both Delany and Brown stipulating that his traveling expenses would be paid—but he declined to respond. Harriet Tubman also did not attend, but is said to have sent at least four fugitive slaves to Chatham for the meetings, and assuredly, whatever the reason for her absence, it was not for lack of interest as some may suppose. The individuals most instrumental in the proceedings after Delany were also emigrants from the United States; James Madison Bell from Cincinnati and Osborne Perry Anderson from West Chester, Pennsylvania. Bell, a twenty-eight-year-old poet,

was part of the circle around the *Provincial Freeman*, and was
employed as a plasterer. He first met Brown when the latter returned
from Iowa, when the elder man presented himself at his door in
Chatham with a letter of introduction from William Howard Day.
Brown stayed several days a guest in Bell's home, thereafter taking
lodging at the Villa Mansion. Bell would be at Brown's side throughout the meetings and would receive all of the "old man's" incoming
mail. Osborne Anderson, likewise, was at the *Provincial Freeman*,
where he was employed as a printer's devil and general assistant. He
immigrated to Canada with a branch of the Shadd family who purchased two farms near Chatham, which Anderson managed for a
time. Anderson later wrote that "John Brown . . . made a profound
impression upon those who saw or became acquainted with him.
Some supposed him to be a staid but modernized Quaker; others, a
solid businessman, from 'somewhere,' and without question a philanthropist. His long white beard, thoughtful and referent brow and
physiognomy, his sturdy measured tread, as he circulated about with
his hands, as portrayed in the best lithographs, under the pendant
coat-skirt of plain brown tweed, with other garments to match,
revived to those honored with his acquaintance and knowing his history, the memory of a Puritan of the most exalted type."[27]

Preparatory to the convention were a series of private talks where
many points arose and were settled. At these gatherings one supposes
Brown's so-called Kansas emigration scheme that Delany recounted as
being broached by Brown in their first meeting was discussed. Delany
remarked in the biography by Rollins that Brown had dubbed this
SPW for the Subterranean Pass Way, to distinguish it from the underground terminating in Canada. But these initials, many will know,
were embossed on a flag beside which Brown stood with his hand
upraised swearing an oath over a decade before. He had in fact entertained many ideas for testing the pertinent issue of black freedom on
United States soil, but the fugitives in question in Delany's observation
would have originated from Brown's Virginia campaign, the real issue
in view, about which Delany could not have been mistaken.

Many of Brown's notions, however, were not accepted out of
hand. One participant recounted in James C. Hamilton's *John Brown
in Canada*:

> During one of the sittings Mr. Jones had the floor, and discussed
> the chances of the success or failure of the slaves rising to support

the plan proposed. . . . Jones expressed fear that he would be disappointed because the slaves did not know enough to rally to his support. The American slaves . . . were different from those of the West India Island of San Domingo, whose successful uprising is a matter of history, as they had there imbibed some of the impetuous character of their French masters, and were not so overawed by white men. "Mr. Brown, no doubt thought," said Mr. Jones, "that I was making an impression on some of the members, if not on him, for he arose suddenly and remarked, 'friend Jones, you will please say no more on that side. There will be a plenty to defend that side of the question.' A general laugh took place."

At one setting Brown's constitution was examined, with Delany serving as chairman, and John Henry Kagi and Osborne Anderson as secretaries. The question of the organization as a political entity was considered, it being raised that since blacks had no rights, they "could have no right to petition and none to sovereignty." Therefore it would be "a mockery to set up a claim as a fundamental right." Delany remarked, "To obviate this, and avoid the charge against them as lawless and unorganized, existing without government, it was proposed that an independent community be established, without the state sovereignty of the compact, similar to the Cherokee Nation of Indians, or the Mormons. To these last named, references were made, as parallel cases, at the time." Also raised was the question as to an opportune time for making the attack, one speaker holding it would be folly to begin while the United States was at peace with other nations, and he advocated that they wait till the country was embroiled in a war with a "first class" foreign adversary. One participant wrote, "Mr. Brown listened to the argument for some time, then slowly arose to his full height, and said: 'I would be the last one to take the advantage of my country in the face of a foreign foe.' He seemed to regard it as a great insult." Some of those at the meetings, too, were not sure Brown's plan would succeed in effecting much at the South. Delany in particular seems to have strongly questioned some of Brown's tenets. According to Richard Realf, Delany, "having objected repeatedly to certain proposed measures, the old captain sprang suddenly to his feet, and exclaimed severely, 'Gentlemen, if Dr. Delany is afraid, don't let him make you all cowards!' Dr. Delany replied immediately to this, courteously, yet decidedly. Said he, 'Captain Brown does not know the man of whom he speaks. There exists no one in whose veins the blood of cowardice courses less freely, and it must not be said, even by John Brown of Osawatomie.' As he concluded, the old man bowed approvingly to him, then arose and made explanations."[28]

In Brown's company were twelve men beside himself, all white except Richardson, and they were joined by an equal number of Canadian refugees in these preparatory meetings, where Delany, Brown, Bell, Anderson, Kagi, and Realf imparted solid oratory. Osborne Anderson wrote:

> The "boys" of the party of "Surveyors," as they were called, were admired of those who knew them, and the subject of curious remark and inquiry by strangers. So many intellectual looking men are seldom in one party, and at the same time, such utter disregard of prevailing custom, or style, in dress and other little convention-alities. Hour after hour they would sit in council, thoughtful, ready; some of them eloquent, all fearless, patient of the fatigues of business; anon, here and there over the "track," and again in the assembly; when the time for relaxation came, sallying forth arm in arm, unshaven, unshorn, and altogether indifferent about it; or one, it may be impressed with the coming responsibility, saunter-ing alone, in earnest thought, apparently indifferent to all outward objects, but ready at a word or sign from the chief to undertake any task.[29]

On the 5th of May invitations went out to some thirty persons in the United States and Canada who had all been apprised of the proceedings, which simply read: "My Dear friend: I have called a quiet convention in this place of true friends of freedom. Your attendance is earnestly requested. Your Friend, John Brown."

Among those answering the summons were William Lambert of Detroit, head of the Vigilance Committee in that city; the Rev. William C. Munroe, who had presided over the "emigration conven-tion" in Cleveland in 1854; he was also from Detroit and would also serve as president of John Brown's Chatham convention. The Rev. Thomas Kinnard from Toronto came, and James H. Harris from Cleveland. One of the strongest individuals arriving was G. J. Reynolds, a coppersmith and Underground Railroad leader from Sandusky, Ohio. James Bell, I. D. Shadd, Thomas Stringer, Thomas Cary, Martin Delany, Alfred Whipple, J. M. Jones, and Isaac Holden were among those attending from Chatham. Many of those coming were fugitive slaves from St. Catharines, Ingersoll, Chatham, and Buxton. When all were assembled there were forty-four persons attending.

While John Brown's Chatham convention has received some comment and curiosity, there is no extended treatment that gives it the attention and emphasis it deserves, save that of DuBois. This is attributable to the fact that none of Brown's most sympathetic supporters or biographers was a participant, but also because it is often viewed as producing little or no appreciable result; and undoubtedly too because it was comprised of fugitive slaves and "emigrationists," a wholly black contingency, aside from Brown and his company. A glimpse into its scenes is afforded, however, by the recollections and testimony of Delany, Jones, Gill, Realf, and Anderson, and to the circumstance that the minutes of the convention were captured, along with John Brown in Virginia, to be scrutinized and published. This exposure led Osborne Anderson to publish them in his *A Voice from Harper's Ferry* in 1861. Beyond that Anderson would offer nothing further, writing only "to give facts in that connection, would only forestall future action, without really benefiting the slaves, or winning over to that sort of work the anti-slavery men who do not favor physical resistance to slavery."[30]

On May 8, a Saturday, the convention was called to order by John J. Jackson. Munroe was chosen president, and on motion of Brown, Kagi was elected secretary. With a growing crowd of the curious gathering outside, Munroe moved they relocate to another building. Giving the explanation that the proceedings were for the purpose of organizing a "Masonic lodge," the assembly walked two blocks to the No. 3 Engine House—a black fire company built by Isaac Holden and his friends—where they would be free from outside scrutiny. Newly situated, Delany called for John Brown.

Anderson wrote in his account:

When the Convention assembled . . . Mr. Brown unfolded his plans and purpose. He regarded slavery as a state of perpetual war against the slave, and was fully impressed with the idea that himself and his friends had the right to take liberty, and to use arms in defending the same. Being a devout Bible Christian, he sustained his views and shaped his plans in conformity to the Bible; and when setting them forth, he quoted freely from the Scripture to sustain his position. He realized and enforced the doctrine of destroying the tree that bringeth forth corrupt fruit. Slavery was to him the corrupt tree, and the duty of every Christian man was to strike down slavery, and to commit its fragments to the flames.

Brown went on to give a general explanation of his plan, and the execution of the project, in view of the convention. Delany and others rose to speak in favor of the plan and project, and both were agreed on unanimously.

Next Brown presented his draft for the organization's charter entitled "Provisional Constitution and Ordinances for the People of the United States," and moved for the reading of the document. Thomas Kinnard objected to its hearing until an oath of secrecy was taken by each member of the convention, whereupon Delany moved that the following oath be taken: "I solemnly affirm that I will not in any way divulge any of the secrets of this convention except to persons entitled to know the same, on pain of forfeiting the respect and protection of this organization." The motion carried and the president administered the oath of obligation.

Many observers mistakenly maintain that the document drawn up by Brown was intended for the overthrow of the United States government, with himself as commander-in-chief. Brown's design to the contrary should be seen as eminently adapted to the political ends he had in view, as he sought to ensure that the actions of this "provisional government" and its "provisional army" adhered to the republican and constitutional forms of the United States. While the wording of the document parallels the national framework in its executive, judicial, legislative, and military offices, it was clearly crafted with a view toward the people whom Brown intended to rally and was meant to subserve their need for direction as a fighting organization, and for civil government within the existing states of the South. The preamble of the document stated:

Whereas slavery, throughout its entire existence in the United States, is none other than a most barbarous, unprovoked, and unjustifiable war of one portion of its citizens upon another portion—the only conditions of which are perpetual imprisonment and hopeless servitude or absolute extermination—in utter disregard and violation of those eternal and self-evident truths set forth in our Declaration of Independence:

Therefore we, citizens of the United States, and the oppressed people who, by a recent decision of the Supreme Court, are declared to have no rights which the white man is bound to respect, together with all other people degraded by the laws thereof, do, for the time being, ordain and establish for ourselves the following Provisional Constitution and Ordinances, the better to protect our persons, property, lives, and liberties, and to govern our actions.

The document consisted of forty-eight articles. One to forty-five were read and adopted without significant discussion. On the reading of the forty-sixth article, Reynolds moved to have it stricken. The article reads: "The foregoing articles shall not be so as in any way to encourage the overthrow of any state government, or the general government of the United States, and looks to no dissolution of the Union, but simply to amendment and repeal, and our flag shall be the same that our fathers fought for under the Revolution."

Reynolds stated he could not fight under the emblem of the United States, that he and other members of the convention had no allegiance to a nation that had robbed and humiliated their race, and that they "already carried their emblem on their backs." Brown, Kagi, Delany, and others rose to advocate the article, and it passed. In his statement Brown said it was a flag that was good enough for the Revolutionary patriots in their war against the tyrants of the old world, and he intended to make it do duty now for the black people. He would not give up the "stars and stripes."

The remaining articles were read and adopted. There were provisions for governance and for removal from office, for the care and trial of prisoners, and punishment for breaches in the rules of war. All terms were to expire after three years. Delany motioned that all supporting the constitution as approved should stand to sign the same. As the paper was presented for signature Brown said, "Now, friend Jones, give us John Hancock bold and strong." There were forty-six signatories, thirteen from Brown's company, including himself, ten blacks from the northern states, with the balance being from the Canadian Dominion. After congratulatory remarks by Kinnard and Delany, the convention adjourned at a quarter to four in the afternoon till six, on the motion of Stevens.

When the members reconvened that evening they were called to order by Delany, on whose nomination Munroe was chosen president, and Kagi secretary. The constitution stipulated the president should summon another convention charged with the election of officers. A committee was then chosen to select candidates for the offices, consisting of Whipple, Bell, Cook, and Kagi. After sitting, the committee reported, and asking leave to sit again, was refused and discharged. On motion of Bell, the convention went to the election of officers. John Brown was nominated for commander-in-chief by Whipple, seconded by Delany, and elected by acclamation. On the nomination of Realf, Kagi was elected by acclamation as secretary of war. On the motion of Brown, the convention adjourned till nine o'clock Monday morning, May 10.

On Monday after the proceedings of Saturday were read and approved, Munroe announced the purpose of the meeting was the further election of officers. Alfred Whipple nominated Thomas Kinnard president. In a speech of length, Kinnard declined. J. W. Loguen was then nominated, but not being present—and it being announced he would not serve if elected—the nomination was withdrawn. Brown moved to postpone the election of president for the time being, which carried. The convention went on to elect A. M. Ellsworth and Osborne Anderson, both of Canada, members of congress, and Richard Realf, secretary of state. The convention adjourned and immediately reassembled for the balloting for treasurer and secretary of the treasury, electing Owen Brown and George Gill, respectively. A resolution was then introduced and carried appointing a committee delegated to fill the remaining offices named in the constitution; these were to include from five to ten members of congress, five supreme court members, a president and vice-president, with all positions unsalaried and for the duration of three years. The convention adjourned. After this last session another committee was organized having a relation to the provisional army in the field; Martin Delany became its president and I. D. Shadd its secretary.

John Brown was well satisfied and he wrote home: "Had a good abolition convention here, from different parts, on the 8th and 10th inst. Constitution slightly amended and adopted, and Society organized." Kagi journeyed to Hamilton with William Howard Day to begin printing up the documents. When the constitution was completed it was a fifteen-page handbound booklet. Among the documents of Brown's authorship also discussed at the Chatham meetings and printed, was *A Declaration of Liberty By the Representatives of the Slave Population of the United States of America*. That declaration, to be released for circulation on July 4, 1858, reads in part: "We therefore, the Representatives of the circumscribed citizens of the United States of America in General Congress assembled, appealing to the supreme Judge of the World, for the rectitude of our intentions, do in the name, & by the authority of the oppressed Citizens of the Slave States, Solemnly publish and Declare; that the Slaves are & of right ought to be as free & independent as the unchangeable Law of God, requires that All Men Shall be."

Martin R. Delany, circa 1865.

Its last sentence reads: "Nature is mourning for its murdered, and Afflicted Children. Hung be the Heavens in scarlet."

As the proceedings in Chatham concluded two issues arose; the first resulted in the delay of the plan, the second in its deferment altogether. By the close of the convention Brown had spent all the cash he had received from his supporters, mostly on travel and accommodations for himself and his men, and on similar expenses for some of the delegates attending from the United States. Bills totaling three hundred dollars were unpaid. But as Brown lingered in Canada awaiting money to cover his receipts, and much more troubling to him, new disclosures by Forbes were coming to his attention. There is no doubt that he was committed to seeing "the thing through," and he wrote Douglass he would soon need all the help he could get. However delay was to prove detrimental, as the possibility of having many of the Canadian refugees on hand would fall precipitously when the plan was not actualized when expected. It would cost Brown some of his own followers as well.

Just as the convention began, word reached him that some of his supporters in Boston were insisting on postponement. Douglass wrote in *Life and Times*: "I think I was the first to be informed of [Forbes'] tactics, and I promptly communicated them to Captain Brown. Through my friend Miss Assing, I found that Forbes had told of Brown's designs to Horace Greeley, and to the government officials at Washington, of which I informed Captain Brown." In a letter sent to Jason Brown intended for his father, Higginson urged the project go forward without delay. He wrote: "Sanborn writes an alarming letter of a certain H.F. who wishes to veto our veteran friend's project entirely. Who the man is I have no conception—but I utterly protest against any postponement. If the thing is postponed, it is postponed forever—for H.F. can do as much harm next year as this. His malice must be in some way put down or outwitted—& after the move is once begun, his plots will be of little importance. I believe that we have gone too far to go back without certain failure, & I believe our friend the veteran will think so too."[31]

Forbes, who had been in New York City since November 1857 with letters of introduction from Douglass, had circulated in abolitionist circles and among the staff at the *Tribune*. Complaining to anyone who would listen, he said he had been deceived by "a vicious man," that he'd given up work in New York to travel to Iowa to drill men, and that he had not been paid for this, and now he was destitute and his family was starving in Paris and he was unable to support them. Obtaining some sympathy and some money, he then decided to travel to Washington to solicit Republican leaders, where he began denouncing Brown's Boston supporters, of whom he'd evidently learned while in the confidence of Dr. James McCune Smith in New York. Approaching Senator Henry Wilson at his desk on the Senate floor, Forbes poured out his story making referral to some well-placed persons in Massachusetts. Wilson now wrote to Dr. Howe warning him confidentially that any arms given to Brown should be removed from his control, surmising only that he intended to use them for some action in Missouri. Wilson wrote: "If they should be used for other purposes, as rumor says they may be, it might be of disadvantage to the men who were induced to contribute to that very foolish movement." Howe quickly replied: "Prompt measures have been taken and will resolutely be followed up to prevent any such monstrous perversion of a trust." Then Forbes sought out Senator Seward in his home, who "found his story incoherent," and concluded he was a confused man soliciting charity.[32]

On May 14 Stearns wrote Brown at Chatham that as chairman of the Massachusetts Kansas Aid Committee he must warn him not to use the arms "for any other purpose and to hold them subject to my order as chairman of said committee." Brown replied: "None of our friends need have any fears in relation to rash steps being taken by us.

As Knowledge is said to be Power, we propose to become possessed of more knowledge. We have many reasons for begging our eastern friends to keep clear of F. personally unless he throws himself upon them. We have those who are thoroughly posted up to be put on his track and we humbly beg to be allowed to do so."[33]

Brown now sent Realf to New York City to get into his country-man's confidence, find out what he knew and secure any documents that may have fallen into his hands. Meanwhile, part of the company—including Stevens, Cook, and Owen Brown—had left Canada and gone to Cleveland, taking day jobs in the countryside. Owen received this from his father as he struggled to avoid defections:

> It seems that all but three have managed to stop their board bills, and I do hope the balance will follow the manlike and noble exam-ple of patience and perseverance set them by the others, instead of being either discouraged or out of humor. The weather is so wet here that no work can be obtained. I have only received $15 from the East, and such has been the effect of the course taken by F., on our Eastern friends, that I have some fears that we shall be com-pelled to delay further action for the present. They urge us to do so, promising us liberal assistance after a while. I am in hourly expectation of help sufficient to pay off our bills here, and to take us on to Cleveland, to see and advise with you, which we shall do at once when we shall get the means. Suppose we do have to defer our direct efforts; shall great and noble minds either indulge in use-less complaint, or fold their arms in discouragement, or sit in idle-ness, when we may at least avoid losing ground? It is in times of difficulty that men show what they are; it is in such times that men mark themselves. Are our difficulties such as to make us give up one of the noblest enterprises in which men ever were engaged?[34]

One of the party decided to use the time to take a little trip to see Reynolds in Sandusky. George Gill left an account of his visit, includ-ed in Hinton's *John Brown and His Men*, that relates much of inter-est. He wrote:

> My object in wishing to see Mr. Reynolds . . . was in regard to a military organization which, I had understood, was in existence among the colored people. He assured me that such was the fact, and that its ramifications extended through most, or nearly all, of

the slave states. He himself, I think, had been through many of the slave states visiting and organizing. He referred me to many references in the Southern papers, telling of this and that favorite slave being killed or found dead. These, he asserted must be taken care of, being the most dangerous element they had to contend with. He also asserted that they were only waiting for Brown, or someone else, to make a successful initiative move when their forces would be put in motion. None but colored persons could be admitted to membership, and, in part to corroborate his assertions, he took me to the room in which they held their meetings and used as their arsenal. He showed me a fine collection of arms. He gave me this under the pledge of secrecy which we gave to each other at the Chatham Convention.[35]

Gill's return to the others was facilitated by Reynolds, who passed him through the organization; first to J. J. Pierce in Milan, who paid his bill overnight at a hotel, and gave him money, then to E. Moore in Norwalk, who also paid for his lodging and purchased his rail fare through to Cleveland. When Gill reached Cleveland he found Stevens and Cook in a hotel; they had been joined, on his way east, by Realf. With Cook talking loosely and making rash avowals to strangers, Gill confided to his comrades the details of his visit to Sandusky. Unnerved by these indiscretions, Realf wrote to an uncle in England about his misgivings, stating one of their number had "disclosed objects to the members of a Secret Society calling itself the 'American Mysteries,' or some other confounded humbug. I suppose it is likely that these people are good men enough but to make a sort of wholesale divulgement of matters at hazard is too steep even for me, who are not by any means over-cautious."[36]

After two weeks' delay Brown received money to settle accounts in Canada, and immediately came on to Cleveland, leaving only Kagi behind in Hamilton to finish printing the documents. In Boston, Brown's supporters were bedeviled by letters from Forbes. Threatening to make their involvement public unless they removed Brown as commander of the projected expedition, to be replaced by himself or someone else, Sanborn, Smith, and Parker concluded the plan must go no further for the moment. Sanborn wrote: "It looks as if the project must, for the present, be deferred, for I find by reading Forbes' epistles to the doctor that he knows (what very few do) that the doctor, Mr. Stearns, and myself are informed of it. How he got this knowledge is a mystery." Only Higginson remained firm, writing to Parker: "I regard any postponement as simply abandoning the project; for if we give up now, at the command or threat of H.F., it will be the same next year. The only way is to circumvent the man somehow (if he cannot be restrained in his malice). When the thing is well started, who cares what he says?"[37]

Absent Higginson, five of the "committee" met in Boston's Revere House on May 24 to consider the situation. Decisions requiring Brown to put off the attack and place the arms under temporary interdict had already been made; the issues remaining were whether he should be required to go to Kansas, where there had been a new spasm of violence, and how much money should be raised for him in the future. Resolving unanimously that Brown should go to Kansas immediately both to reinforce the Free State cause and to blind Forbes, they suggested his Virginia campaign could safely be brought off in the next winter or spring, for which they severally pledged from two to three thousand dollars. They further resolved that henceforth they would not know nor inquire about Brown's plans.

Soon after the meeting at the Revere House, Brown arrived in Boston. Stating his objections to Higginson, he said delay was very discouraging to his men and to those in Canada. The others of the committee were not men of action; they had been intimidated by Wilson's letter and magnified the obstacles. The knowledge Forbes could give of his plan was injurious, for he wished his opponents to underrate him; still the increased terror could counterbalance this, and it would not make much difference. If he had the means he would not lose a day; it would cost him no more than twenty-five dollars apiece to get his men from Ohio. Still, it was essential his backers not think him reckless, and they held the purse. Faced with being cut off from his financial resources, Brown finally acknowledged that he had little choice but to acquiesce.

With all in agreement, Stearns foreclosed on his title to the Sharps rifles and gave them to Brown as a gift with no conditions attached. Brown left Boston on June 3, and after a brief visit to North Elba, returned to Cleveland. The arms were hidden inside a haystack near Ashtabula; then Brown, with his son Owen, Kagi, and Stevens, departed for Kansas. Others in the company would remain in Ohio for a time, some eventually coming out to Kansas. John Cook volunteered or was delegated to take up residence in Harper's Ferry, Virginia. When Realf reached New York City he found Forbes was in

Washington, and the two never met. Realf would travel to England, deserting his leader and his comrades without a word. After lecturing on antislavery and Kansas affairs, he returned to the States in 1860; summoned from Texas to appear before the Select Committee of the United States Senate investigating "the Harper's Ferry Affair," he would denounce Brown for having committed "an intellectual error which had precipitated him upon a cruel and wicked deed."[38]

Realf perhaps never lived down his association with John Brown and his allies, ending his life in 1878 at age forty-four in a hotel room in Oakland, California. Brown often expressed high regard and affection for the young bard who had written,

> Born unto singing
> And a burthen lay mightily on him.

Five

An American Spartacus

I heard him tell one evening . . . the obscure and forgotten stories of Isaac, Denmark Vesey, Nat Turner, and the Cumberland region insurrectionary affairs in South Carolina, Virginia, and Tennessee . . . and he knew the story of Haiti and Jamaica, too, by heart.
—Richard Hinton, *John Brown and His Men*

S INCE THE STRIFE-FILLED DAYS WHEN THE RIGHT OF FREE-STATE settlers to exist had been in doubt in Kansas, the situation had changed dramatically. After winning the majority of seats in territorial election in October, free-state electors came to power in December 1857, when the acting governor called a special session of the Lecompton legislature. George Deitzler, "procurer of Sharp's rifles,"[1] was speedily elected speaker, as Jim Lane, after marching into the proslavery capital at the head of the Lawrence Cornet Band, was named major-general of the territorial militia. But in Washington the president and his cabinet were determined Kansas would only be admitted into the Union under the Lecompton Constitution. An exasperated James Buchanan now called upon his Indian commissioner, who was in the territory and a friend of David Atchison, to become governor, and James Denver was sworn in on December 21, 1857.

While northeastern Kansas was solidly under the banner of "Free-Soil," to the southeast in Bourbon and Linn counties, with its center of power in Fort Scott, the proslavery party held sway. As proslavery Democrats awaited the action of the Congress, ex-Indian agent G. W. Clarke and ex-chief justice of the Iowa Supreme Court "Fiddling" Williams administered justice from the abandoned army outpost. Most of the settlers in the southeast were from Missouri, having resettled there after being driven off claims in the northern counties. Living among them was James Montgomery, who, along with his followers, had decided to ignore Judge Williams's decisions in land disputes in favor of proslavery claimants and had organized for their self-protection. When the United States marshal was ordered to break up Montgomery's "Self-Protection Company," men from Lawrence led by Jim Lane, William A. Phillips, and Preston Plumb hastened to assist them, and the new governor ordered United States dragoons to disperse Lane's militia as well as Clarke's regulators.

During this time a new figure appeared among the proslavery settlers with an abiding hatred for abolitionists: the handsome son of a wealthy Georgia planter, Charles A. Hambleton. As a provocation Hambleton built a substantial log house complete with slave quarters on a property near the settlement called Trading Post just across the Missouri border inside Kansas, and from there he commanded a company of proslavery partisans. Montgomery and his company rode into this stronghold, and smashing barrels of corn whiskey, ordered Hambleton and his men out of Kansas. On May 19, rallying men in Missouri, Hambleton rode back into Kansas seeking vengeance. Seizing a store clerk in Trading Post, and a Baptist missionary and two men conversing with him a half mile off, Hambleton and his men then drove four men from their cabins and seized two more working in their cornfields, and then another on the road. These eleven men were marched to a ravine and stood in a line, while Hambleton's mounted men formed to the side.

One of the prisoners, William Hairgrove, said: "Gentlemen, if you are going to shoot, take good aim." Hambleton ordered: "Make ready, take aim . . . " At this Pate's lieutenant at Black Jack, W. B. Brochett, turned away, saying, "I'll be damned if I'll have anything to do with a goddam piece of business like this. If it was in a fight, I would fire."

Hambleton drew his revolver and fired, as did others. Five men fell dead, five were severely wounded and thought to be dead, and one man lay unhurt feigning death. The murderers rifled their victims' pockets, and separating as they rode so as to leave no tracks, hastened back across the border. This was the so-called Marais des Cygnes Massacre.[2]

On July 1 Brown visited the scene and, broadcasting his defiance, soon arranged to buy land nearby and build a fortification, affording a commanding view across the border into Missouri and of the surrounding country. Eli Snyder, a blacksmith, owned the parcel he selected. Brown sent the following off to Boston:

> I am here with about ten of my men, located on the same quarter section where the terrible murders of the 19th of May were committed. . . . Deserted farms and dwellings lie in all direction for some miles along the line, and the remaining inhabitants watch every appearance of persons moving about, with anxious jealousy and vigilance. Four of the persons wounded or attacked on that occasion are staying with me. The blacksmith Snyder, who fought the murderers, with his brother and sons, are of the number. Old Mr. Hairgrove, who was terribly wounded at the same time, is another. The blacksmith returned here with me and intends to bring his family on to his claim within two or three days. A constant fear of new trouble seems to prevail on both sides of the line, and on both sides are companies of armed men. Any little affair may open the quarrel afresh. Two murders and cases of robbery are reported of late. I have also a man with me who fled from his family and farm in Missouri but a day or two since, his life being threatened on account of being accused of informing Kansas men of the whereabouts of one of the murderers, who was lately taken and brought to this side. I have concealed the fact of my presence pretty much; but it is getting leaked out, and will soon be known to all. As I am not here to seek or secure revenge, I do not mean to be the first to reopen the quarrel. How soon it may be raised against me, I cannot say; nor am I over anxious.[3]

Brown and his men, together with some of his old Kansas company, now reformed as Shubel Morgan's company, his new nom de guerre. In erecting the fortification he undoubtedly thought it in his interest to occupy a strong position, but it is also likely he had in mind giving some practical instruction to his men in building a type of defensive work which, he had, before this, largely theorized. A stream ran onto the property, and the point where it sank back underground was the site of John Brown's fort. The structure was two sto-

ries high, of hewn logs twelve by fourteen feet with portholes inter-
spaced on each floor for firing positions. Earth and stone were
mounded around the perimeter to a height of four feet, making it
both a thorny problem even to assault and impervious to artillery.
After its completion Brown wrote to John Jr., "In Missouri . . . the
idea of having such a neighbour improving a Claim (as was the case)
right on a conspicuous space and in full view for miles around in
Missouri produced a ferment there you can better imagine than I can
describe. Which of the passions most predominated, fear or rage, I do
not pretend to say."4 It was undoubtedly an effective deterrent, as no
marauding proslavery gang dared come near the fortification, particularly as it was now generally known it was the lair of John Brown.

Brown also had gained a valuable ally, often holding counsel with
the Campbellite preacher and partisan fighter James Montgomery.
The two leaders actively sought each other's cooperation and often
appeared in each other's camps. Richard Hinton recorded the elder
man's favorable remarks in regard to Montgomery's activities:
"Captain Montgomery is the only soldier I have met among the
prominent Kansas men. He understands my system of warfare exactly. He is a natural Chieftain, and knows how to lead." Pursued in
April by United States dragoons ordered out against him by Governor
Denver, Montgomery posted his men in good position, killing one soldier and wounding half a dozen, leaving a number of dead horses on
the field. Montgomery's biographer wrote, "When he learned the particulars of this engagement, [John Brown] said the like had not happened before in the Territory, and that the skill with which he conducted the engagement, stamped him as one of the first commanders
of the age."5

Montgomery had been born in Ashtabula County in Ohio's
Western Reserve in 1814. Migrating first to Kentucky, where he mar-
ried, in 1852 he moved with his family to Missouri, settling finally in
Kansas in 1856 when Montgomery took up the free-state cause. Like
John Brown, he was of a religious nature, and would famously command a black regiment in South Carolina during the Civil War.
Franklin Sanborn, who met him in 1860, was surprised to find that
"he had an air of elegance and distinction which I hardly expected.
He was a slender, courteous, person with a gentle cultivated voice and
the manner of a French chevalier." Brown described Montgomery as
a "brave and talented officer," who was both "kind and gentlemanly
. . . and what was infinitely more a lover of Freedom."6

Touring the southeastern counties of the territory, Governor Denver reported to Buchanan's secretary of state, Lewis Cass: "We passed through a country almost depopulated by the depredations of the predatory bands under Montgomery, presenting a scene of desolation such as I never expected to have witnessed in any country inhabited by American citizens." In Washington, with Stephen Douglas fulminating that it was a "trick, a fraud upon the rights of the people," the Senate approved the Lecompton Constitution despite the manifest will of the majority in Kansas. In the House, opposition to its adoption was so intense that in an all-night session members engaged one another in a brawl along sectional lines. Finally Buchanan put out a backhanded inducement, offering a block of public land for new settlement in Kansas. But on August 2 this was turned down when all but two thousand of thirteen thousand votes were cast against it. At the end of the month Brown remarked in a letter to his son, "The Election of the 2nd Inst. passed off quietly on this part of the Line."[7]

As things remained quiet, Brown, John Henry Kagi, Charles Plummer Tidd, and the blacksmith Eli Snyder rode into Missouri on reconnaissance. Running a surveyor's line, Brown and Snyder headed in one direction, while the younger men, pretending to be teachers looking for jobs, went in another. Meeting later on a hill under a clump of trees and looking down to the east through a lens, they spied the Rev. Martin White, the murderer of Frederick Brown, reclining half a mile off while reading a book under a tree. Snyder proposed that he and Brown pay White a visit. "I won't do that," Brown replied. Then Kagi said he and Snyder might go. "Go if you wish but don't hurt a hair of his head," Brown cautioned. As Kagi replied he would only go if he had no instructions, the matter ended. Brown later explained he "would not go an inch to take his life. I do not harbor the feelings of revenge. I act only from a principle, to restore human rights."[8]

By the end of August Brown was ill again, as he had been early in July when he convalesced at the farm of Augustus Wattles. With his renewed illness he completely abandoned his fort, recuperating at the home of his cousin, the Reverend Adair, near the town of Osawatomie. It was during this time at the end of summer that Brown and Kagi met with Richard Hinton for an extended interview. Hinton had figured in Kansas affairs both as a journalist and as a partisan fighter, and since the fall of 1857, on his own testimony, had main-

tained "an important, if irregular correspondence" with Kagi. It was
at Kagi's behest that Hinton began in the previous year "a systematic
investigation of conditions, roads, and topography of the Southeast
. . . under guise of examining railroad routes."[9] His survey was with
reference to Brown's projected attack in Louisiana as a way of relieving pressure on Kansas, a scheme Brown pushed in his agitation upon
his thesis of "carrying the war into Africa," as he phrased it. His
daughter Anne, in Hinton's *John Brown and His Men*, would later disclose the reason this plan had been abandoned in favor of the attack
in Virginia. The slave system there, Brown deemed, had been drawn to
such an extreme pitch of exploitation that any countervailing tendencies to its inhumanity had been repressed. An uprising of Louisiana
slaves, he believed, would erupt in such ferocity, leaping beyond all
possibility of control, as to result in indiscriminate bloodshed.

It was not until the end of August 1858, when Hinton received an
invitation to a private interview in Lawrence, that the Scotsman
learned the true orientation of Brown's plan. After dinner Brown and
his lieutenant had some conversation apart; then saying he wanted to
do some fishing, Kagi asked Hinton to accompany him to the river.
Stopping half way, the two sat on a fence rail as the conversation
turned to Brown's nearer object. In his book, Hinton states Kagi began
by asking him what he supposed to be the plan of John Brown. Hinton
replied that he presumed that to be reference to an attack in Louisiana
and retreat into the southeast states and the Indian Territory of
Oklahoma. He was then given a full account of the meetings in
Chatham, as well as of the organization effected there. The true location of their operations was to be in the mountains of Virginia, said
Kagi; and he went on to describe their mode of operation, sketching a
theory of how the campaign would develop and be received in the
North and in the South. One of the reasons that had induced Kagi,
Hinton wrote, "to go into the enterprise was a full conviction that at
no very distant day forcible efforts for freedom would break out
among the slaves, and that slavery might be more speedily abolished
by such efforts than by any other means. . . . Believing that such a blow
would soon be struck, he wanted to organize it so as to make it more
effectual, and also, by directing and controlling the Negroes, to prevent some of the atrocities that would necessarily arise from the sudden upheaving of such a mass as the Southern slaves."

Given a country admirably adaptable to guerrilla warfare, the
freed slaves were to be armed and organized into companies, headed

by the men Brown had selected and the Canadian recruits that would be sent down. The southern oligarchy, Kagi continued, would become alarmed by the discipline maintained and by the show of organization. At no point was the South more vulnerable, he emphasized, than in its fear of servile insurrection; they would imagine that the whole North was down upon them, as well as their slaves.

After a long exchange on the topic and its ramifications the two young men returned to the house where the discussion continued with Brown, Hinton wrote, "mostly upon his movements, and the use of arms." Hinton recorded these concise utterances made by Brown: "Any resistance, however bloody, is better than the system which makes every seventh woman a concubine"; "A few men in the right and knowing they are right, can overturn a king"; "A ravine is better than a plain. Woods and mountain sides can be held by resolute men against ten times their number."

Brown was also critical of the Kansas free-state leaders. They "acted up to their instincts as politicians," he said. "They thought every man wanted to lead, and therefore supposed I might be in the way of their schemes." Hinton remarked that from an antislavery point of view their policy had been an "abortion." Struck by the word, Brown gazed into the younger man's face "with a peculiar expression in his eyes." "Abortion?" he mused. "Yes, that's the word."

Hinton ends his summary with Brown's remarkable evocation of the storied leader of the Southampton slave revolt: "Nat Turner with fifty men held a portion of Virginia for five weeks. The same number, well organized and armed, can shake the system out of the state." Brown, of course, was referring to the endemic fear in the South of "servile insurrection," a pathological predicament arising out of the relation whereby one group by their color had supremacy in all matters over another. Yet even Herbert Aptheker, a scholar without parallel in his research in African American history, misapprehends Brown's emphatic remark. In his *Nat Turner's Slave Rebellion,* he wrote, "This would indicate that Brown was misinformed about the Turner episode."

It has been maintained that "insurrectionary movements" were of little significance in the advance of the antebellum South; that they neither had an impact on slavery's development as a socioeconomic system nor brought about its formal internment in the course of the war of 1861–65. They were local outbreaks that were quickly suppressed. Indeed, several factors were seen to dampen the tendency toward rebellion among the American slaves: There was a disproportionate use of force for punishment and maintaining order, for one; for another, all means of enlightenment were withheld from them, and they were closely watched and were isolated as much as possible, most slaves living and dying on the plantations on which they were born. With few traveling more than several miles in a lifetime, their view of the outside world was overly constricted and provincial. Moreover, it was difficult, if not impossible, for slaves to build a concerted leadership. There was too ever the probability of betrayal from one or another of their number; nor could one expect an uprising to attract sufficient adherents to become self-sustaining.

John Brown knew all these caveats, yet he also knew what few have scarcely dared to contemplate. In his testimony before the congressional committee investigating Harper's Ferry, Richard Realf related some of the historical studies Brown undertook in contemplating his object:

> For twenty or thirty years, the idea had possessed him like a passion of giving liberty to the slaves; that he made a journey to England, during which he made a tour upon the European continent, inspecting all fortifications, and especially all earthwork forts which he could find, with a view of applying the knowledge thus gained, with modifications and inventions of his own, to a mountain warfare in the United States. He stated that he had read all the books upon insurrectionary warfare, that he could lay his hands on: the Roman warfare, the successful opposition of the Spanish chieftains during the period when Spain was a Roman province— how, with ten thousand men, divided and subdivided into small companies, acting simultaneously, yet separately, they withstood the whole consolidated power of the Roman Empire through a number of years. In addition to this he had become very familiar with the successful warfare waged by Schamyl, the Circassian chief, against the Russians; he had posted himself in relation to the wars of Toussaint L'Ouverture; he had become thoroughly acquainted with the wars in Hayti and the islands round about.[10]

In the inaugural issue of the *Anglo-African Magazine*, December 1859, its editor, Thomas Hamilton, noted to his readers, in introducing "The Confessions of Nat Turner" which he was publishing, that the slave leader and John Brown bore comparison. To Hamilton these comparisons were personal: Both men were idealists who harbored their purposes for years, and followed spiritual impulses; one seeking signs in the air, earth, and heavens, and the other believing he was foreordained in his actions from eternity. In methodology, however, these parallels ceased in Hamilton's opinion. Although both movements were predicated on the issue of a first successful stroke, and both sought to beat up the slave-quarters and arm slaves by seizing an arsenal, Turner saw only "the enfranchisement of one race," while Brown believed "that freedom of the enthralled could only be effected by placing them on an equality with their enslavers."[11]

But don't the comparisons go beyond this? The Turner experience was in fact a critical reference for Brown, so much so that the leader, he held, had earned a place in the American patriotic pantheon on a pedestal next to Washington. Both were individuals whose lives have subsequently come to be viewed through the distorting lenses of apocalyptic events they advanced. The fulcrum of fact however turns here, as elsewhere, on concrete historical conditions and on tendencies traversing narratives that only come to fulfillment in an individual mind after long years of struggle. When Nat Turner, therefore, was interrogated about a contemporaneous revolt in North Carolina, he disavowed knowledge of it. But seeing his interrogator probing his face, hoping to penetrate his thoughts, he replied, "I see sir, you doubt my word; but can you not think the same ideas . . . might prompt others, as well as my self." Far from demonstrating the futility of a slave revolt, Brown held that Nat Turner's Rebellion had demonstrated the possibility of its success; the former coming at the beginning of the South's greatest antebellum development, while the latter undertook his with great calculation near its end.

Midway through the 1820s in the southeastern corner of Virginia a slave had a vision that confirmed him in his deeply held belief he had an important mission to accomplish for *his* people. A few years later in the so-called "Confessions of Nat Turner," that slave, the property of Benjamin Turner, told how he'd seen "white spirits and black spirits engaged in battle"; the sun darkened and thunder rolled through the Heavens, then a voice spoke to him: "Such is your luck, such you are called to see, and let it come rough or smooth, you must surely bare it." Toward the end of August 1831 this vision came to fruition in Southampton County, Virginia.

Seventy years later, remarking on its importance, William S. Drewry, a man holding the pronounced racialist views of his day, nevertheless wrote in his *The Southampton Insurrection*: "It was the forerunner of the great slavery debates which resulted in the abolition of slavery in the United States, and was, indirectly, most instrumental in bringing about this result. Its importance is truly conceived by the old negroes of Southampton and vicinity, who reckon all time from 'Nat's Fray,' or 'Old Nat's War.'"

On August 21, a Sunday, by prearrangement half a dozen slaves gathered in a secluded wood near a farm in Southampton County. They had come together to prepare a dinner and resolve on a plan of revolt. At the appointed hour the leader was the last to arrive. There was a pig roasting and a keg of brandy; Nat Turner could see his four trusted confederates had brought two more recruits, Will and Jack. Jack was "only a tool in the hands of Hark," so Turner did not question him, but he asked Will, "Why are you here?" Will answered, "His life was worth no more than others, and his liberty as dear to him." "Do you mean to obtain it?" asked Turner. He replied he did, or lose his life. This was enough to put him in full confidence.

The slave cabal shared the meat and drink through the day as singly Turner took each man aside to talk. After assuring himself they were committed to the cause, through the evening and into the night they determined on a course of action. Turner told the men they were to begin a fight for liberty as the whites had done years before when they drove the British out. To command them he would assume the nom de guerre General Cargill, and they would fight under a banner with a red cross on a white field, distinguishing themselves by red caps or hats ornamented with red ribbons. Arms, provisions, horses, equipment, money, and other things they would need all would come from their own masters. They would begin that night at the home of Nat's master, and each was sworn not to spare a single white skin, regardless of age or sex, until they had gathered sufficient force and armed and equipped themselves. By beginning with a swift, indiscriminate massacre, Turner knew, they would "create more terror than many battles," and once they had begun in blood there would be no turning back; it would harden the men for the terrible fight ahead. After gathering sufficient force, they would come down on the county seat—then called Jerusalem—to seize the arsenal there. Properly armed, they could begin a policy of sparing women and children, and men who ceased to resist. The war could now begin in earnest as their

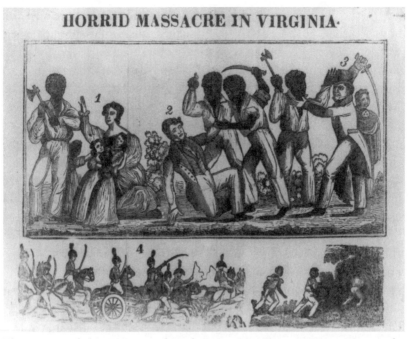

HORRID MASSACRE IN VIRGINIA.

The caption of this 1831 woodcut depicting Nat Turner's insurrection reads: "The Scenes which the above Plate is designed to represent, are—Fig 1, a Mother intreating for the lives of her children.—2. Mr Travis, cruelly murdered by his own Slaves.—3. Mr Barrow, who bravely defended himself until his wife escaped.—4. A comp. of mounted Dragoons in pursuit of the Blacks."

army marched through the country under its banner of liberty, freeing slaves and driving the whites before them. Fully confident they would be able to prevail, if it became necessary, Turner said they would retreat to the Dismal Swamp which could be reached in a day's march, where they could establish themselves, as did runaways, and defy their pursuers.

The plan proceeded unchecked for almost two days; from the initial six recruits, the number became fifteen, then forty. By nine o'clock on the first morning Turner had sixty men variously armed; some carrying axes and scythes, some muskets and other firearms, with a score mounted on horses. As the number increased they were divided, with those mounted advancing as fast the beasts would carry them, while those on foot followed them up.

While leaving a ghastly trail of murder and mutilated corpses, before they could leap beyond the prescribed terrain of depredation, the limitations inherent in the rebellion proved fatal. Meeting a hastily thrown together troop of no more than twenty white men and boys, Turner advanced his men against them. Discharging their weapons the whites began retreating as the blacks pressed forward. Coming over the rise of a hill Turner could see the whites reloading. Discharging their own few arms, the blacks charged against the whites who fired into them, wounding several of Turner's best men. As his men began dispersing in confusion and fear, Turner resolved to gather them together again, and come by another way down on the county seat to the arsenal. But this was to be the end of the rebellion.

Two days after "Nat's war" commenced Virginia's governor John Floyd dispatched a company of cavalry from Richmond with light artillery, having in its care one thousand stand of arms. In addition the federal garrison at Fortress Monroe sent a detachment, and two companies of infantry were dispatched from Norfolk and Portsmouth; while from the Navy Yard in Norfolk, a company of sailors hastened under arms. Three thousand troops were on their way to Southampton, and more were preparing to set out. The Richmond *Constitutional Whig*'s reporter sent this dispatch to his paper: "The Richmond Troop arrived here this morning . . . after a rapid, hot and most fatiguing march. . . . On the road, we met a thousand different reports, no two agreeing. . . . On the route from Petersburg, we found the whole country thoroughly alarmed; every man armed, the dwellings all deserted by the white inhabitants, and the farms most generally left in the possession of the blacks."

Now commenced a massacre that would outdo the slaughter of the previous days as black men were shot down, whether implicated in the rebellion or not, as they appeared. The heads of some of those implicated were cut off and placed on stakes as a warning against further resistance. Armed with only a sword, Nat eluded capture, as the rest of his men were seized or slaughtered. First he lay under a haystack; then he crawled into the hollow of a fallen tree, concealed by brush, never more than half a mile from where he had begun his war for black liberty. In the days and weeks that followed until his capture, there were many sightings of this man of slight stature with thinning hair, large eyes, broad flat nose, and deeply furrowed face. First he was seen in several neighboring Virginia counties; then he was sighted over the border in North Carolina, then in faraway places throughout the South, and finally as far north as Baltimore.

The *Constitutional Whig*'s account continues: "On our arrival at this village, we found Com. Elliot and Col. Worth with 250 U. States troops, from the neighborhood of Old Point, and a considerable militia force. A troop of Horse from Norfolk and one from Prince George, have since arrived. Jerusalem was never so crowded from its foundations; for besides the considerable military force assembled here, the ladies from the adjacent county, to the number of 3 to 400, have sought refuge from the appalling dangers by which they were surrounded."

Nansemond, Isle of Wight, Prince George, and Surry counties were in expectation of "insurrection" on August 26. Arms were appealed for and immediately sent. On the same day in Fredericksburg a "conspiracy" was discovered—appeals were made for arms and granted. The *Richmond Enquirer* reported: "Gen. Eppes writes to the Governor on the 30th that everything was quiet in Southampton, and he thought was likely to continue so—but in consequence of the dispatches received from Nansemond and Isle of Wight, a Council of War has been held, at which it was determined that as a precautionary measure, a strong patrol in these counties ought to be kept up."

On August 31 forty blacks were seized in Nanesmond County; and as August passed into September slaves were reported in a state of insubordination in Northampton County. There were more appeals for arms. On September 16 nine slaves were condemned to hang in Sussex County. On September 20 the governor recorded: "The alarm of the country is great in the counties between this and the Blue Ridge Mountains. I am daily sending them a portion of arms though I know there is no danger as the slaves were never more humble and subdued." On the twenty-second he noted: "This day was spent in giving orders for arms to be distributed to various counties and regiments." Then an intended insurrection was exposed in North Carolina's Duplin, Bladen, and Sampson counties. Rumors swept the country that Wilmington had been burned and half its white inhabitants massacred. Again the entire lower portion of Virginia below the James River was convulsed, spreading southward into North Carolina and north to Maryland's Eastern Shore and into Delaware. Fully five weeks had now passed since Nat Turner set in motion the reign of trepidation.

Henry "Box" Brown, who would enter history later by "mailing" himself inside a crate care of William Still in Philadelphia, lived in Richmond during this period, and wrote in his *Narrative of Henry Box Brown*: "A dark cloud of terrific blackness, seemed to hang over the heads of the whites. So true is it, that 'the wicked flee when no man pursueth.' Great numbers of slaves were locked in the prison, and many were 'half hung,' as it was termed; that is, they were suspended to some limb of a tree, with a rope about their necks so adjusted as not to quite strangle them, and were pelted by men and boys with rotten eggs."

The best minds in Virginia now mobilized around events emanating from Southampton, and they devoted considerable energy to ferreting them out. It seemed to the governor several phases of movement culminated in the recent contagion. First, as he detailed in a letter to South Carolina's governor, James Hamilton Jr., dated November 19, came the influence of the "Yankee population" on the slaves. Though the course was not a direct one, this consisted in making the slaves religious; they learned that "God was no respector of persons . . . that they cannot serve two masters, that the white people rebelled against England to obtain freedom; so have the blacks a right to do." Large assemblages of slaves had been allowed to meet for religious purposes, headed by black preachers, who propagated these ideas. Then, in the governor's view, a more active phase commenced, with the circulation of abolitionist publications, two of which impressed him most forcefully—David Walker's *Appeal* and the *Liberator*, both originating from Boston. These were now placed under ban, and avenues explored to see if their authors could be held accountable and prosecuted. Now the governor wrote: "From all that has come to my knowledge during and since this affair; I am fully convinced that every black preacher, in the whole country east of the Blue Ridge, was in the secret, that the plans as published by these Northern prints were adopted and acted upon by them, that their congregations, as they were called knew nothing of this intended rebellion, except a few leading, and intelligent men, who may have been head men in the church. The mass were prepared by making them aspire to an equal station by such conversation as I related in the first step."[12]

In an ensuing debate before the legislature in the winter of 1831–32 some historians have maintained that Virginia came close to actually abolishing slavery. The truth may be that these debates, occurring in conjunction with the famous tariff controversy, strengthened the hand of slavery in the state, while prefiguring the divisions that came to the fore in Virginia, and in other border states, with the slaveholder-engineered secession in 1860–61. Finally, as the debate began to take on an angry tone, voices were heard saying that it was dangerous; that seeing the whites thus divided would only further agitate the slaves. In the midst of this, the mitigating voice of Thomas Jefferson Randolph, the grandson of his namesake, was raised. In an article for the *Journal of Negro History* in 1920, titled "The Aftermath of Nat Turner's Insurrection," John A. Cromwell summarized his words: "It was the dark, the appalling, the despairing future that had awakened the public mind rather than the Southampton insurrection. He asked whether silence would restore the death like apathy of the negro's mind? It might be wise to let it sleep in its torpor; 'but has not,' he asked, 'its dark chaos been illumined? Does it not move, and feel and think? The hour of the eradication of the evil is advancing, it must come. Whether it is affected by the energy of our minds or by the bloody scenes of Southampton and San Domingo is a tale for future history.'"

All manner of business and industry had come with the continuing arrival of settlers in Kansas—contractors, builders, transporters, outfitters, lumber merchants, and merchants of every kind—making it by 1858 a territory that was, but for its constitution, free of slavery. Horace Greeley, who would travel through on his way to Pike's Peak, Salt Lake City, and the Pacific Coast, would stop off in Osawatomie in the spring of 1859 and be there for the inaugural meeting of the Republican Party. In a dispatch to his paper he wrote of the magnitude of the change, describing the transportation enterprise of Russell, Majors, and Waddell in nearly Whitmanesque terms: "Such acres of wagons! Such pyramids of extra axletrees! Such herds of oxen! Such regiments of drivers and other employees! No one who does not see can realize how vast a business this is, nor how immense its outlay as well as income. I presume the great firm has at this hour two millions of dollars invested in stock, mainly oxen, mules, and wagons. (Last year they employed six thousand teamsters and worked 45,000 oxen.)"[13]

Meanwhile in Kansas, Linn and Bourbon counties continued unaffected by the prosperity, where, under the aegis of the government, Clarke and Williams (the former Indian agent and state supreme court justice) unrelentingly harried free-state settlers. The only hope that the "yoke could be lifted" had come from the resistance of James Montgomery and his followers, and now from John Brown. The months June through December 1859 are often regarded by his biographers as idle months for Brown; indeed, he had been ill part of July and August, and it was his policy, as he said, "to do nothing, when he did not known what to do." But in building a strong fortification near the very ground of the May massacre he had barred further invasions from Missouri, giving free-state settlers in its vicinity greater security. In Montgomery and his partisans he had found allies, with cooperation and coordination between their camps following. When not himself present and active in operations, Brown's two chief lieutenants, Kagi and Stevens, were; Brown, too, was to take two of Montgomery's fighters with him to Harper's Ferry, Jeremiah Anderson and Albert Hazlett.

In October Kagi notified *Tribune* readers: "Captain Brown will now, if necessary, take the field in aid of Captain Montgomery." The signal an offensive had begun came in mid-October when Montgomery led a force to Fort Scott, seizing and ransacking "Fiddling" Williams's courthouse. As payback, a volley was fired into Montgomery's cabin, but he, his wife, and his children all escaped unharmed. At this, Brown proposed that the cabin of Montgomery's mother-in-law be fortified, and he, Stevens, Kagi, and Tidd occupied themselves with this for much of November. On November 13, after Montgomery was indicted for destroying ballot boxes stuffed with "bogus ballots" in the previous January's election, both Brown and Montgomery led their companies to the town of Paris, overturning it looking for indictments and arrest warrants. Now warrants were issued for both leaders, with rewards offered for securing their persons at Fort Scott. Finally to "break up their organization and drive them from the Territory," the sheriff of Linn County and his posse visited "Montgomery's Fort" on November 30, hoping to apprehend Brown. As he was away in Osawatomie, and they did not wish to quarrel with Stevens brandishing a shotgun at the door, they left. The next day the sheriff himself was cornered alone, relieved of his arms, and then sent home. After this Brown worked out a "peace agreement" between Montgomery and his opponents, as the warring factions met at Sugar Mound. In the accord all prisoners held by either side were to be released and all criminal prosecutions against free-state men were to cease; each party was to discontinue acts of "theft or violence against the other on account of their political differences," and all proslavery men forcibly expelled from Kansas were to stay out.

No sooner had the agreement been completed than Williams ordered the jailing of two free-state men at Fort Scott on trumped-up charges. A force larger than any yet seen, seventy-five men, was raised to set them free. That night in council Brown asked for undivided command, promising the assembly he would lay Fort Scott in ashes. When Montgomery rose to oppose this, saying that the only purpose should be to free the prisoners, Brown stalked out, leaving Kagi and Stevens to participate. In a skirmish the next day Fort Scott was captured, and the prisoners being held on the third floor of the Fort Scott Hotel were freed. A storeowner, who was also a deputy marshal, discharged a shotgun into Kagi's breast, but his heavy outer clothing saved him from serious injury. The man was shot and killed, and his store was emptied of merchandise valued at $7,000.

During the fall Brown's nearly constant companion was George Gill. More than thirty years after the death of his chief, in a letter dated July 7, 1893, Gill confided his private thoughts to Richard Hinton regarding the elder man's intolerance of the wishes of others and of his "individuality." He wrote: "All great men have their foibles or what we in our difference from them call their weakness." Intimate acquaintance had demonstrated to Gill that John Brown was very human, with a love of command and adventure, together with firmness and combativeness, which sometimes was given vent in vindictiveness. Brown's "immense egotism coupled with love of approbation and his god idea begot in him a feeling that he was the Moses that was to lead the Exodus of the colored people from their southern taskmasters. Brooding on this, in time he believed that he was God's chosen instrument, and the only one, and that whatever methods he used, God would be his guard and shield." Believing this, Gill wrote, he could brook no rival, and while at first he was very fond of Montgomery, he loved him no more when he found he had thoughts of his own and could not be dictated to. The day after the Fort Scott expedition Brown and Montgomery were to rendezvous, when Brown severely upbraided the commander in front of Gill for carrying along a prisoner. Gill wrote: "Montgomery gave him a trial and he was released by general consent as not meriting punishment. When we returned Brown was furious because the man was not shot. His Calvinism and general organism would have treated Servetus as Calvin did. . . . And yet this very concentration on self commanded the grand advance on American slavery."

Soon after this exchange, Gill was to get a striking illustration of the extraordinary union of messianic belief and fighting antislavery in his chief, when he came across a black man apparently engaged in the earthly business of selling brooms. After each resolved the propriety of making the other his confidant, Gill learned the man was Jim Daniels, who had come across from Missouri looking for help. He, his wife, and his children belonged to an estate that was to be sold at an administrator's sale, and Daniels wished to avoid the certain prospect of seeing his family broken up and sold into the Deep South. They immediately hunted up Brown, who devised a plan to bring the needed assistance. Gill gave an account related in Hinton's *John Brown and His Men*: "I am sure that Brown, in his mind, was just waiting for something to turn up; or, in his way of thinking, was expecting or hoping that God would provide him a basis of action. When this came, he hailed it as heaven sent."

The next night seventeen heavily armed and mounted men crossed into Missouri. Dividing into columns, Brown with his party, guided by Daniels, rode to the farm where he and his family belonged, while Stevens with his party rode toward neighboring farms. The Brown party approached the house of Daniels's proprietor, surrounded it, several entering with drawn revolvers, while others rounded up the animals in the corrals and barns. Less than a mile away Stevens did likewise, telling the startled slave-owner at gunpoint: "We have come after your Negroes and their property. Will you surrender or fight?"

In all, Brown's party freed ten slaves and seized horses, oxen, clothing, boots, bedding, and a wagon that was loaded with provisions consisting of bacon, flour, and meal. Stevens freed but one slave after killing his owner, who had resisted, and also seized property including horses, mules, oxen, and a wagon with provisions. At an early hour the two parties rejoined, taking three white men as hostages, and hastened back into Kansas.

Once across the border the hostages were released with the admonition that they could follow if they wished; while shelter was sought in a ravine from the wind and cold for the animals, the armed men, and the newly freed people—where all would remain throughout the daylight hours. George Gill wrote, referring to the blacks: "to our contrabands, the conditions produced a genial warmth not endorsed by the thermometer."

When night came the entourage resumed its trek, reaching the farm of Augustus Wattles at midnight a day later. Bunked in the loft there with a few of his men, and awakened by the commotion, was James Montgomery, who peered down as Brown ushered the fugitives into the parlor. He exclaimed: "How is this Captain Brown? Whom have you here?" Gesturing broadly with his straw hat, Brown replied: "Allow me to introduce to you a part of my family. Observe, I have carried the war into Africa." Three of the persons presented were men, five were women—one of them, the wife of Jim Daniels, who was pregnant—and three were children, two boys and a girl.

As morning came a covered wagon into which the fugitives were secreted was packed with provisions and tethered to oxen, with Brown and Gill occupying the driver's seat. On Christmas Eve they reached the Adair home near Osawatomie, where they remained throughout Christmas Day in the warmth of the kitchen. In the early hours of the following morning all moved into a roughly hewn abandoned cabin on Pottawatomie Creek. As improvements were hastily made to the dwelling, in this setting was born the first of the John Brown namesakes, John Brown Daniels, attended by a physician of the neighborhood, Dr. J. G. Blunt, who was sympathetic to Brown.

As news of the incursion spread, the Missouri press howled for retribution for "Old Brown's" infamy, while the state's governor, amidst denunciations of robbery and assassination, offered $3,000 for his capture, to which President Buchanan personally added $250 and the Kansas territorial governor an equal amount. With calls for the arrest of both Brown and Montgomery, a five-hundred-man posse assembled to ride up from West Point, as Brown notified his ally they should be in readiness to fight. But Brown's forcible expropriation of slaveholder property also met with severe condemnation in Kansas. One criticism circulating was that Brown intended to leave the territory, and a retaliatory attack would fall upon other heads, not his; another was that the Buchanan administration would now only redouble its effort to impose a proslavery constitution on Kansas. For this censure, Brown retorted: "It is no pleasure to me, an old man, to be living in the saddle, away from home and family, exposing my life, and if the Free State men of Kansas feel that they no longer need me, I will be glad to go." To one of his staunchest friends, Augustus Wattles, Brown said: "I have been at your abolition meetings. Your schemes are perfectly futile. You will not release five slaves in a century. Peaceful emancipation is impossible. The thing has gone beyond that point."[14]

As Montgomery, Brown, and Kagi sat debating the situation at Wattles's home in the early weeks of the new year, Brown wrote his famous "Parallels," published on January 22, 1859, in the *New York Tribune*. Contrasting the reaction of the administration and the law enforcement agencies to the May murders, about which nothing had been done, and his own recent exploit, Brown wrote:

> Now for a comparison, eleven persons are forcibly restored to their natural; & inalienable rights, with but one man killed, and all "Hell is stirred from beneath." It is currently reported that the governor of Missouri has made a requisition upon the governor of Kansas for the delivery of all such as were concerned in the last named "dreadful outrage." The marshal of Kansas is said to be collecting a posse of Missouri (not Kansas) men at West Point in Missouri, a little town about ten miles distant, to "enforce the laws." All Pro-Slavery, conservative Free-State, and doughface men and Administration tools, are filled with holy horror. Consider the two cases, and the action of the administration party.

The publication of Brown's observations and the keen interest the *Tribune* took in Kansas affairs assured that his foray would be well known in the East. "Old Brown Invades Missouri" had been the caption of a lead story on January 6, while Horace Greeley editorialized: "Captain Brown, who had cooperated with Montgomery and whose property had been destroyed and his son murdered in the former wars, did not wait for invasion. He led a party into Bates County, who retorted on the slaveholders of the vicinity the same system of plunder which the Free State people of Kansas have suffered during the recent invasion."

On January 10 Gerrit Smith noted in a letter to his wife: "Do you hear the news from Kansas? Our dear John Brown is invading Missouri and pursuing the policy which he intended to pursue elsewhere." The *Tribune*'s correspondent summed up the results attained: "Some bad may have grown out of the movement, but I have yet to see what it is. Much good has come of it. The bluster of Missouri has lessened. While hundreds of non-slaveholding whites express great indignation at the invasion of their state and boil over with patriotism in public, they privately laugh at the idea of their defending a species of property that is a curse to them and rejoice that certain lordly slaveholders have 'come down to their level.'"

Later Elijah Avey wrote in his book, *The Capture and Execution of John Brown*, "It was reported that the slave population of the two adjacent Missouri counties was diminished from five hundred to fifty within a few weeks, mainly by removal for sale."

But Brown's prescription for resolving the state of war with a more thorough war may have been too strong even for Montgomery. With his support waning among men who had sustained him in the past, he felt a need to distance himself from Brown, and walked into the Territorial Court in Lawrence on January 18 and turned himself in. Freed on $4,000 bail, Montgomery made an appearance at the territorial legislature then in session in Lecompton. The *New York Times* correspondent described the occasion: "Scores were pressing to grasp him by the hand while he 'looked down' upon the heads of those who but a few days before were branding him as the arch-robber. Firm, fearless, erect, he now stood in the same hall where his name had been traduced and vilified. This must have been to him one of the strange vicissitudes of human life."

In this charged atmosphere, on January 20 Brown's ox-drawn wagon lumbered northward over the frozen ground, with George Gill and twelve "contrabands." In parting Brown gave this assessment to Augustus Wattles: "I considered the matter well. You will probably have no more attacks from Missouri. I shall now leave Kansas and probably you will never see me again. I consider it my duty to draw the scene of the excitement to some other part of the country."[15]

Up in Leavenworth there had been uproar when some men tried to kidnap a black barber to sell him into slavery. At a huge rally James Lane boomed: "If you must choose between kidnapping a man into slavery or kidnapping him into freedom, in God's name kidnap him into freedom."[16]

For eighty-one days Brown and his group were to trek before reaching "the land of Canaan," as fugitive slaves called Canada. For nearly a month they had been secreted in a secluded cabin only twenty-five miles from the Missouri border. Armed men were scouring the country for them; slave hunters were on the prowl, spies were everywhere on the roads. Brown's wagon rolled up to the cabin of Major J. B. Abbott on the outskirts of Lawrence four days after leaving Osawatomie. Word had been sent ahead to E. B. Whitman that he needed money, but as he visited Lawrence, Brown was told Whitman was "dissatisfied at your proceedings . . . when you were here before." However, a little money was raised from selling some of the seized property, and the oxen were exchanged for horses.

While in Lawrence, one of the persons Brown met with was William Addison Phillips. A journalist, Phillips was among a select number of men influential in Kansas political circles Brown had sought as allies. Brown now wanted to draw this engagement toward the broader object he had in view—the struggle between slavery and freedom that was pushing the nation toward the precipice of war. Kagi summoned him to a meeting with Brown, saying that he was going away for good and wished to see him. Phillips sent back that he would not come as Brown never took his advice, and he saw no reason to offer any now. Soon however he relented and agreed to come to the hotel where Brown was staying. Ushering Phillips to the room, Kagi posted himself outside to ensure their conversation would not be overheard. This interview, the last of three Phillips conducted with Brown in Kansas, was reported, together with the others, over twenty years later in the *Atlantic Monthly*, December 1879.

Phillips's feeling of ambivalence had perhaps been overcome by his journalist's curiosity. There was a fascination about the "old man" which he had clearly glimpsed, and although he was averse to Brown's more strongly held convictions, Phillips was not of the view that would come to predominate in later decades—that Brown was "fanatical," or a "monomaniac," to use two words often circulating in American scholarship about him. In fact, Phillips would try to placate Brown by suggesting that he and his followers stay in Kansas and take claims; they could keep their arms in self-defense, but otherwise would assure the territory entered upon its statehood a prosperous and free country. Phillips wrote: "He had changed a little. There was in the expression of his face something even more dignified than usual; his eye brighter, and the absorbing and consuming thoughts that were within him seemed to be growing out all over him."

Brown began by reciting the long career of slavery in the United States. "He recalled many circumstances that I had forgotten, or had never heard of," Phillips wrote. When the republic had been founded, he said, the expectation had been that slavery would be shaken off, and this view prevailed for the first twenty-five years of its existence. This was evident in that even before the Constitutional Convention the founders had enacted the Ordinance of 1787 under the Articles of Confederation. Known as the Northwest Ordinance, this had excluded the importation of slaves into the Northwest Territory from which the states of Ohio, Indiana, Michigan, Illinois, and Wisconsin would be formed. In the first Congress, while slavery had been recognized in those territories that became Alabama and

Mississippi, this had only been done after prohibiting the continuance of the Atlantic slave trade after January 1, 1808. Finally as part of the Missouri Compromise of 1820 prohibiting slavery's expansion north of that state's border, Congress had barred slavery in the remaining Louisiana Territory. Compromises such as these were made because the new nation could not have come into and remained in existence otherwise, but the founders had taken care not to use the word "slavery" in the Constitution, believing it incompatible with republican governance and the principles of human equality and liberty.

Although slavery was thought to be on the road to extinction; gradually it had become more profitable and begun to extend and increase itself. This was due to the invention of the cotton gin and the abundance of cheap fertile land. This brought of necessity a more despotic system, and rights once possessed by the slaves had been taken away. The breeding of blacks for sale had likewise become profitable. Little by little the pecuniary interests that rested on slavery enabled proslavery interests to seize the power in the government. Then opinion opposed to slavery in the South was under ban, and an attempt was made to apply that to the northern states. At length the politicians of the South had become mere propagandists for slavery, and the northern politicians "trimmers." When northerners tried to check this alarming growth, the South threatened secession. Now began an era of compromise—"where for peace, men were willing to sacrifice everything for which the republic was founded."[17]

Brown was to remark he had taken little interest in politics since the election of Andrew Jackson, seeing it merely followed the condition of public sentiment on the slavery question. And that sentiment was mainly created by actual collisions between freedom and slavery. Now the point had been reached with the collision in Kansas where nothing but war could settle the question. The South had just missed the opportunity of permanently getting the upper hand. But its aggressions had only temporarily been checked. Phillips noted here he still recalled much of what Brown said verbatim: "It has taken them half a century to get the machinery of government into their hands and they know its significance too well to give it up. They will never peacefully relinquish it. If the Republican party elects its president next year there will be war. The moment they are unable to control they will go out, and as a rival nation alongside they will get the countenance and aid of the European nations, until American republicanism and freedom are overthrown."

Astonished by the thrust of Brown's reasoning, Phillips suggested that surely he was mistaken. The collision that occurred in Kansas would soon die down, and it was in the northern interest to let it do so.

Brown countered: "No, no, the war is not over. It is a treacherous lull before the storm. We are on the eve of one of the greatest wars in history, and I fear slavery will triumph and there will be an end of all aspirations for human freedom. For my part, I drew my sword in Kansas when they attacked us, and I will never sheathe it until this war is over. Our best people do not understand the danger. They are besotted. They have compromised so long that they think principles of right and wrong have no more any power on this earth."

Phillips: "Let us suppose all you say is true. If we keep companies on the one side, they will keep them on the other. Trouble will multiply; there will be collision, which will produce the very state of affairs you depreciate. That would lead to war, and to some extent we should be responsible for it. Better trust events. If there is virtue enough in this people to deserve a free government, they will maintain it."

Brown: "You forget the fearful wrongs that are carried on in the name of government and law."

Phillips: "I do not forget them—I regret them."

Brown: "I regret and will remedy them with all the power that God has given me."

Now Brown went on to parallel his argument with reference to the example of the slave leader Spartacus, who had opposed Rome. Phillips pointed out that the Roman slaves were a warlike people, trained in arms and in warfare; the American slaves were far more domestic and "in all their sufferings they seem to be incapable of resentment or reprisal."

Brown: "You have not studied them right; and you have not studied them long enough. Human nature is the same everywhere." Continuing to point out the mistakes of that ancient insurrection, showing he had given it his best study, Brown said, "Instead of wasting his time in Italy, the leader should have struck at Rome, or if not strong enough for that, escape to the northern provinces to build an army. But Rome's armies were able to swoop down on them to destroy them."

Hearing nothing but more talk of war, Phillips finally bridled, predicting Brown would end by bringing himself and his men into a desperate enterprise, where they would be imprisoned and disgraced.

Brown: "Well, I thought I could get you to understand this, I do not wonder at it. The world is very pleasant to you; but when your household gods are broken, as mine have been, you will see all this more clearly."

As Phillips rose to leave, he said, "Captain, if you thought this, why did you send for me?" Reaching the door he felt Brown's hand on his shoulder. Turning, the old man took Phillips's hands in his, tears showing on his hard bronzed cheeks, as he said: "No, we must not part thus. I wanted to see you and tell you how it appeared to me. With the help of God, I will do what I believe to be best." Holding Phillips's hands firmly in his, Brown leaned forward and kissed him on the cheek. "And," Phillips wrote, "I never saw him again."

Leaving Gill in Lawrence recuperating from frostbite and exhaustion, Brown departed for Topeka on the twenty-fifth with his wagon, with Stevens sharing the driver's seat. Reaching the edge of Topeka, they stayed for a few days in the Sheridan family cabin where new shoes were donated for the "contraband," and food and money obtained. Three days later the party was some fifteen miles north of Topeka, in Holton. The next day, six miles further on, they sought refuge after a snowstorm at the cabin of Abram Fuller. By then it was well known that John Brown was passing through the country and an eighty-man posse led by a United States marshal rode out to apprehend them, even as the territorial governor wired President Buchanan premature confirmation of their success.

As the marshal's advance scouts approached Fuller's cabin, Stevens stepped out to greet them. "Gentlemen," he said, "you look as if you were looking for somebody or something." "Yes," one of them replied, "we think you have some slaves up in that house." "Is that so? Well, come on with me and see." As he reached the door, Stevens opened it just enough to grab a shotgun resting inside. Pointing the weapon at the men, he said: "You want to see slaves, do you? Well, just look up those barrels and see if you can find them."

All three men ran, but as one had a bead drawn on him, he was captured. One of the fugitives, a man named Samuel Harper, later recounted in Hamilton's *John Brown in Canada* what happened next. "Captain Brown went to see the prisoner, and says to him, 'I'll show you what it is to look after slaves, my man.' That frightened the prisoner awful. He was a kind of old fellow and when he heard what the captain said, I suppose he thought he was going to be killed. He began to cry and beg to be let go. The captain only smiled a little bit, and talked some more to him, the next day he was let go." By this time Kagi had come up from Topeka with twenty heavily armed, mounted men. As the marshal's posse waited at a creek crossing ahead, with odds four to one in the marshal's favor, one of the men asked Brown what he proposed to do. "Cross the creek and move north."

"But, Captain," he objected, "the water is high and I doubt if we can get through. There is a much better ford five miles up the creek."

Brown replied: "I intend to travel it straight through and there is no use to talk of turning aside. Those who are afraid may go back. The Lord has marked out a path for me and I intend to follow it. We are ready to move."

Placing a double row of mounted men before the wagon, giving them ample room for maneuver, Brown gave the order to go straight at the marshal's posse. George Gill, who had rejoined the company, wrote: "The scene was ridiculous beyond description." While waiting in ambush, the marshal and his posse found themselves under attack. Seeing horsemen bearing down on them, the entire posse jumped for their horses and fled in disarray. One man could only grab the tail of his horse as it galloped away, and was seen in tow. Samuel Harper's account of what became known as the Battle of the Spurs, continues: "Captain Brown and Kagi and some others chased them, and captured five prisoners. There was a doctor and lawyer amongst them. They all had nice horses. The captain made them get down. Then he told five of us slaves to mount the beasts and we rode them while the white men had to walk. . . . The mud on the roads was away over their ankles. I just tell you it was mighty tough walking, and you can believe those fellows had enough of slave-hunting. The next day the captain let them all go."

On February 1 the party approached the Nebraska border. With the Nemaha River only partially frozen and too high to cross, they waited through a bitterly cold night. By morning the ice was solid enough to bear the weight of a man, but not the team and wagon. Disassembling the wagon, they pushed it across in pieces, and a makeshift bridge of lumber, poles, and brush was laid for the horses. Three days later, after eluding another posse, Brown and his party crossed the Missouri River into Iowa.

Reaching Tabor on February 4 they found the news of their exploit had preceded them. The fugitives would be sheltered in the schoolhouse of a local church and fed by member families in the coming week; but there was consternation in regard to their benefactors. At Sunday service a clergyman refused to entertain Brown's petition to the pulpit asking for "public thanksgiving to Almighty God in behalf of himself, & company; and of their rescued captives," because they had killed a man and taken horses and other property. And on the following evening, as Brown was giving a recital at a public meeting, he interrupted his remarks when a slaveholder from St. Joseph, Missouri, entered. Brown requested the man withdraw, but the sense of the meeting was to the contrary. As Brown walked out he remarked to Tidd, "There are some there who would give us a halter for our pains. We had best look to our arms and horses." The meeting then debated and passed a resolution of support for the fugitives, but stipulated, "We have no sympathy with those who go to slave states to entice away slaves and take property or life when necessary to attain that end."[18]

The caravan left Tabor on February 11. Traveling twenty-five miles a day, on the eighteenth they were in Des Moines, where Kagi paid a visit to the office of the *Register*, giving its editor an account of Brown's Missouri exploit, with Brown following up with a letter. Two days later they reached the town of Grinnell, where Brown knocked at the door of Josiah Bushnell Grinnell, an abolitionist and the founder of the town and college bearing his name. Brown began: "This cannot be a social visit. I am that terrible Brown of whom you have heard." As he explained that he needed a place to stay for himself and his companions, Grinnell forthwith threw open the door to his parlor, henceforth called the Liberty Room, saying, "This room is at your service and you can occupy the stalls at the barn which are not taken. Our hotel will be as safe as any place for part of your company."[19]

The contrast between the receptions in Tabor and in Grinnell was striking, and Brown exalted. On successive nights there were full-house meetings at which Brown and Kagi spoke where they "were loudly cheered; & fully indorsed," as Brown wrote. Contributions in clothing, food, and cash were raised for the fugitives, and J. B. Grinnell, now called "John Brown" Grinnell, arranged for rail conveyance to Chicago for them, from the railhead at West Liberty, Iowa.

By the 25th the party had moved on to Springdale, where Brown's Quaker friends extended their hospitality. Gill would stay here with his family to recover from an inflammation in his joints; he never saw his companions again. As they were enjoying their reunion in

Springdale, word reached there that a posse was being raised in a nearby town to apprehend the fugitives. Brown and his entourage quickly traveled the seven miles to West Liberty and on March 9 boarded an unbilled boxcar for Chicago. The following morning as the train reached Chicago, it halted a half-mile from the depot where the car with the fugitives was detached and pushed to a side track where its occupants were released. Brown was soon in touch with Allen Pinkerton, the famed detective, who showed no reticence whatever in escorting the fugitives to the mill owned by Henry O. Wagner, who put out a sign "Closed for Repairs," while Brown went to the home of his friend, John Jones. That afternoon Pinkerton solicited $400 from C. G. Hammond, superintendent of the Michigan Central Railroad, to hire a freshly provisioned boxcar to Detroit.

Meanwhile Brown had boarded an earlier train after telegraphing Frederick Douglass, whose whereabouts he had learned from Jones, requesting that they meet in Detroit. On the afternoon of March 12 in a blizzard the fugitives arrived at the wharf in Detroit to await embarkation for Windsor. Shortly before boarding the ferry John Brown arrived to see them off. Said he: "Lord, permit Thy servant to die in peace; for mine eyes have seen Thy salvation! I could not brook the thought that any ill should befall you—least of all, that you should be taken back to slavery. The arm of Jehovah protected us."[20]

———

During the winter of 1858–59, Douglass, together with John Jones and H. Ford Douglass, had been making an intensive lecturing tour through Illinois, Wisconsin, and Michigan. Meeting with a generally positive, even an enthusiastic reception, Douglass was to write of "Our Recent Western Tour" in his April *Monthly*: "a Negro lecturer" is "an excellent thermometer of the state of public opinion on the subject of slavery." He was much better than a white antislavery lecturer, because "a hated opinion is not always in sight—a hated color is. . . . The Negro is the test of American civilization." At the conclusion of the tour Douglass was to give special notice in his paper to one of his co-lecturers, writing: "He has that quality without which all speech is vain—earnestness. He throws his whole soul into what he says, and all he says. His person is fine, his voice musical, and his gestures natural and graceful." This acknowledgment was offered, however, "not as incense offered to vanity," but that he might better serve the cause. "We call upon H. Ford Douglass to put himself unreservedly into the lecturing field, not upon the platform of an African Civilization Society."

Frederick Douglass rejected the necessity of seeking, as did others, a "nationality" far from the shores where the principal struggle must take place. He retorted: "We have an African nation on our bodies. . . . We are contending not for the rights of color, but for the rights of men." Douglass called upon H. Ford Douglass to "stand upon the platform of Radical Abolitionism and go to the people of the country with a tangible issue." What better issue could be made he advised than to return to the state of Illinois and agitate for the repeal of its Black Law, a fight being led there by John Jones. "All that is malignant and slaveholding in that State, clings around and supports these atrocious laws. The immediate repeal of these cruel enactments, (which the white [Stephen A.] Douglas is endeavoring to sustain,) should be the demand of the black Douglass of Chicago."

It was while Douglass was completing his tour of Michigan— where he visited Battle Creek, Marshall, Albion, Jackson, and Ann Arbor—that he received the telegram from John Brown, then with John Jones in Chicago, summoning him to a meeting in Detroit. It was for this reason an appearance at Detroit's City Hall was added to Douglass's schedule, and he arrived in the city on March 12. The *Detroit Free Press* was to marvel as to "why Fred Douglass should have come to so inhospitable a region." The reason, to which they were not privy, was a closed-door meeting of selected persons at the home of William Webb, whose guests included William Lambert, William C. Munroe, George DeBaptista, at least three other black Detroiters, and John Brown.

While few details allow for more than a suggestive explication of the get-together, one can surmise it had an earnest, if congratulatory air. Among other things, Brown's recent exploit was discussed, its implications and its incidents narrated. As far as Brown's projected campaign was a concern, the fact that several of the participants had been at his Chatham Convention the previous year indicates it was discussed. It is known that DeBaptista, who had taken over leadership of the Detroit Vigilance Committee from Lambert, questioned the efficacy of Brown's plan. In its stead he is said to have suggested a program of bombings across the South as a way to throw the slave oligarchy off balance and raise a national outcry. Whatever the resolution on the issue, Brown surely demurred from being connected with it on humanitarian grounds. There is indication too that dis-

paraging words were passed between Brown and Douglass, some supposing that Brown challenged Douglass's courage when he opposed
certain measures. But the two would be seen working together in the
coming months, and any friction that may have been displayed in
Webb's home likely had more to do with Brown's aversion to merely
talking abolitionism, than any reservations Douglass may have
expressed. Shortly after their meeting in Detroit, in the columns of his
April *Monthly*, Douglass was to avow his confidence in Brown. The
basis of his idea was that "the enslavement of the humblest human
being is an act of injustice and wrong, for which Almighty God will
hold all mankind responsible; that a case of the kind is one in which
every human being is solemnly bound to interfere; and that he who
has the power to do so, and fails to improve it, is involved in the guilt
of the original crime. He takes this to be sound morality, and sound
Christianity, and we think him not far from the right."

Uppermost on Brown's mind then—and this is another subject that
likely was broached at William Webb's house—was not only the gathering of additional recruits, but above all the need to enlist a recognized black leader of demonstrated ability who would not shrink
from the dangers his Virginia campaign would entail. The setback in
regard to the Canadian refugees had to be addressed, and Brown
must have thought to rely on Harriet Tubman and Martin Delany in
the coming weeks for this. But there had been no contact with either
of them, and Delany—who had corresponded with Kagi in Kansas,
assuring him, in a letter dated August 16, 1858, of his and other
Chathamites continued support—had other priorities. He was
preparing just then to embark to Africa on his Niger Valley expedition, commissioned by the 1854 emigration convention in Cleveland.
Others, who had been at the Chatham convention, and to whom
Brown could have looked, had also turned their attention elsewhere.
Isaac Holden and James Munroe Jones would be traveling to the
Canadian and U.S. Pacific coasts; William Howard Day would be
sailing for Dublin, Ireland, and Reverend Munroe was to set sail for
Africa on the same ship with Delany in May, going under church aus-
pices to Liberia.

Others to whom Brown may have looked were Gloucester in
Brooklyn, Loguen in Syracuse, and Garnet in New York City. But
Garnet had founded his African Civilization Society in 1858, and was
thoroughly engaged in controversy just then. Garnet and Gloucester,
moreover, had been skeptical; counseling that they did not think the

time was ripe, that slaves were not sufficiently aware of their rights to respond in the way Brown predicted, nor were northern blacks sufficiently prepared, as they had been shut out in consequence of the discrimination against them "from both the means and the intelligence necessary," as John S. Rock phrased it.[21]

As a result the office of president in Brown's provisional constitution had been left vacant. The office undoubtedly called for an individual with organizational abilities, with extensive contacts, who would be able to promote and inspire cooperation between the various societies, associations, and individuals that could be looked to for support. Perhaps none of these, excepting Delany and Tubman (whose commitment remained unshaken), had the necessary military flair for Brown's enterprise.

Brown next arrived in Cleveland where the Oberlin-Wellington rescue trial was under way. The case originated in Wellington, Ohio, on September 13, 1858, when a Kentucky slaveholder, searching for a fugitive in nearby Oberlin, recognized a man who had been the slave of his neighbor in Mason County, Kentucky. He obtained the power of attorney to act on the owner's behalf, and a complaint was taken to a commissioner who issued a warrant, and a U.S. marshal was sent to Oberlin.

As it would be difficult to effect the man's capture there, a plan was devised whereby he would be lured to Wellington on a subterfuge. Offered a job digging potatoes for a prominent Democrat farmer, the alleged fugitive, John Price, was seized by four men on his way to perform the work and confined in a local tavern to await a train to Columbus. Word that Price had been seized spread "like a flash of lightning," and people were soon converging by every available means upon the building. After two hours of wrangling with his captors, the secretary of the Ohio Anti-Slavery Society and active Underground Railroad operative Charles H. Langston—the younger brother of the abolitionist John Mercer Langston—led a charge from the growing crowd, freeing the fugitive. Spirited away in a wagon driven by Simeon Bushnell to Cleveland to await a ferry to Canada, Price would be accompanied on his journey to freedom by a pistol-bearing young man named John Copeland.

An outraged Buchanan administration decided to make a signal example of this "villainy," and thirty-seven warrants were issued against residents of Oberlin and Wellington, including professors, clergy, students, free blacks, and fugitive slaves alike.

Twenty-three of those charged refused bail and were confined en masse in Cleveland to await trial. As delegations began arriving from all over the state to show support, the defendants became the object of growing attention and sympathy. Demonstrators marched in ranks around the block where the jail stood, carrying banners, as those inside made speeches from cell windows. As the trial was set to begin, twelve thousand persons rallied to be addressed by Governor Salmon P. Chase and Congressman Joshua R. Giddings.

When Brown arrived on March 20 he took a room at a hotel only four blocks from the U.S. marshal's office, making no effort to conceal his identity, even as posters were being displayed calling for his arrest. Several days later, broadcasting his defiance, Brown stood before an auction to raise money by the sale of two horses and a mule, despite their "questionable title," and they brought good prices. A short time after this a public meeting was called where Brown and Kagi spoke, drawing an audience of fifty at a quarter-dollar admission. The city editor of the *Plain Dealer*, Charles Farrar Browne, the up-and-coming humorist writing under the pseudonym Artemus Ward, wrote for his paper that Brown was "a medium-sized, compactly built and wiry man, and as quick as a cat in his movements. His hair is of salt and pepper hue and as stiff as bristles. He has a long, waving, milk-white goatee, which gives him a somewhat patriarchal appearance. His eyes are gray and sharp. A man of pluck is Brown. You may bet on that. He shows it in his walk, talk and actions. He must be raising sixty and yet we believe he could lick a yard full of wild cats before breakfast and without taking off his coat."[22]

In his remarks that night, as the reporter for the *Cleveland Leader* noted, Brown called attention to the fact that he was "an outlaw, the governor of Missouri having offered a reward of $3000, and James Buchanan $250, for him. He quietly remarked, parenthetically, that 'John Brown' would give two dollars and fifty cents for the safe delivery of the body of James Buchanan in any jail of the free states. He would never submit to an arrest, as he had nothing to gain from submission; but he should settle all question on the spot if any attempt was made to take him."

The story Brown related of his Missouri incursion in the course of his talk was "refreshingly cool," remarked Ward in his column. Brown "would make his jolly fortune by letting himself out as an Ice Cream Freezer," he suggested.

In the audience that evening was a man from Oberlin named Lewis Sheridan Leary, the uncle of John Copeland, who was being held in the jail awaiting the beginning of the trials with the other rescuers. Like Copeland, who had come to Oberlin to enter its university's preparatory school, Leary, a saddle and harness maker by trade, was originally from North Carolina. This uncle and nephew by relation were first among a community where blacks bore an attitude toward whites, a contemporary remarked, that said "touch me if you dare"; and both quickly developed a keen interest in John Brown.

The first defendant to be tried in the Oberlin-Wellington trials was Simeon Bushnell. He was adjudged guilty and sentenced to sixty days in the county jail, with a fine of six hundred dollars and ordered to pay the cost of his prosecution. Charles Langston was the next up. His speech to the court—"Should Colored Men Be Subjected to the Penalties of the Fugitive Slave Law?"—had a powerful effect, both in the court and in wider circulation. Printed in its entirety in both the *Cleveland Leader* and the *Columbus State Journal*, it was again published by the *Anglo-African*, where it appeared together with his brother's account of the entire episode, then circulated as a pamphlet. Langston was also adjudged guilty, but on the strength of his statement the judge meted out a sentence less stringent than previously, sentencing him to twenty days in the county jail with a fine of one hundred dollars plus the cost of prosecution.

As Charles Plummer Tidd and Kagi together with others from Cleveland and Oberlin planned a rescue of those held for trial, the state of Ohio relieved them of its necessity by arresting the four Kentucky men who had seized John Price, as they appeared to testify. After this intervention both sets of prisoners would be freed in June as settlement of the case, with only Bushnell and Langston having been tried and convicted—recognition that the Fugitive Slave Law had been broken in the Western Reserve.

Brown was in Cleveland for only ten days and certainly did not meet with Charles Langston, but was surely familiar with his eloquent appeal for justice to the court. Kagi and others of his party, however, did become Langston's intimate; Kagi reporting on the trials as a correspondent to the *New York Tribune* and the *Cleveland Leader*. While in Cleveland, Brown received an invitation from Joshua Giddings to speak at the Congregational Church in Jefferson where Giddings was a member. Receiving donations from the congregation afterward, including money from Giddings, and an invitation to his house for dinner, Brown undoubtedly considered fortuitous this contact with the leader of antislavery opinion in the House of Representatives, which together with the addition of Leary, Copeland, and Langston as adherents, argued well for his coming campaign.

Before leaving Ohio, Brown stopped for a conference with John Jr., in West Andover. His cache of arms, shipped from Springdale, Iowa, in February 1858, had first been stored in a furniture warehouse in nearby Cherry Valley, then after the Chatham convention moved at Brown's urging by his son, where none "but the Keeper & you will know where to find them." Some had been concealed inside a haymow in Wayne, and the remainder in another farm building. They had been cleaned and oiled, and were ready for shipment.

Six

A Setting Possessed
of Imposing Grandeur

I surely would be a prophet, as the Lord had shewn me things that had happened before my birth. And my father and mother strengthened me in this my first impression, saying in my presence, I was intended for some great purpose.

—Nat Turner, "Confessions"

T OWARD THE SECOND WEEK OF APRIL 1859 HORACE MCGUIRE, employed by Frederick Douglass in his printing office at Rochester, later remembered "a tall, white man, with shaggy whiskers, rather unkempt, a keen piercing eye, and a restlessness of manner," called asking for his employer. The man gave the appearance of one whose "interview was by appointment," and when Douglass returned, McGuire observed, "the greeting between the white man and the former slave was very cordial" and they "talked freely" and "with great earnestness." Biographers that have noted this meeting often infer it was a stop of several hours only, while in point of fact it was at this time that Douglass arranged for Brown's appearance at Rochester City Hall, giving notice of it his *Monthly*, published April 15 under the heading "Old Brown in Rochester." Upbraiding the self-professed "Republicans" of Rochester for staying away, Douglass wrote, "Even our newly appointed Republican janitor ran off with the key to the bell of the City Hall, and refused to ring it on the occasion! Shame upon his little soul, and upon the little souls who sustained him in his conduct."

While Douglass offers no substantive comment on their meeting, merely remarking in his *Life and Times* that he had numerous visits from Brown during his travels to and from Kansas, it is important in demonstrating the level of his interest and commitment at that time. During this visit Douglass introduced Brown to an individual he intended as his contribution to the recruiting for the Virginia campaign, who was then a boarder in the Douglass home. He was a fugitive slave from Charleston, South Carolina, who had arrived in Rochester after escaping aboard a sailing ship in 1856, after the death of his wife, leaving a son in slavery. He had lived for several years in St. Catharines where he was engaged as a house servant and waiter. Returning to Rochester in 1858, he proposed to establish himself in a business as a clothes cleaner, having a card printed stating his work would be done "in a manner to suit the most fastidious and on cheaper terms than anyone else." The man's name had been Esau Brown, which he soon exchanged for another—Shields Green. A full-blooded African, he was twenty-four years old, of slight stature but well built, possessing a self-confident bearing and styling himself in dress and action a "Zouave." Reputedly in the lineage of an African prince, he referred to himself as "Emperor." One of his contemporaries, Lucy Coleman, wrote that the "overseer's lash had cut deeply into his soul." His host described him as "a man of few words," with speech that was singularly broken, "but his courage and self-respect made him quite a dignified character." The convergence between the elderly white and the young black man could not have been greater; Douglass concluded. "John Brown saw at once what 'stuff' Green was made of, and confided to him his plans and purposes. Green easily believed in Brown, and promised to go with him whenever he should be ready to move."[1]

Still it could be asked—just how far Douglass himself was willing to go to sustain John Brown? For two decades he had participated in any number of venues for antislavery action; he had been ever ready to publish, to speak, to combine or to conspire with any individual, society, or party, as he would say, "to head off, hem in, and dam up the desolating tide of slavery." But it was a sad fact, Douglass wrote later that summer, "that in the hands of all these societies and committees, nearly all our anti-slavery instrumentalities have disappeared. . . . The Radical Abolition Society . . . was built on a faultless plan, but where is that Society to-day? Where is its committee? Where its paper, its lecturers and patrons? All gone!" He was therefore opting, he wrote, "to work for the present on the plan of individualism, uttering our word for freedom and justice, wherever we may find ears to hear, and writing our thoughts for whoever will read them." It is on this ground accordingly one must look for concurrence between Brown and Douglass, noting that the latter supplied the former with a recruit, convened meetings and collected money for him, and also had facilitated his contact with Harriet Tubman—all for an attack upon slavery "with the weapons precisely adapted to bring it to the death."[2]

The orientation of Douglass in relation to Brown, and his solidarity is perhaps exhibited in a letter Jeremiah Anderson, who was traveling with Brown as his adjunct, wrote to his brother on June 17, 1859 (in Hinton's biography), announcing unequivocally, "Douglass is to be one of us." This was also an assessment in which others in the Brown family later concurred, with Anne Brown remarking (also in Hinton) her father saying, "Douglass is to be one of us, even unto death."

Three days later Brown and Anderson were in Peterboro at the estate of Gerrit Smith, a journey requiring only a few hours, where they would remain three days. In the evening Brown addressed a small gathering including Smith and his wife, Edwin Morton (employed as a tutor by Smith), and Harriet Tubman, who often traveled across New York from her home in Auburn to St. Catharines, and assuredly would have received word from Douglass that Brown would be at Smith's. After Brown spoke, with an eloquence that moved both Smith and his wife to tears, the philanthropist rose to hail Brown as "the man in all the world I think most truly Christian," and he wrote a pledge for a $400 contribution. Morton later wrote to Franklin Sanborn that Brown had been "tremendous," and that he no longer had any doubts as to the soundness of his course, adding, "I suppose you know where this matter is to be adjudicated. Harriet Tubman suggested the 4th of July as a good time to 'raise the mill.'"[3]

From April 9 to May 5 Brown was with his wife and younger daughters in North Elba, where he would remain mostly convalescing from illness. May 7 to June 2 found him once again in Boston, where Harriet Tubman had also gone. Before arriving in Boston Brown visited Sanborn in Concord, where he delivered an address. A. Bronson Alcott wrote in his diary: "Our best people listen to his words— Emerson, Thoreau, Judge Hoar, my wife; and some of them contribute something in aid of his plans without asking particulars. . . . I have a few words with him after his speech, and find him superior to legal traditions, and a disciple of the right in ideality and the affairs of the state. He is Sanborn's guest, and stays a day only. A young man named Anderson accompanies him. They go armed, I am told, and will defend themselves, if necessary."[4]

In this speech, while not revealing his plans, Brown did not conceal from his auditors "his readiness to strike a blow for freedom at the proper moment." Many could infer his intention when he asserted it was right to repeat such incursions as his Missouri raid as the opportunity arose, and that it was right to take property and even life in the process. Sanborn noted the audience winced at these remarks, but availed him good applause as he concluded, and contributed something that he might continue his activities.

While in Concord Brown broached the subject to Sanborn that he undertake a journey to Canada for him, the scholar demurring because of his obligations. In a letter to Thomas Wentworth Higginson, Sanborn in turn wrote that Brown was "desirous of getting someone to go to Canada and collect recruits, with H. Tubman, or alone as the case may be & urged me to go. . . . Last year he engaged some persons and heard of others, but he does not want to lose time by going himself now. I suggested you to him. Now is the time to help the movement, if ever, for within the next two months the experiment will be made."[5]

Sanborn already had notified Higginson that Brown informed Smith he would be coming east with some new men "to set his mill in operation." Sanborn suggested, "As a reward for what he has done, perhaps money might be raised for him." But Higginson's ardor had cooled from the previous year and he would recall in his *Cheerful Yesterdays*—"It all began to seem rather chimerical." In his reply to Sanborn at the time apparently referring to Brown's raid in Missouri, he wrote, "It is hard for me to solicit money for another retreat."[6]

Beginning his visit to Boston the next day, Brown was again to call on Amos Lawrence, who had also turned decidedly cool to him. Lawrence wrote in his diary: "He has been stealing Negroes and running them off from Missouri. He has a monomania on that subject, I think, and would be hanged if he were taken in a slave state. . . . He and his companion both have the fever and ague, somewhat, probably, a righteous visitation for their fanaticism."

Samuel Gridley Howe and George Stearns conveyed Brown to dinner at the Bird Club where he met Senator Henry Wilson. Remarking that he understood the senator did not approve of his course, Wilson offered that he did not, and that if he had gone into Missouri two years before there would have been a retaliatory invasion with great resultant bloodshed. With a scathing look Brown replied that he thought he acted rightly and that it had exercised a salutary influence. Howe also introduced Brown to John Murray Forbes, a wealthy clipper trader and railroad financier. Invited to visit his home in Milton, Brown held the businessman and a few of his friends up past midnight "with his glittering eye" and talk of the coming war between North and South. Forbes contributed $100 for Brown's "past EXTRAVAGANCES, and none for his future." At another meeting, John A. Andrew (soon to be elected governor of Massachusetts) contributed $25 for Brown and noted, "The old gentleman in conversation scarcely regarded other people, was entirely self-possessed and appeared to have no emotion of any sort but was entirely absorbed in an idea which pre-occupied him and put him in a position transcending ordinary thought and ordinary reason."[7]

Harriet Tubman had come to Boston in the spring of 1859 primarily to arrange for the security of her parents, so she could go on to do "practical work" with John Brown. It was at this time, that the woman whose activity, as Douglass would write, had been witnessed only "by the midnight sky and the silent stars," began to appear on the stage publicly, leaving deep impressions on abolitionist circles and forming lasting friendships with many persons. Sanborn wrote: "Pains were taken to secure her the attention to which her great services to humanity entitled her, and she left New England with a handsome sum of money towards payment of her debt to Mr. Seward."[8]

Wendell Phillips wrote of being called on by two distinguished visitors: "The last time I ever saw John Brown was under my own roof, as he brought Harriet Tubman to me, saying 'Mr. Phillips, I bring you one of the best and bravest persons on this continent—General Tubman as we call her.'" Brown went on to recount her labors and sacrifices on behalf of her people, asserting, "she was a better officer than most whom he had seen, and could command an army as successfully as she had led her small parties of fugitives." After her visit Higginson wrote his mother: "We have had the greatest heroine of the age here, who has been back eight times secretly and brought out in all sixty slaves with her, including all her own family, besides aiding many more in other ways to escape. Her tales of adventure are extraordinary. I have known her for some time and mentioned her in speeches once or twice—the slaves call her Moses. She has had a reward of twelve thousand dollars offered for her in Maryland and will probably be burned alive whenever she is caught, which she probably will be, first or last, as she is going again. She is jet black and cannot read or write, only talk besides acting."[9]

Since February Brown had been preparing his financial backers, or the "secret six" as they were soon to be called, to begin collecting the money promised after the Chatham Convention, but wrote Kagi back in Ohio that the fund-raising "was a delicate and very difficult matter," further complicated by the departure of Theodore Parker. Ill with consumption and no longer able to take New England's winters, he had gone to Cuba in December 1858 accompanied by Dr. Howe. After two months he would sail for Rome and, at age fifty, die in Florence in 1860. Parker's interest in Brown's crusade would continue however, as he wrote Sanborn from Rome: "Tell me how our little speculation in wool goes on, and what dividend accrues there from."[10] But he had to drop all active participation and could contribute nothing financially. Howe, for his part, began to take a more nuanced stance regarding slavery after stopping off on his way home from Cuba at Wade Hampton's plantation in South Carolina. His interest in Brown's sanguinary outlook was diminished after experiencing the gracious hospitality of his host; Howe himself explaining he felt he had to relinquish his stake in the enterprise after envisioning its consequences for Hampton and his family and others of their ilk.

The amount finally raised for Brown came to something over $2,000; Stearns came in with just over $1,200, with $750 coming from Smith, while Higginson handed in a mere $20, with similar amounts coming from others. All in all Brown probably marked up his visit to Boston a success. Sauntering down Court Street oblivious to the glances of passersby, Brown nonchalantly peeled an apple with his jackknife and talked with his companion "Jerry" Anderson. The young man conveyed his excitement at hearing a noted antislavery orator (Phillips), whom he thought had rendered a particularly apt phrase. Temperamentally underimpressed, as the elder man heard it, he replied, "I suppose I ought to say as the boys say—O shit!"[11]

Leaving his supporters in Boston "much in the dark concerning his destination and designs for the coming months," Brown next appeared at six in the evening on June 3 at the shop of Charles Blair in Collinsville, Connecticut. Reminding the blacksmith who he was, Brown said he had come to fulfill the contract they had made in the previous year. "Mr. Brown," said Blair, "the contract I considered forfeited, and I have business of a different kind now. I do not see how I can do it."

"Well, I want to make you perfectly good in the matter. I do not want you to lose a cent," declared Brown.

"Why not take the steel and handles just as they are?" asked Blair.

"No, they are not good for anything as they are."

"What good can they be when they are finished? Kansas matters are all settled."

"They might be of some use if they are finished up. I might be able to dispose of them."

Blair agreed, and when paid the remaining $450 he said he would "find a man in the vicinity to do the work."[12] The next day Brown came again, handing Blair $50 and a $100 check; he mailed the balance in a check on June 7 from Troy, New York.

On June 9 Brown was in North Elba for a week for a last reunion with his wife and young daughters. Then, traveling across New York he arrived on June 18 in West Andover, Ohio, for a last consultation with John Jr. As the senior Brown would go about setting up his "southern headquarters," his eldest son would become his indispensable liaison, shipping arms, collecting money, maintaining contacts with supporters, and gathering and forwarding recruits. Setting out from Akron with two of his sons, Owen and Oliver, and Jeremiah Anderson, Brown notified Kagi that he was on the way to the Ohio River.

Harper's Ferry in 1859 was a bustling center of commerce and industry, with a population of three thousand persons. Two rail lines ran through it; the Baltimore and Ohio, connecting Baltimore and Washington to Ohio west of the mountain ranges, and the Winchester and Potomac, running some thirty miles into Virginia's Shenandoah Valley. The railroads converged in a Y-junction near the confluence of the rivers, crossing the Potomac River to Maryland on a single-truss iron suspension bridge. This bridge was covered and served other traffic, while a second covered bridge forded the Shenandoah a quarter of a mile upstream for wagon traffic. The town occupies a narrow spit of land at the convergence of the rivers, the Potomac flowing out of the west and the Shenandoah flowing south to north out of the valley named for it. Settled beneath two towering ridges called Maryland and Loudoun Heights, Harper's Ferry rises beyond the point onto a great hill known as Bolivar Heights. Three main streets traverse the town; Shenandoah and Potomac streets parallel the respective rivers, and High Street runs up to Bolivar Heights.

Travelers passing through the gorge have long noted its scenic beauty and spoken of the surrounding heights as possessed of imposing grandeur. In the early years of the Republic, Eli Whitney, together with another Connecticut Yankee, had come there to establish a rifle manufactory situated on the Shenandoah River called Hall's Rifle Works. Developing the production of arms on the principle of interchangeable parts, Whitney became wealthy. The United States Congress, too, at the urging of the first president had established a national arms manufactory and arsenal here, which by the 1850s was a substantial works. Divided by a single street, two orderly rows of multistoried buildings ran for two thousand feet along the Potomac; with stocking and machine shops, smith and forging shops, annealing and carpentry shops, mills for sawing and grinding, and warehouses and storehouses. Fronting the government grounds was an open area called the Ferry Lot, across which on the point were some commercial establishments, the train depot, and a hotel called the Wager House. Beside these, near the Shenandoah River, with the Winchester and Harper's Ferry Railroad Depot behind it, was the enclosure of Arsenal Square, with the superintendent's office and large and small arsenal buildings. These buildings housed at any given time between one and two hundred thousand stand of arms.

Brown and his companions would have traveled by rail through Pittsburgh to Harrisburg, before traveling on to Chambersburg, where lodgings were taken for the night by "Isaac Smith & Sons." The next day they reached Hagerstown, Maryland, where they again took lodgings. At eight in the morning on July 3 the quartet arrived at Harper's Ferry. Surveying the town and the imposing view of the surrounding country, Brown and party would have passed over the Potomac Bridge to Maryland to begin inquiring about land for sale or rent. Before they had gone far, they encountered a man who owned a farm in Maryland, and they fell into conversation where he suggested that "Mr. Smith" could see about renting a farm owned by the heirs of a Dr. Kennedy, three miles to the north of Harper's Ferry and

The Kennedy farmhouse sketched in 1859.

a mile or two east the Potomac River on the Boonesborough Pike. Three hundred yards from the road, they would see a farmhouse, he said, and on the other side a cabin. Within a day or two of learning of the place, Brown rented the buildings and land for $35, until March 1, 1860, or for nine months.

Once this preparatory work was completed, Oliver was sent back to North Elba to bring back his sixteen-year-old sister, Anne, and his seventeen-year-old wife, Martha, who was then pregnant with their child. The young women would provide the farm with a setting of normality to outside curiosity, also providing sustenance and comforts for the men as they arrived. Notified of these developments, John Jr. also received $100 from his father to cover traveling expenses and was instructed to "hold back Whipple & Co" till the quarters were ready. John Brown's agent was also told to "be in readiness to make the journey through the country northward," meaning to New York and to Canada. Meanwhile Kagi, who would use the name "John Henrie," had arrived in Chambersburg, fifty miles north of Harper's Ferry, with instructions to take lodgings at an inn on East King Street, of which Mrs. Mary Ritner was the proprietor. He was also instructed to make the acquaintance of Mr. Henry Watson "and his reliable friends," the Underground operatives in that Pennsylvania town, but that he was in no way to appear "fast" with them.

As Brown completed these arrangements he could at last turn his attention to the formidable task at hand. Of necessity in the spring his primary focus was on the funds that had been promised him, without which he would not be able to do anything. Now the men from his

Kansas Company and those recruited along the way could be contacted by mail, and if they needed assistance, money could be forwarded.
But Brown had been out of touch with the crucial Canadian sector;
moreover, some of the principals in Chatham who had attended the
convention in the previous year were no longer on the scene. Now
John Jr. was poised to begin an important journey to renew and vital-
ize these contacts, and would look to the assistance of Harriet
Tubman and others. Meanwhile on July 22 Brown reported to his
wife that "Oliver, Martha & Anne all got in safe," and now the men
could begin moving into the Kennedy farm.[13] John Jr. shipped fifteen
crates to Chambersburg from Ohio via Pittsburgh and Harrisburg, as
Kagi sent out orders from Chambersburg to various quarters for the
men to report to duty.

The Kennedy farm was an excellent location in every respect. Kagi
received and dispatched freight and men as they arrived, Oliver and
Owen Brown received anything and anyone sent down, while to
maintain communications, John Brown often traveled between the
two locations, seeing "that matters might be arranged in due season."
September 1 had originally been designated as the date to begin operations; but the time-consuming and cumbersome arrangements and
the slowness with which some of the men were being mustered, and
the crucial work that remained to be done in Canada, soon led to the
realization that more time would be required. On July 27 Brown sent
notice to North Elba that his son Watson and one of the Thompson
boys, Dauphin, could delay their departure "to allow more time for
haying."[14]

By early August the number of those taking residence at the
Kennedy farm had grown considerably. Besides the leader and the
two teenage girls, there were three of Brown's sons, Owen, Watson,
and Oliver, William Thompson, two Quaker brothers from
Springdale, Edwin and Barclay Coppoc, in addition to Jeremiah
Anderson and Charles Plummer Tidd, who were soon to be joined by
Albert Hazlett, William Leeman, and Stewart Taylor. Aaron Stevens
had taken up station in Hagerstown to assist in the forwarding of
men and in communications between Chambersburg and "headquarters." With matters progressing, Brown was distressed by the reluctance of some of the men to come forward. George Gill was one of
these, prompting Brown to write Kagi, "I hope George G. will so far
redeem himself as to try; & do his duty after all. I shall rejoice over
'one that repenteth.'"[15]

In his diary John Brown recorded making eight trips between the Kennedy farm and Chambersburg; riding either on a mule or in a wagon, sometimes taking overnight lodging at the Union Hotel in Greencastle. These trips were accomplished mostly at night, and while in the Pennsylvania town he would stay for several days. It was in early August during one of these journeys that Brown revealed to his son Owen, who rode with him in the wagon, what only Stevens and Kagi knew—that the initial objective of the company would be the seizure of the government works at Harper's Ferry. Having frequented the town, Owen recoiled, easily apprehending the dangers, and would later recall he told his father, "You are walking straight into the arms of the enemies as Napoleon did when he entered Moscow." When they returned to the farm and a meeting was held to acquaint the rest of the company with the new plan, there was immediate dissension. The younger Browns, Oliver and Watson, and Charles Plummer Tidd, in particular, were opposed to it; bluntly put, it appeared to them they would be committing suicide. Brown answered that even in the event of their death it would be a gain: "We have only one life to live and once to die, and if we lose our lives, it will perhaps do more for the cause than any other way." The quarrel was so great it threatened to break up the camp. Finally to calm the dissension, Brown relented, saying if that was the way they felt he would resign as commander and follow another proposal if they had one. Tidd was so upset he went to stay for a few days with Cook at Harper's Ferry. However since they were an oath-bound company, they all finally agreed to follow the elder man's leadership on the stipulation that both bridges in Harper's Ferry be fired. Owen drafted the following note acknowledging their consent: "Dear Sir—We all agree to sustain your decisions until you have proved incompetent, and many of us will adhere to your decisions so long as you will."[16]

Toward the end of June, as it happened, Harriet Tubman would again be in Boston, and on July 4 was introduced to a gathering in Framingham of the Massachusetts Anti-Slavery Society by its just elected president for the coming year, Thomas Wentworth Higginson. As reported, the president presented "to the audience a conductor on the Underground Railroad, who, having first transformed herself from a chattel into a human being, had since transformed sixty other chattels into other human beings, by her own personal efforts." Speaking briefly "in a style of quaint simplicity," she aroused such interest in her hearers that later, in his *Cheerful Yesterdays*, Higginson was to write: "On the anti-slavery platform where I was reared, I cannot remember one real poor speaker; as Emerson said, 'eloquence was dog-cheap there. . . .' I know that my own teachers were the slave women who came shyly before the audience. . . . We learned to speak because their presence made silence impossible."

On August 1 a meeting of the New England Colored Citizens convened in Boston and was addressed by the Rev. J. W. Loguen and, according to an account in the *Liberator,* by one "Harriet Garrison." The temporary bequeathal of the surname of its editor, no doubt, signaled her acceptance into the antislavery family, but was also a precaution as Tubman was introduced as "one of the most successful conductors on the Underground Railroad." She made a short speech denouncing the colonization movement by telling the story of a man who sowed onions and garlic on his land, thereby hoping to increase its dairy production. When he found his butter too strong and unable to bring a price, he concluded, she said, to sow clover instead. But the wind had blown the onions and garlic all over the field. "Just so, she stated, the white people had got the Negroes here to do their drudgery, and now they were trying to root them out and ship them to Africa. 'But,' said she, 'they can't do it, we're rooted here, and they can't pull us up.' She was much applauded."

Shortly after this Tubman dropped out of sight and her whereabouts were unknown for many weeks. Given to bouts of illness brought on by years of exposure, and suffering from frequent spells of unconsciousness, caused by being hit in her head with a heavy object by an overseer as a girl, she had gone to convalesce at the home of a friend in New Bedford. Although few have looked beyond the simple assertion that there was a profound bond between Brown and Tubman, there has never been any doubt about her determination to join his campaign in the Virginia mountains. In his *John Brown,* DuBois writes of Tubman as having "wild, half-mystic ways with dreams, rhapsodies and trances." She laid great stress on her nocturnal visitations, and just before meeting Brown in Canada had a recurring dream where she saw the head of a snake rising among rocks and bushes in a rugged country. As it rose it became the head of an old man with a long white beard gazing at her "wishful like" as she related in Sarah Bradford's *Harriet, the Moses of Her People,* looking at her "jes as if he war gwine to speak to me." Then two younger heads rose up beside it, and as she stood wondering at the appearance a crowd of men rushed up to beat down the younger heads, and then the head of the old man who had continued looking at her. As the dream repeated over several nights, she could not guess at its meaning until Brown appeared in St Catharines, and it was only when the news of the raid at Harper's Ferry broke that she knew that the two other heads she had seen were those of two of his sons, Oliver and Watson.

After arranging for the arms shipment to "headquarters," John Jr. awaited the order, contingent upon the state of readiness at the farm, to begin his "Northern tour." The key to the full development of his Virginia campaign now lay, in Brown's opinion, with the Canadian recruits, and with Douglass and Tubman. Accordingly the important work undertaken by his son, in its scope and bearing on the plan, and on its sequel, is indispensable to an understanding of that action and its outcome. Shortly after the first week of August John Jr. arrived in Rochester at the home of Frederick Douglass. Finding Douglass away in Niagara Falls, he awaited his return. The next day Douglass did so, and John Jr. informed Kagi in a letter dated August 11: "I spent remainder of day and evening with him and Mr. E. Morton, with whom friend Isaac is acquainted." Morton was then on an extended visit with Douglass, and was undoubtedly engaged in more than social obligations. "He was much pleased to hear from you;" the letter continued;"was anxious for a copy of that letter of instructions to show your friend at 'Pr.,' who, Mr. M. says, has his whole soul absorbed in this matter."[17]

Said instructions, in part, were how to reach Brown by mail, namely by addressing an envelope to Mrs. Ritner in Chambersburg, Pennsylvania, and placing another inside marked "H. K." Referring broadly to his discussion with Douglass, John Jr. wrote: "The friend at Rochester will set out to make you a visit in a few days. He will be accompanied by that 'other young man,' and also, if it can be brought around, by the woman that the Syracuse friend could tell me of. The son will probably remain back for a while." The fact that Douglass's son Lewis was being considered for active service with Brown shows a level of commitment on his part beyond what is usually acknowledged. Douglass had also volunteered that "'the woman' . . . whose services might prove invaluable, had better be helped on." John Jr. added: "If alive and well, you will see him ere long. I found him in rather low spirits; left him in high."[18]

On August 11 John Jr. reported to Kagi from Syracuse, where he had gone to meet with Loguen and Tubman—that he and "also said woman" had gone to Boston. Informed by Loguen's wife that he was expected to visit Canada soon, and "would contrive to go immediately," Brown resolved to go on to Boston. In his letter to Kagi he added, "Morton says our particular friend Mr. Sanborn, in that city, is especially anxious to hear from you; has his heart and hand engaged in the cause. Shall try and find him. . . . I leave this evening on the 11:35 train from here; shall return as soon as possible to make my visit at Chatham."[19]

The next morning, a Friday, arriving in Boston, John Jr. soon found the Reverend Loguen, who informed him that Tubman was not known to be in the city, nor was anything known of her present whereabouts. As for going to Canada, his engagements were such that he could not possibly leave till the end of the following week. It was not until the next Tuesday that the younger Brown posted a letter to Kagi informing him of this, and that he had decided in the meanwhile to improve the time by "making acquaintance of those staunch friends of our friend Isaac."[20]

First he called upon Dr. Howe, who although he had no letter of introduction, received him cordially. In his letter John Jr. wrote: "[Howe] gave me a letter to the friend who does business on Mills Street. Went with him to his home in Medford, and took dinner. The last word he said to me was, 'Tell friend Isaac that we have the fullest confidence in his endeavor, whatever may be the result.' I have met no man in whom I think more implicit reliance may be placed. He views matters from the standpoint of reason and principle, and I think his firmness is unshakable."[21]

One of the strongest individuals John Jr. met with, and who was to make a substantial contribution was Lewis Hayden, the head of Boston's Vigilance Committee and Underground operations. Hayden had seen all of his family separated and sold, and twice stood on the auction block, while his mother descended into madness. At age thirty-three, with his wife and son, he fled from Kentucky to Canada. Moving thence to Detroit, he became a prominent member of that city's growing black community, building a church and a school; in the early 1850s he moved to Boston where he established a clothing store and played an outstanding role in Beacon Hill's African American community. He would be instrumental in locating the elusive Harriet Tubman, and also raising money and recruits to be sent to the elder Brown in Virginia.

Traveling to Concord during this interval, John Jr. found that Sanborn was away on a brief trip to Springfield. He wrote: "The others here will, however, communicate with him. They are all, in short, very much gratified and have had their faith and hopes strengthened. Found a number of earnest and warm friends, whose sympathies and theories do not exactly harmonize; but in spite of them selves their hearts will lead their heads."[22] On his return, Sanborn too was to undertake a new round of fund-raising, guaranteeing that an additional $300 would reach Brown via Kagi by the end of the month.

On August 20 Brown was back in Chambersburg to arrange for the transfer of the crated arms that had arrived from Ohio, and just as importantly to keep his appointment with Frederick Douglass. Joining Kagi at the boardinghouse of Mrs. Mary Ritner on East King Street, he was in touch with Henry Watson as well, confiding to him details of his expected meeting, information likely disclosed to Joseph Winters, a known confidant of Brown and a leader of the Underground Railroad operations in Chambersburg and prominent in its black community. The old stone quarry selected for the rendezvous was situated on the south side of the town near a bend in Conococheague Creek before it meanders northward, only a short walk from central Chambersburg. At the scheduled hour no doubt Brown and Kagi were close enough to hear the train as it pulled into the depot from the east, among whose passengers would be the famed abolitionist and his companion, a vigorous young man with sharply drawn African features.

While it was not entirely unexpected that Douglass should come to south-central Pennsylvania, it was unusual that he would appear unannounced, and before he could clear the platform, those who recognized him expressed surprise. Given a cordial welcome, he resolved the propriety of his being there by saying he was on personal business, and when offered the invitation to speak at a public meeting to be held in Chambersburg that evening, agreed to make an appearance. The duo then made their way to the barbershop of Henry Watson, who, though busily engaged, pointed out the way where Brown and Kagi waited. Although far from being the only antislavery action Douglass was prepared to embrace, in his view Brown's movement offered a more "zealous and laborious self-sacrificing spirit" lacking in other quarters. Not only was he remarkable in his stewardship of men; he also had heart enough to set his plan in motion while seeking to establish cooperation among abolitionists. In this spirit he was prepared to answer John Brown's summons, and to do what he could in assuring his success. But despite his deep respect for the project, Douglass was never disposed, as some were, to subordinate himself completely to it. He appears to have merely been prepared to offer support, as he said, "in an individual way"; and while his commitment was considerable, he was also planning a trip for the fall and winter of 1859 to the British Isles and France, and was contracting for a number of lectures in Philadelphia, Boston, and Rhode Island, before his departure, a departure only to be hastened by what was to transpire. Brown however, it must be stipulated, evidenced a great deal of confidence in and respect for Douglass due to his past experiences with him that could only have served as the basis for summoning him to Chambersburg.

In his *Life and Times*, Douglass wrote:

> I approached the old quarry very cautiously, for John Brown was generally well armed, and regarded strangers with suspicion. . . . As I came near . . . he . . . soon recognized me, and received me cordially. He had in his hand when I met him a fishing-tackle, with which he had apparently been fishing in a stream hard by, but I saw no fish, and did not suppose that he cared much for his "fisherman's luck.". . . He looked every way like a man of the neighborhood, and was as much at home as any of the farmers around there. His hat was old and storm-beaten, and his clothing was about the color of the stone-quarry itself. . . . His face wore an anxious expression, and he was much worn by thought and exposure. I felt I was on a dangerous mission, and was as little desirous of discovery as himself.

After cordial greetings, propriety perhaps dictated that Douglass hand Brown a letter of which he was the recipient. He and Green had stopped the previous night in Brooklyn where they had been the guests of the Gloucesters. Elizabeth Gloucester had written expressing her best wishes for Brown's cause and for his welfare, enclosing a ten-dollar contribution. The four men—John Brown, John Henry Kagi, Frederick Douglass, and Shields Green—now settled down among the remains and castaways of the old quarry with the wagon loaded with Brown's "mining tools" hard by, to discuss the enterprise soon to be undertaken.

Even as only the barest hint of the exchange may be recoverable, a careful reading of Douglass's autobiographical and other writings, and those of his contemporaries, may still bring out what must be considered key. While many of the elements can only be the object of conjecture, it is known Douglass was conversant with an earlier version of Brown's plan, as he was aware he had something in mind concerning Harper's Ferry. Soon, however, he would be surprised to hear that Brown had renounced this old plan in favor of a bold new strategy that called for the seizure of the United States arsenal at the outset. The two principals would spend many hours of two days in deliberation, with Green and Kagi remaining, on Douglass's telling in his autobiography, "for the most part silent listeners." The bond between the two clearly was an affiliation more than friendship, and they argued from the fullness of their hearts, as of their minds—surely one of the most significant encounters known to the American scene.

The idea behind the selection of Harper's Ferry for an opening salvo was twofold. Foremost was to give southern "firebrands" the pretext they sought to carry out their threat of secession; this "far more than any other" was the idea behind it, Salmon Brown was to relate (in Oswald Garrison Villard's *John Brown: Fifty Years After*). "All writers," Brown's son Salmon said, had "failed heretofore to bring out this far-reaching idea to the extent it merits."[23] Once a complete schism had taken place, the predicate was, the North would be compelled to whip the South back into the Union without slavery. And as its corollary, it was Brown's intention to anticipate this rebellion by using Harper's Ferry as "a trumpet to rally the slaves to his standard," as Douglass phrased it, as a signal to be heard by slaves across the South that "friends" in the North were prepared to intervene on their behalf and inspire them to rally. At one blow, the political platforms upon which the country tottered would come crashing down and a war begun for the liberation of the slaves.

Brown knew, as few others did, how slavery had come to threaten the American republic and the very basis of its liberty, and he was certain that the coming war would be of such scope, bringing such forces and hitherto unimagined destructive powers to battlefields, as to redden rivers with blood. This goes a long way to explaining the intractability of what he was proposing. He had "unsheathed his sword," as he said, to forestall for the country the necessity of going through the war he had seen prefigured in the events of the day—a war issuing on the requirements of the collisions of great armies, over great distances. He was aware of changes in arms production, the extent and use of rail power, of steam, of the telegraph, and how they transformed war. And he was aware of America's far-flung geography; of its rivers and mountains, its towns and cities, and its populations. The war he was proposing was the inverse of that which was to be fought—his was a revolutionary war.

As well as anyone Douglass understood that the sectional contest was ultimately one for dominion. But for over a decade he had argued against abolitionists advocating the dissolution of the Union as a way to disenthrall the republic from slavery. He took the position espoused by Gerrit Smith and others—that the Congress had the right to legislate on slavery, and, as John Quincy Adams had argued, by its war powers under the Constitution could abolish it. Certainly Douglass put considerable effort along these lines in trying to dissuade Brown, believing, contrary to his expectation, his gambit could only lead to the opposite state of affairs than he sought. It was bound to misfire, said Douglass, and would sacrifice the lives of everyone engaged in it. The battle for freedom must be fought within the Union, he argued; it would only be the power of the federal government after all, supported by its armed forces that could deal with the issue on the scale proposed. Beginning on such a rousing blow would not help their cause, and it would do nothing to help the slave. Brown would be discredited and the South given the excuse, not for secession, but for the complete suppression of all antislavery opinion throughout the northern states. Assuredly too, he suggested, in attempting to mount such a scheme, however unlikely its success, an indiscriminate slaughter of the male portion of the slaves would commence.

Clearly Brown felt none of Douglass's scruples, believing a resounding blow just the thing to wake the northern people out of their torpor on the subject of slavery. He expected, as Richard Realf would later relate before the Mason Committee, convened in 1860 to investigate the Harper's Ferry raid, "that all the free negroes in the Northern states would immediately flock to his standard";[24] and as Brown called on the aroused strength of this black nation, it would create among the onlooking whites the sympathy they would need to gain their support. Any slaughter of the kind Douglass feared, once begun, Brown likely held, would cure itself. In the sight of the American people and of the world, the slaveholders could proceed only to a limited extent in this shameless conduct. In Brown's view it was too late to destroy and erase the South's peculiar institution by means less harmful than the evil itself. The slaveholders not only lived south, to the exclusion of all others they were "the South" itself; they were the only active power there. They could not be "talked down," and would not consent to be hemmed in by political means alone. No one in the country better understood the nature of the business they were engaged in than these men. They were broadcasting it in the powder they were buying and in the arms they were stealing. War was at that very moment being contemplated in the cabinet of President Buchanan. Out of his success at Harper's Ferry, Brown was convinced, would come freedom for the enslaved; and since slavery could only end in blood, there could be no better time to end it than now.

Douglass wrote: "He described the place as to its means of defense, and how impossible it would be to dislodge him if once in possession. Of course I was no match for him in such matters, but I told him, and these were my words, that all his arguments, and all his descriptions of the place, convinced me that he was going into a perfect steel trap, and once in he would never get out alive. . . . He was not to be shaken by anything I could say, but treated my views respectfully, replying that even if surrounded he would find means for cutting his way out."[25]

Brown's plan called for a party to cross the Potomac Bridge at night, approaching from the west on the Maryland side of the river along the Chesapeake and Ohio Canal. The telegraph wire would be cut and the bridge seized, detaining the watchman on his regular rounds. The telegraph wire on the Virginia side would also be cut. Crossing the Ferry Lot to the armory gate, the watchman would be called out and compelled to open it, or it would be broken open. The two buildings fronting the government works overlooking the Ferry Lot were to be seized—the enginehouse and the paymaster's office. On Shenandoah Street, across the Lot, the arsenal would be invested—the watchman there also detained—and behind it the bridge to Loudoun Heights. Finally Hall's Rifle Works would be seized, and its watchman, along with the arsenal watchman escorted to the enginehouse in the armory. These watchmen were the only guards.

All the gaslights in the lower part of town would be extinguished, and to assure unimpeded access and communication among the points held sentries and a roving picket would be established. Trains arriving during the night would be stopped and prevented from passing, and the tracks torn up on each side of the river to ensure it.

Incendiary devices would be placed on the bridges so that they could be fired if that became advisable. By beginning at night the raiders would minimize the possibility of encountering many people, and the town would quickly be secured before alarm could be given. As people awakened and the day began, unaware of what was happening, they would find the government works and their streets under control of armed men, who would have, as well, a monopoly on all the arms in the neighborhood. They would be dissuaded from opposing Brown and hold to their houses, or else flee the town. Those who did not, and who sought to interfere, would, for their own safety, be held prisoner for a time. At the outset "without the snapping of a gun," said Brown, they would have possession of the government property and the rifle factory, and the points of entrance and exit from the town.

When the Ferry was secure and before daybreak a second phase of the plan would have begun. A party was to proceed up Bolivar Heights where selected slaveholding estates were to be visited. Slaves would be freed and brought back to the Ferry along with their masters, together with other requisitioned property, horses, and wagons. A similar process would be executed in Maryland, seizing selected slaveholders and their property and bringing them to the Ferry to be held with the others. These prominent slaveholding citizens would be the very persons, said Brown, who could be expected to lead any resistance, and those who wanted to oppose him would be restrained so as not to harm them. In the worst case, too, he would be able to dictate the terms of his withdrawal by the influence and prominence of these men. His intention, moreover, was not to harm them, but to make a signal example of them by arranging with their "friends" for their exchange for able-bodied male slaves.

The dangers posed by Brown's radical inversion of priorities were palpable, and Douglass clearly recoiled at the suggestion of so daring an act. One of the difficulties he foresaw was presented by the terrain itself. Constricted by the natural setting and by the limited points of entrance and of egress and by the relatively large population, Brown, however securely in control, faced the obvious risk of being bottled up. One band of their party could become isolated from another, and if any casualties were sustained it could be catastrophic to them. In addition, there were any number of towns within a very short radius and the country was well settled. The various militia companies maintained in every community could be assembled against them while they were still at the Ferry. Then, too, Harper's Ferry was serviced by two rail lines, one of them connecting to a major city, Baltimore—scarcely an hour away—and beyond that the nation's capital. On the first news of uprising, troops would be dispatched to deal with them. If this happened, Douglass predicted, those opposing them in the neighborhood would become an enraged mob, and attempt to seize them, rather than let them escape, and lynch law would ensue. Brown was thrusting himself and his men, said Douglass, into dangers, and worse, that may be easy to get into, but impossible to get out of. He was walking into a perfect cul-de-sac.

Indeed, some observers have cited Harper's Ferry as evidence Brown lost his bearings in an "overlong contemplation" of the issue, and that it was an illustration of his military ineptitude. But these reservations fail to appreciate Brown's true resources. For his part, Brown thought the militia companies would not be eager to converge on Harper's Ferry until they had looked to their own neighborhoods. News that men were in possession of the armory, and were freeing and arming slaves, would alarm and disconcert them. The militia companies weren't a serious military threat either; they were both poorly armed and poorly trained. Even if they did manage to converge on the Ferry they would only come with outdated muskets, and would quickly be dissuaded by the disproportion of force. Neither did Brown think federal troops would immediately be a factor. The governor of Virginia would have to call upon the president for them, and would first want to see how the militia fared against Brown. In any case, since he didn't intend to remain at Harper's Ferry, they would be delayed, as all means of communication would be severed, the rail lines broken, and the bridges burned. But even in the event they were surrounded, Brown assured Douglass he would have ample means to cut his way out.

Douglass places his various reminisces within a curious template, which serves to obscure rather than reveal the true extent of his commitment prior to his last meeting with John Brown. Of his first meeting with Brown, Douglass said much later, "From 8 o'clock in the evening till 3 in the morning, Capt. Brown and I sat face to face, he arguing in favor of his plan, and I finding all the objections I could against it." And of the meeting in the stone quarry he wrote, "Our talk was long and earnest; we spent the most of Saturday and a part of Sunday in this debate . . . he for striking a blow which should instantly rouse the country, and I for the policy of gradually and unaccountably drawing off the slaves to the mountains, as at first suggested and proposed by him." Of that earlier meeting, too, Douglass was to recall Brown saying he "had been watching and waiting" for the heads of collaborators such as he would need from among his race to "pop-up," as it were, above the surface of the water.[26] In August 1859, at age forty-two, Frederick Douglass was a man of considerable bearing and among the most capable of men; at the height of his pre-Civil War fame, he was a veritable Spartacus of his people as evidenced in his untiring oratorical combat and commitment to all things pertaining to antislavery agitation. Brown had summoned him not to elicit his further understanding prior to beginning "his work," but because he wanted Douglass to join him in it as co-leader. This is the aspect of the drama passed over without remark by Douglass, hidden beneath the wavering conceit of a rhetorical device. Brown himself was possessed of complete composure and confidence, and this not just as is often supposed because he had internally wrestled with his God, but more especially because he was thoroughly grounded in and responded to the American condition in relation to its racial division, of which he was profoundly aware. He wanted Douglass to appear with him at Harper's Ferry; his success, Brown now believed was contingent upon it. If Douglass be for him, who could be against him?

The foregoing sketch has been culled from a careful parsing of sources widely available. But the parameters of their encounter can be sketched more broadly still as suggested by three trenchant contemporaneous presentations: Wendell Phillips's speech "The Argument for Disunion," an article by Karl Marx entitled "The American Civil War," and Ralph Waldo Emerson's essay "American Civilization."

The strength of the so-called slave power was comprised of three elements or instrumentalities, said Phillips. The first came in the omnipotence of money: with two thousand million dollars invested in slaves, the South accrued the ability to draw into its reach the sympathy of all other large capital. A second strand making this power came in the "three-fifths" clause in the Constitution, allowing three or four large planters of South Carolina "riding leisurely to the polls, and throwing in their visiting cards for ballots," to "blot out the entire influence of [a] New England town in the Federal government." A third strand came in "the potent and baleful prejudice of color."

With no more than fifty thousand of the three hundred thousand slaveholders set up as planters, the ruling element of the South lived as an irresistible aristocracy in a population of five million whites and four million black slaves. To these were added another two thousand state officials, making a complete proslavery party that had gained ascendancy in the administrations of Pierce and Buchanan, exercising the immense powers of the government on behalf of slavery.

As a geographical entity its hegemony extended from the mid-Atlantic states down the seaboard through the Old South, pushing westward through the tier of states known as the Deep South; touching thence on the Gulf of Mexico, it ranged up the Mississippi River Valley, claiming all land to the northern borders of Kentucky and Missouri; extending thereafter, on the western bank of the Mississippi, into the rich cotton country in Arkansas, and claiming the river valleys of eastern Texas. The South drew in fully fifteen states; with an extensive shoreline, it encompassed the broad middle region called the border states, as well as a part of the Southwest. Each of these regions was as different from the other as they were from any of the northern states—in geography and latitude, in the admixture of the emigrant European stock, in resources and the various pursuits of industry. What gave this expanse its cohesion was the recognition by law and custom of the ownership of black-skinned human chattel.

"The South . . . is neither a territory strictly detached from the North geographically, nor a moral unity," Marx contended. "It is not a country at all, but a battle slogan." Citing the "decisive" importance of a strictly "numerical proportion" of black to white in the temperamental disposition of the South, in an article for *Die Presse*, November 7, 1861, "The Civil War in the United States," Marx elaborated: "The soul of the whole secession movement is South Carolina. It has 402,541 slaves and 301,271 free men. Mississippi . . . comes second. It has 436,696 slaves and 354,699 free men. Alabama comes third, with 435,132 slaves and 529,164 free men."

As Brown was continually reiterating, nowhere is the South more vulnerable than in its fear of servile insurrection. A successful issue at Harper's Ferry would recoil throughout the region, and the slaveholders would imagine the whole of the North and all of their slaves were down upon them pell-mell. "If I could conquer Virginia," Brown was to say, "the balance of the southern states would nearly conquer themselves, there being such a large number of slaves in them."[27]

Virginia held the largest numbers of this captive labor. Ranging through two principal areas of the South, starting at the head of the Blue Ridge, a continuous chain of counties extended clear to Virginia's southern border where black predominated over white. Together with the northern counties of North Carolina and the eastern counties of Maryland and its eastern shore—this comprised the first of slavery's great cantons. The other, extending from the Atlantic shore islands through South Carolina and mid-Georgia, taking in a large swath of north-central Florida to its panhandle, thence ranging through mid-Alabama down to the gulf and embracing the whole of the southern Mississippi River Valley, found its western extent in the river valleys of eastern Texas.

Brown and Kagi had devoted considerable effort in plotting out these "black belts" on a county-by-county basis in a series of large maps. Douglass was familiar with them and would have been cognizant of their import; noting such details as census figures, indicating many of the large plantations where masses of slaves were held, as well as marking out the mountain ranges, rivers, and railroad lines and many of the swamps and places of refuge, and also designating some of the places to be attacked as the campaign took hold and developed. This campaign had been projected to extend down the mountainous line in Virginia into North and South Carolina, and westward into Tennessee.

The South's aim really, Emerson contended in his essay "American Civilization," was not the dissolution of the national government, but its reorganization on the basis of slavery. The motive of its political leaders in threatening secession was clear—by withdrawing from the authority of the Union and uniting the seceded states in a Confederacy, they sought to secure the border states and then lay claim to the entire territory of the United States from the old line of the Missouri Compromise to the Pacific Ocean. They had set themselves on this course because their goals were no longer amenable under the Union. Since a large part of their claim was under control of that Union and would have to be wrested from it, this necessitated on their part "a war of conquest." If they did anything other than wage war on these terms they would be relinquishing their capacity to continue with their dominion in the South and defeat the purpose of secession itself. Peaceable secession was impossible, Brown contended—because the divided sentiment of the border states made it so, as well as the insatiable South. Once in possession of New Orleans and Charleston and Richmond, slaveholders would demand St. Louis and Baltimore. If they got these, they would insist on Washington. Once in Washington they would assume the army and navy, and, through these, Philadelphia, New York, and Boston. This contest would force the North, which would fight for the survival of the Union, to adopt on its banner the formal internment of slavery.

Brown's intent was to anticipate this rebellion upon the soil where it originated, calling the slaves into service to march as a liberating army into the South. Sanborn was to remark, "In Kansas his bold policy had succeeded against the pro-slavery administration headed in its military department by Jefferson Davis. Now Brown hoped it might also succeed in the slave states."[28] What John Brown did was "pure pedagogy," Phillips would remark, saying substantively to the northerners: you are going to have to fight the slave power one way or another; this is how to do it!

Many observers have pointed out that the location Brown chose had a relatively sparse population in slaves, and in northern Virginia slavery was comparatively benign, evidencing contacts and behaviors between the races of a more positive relation. It is often said slaves were neither ready nor willing to take the arms and the freedom Brown was offering, with more than a few holding he was entirely deluded on the matter, gaining all of his information, some have supposed, from his reading of abolitionist literature. But Brown did not imagine that men who had previously been peacefully at labor, other than the very brave or more desperate, would join a fight whose outcome was still in doubt. Nor would it have been prudent for him to expect it. To Brown's way of thinking, he was only expecting enough men to man the freight in arms and equipment over the rough terrain, and to begin building the basis for the army he projected. These men would form the basis around which other slaves could later be drawn. With the success of the initial stroke, though, he expected a series of rapidly cascading events that would net him large numbers.

Twenty-five miles below Maryland's border begins a cordillera that broadens and rises reaching heights of three thousand feet and more; within a thirty-mile radius of Harper's Ferry were twenty thousand slaves, five thousand of whom were able-bodied males. Brown's intention was to seize the government's armory and arsenal as a "trumpet call," as he said, to attentive ears— not, however, to wait for them as is usually supposed—but to begin a march within the angle projected by the rivers toward the Blue Ridge Mountains and

the slave-filled counties abutting its eastern slopes. Sweeping down
this line they expected to readily gather up all the able-bodied males
they could draw. This "retreat," over a largely hilly and forested terrain, would bring them across the Rappahannock River below Front
Royal into the mountain wilderness. If troops should follow them,
which is what Brown expected, their best fighters could bring up the
rear, and with the aid of good positions they could cut the chase out
of all who pursued. When they had fallen sufficiently into the mountain fastness, Brown undoubtedly conveyed to Douglass—reiterating
his old plan—they would set about building a system of fortifications
to act as draws to the enemy, and as warrens for the freed slaves, so
situated that if one were carried, another could be fallen back on, rendering them practically unconquerable. These warrens too would
serve to minimize and mitigate many of the unwieldy circumstances
attendant upon mass upheaval, introducing from the outset a controlling organization.

A plan had been devised to divide and subdivide the freed slaves
into units, to operate separately or together, commanded by the men
commissioned pursuant to the meetings at Chatham. Seven men were
to comprise a band under a corporal; two bands would comprise a
section under a sergeant; two sections, a platoon under a lieutenant;
two platoons, a company under a captain; and finally, five companies
would comprise a brigade. Operations on this model were to be
extended down the line of mountainous country, reaching in time into
the Carolinas and extending west into Tennessee. The organization
effected in Canada, and his friends and supporters, Brown suggested,
could establish headquarters in Harrisburg, within easy communication with Philadelphia, New York, and Boston, and with the lower
counties of central Pennsylvania. While the campaign developed, the
corridor to the north whence they had come was to be kept open as
much as possible, both to maintain communications and the possibility of continuing support. In addition, if it became advisable, they
could retreat up this route, sending freed slaves to safety in the North.
And so repeat the experiment in another locality.

The documents produced at the Chatham Convention—the provisional constitution and the declaration of independence—were to be
distributed, and Brown had selected a photograph of himself as commander-in-chief of the Provisional army to be circulated as well. As
operations took hold, he envisioned the white population would peel
off the abutting areas, and they would extend their control over them.

With growing success a free territory would be created within Virginia, and if possible within the jurisdiction of several southern states, where a complete civil government would be instituted for the freed slaves, with schools and churches. An administration which in time would become the basis for coalescing *Africa* as a political entity within and under the United States, as Delany suggested, as the Cherokee Nation of Indians and the Mormons had been recognized. In this new situation the government would be impressed to begin proceedings whereby slavery would be formally interred under the United States.

Douglass must have felt deeply apprehensive, if not overwhelmed, when he and Green left the stone quarry in the late afternoon of August 20. But couldn't he also have felt a little buoyed—something new needed to be tried. Just then Douglass was giving much thought to the state of the antislavery cause, and to the corresponding progress of slavery.

He could well have pondered these, as he gave attention to what he would say that evening in Chambersburg's Franklin Hall. The prospect of a nearing contestation on the very ground of slavery's existence in the United States could have made him somber, and his speech might have taken the form of an exposition of the hour and a denunciation of the Slave Power. His platform after all was in a town that had been true in its devotion to the fugitive, and he could almost speak to slaveholders themselves by his proximity to the Maryland border.

That night, after what he had heard, Douglass undoubtedly cast his thoughts on the war John Brown was about to commence, mindful of his own relation to that war. On that night, as he surveyed the situation, it was evident that the previous twenty years had brought a marked change in the condition of the antislavery lecturer. Once he and his colleagues had been met with mobs with rotten eggs and brickbats, but now the country had been rocked from end to end by the strength of the faithfulness of abolitionists to the freedom of the slave. Yet slaveholders were still coolly estimating the value of their victories and congratulating themselves upon their security. It was impossible to disguise the fact that slavery had made great progress and had riveted itself more firmly in the Southern mind and heart, and the whole moral atmosphere of the South had undergone a decided change for the worse. These were thoughts Douglass penned in his *Monthly* that appeared that August in 1859.

Douglass may not have acknowledged the old man known as Isaac Smith in his coming and going at Franklin Hall that night; Brown may not have thought it prudent. But Brown was there, as he may have come in and departed alone, or sat near and conversed with his confidants. After the meeting, as Douglass took his leave and walked to his lodging at the home of Henry Watson, he could not have been cognizant that away in the night were dozens of names studded throughout the far-flung landscape of the American scene—names that in a few years would be lifted to immortality by a dignity only death could bestow upon them; names that would be pronounced like anthems on the lips of those touched by what was to happen in or near them.

On that night the names of places that could have reached Douglass would have brought no special significance to his ear. He could not have imagined on that night that the Emancipation Proclamation promulgated by a president of the United States named Abraham Lincoln was less than three years away. In the morning he would return to his meeting in the stone quarry while in a few years more he would be called upon to help raise black recruits to fill regiments for the Union army—troops that would march into the South carrying the banner of the Union and singing "John Brown's Body."

Salmon Brown was to remark that his father and elder brother could sit and discuss the issues contiguous to slavery "by the hour." Certainly Brown did the same with his many other confidants, including Frederick Douglass. As was noted, Douglass conflated details of his first meeting with Brown in 1847 and Brown's stay in his home a decade later. He had also observed after that initial meeting his remarks on the lecturing circuit began to take on "a color more and more tinged" by his friend's "strong impressions."[29] But as is suggested it was after Harper's Ferry and Brown's death—in his speeches in the years 1859–62—that traces of the talk passing between them in those August hours can also be found.

Brown, it is well known, was given to a penchant for urging blacks in the strongest terms to a defense of their liberty, as Douglass knew too well the proscriptions facing them in both sections of the nation. Accordingly the following exchange has been culled from Douglass's words, in composition with Brown's own, regarding aspects brought together in John Brown's Harper's Ferry plan.[30]

Brown: The men I have with me, Douglass, stand ready to peril everything at the first opportunity for a fight. Neither I, nor my men, count our lives as anything outside this fight when weighted against the freedom of millions. There is a latent element in the national character, which, if fairly called into action, will sweep anything down in its course. The American people admire courage displayed in defense of liberty, and will catch the flame of sympathy from the sparks of its heroic fire. This trait has been long manifest in the reception of the patriots who have been cast upon this country's shores from the wrecks of European revolutions. Call the servile population of the South to arms, and inspire them to fight a few battles for freedom, and the mere animal instincts and sympathies of this people will do more for them than has been accomplished by a quarter of a century of oratorical philanthropy.

Douglass: The Anglo-Saxon, Teutonic and Celtic races have utterly failed in the magnanimity and philanthropy necessary to promote the rights of another and weaker race than themselves. As a nation America is bound as by a spell of enchantment to slavery. The attempt to reconcile slavery with freedom in this country has dethroned logic and converted statesmanship into stultified imbecility. For the non-slaveholding whites of the South the highest ambition is to be able to own and flog a Negro. They are in the utmost dread of the slave. They furnish the overseers, the drivers, the patrols, the slave hunters and are in their sphere, as completely the tools of the slaveholders, as the slaves themselves. They are such because they can't be otherwise. The whites of the North have no adequate idea of the power of these master spirits of the South; and yet the fact is they are under the same influence. The European races have drunk deep of the poisoned cup of slaveholding malignity, and only after they have been made to experience a little more of the savage barbarism of slavery will they be willing to make war upon it.

Brown: We can never cease to regret that an appeal to the higher and better elements of human nature is in this case, so barren of fitting response. But it is so, and until this people has passed through several generations of humanitarian culture, so it will be. Outside philanthropy never disenthralled any people. Heaven cannot help the Negro but by moving him to help himself. It required a Spartacus to arouse the servile population of Italy and defeat some of the most powerful armies of Rome.

Douglass: The Negro can do much, but he cannot hope to whip two sections of the country at once. I know well the proscription of each on the condition of the Negro in America. I long for the end of my people's bondage, and would give all I possess to witness the great jubilee, but I cannot see that it will come out of this attack. Not even the allowance that the whites of the North will wink at a John Brown movement could induce me to advocate it. I am sick of seeing mere isolated, extemporaneous insurrections, the only result of which is the shooting and hanging of a few brave men who take part in them—and not being willing to take the chances of such an insurrection myself I cannot advise any one else to take part in them. The time is not right for such a war. The political uprising of the North against slavery has only risen to the level of being negatively antislavery; it now opposes the political power of slavery in the national government, and will one day arrest the spread of the system, humble its power and defeat its plans for giving any further guarantee of permanence. All this is desirable, but it still leaves the great work of abolishing slavery to be accomplished in the future.

Brown: If the Republican Party elects its candidate next year he will enter Washington upon his peril, the way a fugitive slave enters the North. But the result will show them merely to be the continuation of the Pierce and Buchanan administrations. They will bend a knee to slavery, and be indebted to the South for their law and their gospel.

Douglass: Your attack will only create a more active resistance by the party of slavery to the cause of freedom and its advocates. What is wanted is an antislavery government first —in harmony with antislavery speech. For this the ballot is needed, and if that is not to be heard. . . then the bullet.

Brown: Whether the slaveholder is in the cotton states, the slave breeding states or the border states—one is as bad as another, whether in South Carolina, Virginia, Missouri or Kentucky. In every state where they hold the reins of government they will take sides openly. They know that if the government is a miserable and contemptible failure—then that government must meet them in the field and put them down, or itself be put down. They are all traitors to the government and the constitution, and are only waiting to spring up by the heat of surrounding treason. When Virginia is a free state, Maryland cannot be a slave state. This conspiracy must stand together or fall together. Strike it at either extreme—either on the head or at the heel, and it dies.

Finally it cannot be wholly conjectural that Brown asked Douglass to prepare and publish a call upon the Negro in both the North and in the South, with words that could not differ appreciably from these.

Brown: There are men out there, Douglass, who only wait to be brought into this war. From Rochester and Auburn, from Syracuse and Ithaca, from Troy and Albany, could be drawn an hundred men. Your own son is one of these. Philadelphia alone has a hundred, as do New York and Boston. In the west—in Buffalo and Cleveland and Chicago—could be added five hundred to the cause. Add to these the hundreds in Canada West and the thousands of the South, and I have no hesitation in saying that ten thousand men might be raised in the next thirty days to march through and through the South. One black regiment alone, in such a war, would be the full equal of two white ones. The very fact of color in this case would be more terrible than powder and balls.

Douglass sums up everything of this meeting with Brown in August 1859 in a succinct paragraph in his speech at Storer College in 1881:

> Capt. Brown summoned me to meet him in an old stone quarry. . . near the town of Chambersburg. . . . . His arms and ammunition were stored in that town and were to be moved on to Harper's Ferry. In company with Shields Green I obeyed the summons, and prompt to the hour we met the dear old man, with Kagi his secretary, at the appointed place. Our meeting was in some sense a council of war. We spent the Saturday and succeeding Sunday in conference on the question, whether the desperate step should then be taken, or the old plan as already described should be carried out. He was for boldly striking Harper's Ferry at once and running the risk of getting into the mountains afterwards. I was for avoiding Harper's Ferry altogether. Shields Green and Mr. Kagi remained silent listeners throughout. It is needless to repeat here what was said, after what has happened. Suffice it, that after all I could say, I saw that my old friend had resolved on his course and that it was idle to parley.

And like those silent listeners history itself has been mute; but this summons to a council of war, the details of which it is needless to repeat, after what has happened, possesses an eloquence of its own.

Another eventful twenty-two years would pass before Douglass was to give the account in the speech cited above that touches upon his last moments with John Brown. It was concerning something, as he tells it, taking place during Brown's stay with him in February 1858. An article appeared in the newspapers at the time in connection with the Sepoy War in India that caught Brown's attention:

> A Scotch missionary and his family were in the hands of the enemy and were to be massacred the next morning. During the night, when they had given up every hope of rescue, suddenly the wife insisted that relief would come. Placing her ear close to the ground she declared she heard the Slogan—the Scotch war song. For long hours in the night no member of the family could hear the advancing music but herself. "Dinna ye hear it? Dinna ye hear it?" she would say, but they could not hear it.
>
> As the morning slowly dawned a Scotch regiment was found encamped indeed about them, and they were saved from the threatened slaughter. This circumstance, coming at such a time, gave Capt. Brown a new word of cheer. He would come to the table fairly illuminated, saying that he had heard the Slogan, and would add, "Dinna ye hear it?" "Dinna ye hear it?" Alas! Like the Scotch missionary I was obliged to say "No."

Through long hours Douglass tried to moderate Brown's course, seeking a return to an undertaking originally proposed by Brown, to which he had pledged further aid. Some have suggested there were sharply bitter words exchanged between the two men. Brown is said by one account to have accused Douglass of cowardice, "of being afraid to face a gun." Although their differences may not have devolved in quite this acrimonious fashion, one can see Brown rising to his feet, exasperated at not being able to win Douglass over. His friend was reluctant just at the moment decisive action was called for. One can hear Brown, as one writer has, bitterly accusing Douglass of becoming "soft,"—he had already begun to develop his paunch—of enjoying overmuch "the limelight" that his position had brought him.

And when Douglass found he had exhausted himself trying to move Brown to renounce his new plan in favor of the old one, he too rose, announcing he could not countenance such an action. Expressing his "astonishment" that Brown "could rest on a reed so weak and broken," that he could think he would be able to guarantee his safety and that of his men by the fact he would retain a number of hostages from among Virginia's citizens, he warned flatly "that Virginia would blow him and his hostages sky-high rather than that he should hold Harper's Ferry an hour."

Feeling he'd fully justified himself, Douglass turned to Green and said: "Now Shields, you have heard our discussion. If in view of it, you do not wish to stay, you have but to say so, and you can go back with me."

As is known, Brown put his arm around Douglass in a most intimate way, and implored, "Come with me, Douglass; I will defend you with my life. . . " There is perhaps no ambiguity in his telling of this moment, when, tears glistening on his cheeks, Brown said poignantly, "I want you for a special purpose. When I strike, the bees shall begin to swarm and I shall want you to help me to hive them." But Douglass only allowed, "Discretion or cowardice had made me proof against the dear old man's eloquence—perhaps it was something of both which determined my course. When about to leave I asked Green what he had decided to do, and was surprised by his coolly saying, in his broken way, 'I b'leve I'll go wid de ole man.' Here we separated—they to go to Harper's Ferry, I to Rochester."[31]

Seven

The Battle of Harper's Ferry

All through the conflict, up and down
Marched Uncle Tom and Old John Brown,
One ghost, one form ideal;
And which was false and which was true,
And which was mightier of the two,
The wisest sibyl never knew,
For both alike were real.
—Oliver Wendell Holmes,
"Two Poems to Harriet Beecher Stowe"

WITH JOHN BROWN JR. AND J. W. LOGUEN ON THEIR WAY TO Ontario by August 18, the prospects for organizing the recruits in Canada began to brighten. In the latter part of August through the first week of September they visited St. Catharines, Hamilton, London, Chatham, Buxton, and Windsor, and on the U.S. side of the border, Detroit, Sandusky, and Cleveland. In each of these locations an auxiliary named the League of Liberty was formed to ensure a steady issue of recruits for Brown's campaign. Thomas Cary became its chairman, its corresponding secretary I. D. Shadd, its secretary James M. Bell, and William Lambert the treasurer. On September 8, back in West Andover, John Jr. sent the following to John Henry Kagi: "I yesterday received yours of September 2, and I not only hasten to reply, but to lay its contents before those who are interested. . . . Through those associations which I formed in Canada, I am able to reach each individual member at the shortest notice by letter. . . . I hope we shall be able to get on in season some of those old miners of whom I wrote you. Shall strain every nerve to accomplish this. . . . You may be assured that what you say to me will reach those who may be benefited thereby, and those who would take stock in the shortest possible time, so don't fail to keep me posted."

It is evident from a letter dated the previous week and posted from Sandusky that John Jr. had been directed to a man living in Buxton with whom his father had become acquainted in 1858. Living only ten miles from Chatham, he was not among those listed as attending Brown's convention, although assuredly he knew of it—his name was William Parker, formerly of Christiana, Pennsylvania. He was about five years younger than Frederick Douglass, and their lives took a parallel course, both escaping from slavery in Maryland—Parker about age seventeen, and Douglass about twenty-one. Both were to make great efforts to throw off the mental shackles bred in them from servitude, becoming leaders "among their people"; and just as the one became a confident of John Brown, so had the other.

Born about 1822, Parker belonged to a planter in Anne Arundel County near Annapolis. By 1839 he was in Baltimore, from whence Parker and his brother set off together on the York Pike to Pennsylvania. The brothers reached Wrightsville on the western bank of the Susquehanna River, and by crossing the bridge to Columbia left "the empire of slavery." Safe haven was found where many fugitives passing that way found it—at the first house at the end of the bridge, that of the lumber merchant and Underground operative William Whipper. For the next twelve years Parker lived and worked in the central Lancaster County area, coming finally to reside in Christiana. As he narrated in "The Freedman's Story," published in the *Atlantic Monthly* in 1866: "I thought of my fellow-servants left behind, bound in the chains of slavery—and I was free! I thought, that, if I had the power, they should soon be as free as I was; and I formed a resolution that I would assist in liberating everyone within my reach, at the risk of my own life, and that I would devise some plan for their liberation."

Blacks in the vicinity of Christiana were in perpetual fear of marauding slave-catchers. Every two or three weeks there would be an incident where doors were broken in and homes entered, "and when refused admission, or when a manly and determined spirit was shown, they would present pistols and knock down men and women indiscriminately."[1] Vowing never to be taken, with six or seven others Parker formed "an organization of mutual protection" in the 1840s. It was in the winter of 1843 that William Lloyd Garrison and Frederick Douglass visited as lecturers and found Parker in the audience.

By 1851, with the bearing of a proven leader, Parker was dubbed "the preacher" and his home became a center for meetings held nearly every Sunday. During the uproar following passage of the Fugitive Slave Law that year, a slaveholder from Maryland named Edward Gorsuch, together with his son and a U.S. marshal and his posse, invaded Parker's home in search of two fugitives. Parker's wife, Eliza, summoning help, blew a horn from a second-floor window. Nearly a score of armed blacks, with some Quakers, soon gathered to confront the government-warranted slave-catchers. A melee ensued in which Gorsuch, hysterically insisting that he must have "his property," was shot and killed by Parker, and the marshal and his posse dispersed.

Fleeing to Canada, the Parkers along with two others were received in Rochester at the home of Frederick Douglass. In *Life and Times* Douglass wrote that the news of the resistance at Christiana reached him simultaneously with the arrival of the men, and they "were thus almost in advance of the lightning, and much in advance of probable pursuit."

Dispatching Julia Griffiths, his indispensable assistant and intellectual coadjutor at his paper, to the landing three miles away on the Genesee River to ascertain if a steamer was leaving for any port in Canada that night, Douglass remained in his home "anxious hours" guarding his "tired, dust-covered, and sleeping guests." It happened that a steamer would embark for Toronto that night, but this did not completely relieve Douglass; between his house and the landing or at the landing itself they might meet with trouble. He wrote:

As patiently as I could, I waited for the shades of night to come on, and then put the men in my "Democrat carriage," and started for the landing on the Genesee. It was an exciting ride, and somewhat speedy withal. We reached the boat at least fifteen minutes before the time of its departure, and that without remark or molestation. But those fifteen minutes seemed much longer than usual. I remained on board till the order to haul in the gangplank was given; I shook hands with my friends, and received from Parker the

revolver that fell from the hand of Gorsuch when he died, presented now as a token of gratitude and a memento of the battle for liberty at Christiana.

With cognizance of the assistance Douglass rendered Brown in
regard to Harriet Tubman and Shields Green, it is likely the two also
discussed the prospect of enlisting Parker; indeed, in light of Brown's
interest in arms, one can presume that he had seen Gorsuch's pistol
and handled it with particular interest. On August 27 John Jr. had
written Kagi as referenced above, revealing his meeting with William
Parker. He wrote:

> At "B—n" I found the man, the leading spirit in that "affair,"
> which you, Henrie, referred to. On Thursday night last, I went
> with him on foot 12 miles; much of the way through new paths,
> and sought out in "the bush" some of the choicest. Had a meeting
> after 1 o'clock at night at his home. He has a wife and 5 children;
> all small, and they are living poorly indeed, "roughing it in the
> bush," but his wife is a heroine, and he will be on hand as soon as
> his family can be provided for. He owes about $30; says that a
> hundred additional would enable him to leave them comfortable
> for a good while. After viewing him in all points which I am capa-
> ble of, I have to say that I think him worth in our market as much
> as two or three hundred average men, and even at this rate I should
> rate him too low. For physical capacity, for practical judgment, for
> courage and moral tone, for energy and force of will, for experi-
> ence that would not only enable him to meet difficulty, but give
> confidence to overcome it, I should have to go a long way to find
> his equal, and in my judgment, [he] would be a cheap acquisition
> at almost any price. I shall individually make a strenuous effort to
> raise the means to send him on.[2]

A few days after Douglass had returned to Rochester, Owen
Brown and Shields Green set out on foot for the Kennedy farm.
Richard Hinton noted, "The section of Pennsylvania over which they
passed was then more dangerous to them than the neighborhood of
Harper's Ferry itself, 'hunting niggers' was a regular occupation at
that date."[3] As it happened, four men intent on catching slaves spotted the two, whereupon the intended prey fled into a wood, while the
slave-catchers returned whence they had come for more help. Coming

to a river, Brown took Green, who could not swim, upon his back, and since they were headed south and not north, they escaped.

Green was the first black to arrive at "headquarters." A second black, Dangerfield Newby, was recruited several days later; he was a freeman from the neighborhood, born about 1825 in Fauquier County to a slave mother, with a Scotsman as his father/owner. Together with his brothers, Gabriel and James, he was freed by his father and taken to Ohio. The brothers later returned to Virginia where they lived and worked as freemen, Dangerfield as a blacksmith. He had married, and with his wife, a slave woman named Harriet living in Warrenton twenty miles farther south, had six children. The letters of Harriet to her husband are revelatory of the personal stake he must have felt in joining Brown and his growing company. She wrote about this time that he should come and buy her and their children as soon as possible, as was their plan, "for if you do not get me somebody else will." "Oh, dear Dangerfield, come this fall, without fail, money or no money, I want to see you so much, that is the one bright hope I have before me."[4]

It is often assumed that Brown had no contact with, nor had given advance notice of his intentions, to blacks in either Virginia or Maryland. While this was largely so, no doubt, to avoid disclosure, Newby surely provided Brown with information about the feeling and condition of blacks in the neighborhood, and there is suggestion that his brothers were in his confidence. The manner in which Newby came to this association may not be discernible, although it wouldn't have been unlikely, given the unseen contacts that existed, that he was a referral of Joseph Winters and his "reliable friends" in Chambersburg. Newby would remain in the community and join the company on the night of the march to Harper's Ferry. Brown was so satisfied with the way "the business" was progressing that he wrote exaltedly, "The fields whiten unto harvest"; beckoning to a few choice correspondents suggestively, "Your friends at headquarters want you at their elbow." Although the networks put in motion in preparation for the Harper's Ferry raid have eluded almost all scrutiny, passing silhouettes are still discernible, and their movement, perceptible in this retrospective glance, brings out their deeper context and meaning.

At the end of August, Sanborn wrote Brown from Springfield of his activities, including the following crucial information: "Harriet Tubman is probably in New Bedford sick. She has staid in N.E. a long time, and been a kind of missionary." During the same week the

Syracuse Herald published a letter written by Gerrit Smith, where he announced he would not preside that year at the anniversary of the "Jerry Rescue," a celebration at which he had heretofore been an aficionado. He now considered commemorative exercises of the kind "shameless" and "pernicious hypocrisy." Then in a speech in September he placed himself on the same bedrock on which Brown stood, declaring slavery could no longer be ended by peaceable means, and that "the movement to abolish American slavery is a failure." To those who hoped to repeat the example in America of the British emancipation in the West Indies, Smith pointed out the analogy was imperfect: "England was not debauched and ruled by her slavery—but American slavery has left scarcely one sound spot in American character; and it is, confessedly, the ruler of America." He warned, "For insurrection, we may look any year, any month, and day. A terrible remedy for a terrible wrong." In the previous winter at a meeting of the Massachusetts Anti-Slavery Society in Boston, Richard Hinton had addressed the gathering, reading a greeting from John Brown. In his letter Brown proclaimed that "a new era in the anti-slavery movement" had dawned, and by this Hinton explained Brown meant "The rifle shot that laid low the first victim in Kansas, has rung the death-knell of slavery on this continent. . . . The terrible Logic of History teaches plainly that no great wrong was ever cleansed without blood." It was Kansas that had demonstrated, Hinton said, "the mode and manner by which the most vulnerable point of slavery, that of insurrection may be reached" and it "also educated men for the work." Hinton concluded, "For one believing in the right of resistance for myself, I extend the same to my African brother and stand ready at any time to aid in the overthrow of slavery by any and all means—rifle or revolver, the dagger or torch."[5]

In June in Kansas the Republican Party had convened a meeting in Osawatomie, to unite the disparate Democrat, free-soil, and antislavery voters under a single canopy. The gathering extended a cordial welcome to Horace Greeley, then on his famous western tour, treating him to a parade with the marchers wearing "Tribunes" in their hats. Then in July, in Missouri Dr. John Doy and his accomplices from Kansas were sentenced to penitentiary time for conducting an Underground Railroad train, concurrent with Brown's celebrated demonstration in the previous winter. The entire party had been captured by a federal posse out looking for Brown and taken to Missouri, where the fugitives were sold into slavery and the white men held in jail for trial. Upon sentencing, but before they were incarcerated, a band of Kansas men crossed the border and freed them. On their return to Kansas, all the men—the rescued and the rescuers—proudly posed for photographs, complete with arms. As news of the incident filtered east and was erroneously attributed to Brown, John Jr. wrote Kagi: "By the way, I notice through the Cleveland Leader that Old Brown is again figuring in Kansas. Well, every dog must have his day and he will no doubt find the end of his tether. Did you ever know of such a highhanded piece of business? However, it is just like him. The Black Republicans, some of them, may wink at such things, but I tell you . . . he is too salt a dose for many of them to swallow and I can already see symptoms of division in their ranks. We are bound to roll up a good stiff majority for our side this fall."[6]

With Harriet Tubman's whereabouts unknown, and Douglass's refusal to take a direct hand after his meeting with Brown in Chambersburg, a crucial sector from which manpower was expected—New York and Pennsylvania—was underrepresented. To redress this, a meeting of select persons was arranged sometime during the third week of September at an undisclosed location; a meeting which Anne Brown would refer to (in Hinton) as constituting the "missing link" in her father's movements. Although this meeting has escaped comprehensive notice in the historical record, it is likely to have taken place in a village called Mont Alto on the western slope of South Mountain, about twelve miles east of Chambersburg. This conjecture can be based upon Brown's connections there; he reportedly was a worshiper in Mont Alto at the Emmanuel Chapel, where he set up "Sunday school classes for Negro children," and is supposed to have contracted some work in an iron mill near there for the South Mountain Railroad. One or more buildings in or near the village may have been at his disposal. Osborne Anderson indicates that some unspecified "friend" was sent down to the Kennedy farm to accompany Shields Green to the mysterious location, "whereupon a meeting of Capt. Brown, Kagi and other distinguished persons, convened for consultation." Who were the men, and how many were attending? Anderson represents them as "distinguished," opening the likelihood that some were known leaders, evidently from Philadelphia, with others probably coming from Chambersburg, and it can be wondered, other locals in central Pennsylvania such as Mercersburg with its African American community tracing its roots to the eighteenth century. This meeting clearly was crucial to the maturation of Brown's plans, and although its impact on what ultimately transpired

owever, a few laborers may be looked for as certain. I would like
o hear of your congregation numbering more than "15 and 2" to
ommence a good revival; still our few will be adding strength to
e good work.7

e "hand" referred to was Osborne Anderson, arriving in
hbersburg on the eleven o'clock train on the morning of
mber 16. He had been sent under the auspices of the *Provincial*
an, whose editor, I. D. Shadd, thought it obligatory his office be
ented. Accordingly lots had been drawn, the distinction falling
derson. He was the only member of Brown's party to write a
nd account—*A Voice from Harper's Ferry*—and remarked on
ival he was "surprised" to find that all but a small part of the
ad been removed from Chambersburg to the Kennedy farm.
ould indicate that a date to begin operations had been select-
tingent upon the ill-omened meeting in Mont Alto. That date
tober 25.

ral days after this, a letter posted from Boston reached Kagi
wis Hayden: "My dear sir—I received your very kind letter,
ild state that I have sent a note to Harriet [Tubman] request-
o come to Boston, saying to her in the note that she must
ht on, which I think she will do, and when she does come I
will find some way to send her on. I have seen our friend at
he is a true man. I have not yet said anything to any body
m. I do not think it is wise for me to do so. I shall, therefore,
rriet comes send for our Concord friend, who will attend to
r. Have you all the hands you wish? Write soon./Yours, L.

n wrote Thomas Wentworth Higginson on September 14
an was "to be sent forward soon," but then says later she
t heard from. Meanwhile Lewis Hayden had been raising
l recruits, one of whom was to be on hand. This was
Ierriam, a young white man of a New England aristocrat-
irned abolitionist crusader. Hayden met him in Boston and
words said, "I want five hundred dollars and must have
h Merriam replied: "If you have a good cause, you shall
iayden then told him what he had learned from John
iat his father was preparing to lead men into the Virginia
ind needed money. Merriam replied, "If you tell me John
re, you can have my money and me along with it."9

has never been weighted, it was undoubtedly gre
attendant led John Jr. and Hinton in later years
had acted as a "representative" for Douglass. Bi
cation of this from Douglass. In the letter s
signed by a number of "colored men," he only n
"I never knew how they came to send it, but
been prompted by Kagi who was with Brow
would not go to Harper's Ferry." That letter,
and posted from Philadelphia—found in Sa
Letters of John Brown and in Douglass's *Li*
follows, with all the signatories omitted in
Douglass only remarking that he received th
number of colored men."

> Dear Sir—The Undersigned feel it to be of
> our class be properly represented in a con
> away [near] Chambersburg, in this state. \
> of all others to represent us; and we sever
> in case you will come right on we will se
> ed for during your absence, or until
> Answer to us and John Henrie, Esq.
> once. We are ready to make you a rem
> now quite a number of good but not
> tives collected. Some of our members

It has been intimated that Brown
and it was a profound disappointment
ly lessened his prospect of obtaining
the time of this meeting, Kagi received
M. Bell in Chatham:

> Dear Sir—Yours came to hand las
> night, and will be found an effici
> to be at work as a missionary to
> will start in a few days. Another
> not with him. More laborers ma
> sure." Alexander has received
> tions have come to hand, so far.
> work as he agreed. I fear he v
> Dull times affect missionary m

Char
Septe
Freer
repres
to An
firstha
his ar
arms
This w
ed, con
was Oc
Seve
from Lo
and wo
ing her
come rig
think we
Concord
except hi
when Ha
the matte
H."8
Sanbo
that Tubn
still was n
money an
Francis J.
ic family, t
after a few
it." To whi
have it." F
Brown Jr., t
Mountains
Brown is th

Richard Hinton received the following from Kagi: "I have to-day written Redpath and Merriam respecting our Nicaragua Emigration, and wishing them to meet me in Boston at an early day. . . . I wrote them in care of Francis Jackson. I need not say that I would like to see you also at that time, which I am now unable to name."[10]

Osborne Perry Anderson, circa 1865.

Brown's efforts to have an important black leader on hand also included attempts to enlist J. W. Loguen and a principal in the Oberlin-Wellington rescue trials, Charles Langston. In a letter dated the previous May, Brown had written suggestively to Loguen, "I will just whisper in your private ear that I have no doubt you will soon have a call from God to minister at a different location." But John Jr. was to confide to Kagi that Loguen's "heart was only passively in the cause" and that he was "too fat" for the arduous undertaking. Charles Langston, of considerably slighter stature, had been ill, and moreover John Jr. reported he had become "discouraged about the mining business," believing there were too few hands. "Physical weakness is his fault," he concluded. James H. Harris, who had attended the Chatham Convention, had been working with Langston in regard to recruitment for Brown but had met with disappointment. A letter from his hand dated August 22, reached Kagi stating: "I am disgusted with myself and the whole Negro set, God dam em!"[11]

On September 20 Osborne Anderson related, "Capt. Brown, Watson Brown, Kagi, myself, and several friends, held another meeting."[12] It is worth remarking parenthetically that these "several friends" were likely men from Chambersburg; it is also evident Anderson was to assume an increasingly prominent role in the subsequent weeks, as Brown saw him as the most promising candidate then on hand to provide the leadership necessary for the expected black conscripts. He was to be dubbed by those at the farm "Chatham Anderson," no doubt to distinguish him from Jeremiah Anderson, and he would have the distinction of chairing the council in preparation for the engagement at Harper's Ferry.

John Brown Jr. sent the following to "Friend Henri," dated September 27:

Since I became aware that you intended opening the mines before spring, I have spared no pains, and have strained every nerve to get hands forward in season. I do not, therefore, feel blame for any error in respect to time. I had before never heard anything else than that the spring was the favorable time, unless uncontrollable circumstances should otherwise compel! At this distance I am not prepared to judge, but take for granted that wisdom, or perhaps necessity dictates the change of programme. Immediately on receipt of your urgent communications I have dispatched copies where they would be most likely to avail anything, and have devoted, and am still devoting my whole time to forming associations for the purpose of aiding. There will be a meeting of stockholders at my house this eve; a distinguished gentleman from New Hampshire, who is anxious to invest, will be present. Whether it is best for me to come to you now or not I cannot say; but suppose it will be impossible for me to remain here when you are actually realizing your brightest prospects. When in C., and in all other places, I have at all times urged all hands to go on at once, since necessity might render their presence an imperative want at any time.[13]

Toward the end of September, after his introduction to the preparations in Chambersburg, Osborne Anderson started out for the Kennedy farm. Walking alone at night, he reached the Maryland border and found John Brown waiting in his wagon. Riding the remainder of the journey, they reached the Kennedy farm about daybreak. The house where they made their "headquarters" was rough hewn, of log construction; kitchen, parlor, and dining room below, above a spacious attic that served as bunk room, storehouse, and drill room. The men were under strict instructions to be as discreet as possible and not to be seen around the yard; with chores delegated between one and another of the men, while Anne Brown and Mary Thompson did the cooking and superintended. With the arrival of Anderson, as previously with Shields Green, the comings and goings at the farm became even more problematic, as the presence of black men would raise suspicions that the premises were being used as a station on the Underground Railroad.

During the weeks that the men were in confined quarters they occupied themselves with regular study of Forbes's *Manual* and a ridge drill under Stevens. At other times there were discussions of antislavery and reform issues, and they read Paine's *Age of Reason*, or applied a preparation for bronzing rifle barrels obtained by Cook from an armory workman. Anderson wrote: "But when our resources became pretty well exhausted, the *ennui* from confinement, imposed silence, etc., would make the men almost desperate. At such times, neither slavery nor slaveholders were discussed mincingly. We were while the ladies remained, often relieved of much of the dullness growing out of restraint by their kindness. As we could not circulate freely, they would bring in wild fruit and flowers from the woods and fields."

At night the men were able to emerge to ramble and enjoy their solitude in the mountain air. Anderson continued, "There was no milk and water sentimentality—no offensive contempt for the Negro, while working in his cause; the pulsations of each and every heart beat in harmony for the suffering and pleading slave." John Brown loved the fullest expression of opinion, he wrote, and when a subject "was being severely scrutinized" he "would be one of the most interested and earnest hearers." It was gratifying, said the elder man, to see "young men grapple with moral and other important questions . . . it was evidence of self-sustaining power." Anderson concluded, "I thank God that I have been permitted to realize to its furthest, fullest extent, the moral, mental, physical, social harmony of an Anti-Slavery family, carrying out to the letter the principles of its antetype, the Anti-Slavery cause."[14]

As the decisive hour approached— and Anne and Mary prepared to leave, to be accompanied as far as Troy, New York, by Oliver Brown—all the "inmates" were drawn from the attic by a curious "noise" in the parlor. Shields Green, the self-styled "Emperor," was giving the women his heartfelt farewell, said by Anne Brown to be an outlandish conglomeration of ill-fitting words, piled up no doubt from all that he had gleaned from the talk of the previous weeks. It was however a fitting tribute, she allowed, from a man with such distinctive attributes.

On October 6 with issues of importance remaining, Brown and Kagi journeyed to Philadelphia where they would remain for three days. DuBois remarks, "[Thomas] Dorsey the caterer with whom he stayed, at 1221 Locust Street, is said to have given him $300." Upon their return they met Francis Merriam in Chambersburg, whom Lewis Hayden had forwarded to Sanborn. Sanborn in turn sent him on to Higginson with a note that included "Perhaps you will have a message for the Shepherd." Of slight frame and build, Merriam to all appearance was unfit for a soldier's life, and in the bargain had but one eye. The other was glass. Higginson had queried Sanborn about him, who replied, "I consider him about as fit in this enterprise as the Devil is to keep a powder house, but everything has its use & must be put to it if possible." Merriam had withal a purse of six hundred dollars in gold pieces, and he was dispatched directly for Baltimore with instruction to buy a large quantity of percussion caps. When they returned to the Kennedy farm a meeting of those present was called where Brown acquainted all with what had transpired in Philadelphia. Anderson indicates the purposes touched upon, their significance, and the deep pathos felt by all as during his recital Brown wept. Anderson wrote: "How affected by, and affecting the main features of the enterprise, we at the farm knew well after their return, as the old Captain, in the fullness of his overflowing, saddened heart, detailed point after point of interest. God bless the old veteran, who could and did chase a thousand in life, and defied more than ten thousand by the moral solemnity of his death!"[15]

It has been remarked that Brown's Provisional army was led by whites, with blacks filling in the lower ranks, while in fact, as has been seen, he sought to establish a fully collaborative relationship with blacks taking positions of co-leadership. Anderson addressed this issue:

> It has been a matter of inquiry, even among friends, why colored men were not commissioned by John Brown to act as captains, lieutenants, &c. I reply, with the knowledge that men in the movement now living will confirm it, that John Brown did offer the captaincy, and other military positions, to colored men equally with others, but a want of acquaintance with military tactics was the invariable excuse. Holding a civil position, as we termed it, I declined a captain's commission tendered by the brave old man, as better suited to those more experienced; and as I was willing to give my life to the cause, trusting to experience and fidelity to make me more worthy, my excuse was accepted. The same must be said of other colored men . . . who proved their worthiness by their able defense of freedom at the Ferry.[16]

Earlier that week Leary and Copeland had received word to come forward. "Without tools" of their own, they were helped along by two Oberlin professors, Ralph and Samuel Plumb. On October 10 they were in Cleveland at the home of the Sturtevants, supporters of John Brown, and while there met with Charles Langston and James H. Harris, who would not be coming with them, but in any event wished them godspeed. Traveling via Pittsburgh and Harrisburg, the two men arrived in Chambersburg on October 12 and stayed in the home of Henry Watson. Occupying the next day and part of the night traveling to Maryland, they reached the farm as the sun rose on the morning of the 14th. Meanwhile, George Stearns had been in consultation with Lewis Hayden and other black Bostonians. Harriet Tubman was feeling well enough and Hayden had written Gerrit Smith requesting and receiving money for her expenses. By October 15 she was in New York City in the company of four recruits headed for Chambersburg. That same day, Richard Hinton reached Hagerstown to be met with a letter and money instructing him to return to Chambersburg to hire a horse and wagon to carry a quantity of arms still in the town. Hinton arrived too late to effect this and lodged the night with Henry Watson. Elsewhere men were on the roads—George Gill among them. He was to write later regarding this, "I had been in correspondence with Kagi and knew the exact time to be on hand and was on my way to the cars when the thrilling news came that the blow had been struck. Of course, I went no further." Charles Moffet and Luke Parsons received a letter summoning them on the same day the news of the raid reached them. Moffet wrote: "I know positively of my own knowledge and from men's mouths that there were from one to five hundred men on the road when the news burst."[17]

Kagi expected to maintain correspondence so far as possible with three persons. One was John Brown Jr., as their general agent in the North; the others were Charles A. Dana of the *Tribune* and William A. Phillips of Lawrence, Kansas. In his last letter to John Jr., dated October 10, Kagi wrote:

> We shall not be able to receive any thing from you after to-day. It will not do for any one to try to find us now. You must by all means keep back the men you talked of sending and furnish them work to live upon until you receive further instructions. Any one arriving here after to-day and trying to join us, would be trying a very hazardous and foolish experiment. They must keep off the border until we open the way clear up to the line from the south.

Until then, it will be just as dangerous here as on the other side, in fact more so; for, there will be protection also, but not here. It will not do to write to Harper's Ferry. It will never get there—would do no good if it did. . . . We will try to communicate with you as soon as possible after we strike, but it may not be possible for us to do so soon. If we succeed in getting news from outside our own district it will be quite satisfactory, but we have not the most distant hope that it will be possible for us to receive recruits for weeks, or quite likely months to come. We must first make a complete and undisputably open road to the free states. That will require both labor and time.

The letter concluded:

This is just the right time. The year's crops have been good, and they are now perfectly housed, and in the best condition for use. The mood is just right. Slaves are discontented at this season more than at any other, the reasons for which reflection will show you. We can't live longer without money—we couldn't get along much longer without being exposed. A great revival is going on and has its advantages. Under its influence, people who are commonly barely unfavorable to slavery under religious excitement in meetings speak boldly against it. In addition to this and as a stimulant to the religious feeling, a fine slave man near our headquarters, hung himself a few days ago because his master sold his wife away from him. This also arouses the slaves. There are more reasons which I could give, but I have not time.[18]

Many of those who could be expected but were still coming forward, had been informed they must be on hand by October 25. But after one or more of the blacks had been seen by a prying neighbor, suspicions had been aroused that the Kennedy farm was being used as a station on the Underground Railroad. During the second week of October information was received from outside that a search was to be instituted of the property and the buildings on the following Monday, the 17th. To allay this danger Brown decided to seize Harper's Ferry on the night of the 16th. With the date changed, hope that others would arrive in time had to be abandoned. This was to make the raid far more hazardous and problematic that it would have been.

Twenty years later Douglass essayed the enduring significance of Harper's Ferry in his speech at Storer College: "In all the thirty years conflict with slavery, if we except the late tremendous war, there is no subject which in its interest and importance will be remembered longer, or will form a more thrilling chapter in American history than this strange, wild, bloody and mournful drama." But in history as to reason, Douglass proposed, nothing is reaped that has not been sown, and "the bloody harvest of Harper's Ferry was ripened by the heat and moisture of merciless bondage of more than two hundred years. That startling cry of alarm on the banks of the Potomac was but the answering back of the avenging angel."[19]

Saturday, October 15, was a busy day for all hands. Besides two hundred Sharps rifles and an equal number of revolvers and all the ammunition for these, there were 950 pikes, one small cannon mounted on a swivel, ten kegs of gunpowder, and a number of sabers, bayonets, picks, shovels, axes, and torches. Further supplies included four large tents, blankets and clothing, field glasses, surgical equipment, and sundry items including boatswain's whistles. The next morning John Brown rose earlier than usual and called his men down for worship. Reading a passage from the Bible applicable to the condition of the slaves and on the duty of all to assist them, he then offered up his prayer for the liberation of the bondsmen. Anderson wrote: "The services were impressive beyond expression. Every man there assembled seemed to respond from the depths of his soul, and throughout the entire day, a deep solemnity pervaded the place."[20]

After breakfast, roll was called. With a sentinel posted outside the door, the men listened to preparatory remarks to a council to assemble at 10 o'clock, chaired by Anderson. In the afternoon eleven orders were issued pertinent to the seizure of Harper's Ferry. In the evening Brown gave the men final instructions, ending with this: "And now, gentlemen, let me impress this one thing upon your minds. You all know how dear life is to you, and how dear life is to your friends; and remembering that, consider that the lives of others are as dear to them as yours are to you; do not, therefore, take the life of any one if you can possibly avoid it; but if it is necessary to take life in order to save your own, then make sure work of it."[21]

Biographers and historians of Brown and the war he heralded have written accounts and commentary on what occurred at Harper's

Ferry, all varying in details as well as in important phases of the action; this assessment will rely on a careful cross-referencing of these sources against eyewitness accounts. Eighteen men made the march, thirteen white and five black. Some of them had fought for the free-state cause in Kansas; some carried arms into battle for the first time; only one, Aaron Stevens, was a professional soldier. Besides their leader, who was nearing sixty, only two had passed thirty years. Each man knew, however, the significance behind their march; a few understood with certainty that they were going to their deaths. But each equally knew he was about to strike a blow that would transform the politics of the nation whose ultimate victory would mean the death of slavery.

On the cold, rainy evening of October 16 between 8 and 9 o'clock, John Brown gave the order for his men to get on their arms and proceed to the Ferry. Each man tucked a bowie knife and a pair of revolvers in his belt, put on a cartridge belt with forty rounds, and pulled on a heavy woolen gray shawl for protection from the weather and to conceal the arms. Finally each man picked up a Sharps rifle. A horse and wagon were brought up loaded with a bundle of pikes, some extra Sharps rifles and ammunition, torches and incendiary devices, crowbars, and a sledgehammer. Brown climbed into the driver's seat. A procession was formed with Cook and Tidd leading, while the others fell in at measured intervals two by two behind the wagon. Owen Brown, Barclay Coppoc, and Francis Merriam, delegated to guard the remaining arms and equipment until they were called for, now stepped forward to take leave of the others.

At half past 10 o'clock without being seen by a single person, the cloaked invaders reached the abutment of the Potomac Bridge where Cook climbed the telegraph pole and cut the wire. Kagi and Stevens proceeded onto the bridge, approaching the watchman at the opposite end, where they detained him. Then Brown entered the bridge in his wagon followed by the company in double file. Watson Brown and Stewart Taylor took positions assigned them, standing on opposite sides at the entrance of the bridge, the rest of the party proceeding across the nine-hundred-foot span.

Now Stevens ordered to make ready to take the town and each man fastened his cartridge belt outside his tunic. Leaving the bridge they turned the corner at the train depot and approached the armory gate at sixty paces. The gate was chained and locked with the watchman in his room adjacent to the enginehouse. He was called out and Sharps rifles were thrust against his breast.

From left to right: Albert Hazlett, Oliver Brown, William Thompson.

"Open the gate or give us the Key!" He refused. Within moments the lock was wrenched open, the watchman taken prisoner, and the enginehouse invested. The building, of brick construction, consisted of two rooms, divided by a brick partition; the fire engine room proper and a watchman's room. It was in the latter that Jeremiah Anderson and Adolphus Thompson were detailed to hold the detainees.

Stevens now arranged the men to take possession of their remaining objectives. Albert Hazlett and Edwin Coppoc were to hold the arsenal for the time being; Oliver Brown and William Thompson were to post themselves at the wagon bridge across the Shenandoah, on opposite sides. The last building to be taken was Hall's Rifle Works on the bank of the Shenandoah. The watchman there was to be taken prisoner and brought to the government grounds to be detained with the others, while Kagi and Copeland took their posts, with orders to hold the works until reinforcements were sent.

By this time all the gaslights in the lower part of town had been extinguished and the connecting streets secured. It was then that a protracted camp meeting at a Methodist church let out, and as a number of people were returning to their homes additional prisoners were escorted to the enginehouse. Inside the watchman's room Brown offered the detainees a brief summary of his purpose. "I came here from Kansas, and this is a slave state. I want to free all the Negroes in this state. I have possession now of the United States Armory, and if the citizens interfere with me I must only burn the town and have blood."[22]

It was just before midnight when Brown called Stevens to his side, together with those who had no assigned posts. Now the second

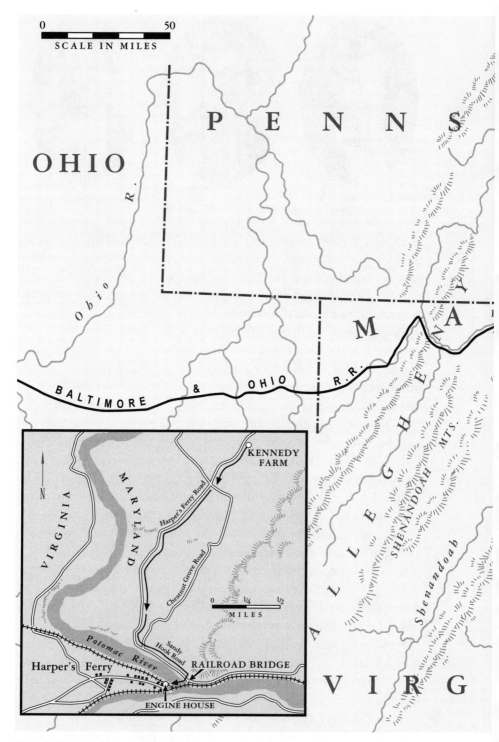

Map 3. Harper's Ferry and Its Vicinity.

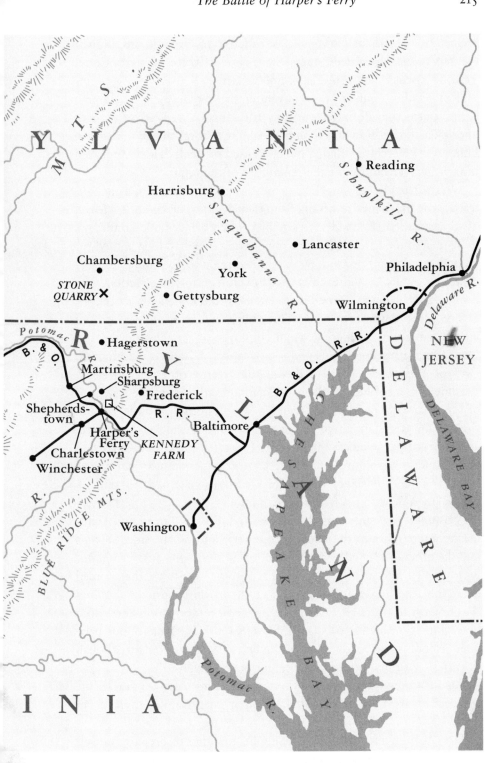

phase was begun as Stevens was ordered to proceed up Bolivar Heights to predesignated estates. Stevens selected for his party Cook, Tidd, Osborne Anderson, Leary, and Green.

Just as this council was convening the replacement watchman came onto the Potomac Bridge for duty. As Watson Brown approached, the brawling Irishman caught him on the chin, decking him. As the watchman fled, a bullet creased his scalp. He disappeared into the Wager House for the remainder of the night.

On the road to Bolivar Heights Stevens and his party soon encountered some black men returning from their Sunday, or "French," leave, as it was called. By Anderson's telling, after being made confidants these men immediately agreed to join, saying they had long waited for an opportunity of this kind. Stevens asked them to go around and circulate the news "when each started off in a different direction. The result was that many colored men gathered to the scene of action."[23]

For several weeks prior, Cook, whose wife was expecting, had been augmenting his income as an employee of the Chesapeake and Ohio Canal by selling maps and a biography of George Washington. One of the places he visited was the estate of Col. Lewis W. Washington, a nephew of the great man and adjunct to the governor of Virginia, who, perhaps aware of Cook's reputation as a marksman, showed the young man two pistols presented to his uncle by Lafayette and a sword presented to the same by Frederick the Great. Cook had also taken the time during these weeks to conduct an informal census of the number of slaves along the Charlestown Road, to settle a wager, he said, with his friend Isaac Smith.

Washington's estate was the first visited. Placing Green and Leary on the approaches to the house in the front and on the side, Stevens and the others knocked at the front door. With no one answering, but with women and girls appearing at the upper windows, they forced entry into the house and began a search for its proprietor. Finding Washington on the first floor as he opened the door to his bedroom beseeching that his life be spared, Stevens replied: "You are our prisoner. You must get ready to go to the Ferry. We have come to abolish slavery in Virginia, not to take life but in self-defense."

Again Washington pleaded, "You can have my slaves, if you let me remain." "No," Stevens insisted, "You must go along too; so get ready."

Stevens then left the house, leaving his prisoner to prepare himself under guard. In a short time three of Washington's male slaves then

present were gathered together and an explanation for their action offered; one of them,
Washington's coachman, Jim, reportedly
joining "with a good will." Harnessing a
two-horse carriage and four-horse wagon,
they were quickly brought around to the
front of the house. As Stevens returned to the
house Washington continued to plead he be
allowed to remain and went to his sideboard
to take out whisky, offering drinks. Now
Stevens carried out an express order from
John Brown that Washington surrender all
the arms in his possession and in particular

Lewis W. Washington in
1861.

that he present the sword given the late president by Frederick the
Great and the set of pistols from Lafayette. When Washington
brought these out, Stevens ordered Osborne Anderson to step forward and accept them, Brown having remarked at the preparatory
meeting, "Anderson being a colored man, and colored men being
only things in the South, it is proper that the South be taught a lesson
on this point." Handing the weapons over, Washington "appeared to
be taken aback," wrote Anderson.[24]

The party now rode in the wagons to another estate, but with the
wife recently widowed, and being told that there were only women
and girls living there, they drove on. Appearing next at the Allstadt
estate, they announced their presence by crashing a fence rail through
his front door. Compelling John Allstadt and his eighteen-year-old
son to join their neighbor in his carriage, six of his male slaves were
induced to join the others in the big wagon. Before moving on, another of Washington's slaves who had been away but heard about the
excitement, ran up to join the others. He was a strapping youth, an
eighteen-year-old named Phil.

On their descent from the Heights, Anderson wrote, the party
stopped at the home of an "old colored lady . . . a little way from the
town, [who] had a good time over the message we took her. This liberating the slaves was the very thing she had longed for, prayed for,
and dreamed about, time and again; and her heart was full of rejoicing over the fulfillment of a prophecy which had been her faith for
long years. 'God bless you! God bless you!' she cried, kissing the men
at her door and offering a prayer for their success."[25]

At 1:15 A.M. a train bound for Baltimore from Cincinnati and Wheeling pulled into Harper's Ferry. The conductor, conferring with the hotel clerk and the absconding watchman, and hearing talk of armed men on the bridge, ordered his baggage master to investigate. Confronted with Sharps rifles, the baggage master hastily fled to the safety of the hotel, while the engineer hurriedly climbed into his cab, moving the train back a hundred feet. The station porter, Sherwood Hayward, a black man, who had also ventured out, when ordered to halt, turned to flee. He was mortally wounded by a bullet entering his back and exiting near his heart. At this shooting, the train's passengers began disembarking in pandemonium, seeking shelter in the hotel and train depot. The conductor who had remained at the bridge's entrance was detained and escorted into the armory. It was with this man, Andrew Phelps, that the initial reports on the "Harper's Ferry Insurrection" were based—as he telegraphed later that morning to his superiors that the government works were teeming with insurgents, magnified into several hundred abolitionists, black and white. They were arming "Negroes," his message stated, and sending wagons with arms into the hills and were firing the town.

As Hayward was shot, Dr. John Starry, who had a room in the lower part of town, awakened and came out to see what was going on. Finding the wounded porter near death, he rendered the assistance he could, and dressed the watchman's scalp. It was about 4:30 in the morning when Stevens and his party returned from the Heights, depositing their prisoners in the watchman's room with the others. A short time later the nine slaves along with a number of other blacks coming on the scene were drawn up in a semicircle in front of the enginehouse and armed at Brown's order by Osborne Anderson. Some were given pikes, while a few were variously armed; Jim, Washington's coachman armed himself, it was reported, with a pistol and a supply of balls, and William Leeman loaded a double-barreled shotgun taken from Washington and placed it in the hands of "an elderly slave man." Four freed slaves were posted as guards over their masters and the other prisoners in the watchman's room of the enginehouse.

After warming themselves by a stove in the enginehouse, Cook, Tidd, and Leeman were ordered by Brown to take the heavy wagon with a number of freed slaves (Anderson says fourteen, others report four and five) into Maryland. There they were to take another slaveholder prisoner whom Leeman and Cook would detain at his home, while the others began the removal and forwarding of the arms and

equipment at the farm. The previous day Brown selected an intermediate point as a temporary depot—a one-room schoolhouse three-
quarters of a mile from the Ferry on the opposite side of the Potomac,
where the Boonsborough Pike meets the river at a right angle.

As the wagon rumbled into the darkness, Dr. Starry peered after it.
He would now rouse some boys he knew, and send them to
Charlestown with the urgent message that "abolitionists" had seized
the government armory and were arming "Negroes." Then he went
off to rouse the acting superintendent of the armory, A. M. Kitzmiller.
At this moment John Brown sought out Washington and Allstadt,
inviting them into the enginehouse with him. Addressing Washington,
he said:

> I presume that you are Mr. Washington. You will find a fire in
> here, sir. It is rather cool this morning. I think, after a while, pos-
> sibly, I shall be enabled to release you, but only on the condition
> of getting your friends to send a Negro man as a ransom. I shall be
> very attentive to you sir, for I may get the worst of it in my first
> encounter, and if so, your life is worth as much as mine. I shall be
> very particular to pay attention to you. My particular reason for
> taking you first was that, as the aide to the governor of Virginia, I
> knew you would have been a troublesome customer to me; and,
> apart from that, I wanted you particularly for the moral effect it
> would give our cause having one of your name as prisoner.[26]

Few in Harper's Ferry were yet aware of what was taking place,
the residential neighborhoods being for the most part remote from
the scene of action. During the four hours the train sat near the
Potomac Bridge, Brown sent two messages that the passengers and
crew should prepare to depart, but the conductor, fearing the bridge
was mined, would not order it out. The engineer would later report
as many as three hundred blacks gathered around the cars in these
early hours, angrily excoriating those inside, saying they'd been slaves
long enough. As the sun began to rise Brown came out of the armory
to assure Phillips it was safe for the train to pass, saying, "You need
have no fears, I will walk ahead of you just ahead of the engine."[27]

Just before six, James Darrell, the bellringer at the armory, came
out of his house and walked down Shenandoah Street, as Walter
Kemp, a bartender at the Wager House, was also headed to work.
Dangerfield Newby, six feet two inches and large framed, encoun-

tered the bellringer, whom he ordered to halt, saying he could not go about his business that day. Darrell swung his lantern at Newby, but was further dissuaded by the appearance of other pickets who came up, conveying him and the bartender to the watchman's room to sit out the morning. Now the train was passing slowly over the bridge and soon would speed toward Baltimore with its panic-stricken passengers. In parting Brown remarked to the conductor, "there must be no more trains through Harper's Ferry, east or west."[28]

At 6:30 the acting superintendent of the armory appeared at the armory gate and was taken inside. Nearby the paymaster at the armory, J. E. P. Daingerfield, also approached and was stopped by three pickets.

"What does this mean?" he asked.

"Nothing, only we've taken possession of the government works," said one.

"You talk like a crazy man."

"Not so crazy as you think. You'll soon see."

Seeing arms for the first time as the men drew back their shawls, Daingerfield said, "I think I'll return home."

"No, you're our prisoner."

As he arrived at the government works, Daingerfield was to report decades later, he "saw what indeed looked like war. There were negroes armed with pikes, and sentinels with muskets all around." Brown ushered the prisoners into the room with Colonel Washington and Mr. Allstadt. Asked what was his object? Brown is said to have replied "to free the Negroes of Virginia. By noon there will be fifteen hundred men with us, ready armed."[29] This last expression was one that Andrew Phillips also recalled Brown saying as he warned no more trains would be allowed through Harper's Ferry.

On the hill Dr. Starry was having the Lutheran church's bell rung, and compelling a search for townsmen with guns. He was soon to conclude not much could be done. Around seven on High Street a man did come out of his home with a shotgun and had his hat shot off. Thomas Boerly, a grocer and prominent citizen of the Ferry, crept up toward the armory gate with a similar weapon looking for a target, shooting at Edwin Coppoc standing sentinel. Stevens ordered the old man armed earlier by Leeman to arrest him. Anderson wrote: "The old man ordered him to halt, which he refused to do when instantly the terrible load was discharged into him and he fell, and expired without a struggle."[30]

Elsewhere another armory worker, Joseph Barry, argued against his arrest and soon had four or five pickets surrounding him. Bolting down an alley among the workmen's cottages along Potomac Street, a slave woman named Hannah stepped out spreading her arms, beseeching he not be killed. This intervention gave Barry his opportunity to escape.

Describing the scene on that morning as "a time of stirring and exciting events," Osborne Anderson wrote:

> In consequence of the movements of the night before, we were prepared for commotion and tumult, but certainly not for more than we beheld around us. Gray dawn and yet brighter daylight revealed great confusion, and as the sun arose, the panic spread like wild-fire. Men, women and children could be seen leaving their homes in every direction; some seeking refuge among residents, and in quarters further away, others climbing up the hillsides, and hurrying off in various directions, evidently impelled by a sudden fear, which was plainly visible in their countenances or in their movements.
>
> Capt. Brown was all activity, though I could not help thinking that at times he appeared somewhat puzzled. He ordered Sherrard Lewis Leary, and four slaves, and a free man belonging to the neighborhood, to join John Henry Kagi and John Copeland at the rifle factory, which they immediately did.[31]

At this hour John Brown was in control of the situation, but elsewhere events were occurring that would soon recoil upon him. On the train racing through the countryside toward Baltimore, first one passenger then several penned messages dropped from windows giving a frenzied view of what they'd seen in Harper's Ferry. This was further amplified when the relay station at Monocracy was reached and the wires began carrying the news to Baltimore and Washington and beyond of

FRIGHTENING INTELLIGENCE—SERVILE INSURRECTION BELOW THE MARYLAND BORDER

As Dr. Starry rode up from the Ferry he found Charlestown rent by rumors of fire and blood; as if an invading army was about to sweep upon them, church bells were ringing and drums rolled as militia companies began assembling. Meanwhile in Maryland on the

Sharpsburg Pike a slaveholding farmer named Terence Byrne encountered Cook, Tidd, and Leeman with the wagon brimming with black men. Byrne knew Cook and what he saw must have piqued his curiosity, but he was quickly to learn that he was their prisoner. Brown's order had been that Cook and Leeman should hold Byrne while Tidd went to the Kennedy farm with the wagon and the blacks. When the wagon was loaded they were to return to pick up Byrne together with his male slaves, who would be escorted by Leeman to the Ferry. The arms and supplies were to be deposited in the schoolhouse where Cook and several of the blacks would remain as guards. Tidd and the others were to return to the farm for the remaining equipment, when Owen Brown, Barclay Coppoc, and Merriam would come down with them. With sufficient dispatch Brown must have been expecting this would all have been done at the latest by mid-morning.

By 7:30 there were some forty prisoners confined in the enginehouse, mostly armory officials and workers, in addition to the slaveholder hostages. Three men had been shot and were either dead or dying, as sporadic gunfire continued to interrupt what would have been a normal workday. In the Wager House and Galt House saloon men were milling, intently watching the goings on in the government works opposite them. The compass of the action formed two interconnecting triangles, given distinction by the points seized. The first was from the Potomac Bridge to the armory gate and the enginehouse, with another tangent being from there to the arsenal—an area where the train depot, Wager House Hotel, and Galt Saloon were situated. The second triad was defined by the rear of the arsenal to the Shenandoah Bridge, then to the rifle factory. Both were in the lowlying part of town near the confluence of the rivers.

Unassailable for the moment, Brown waited for his men to complete their work in Maryland, and for a time a lull marked by a curious atmosphere of courtesy settled in. It was at this time Brown asked the bartender at the Wager House who was his prisoner to call upon his manager for breakfast for forty-five men. Before the hour was out "colored waiters" delivered plates of ham and eggs, although neither Brown nor his two hostages would partake of a share. To calm fears among armory officials and workers, a few were allowed to return home under escort to reassure their wives and family of their safety. During this interlude, Joseph Brewer, an official of the armory and nominally a prisoner, circulated among the citizens, asking them not to shoot for the sake of the hostages. This seemingly formless waiting continued to 9 o'clock, when a warehouse containing muskets removed from the arsenal when there was danger from flooding was pointed out by an armory worker, and broken into. A few of the arms were distributed and some men began a desultory fire from Camp Hill down on the government works. But with little ammunition their effect was negligible.

As 10 o'clock neared, the watchman who had had his scalp grazed during the night came out of the Wager House to get water for the dying porter. As he did, William Thompson approached from his post for a drink, asking him to take water also for Oliver Brown and Newby on the Shenandoah Bridge, which he did. At this time Brown also advised Washington and the others that they could exercise their legs, if they desired, outside the enginehouse. But given the occasional shots being fired from the hill, they declined. As 11 o'clock approached Brown called William Thompson to his side, telling him to ride into Maryland to communicate with those on the other side of the Potomac that they were in control of the Ferry; and undoubtedly, too, to hasten them in their work. About this time a train from the east drew up at Sandy Hook just beyond the Potomac Bridge on the Maryland side, and the conductor and brakeman who ventured out on the bridge were taken in tow to Brown. The train would not be allowed to pass through the juncture, they were told. By now several trains were also waiting in the west.

Various writers have noted that his success during the night and in the early morning seemed to have "a soporific effect" on Brown, that he seemed "constitutionally unable to formulate a decisive plan of action." Hermann E. von Holst wrote: "It is as though during those decisive hours a thick veil had fallen over the eyes that were wont to see so clearly." John Brown, Charles Robinson said caustically years later, "demonstrated his wonderful generalship by committing suicide at Harper's Ferry."[32] Brown is said to have failed because he put himself across the river from his supplies and part of his force without being able to assure that the bridge over the Potomac could be held. He was also said to have been mistaken in believing he could hold a town that size while keeping the local population at bay, and not concern himself that they might receive reinforcements from surrounding towns. Brown's conduct, to the contrary, shows that he felt secure and must have expected to be able to withdraw from the town before measures being prepared against him could be taken. He was hardly unaware that he was stirring up "a mighty passel of enemies," as it had been his design to strike in such a way as to inspire maximal alarm. These hours are problematic to most observers because they have never properly understood them. That Brown seems to have remarked he was expecting to be joined by as many as fifteen hundred slaves is puzzling. How long was he prepared to wait; how would they get there? Wouldn't such a mass, by its nature unwieldy, degenerate into anarchy? But this statement may only have been part of Brown's design to "agitate," as Salmon Brown was to remark years later. Whatever its import, he was awaiting the arrival of his own arms, enough to arm fifteen hundred men all counted.

Given to making wry comments, Brown perhaps intended this meaning. But all commentary falters here, falling on the facile notion that Brown was quixotic in his expectations. Indeed it is often concluded that Brown hadn't concerned himself with what he would do after he was "inside" the armory, that he thought he could merely improvise his tactics and be guided by events.

Finally it is often assumed that the attack failed for want of participation on the part of the people it was designed to free, and this is given as proof that the whole enterprise was mistaken in its estimation of American slavery. Most commentators maintain that Brown couldn't have picked a better place to demonstrate the loyalty of the slaves. One slave owner from Lynchburg remarked: "If there were any danger at all, I would go to my plantation for a body guard of my slaves." To the contrary, Anderson wrote, "Captain Brown . . . was surprised, and pleased by the promptitude with which they volunteered, and with their manly bearing at the scene of violence. [He remarked] that he was agreeably disappointed in the behavior of the slaves; for he did not expect one out of ten to be willing to fight." DuBois estimated Brown had between twenty-five and fifty slaves and free blacks cooperating with him. Anderson said, "Many colored men gathered to the scene of action," and "a number were armed." An early press report highlighted "Anderson as ringleader of the Negroes," after blacks were drawn up in front of the enginehouse, and Brown ordered them armed by Anderson.[33] Four slaves and a free man were ordered by Brown to reinforce Kagi in the rifle factory; according to witnesses another four were seen on the Potomac Bridge, and several reinforced the men in the arsenal. Blacks, in addition, stood as sentinels with Brown's men at the armory gate, while another four stood guard over their masters in the enginehouse. To varying degrees of effectiveness all were seen cooperating with Brown, as many more undoubtedly waited to see how the issue would be decided. The fact is, and it has been so difficult for so many to see—Brown was staging a demonstration at Harper's Ferry, as the sequel to what had occurred under Nat Turner, of armed blacks openly defending their liberty in a slaveholding community.

Frederick Douglass years later stated, "When John Brown proclaimed emancipation to the slaves of Maryland and Virginia he added to his war power the force of a moral earthquake. Virginia felt all her strong-ribbed mountains shake under the heavy tread of armed insurgents. Of his army of nineteen her conscience made an army of nineteen hundred." Then further on, Douglass said, "Conscious of her guilt and therefore full of suspicion, sleeping on pistols for pillows, startled at every unusual sound, constantly fearing and expecting a repetition of the Nat Turner insurrection, she at once understood the meaning, if not the magnitude of the affair."[34]

That morning as William Thompson rode across the Potomac Bridge he encountered Leeman escorting Terrance Byrne and his brother to the Ferry. Leeman asked how the people of Harper's Ferry were taking it all. "Pretty well," Thompson answered. "They're more frightened than hurt." But up on Bolivar Heights a group of "indifferently armed" men could see that the invaders were vulnerable from an attack on the opposite side of the bridges. If they crossed the rivers upstream, they could come down and drive them off. Shortly after this Kagi's position in the rifle factory began to come under fire, and he sent an urgent message to the armory apprising Brown of the circumstance. Leary carried his note, which read: "Get over the bridge and into the hills. Do not delay. Our purpose is accomplished. The blacks will respond. Pray you not remain here."[35]

Seeing the maneuvers against him, Brown said, "We will hold on to our three positions, if they are unwilling to come to terms, and die like men." Stevens ordered Leary, "Tell Kagi to remain firm."[36]

In Washington and elsewhere crowds began congregating at telegraph offices, where reports were coming in that told of an outbreak of alarming proportions. By mid-morning Secretary of War John Floyd was handed a message by his chief signal officer from the president of the Baltimore & Ohio Railroad saying that armed men were in control of the United States Armory at Harper's Ferry, they were stopping trains, and they were inciting slaves to join them. An urgent wire reached President Buchanan and another reached Virginia's Governor Henry A. Wise, saying that hundreds of abolitionists, black and white, had rallied a large number of slaves, and held prominent Virginia citizens hostage; the telegraph wires were cut, and the town was being fired and arms sent into the hills.

By 6:30 that morning two companies of militia had begun to muster at ten miles distance from Harper's Ferry in Charlestown. In Martinsburg, twenty miles to the west, another company would begin to muster later that morning. Before noon President Buchanan was meeting with his Secretary of War, ordering him to send troops to aid in suppressing the insurrection. Orders were telegraphed to Fort Monroe for three artillery companies to move on the next train. Already a contingent of 99 marines was marching up from the Navy Yard in Washington and the United States had accepted the proffered services of a militia company from Frederick, Maryland. Before the end of the morning the entire eastern portion of Virginia was alerted of an insurrection below the Maryland border, and Washington began hearing calls to protect the Congress and the president and the residents of the city. Two hundred muskets with ammunition were brought up from an arsenal to City Hall, and militia units were mustered there and in nearby Alexandria. Still the president of the B&O cautioned these might not be enough to suppress the insurrection. "This is a moment of the greatest peril," he concluded.[37]

After dispatching Thompson, Brown began to consider a proposition for the release of his prisoners in preparation for the arrival of his men with the supplies. He passed from the armory to the bridge to the arsenal encouraging his men, "Hold on a little longer, boys, until I get matters arranged with the prisoners." Osborne Anderson wrote: "This tardiness on the part of our brave leader was sensibly felt to be an omen of evil by some of us, and was eventually the cause of our defeat."[38]

The first company of militia arriving by train from Charlestown and disembarking on Bolivar Heights—the Jefferson Guards—took up the strategy proposed by Harper's Ferry men only moments before. Their commander divided them in two; one part was to go west of the town to a location called Old Furnace, where unseen they could cross the Potomac in boats; the other part meanwhile was to work its way down through the upper neighborhood lateral to the rifle works, seize the Shenandoah Bridge, and come up on the rear of the arsenal.

As noon neared, the Jefferson Guard approached the Potomac Bridge along the towpath of the C&O Canal. Spotting them, Brown exclaimed, "Here come our men." Stevens quickly corrected him, "Captain, I'm afraid those aren't our men." When the militia reached the bridge they formed into marching order to cross it. Fastening Washington's famous sword to his side with which to command his men, Brown said, "The troops are on the bridge, coming into town. We will give them a warm reception." Leading men out of the armory and calling others out of the arsenal, Brown passed among them. "Men, be cool! Don't waste your powder and shot! Take aim, and make every shot count! The troops will look for us to retreat on their first appearance; be careful to shoot first."[39]

As the Jefferson Guard came out of the bridge and began approaching, Brown shouted, "Let go upon them." Osborne Anderson wrote of the effect of their repeated volleys:

From marching in solid martial columns, they became scattered. Some hastened to seize upon and bear up the wounded and dying—several lay dead upon the ground. They seemed not to realize, at first, that we would fire upon them, but evidently expected we would be driven out by them without firing. Capt. Brown seemed fully to understand the matter, and hence, very properly in our defense, undertook to forestall their movements. The consequence of their unexpected reception was, after leaving several of their dead on the field, they beat a confused retreat into the bridge, and there stayed under cover until reinforcements came to the Ferry.

Just as Brown was ordering his men back to their positions, George W. Turner, a prominent slaveowner who had come to the Ferry to see about his friends, leaned on a fence rail on High Street as he took aim. Seeing him, Dangerfield Newby brought up his rifle and fired. Hitting Turner in the shoulder, the bullet exited from his neck, killing him.

Across the street in an overlooking second-floor window, a man took aim at Newby with a musket loaded with a six-inch spike for want of a ball. As Newby came into view the weapon was discharged, striking him in the throat with such force that his body spun like a man who'd been pole-axed, opening a wound from ear to ear. In the next instant, Shields Green, whom witnesses described as firing "rapidly and diligently," shot at the assailant as he withdrew from the window.

Newby was the first of Brown's company to die, while two of the freed slaves were also slain in this fight outside the armory gate. Simultaneous with this, the rifle factory was now being beleaguered from behind boulders and trees. Kagi sent a second message apprising Brown of the situation. Jeremiah Anderson carried Brown's instructions: "Hold out a few minutes longer, when we [will] all evacuate the place."[40] Anderson was mortally wounded, but managed to return to the enginehouse.

Now both bridges were lost and the three points still held were isolated and under siege, as militia from the surrounding country continued to pour in. Moments before this Green had been sent from the enginehouse to the arsenal with a message. Osborne Anderson and Albert Hazlett, who had remained there, could see that a disastrous general encounter was about to ensue. Anderson told Green he'd better remain with them, where, since the enemy's focus was on the enginehouse and the prisoners, they might have an opportunity to escape. Green turned and looked in the direction whence he had come.

"You think der's no chance, Osborne?"

"Not one," came the reply.

"And de old Captain can't get away?"

"No," said both men.

With a long look toward the enginehouse and its handful of defenders, Green echoed his utterance several months previous in the stone quarry in Chambersburg: "Well, I guess I'll go back to de old man."[41]

Seeing that his situation was dire, Brown made selection among his prisoners, taking Washington, Allstadt and his son, Bryne, and a half dozen armory officials from the watchman's room into the enginehouse proper, henceforth known as John Brown's Fort. Also in the enginehouse were Stevens, Oliver and Watson Brown, Edwin Coppoc, Stewart Taylor, William Leeman, Jeremiah Anderson (who was dying), Shields Green, and seven of the freed slaves. Ordering his men to make it as strong as possible against assault, Brown said to Washington's slave Phil: "You're a pretty stout-looking fellow. Can't you strike holes through the wall for me?" Phil took up a mason's pick and chopped four holes for firing portals. Once this was done Brown addressed his hostages: "Gentlemen, perhaps you wonder why I have selected you from the others. It is because I believe you to be

The engine house, or "John Brown's fort," in its original location at Harper's Ferry.

the most influential; and I have only to say now, that you will have to share precisely the same fate that your friends extend to my men."[42]

As this was happening William Thompson approached the Potomac Bridge and was captured by railroad "tonnage men" who had come in from Martinsburg. He was bound and hurried across the bridge to be held in a room under guard in the Wager House. Seeing the advantage change favorably to themselves, near pandemonium broke out among the Virginians; many had been drinking in the saloon and hotel for over an hour, and they became emboldened. By now a general assault on the rifle factory had ensued, with more than forty men raining shot of such intensity that Kagi and the others could only seek cover behind machinery. Kagi ordered a brisk fire and retreat to one of the buildings nearest the river. Again the firing became too intense, and he ordered a last retreat across the Shenandoah River.

As the men waded the shallow water Virginians crowded the shore sending a hail of shot after them. A ball to the head killed Kagi before he reached midstream. Four others—Leary, Copeland, and two freed slaves—got to a large rock and returned fire. Leary and one slave were mortally wounded; the other slave was killed outright. Copeland stood alone unharmed. A Virginian waded out with raised pistol. When within yards he snapped his gun, as did Copeland, but neither gun fired. Copeland was captured alive.

With the siege degenerating, to avoid needless carnage Brown asked his son Watson to go out under flag of truce to ask for a cease-fire. After only a few steps with a raised white cloth Watson was brought down by a mortal shot to the pit of his stomach. He managed to crawl back into the enginehouse.

During this shooting William Leeman tried to make an escape. Passing behind the armory buildings he crossed the railroad tracks and began fording the river. Militiamen on both banks spotted him and riddled the water with shot. As Leeman sought refuge on a rock, a Virginian waded out, discharging his gun point blank into his face. After this death some men came out of the saloon to where Newby's body was laid, kicking and dragging it up Shenandoah Street. Stripping Newby of his cowhide boots, they cut off his ears.

A volunteer company of railroad workers now decided it was time to make an assault from inside the armory on the enginehouse. Forming their skirmish line and advancing along the railroad tracks, they were within a hundred yards when some began firing as others began to make a dash. In the next instant they were stopped by galling fire from inside the fort. With half a dozen wounded, one of them severely, the Martinsburg volunteers retreated, as the prisoners left unguarded in the watchman's room took the opportunity to escape.

Between 2 and 3 P.M. rain began to fall, dampening the fight. When it subsided the firing began again more intensely. The acting superintendent of the armory, Kitzmiller, now offered to go out and try to end the fighting; Brown accepted, asking his second in command, Stevens, to accompany him. Carrying colors of truce with the official at his side, after a few paces Stevens was badly wounded in the neck and shoulder by shots coming from inside the Galt House. As Kitzmiller scurried to safety, Joseph Brewer came out of the enginehouse, took Stevens in his arms, and carried him whence his superior had just fled. Stevens was placed on a bed in the room with Thompson, and Brewer, in a gesture of conciliation, returned to the enginehouse.

Across the Ferry Lot, Edwin Coppoc could see a figure ducking around buildings and popping up between freight cars. As the man climbed the water tower for a bird's-eye view, Coppoc said, "If he keeps peeking, I'm going to shoot."[43] It was Fontaine Beckham, mayor of the town and for many years the agent of the B&O, who was unarmed but his movements looked suspicious to men under fire. As Beckham reappeared, Coppoc shot, missing him. A second shot killed the man.

A contemporray drawing of Harper's Ferry at the time of John Brown's raid showing local militia moving toward and firing into the town.

As news of Beckham's killing circulated, a group of well-liquored men stormed into the Wager House seizing William Thompson despite the pleading of the proprietor's daughter. Dragging him from the room into the street with the intent of lynching him, they pushed him toward the abutment of the Potomac Bridge. Finding no rope they ineffectually attempted to tie handkerchiefs together, but finally drew their pistols. Facing the mob, Thompson said, "Though you may take my life, eighty thousand will rise to avenge me and give liberty to the slaves."[44] As he was shot his body fell through the trestle forty feet into the shallow river and the rocks below. His assailants returned to the room where Stevens was held, but he was in such agony they left him, thinking he would be dead in hours. To conclude their reprisal, they returned to fire their pistols into William Thompson's broken body.

It was around 4:30 when reporters from Baltimore strolled into town. Making their way to the room where Aaron Stevens was held, he told them the raiders' sole purpose in coming to Harper's Ferry was to free slaves, and for the first time revealed the number of their party. Then the reporters made their way to a copper shop on the bank of the Shenandoah where Leary lay bleeding unattended. Asking the reporters to write his wife, Leary said, "I am ready to die."[45]

There were now over three hundred militia troops in Harper's Ferry; they and several hundred enraged men held Brown and his men with no possibility of escape. At the enginehouse door, catching sight of a gun raised against him, Oliver Brown brought his rifle up to fire. Before he could get off, he fell mortally wounded. Now two of John Brown's sons lay dying on the enginehouse floor.

As twilight came, shots were heard whistling through the air above the heads of militia troops standing on High Street, with the congressional representative from the district, Alexander Boteler, standing by. It was John Cook, who had climbed a tree part way up Maryland Heights, to see if he could bring assistance to his besieged comrades. The militia returned his fire, severing a branch upon which he rested, and Cook came crashing to the ground.

Cook had come up from the schoolhouse with two freed slaves, leaving Tidd and his party behind. Finding that things had gone badly, Cook returned to the schoolhouse to find it abandoned, with all the arms inside. In his absence, Tidd had gone back to the farm for Owen Brown and the others who would then come down along the towpath in the gathering darkness. Encountering a black woman, she told them what had happened: That all their men in the Ferry were surrounded and many were killed; the "old man" himself had been killed about 4 o'clock, she said.

Had Cook and Tidd and Owen Brown come up when they must have been expected—around noon—things probably would have turned out differently. But further fighting now was pointless. Realizing they could do nothing more, Owen Brown burst into tears. They would return to the farm for the food and provisions still there, hide in the woods for several hours to rest, and then flee to the north. All would escape but Cook, who was captured near Mont Alto nine days later.

Troops would continue pouring in for many hours—from Shepardstown, Martinsburg, and Winchester in Virginia, and from Frederick and Baltimore in Maryland. Governor Henry A. Wise was hastening with militia companies from Richmond; a company of United States Marines commanded by Colonel Robert E. Lee hastened from Washington. By evening there would be over eight hundred militia troops in Harper's Ferry.

At this hour Albert Hazlett and Osborne Anderson, who had remained hidden on the second floor of the arsenal, seized the opportunity to escape. As night fell, with new men continuing to enter town, new faces wouldn't arouse inordinate suspicion. In addition, the town had been disorderly; there was plentiful "bonded Bourbon" on hand, and "cheap redeye." The two came out from the rear of the building and walked briskly away from the scene of everyone's attention along the railroad tracks running parallel to the Shenandoah River. Soon they were on Bolivar Heights and crawled under cover of bushes.

After several hours, with no one approaching, they felt they could make good their escape. Passing to the opposite side of the hill, they made their way down through the town and walked along the Potomac River. Finally stumbling upon a flat-bottomed boat whose oars had been left in it, they crossed the river, hurrying along the towpath of the canal, past the schoolhouse and up into the woods. In two days they would cross the Pennsylvania border. Hazlett, succumbing to hunger, would be captured near Carlisle by federal marshals and remanded to Virginia, along with Cook. Anderson would make good his escape, traveling by rail through Philadelphia and New York City to Rochester, seeking refuge in the home of Frederick Douglass, before the two fled to Canada, just hours ahead of pursuing federal marshals.

The Harper's Ferry raid, as it's been called, was the first national news story in an age that had seen the influence of newspapers vastly increase—an age of steam, iron horses, and electric wires. But there have long been troubling issues as to Brown's intentions which no observer has been able to resolve, many merely assuming he had a confused and ill-thought-out plan. The secret of the extraordinary collapse of the foray at a time when victory was within grasp, speculated DuBois, lay in the fact that it took eleven hours to move two wagon loads less than five miles. Whatever the reason for this delay is *the* explanation, he conjectured, for the defeat and capture of John Brown at Harper's Ferry. Tidd, who was nominally in charge, when in Boston the following year disclosed to Higginson that several of the freed slaves had gone off unsuccessfully to recruit their "friends," while they had waited for them. This delay, however, while their comrades held the way to the mountains open, was inexcusable, and DuBois speculates further that Tidd was being recalcitrant because he had disagreed with the plan.

But does this really explain all that was entailed in the raid, of all that had been planned to happen? The next day, in reply to a query from Virginia's senator James Mason, who lived in Winchester, Brown laid blame for the debacle to his own account. He said: "It is by my own folly that I have been taken. I could easily have saved myself from it, had I exercised my own better judgment rather than yielded to my feelings."[46]

He had allowed himself "to be surrounded by a force by being too tardy," he explained. "I should have gone away; but I had thirty odd prisoners, whose wives and daughters were in tears for their safety, and I felt for them. Besides, I wanted to allay the fears of those who believed we came here to burn and kill. For this reason I allowed the train to cross the bridge, and gave them full liberty to pass on."[47]

Clearly resolution of the issue must lie, at least in part, in consideration of the role of Frederick Douglass, or rather of the proposition he refused. In his writings Douglass was almost entirely circumspect on this point. The measure was never encouraged by his "word" or by his "vote," he maintained.

> At any time or place, my wisdom or my cowardice has not only kept me from Harper's Ferry, but has equally kept me from making any promise to go there. I desire to be quite emphatic here, for of all guilty men, he is the guiltiest who lures his fellowmen to an undertaking of this sort, under promise of assistance which he afterwards fails to render. I therefore declare that there is no man living, and no man dead, who, if living, could truthfully say that I ever promised him, or anybody else, either conditionally, or otherwise, that I would be present in person at the Harper's Ferry insurrection. My field of labor for the abolition of slavery has not extended to an attack upon the United States arsenal. In the teeth of the documents already published and of those which may hereafter be published, I affirm that no man connected with that insurrection, from its noble and heroic leader down, can connect my name with a single broken promise of any sort whatever. So much I deem it proper to say negatively. The time for a full statement of what I know and of all I know of this desperate but sublimely disinterested effort to emancipate the slaves of Maryland and Virginia from their cruel taskmasters, has not yet come, and may never come.[48]

The reason for Douglass's forceful disavowal was the confession of John Cook, where the unfortunate prisoner stated Douglass had failed to make good on his pledge to furnish additional men and that he had himself "promised to be present at the Harper's Ferry insurrection." Douglass wrote in a letter to the *Rochester Democrat and American*, October 31, 1859: "Mr. Cook may be perfectly right in denouncing me as a coward—I have not one word to say in defense or vindication of my character for courage; I have always been more distinguished for running than fighting, and, tried by the Harper's Ferry insurrection test, I am most miserably deficient in courage, even more so than Cook when he deserted his brave old captain and fled to the mountains."

One of the difficulties Brown attempted to resolve by beginning at Harper's Ferry was that of introducing a significant body of men into Virginia's Blue Ridge Mountains; another was to give the movement sufficient momentum. These two purposes were co-determinate. The initial stroke was designed to inspire confidence in the slaves, while exciting a corresponding debilitating fear in slaveholders. The seizure of the government armory and the capture of prominent hostages, while freeing and arming slaves, was the prelude to moving arms and supplies to the country beyond. When the first movement of material from the Kennedy farm to the schoolhouse was completed it was nearly ten o'clock. Thompson was sent out to say that they were in control of the town. He passed Leeman on his way to the Ferry with Byrne, and met Cook at the schoolhouse, but probably, as DuBois surmises, did not overtake Tidd, who would not return with the second load till nearly 4 o'clock.

The first part of the movement had been a success. Anderson wrote: "That hundreds of slaves were ready and would have joined in the work had Captain Brown's sympathies not been aroused in favor of the families of his prisoners, and that a very different result would have been seen, in consequence, there is no question." But there had not been coordination between the two phases. Yet isn't it also possible that some persons were expected who failed to appear? The success of the plan clearly depended on the party seizing the Ferry being "seconded," as Anderson phrased it, "from another quarter." The fatal flaw may have been that Brown was forced to begin a week earlier than planned—on October 16 instead of October 25—and word of this change did not reach those expected to come forward. Frederick Douglass adamantly protested against the plan on grounds that it rested on a supposition completely unsubstantiated: that forces concentrating against them would not attack while Brown held hostages. While Brown never deviated in his conviction, it was no little matter for him to ask Douglass to appear in person at Harper's Ferry, and he, of course, did not intend to thrust his friend into dangers he was scarcely prepared for. But an onlooking nation would ask: if Douglass was with him, who could be against him?

It has been seen that there was a body of men in Philadelphia who had pledged they would participate under Douglass's leadership. Curiously this was to find its echo in July 1863, just as Douglass was preparing to travel to Philadelphia for a speech at National Hall urging enlistment of the city's black men in the Union army. An article appeared in Thomas Hamilton's *Anglo-African* by a resident of that city named Parker T. Smith. Citing an "influential friend," Smith stated there were men in Philadelphia who were eager to form a company "if only" Douglass himself was willing to step up and lead them. Noting Douglass's frequent mention of his two sons in the Massachusetts 54th Regiment, the writer added, "no man's sons can work out his political salvation."

Toward the end of July Douglass sent his response to the magazine, and published it in his *Monthly* that August. He wrote querulously:

Let me tell the said Mr. Smith, if you please, that when he or his influential friend, of whom he speaks, shall have furnished any considerable evidence of his ability to fill my place at the North, he will have done something to convince me that I ought to assume the position he assigns me in the army at the South. I certainly have a pretty high sense of my importance, but Mr. Smith carries it a peg higher when he represents my not enlisting as being the cause of hesitation in his influential friend and others. According to him, there are numerous fighting men in Philadelphia, burning to go to battle, who are only kept back from deeds of valor, because I do not lead them. This is very strange. Whence came this general confidence in me, as a warrior? When have I been heard of as a military man? How happened it that among all the fighting material of Philadelphia, of which Mr. Smith speaks, not one man can be found, who could raise a company of these eager warriors?

For many of Brown's biographers, one of the confusing issues in the movements of that day has been the placement of the arms in the schoolhouse on the Maryland side of the Potomac a mile from the

Ferry. Some have thought this signaled that Brown intended retreat into Maryland after freeing slaves. It has also been remarked that arms were left behind in Chambersburg, which has often been attributed to an oversight that Richard Hinton was sent to rectify at the last moment. There is indication, too, although it has never been divulged, that men from Chambersburg were expected. There was a letter dated October 21 by William A. Boyle, a physician from the town, informing Governor Wise that "a portion of our Negroes knew of it and were expected to join in it."[49]

No one has ever surmised this missing element: having seized Harper's Ferry, with slaves and others rallying to his standard, John Brown planned another demonstration—that of none other than Frederick Douglass himself coming up with the body of men named to march into the Ferry. Brown was clearly expecting a party three or four times the size that went with him. With fifty to seventy-five men, instead of nineteen, the inundation of militia companies could have been entirely forestalled. The fact that Brown envisioned an exchange of his slaveholder hostages "for able bodied male slaves" suggests a formal setting, where, he also intended, after putting them into service at Harper's Ferry, returning the weapons of President George Washington to their proprietor after they had been surrendered into the hands of a black man. All this, highly symbolic, was intended to be duly noted in the direction of the South, when before a sizable assembly, a thunderbolt would be delivered to roll across the nation, as, standing within the grounds of the United States Armory, Frederick Douglass would proclaim emancipation in Maryland and Virginia. Finally, Brown planned for the bell in the cupola of the enginehouse to be rung, summoning slaves and others from the countryside. A culmination implicit in the movements on that day that has never appeared in accounts of the Harper's Ferry raid.

Even before he began, Brown's expectations had almost come to naught. Yet he hurried on, certain that although his blow "may be but like the final struggle of Samson," it was to be one from which American slavery could never recover. He never wavered from this belief despite his situation, and was to write in his captivity: "As I believe most firmly that God reigns, I cannot believe that anything I have done, suffered, or may yet suffer, will be lost to the cause of God or of humanity. And before I began my work at Harper's Ferry, I felt assured that in the worst event it would certainly pay. I often expressed that belief; and can now see no possible cause to alter my mind. I am not as yet, in the main, at all disappointed. I have been a good deal disappointed as it regards myself in not keeping up to my own plans; but I now feel entirely reconciled to that, even—for God's plan was infinitely better, no doubt, or I should have kept to my own."[50]

Some of the others who went with him also had no compunction about mere "personal defeat." Hinton, who just missed being present at Harper's Ferry, only conjectured in his book that Kagi had sent him back to Chambersburg to save his life.

In the decade after the war Hinton would serve as the secretary of the Washington, D.C. branch of the International Workingmen's Association, and in May 1871 in an article appearing in the *Atlantic Monthly* lauded the *communards* of Paris for giving "form and substance to a purpose so grand, a spirit so comprehensive," and Marx was to conclude "the triumph of the American Republic over slavery has given such impetus to all radical agitation in Europe." Harper's Ferry and the revolutionary Commune were the counterpoints on two continents around which entire historic tendencies converged.

Eight

Year of Meteors

I would sing in words retrospective some of your
Deeds and signs,
I would sing of your contest for the 19th Presidentidad,
I would sing how an old man, tall with white hair,
Mounted the scaffold in Virginia . . .
—Walt Whitman, "Year of Meteors, 1859 '60,"
Leaves of Grass

"ON THE EVENING WHEN THE NEWS CAME THAT JOHN BROWN had taken and was then holding the town of Harper's Ferry, it so happened that I was speaking to a large audience in National Hall, Philadelphia. The announcement came upon us with the startling effect of an earthquake. It was something to make the boldest hold his breath," wrote Frederick Douglass in his autobiography. Soon upon this came report that Brown had fortified and was holding a small enginehouse with two or three of his men, and just as fast came news that United States troops had made a breach in the building and captured and mortally wounded Brown. Visiting Richmond, Virginia, at the time of his friend's debacle, Alexander M. Ross of Toronto later recalled the scene on the night of October 17: "Crowds of rough, excited men filled with whiskey and wickedness stood for hours together through the night in front of the offices of the

Enquirer listening to reports as they were announced within. When news of Brown's defeat and capture was read from the window, the vast crowd set up a demonic yell of delight, which to me sounded like a death-knell to all my hopes for the freedom of the enslaved."[1]

In the enginehouse that night John Brown had waited for the inevitable to transpire, attended to his dying sons, and spoke at various times with his prisoners. With him were four of his men, three of them unwounded, seven freed slaves, and ten hostages. Eight of his men had been killed in the fighting and two were captives—Stevens and Copeland—and six had escaped, with two of these later captured. At least seven of the freed slaves who had taken arms had been killed, and many of the others left the scene as things started to go badly. The people of Harper's Ferry and the Virginia and Maryland militia companies suffered more than the four officially reported killed, with many wounded, but these numbers were to be quickly suppressed.

Meanwhile, the telegraph had been repaired and Governor Wise of Virginia sent his orders to the militia companies establishing a command. Just after midnight Colonel Lee arrived with ninety-nine United States Marines, taking possession of the armory, establishing order in the town, and closing the saloons. The rain, which had been intermittent, had stopped; now a river fog slid into the town. At an early hour in the morning Oliver Brown died. His father went over to his body, removing his cartridge belt and straightening his limbs. Then he said to one of the prisoners: "This is the third son I have lost in this cause." That man said to him: "You are as brave as any man I know, and as sensible on all other subjects as any, but I cannot go along with your feeling on the blacks."[2]

After Robert E. Lee secured the armory, he reconnoitered the enginehouse in anticipation of a bloody fight. Then he sent his adjunct, J. E. B. Stuart, to the building's entrance to notify the defenders of his arrival. One of Brown's prisoners, the paymaster of the armory John E. P. Daingerfield, later wrote what he witnessed of the encounter, which appeared in June 1885 in *Century* magazine:

When Stuart was admitted, and a light brought, he exclaimed, "Why, aren't you old Osawatomie Brown of Kansas, whom I once had there as my prisoner?" "Yes," was the answer, "but you did not keep me." This was the first intimation we had of Brown's real name. When Colonel Lee advised Brown to trust to the clemency of the government, Brown responded that he knew what that meant—a rope for his men and himself, adding "I prefer to die just here." Stuart told him he would return at early morning for his final reply, and left him. When he had gone Brown at once proceeded to barricade the doors, windows etc., endeavoring to make the place as strong as possible. All this time no one of Brown's men showed the slightest fear, but calmly awaited the attack, selecting the best situation to fire from, and arranging their guns and pistols so that a fresh one could be taken up as soon as one discharged.

Besides the marines and the men in the enginehouse, there were many hundreds of armed men in Harper's Ferry. At dawn at the sound of marching the town's people came out, filling the streets, some climbing atop their roofs for a view, and standing on the stone trestle of the Potomac Bridge. Two thousand people drew up to see the assault on the enginehouse. Before commencing, Colonel Lee went to the side of the commandant of the Maryland militia companies, offering him the honor of beginning the assault with his troops. He declined. Lee turned to the ranking officer of the Virginia militias, offering the same; he too declined. Then the future commander of the Army of Northern Virginia turned to Lieutenant Green of the marines and asked if he wished "the honor of taking those men out." He accepted.

Twelve volunteers were called for from among the marines, three of whom were handed sledgehammers to beat the doors in. Lee instructed the storming party to use bayonets only, to distinguish between the prisoners and the defenders, and to ignore the blacks unless they showed signs of fight. It was 7 o'clock in the morning, Tuesday, October 18, 1859, when "Jeb" Stuart, wearing the regalia for which he became distinguished—the ostrich plumed hat—stepped forward. The waving of the hat was the signal to begin.

The marines pounded the doors, but to no effect. Then dropping their hammers, they picked up a heavy ladder that lay at the side of the building to use as a battering ram. The first blows did not tell. Finally a blow splintered the boards making a breach large enough for a man to squeeze through. Lieutenant Green and Major Russell of the marines slipped through; the next two marines to force through were mortally wounded. Daingerfield ends his account this way:

> The engine rolled partially back, making a small aperture, through which Lieutenant Green of the marines forced his way, jumped on

top of the engine, and stood, amidst a shower of balls, looking for
John Brown, when he saw Brown, he sprang about twelve feet at
him, giving an under thrust of his sword, striking Brown about
midway the body, and raising him completely from the ground.
Brown fell forward, with his head between his knees, while Green
struck him several times over the head, and, as I then supposed
split his skull at every stroke. I was not two feet from Brown at the
time. Of course, I got out as soon as possible, and did not know
till some time later that Brown was not killed. It seems that Green's
sword, in making the thrust, struck Brown's belt and did not penetrate the body, the sword was bent double. The reason that
Brown was not killed when struck on the head was that Green was
holding his sword in the middle, striking with the hilt, and making
only scalp wounds.

Two of Brown's men were run through in the assault, while two
were captured unharmed. John Brown and his son Watson were
brought out and laid on the grass. Then the bodies of Dauphin
Thompson, Oliver Brown, and Stuart Taylor were brought out.
Edwin Coppoc and Shields Green were brought out under the grip of
the marines, and the blacks were brought out and stood to the side.
Now the prisoners emerged. Blankets were brought for the prostrate
men and physicians summoned to attend them. Now Lee ordered
Brown and the wounded be moved indoors into the paymaster's
office in the armory building. Stevens was carried from the hotel and
laid beside his commander.

On the afternoon of October 18 Governor Henry A. Wise and
Senator James M. Mason arrived in Harper's Ferry and hastened to
the room where the wounded insurgents were held. With them were
representative Clement C. Vallandigham of Ohio, as well as other
officials and officers along with newspaper reporters. They were to
interview Brown for over three hours. Governor Wise bent over the
matted and bloodstained body on the floor.

"Who are you?" he asked.

"My name is John Brown; I have been well known as old John
Brown of Kansas. Two of my sons were killed here today, and I'm
dying too. I came here to liberate slaves, and was to receive no
reward. I have acted from a sense of duty, and am content to await
my fate; but I think the crowd have treated me badly. I am an old
man. Yesterday I could have killed whom I chose; but I had no desire

to kill any person, and would not have killed a man had they not tried to kill me and my men. I could have sacked and burned the town, but did not. I have treated persons whom I took as hostages kindly, and I appeal to them for the truth of what I say. If I had succeeded in running off slaves this time, I could have raised twenty times as many men as I have now, for a similar expedition. But I have failed."

Senator Mason continued the questioning.

"Can you tell us who furnished money for your expedition?"

"I furnished most of it myself; I cannot implicate others. It was by my own folly that I have been taken. I could easily have saved myself from it, had I exercised my own better judgment rather than yielded to my feelings."

Next into this discourse entered Representative Vallandigham.

"Mr. Brown," he snapped, "who sent you here?"

"No man sent me here; it was my own prompting and that of my Maker, or that of the Devil—which ever you please to ascribe it to. I acknowledge no master in human form."

Mason broke in—

"What was your object in coming?"

"We came to free the slaves, and only that."

One of the militia officers wanted to know how many men in all had held them at bay?

"I came to Virginia with eighteen men only, besides myself."

"What in the world did you suppose you could do here in Virginia with that amount of men?"

"Young man, I do not wish to discuss that question here."

"You could not do anything."

"Well, perhaps your ideas and mine on military subjects would differ materially."

Someone asked—

"Do you consider this a religious movement?"

"It is, in my opinion, the greatest service a man can render to God."

"Do you consider yourself an instrument in the hands of Providence?"

"I do."

"Upon what principle do you justify your acts?"

"Upon the golden rule. I pity the poor in bondage that have none to help them. That is why I came here, it is not to gratify any personal animosity, or feeling of revenge, or vindictive spirit. It is my sympathy with the oppressed and the wronged, that are as good as you, and as precious in the sight of God."

"Certainly. But why take the slaves against their will?"

"I never did."

"Why did you do it secretly?"

"Because I thought that necessary to success, and for no other reason."

Vallandigham questioned him closely on his activities in Ohio the previous spring, showing he had followed in some manner on his passing.

"When in Cleveland, did you attend the Fugitive Slave law convention there?"

"No. I was there about the time of the sitting of the court to try the Oberlin rescuers. I spoke there publicly on that subject; on the Fugitive Slave Law and my own rescue. Of course, so far as I had any influence at all, I was supposed to justify the Oberlin people for rescuing the slave, because I have myself forcibly taken slaves from bondage. I was concerned in taking eleven slaves from Missouri to Canada last winter. I think I spoke in Cleveland before the convention. I do not know that I had conversation with any of the Oberlin rescuers. I was sick part of the time I was in Ohio with the ague, in Ashtabula County."

Finally Brown broke off the line of questioning: "I want you to understand, gentlemen, and (to the reporter from the 'Herald') you may report that—I want you to understand that I respect the rights of the poorest and weakest of the colored people, oppressed by the slave system, just as much as I do those of the wealthy and powerful. This is the idea that has moved me, and that alone. We expected no reward except the satisfaction of endeavoring to do for those in distress and greatly oppressed as we would be done by. The cry of distress of the oppressed is my reason, and the only thing that prompted me to come here."

One Virginia upstart shot in—

"Brown. Suppose you had every nigger in the United States, what would you do with them?"

(In a loud voice) "Set them free."

"Your intention was to carry them off and free them?"

"Not at all."

Another bystander came in—

"To set them free would sacrifice the life of every man in this country."

"I do not think so."

"I know it. I think you are fanatical."

"And I think you are fanatical. 'Whom the gods would destroy they first make mad,' and you are mad."

Toward the end of the interview, seeing they were making no headway on their own terms, Governor Wise interrupted: "Mr. Brown, the silver of your hair is reddened by the blood of crime, and you should eschew these hard words and think upon eternity. You are suffering from wounds, perhaps fatal; and should you escape death from these causes, you must submit to a trial which may involve death. Your confessions justify the presumption that you will be found guilty; and even now you are committing a felony under the laws of Virginia, by uttering sentiments like these. It is better you should turn your attention to your eternal future than be dealing in denunciations which can only injure you."

Brown replied: "Governor, I have from all appearances not more than fifteen or twenty years the start of you in the journey to that eternity of which you kindly warn me; and whether my time here shall be fifteen months, or fifteen days, or fifteen hours, I am equally prepared to go. There is an eternity behind and an eternity before; and this little speck in the center, however long, is but comparatively a minute. The difference between your tenure and mine is trifling, and I therefore tell you to be prepared. I am prepared. You have a heavy responsibility, and it behooves you to prepare more than it does me."[3]

This was the only interrogation Brown submitted to. Said Wendell Phillips, "Having taken possession of Harper's Ferry, he began to edit the *New York Tribune* and the *New York Herald* for the next three weeks." Already Brown seemed to know he was addressing a national audience, answering all questions forthrightly, but from the beginning, despite the weighty experience of his interrogators, John Brown commanded the interview. His name now would no longer be associated with a particular locality or with events in a section of the country, as Governor Wise unwittingly set the stage for the protagonist hero to show the nation the way of deliverance.

Wise arrived at Harper's Ferry about one o'clock on Tuesday, October 18, almost five hours after the insurgents had been subdued by the marines and Colonel Lee.

If he had arrived before their surrender, Wise explained in his message to the legislature on December 5, he "would have proclaimed martial law," stormed the enginehouse "in the quickest time possible," given no quarter, tried any survivors by court-martial, and shot the condemned on the spot. But he had been too late, and found he was bound to protect the prisoners as a mob gathered demanding they be lynched. Soon he had secured Brown's carpetbag, finding numerous letters and documents implicating many of his supporters. Later that day Douglass received a telegram warning him that he and others were to be arrested. Urged by friends to leave Philadelphia on the next train, Douglass hastened with an escort to the Walnut Street Ferry in expectation of momentary arrest if he did not depart.

Crossing to Camden, he reached New York City that night and went at once to the Barclay Street Ferry. Crossing the Hudson to Hoboken, he went directly to the home of Mrs. Marks on Washington Street and there spent a restless night.

The next morning, across the Hudson River, a *New York Herald* headline announced "Gerrit Smith, Joshua Giddings, Fred Douglass and Other Abolitionists and Republicans Implicated," and a search would be made "for papers as well as persons" and that the government would bring to punishment all that were connected with the "outrage." Douglass had not, to be sure, gone all the way with John Brown, but he knew Virginia could prove he had on numerous occasions met with him; that he had supplied him with a recruit and financial aid, and had given his countenance in other ways. Upon proper requisition Douglass expected to be delivered up for trial in Virginia with Brown, and in the present climate he too would assuredly be found guilty and executed. What is more, Douglass had numerous letters from Brown in his desk in Rochester where he also kept a copy of his provisional constitution.

To prevent these papers from falling into the hands of his enemies, Douglass dictated a note to his friend, Miss Ottilia Assing, addressed to the telegraph operator in Rochester, also a friend, who would understand its meaning. "B. F. Blackwell, Esq.: Tell Lewis to secure all the important papers in my high desk." Years later Douglass was to write, "The mark of the chisel with which the desk was opened is still on the drawer, and is one of the traces of the John Brown raid."[4]

Douglass saw his future compounded with uncertainty. He had not yet finished his schedule of speaking engagements in the run up to his trip to England, but now the speaking tour had to be abandoned and his travel abroad delayed. To stay in Hoboken was out of the question. But should he go to Rochester, where it would be difficult for the authorities to take him? By traveling there he would

expose himself to arrest on the road. Which
route to travel? New York City was as
incensed against John Brown as many of the
cities of the South. Finally Douglass's friends
decided he should leave Hoboken for
Paterson at night, and by private conveyance. There he could board the Erie
Railroad, reaching Rochester in the shortest
possible time.

On the morning of October 20, in agree-
ment with President Buchanan, Wise took
John Brown, Shields Green, John Copeland,
Edwin Coppoc, and Aaron Stevens under the

Henry A. Wise.

jurisdiction of Virginia—Watson Brown died at noon on the 19th—
and conveyed them under marine guard to the jail in Charlestown,
there to be "proceeded against regularly by the civil authority, under
process of both state and federal governments." He deigned not to
remove the prisoners further into the interior because he was determined, he said, to show he had no apprehension as to their rescue,
and if the jail of Jefferson County had been on the state line they
would have been kept there.

Wise was a Democrat who had attained the governorship of
Virginia, with aspirations for national office, in the election of 1856,
swearing if John C. Frémont won the presidency he would raise fifty
thousand men of the South to march under arms into the North to
make a revolution. Tall with a slim physique, Wise was fifty-three at
the time, a great "talker and chewer" of whom it was said he "rioted
in the eccentricities of his genius." He had a flair withal for public
speaking. Seeking to dramatize the affair at Harper's Ferry to defend
slavery, he said it was no result of ordinary crime, "but it was an
extraordinary and actual invasion, by a sectional organization, specially upon slaveholders and upon their property in Negro slaves."
On the evening of his arrival at the Ferry, in a stump speech outside
the Wager House he denounced the residents of the town as "a flock
of sheep," and said he was ready to weep when he heard "the town
was taken in ten minutes by twelve men." His antagonist he portrayed as "a broken winged hawk, lying on his back, with a fearless
eye, and his talons set for further fight, if need be."[5]

The governor appointed Charlestown lawyer Andrew Hunter special prosecutor, and on October 26 the grand jury, then sitting,

brought indictments against the prisoners for murder, treason, and exciting slaves to rebel. Hunter and Wise were in agreement that the defendants should be "arraigned, tried, found guilty, sentenced and hung," all within ten days if possible. Wise was determined to demonstrate that Brown could only have proceeded with his plans with extensive financial and material backing in the North, and sought and brought indictments against supporters revealed in the captured correspondence.

Upon Douglass's arrival in Rochester a tried friend, accompanied by New York's lieutenant governor, who lived in Rochester, paid a visit to him. Assuring Douglass there were many in Rochester who would act to prevent his arrest and stop any process against him, they also stated that the governor would surely honor a request from Virginia for his extradition. To prevent bloodshed they urged he flee the state for Canada. Heeding this advice, Douglass fled with Osborne Anderson as soon as he arrived. Later in his autobiography, Douglass noted that federal marshals knocked at his door within six hours of his departure, armed with warrants for his arrest for murder and robbery and for inciting servile insurrection in Virginia. Within a week of this, the district attorney for western New York and other officials alighted in Rochester making inquiries about Frederick Douglass, while Governor Wise hired a detective in Virginia to circulate under the code name "C. Camp," to determine his whereabouts.

Just as Douglass arrived in Canada, an account was published of his supposed involvement in the Harper's Ferry affair based upon John Cook's statements to his Virginia interrogators. Douglass took this as an opportunity to explain his actions in the *Rochester Democrat and American*, October 31, and reprinted in the *New York Herald*, November 4. Denying Cook's allegation that he had promised to be present at Harper's Ferry and bring with him a large contingent, Douglass passed on Cook's attribution of his failure to appear to "cowardice." He wrote: "Having no acquaintance whatever with Mr. Cook, and never having exchanged a word with him about the Harper's Ferry insurrection, I am disposed to doubt if he could have used the language concerning me which the wires attribute to him. The lightning, when speaking for itself is among the most direct, reliable and truthful of things, but when speaking of the terror-stricken slave-holders at Harper's Ferry, it has been made the swiftest of liars."

By the act of slaveholding, owners "voluntarily placed themselves beyond the laws of justice and honor," Douglass wrote, and it could never be wrong to rise up and strike them down. If anyone was inclined to think less of him for this sentiment, or because he had been aware of what was about to take place and did not assume the character of an informer—so be it. "Let every man work for the abolition of slavery in his own way. I would help all and hinder none," Douglass affirmed. He had no apology for keeping out of the way of the United States marshals: "A government, recognizing the validity of the Dred Scott decision at such a time as this, is not likely to have any very charitable feelings toward me. . . . I have quite insuperable objections to being caught by the hounds of Mr. Buchanan, and 'bagged' by Gov. Wise. For this appears to be the arrangement. Buchanan does the fighting and hunting, and Wise 'bags' the game."

Wise later took note of Douglass's statement, remarking, "This negro had published his proclamation against Governor Wise. . . . He had said then he has no idea of going back to New York or Pennsylvania for fear that Governor Wise, through federal agents, will bag him. I will never put my hemp in the form of a bag for him, it will be in the shape of a rope."[6]

In Concord, Massachusetts, Franklin Sanborn, consulting with his lawyer, was informed that he might be extradited and tried with Brown. One day after Frederick Douglass fled for Canada, Sanborn took a steamer to Portland, Maine, and there would remain until Brown was executed. Two days after Sanborn left Concord, Dr. Howe and George Stearns fled Boston for Quebec. In New York, when Gerrit Smith was asked by a reporter what his connection had been to Brown, he found it hard to remember just who he was, and would subsequently seek to destroy evidence of his dealing and friendship with him. Wracked with fear and growing incoherent, he committed himself to the New York State Asylum for the Insane at Utica.

With Theodore Parker dying in Rome, only Thomas Wentworth Higginson of all the "secret six" would remain in the States, although his involvement had not yet become public. In exile Stearns would remain silent, although as perplexed as the others; while Howe, and to a lesser degree Sanborn, engaged in recriminations and denial. Howe complained that the event had been "unforeseen and unexpected" by him, and neither his acquaintance with nor knowledge of Brown's character allowed him to "reconcile" it. "It is still to me a mystery and a marvel," he wrote. To Higginson he wrote: "It is true that I ought to have expected an onslaught somewhere but the point is that I did not expect anything like what happened." Howe's back-tracking caught the contempt of Higginson: "Since language was first invented to conceal thoughts, there has been no more skillful combinations of words." And to Sanborn he wrote: "Sanborn, is there no such thing as honor among confederates? . . . Can your moral sense be so sophisticated to justify holding one's tongue in the face of this lying—and lying under the meanest of circumstances—to save ourselves from all share in even the reprobation of society when the nobler man whom we have provoked on to danger is the scapegoat of that reprobation—& the gallows too?"[7]

Meanwhile, advised that Douglass was somewhere in the state of Michigan, Governor Wise made requisition upon the governor there for his surrender. It was not out of the question that he could be kidnapped from Canada and taken to Virginia, so what he had looked forward to as a pleasure was now a necessity. On November 12 Douglass took passage on a steamer from Quebec bound for England. Douglass wrote: "I could but feel that I was going into exile, perhaps for life. Slavery seemed to be at the very top of its power. . . . Nobody could then see that in the short space of four years this power would be broken and the slave system destroyed. So I started on my voyage with feelings far from cheerful. No one who has not himself been compelled to leave his home and country and go into permanent banishment can well imagine the state of mind and heart which such a condition brings."[8]

Learning of Douglass's escape with disappointment, Wise was to give a somewhat grandiloquent venting of his frustration in Richmond on December 26 in his speech "The Question of the Day": "Fred Douglass says that he is bound for England. Let him. Oh! If I had had one good, long, black, low, rakish, well-armed steamer in Hampton Roads, I would have placed her on the Newfoundland Banks with orders that if she found a British packet with that negro on board to take him. (Tremendous applause.) And by the eternal gods he should have been taken—taken with the very particular instructions not to hang him before I had the privilege of seeing him well hung. (Laughter and applause.)"[9]

The trials of the defendants began on October 27, with John Brown's the first one up. Convicted on November 2, he was sentenced to hang in one month, on December 2, 1859. Albert Hazlett and John Cook were brought in after their capture and tried in succession with the others. Each was found guilty of murder and treason and of exciting slaves to insurrection; each was sentenced to the scaffold. All sentences were duly carried out.

Throughout his trial Brown lay on a cot, slowly healing his wounds. The pain in his left kidney was especially severe, making his ordeal a miserable one. His trial was cut and dried. In came Colonel Washington, Allstadt, and a few other witnesses who told their stories. Motions were made for postponement on account of Brown's wounds, or until his lawyer could arrive from the North—a young man named Hoyt sent by Higginson to join Brown's court-appointed attorneys. The court refused postponement each time.

At the outset of the trial one of Brown's court-appointed attorneys presented a letter he had received from Akron, Ohio, alleging insanity ran in Brown's family, offering the letter as justification for declaring the prisoner innocent by insanity. Brown rose from his cot to speak: "I look upon it as a miserable artifice and pretext. I view it with contempt more than otherwise. As I remarked to Mr. Green, insane persons, so far as my experience goes, have but little ability to judge their own sanity; and if I am insane, of course, I should think I know more than all the rest of the world. But I do not think so. I am perfectly unconscious of insanity, and reject, so far as I am capable, any attempt to interfere in my behalf on that score."[10] This ended the matter.

As the witnesses told their stories Brown instructed his counsel:

"We gave numerous prisoners perfect liberty.

Get all their names.

We allowed numerous other prisoners to visit their families to quiet their fears.

Get all their names.

We allowed the conductor to pass his train over the bridge with all his passengers, I myself crossing the bridge with him, and assuring all the passengers of their perfect safety.

Get the conductor's name, and the names of the passengers, so far as may be.

Our orders, from the first and throughout, were, that no unarmed person should be injured, under any circumstances.

Prove that by ALL the prisoners.

We committed no destruction or waste of property.

Prove that."[11]

Brown often appeared to sleep, and nearly always remained in perfect calm. He began to cast a spell on his onlookers, who included not only the five hundred spectators crowded into the court, but numerous reporters from the major Eastern newspapers, and from Baltimore, Richmond, and Washington. "His confinement has not at all tamed the daring of his spirit," wired the correspondent of the *New York Herald*.[12]

On the second day of the trial the man who had been the leader of the mob that seized and murdered William Thompson was called as a witness. The son of Andrew Hunter, the prosecuting attorney, he had come in with the Jefferson Guard, but had gone off to fight on his "own hook," as he said. John Brown sat up on his cot and raised himself to his feet. Looking directly into the young man's eyes, he said: "Tell us the details. Hold nothing back." As the witness began to tell his story Brown laid down and drew his blanket over his shoulder with a groan.

After several more witnesses, Brown rose again. "May it please the court, I discover that notwithstanding the assurances I have received of a fair trial, nothing like a fair trial is to be given me." He denounced his counsel for not having subpoenaed witnesses, and complained that all his money had been taken from him, so he could get "nobody to do any errand" on his behalf. The rest of the proceedings he was quiescent and took no part. Finally the jury filed out to deliberate. Forty-five minutes later they filed back into the jury box. Brown sat up on his cot. "The jury find the prisoner guilty of treason in advising and conspiring with slaves and others to rebel, and of murder in the first degree."[13]

Brown was sent back to his cell while the second defendant, Edwin Coppoc, was arraigned and his jury impaneled. The trial had been completed just four days after the arraignment. The trials of Coppoc, Cook, Green, and Copeland were all over by November 9. Hazlett's trial would be delayed because of uncertainty as to his identity—he purported to be William Harrison and had never been at Harper's Ferry—while Brown and the others would maintain to the end they did not know him. Stevens's trial, too, was delayed on account of the severity of his wounds, but also because Wise and Hunter contemplated a federal trial for him to pursue indictments against some of Brown's supporters. But this tack was dropped when, on December 15, the United States Senate appointed to investigate the "Harper's

Ferry Raid" the Mason Committee, which subsequently subpoenaed many of those involved.

The first northerners' reactions to the seizure of Harper's Ferry had been dismay and bewilderment—the work of a "madman" or "fanatic"—and but for the loss of life it seemed ridiculous. Yet the determination to conclude the proceedings against Brown in the shortest possible time, providing a mere façade of a fair trial, began to recoil upon Virginia. In Lawrence, as Brown's trial concluded, the *Kansas Republican* editorialized: "We defy an instance to be shown in a civilized community where a prisoner has been forced to trial for his life, when so disabled from his ghastly wounds as to be unable even to sit up during the proceedings and compelled to be carried to the judgment hall upon a litter."

With the people in the northern counties of Virginia intensely excited, and with the accompanying military measures, the trials took on an inquisitorial air, as day after day, with public interest unabated, reports continued to fill the papers. The "raid" now began to appear more plausible, for it was seen that the South had blanched with terror as nineteen men held a town of three thousand on its northern border and fought the militia companies sent to aid them to a standstill.

One of the first to speak in Brown's favor was Henry David Thoreau, whose *A Plea for Captain John Brown* is one of the outstanding statements of the age. As Brown's trial was under way on October 30, Thoreau sent a messenger from house to house in Concord notifying his neighbors he was going to speak in the Unitarian Vestry on John Brown that evening. A note came back from a "Republican" friend that such a speech might be premature. "You misunderstand," countered the naturalist and philosopher, "I did not send to you for advice but to announce that I would speak."[14]

He had called his audience together "to correct the tone and the statements of the newspapers," Thoreau said, and to express his sympathy and admiration for Brown and his companions.

> There are at least as many as two or three individuals to a town throughout the North who think much as the present speaker does about him and his enterprise. I do not hesitate to say that they are an important and growing party. We aspire to be something more than stupid and timid chattels, pretending to read history and our Bibles, but desecrating every house and every day we breathe in. Perhaps anxious politicians may prove that only seventeen white

men and five Negroes were concerned in the late enterprise; but their very anxiety to prove this might suggest to themselves that all is not told. Why do they still dodge the truth? They are so anxious because of a dim consciousness of the fact, which they do not distinctly face, that at least a million of the free inhabitants of the United States would have rejoiced if it had succeeded.

Who then were Brown's constituents? asked Thoreau. He answered:

> If you read his words understandingly you will find out. In his case there is no idle eloquence, no made, nor maiden, speech, no complements to the oppressor. Truth is his inspirer, and earnestness the polisher of his sentences. He could afford to lose his Sharpe's rifles, while he retained his faculty of speech, a Sharpe's rifle of infinitely surer and longer range.
>
> And the *New York Herald* reports the conversation "verbatim"! It does not know of what undying words it is made the vehicle.[15]

Two nights later in Boston, the same address was heard as "the fifth lecture of the Fraternity Course," originally scheduled for an appearance of Frederick Douglass. "The reason I am here, is the reason Douglass is not," Thoreau began in Boston.

On the morning of November 2, John Brown walked to the court wearing manacles between double rows of troops, dividing a crowd straining to see the prisoner. When the court reconvened the clerk asked: "Is there any reason why sentence should not now be passed upon you?" Surprised by the abruptness of the passing of sentence, and not having yet taken time to compose his statement, Brown, nevertheless, rose from his cot and stood before the Court. He began:

> I have, may it please the Court, a few words to say.
>
> In the first place, I deny everything but what I have all along admitted—the design on my part to free the slaves. I intended certainly to have made a clean thing of that matter, as I did last winter, when I went into Missouri and there took slaves without the snapping of a gun on either side, moved them through the country, and finally left them in Canada. I designed to have done the same thing again, on a larger scale. That was all I intended. I never did intend murder, or treason, or the destruction of property, or to

excite or incite slaves to rebellion, or to make insurrection. I have another objection, and that is, it is unjust that I should suffer such a penalty. Had I interfered in the manner which I admit, and which I admit has been fairly proved (for I admire the truthfulness and candor of the greater portion of the witnesses who have testified in the case), had I so interfered in behalf of the rich, the powerful, the intelligent, the so-called great, or in behalf of their friends—either father, mother, brothers, sisters, wife, or children, or any of that class—and suffered and sacrificed what I have in this interference, it would have been all right; and every man in the court would have deemed it an act worthy of reward rather than punishment.

This court acknowledges, as I suppose, the validity of the law of God. I see a book kissed here which I suppose to be the Bible, or at least the New Testament. That teaches me that all things whatsoever I would that men should do to me, I should do even so to them. It teaches me, further, to "remember them that are in bonds, as bound to them." I endeavored to act up to that instruction. I say, I am as yet too young to understand that God is any respecter of persons. I believe that to have interfered as I have done—as I have always freely admitted I have done—in behalf of His despised poor, was not wrong, but right. Now, if it is deemed necessary that I should forfeit my life for the furtherance of the ends of justice, and mingle my blood further with the blood of my children and with the blood of millions in this slave country whose rights are disregarded by wicked, cruel, and unjust enactments—I submit; so let it be done!

Let me say one word further.

I feel entirely satisfied with the treatment I have received on my trial. Considering all the circumstances, it has been more generous than I expected. But I feel no consciousness of guilt. I have stated from the first what was my intention, and what was not. I never have had any design against the life of any person, nor any disposition to commit treason, or excite slaves to rebel, or make any general insurrection. I never encouraged any man to do so, but always discouraged any idea of that kind.

Let me say, also, a word in regard to the statements made by some of those connected with me. I hear it has been stated by some of them that I have induced them to join me. But the contrary is true. I do not say this to injure them, but as regretting their weakness. There is not one of them but joined me of his own accord, and the greater part of them at their own expense. A number of them I never saw, and never had a word of conversation with, till the day they came to me; and that was for the purpose I have stated.

Now, I have done.[16]

On the day Brown's trial began, haymows on a farm outside Charlestown were set on fire, taking with them several outbuildings. This was followed by more fires, including the destruction of the "big house" on the estate of George W. Turner, the slaveholder killed at Harper's Ferry who was said to have been hated by "his Negroes." Three of his jurors had property burned following Brown's conviction; but still more disconcerting were rumors circulating of a rescue of Old Brown, and of preparations for further armed invasions from the North. It was said five hundred men, who had fought with Brown and had sworn not to let him hang without spilling more blood on Virginia's soil, were coming forward from Kansas to effect his deliverance. Then a letter was received by the special prosecutor from the United States marshal in Cleveland stating he had positive evidence that a thousand men were gathering arms in Ohio for a march on Charlestown. Governor Wise easily believed these rumors, seeing evidence of sinister import behind every shadow. Ordering his militia officers to extend their vigilance and to make full use of their troops, he sent an additional two hundred muskets to Charlestown together with ammunition. Guards were stationed at the jail and across the street at the courthouse, the railroad was placed under guard, as an armed circle was drawn around the center of town. On the second day of trial loaded cannon were placed at the front corners of the building, aimed at the building itself, and additional troops were placed at the approaches to the town. The curious and sightseers began making the trek from Winchester and from other parts of Virginia, drawing with them hawkers and peddlers; while Edwin Ruffin, the agriculturist and sage of the South, came to observe the trials. With its streets turned into a parade ground, travel to and from Charlestown was hindered; and as tempers flared farmers began to refuse to bring in produce. Wise wrote to Andrew Hunter: "Information from every quarter leads to the conviction that there is an organized plan to harass our whole slave border at every point. Day is the very time to commit arson with the best chance against detection. No light shines, nor smoke shows in daylight before the flame is off and up past putting out. . . . I tell you those Devils are trained in all the arts of predatory war. They come one by one, two by two, in open day, and make you stare that the thing be attempted as it was done."[17]

In New York City on the Sunday before Brown's trial began, Henry Highland Garnet told his congregation it was the duty of freedom-loving people "to affirm the rightness" of the raid. The only right of slavery was the right to die, he avowed. But in a later speech he suggested if Brown had spent twenty thousand dollars, as the papers reported, he only needed twenty. "All that was needed was a box of matches in the pocket of every slave, and then slavery would be set right." Night after night "the heavens are illuminated by the lurid glare of burning property," it was reported, as barns, stables, and haystacks were set afire. In addition several masters had been attacked and beaten by their servants. One journalist from the North noted that blacks in the vicinity of Harper's Ferry seemed eager to learn "every particular" of the raid, and what one found out, others soon knew. He wrote: "They have a pretty effective and secret Free Masonry among them." Ten days after his trial Brown wrote his wife, the fires "are almost of daily and nightly occurrence in this immediate neighborhood."[18]

In mid-November there was another alarm at Harper's Ferry, as Superintendent Kitzmiller detailed intelligence he had of a new raid planned upon the arsenal to obtain arms for Brown's rescue. The governor wired Petersburg for troops, and ordered others up from Richmond—four hundred men in all—to occupy Harper's Ferry anew. In Charlestown a colonel of the militia telegraphed Richmond: "We are ready for them. If attack be made, the prisoners will be shot by the inside guards." Then Wise ordered officers in Berkeley, Jefferson, and Frederick counties to be ready, as he entrained for Harper's Ferry himself amid reports that a large body was approaching from Wheeling. As he arrived, Wise wired his commander in Charlestown: "Be cautious. Commit no mistake tonight. Men will march tomorrow morning."[19] A reporter from the *Baltimore American* wrote of the subsequent scene: "Even the citizens of the town cannot pass through the suburbs without being arrested and carried off to headquarters. Persons coming into the town have to be detained an hour or more, and then marched by an armed guard to the presence of the military authorities to give an account of themselves. On leaving they have to obtain a pass, and run the gauntlet of a dozen sentinels to return to their homes again."

These precautions did not prevent Brown's young associate from Kansas, Charles Lenhart, from coming to Charlestown and loudly and bitterly denouncing abolitionism, Brown, and his confederates. He soon obtained an appointment as a guard inside the jail.

A group of German immigrants living in New York City were interested in freeing Brown and the other prisoners. But any plan, of course, needed financial backing and faced daunting logistical obstacles. Discussions were held where Higginson became the center, and sent the "boy lawyer," George H. Hoyt, to explore the possibility of rescue with the captives and to assist Brown's defense. Hoyt, however, was generally regarded as a spy by the Virginians and closely watched. There was the suggestion, with some exploration of it by the "plotters," to put a swift boat on the James River and kidnap Henry Wise from Richmond, then spirit him out to sea. The vessel would, it was proposed, seek port in Philadelphia, New York, or Boston, holding the governor hostage for the five jailed men in Charlestown. In Ohio there were also some preliminary explorations of the possibility of a rescue centered around John Brown Jr., with hundreds of conspirators known as the "Black String"; and in Kansas James Montgomery and others prepared to come east. But Montgomery would not do so until after Brown's execution, actually being on the ground by the date of Aaron Stevens's execution, delayed to March 16, 1860. Charles Lenhart, though, was nearly able to effect an escape by Cook and Coppoc on the day before their execution, foiled only by a failure of coordination.

For his part, Brown had determined that his body was worth "infinitely" more dead than alive, and let it be known unequivocally that no attempt at his rescue should be made. Besides the moral and political effect he saw his sacrifice would have, a bond had developed between him and his jailer, John Avis, who became solicitous of his needs, and who, it turned out, had been a boyhood friend of Martin Delany when he lived in Charlestown.

In his message to the Virginia legislature Governor Wise was to justify his extraordinary precautions: "These reports and rumors, from so many sources, of every character and form, so simultaneous, from places so far apart, at the same time, from persons so unlike in evidence of education, could be from no conspiracy to hoax: but I relied not so much upon them as upon the earnest continued general appeal of sympathizers with the crimes. It was impossible for so much of such sympathy to exist without exciting bad men to action of rescue or revenge. On this I acted."[20]

After sending 563 troops from Richmond, Alexandria, Petersburg, and Fauquier County, a force totaling 560 men was called from various other Virginia localities, including a howitzer corps from the Virginia Military Institute. Then "compelled by apprehension of the most unparalleled border war," Wise ordered, so far as it could be, the entire border east of the Piedmont placed under guard, from Point of Rocks to Alexandria. John Brown's invasion after all had been the work of a sectional organization, specifically upon slaveholders and their slave property. The governor told the legislature: "I believe in truth that the very policy of the prime promoters of this apparently mad movement is purely tentative: to try whether we will face the danger which is now sealed in blood. . . . [Our] safety and national peace will be best secured by a direct settlement at once—the sooner the better."

The most irritating feature of the movement, said Wise, had been that the seat of its origin was in a British Dominion, "which furnish asylums for our fugitives . . . and sends them and their hired outlaws back upon us from depots and rendezvous in bordering states." When the fight came with the "enemies of the constitution and the Union," he declared, he would not be content to meet them upon ground south of Mason and Dixon's line, nor would he confine it to ground north of that line in the Union; he would carry the war into Canada. "Let her furnish any longer an asylum for Fred Douglass to fly to; let her furnish a depot for the provisional government of John Brown at Chatham."

It has been defiantly proclaimed aloud, said the governor, referring to a speech by Wendell Phillips in Boston; "Insurrection is the lesson of the hour. . . . that a slave has the right to throttle his master, and if he makes his way over dead bodies from Virginia to Canada, he has a right to do so." There was no danger to Virginia from their slaves or "colored people," the governor asserted, but that fact could not be taken as solace because therein lay the root of their danger. "Masters in the border counties now hold their slaves by sufferance. . . . The slave could fly to John Brown much easier than he could come and take him. The slaves at will can liberate themselves by running away. . . . We must, then acknowledge and act on the fact that present relations between the states cannot be permitted longer to exist without abolishing slavery throughout the United States or compelling us to defend it by force of arms."

Writing before his departure for England from the Canadian province so galling to the Virginia governor, Frederick Douglass addressed the slaveholder-advanced assertion that the slaves did not take up arms against their masters in an article in his November 1859 *Monthly* entitled "To My American Readers and Friends":

> The slaveholders of Virginia and of the South generally, are endeavoring to make the impression that the Negroes summoned to the standard of freedom by John Brown, viewed the effort to emancipate them with indifference. An eyewitness, and a prominent actor in the transactions at Harper's Ferry, now at my side, [Osborne Anderson] tells me that this is grossly aside from the truth. But even if the contrary were shown, it would afford small comfort to the slaveholders. The slaves were sensible enough not to shout before they gained the prize, and their conduct was creditable to their wisdom. The brief space allowed them in freedom, was not sufficient to bring home to them in its fullness the real significance of the occasion. All the efforts to disparage the valor of the colored insurgents are grounded in the fears of the slaveholders, not in the facts of the action.

But all trace of individuality among the blacks joining in the fight at Harper's Ferry had been wiped from the received history. However, there remains Phil, the youth who ran to join the others at the Allstadt estate, and who chopped the holes in the enginehouse walls at Brown's request. Wounded in the assault of the enginehouse, he was said by Virginia newspapers to have "died of fright," but more likely from pneumonia, in the Charlestown jail a week later. And there was Washington's coachman, Jim, who fled into the Shenandoah River where he was said to have "drowned" in its shallow currents. And there was the "elderly slave man" into whose hands Leeman placed a shotgun, with which he killed the grocer, Boerly, as he crept up to discharge his weapon at a sentinel standing at the armory gate. There were two slaves with Cook when he began his ineffectual enfilade, seen by various witnesses offering a fight before being killed. And seven bodies not attributed to the original invaders were disposed of with the others.[21]

Was Harper's Ferry a defeat or a victory? It is true that John Brown failed to free a single slave; but out of this failure a new perception dawned. Wasn't Brown right after all, right regardless of what one might think of his methods? The Boston lawyer and future wartime governor of Massachusetts asked that question, as did hundreds of thousands throughout the North. "'Capt. Brown's expedition was a failure,' I hear it said," Theodore Parker wrote from Rome. "I am not quite sure of that. True, it kills fifteen men by sword and shot, and four or five men by the gallows. But it shows the weakness of the greatest slave State in America, the worthlessness of her soldiery, the utter fear which slavery genders in the bosoms of the masters. Think of the condition of the city of Washington while Brown was at work!" Just over a month after the raid the *New York Independent* in a remarkable editorial concluded: "Not John Brown but Slavery will be gibbeted when he hangs upon the gallows. Slavery itself will receive the scorn and execration it has invoked for him. When John Brown is executed, it will be seen that he has done his work more effectively than if he had succeeded in running off a few hundred slaves. The terror by night that rules in every household on her soil, drawing sleep from mothers and children, the anxieties and fears that for months to come will burden her population, the spirit of revenge—all these will make the cost of slavery to Virginia greater than she can bear."

During the last six weeks of John Brown's life, remarked more than one writer, his reknown was meteorlike. Strange it was to watch it work, DuBois wrote—"To be sure, the nation had long been thinking over the problem of the black man, but never before had its attention been held by such deep dramatic and personal interest." His speech to the court was recognized as masterful, and taken together with more than one hundred letters in his final weeks written to supporters and sympathizers, to friends and family, most of which found their way into print—they constitute a legacy without equal in American letters. "Where is our professor of belles-lettres, or of logic and rhetoric, that can write so well?" asked Thoreau. He could only fight with words now —and his magnanimity of spirit impressed and persuaded a growing audience, as the courage and the joy he evinced facing death exalted him.

Sharing a cell with Stevens, with his feet manacled, Brown was allowed two visitors a day; and there were a steady stream of them. Henry Clay Pate came; as did Samuel C. Pomeroy of the Emigrant Aid Society from Kansas. Lydia Maria Child was allowed to come to nurse his wounds, and Rebecca Spring of Perth Amboy, New Jersey, came to sit quietly with him. Judge Russell and his wife came from Boston, and there were a number of Southerners whom he received patiently, including Governor Wise. Visited by several southern clergy, Brown wrote: "There are no ministers of Christ here. These ministers who profess to be Christian and hold slaves or advocate slavery, I cannot abide them. My knees will not bend in prayer with them while their hands are stained with the blood of souls." To one clergyman he said, "I of course respect you as a gentleman, but as a heathen gentleman."[22]

Alexander Ross, leaving Richmond, attempted to reach Charlestown to see Brown before his execution, but was prevented from traveling beyond Harper's Ferry. Back in Richmond he obtained an interview with Wise, who lashed out at the abolitionist: "I am wise enough to understand your object in wishing to go to Charlestown, and I dare you to go. If you attempt it I will have you shot. It is such men as you who have urged Brown to make his crazy attack. I would like to hang a dozen of you leading Abolitionists." Wise handed him a paper ordering him to leave the state "within 24 hours."[23]

Brown wrote and read constantly, while sharing some talk with Stevens and his jailer, Avis. His reading included Carlyle on the French Revolution and a book about Toussaint L'Ouverture, whom he told one of his guards he ranked as one of the world's greatest men, the equal of a Socrates, a Luther, or a John Hampden.

In a letter to Thomas M. Musgrove, dated Nov. 17, 1859, he wrote: "Men cannot imprison or chain or hang the soul. I go joyfully in behalf of millions that have no rights that the "great and glorious"; "this Christian Republic" "is bound to respect." "Strange change in morals, political as well as Christian, since 1776. I look forward to other changes in God's good time, believing that 'the fashion of this world passeth' away."

To the Reverend Dr. Humphrey, Brown wrote: "For many years I have felt a strong impression that God has given me power and faculties, unworthy as I was, that He intended to use for a similar purpose. This most unmerited honor He has seen fit to bestow. . . . The scaffold has few terrors for me. God has often covered my head on the day of battle and granted me many times deliverances that were almost so miraculous that I can scarce realize their truth, and now when it seems quite certain that He intends to use me in a different way, shall I not most cheerfully go?"

Drawing a parallel between his failure to adhere to his original plan and a biblical story, to the Reverend H. L. Vaill he wrote: "Had Samson kept to his determination of not telling Delilah where his great strength lay, he would probably have never over turned the house. I did not tell Delilah; but I was induced to act very contrary to my better judgment; & I have lost my two noble boys; & other friends, if not my two eyes."

To his sisters, Mary and Martha: "I feel astonished that one so rude and unworthy as I should ever be suffered to have a place anyhow or anywhere among the least of all who when they come to die (as all must) were permitted to pay the debt of nature in defense of the right of God's eternal truth . . . it is even so I am now shedding tears but they are no longer tears of grief or sorrow. I trust I have nearly DONE with those. I am weeping for joy."

Among the many letters Brown received, this came from Thaddeus Hyatt: "Your courage, my brother, challenges the admiration of men! Your faith, the admiration of Angels! Be steadfast to the end!"

From Kendalville, Indiana, came a letter from the pen of the poet Frances Ellen Watkins: "I thank you, that you have been brave enough to reach out your hands to the crushed and blighted of my race. You have rocked the bloody Bastille; and I hope that from your sad fate great good may arise to the cause of freedom. Already from your prison has come a shout of triumph against the giant sin of our country."

He was so fully possessed of self-composure, repeatedly describing his state of mind as "cheerful," that a wide public was impressed. To his family he advised they be resigned and composed of mind as well, writing on November 8: "I can trust God with both the time and manner of my death, believing, as I now do, that for me at this time to seal my testimony for God and humanity with my blood will do vastly more toward advancing the cause I have earnestly endeavored to promote, than all I have done in my life before."

Perceiving the change in his circumstances and the opportunity it afforded him, on November 11 he again wrote to his wife: "I have been whipped, as the saying goes, but I am sure I can recover all the lost capital occasioned by that disaster by only hanging a few moments by the neck, and I feel quite determined to make the utmost possible out of a defeat. I am daily and hourly striving to gather up what little I may from the wreck."

In several letters he took up the education of his daughters, the welfare of his wife and family, receiving unsolicited donations, which he sent in checks payable to them. His words respecting the education of his daughters, Thoreau wrote, "deserve to be framed and hung over every mantle piece in the land. Compare this earnest wisdom with that of Poor Richard."

In his last letter Brown wrote: "Pure & undefiled religion before God & the Father is as I understand it: an active (not a dormant) principle." In another letter he called Christ "the great captain of Liberty," who had taken the sword of steel from him and given the sword of the spirit. Had he not once armed Peter, "and there continued it so long as he saw best." "I wish you could know with what cheerfulness I am now wielding the 'sword of the spirit' on the right hand and on the left," he wrote to one correspondent. Above all Brown kept reiterating his message that blacks had a right to enjoy unfettered freedom in the United States and that it was a Christian as well as a political duty of a citizen to help them to it. He also repeatedly expressed his opinion that his "disaster" at Harper's Ferry was due to his own "folly," and not to any other cause. He wrote a correspondent: "Of this you have no proper means to judging, not being on the ground, or a practical soldier. I will only add, that it was in yielding to my feelings of humanity (if I ever exercised such a feeling), in leaving my proper place and mingling with my prisoners to quiet their fears, that occasioned our being caught."

All of Brown's correspondence was, of course, monitored, and he was very circumspect about his actual plans, but he tried to make its main features clear. In his conversations with Governor Wise he repelled the idea that he had any intention to run slaves off, avowing rather his purpose was to arm them to fight by his side in defense of their freedom. The governor reported, "If assailed by their owners, or any one else; he said his purpose especially was to war upon the slaveholders, and to levy upon their other property to pay the expense of emancipating their slaves." He acknowledged further, said Wise, "that he expected to be joined by the slaves and by numerous white persons from many of the slave as well as the free states." Asked from whence he expected assistance once setting things in motion, Brown replied, "From more than you'd believe if I should name them all, but I expected more from Virginia, Tennessee and the Carolinas than from any others."

That there was a disagreement between his statements to the governor and that given at the time he received his sentence was brought to Brown's attention. He addressed the issue in a letter to his prosecutor, Andrew Hunter: "There need be no such confliction, and a few words of explanation will . . . be quite sufficient. I had given governor Wise a full and particular account of that, and when called in court to say whether I had anything further to urge, I was taken wholly by surprise, as I did not expect my sentence before the others. In the hurry of the moment, I forgot much that I had before intended to say, and did not consider the full bearing of what I then said. I intended to convey this idea, that it was my object to place the slaves in a condition to defend their liberties, if they would, without any bloodshed, but not that I intended to run them out of the slave states."

Brown said little else concerning his plans, but in the little said, it is clear something had gone wrong. To the young lawyer, Hoyt, he remarked that his company had performed with valor, but from a military point of view they had been sorely lacking. Among the many visitors at the jail at Charlestown to see Brown during his confinement were Judge Russell and his wife, and in a later year Mrs. Russell related to Katherine Mayo, a researcher for Oswald Garrison Villard: "To my husband he said but little of the raid, yet in that little it was evident that something had gone very wrong—that something had been done that he had expressly forbidden, or which was against his will. He had no fondness for Fred Douglass. Once I heard him say to my husband, of some defeated plan, some great opportunity lost, 'That we owe to the famous Mr. Frederick Douglass!' and he shut his mouth in a way he had when he thought no good."[24]

The last indication of Brown's attitude regarding his friend and confidant came from Alexander Boteler, the member of Congress from the district who was among those who interviewed Brown after his capture. He had the opportunity to look at some of Brown's correspondence and papers and saw a page on which Brown had written two columns of names under the headings, "reliable" and "unreliable." Frederick Douglass headed the latter column.

Virginia now saw the largest military gathering since Cornwallis surrendered to Washington at Yorktown. On November 26 Major-General Taliaferro issued a proclamation pursuant the governor: From that day, he declared, to the following Friday, the day of John

Brown's execution, any "STRANGERS found within the county of
Jefferson and counties adjacent, having no known or satisfactory
account of themselves will be at once arrested." Further the document
stated that any strangers or "parties approaching under the pretext of
being present at the execution" would be met by military force. No
women or children would be allowed to come near the grounds, and
all citizens were urged to remain in their houses. "Keep full guard on
the line of frontier from Martinsburg to Harper's Ferry," the governor instructed his commander. Wise also importuned President
Buchanan to send federal troops, and forthwith four companies were
dispatched from Fort Monroe to Harper's Ferry under Colonel Lee to
protect United States property.

Governors Packer of Pennsylvania and Gist of South Carolina also
tendered an offer to send troops, but were refused. The president of
the Baltimore and Ohio Railroad ordered that unless a purchaser had
a certificate of approval from an authorized officer of the road, no
tickets would be sold for Harper's Ferry; if trains were seized engineers were instructed to run the train down the first embankment
they came to.

John Brown's wife, Mary, from the first had been wanting to come
to Charlestown to be with her husband, but he urged her repeatedly
not to "come on for the present," reasoning it would use up her
scanty means, while there was "but little more of the romantic about
trying to relieve poor 'niggers,' as helping a widow"; then again she
would become an object of curiosity and "a gazing stock throughout
the whole journey." More sympathy would reach her if she stayed at
home, yet again "the pains of a final separation" would dearly buy
the comfort it might afford them.[25]

At the end of October, despite Brown's wishes, Higginson had
gone to North Elba and by November 4 was escorting Mary Brown
south. On November 9, accompanied by friends, she was refused permission to travel beyond Baltimore. For the next two and a half
weeks she divided her time between the homes of the abolitionists
Rebecca Spring of Perth Amboy and Mr. J. Miller McKim and his
wife of Philadelphia. Finally, at the end of November she gained the
consent of Governor Wise and the agreement of her husband, arriving at Harper's Ferry accompanied by the McKims and the young
Philadelphia lawyer Hector Hyndale. Authorities in Richmond
telegraphed: "Detain Mrs. Brown at Harper's Ferry until further
orders with the lady and two gentlemen who accompany her and

watch them." On the afternoon of the following day she was allowed to proceed alone
under escort of an officer and a sergeant and
eight mounted soldiers; an arrival that had
been preceded by the placing of three brass
cannon, in addition to two already in place
facing the jail, inside a hollow square of a
thousand troops with fixed bayonets.

The last meeting of husband and wife was
brief, occupying the early hours of the
evening, including a dinner shared in the
apartment of Brown's jailer, John Avis. They
embraced, and their initial conversation
went accordingly:

Mary Brown.

"My dear husband, it is a hard fate."

"Well, well, cheer up, cheer up Mary. We must bear it the best
manner we can. I believe it is all for the best."

"Our poor children—God help them."

"Those that are dead to this world are angels in another. To those
still living, tell them their father died without a single regret for the
course he has pursued."[26]

Brown urged his wife to see that his body, together with those of
their two sons, and the two Thompsons, be burned and the ashes collected and carried to North Elba for burial. But Virginia had already
made it plain to her that none of the bodies would be released but
that of her husband—and she would not hear of cremation. After
they talked over pertinent family matters and shared a meal, the commanding officer entered and said that Mrs. Brown must depart.

Mary Brown reached Harper's Ferry at 9 P.M., showing the effects
of grief and exhausted by all the circumstances attending. Brown is
said to have slept peacefully his final night, having written Judge
Tilden of Cleveland: "I fancy myself entirely composed and that my
sleep in particular is as sweet as that of a healthy joyous little infant.
I pray God that He will grant me a continuance of the same calm but
delightful dream until I know of those realities which eyes have not
seen and which ears have not heard."

Concerning his last rites and burial services, he'd written the wife
of George Stearns: "I have asked to be spared from having any mock;
or hypocritical prayers made over me, when I am publicly murdered:
& that my only religious attendants be poor little, dirty, ragged, bare-

headed, & barefooted slave boys; & girls; led by some old gray headed slave mother. Farewell. Farewell."27

The day opened beautifully on December 2, the reporter of the
Washington Star noted; the eastern sky "glided with all the colors of
the rainbow," the air "deliciously soft and balmy." Brown rose as
early as usual; first he finished a project he'd begun several days
before, marking out passages in his Bible that had influenced him the
most by underlining them in pencil. In a final letter to his wife he
instructed how the gravestone of his grandfather, transported from
Connecticut to North Elba, should be inscribed. Along with his
grandfather, Captain John Brown, were to appear his own name with
those of his three slain sons, Frederick, Watson, and Oliver, together
with birth and death dates for each. Then there was some more correspondence during which Brown penned two sentences, which he
was to hand to one of his guards on the way to his execution. It read:
"I John Brown am now quite certain that the crimes of this guilty
land: will never be purged away; but with Blood. I had as I now think:
vainly flattered myself that without very much bloodshed; it might be
done."28

After having Andrew Hunter write a codicil to his will leaving
John Avis his Sharps rifle, Brown handed out his books to his guards
and bid them farewell. At 10:30 he asked to say good-bye to his men,
who, excepting his cellmate Stevens, were on the second floor. To
Cook and Coppoc, who shared a cell, he expressed his displeasure at
the statements they had made; telling Cook he could not forgive him
his "confession." He shook hands with both, and handed Coppoc a
quarter as a remembrance, and said, "Farewell." To Copeland and
Green he handed a quarter each and shook their hands warmly, saying in parting, "If you must die, die like men." Albert Hazlett in an
adjoining cell he ignored.

Returning to Stevens, he said, "I am here to bid you farewell as the
others." Overcome with emotion, Stevens said: "I feel in my soul,
Captain, that you are going to a better world." Brown: "Yes, yes, but
stand up like a man. No flinching now. Farewell, farewell."

Wearing a black cassimere suit with his arms bound at his side,
Brown came out of the jail and descended the stairs, his white cotton
shirt opened at the collar, with white woolen socks and red carpet
slippers, with a slouch hat upturned in the front. At either side stood
John Avis and Sheriff Campbell, who were immediately surrounded
by the Petersburg Greys opening ranks to permit their passage. The

prisoner was assisted into a farm wagon bearing his coffin and took a seat on its forward end. As he rode to the execution ground with four companies of mounted military escort, the parade entered two hollow concentric squares of troops, one thousand in all and beyond these as far as anyone could see were cavalry and squads of sentinels. Notables within this military array were a professor of the Virginia Military Institute, Thomas J. Jackson, later to win the sobriquet "Stonewall," and in company F from Richmond, the younger brother of the tragedian Edwin Booth, John Wilkes Booth, who himself had left the theater immediately after a performance, hastening to join his company.

When the wagon reached the execution site Brown fairly leapt from his perch and was the first to ascend the stairs to the gallows, politely bowing to Andrew Hunter and other officials. As he stood over the trap a white hood was quickly placed over his head to prevent him from making any speech, which, at any rate, he had not intended. Standing completely motionless except for the breeze ruffling his clothing, the entire entourage waited fifteen minutes while soldiers maneuvered into position. Finally the trap was sprung and Brown's knees fell level with the scaffold and his pinioned arms shot back nearly horizontal behind him, his strong, hard hands clenched tight. After a moment a mounted officer in front of the scaffold called out: "So perish all such enemies of Virginia! All such enemies of the Union! All such enemies of the human race!"

After twenty-five minutes a physician gave his opinion that all life had been extinguished from the body, and after forty minutes it was cut down and placed in the coffin.

Several uniformed companies escorted John Brown's body back to the jail, and by 6 P.M. it was delivered to his wife at Harper's Ferry.

Meetings for prayer and protest had convened throughout northern cities and towns in a solemn and funereal atmosphere as Brown mounted the scaffold in Virginia. In all of these gatherings, blacks were seen beside whites, and in many cases blacks organized and conducted meetings like the one at New York City's Shiloh Church, of which Henry Highland Garnet was pastor. In a eulogy that morning Garnet had said, "Today John Brown is to offer up his life a sacrifice for the sake of justice and equal human rights. Henceforth the Second day of December will be called 'Martyr's Day.'" A meeting drew nearly four thousand persons at Philadelphia's National Hall, where a roster of prominent abolitionists including Lucretia Mott addressed the gathering. Finally Robert Purvis took the stand to convey the momentous fact that many were acknowledging: John Brown's death would work out to the salvation of the slave in America. He was Christ's apostle of liberty in the nineteenth century, said Purvis. "The coward fiends of Virginia have sowed the winds, to gather, in the coming wrath of God, the whirlwind." At this statement pandemonium erupted in the auditorium, orchestrated with catcalls and hisses by a contingent of Southern medical students. Fearing violence, police brought the meeting to an end and closed the hall.[29]

Elsewhere in the city meetings continued without interruption, many participants in Philadelphia and other places wearing crepe armbands with a rosette, or with Brown's portrait. In New York City, in Boston, and in Philadelphia all black-owned businesses were closed, and in Worcester blacks quit work between 11 A.M. and 3 P.M. In New York the *Herald* reported: "Dark barbers darkly doomed their razors to one day of rest; at dinner waiters were abstracted; to get one's boots blacked was difficult; to have a carpet shaken impossible."

It was almost universally seen by northern blacks that John Brown was a true friend who had proven it by giving his life. What is more, the event of his death had opened a wedge for the overthrow of slavery. Garnet put it this way before his congregation: "The nation needed to see a picture of the future of slavery and its end and methinks God has been pleased to draw it in crimson lines. Americans, Patriots, Christians, Tyrants, look upon it, and be instructed. Is it not a singular coincidence that in Virginia, the very soil which African slavery in this part of the New World commenced its reign of terror, the system should receive its first damaging blow?"

In Boston the largest meeting in the history of Tremont Temple gathered, with three thousand persons remaining outside. James Freeman Clarke and William Lloyd Garrison addressed the standing room only gathering, other speakers included J. Stella Martin, pastor of the Joy Street Baptist Church. Martin said the difference between John Brown and the revolutionary patriots lay only in that Brown designed to help black people to their rights. "Though his body falls, the spirit of slavery and despotism falls with it," he said.[30]

Many, like Garrison, were recognizing that a new phase of the antislavery struggle had been ushered in. For three decades the most outstanding representative of the "moral suasion school," editor of the *Liberator,* and principal in the movement, Garrison declared in that meeting at Tremont Temple: "I am a non-resistant and I not only desire, but have labored unremittingly to effect, the peaceful abolition of slavery, by an appeal to the reason and conscience of the slaveholder; yet, as a peace man—an 'ultra' peace man—I am prepared to say 'success to every slave insurrection at the South, and every slave country.'"[31]

Elsewhere in Boston a continuous prayer meeting was held at the Twelfth Baptist Church, whose pastor, Leonard Grimes, was a fugitive slave, and where Charles Lenox Remond delivered an address. In Worcester a meeting was held under the auspices of the Anti-Slavery and Temperance Society of Colored Citizens; and blacks in Hartford, New Haven, and New Bedford convened meetings. Providence, Rhode Island, heard an address by William Wells Brown titled "The Heroes of Insurrection." John Brown and his little band, Wells said, although dashing themselves to a bloody death, "shook the prison-walls from summit to foundation, and shot wild alarm into every tyrant-heart in all the slave-land." Let no one, he continued, who glories in revolutionary struggles for freedom "deny the right of the American bondsman to imitate their high example."[32]

In Pittsburgh a meeting was addressed by George Vashon; and in Albany, after a hundred-gun salute to Brown from the city, at the Wesleyan Church William Watkins announced, "Black men, be of good courage, for the day of liberty draws near," and he challenged anyone in his audience to rise and say Brown was not right. In Chatham five prayer meetings were held and two members of John Brown's convention there, James Bell and James Harris, delivered eulogies. At a memorial meeting at St. Lawrence Hall, Toronto, Thomas Kinnard spoke. In Cleveland two thousand persons assembled at Melodeon Hall, Charles Langston presiding; while a meeting held at the Second Baptist Church in Detroit was called to order by William Lambert. Following him, three clergy spoke on the Christian virtues of John Brown, and then there was a presentation by the Brown Liberty Songsters. When the president of Detroit's all-black Old John Brown Liberty League assumed the chair, a declaration of sentiments was read by Lambert that stated what many already understood—that Brown's effort had "put the liberty ball in motion," and that it would not stop until slavery was abolished in the United States.[33]

That night John Brown's widow accompanied his body to Baltimore and on the following day to Philadelphia. At the station at

Broad and Pierce streets a large crowd had gathered including a delegation of fifty from the Shiloh Baptist Colored Church, who were not allowed admittance to the station. In view of the excitement and the divided state of the public, said the mayor, the coffin should not be unloaded in Philadelphia. As the station was cleared, the coffin was driven to the Walnut Street wharf for a steamer to New York City.

In New York the body was taken to a funeral parlor on the Bowery for embalming, where a line queued to pay respects. On Monday at two in the afternoon the funeral cortege, which included the McKims joined by Wendell Phillips, reached Troy, staying overnight at the America House. The next morning they were in Rutland, Vermont, then on to Vergennes for passage across Lake Champlain. In many of the towns through which they passed bells tolled and people came out to murmur condolences to the widow. At Elizabethtown an honor guard watched over the coffin till dawn at the courthouse.

After another day's travel, after sundown in bitter cold, on December 7 the group reached North Elba, where lanterns had been lit as the Brown and Thompson families waited. With four widows present, the funeral service was held in the early afternoon on December 8, as Brown's mortal remains were buried next to a huge boulder not far from his dooryard. Joshua Young of Burlington, Vermont, offered prayers, and hymns were sung led by the Epps family, the Browns' black neighbors. Ruth Brown Thompson wrote: "It seemed as though my dear father was with us in spirit and joined in the chorus of his favorite hymn."

> Blow yet trumpet, blow—
> The gladly solemn sound;
> Let all the nations know,
> To earth's remotest bound,
> The year of jubilee has come.

Phillips delivered his matchless eulogy to an intimate gathering from inside the house:

Harper's Ferry was no single hour, standing alone—taken out from a common life—it was the flowering out of fifty years of single-hearted devotion. He must have lived wholly for one great idea, when these who owe their being to him, and those whom love has joined to the circle, group so harmoniously around him, each accepting serenely his or her part.

I feel honored to stand under such a roof. Hereafter you will tell children standing at your knee, "I saw John Brown buried—I sat under his roof." Thank God for such a master. . . . What lesson shall those lips teach us? Before that still, calm brow let us take a new baptism. How can we stand here without a fresh and utter consecration? These tears! How shall we dare even to offer consolation? Only lips fresh from such a vow have the right to mingle their words with your tears.[34]

This strange and startling passage of John Brown—from the night of October 16 through his trial and imprisonment and hanging in Virginia to his burial at his home in the Adirondack Mountains—had "changed the thoughts of millions," said Phillips. And now a million hearts guarded his words.

Thoreau wrote: "What a transit was that of his horizontal body alone, but just cut down from the gallows-tree! We read that at such a time it passed through Philadelphia, and by Saturday night had reached New York. Thus like a meteor it shot through the Union from the Southern regions toward the North! No such freight had the cars borne since they carried him Southward alive."[35]

Nine

1860

Old Brown,
John Brown,
Osawatomie Brown.
He'll trouble them more when his Coffin's nailed down!
—Diary of J. S. Reader of Kansas

F REDERICK DOUGLASS'S VOYAGE OUT OF QUEBEC WAS BY NORTH
passage in weather that was cold, dark, and stormy. But after
"fourteen long days," he later wrote, "I gratefully found myself upon
the soil of Great Britain, beyond the reach of Buchanan's power and
Virginia's prisons."[1] He arrived in Liverpool November 26, 1859, six
days before John Brown's execution.

Douglass's account of his mood and the circumstances of his
departure was colored by the years that passed between those events
and the telling. But clearly among the things weighing heavily upon
him was the fact that he had fled, leaving the impression he had failed
to meet his commitments to his condemned friend, not to mention his
separation and continuing obligations to his family. He had left, too,
at a crucial moment, when the entire antislavery enterprise seemed to
be in doubt and slaveholder influence in the national government
appeared at its zenith; moreover, his paper was dwindling in subscriptions, its survival uncertain.

But adversity was nothing new to him. As he stepped to the docks he found a warm reception awaiting him, and was again welcomed into the ranks of the British antislavery cause, quickly renewing many an old acquaintance. Friends were eager to use his reputation and talents to aid in a revival of their own public, and he immediately began speaking on American slavery in different parts of the country. The fact that he was a fugitive would guarantee an audience, and Douglass was called upon to give an account of the men who had boldly tried to free slaves and lost their lives in the attempt. He wrote: "My own relation to the affair was a subject of much interest, as was the fact of my presence there being in some sense to elude the demands of Governor Wise."[2]

Many of the notable abolitionist orators in Britain were conversant with, if not adherents to, the Garrisonian doctrines of the American school of antislavery, and they sounded Douglass on the question of the right of the slave to gain freedom by a resort to arms. To allay apprehensions, Douglass gave assurances that his advocacy of the cause of the slave while in England would have no reference whatever to any plan involving arms; but added, in a letter to Helen Boucaster, secretary of the Sheffield Anti-Slavery Association: "The slave is a victim of a constant insurrection, by which his blood is drawn out drop by drop! It may not be altogether impartial to lay down the rule of submission to him, too sternly, especially since he has submitted already two hundred years."[3]

On February 27, 1860, Douglass's much-esteemed friend, the British abolitionist George Thompson, broadly assailed Douglass in Glasgow before the two had a chance to meet, attacking his position on the U.S. Constitution. Noting that Douglass had at one time also espoused the Garrisonian view, Thompson's strongest point was that the practice of the U.S. government was proslavery, and this was convincing proof as to the proslavery character of the Constitution. In the following month, when in Glasgow, Douglass was given the opportunity to answer his critic when he appeared at Wakefield, sharing the platform with Sarah Parker Remond, sister of Charles Lennox Remond. It was his or any man's right to change their opinion, said Douglass, but on the subject of the Constitution, it would be seen to afford slavery no protection when it ceased to be administered by slaveholders, once there was a will in the American people to abolish slavery.

There was "no word, no syllable in the Constitution to forbid that result," he reiterated.[4] That which gave the Constitution the proslavery interpretation it would not have had was the invention of the cotton gin, which had raised the value of a slave from two hundred to eight hundred and at last a thousand dollars.

When the 36th Congress convened on December 5, 1859, many House members came armed with pistols and bowie knives; John Brown's call to the slaves to join him in the fight for freedom had plunged like a dagger into the Southern psyche, provoking fear and rage. The success of the Republicans in the November elections, too, had spelled out unmistakably that Harper's Ferry had only been a harbinger in a long-expected drama, one that had roiled in the imagination of whites of the South for decades. The Republicans now held a plurality, and with Democrats split over slavery in Kansas, and their leading politician, Stephen Douglas, discredited, the small Know-Nothing Party (the American Party) held the balance of power. When John Sherman of Ohio was nominated as Speaker of the House, a fierce debate broke out between the camps, threatening violence on the House floor. The ostensible reason for this was Hinton Rowan Helper's book *The Impending Crisis*.

Helper was a North Carolinian, a slaveholder's son who did not himself own slaves. His book, published in 1857, was a bitter indictment of slavery as an impediment to all development of the South. While still imbibing the racialist views of the white southerner, southern politicians labeled the book incendiary; copies were burned in North Carolina, and its circulation there and in many southern states was banned. As the book came to the attention of Horace Greeley, the *New York Tribune* ran an unheard of eight-column review, and for the 1859 election cycle Republicans raised money to publish an abridged edition as a campaign pamphlet. Sixty-eight Republican congressmen had endorsed the book in a circular; one of them was John Sherman.

Then on December 15—the day before John Cook, Edwin Coppock, John Copeland, and Shields Green were hanged in Charlestown for their roles in the raid—the Senate appointed a five-man committee "to inquire into the late invasion and seizure of the public property at Harper's Ferry;" James Mason was its chairman, and Jefferson Davis joined him in its work. The committee was impaneled to find whether, and to what extent "any citizens of the United States not present were implicated therein, or accessory thereto, by contributions of money, arms, munitions, or otherwise." The committee's task was to spell out implications for the security of "Southern Institutions," and to draw inferences implicating Republican politicians. Although not a judicial investigation, it had the power to subpoena, and was to report on whether any legislation was required. The names Frederick Douglass, Gerrit Smith, George Stearns, Samuel Howe, Theodore Parker, and Franklin Sanborn had quickly come to public notice in relation to Harper's Ferry. But just as quickly there was information by implication that senior Republicans had advance notice; these were Senators Seward and Wilson, through their meetings with Hugh Forbes in Washington, and, through his meetings and correspondence with John Brown, Representative Joshua Giddings. As the inquiry began, knowledge of Brown's supporters and contacts widened, bringing to the committee's attention Lewis Hayden, George De Baptiste, John Brown Jr., Richard Hinton, Richard Realf, James Redpath, Thaddeus Hyatt, and others.

Among those ordered to appear were Hayden and De Baptiste, on January 23 and 27, respectively. Before serving the summons, the United States marshal at Detroit wired Mason: "This De Baptiste is a Negro, though a smart and intelligent fellow." Learning also of Hayden's identity, the marshals in Detroit and Boston were ordered to withhold the summonses, the senators preferring to ignore the role of blacks, except for Frederick Douglass, who it was understood was beyond their reach. Sanborn vowed not to appear before the committee, fleeing to Canada before being summoned; and Thaddeus Hyatt, when summoned, also refused to speak and was imprisoned. Although eager to speak his mind before the senators, Thomas Wentworth Higginson received no summons. John Brown Jr., Mason wrote in his report, "at first evaded the process of the Senate, and afterwards, with a number of other persons, armed themselves to prevent arrest."[5] However, others summoned to testify, including Stearns, Howe, and Realf, did appear, and by adroitly side-stepping the committee's chief concern, for the most part managed not to give any substantive information.

Taken to task by Higginson for again not standing his ground, Sanborn responded: "There are a thousand better ways to spending a year in warfare against slavery than by lying in a Washington prison. . . . Some of us are so fond of charging bayonets that for fault of any enemy we rush upon our friends . . . do not tell what you know to the enemies of the cause, I implore you."[6]

Stearns, testifying on February 24, 1860, denied he had foreknowledge of Brown's intention to seize the arsenal at Harper's Ferry, but admitted he had provided him with money. Why? he was asked.

"Well, my object in giving him the money was because I considered that as long as Kansas was not a free state, John Brown might be a useful man there. That was my object. Another was a very high personal respect for him. Knowing that the man had an idea and that he was engaged in a work that I believed to be a righteous one, I gave him money to enable him to live or do whatever he thought was right."

Had not John Brown Jr. conferred with him in his home only two months before the raid, and didn't he then discuss his father's plans with him? That had not occurred, Stearns assured the committee, but remarked that upon Brown's sons visit he "was particularly struck by the fact that he inquired about some bas-reliefs that I had put on the wall. He looked at the garden and picked one or two flowers and asked me that he might take them home to his wife. I told him he might take as many as he chose. In a few minutes I found that he was holding them up and contrasting the colors, what not a man in five hundred would do. I was struck particularly with the natural love he showed not only for art but for nature. That was all that occurred at that time."

Asked if he disapproved of the raid, Stearns replied: "I should have disapproved of it if I had known of it, but I have since changed my opinion. I believe John Brown to be the representative man of this century, as Washington was of the last—the Harper's Ferry affair and the capacity shown by the Italians for self-government, the great events of the age. One will free Europe, and the other America."[7]

Three weeks later, on March, 16, the last two Harper's Ferry raiders in custody, Aaron Stevens and Albert Hazlett, were hanged in Charlestown.

Elected or governmental officials like Wilson, Giddings, and Secretary of War John Floyd were not subpoenaed, but were invited to come before the committee after some witness or other brought up their names, and to give rebuttal. The committee was particularly interested in the testimony of Joshua Giddings. Senator Davis directed the questioning, asking the witness to expound on the meaning of "higher law," the exposition of which he had devoted numerous lectures, and a phrase often uttered by Senator Seward and other

Republicans. Giddings replied: "What I mean by the higher law is, that power which for the last two centuries has been proclaimed by the philosophers and jurists and statesmen of Germany, Europe, and the United States, called, in other words, the law of nature; by which we suppose that God, in giving man his existence, gave him the right to exist; the right to breathe vital air; the right to enjoy the light of the sun; to drink the waters of the earth; to unfold his moral nature; to learn the laws that control his moral and physical being; to bring himself into harmony with those laws, and enjoy that happiness which is consequent on such obedience."

Davis: "In your lectures, was the theory of that law applied to the condition of African slavery in the United States?"

Giddings: "Unquestionably, to all. Wherever a human soul exists, that law applies. I mean by the term 'soul,' that immortal principle in man that exists hereafter, which is called the human soul; and wherever such soul exists to sustain life, obey the law of his Creator, and enjoy heaven or happiness."

Davis: "Did you, in inculcating, by popular lectures, the doctrine of a law higher than that of the social compact, make your application exclusively to Negro slaves, or did you also include minors, convicts, and lunatics, who might be restrained of their liberty by the laws of the land?"

Giddings: "Permit me . . . with all due deference, to suggest, so that I may understand you, do you intend to inquire whether those lectures would indicate whether your slaves had a right at all time to their liberty?"

Davis: "I will put the question in that form if you like it."

Giddings: "My lectures, in all instances, would indicate the right of every human soul in the enjoyment of reason, while he is charged with no crime or offense, to maintain his life, his liberty, the pursuit of his own happiness; that this has reference to the enslaved of all states as it had reference to our own people while by the Algerines in Africa."

Davis: "Then the next question is, whether the same right was asserted for minors and apprentices, being men in good reason, yet restrained of their liberty by the laws of the land."

Giddings: "I will answer at once that the proposition or comparison is conflicting with the dictates of truth. The minor is, from the law of nature, under the restraints of parental affection for the purposes of nurture, of education, of preparing him to secure and maintain the very rights to which I refer."[8]

For fully two months and forty-four ballots the House remained deadlocked on the election of majority speaker. During this time South Carolina's governor Gist notified one of his state's representatives: "If . . . you upon consultation decide to make the issue of force in Washington, write or telegraph me, and I will have a regiment in or near Washington in the shortest possible time." But before this resort was called for, the impasse was broken when Sherman withdrew and Republicans nominated New Jersey representative William Pennington, a supporter of the Fugitive Slave Law, who was duly elected.

One day after the pandemonium in the House was allayed, Jefferson Davis introduced resolutions in the Senate for acceptance of a straight slave code in the Territories. It was not expected the Senate would pass them; those who supported them looked to their adoption into the Democratic platform in the nominating convention in Charleston that spring. These affirmed that neither Congress nor a territorial legislature could "impair the constitutional right of any citizen of the United States to take his slave property into the common territories. . . . It is the duty of the federal government there to afford, for that as for other species of property, the needful protection." Aiding fugitive slaves and interfering with enforcement of the Fugitive Slave Law, the resolutions stated, were acts "hostile in character, subversive of the Constitution and revolutionary in their effect."[9] As the Senate poised for a vote on the resolutions, their ulterior motive was clearly pointed out by Davis on May 7, in a speech titled "Relations of the States." Of John Brown, Davis said, "It was only last fall that an open act of treason was committed by men who were sustained by arms and money raised by extensive combinations among the non-slaveholding states to carry treasonable war against the state of Virginia." And of the Republicans: "The power of resistance consists, in no small degree, in meeting the enemy at the outer gate. I can speak for myself—having no right to speak for others—and do say that if I belonged to a party organized on the basis of making war on any section or interest in the United States, if I know myself, I would instantly quit it. We of the South have made no war upon the North. We have asked no discrimination in our favor. We claim but to have the Constitution fairly and equally administered."

Edmund Ruffin, who came away with a number of John Brown's pikes as souvenirs, began mailing them to friends throughout the South with the label "Sample of the favors designed for us by our

Northern Brethren." South Carolina's Gist received one. Soon after his capture, too, Brown's maps—seven of them—began to be reported in the southern papers. People in many localities began to apprise the fact that their communities had been marked out as points of attack. The *Savannah Republican* reported on a number of incendiary fires in Talbot County in southwest Georgia, an area marked on the maps, as evidence that Brown's agents had passed through. On October 26, 1859, in a letter to the *Mercury*, a slaveholder in Charleston urged that recalcitrant elements in Southern society be exposed: "We are not alarmists; neither are we agitated or surprised at the recent disturbances in Virginia; but we would speak the word of caution to our citizens. We would have them examine our community and punish its lawbreakers. Neither talents, social position or wealth should screen them from public exposure and denunciation."

When the State Assembly convened, the *Mercury* charged them "to speak out in terms not to be mistaken. We look for stringent measures calculated to put down these concerted emissaries of mischief—to award them full justice—that is, a speedy trial—quick execution." With the press calling for action, vigilante activity intensified across the South, a movement that quickly became politically significant as slaveholders assumed the reins. Patrols, based on membership in state militia companies, charged every white man with responsibly for good order around the plantations. These barred assemblies of slaves, inspected slave quarters for weapons and for stolen or illegal property, and arrested slaves without proper passes throughout the countryside. Dozens of supposed abolitionist agents and sympathizers were drummed out of communities; while political leaders denounced abolitionism for having split the national churches, stolen slaves, incited insurrection, excluded the southern people from the common national territories, and now, to complete their purposes, organized a political party to capture the federal government. John Brown, the indictment went, had merely shown what awaited the entire South once men like Seward, Sumner, Wilson, and Giddings gained control of the government.

The day after Davis's speech on "Relations of the States," he again took the floor to attack the leading politicians of the Republican Party and its presumed presidential candidate that year, Seward. That senator, he advised his colleagues, had advance knowledge of Harper's Ferry, and "like John Brown, deserves, I think, the gallows, for his participation in it."[10] Into this growing welter and onto the speaker's platform at New York's Cooper Institute stepped the Hon. Abraham Lincoln. He was as yet little known in national politics, but had served in the Congress some years before in the Whig Party.

Now a Republican, he had made an impressive showing in the debates with Stephen A. Douglas, although he lost the race for the United States Senate to Douglas in 1858. His speech on February 27 was to bring him for the first time before the eyes and ears of the eastern leaders of the Republican Party, and was an important one, as it would give him, through the New York press, a national forum. In condemning Brown's actions and denying Republican complicity in them he merely repeated what many of the party's spokesmen were saying. Seward said as much when he declared it was "an act of sedition and treason and criminal in just the extent that it affected the public peace and was destructive of human happiness and life." But Lincoln displayed something more in his approach when he denounced the "slander" that Brown was the natural outgrowth of Republican "doctrines":

> You [the South] never dealt fairly by us in relation to this affair. When it occurred, some important State elections were near at hand, and you were in evident glee with the belief that, by charging the blame upon us, you could get an advantage of us in those elections. The elections came, and your expectations were not quite fulfilled. Every Republican man knew that, as to himself at least, your charge was a slander, and he was not much inclined by it to cast his vote in your favor. Republican doctrines and declarations are accompanied with a continual protest against any interference whatever with your slaves, or with you about your slaves. Surely, this does not encourage them to revolt.[11]

Notwithstanding Lincoln's statement, weeks earlier on the floor of the Senate Jefferson Davis held that he found all speculation as to whether "our servants" would rebel or nor exceedingly offensive. Of prime consideration to this Mississippi senator and his fellow slaveholders in assessing the value of the Union was whether and the extent to which it facilitated their control of their slaves. Southerners had displayed no lack of confidence before Harper's Ferry in their ability to protect their slave property within the constitutional framework of the Union as it had been practiced. Now that had changed; for the first time they were faced with the crisis of having two billion dollars invested in slaves swept away. They were faced, too, with the daunting setback of seeing their vaunted political power reduced to permanent impotence, while facing the horrors posed to them by racial equality and impending racial conflict. "The negro-stealing mobs of the North" only proved that neither Constitution nor enactments of Congress, nor decisions of the Supreme Court could be relied upon to protect "Southern Property."[12]

Just as Brown had been carrying forward his work in 1858, the eccentric Boston lawyer and abolitionist Lysander Spooner had written and printed a circular to be distributed throughout the South calling for slave insurrection. On one side of his sheet he addressed "The Non-Slaveholders of the South," calling upon them, for their own advancement, to throw off the domination of the slaveholders. The other side was presented under the heading "A Plan for the Abolition of Slavery." In his argument Spooner called for the formation of "Leagues of Freedom" to raise money and arms to liberate the slaves; then a campaign was to be initiated with the purpose of destroying slaveholder property and morale. Some copies of this circular were sent to selected persons via the mail in both the North and the South. Theodore Parker saw a longer elaboration upon Spooner's theme and wrote him: "Your paper is very well thought & expressed, as indeed are all your writings. If it were widely circulated at the South, it would strike a panic terror into those men, whose 2,000,000 is invested neither in land nor things. But I think you can't get a Corporal's Guard to carry your plan into execution. When I am well enough I will come & talk with you about it."[13]

Higginson, too, saw it and commented: "The increase of interest in the subject of Slave Insurrection is one of the most important signs of the time." In continuing he said, "within a few years, the phase of the subject will urge itself on general attention, and the root of the matter be thus reached. I think that this will be done by the action of the slaves themselves, in certain localities, with the aid of secret co-operation from whites."

His criticism of Spooner's plan, however, was a cardinal one: "In Revolutions the practical ends always comes first & the theory afterwards; just as ours others, long after the Battle of Bunker Hill, still disavowed the thought of separation—and honestly." For each man who could support Spooner's proposition, "there are ten who would applaud it, when it actually came to the point. People's hearts go faster than their heads. . . . In place therefore of forming a society or otherwise propounding insurrection as a plan, my wish would be assure it as a fact." He could give assurances, were he free, "that what I say means something, & that other influences than these of which you speak are even now working to the same end."[14]

John Brown himself spoke privately with Spooner when in Boston in May 1859, urging that he curtail further circulation of his proposal so as not to publicize in advance the "practical demonstration" Brown was preparing.

Several weeks after Brown's execution, slaves in Bolivar in southwestern Missouri, armed only with sticks and stones, attacked their master and other whites. The *Missouri Democrat*, in a report subsequently carried in the *New York Tribune*, told of a mounted company "ranging the woods in search of negroes. The owner of some rebellious slaves was badly wounded, and only saved himself by flight. The greatest excitement prevailed, and every man was armed and prepared for a more serious attack." This dispatch caught the critical eye of a London-based reader who nearly leapt at the news. In a letter to Frederick Engels dated January 11, the day after Jefferson Davis's remark to the Senate to the contrary, Karl Marx wrote: "In my opinion the biggest things that are happening in the world today are on the one hand the movement of the slaves in America started by the death of John Brown, and on the other the movement of the serfs in Russia. . . . I have just seen in the *Tribune* that there has been a fresh rising of slaves in Missouri, naturally suppressed. But the signal has now been given."[15]

Some week's later Marx's correspondent replied: "Your opinion of the significance of the slave movement in America and Russia is now confirmed. The Harper's Ferry affair with its aftermath in Missouri bears its fruit; the free Negroes in the South are everywhere hunted out of the states, and I have just read in the first New York cotton report . . . that the planters have hurried their cotton on to the ports in order to guard against any probable consequences arising out of the Harper's Ferry affair."[16]

That the South was clamoring terribly against "Northern emissaries" and "Black Republicans" exciting servile war is duly noted by American scholarship, but its corollary has hardly been grasped. Herbert Aptheker has pointed out that "to draw the lesson [from

John Brown's] failure that the slaves were docile, as has so often been done, is absurd. And it would be absurd even if one did not have the record of the bitter struggle of the Negro people against slavery." Aptheker's chronicle of these ongoing phenomena in *American Negro Slave Revolts*, notes that in July 1860 the *Austin State Gazette* reported that for eight weeks fires of breathtaking dimension swept many towns and settlements across north Texas. Texas senator Louis T. Wigfall was reported saying in the *New York Herald*, September 21, 1860: "We say to those States [in the North] that you shall not . . . permit men to go there and excite your citizens to make John Brown raids or bring fire and strychnine within the limits of the State to which I owe my allegiance. You shall not publish newspapers and pamphlets to excite our slaves to insurrection. . . . We will have peace . . . [or] withdraw from the Union." He then revealed, the *Herald* wrote: "An association called 'the Mystic Red' was entered into by members of the Methodist Church North and the John-Brown men; and their purpose was to carry out the irrepressible conflict, to burn up the mills, to bring free-soil northern capital in, and thus get possession of Texas."

In August 1860, a "conspiracy" among slaves was reported crushed in Talladega County, Alabama, followed by reports of disturbance in Pine Level in Montgomery County. On the 29th of that month in Georgia the *Columbus Sun* reported: "By a private letter from Upper Georgia, we learn that an insurrectionary plot has been discovered among the negroes in the vicinity of Dalton and Marietta and great excitement was occasioned by it, and still prevails." The plan, involving slaves from Whitfield, Cobb, and Floyd counties, called for Dalton to be set ablaze and a train captured and crashed in Marietta. In Mississippi, trouble was reported in Winston County in September, and then near Aberdeen. In late October a plot was heard involving blacks in Norfolk and Princess Anne County in Virginia, followed by reports of trouble below there in Currituck and Washington counties, North Carolina; and in November in Crawford County, south of Macon, Georgia, and north of there in Habersham County. In December, returning to Alabama, there were disturbances surrounding Montgomery, reverberating in Pine Level, Autaugaville, Prattville, and Haynesville. The *Montgomery Advertiser* reported on December 13: "We have found out a deep laid plan among the negroes of our neighborhood, and from what we can find out from our negroes, it is general all over the country. . . . We hear some startling facts. They have gone far enough in the plot to divide our estates, mules, lands, and household furniture."

That slaves had perfected the art of communicating among themselves and realized a considerable organization has been intimated in many sources, but the true extent and ramification of this has never fully been understood. James R. Gilmore, a freelance journalist who wrote under the name Edmund Kirke, and who became a close associate of Lincoln during the war, in December 1860 had conversation in South Carolina with a slave leader named Scipio, and found that "there exists among the blacks a secret and wide-spread organization of a Masonic character, having a grip, password, and oath. It has various grades of leaders, who are competent and earnest men and its ultimate object is FREEDOM." The organization was established for "RIGHT and JUSTICE," said Scipio, and that the South was bound to be defeated "cause you see dey'll fight wid only one hand. When dey fight de Norf wid de right hand, dey'll hev to hold de nigga wid de leff."17

Into the shadow of things done and the effects of those still to come, no one achieved the distinction to match the Bostonian abolitionist and reform orator Wendell Phillips. For years Phillips had been one of the coadjutors of William Lloyd Garrison, whose slogan "immediate emancipation" sounded the clarion for thirty years of agitation. But as a new hour struck—high noon, as Garrison remarked—Phillips transcended the restrictions of Garrisonian sectarianism and with the legacy of John Brown gained new prominence. Phillips first spoke on John Brown on November 1, 1859, at the Reverend Beecher's Plymouth Church in Brooklyn. Then on December 2, the day of Brown's hanging, he spoke at Boston's Tremont Temple, followed by his eulogy at John Brown's graveside on December 8. Finally, as proslavery toughs hired to prevent him from speaking harangued and mobbed him, he spoke in New York City on December 15, followed by appearances on Staten Island and in Troy, New York, where mobs again tried and failed to silence him. One of his biographers wrote: "Taking his text from the insults of his enemies [he] hurled defiance back in their teeth . . . and fairly quelled the rioters by his courage, address, and personal magnetism."

Phillips's keynote came in the form of the question "The Lesson of the Hour?" He answered: "I think the lesson of the hour is insurrection. Insurrection of thought always precedes the insurrection of

arms." The insurrection of thought had been carried on for thirty
years and now it was time for a new phase in the struggle, Phillips
avowed: "I value this element that Brown has introduced into
American politics." Then: "Virginia did not tremble at an old gray-
headed man at Harper's Ferry; they trembled at a John Brown in
every man's conscience."[18]

Higginson described the basis of Phillips's oratory:

> It was essentially conversational—the conversational raised to its
> highest power. Perhaps no orator ever spoke with so little appar-
> ent effort or began so entirely on the plane of his average hearers.
> It was as if he simply repeated, in a little louder tone, what he had
> just been saying to some familiar friend at his elbow. The effect
> was absolutely disarming. Those accustomed to spread-eagle elo-
> quence felt, perhaps, a slight sense of disappointment. Could this
> easy, effortless man be Wendell Phillips? But he held them by his
> very quietness; it did not seem to have occurred to him to doubt
> his power to hold them. . . . Then, as the argument went on, the
> voice grew deeper, the action more animated, and the sentences
> came in a long, sonorous swell, still easy and graceful, but power-
> ful as the soft stretching of a tiger's paw.[19]

As delegates to the nominating convention for the Democratic Party
convened in Charleston, South Carolina, April 23, 1860, every one of
them knew in his heart Harper's Ferry had only been a foretaste of
what lay ahead if "Black" Republicans were successful in that fall's
presidential election. Although the speeches of Seward and Lincoln
had conclusively demonstrated to the South that the goal of the
Republican Party was total abolition, it was John Brown who convinced them that the election was to spell the final verdict on slavery.
In contrast, a victory putting Stephen Douglas at the helm would be
little better than a defeat, the choice between "squatter sovereignty"
and "irrepressible conflict," it was said, being one of no difference.
For the South the nomination of an acceptable candidate with a plank
favorable to slavery, therefore, necessitated a split from the northern
Democrats. By presenting a solid front the South could force a settlement upon the North, proclaiming the terms upon which the Union
could be maintained and the nightmare of abolitionism vanquished.

Two persons were to be preeminent in the political engineering to ensure a favorable outcome at the convention—R. B. Rhett Jr., editor of the *Charleston Mercury*, and William Lowndes Yancey from Alabama, prince of the so-called "fire-eaters." States with delegations amenable to their goals, they calculated, were South Carolina, Alabama, and Mississippi. In his paper Rhett had called for men "having both nerve and self-sacrificing patriotism," to head the movement, and by shaping its course, they would control and compel "their inferior contemporaries"; while Yancey pledged to split the party if it would not adopt the strict slave code for the territories introduced in the Senate in February by Jefferson Davis, and uphold it as the party platform.

On January 29, Rhett editorialized: "So long as the Democratic Party, as a 'national' organization, exists in power at the South, and so long as our public men trim their sails with an eye to either its favor or enmity, just so long need we hope for no Southern action for our disenthrallment and security. The South must dissever itself from the rotten Northern element."

In his speech before the convention to hammer through the slavery plank, Yancey lambasted the northern delegates for treating slaves as less than an equal under the Constitution and slavery as a moral evil. As if the ghost of John Brown was at his side, he nearly stammered out his pathology: "Ours is the property invaded. Ours are the institutions which are at stake; ours is the peace that is to be destroyed; ours is the property that is to be destroyed; ours is the honor at stake—the honor of children, the honor of families, the lives, perhaps of all—all of which rests upon what your course may ultimately make a great heaving volcano of passion and crime, if you are enabled to consummate your designs. Bear with us then, if we stand sternly here upon what is yet that dormant volcano, and say we yield no position here until we are convinced we are wrong."[20]

At this point George A. Pugh from the Ohio delegation, a Douglas supporter, rose to make a reply that sent all delegates into uproar: "Gentlemen of the South, you mistake us—you mistake us—we will not do it." As the proceedings degenerated into pandemonium the chairman, Caleb Cushing, did the only thing he could to restore order, he brought his gavel down and announced the convention adjourned until morning.

On the following morning the convention voted by the narrowest margin to recommit the resolutions for a slavery plank to the platform committee. In the afternoon they came back, only slightly modified. After more wrangling accomplished nothing, the convention adjourned, to give way to meetings conducted in hotel lobbies and barrooms. That Monday, April 30, the convention adopted the Douglas platform with a margin of less than thirty votes, a reaffirmation of Cincinnati's Democratic platform four years before that had only tepidly dealt upon the slavery issue. Now the delegates of the cotton states withdrew. The correspondent to the *Richmond Dispatch* recalled the scene, writing on May 5: "There was no swagger, no bluster. There were no threats, no denunciations. The language employed by the representatives of these seven independent sovereignties was as dignified as it was feeling, and as courteous as it was either. As one followed another in quick succession, one could see the entire crowd quiver as under a heavy blow. Every man seemed to look anxiously at his neighbor as if inquiring what is going to happen next. Down many a manly cheek did I see flow tears of heart felt sorrow."

Fifty delegates bolted and reassembled in another hall that night to hear Yancey announce they would form the "constitutional Democratic convention." The delegates remaining in the original convention proceeded to the selection of a candidate, where Douglas was the front-runner. With his enemies withdrawn, it seemed as if he would receive the nomination, but Caleb Cushing ruled the nominee must get two-thirds of all votes originally accredited to the convention; a ruling, it became apparent, that would yield no winner. The convention went through fifty-seven ballots, finally agreeing to try again in Baltimore on June 18. With the issue of the viability of Douglas's candidacy undecided, the secessionists could do nothing more than agree to meet in Richmond on June 11.

The convention in Charleston, said the *Dispatch* in an article on May 7, had not faltered over the platform, but at the selection of a man. The paper consoled its readers: "After all, the public have not much faith in any platforms, except such as Gov. Wise constructed for John Brown and those other distinguished members of the Republican party who called a Convention and nominated a ticket in Virginia last fall."

Earlier in April, after President James Buchanan's minister to Great Britain had refused him a passport to travel in the Continent on the ground that he was not a United States citizen, Frederick Douglass obtained one from the French ambassador in London. But as Douglass prepared to visit France, he received news that would compel him to return home—word of the tragic death of his ten-year-old daughter, Anne. Against the advice of his British antislavery friends and acting on impulse, he boarded the first outgoing steamer for Portland, Maine; and after a rough seventeen-day voyage, then traveling through Canada, he reached his grieving family in Rochester. "The light" of his home had become ill, he learned, and could not recover because of the distress she felt at her father's prolonged absence and because of her despondency over the death of John Brown, to whom she had been a favorite and on whose knee she often sat only a year before.

As Douglass returned, with the Mason Committee still empaneled, he had to remain vigilant lest he be arrested. But he soon saw that great changes had been worked in the public's mind in his absence. John Brown's "defeat," Douglass wrote, "was already assuming the form and pressure of victory, and his death was giving new life and power to the principles of justice and liberty."[21] Daily Douglass was at his office, and openly walking the streets. It was evident he would be able to proceed untroubled. To be sure, it would not do for him to enter Washington, or any city in proximity to a slave state, where even a white man with the slightest abolitionist sympathies was now in danger, but in much of the North and in Rochester reliable friends would be able to get up men to resist his being taken. At his paper Douglass found receipts had fallen well behind expenses, so he was compelled to contemplate its discontinuance, deciding, yet again, that he must shoulder on, and he went about raising new subscriptions and collecting those outstanding. It was his intention, in agreement with his British friends, after setting his paper on a sound basis that he would return to England as early as September.

Engagements had been booked and Douglass expected to resume his tour, with all of Ireland and the south of England to be visited. Great Britain would have a great influence upon the issue of slavery in the United States, and it was better, reasoned Douglass, that it be weighted against slavery. For that the workers of Manchester must know something about it besides cotton, and there was no better way than for a representative of his race to stand before them and speak.

Just as the Democrats met in Charleston, Gerrit Smith called Harriet Tubman to his home in Peterboro to urge her to attend the gathering of the American Anti-Slavery Society in Boston that May.

She had been spending some quiet months at the home of her parents, just down the road from the William H. Seward residence, in Auburn. Recuperated in health, she agreed to Smith's suggestion and, just as the Democrats were on the verge of their historic split, stopped over en route at Troy. It was there that the fugitive slave case of Charles Nalle was being heard.

As the case unfolded, the city of Troy itself was in tumult, and businesses were closed as those both for and against the process poured into the town center below the commissioner's office. To gain admittance to the hearing Tubman employed a ruse she had used in some of her rescue work. Making her way to the head of the pushing and shouting crowd, she appeared as a crouched and wizened old woman, her face concealed by a shawl drawn over her head. Carrying a nearly empty basket in front of her, she gently pulled on the guard's lapel. Allowed entrance with two stout black women supporting her, she climbed the stairs in the two-story building as the crowd outside began to call for a rescue of Nalle, while others urged calm and "law and order." One of the men on the street, a black named William Henry, began preparing the crowd for action. The *Troy Whig* reported his words in an article on April 27: "There is a fugitive in that office. Pretty soon you will see him come forth. He's going to be taken down South, and you'll have a chance to see him. He is to be taken to the depot, to go to Virginia in the first train."

In the commissioner's office, evidence against the accused was being given by his very brother, together with the agent of their Virginia master. As he realized he was becoming a *cause célèbre*, Nalle looked out the window at his would-be-rescuers. When the decision against him was announced, he swiftly walked to the window, opened it, and was out on the ledge, positioning himself to drop to the pavement. The *Whig* reported: "The crowd at this time numbered nearly a thousand persons. Many of them black, and a good share were of the female sex. They blocked up State Street from First Street to the alley, and kept surging to and fro."

But Nalle was soon clasped by the guards and hauled back inside, now a helpless prisoner. As the two brothers glowered at each other, there was a lull in the proceedings as the attorney for the fugitive hastened to a nearby court to appeal the judgment. Suddenly from the street came the cry: "We will buy his freedom. What is his master's price?"

"Twelve hundred dollars," shouted the master's agent from the opened window.

A flurry of notes changed hands as a collection was taken to meet the owner's price.

Soon a shout went up, "We have raised twelve hundred dollars!"

Avarice taunted back, "Fifteen hundred dollars!"

With this slap in its face, a man from the crowd called back, "Two hundred dollars for his freedom but not one cent for his master!"

The lawyer returned to the commissioner's office with an order that the fugitive be brought immediately for an appearance before a judge of the New York Supreme Court. With no alternative the commissioner ordered the prisoner manacled, and he was guided to the door between several guards. Seizing the moment, Tubman dropped her ruse and dashed to the opened window, calling below, "Here he comes! Take him!"

Hurrying to the street, she instructed the throng: "This man shall not go back to slavery! Take him, friends! Drag him to the river! Drown him! But don't let them take him back!" With this Tubman locked her arms around Nalle's manacled wrists, charging headlong as clubs rained down and pistols discharged. The prearranged understanding of the rescuers was to steer him toward the river where a waiting skiff would cross to Albany County. In the melee, the throng brought the prisoner to the shore and he was carried to the other side. The *Whig* related: "Then there was another rush for the steam ferry boat; which carried four hundred persons and left as many more—a few of the latter being dowsed in their efforts to get on the boat."

On the opposite shore police seized a nearly unconscious Nalle and rushed him to the Police Justices Office, the rescue temporarily coming to naught. But soon men were clamoring outside the barricaded door where the fugitive was held, as others threw stones against the building, "shouting and execrating the officers." One determined rescuer finally concluded: "They can only kill a dozen of us—come on!" The *Whig*'s account reported: "At last the door was pulled open by an immense Negro and in a moment he was felled by the hatchet in the hands of Deputy Sheriff Morrison; but the body of the fallen man blocked up the door so that it could not be shut." Following Harriet, several black women scrambled over the man's body and brought Nalle out, "when he was put on the first wagon passing, headed west."

Harriet Tubman appeared in Boston in May. It was at this time that she was lifted out of the semi-obscurity in which she operated into the homes of the likes of Garrison, Ralph Waldo Emerson, Bronson Alcott, and Whitney. Franklin Sanborn ushered her from one eminence to another, where she occupied parlors before audiences regarding her with rapt attention; she was the black nation itself. Besides being a "natural spinner of tales," she could readily demonstrate the folk dances of her home, or sing songs few outside the South had heard; to Tubman, moreover, the moral censure of slavery meant nothing if it did not arouse action, and it was to this keen sense that she spoke in Boston's antislavery circles. Her reputation as "the Moses of her people" was well known, as was the fact that she had been in confidence with John Brown and knew the details of his plans.

"It was not John Brown that died at Charlestown. It was Christ— it was the Savior of our people," was a saying of hers widely quoted. Sanborn was to write to Sarah Bradford after the war: "The first time she came to my house, in Concord, after that tragedy, she was shown into a room in the evening, where Brackett's bust of John Brown was standing. The sight of it, which was new to her, threw her into a sort of ecstasy of sorrow and admiration, and she went on in her rhapsodical way to pronounce his apotheosis." Well ahead of what many could accept, she knew how things stood, and remarked to Sanborn, "They may say, 'Peace, Peace!' as much as they like; I know there's going to be war!"[22]

On May 10 the Tenth National Women's Rights Convention convened in Boston to demand "an even platform with proud man himself." The movement for women's rights in America, had, of course, been contemporaneous for over a decade with the struggle for black freedom, and neither would be fully realized, many understood, without the success of the other. At this assembly, together with Lucy Stone, Elizabeth Cady Stanton, Lydia Maria Child, Abbey Kelly Foster, Sallie Holley, both Wendell Phillips and Richard Hinton spoke. Following Phillips, Hinton remarked to the convention that he had observed the ideas of women's rights were moving west from where he'd just come "with seven-leagued boots."[23]

As higher law clashed with the law of the preservation of slavery, William H. Seward, his eye on the nominating convention of the Republican Party, and under direct attack for his supposed complicity in Harper's Ferry, began to prevaricate. Late in February in a speech to the Senate on the admission of Kansas to the Union, he scrupulously avoided reference to doctrines he had uttered, and in a direct bid to appease his enemies that could only have elicited scorn from them, the leading Republican said: "Differences of opinion, even upon the subject of slavery, are with us political, not social and personal differences. There is not one disunionist or disloyalist among us all. We are altogether unconscious of any process of dissolution going on among us or around us. We have never been more patient, and never loved the representatives of other sections more than now."[24]

With this obsequiousness, Seward was determined to stoop even further, and in the run-up to the convention delivered a major address where he repudiated John Brown, calling his execution "necessary and just." As his presidential aspirations grew, the enthusiasm of many Republicans lessened for him, and would dim even further as Wendell Phillips stood forth in a "public excoriation" of him. The way had been prepared for his candidacy to be slapped down.

In Chicago where the convention was set to convene on May 16 in a gigantic wood structure built especially for the occasion called the "Wigwam," hordes poured in from the Northwest, as trainloads of delegates and supporters arrived from the East to see the West for the first time. It was to be a gathering unequaled in numbers at that time, and one that no participant would ever forget. As party strategists surveyed the situation, the problem was to carry as many of the free states in order to win the election. California and Oregon were lost to them, as was perhaps New Jersey; this meant they would have to win in states where the Republican Party had lost in 1856— Pennsylvania and either Indiana or Illinois. Seward was the only candidate with a national reputation, but his standing as a radical on slavery undercut his support in the Northwest, just as his backtracking before the convention had hurt him with the radicals. Seward's support going into the convention, with the management of Thurlow Weed, was strong, but the New York machine also tainted him with the issue of corruption. While he had come to the convention hoping for a first-ballot nomination, at the opening gavel the pragmatic and the doubtful were combining in a stop-Seward movement.

The platform of the party softened the antislavery stance from what it had been in 1856, affirming the right of the territories to be free of slavery and demanding the admission of Kansas as a free state, but emphasizing that they meant no interference with slavery where it existed. There was no mention of the Fugitive Slave Law, of Dred Scott, nor of the abolition of slavery in the District of Columbia, while the John Brown "raid" was condemned as "the gravest of crimes." Representative J. R. Giddings proposed an amendment to a number of resolutions reaffirming the Declaration of Independence; but as it was voted down, Giddings withdrew from the convention. The amendment was then taken up by George W. Curtis, who asked the delegates if they were prepared to go to the country as the party which voted down the principle "that all men are created equal, endowed by their creator with certain inalienable rights." The amendment was passed to hurrahs of approval. The platform also pledged to support the Homestead Act, internal improvements, a transcontinental railroad, and tariffs, particularly for Pennsylvania's ironmakers.[25]

By the time balloting began, Lincoln was emerging as Seward's chief rival. Personifying the "free-labor" ideology, Lincoln was known as a moderate, a former antislavery Whig in a party largely made of the same. What is more, Lincoln came from the Northwest, the fastest-growing region of the Union, which had just demonstrated its burgeoning power in Charleston. As the first ballot was counted, the front runner fell sixty votes short of the nomination, while Lincoln made an impressive showing. On the second ballot Lincoln drew almost even. With the convention electric with excitement, on the third ballot Lincoln was within two votes of the nomination, whereupon the chairman of the Ohio delegation announced a change of four votes in favor his candidacy. To the forty thousand people in and around the "Wigwam," that was a moment indelibly etched in memory.

On May 27 the New England Anti-Slavery Conference opened with many of the leading abolitionists attending. By that time news of Theodore Parker's death in Florence, Italy, had crossed the Atlantic, and as John Brown was eulogized, his passing was also duly mourned. Among the speakers at the conference were Garrison, Robert Purvis, Susan B. Anthony, Charles Lennox Remond, Lydia Maria Child, Samuel J. May, and Wendell Phillips. The stenographer James Yerrington caught its fleeting moments for posterity in shorthand, as Harriet Tubman tracked speaker after speaker, each confirming her in what she already knew—that the day of jubilee was near. When Wendell Phillips rose to speak, striding to the platform, those in the auditorium sensed he was now moving in history's wake.

Seward's fall had just shown his potency, and this was his first major speech since that event.

In his lengthy oration Phillips said the late convention in Chicago that brought forth a man of the Northwest, had been, too, the coming out for that region. It was also the affirmation of something else: of the slave's right to rise in his own defense. The time had arrived, said Phillips, when the slaves would strike out for their freedom. Then referring to Seward's pre-convention avowal that John Brown had been justly hung, to sustained applause, Phillips said: "I thank God therefore, that William H. Seward was rejected after making such a speech. It is a good sign that far off in the Northwest there is a leaven of that spirit, that looks upon the Negro as a nation, with the right to take arms into its hands and summon its friends to its side, and that looks upon the gibbet of John Brown, not as the scaffold of a felon, but as the cross of a martyr."[26]

Not present at the conference, but working almost in tandem with its sentiments, Frederick Douglass prepared the next issue of his *Monthly* in his office in Rochester. His own hopes had lain with a Seward nomination, but now he editorialized under the title "The Chicago Nominations": "The road to the Presidency does not lead through the swamps of compromise and concessions any longer, and Mr. Seward ought to have made that discovery, before John Brown frightened him into making his last great speech. In that speech he stooped quite too low for his future fame, and lost the prize that tempted the stoop after all. He had far better have lost it while standing erect."

Lincoln was a man of untried abilities, said Douglass, whose political history was "too meager to form a basis on which to judge of his future." On the positive side, he was of untarnished character, stood at the head of the bar in his state, and had firmness of will. His party had placed a faith in him "before his greatness was ripe," but when elected it would no longer be perilous for him to develop abilities the occasion demanded if he were to succeed. Once the South's expansive ambitions had been curbed and it no longer wielded the reins of power, the four million bondsmen in the South would remain; a Republican victory in November was only the prelude to the struggle for total abolition, Douglass predicted.

The week before the Republican convention former Whig Party conservatives held a convention under a banner that pledged "to recognize no political principle other than the Constitution . . . the

Union . . . and the Enforcement of the Laws." This Constitutional Union Party nominated John Bell of Tennessee for president and the venerable Edward Everett of Massachusetts for vice president. Theirs was a spoiling strategy. Not expecting to win the election themselves, they did expect to draw votes from the lower North in hope of denying a Republican electoral majority.

Wendell Phillips.

In Rochester, Douglass was at work every day on the forthcoming issue of his paper, and just as the New England meetings concluded he published a letter "To My British Anti-Slavery Friends." He wrote: "I have nothing to add on the present aspect of anti-slavery affairs here to what will be found in the other columns of the *Monthly*, except to say, that I have never known the slaves to be escaping from slavery more rapidly than during the several weeks I have been at home. Ten have found food, shelter, counsel and comfort under my roof since I came home, and have been duly forwarded where they are beyond the reach of the slave-hunter. God speed the year of Jubilee the wide world o'er!"[27]

On June 15 the Mason Committee asked to be discharged and ended its investigation, despite having failed to fulfill the responsibility for which it was appointed. Frederick Douglass wrote years later in his *Life and Times*: "I have never been able to account satisfactorily for the sudden abandonment of this investigation on any other ground than that the men engaged in it expected soon to be in rebellion themselves, and that, not in rebellion for liberty, like that of John Brown, but a rebellion for slavery, and that they saw that by using their senatorial power in search of rebels they might be whetting a knife for their own throats."

One of the resolutions adopted by the Senate was that the committee "report whether and what legislation may . . . be necessary on the part of the United States for the future preservation of the peace of the country." Mason's report stated that "after much consideration, [the committee was] not prepared to suggest any legislation," as it would be incapable of preventing like occurrences in the future. The articles of the Constitution pertaining—Section 8 of Article One and Section 4 of Article Four, delegating to Congress the power to suppress insurrections, guarantee a Republican form of government, and to repel invasions—were deemed not applicable to the case because John Brown had sanction by neither a public nor a political authority, which Mason's report suggested was meant in the Constitutional provisions for congressional action. Brown had affirmed that he, and he alone, was responsible for all that had happened, that he had many friends but no instigators, and the committee had brought very little to light that proved otherwise.

Mason's committee was cast out of the way, moreover, because there was bigger work to be done; on June 18 the Democratic nominating convention reconvened in Baltimore. Murat Halstead, editor of the *Cincinnati Commercial*, who covered all four of the conventions that year, chronicling them in his *Caucuses of 1860*, provided a startlingly telegraphic appraisal, reporting: "The Democracy of the Northwest rose out of the status of serfdom. There was servile insurrection, with attendant horrors, and Baltimore became a political St. Domingo."

As the Democrats assembled for the second time to nominate candidates, the bolters at Charleston had been meeting in Richmond with delegations from the upper South and were poised to make a bid for readmission to the convention. A delegate from South Carolina, William Preston, scorning the axiom "Cotton is King," affirmed: "Slavery is our King; Slavery is our Truth; Slavery is our Divine Right." Now he looked to a reunification at Baltimore on a basis satisfactory to the South—a slave-code platform with an acceptable candidate. But the radical leaders were determined to rule or ruin. Joined by the delegates from the upper South who had promised solidarity with them, they would walk out again, forming a new party.

As the convention came to order Caleb Cushing looked dignified ensconced in the chair, dressed in a blue coat with brass buttons. Stephen Douglas's supporters had organized delegations to replace those who had left in Charleston, and now there was to be a fractious credentials fight. After three days the convention finally proposed to seat some of the delegates from both sides, but this left many in Richmond on the sidelines. Now the southern delegations sought a permanent disruption of the national party and walked out, taking a third of the party. This faction quickly convened at the Maryland Institute Hall with Caleb Cushing once more presiding. Buchanan's vice president, John C. Breckinridge of Kentucky, was nominated for president and mounted atop a slave-code stand. The "loyalists" of the sundered party meanwhile nominated Douglas for president.

Delegates of the two camps were alternately stunned and elated, with few understanding the breakup's ultimate consequence—nothing now was left to hold down the fierce antagonisms of the two competing sections. Commenting on this outcome in his *Monthly*, in August 1860, Frederick Douglass wrote:

> With a name altogether attractive to the masses, and a long period of uninterrupted strength and prosperity, its leaders and managers, like all others thus conditioned, began to think the party immortal. In every division of opinion, in all the contests of fractions, the quarrels of greedy and aspiring candidates, hitherto the party has at the trial hour been found untied and strong. The Whig party crumbled under the sturdy blows of the Abolitionists, and went to its own place, having outlived its usefulness eight years ago; but the Democratic party stood firm and united, impressing its enemies, as well as its friends, with the idea of its firmness and indivisibility. The illusion is now dispelled. Babylon has fallen.

Just as Douglass's pen was sketching the ramifications of the momentous event in the month of June, from James Redpath came an invitation to attend a meeting of "friends" at Brown's grave on the 4th of July. Douglass replied his heart was with them, but it would not be possible for him to break away from his duties on such short notice. Had the invitation reached his desk but a day or two earlier, he would have "very gladly" come "to do honor to the memory of one whom I regard as the man of the nineteenth century." It had been among his highest privileges, Douglass wrote, "to have been acquainted with John Brown, shared his counsels, enjoyed his confidence, and sympathized with the great objects of his life and death."[28]

Several days later in a letter to William Still, Douglass confided his deepest longing with the Underground Railroad leader "for the end of my people's bondage, and would give all I possess to witness the great jubilee." The signs on the road to the downfall of slavery were numerous, but he recalled, twenty years before it did really seem too to be hastening to its death, only to become entrenched in a deeper abyss. Douglass remarked, "I will walk by faith, not by right, for all ground of hope founded on external appearance, have thus far signally failed and broken down under me."[29]

Douglass began to place more emphasis than ever on the efficacy and necessity of revolutionary violence on the part of the slaves. He wrote Redpath: "The eight-and-forty hours of John Brown's school in Virginia, taught the slaves more than they could have otherwise learned in a half-century." Though it seemed that Brown's effort had yielded little, Douglass affirmed it had "done more to upset the logic and shake the security of slavery, than all other efforts in that direction for twenty years."[30] Repeatedly striking this theme canvassed by Brown, in letters, in his press, and in speeches—in the August issue of the *Monthly*, "The Prospect for the Future," Douglass wrote: "Outside philanthropy never disenthralled any people. It required a Spartacus, himself a Roman slave and gladiator, to arouse the servile population of Italy, and defeat some of the most powerful armies of Rome, at the head of an army of slaves; and the slaves of America await the advent of an African Spartacus." That very year had seen Giuseppe Garibaldi with his thousand "red shirts" lead a successful expedition to Sicily and Naples, overthrowing the Bourbon monarchy. Even law-and-order conservatives and slaveholders in America had greeted this with hosannas. Douglass asked: "Why should we shout when a tyrant is driven from his throne by Garibaldi's bayonets, and shudder and cry peace at the thought that the American slave may one day learn the use of bayonets also?"

After a four-year absence Charles Sumner returned to the Senate floor on June 4, 1860. Massachusetts voters had retained him as their senator despite his incapacity following his beating in the Senate chamber, and his speech titled "The Barbarism of Slavery" was to again place him in the forefront of the Republican Party and draw great appreciation for his courageous stand. The speech was in opposition to the bill providing for the admission of Kansas to the Union with slave-code provisions, and Sumner took occasion to deliver a thorough refutation of the defenses of slavery, declaring the total social and political equality of blacks and whites. Analyzing the slave codes and the present conduct of slave owners, Sumner said slavery was a "Upas tree with all its gigantic poison," and he expounded the most systematic and careful statement against slavery at that crucial time within the Senate. Immediately hastening to rally to Sumner's standard, William Still wrote, "You have effectually laid the axe at the root of the tree"; from Chicago, H. O. Wagoner wrote to offer his heartfelt thanks in "behalf of seven or eight thousand colored people of the State of Illinois." Should the slaves be made to comprehend the speech and the circumstances under which it was given, "it would thrill their very souls with emotions of joy unspeakable." From John S. Rock came: "Your immortal speech has sent a thrill of joy to all lovers of Freedom everywhere." Frederick Douglass printed the speech in its entirety in his July *Monthly*, and only a few days after Sumner's effort, he penned these words: "I wish I could tell you how deeply grateful I am to you, and to God, for the speech you have been able to make. . . . You spoke to the Senate and the nation, but you have another and a mightier audience. The civilized world will hear you." The Massachusetts legislature endorsed the speech in a resolution and thanked Sumner, declaring, "the stern morality of that speech, its logic and its power, command our entire admiration, and that it expresses with fidelity the sentiments of Massachusetts upon the question therein discussed."[31]

Into this turmoil that absorbed and roiled party and nation, Frederick Douglass threw himself, as he wrote, "with firmer faith and more ardent hope than ever before." He did so not by going on the stump or editorializing in support of the Republican Party and its candidate; rather, as he saw it, out of the campaign and election must either come a strengthening or a weakening of "the abolition element in the country, and it [was] for the Republicans to say which it shall be." There was no denying, after all, that the "vital element" for its support was in antislavery opinion. But at the convention in Chicago and since then there had been a concerted effort to make the party appear as "indifferent" on the question. In regard to the Fugitive Slave Law, the party occupied the same position as did the Whigs in 1852—they would not repeal the law and would do nothing to lessen its efficiency; the South, said Lincoln, was entitled to assistance in recapturing its fugitives.

But this newfound opposition to all elements of antislavery agitation, said Douglass, was "a fatal trick upon the very life of the party itself." The vitality it had gained had been brought by that element "which sees a brother in the blackest slave. . . . All else is weak and standing alone is worthless."

That summer, in the *Liberator*, Wendell Phillips bitterly denounced the Republican presidential candidate as "the Slave Hound of Illinois"; drawing attention to the fact that on January 10, 1849, as a Whig congressman, Lincoln had introduced a bill requiring municipal officials in Washington and Georgetown "to arrest and deliver up to their owners all fugitive slaves escaping into said District." True,

Phillips wrote, Lincoln's bill was offered as a concession for a proposal abolishing both the slave trade and slavery in the capital, but he argued, "It is the nature of the compromise with which I find fault. . . . Some things are too sacred to be made counters of, to be traded in or compromised away. My charge is that in 1849 Mr. Lincoln did not know that slave-hunting was one of these."

On July 2 Douglass wrote to Gerrit Smith thanking him for two contributions to his paper and discussing the prospects for its continuance. He wrote he was still planning to return to England in September, saying: "I cannot support Lincoln, but whether there is life enough in the Abolitionists to name a candidate, I cannot say." The Radical Abolition Party was nearly moribund, but he, Smith, and others would give their attention in the interim before the November election to see what could be done in reviving it. After all, the Liberty Party with a handful of members had been the source of the Republican Party; they too could have importance outweighing their prospects and circumstances. But Douglass was far from optimistic from a radical abolition prospective; its adherents had "reached a point of weary hopelessness." True, the work of educating the public had been a success, but that work had borne no practical results. The public seems to regard the whole struggle against slavery, Douglass remarked, "as a grand operatic performance, of which they are the spectators." Yet despite "doubt and gloom" as to the ultimate outcome and despite Republican attempts to dampen their antislavery appeal—one thing was certain: "no professions of loyalty to the South, no pledges to carry out that foul and merciless abomination of the Fugitive Slave Law, no expression of contempt for the rights of Negroes, no bowing or cringing to the popular prejudice against color, will win for the Republican party the support of genuine proslavery men, or avert from the party the odium of being the advocate and defender of the Negro as a man and a member of society."[32]

On August 29 the Radical Abolition Party met in Syracuse, nominating Gerrit Smith for president, and choosing Douglass as an elector. It was just prior to this that Stephen S. Foster and John Pierpoint sent out a call for a political antislavery convention to convene in Worcester, "to consider the propriety of organizing a Political Party upon an Anti-Slavery interpretation of the U.S. Constitution, with the avowed purpose of abolishing slavery in the states, as well as the territories of the Union." Douglass was "in half and half condition about attending," lest the Garrisonians present snub him and their disagreements muddy the proceedings, he professed in a letter on August 25, responding to an invitation from Elizabeth Cady Stanton. But as that convention neared Douglass decided he must attend and would serve on its executive committee, reasoning that with the abysmal Republican platform on slavery, the scattered elements of antislavery in the country might be rekindled and united to "produce one solid abolition organization."[33]

With opportunities for significant agitation renewing, Douglass notified his public and his friends in England that he had given up on the idea of returning to complete his tour. "He who speaks now," he wrote in his September *Monthly*—"may have an audience. . . . Mind is active; opinions and principles clash; truth with error meets in stern debate before all people." An additional reason given for not going abroad just then was a proposition before the people of New York State for "Colored Suffrage"; Douglass believed that "he who has any influence should remain and exert it in bringing the State to the great measure of justice now proposed."

The presidential campaign of 1860 was to be one of a thousand voices with few hearing what the other was saying. But throughout the North, in New England and the Mid-Atlantic, the Middle West and Northwest, young Republican clubs dubbed Wide-Awakes were casting portentous shadows, giving Lincoln's campaign the appearance of being a movement. Assuming a quasi-military order, wearing oilcloth capes and bearing smoking torches, some carrying rails— since Lincoln was "the rail-splitter"—they marched in towns and cities alike from Maine to Minnesota. First appearing outside Cooper Institute in March when Lincoln spoke, they began to convey the idea of a candidacy with huge popular support. The largest political demonstration to occur in the country up to that time took place August 8, the *Springfield State Journal* reported, when groups from all parts of Illinois detrained by the hundreds in Springfield to celebrate Lincoln's nomination. It was "a veritable political earthquake," the paper observed.

The leaders of the South were being driven under a sharp compulsion. "If Lincoln is elected—what then?" South Carolina congressman Lawrence M. Keitt posed in a letter to James Hammond on September 10. This interrogative was made all the more affecting by the fact that slaves had slain Keitt's brother in February on his Florida plantation. Keitt wrote: "I am in earnest, I'd cut loose through fire and blood if necessary—See—poison in the wells of Texas and fire for the houses in Alabama—Our Negroes are being enlisted in politics—With poison and fire how can we stand? I confess this new feature alarms me more than even everything in the past. If Northern men get access to our negroes to advise poison and torch we must prevent it at every hazard. The future will not 'down' because we are blind."[34]

The split in Charleston, with cotton-state delegates opting for an independent proslavery ticket in Baltimore, was a deliberate action taken, not to win the election, but to lose it. The calculation was that a Republican victory would culminate in the secession of all the slave-holding states of the Union; but in the summer and fall of 1860, it did not suit their goal to go out immediately. They could not justify it while the Buchanan administration, which had been friendly to their interests, was still in office. And, moreover, many of its officials would be leaders in the future Confederate government, and many were not yet ready to resign. It did not go well, either, because they were the ones who most often held up the argument of the inviolability of the Constitution and its processes. The months ahead would give them time to firm up sentiment, solidify their movement, collect arms and munitions, and ready their cotton for export.

That September Charleston-based secessionists formed the 1860 Association, and by the end of October it had become a veritable font of pamphlets, publishing such titles as "The South Alone Should Govern the South," and "African Slavery Should be Controlled by those Only Who are Friendly to It." Another was "The Doom of Slavery in the Union: Its Safety Out of It."

The agitators proclaimed that abolition at the hands of the federal government, with the resulting political, social, and sexual equality of the races, was inevitable with a "Black" Republican victory. The financial basis of the South—the rising value of land and crops and black skins—would be destroyed and the freedmen reduced to crime and idleness. The only alternative to a second "John Pike Brown" was withdrawal from the Union. Accordingly, the association became the center for military and psychological grounding for disunion, establishing throughout the cotton and border slave states committees of correspondence on the model of those established during the revolution by patriots. Newspaper editorialists received hundreds of letters detailing "a settled and widely-extended purpose to break up the Union."

As the fear sweeping the South rose to a fever pitch, Lincoln was urged to make a mollifying statement. He asked: "What is it I could say which would quiet alarm? Is it that no interference by the government is intended? I have said this often already, that a repetition of it is but mockery, bearing an appearance of weakness." Speaking in St. Paul, Minnesota, Seward ridiculed the effort. "Who's afraid?" he asked to laughter and cries of "no one." "Nobody's afraid; nobody can be bought." In this speech reprinted in pamphlet form and distributed in every northern state, Seward declared: "The man is born today who will live to see the American Union, the American people—the whole of them—coming into the harmonious understanding that this is the land of free men—for free men—that it is the land for the white man; and that whatever elements there are to disturb its present peace or irritate the passions of its possessors will in the end—and that end will come before long—pass away, without capacity in any way to disturb the harmony of or endanger, this great Union."[35]

The *Charleston Mercury* in July surmised Seward had been rejected in Chicago because "he was disposed to temporize"; Lincoln on the other hand was "the beau ideal of a relentless, dogged free-soil border ruffian. . . . a vulgar mobocratic and a Southern hater."

At the same time the *New Orleans Crescent* declared that every vote cast for Lincoln would be "a deliberate, cold-blooded insult and outrage" against Southern honor. John J. Crittenden, senator from Kentucky, denounced the "profound fanaticism" that would "think it [its] duty to destroy . . . the white man, in order that the black might be free."

Ever a realist, Stephen Douglas had been compelled to accept the northern interpretation of his "squatter sovereign" doctrine, but as a savvy political insider he remained a formidable candidate. After Republicans swept in crucial state elections in October, Douglas confided to his secretary: "Mr. Lincoln is the next President. We must try to save the Union. I will go South." Breaking with tradition he personally campaigned throughout the states, except in California and Oregon, carrying the message that he was the only national candidate. But Douglas's standing with the South had slid further when on August 25 in Norfolk he was asked from the audience for his stand on disunion. In weakened health and hoarse voice he invoked the name of Andrew Jackson to repudiate the idea of peaceable acquiescence to secession.

"A frightful number of patriots are modestly consenting to assume the burden of Presidential honors," Frederick Douglass observed in a speech titled "The Presidential Campaign of 1860," on August 1. "Instead of five loaves and two fishes—the usual number of political principles—we have five parties and no principles in the present canvass. And yet, since the organization of the Government there has been no election so exciting and interesting as this. The elements are everywhere deeply stirred, and nowhere are they more deeply stirred than at the South. Our political philosophers call the present contest a sectional strife; as if there could be conscious antagonism between two pieces of land not even separated by a stream of fresh water. . . . The single bone of contention between all parties," declared Douglass, was slavery; and every man would go to the election either to help or to hinder slavery, or with the idea of neither helping nor hindering slavery.

In both instances, it was the only interest. What was the difference between Stephen A. Douglas and Breckinridge? It was that one believed the Supreme Court with *Dred Scott* had decided that slaveholders had the right to carry slaves into the territories and that the Congress had no power to prohibit the relation of master and slave; while the other did not believe the Court had so decided, but declared he would abide by such a decision if it were made. Both served the same master, said Douglass. One thought himself already sent, while the other held "himself ready upon the moment of receiving orders." Nothing need be said of the Bell and Everett ticket, in Douglass's view, because "a party without any opinions need have no opinion expressed of it." Although the Republican Party was far from being an abolition party, Douglass did "not accord with those who prefer" its defeat "from fear that it [would] serve slavery as faithfully as the Democratic party, or either branch of it." To do that, Douglass reiterated, "would be to cut its own thread of existence."

On election day, with only 40 percent of the national vote, but with 54 percent in the North, Lincoln claimed 180 electoral votes to win the presidency. Douglas won three electors in New Jersey but carried only in Missouri. Bell won in Virginia, Kentucky, and Tennessee, while Breckinridge carried in the rest of the South. At a cheering gathering in Boston, Wendell Phillips spoke:

Ladies and Gentlemen: If the telegraph speaks truth, for the first time in our history the slave has chosen a President of the United States. We have passed the Rubicon, for Mr. Lincoln rules today as much as he will after the 4th of March. It is the moral effect of this victory, not anything which his administration can or will probably do, that gives value to this success. Not an Abolitionist, hardly an anti-slavery man, Mr. Lincoln consents to represent an anti-slavery idea. A pawn on the political chessboard, his value is in his position; with fair effort, we may soon change him for knight, bishop, or queen, and sweep the board. . . . In 1760 what rebels felt, James Otis spoke, George Washington achieved, and Everett praises today. The same routine will go on. What fanatics now feel, Garrison prints, Lincoln will achieve. . . . You see exactly what my hopes rest upon. Growth! The Republican party have undertaken a problem the solution of which will force them to our position.

As Lincoln's election was being tallied, South Carolina senators James Chesnut Jr. and James H. Hammond resigned from the United States Senate. A few days before a convention had been called by the state's legislature "to consider [South Carolina's] relation to the Northern States and to the United States Government," while simultaneously conventions were called in Mississippi, Alabama, and Georgia. Most observers in the North regarded the culminations of "fire-eating" rhetoric as a political bluff, a device to frighten them into granting favorable concessions to the South. Lincoln was reported saying that he "believed that when the leaders saw their efforts in that direction unavailing, the tumult would subside."[36]

Immediately after the election, parties who had lost in the vote made an effort to compel the victors to abandon the principle upon which they were elected. On December 3 Buchanan delivered his last message to the Congress, where he gave a truckling recital of the defeated Breckinridge platform as a way to salvage the Union. It was the North, he said, and specifically the Republicans, who were to blame for "the incessant and violent agitation of the slavery question" that had "produced its national effects" in the secession movement. This opposition must be stopped, the Fugitive Slave Law must be enforced, and he urged the repeal of "unconstitutional and obnoxious" personal liberty laws in the North. Advocating the adoption of an amendment to the Constitution protecting slavery in the territories, Buchanan also advised the acquisition of Cuba as a new slave state to the Union. He had begun, however, by asserting the Union was not "a mere voluntary association of States," but was intended to be perpetual; but confessed he also felt Congress could do nothing about dissolution, ending with an appeal to the nation for "prayer and fasting."[37]

Paralleling the president's message, committees were appointed in both chambers of Congress to see what could be done to calm the South. In the Senate a committee of thirteen was formed that included Seward, Benjamin Wade, Stephen Douglas, Robert Toombs, Jefferson Davis, and John J. Crittenden. This committee would offer the so-called Crittenden Compromise as an amendment to the Constitution, declaring slavery perpetually free of interference from the national government, reinstating the prohibition on slavery north of the line 36 degrees 30 minutes, but providing for its spread south of the line. The compromise would also prohibit Congress from abolishing slavery in the forts, arsenals, and naval bases of the federal government, while allowing its abolition in the District of Columbia provided slavery was first abolished in both Virginia and Maryland. It further prohibited interference in the interstate slave trade and would provide compensation for slaveholders who "lost" property escaping into the North.

On November 11, in an editorial in the *New York Tribune*, Horace Greeley cautioned against using force to maintain the Union. He wrote: "If the Cotton States shall become satisfied that they can do better out of the Union than in it, we insist on letting them go in peace. . . . When ever a considerable section of our Union shall deliberately resolve to go out, we shall resist all coercive measures designed to keep it in. We hope never to live in a republic whereof one section is pinned to the residue by bayonets."

Thurlow Weed, in the *Albany Evening Journal*, came out in a similar fashion, suggesting by implication that Seward too favored peaceful secession to war. Frederick Douglass wrote: "Some of these peace propositions would have been shocking to the last degree to the moral sense of the North, had not fear for the safety of the Union overwhelmed all moral conviction. Such men as William H. Seward, Charles Francis Adams, Henry B. Anthony, Joshua R. Giddings, and others—men whose courage had been equal to all other emergencies—bent before this southern storm, and were ready to purchase peace at any price. Those who had stimulated the courage of the North before the election, and had shouted 'Who's afraid?' were now shaking in their shoes with apprehension and dread."[38]

To convince the South they had nothing to fear by remaining in the Union, and that their property relations would not be interfered with, northern conservatives launched a movement to repeal the "Personal

Liberty Laws" which barred, in some instances, state officials from enforcing the Fugitive Slave Law, and in others prohibited state jails from holding fugitive slaves, or otherwise protected free blacks from kidnapping and guaranteed fugitives due process. Prominent merchants, who had large uncollected debts with the South and held mortgages on slaves, took it upon themselves through hirelings to suppress abolitionist meetings wherever they occurred. The first to be attacked was a meeting in Boston's Tremont Temple on December 3 commemorating the first anniversary of John Brown's death. Calling abolitionists together to consider "The Best Method to Abolish Slavery," Frederick Douglass traveled five hundred miles from Rochester to be present, while John Brown Jr. traveled even farther, from Ohio. Arriving at Syracuse Station and met by Samuel J. May, Brown's oldest surviving son disclosed details "of a conspiracy against the lives of prominent colored men and abolitionists."[39] The plot's leaders had vowed that blood must be shed to appease the honor of the South and restore peaceful relations between the two sections of the country.

The meeting convened with a full hall and included some of Boston's leading businessmen bent on provoking action to break up the assembly and attack its principals. As Frederick Douglass took the stand he was greeted by a chorus of catcalls, boos, and epithets. He had come, he began, to advocate the best way of abolishing slavery, a new way that had been demonstrated by the man whom they were there to remember: he had come to advocate "the John Brown way." That became the signal for the assault to begin. As those there to disrupt the meeting rose, a fierce fight broke out where many in the audience were physically assaulted and roughly pushed toward the exits. On the front steps of Tremont Temple, Franklin Sanborn was collared and hurled onto the pavement. Surrounded by assailants, several witnesses would relate, Frederick Douglass fought "like a trained pugilist."[40]

The next day the *Boston Courier* crowed that the parties that "got up" the meeting "were all irresponsible," and that, as an added affront, "a large number were Negroes." Glorying that the police had closed the hall after Boston's "citizens" drove the abolitionists from it, editorials in many papers now called for demonstrations wherever abolitionists gathered. Following the example in Boston, meetings would be broken up or prevented in New York, Philadelphia, Albany, Troy, Buffalo, Cincinnati, and Chicago.

That evening at Boston's Joy Street Baptist Church the meeting

"Expulsion of Negroes and Abolitionists from Tremont Temple, Boston, Massachusetts, on December 3, 1860."

reassembled, now joined by Wendell Phillips, who had not been present earlier. Addressing that fact, Douglass remarked: "I am sorry that Mr. Phillips was not there to look that Fay in the face. (Hear!) I believe that he, and a few Abolitionists like him in the city of Boston, well-known, honorable men, esteemed among their fellow citizens— had they been there to help us take the initiatory steps in the organization of that meeting, we might, perhaps, have been broken up, but it would have been a greater struggle, certainly, than that which it cost to break up the meeting this morning. (Applause.)"[41]

Beginning again from his interrupted speech: For twenty-five years, said Douglass, the moral and social means of opposing slavery had been given prominence. But far from winning over the heart and conscience of the slaveholder, the opposite had happened; they had only become harder, bolder, and more maddened. "Now what remains? What remains?" cried Douglass. Now he stood and announced for the "John Brown way":

We must, as John Brown, Jr.—thank God that he lives and is with us to-night! (Applause)—we must, as John Brown, Jr., has taught us this evening reach the slave-holder's conscience through his fear of personal danger. We must make him feel that there is death in the air about him, that there is death in the pot before him, that there is death all around him. We must do this in some way.

I know that all hope of general insurrection is vain. We do not need a general insurrection to bring about this result. We need the fact to be known in the Southern States generally, that there is liberty in yonder mountains planted by John Brown.

Signifying that he now embraced the opportunity he had let slip, he went on using words that could have been spoken by Brown:

The slaveholders have but to know, and they do now know, but will be made to know it ever more certainly before long—that from the Alleghanies, from the state of Pennsylvania, there is a vast broken country extending clear down into the very heart of Alabama—mountains flung there by the hand and the providence of God for the protection of Liberty—(Cheers)—mountains where there are rocks, and ravines, and fastnesses, dens and caves, ten thousand Sabastapols, piled up by the hand of the living God, where one man for defense will be as good as a hundred for attack. There let them learn that there are men hid in those vastnesses, who will sally out upon them and conduct their slaves from the chains and fetters in which they are now bound, to breathe the free air of liberty upon those mountains. Let, I say only a thousand men be scattered in those hills, and slavery is dead. It cannot live in the presence of such a danger. Such a state of things would put an end to the planting of cotton; it would put an end not only to planting cotton, but to planting anything in that region.[42]

Douglass remained in Boston after the attack upon the meeting at Tremont Temple, so "mortifying and disgraceful" had been that act that he vowed to speak again in the city.

The following Monday in the Music Hall he delivered "A Plea for Free Speech in Boston." The scene had been an instructive one, said Douglass. Men referred to as gentlemen, led by some of the wealthiest and most respected merchants of Boston, had invaded, insulted, and captured the meeting hall, whereupon it was broken up and persons assailed—and this despite pleas to the mayor and the police for protection. Now, he noted, the friends of freedom were of two voices, denouncing the mob while regretting the meeting: "We are told that the meeting was ill-timed, and the parties to it unwise. . . . Why, what is the matter with us? Are we going to palliate and excuse a palpable and flagrant outrage on the right of free speech, by implying that only a particular description of persons should exercise that right? . . . until the right is accorded to the humblest as freely as to the most exalted citizen, the government of Boston is but an empty name, and its freedom a mockery."[43]

But this was not the end of mob action in Boston. Because of his special prominence and effectiveness as a public orator, Phillips was to be singled out for physical assault, if not worse, by persons sworn to prevent him from speaking. A few days after the Music Hall meeting, Phillips was invited to occupy Theodore Parker's pulpit, selecting for his theme "Mobs and Education." Reviling the conduct of the mob, amid hisses and boos, the *Courier* reported the speech "a mass of poisonous and malignant trash—a thorough jail delivery of bad temper, vituperation, and hatred." Friends hastened to him afterward, congratulating Phillips for his courage in the face of the threats. As he came out into the street to a taunting crowd of nearly a thousand, Phillips walked behind a cordon of young men armed with pistols, and he too was armed.

"There he is!" someone shouted. "Down with the Abolitionists."[44] As the crowd surged toward him they were forced to give way by the volunteer bodyguard and two hundred police officers. The mob followed to Phillips's house on Essex Street, and as they were ordered to disperse three cheers went up for Phillips. For weeks the volunteer bodyguard watched the house night and day; and as many as forty young men would accompany Phillips when he spoke in Boston. These were German-American youths of the Turn Verein, whose leader, Karl Heinzen, publisher of *Der Pionier*, was a determined champion of free speech.

In the summer and fall of 1860 Harriet Tubman divided her time between New England and New York State. As the election approached she was determined to make another trip south to rescue whomever she could. With money donated by abolitionist friends she arrived in late November on Maryland's Eastern Shore in a neighborhood familiar to her. Traces of this furtive activity were indelibly drawn in a letter of Thomas Garrett to William Still, December 1, 1860.

Respected friend:—William Still:—I write to let thee know that Harriet Tubman is again in these parts. She arrived last evening from one of her trips of mercy to God's poor, bringing two men with her as far as New Castle [Delaware]. I agreed to pay a man last evening, to pilot them on their way to Chester County [Pennsylvania]; the wife of one of the men, with two or three children, was left some thirty miles below, and I gave Harriet ten dollars, to hire a man with a carriage, to take them to Chester County. She said a man had offered for that sum, to bring them on. I shall be very uneasy about them, till I hear they are safe. There is now much more risk on the road, till they arrive here, than there has been for several months past, as we find that some poor, worthless wretches are constantly on the look out on two roads, that they cannot well avoid more especially with carriage, yet as it is Harriet who seems to have had a special angel to guard her on her journey of mercy, I have hope."[45]

Tubman arrived in New York City with her party just as mobs were directing their fury against abolitionists, and Seward and other politicians in Washington were "down on their knees" begging that the South not secede. It was evident to her friends that Tubman's increasing prominence would make her especially a target for the wrath of this proslavery vengeance. Franklin Sanborn was to write in the *Commonwealth*, on July 17, 1863: "Those anxious months, when darkness settled over our political prospects, were viewed by all classes with deep forebodings and by none more so than those who, like Harriet, had rendered themselves obnoxious to the supporters of slavery by running off so many of their race from its dominions. Fear for her personal safety caused Harriet's friends to hurry her off to Canada, sorely against her will."

Before being whisked away she stayed at the home of Rev. Henry Highland Garnet in New York City. He later related a story of how she came down to the breakfast table one morning singing in rapture—"My people are Free! My people are Free!"

"Oh, Harriet! Harriet!" interrupted Garnet. "You've come to torment us before the time; do cease this noise! My grandchildren may see the day of the emancipation of our people, but you and I will never see it."

"I tell you, sir, you'll see it, and you'll see it soon. My people are Free! My people are Free!"[46]

Sarah Bradford was to write in *Harriet, Moses of Her People*: "When, three years later, President Lincoln's proclamation of emancipation was given forth, and there was a great jubilee among the friends of the slaves, Tubman was continually asked, 'Why do you not join with the rest in their rejoicing?' 'Oh,' she answered, 'I had my jubilee three years ago. I rejoiced all I could den; I can't rejoice no more.'"

Ten

Terrible Swift Sword

The land is now to weep and howl, amid ten thousand desolations brought upon it by the sins of two centuries against millions on both sides of eternity.

—Nemesis, *Douglass' Monthly*, May 1861

A S THE CONVENTION PRONOUNCING A "REPUBLIC OF SOUTH Carolina" finished its business, political leaders and editorialists turned their attention to a problem they need consider—the forts in Charleston Harbor that until now had been under United States authority. These included Fort Moultrie on Sullivan Island, the largely abandoned Fort Johnson on James Island, and, still under construction in the harbor entrance, Fort Sumter. These were now deemed the property of the "new nation," and accordingly three commissioners were dispatched to Washington to negotiate their transfer, along with all other property of which the United States held title.

On December 24, 1860, the convention produced two documents meant to justify and secure further adherents to secession. One bore the title "Declaration of Immediate Causes," asserting the Constitution had been undermined by states assuming "the right of deciding upon the propriety of our domestic institutions; and have denied the rights of property established in fifteen of the States and recognized by the Constitution." The second, addressing "The People of the Slave-holding States of the United States," asserted the government was a despotism, and its Constitution a failed experiment, while South Carolina was   "in the van of the great controversy between the Northern and Southern States. . . . Compromise after compromise, formed by your concessions, have been trampled under foot by your Northern confederates. . . . United together, we must be the most independent as we are the most important of the nations of the world. United together, and we require no other instrument to conquer peace than our beneficent productions."

On the day following the arrival of South Carolina's commissioners in Washington, President Buchanan received a telegram stating that the commander at Fort Moultrie, a major of artillery named Robert Anderson, had spiked its guns and moved his garrison to Fort Sumter, considering it more defensible. On December 27 and 28, two meetings were held with southern leaders in Congress, the president and the commissioners attending, where Buchanan was told he had broken his word. "And now, Mr. President," said Jefferson Davis, "you are surrounded with blood and dishonor on all sides."[1]

Giving way in near panic, Buchanan drafted a paper ordering Anderson and his garrison back to Moultrie. When this was presented to Jeremiah S. Black, the secretary of state, he and Attorney General Edwin Stanton, both new to the cabinet, threatened to resign. Buchanan handed the paper over to his chief cabinet minister, telling him to alter it as he saw fit. Anderson would remain at Sumter. South Carolina, however, was not prepared for a shooting war, and as Major Anderson stood firm, a stalemate ensued. The garrison was still supplied in its daily needs from the Charleston markets, and South Carolina continued to build batteries to bear on the fort when the time was ripe, as South Carolina militia occupied Fort Moultrie and other military installations in the harbor, and soon would seize the federal arsenal in Charleston.

Still in the Senate in Washington, Jefferson Davis advised Governor Pickens of South Carolina to exercise caution: "The little garrison in its present position presses on nothing but a point of pride & to you I need not say that war is made up of real elements. It is a physical problem from the solution of which we must need exclude all sentiment. I hope we shall soon have a Southern Confederacy, shall soon be ready to do all which interest or even pride demands."[2] Following Davis's letter, secession became an accomplished fact in his home state, followed by Florida and Alabama, and in the succeeding weeks by Georgia, Louisiana, and Texas. Federal forts and arsenals were seized in all of these without the shedding of blood, including Fort Morgan in Mobile Bay, Forts Jackson, St. Philip, and Pike near New Orleans, and Barrancas and McRae in Florida. Besides Sumter, the only other coastal fortification not taken was Fort Pickens in Florida, where a siege was also being prepared. The arsenals at Mount Vernon in Alabama and at Baton Rouge in Louisiana were among those seized.

On January 21, as he prepared to leave Washington, a gaunt and pallid Jefferson Davis stood before a muted United States Senate to deliver his resignation speech. Oddly echoing the consolatory words of Seward in the previous year, his words did not seem his own, Davis confided to a friend, "but rather leaves torn from the book of fate." Only now was a bill brought before Congress that was to resolve the central dispute of the sectional controversy in the previous years—in whose name blood had flowed and argument raged, and from which one man had come away determined to strike at the very heart of the rebellion he saw nurturing on cotton and black skins—as Kansas was admitted to the Union a free state.

In the border slave states remaining in the Union, secession was not carrying the day, Unionists in them believing they had not exhausted efforts for a solution within it. In North Carolina delegates rejected a call for a secession convention and its governor returned two forts to the federal government that had been seized by its citizens. In Tennessee, the legislature also balked at secession, calling instead for amendments to the Constitution to protect "Southern Rights." Kentucky was of like mind, opting for a national convention, and voicing support for compromise on the basis of Senator Crittenden's proposals for settlement. Maryland's governor, too, refused to call for secession, while Delaware expressed "unqualified disapproval for the remedy." In Missouri, the legislature summoned a convention for February 18, but stipulated secession ordinances would have to be ratified by statewide vote. Its neighbor Arkansas was not nearly as tentative, calling an election for the same day, as its armed citizens had already taken possession of the federal arsenal at Little Rock and expelled the United States troops across the border. The key to the ultimate success of a "Southern Confederacy," however, depended on the most populous slave state, Virginia, with deep sympathies for secession but with a strong Unionist sentiment in its western portion beyond the mountains. As Virginia's governor Letcher wavered, seeing an ominous future, he hoped a national convention, minus New England, could solve the problem. The same split in sentiment was manifest in eastern Tennessee and in western North Carolina, and was also a factor in the hill country of South Carolina, Georgia, and Alabama, on whose rocks secession faltered.

Constituting an area hardly tenable as a defensible nation, only seven of fifteen slave states had gone out. On its northern frontier was a large mountainous country offering an inviting haven for escaping slaves, with many avenues, as well, to support invasion from the north, while also possessing an extensive coastline on both the Atlantic and the Gulf of Mexico difficult to defend. But these vulnerabilities served rather to stiffen the resolve of the leaders of secession, who chose Montgomery, Alabama, as the capital of what they proposed would be a powerful and prestigious nation. On February 4 thirty-seven delegates, with one delayed, from six of the seceded states (minus Texas), settled into one or the other of Montgomery's two hotels, meeting in secret sessions to create a government for the Confederate States of America. At the Capitol Building, Howell Cobb was chosen president of the convention. On February 8 a provisional constitution was adopted, a provisional president named on the following day, and within a week, as the convention became a congress, a government was up and running. All this, done with the utmost speed so as to be in place before the inauguration of the new president at Washington on March 4, and to present a face of legitimacy to the world. Jefferson Davis, who had taken no part in the deliberations and who would have preferred the rank of major-general in the Confederate Army, would assume the presidency, and Alexander Stephens of Georgia, who had been a Douglas supporter in the late political canvas, would be vice president.

Davis reached Montgomery on February 16. His words to the delegation that met him at the station left no doubt as to the determination with which he would face his "country's" prospects; vowing that its foe to the north would "smell Southern powder and feel Southern steel if coercion is persisted in."[3] That evening on a balcony of the Exchange Hotel, with William L. Yancey at his side, Davis painted a more exalted image. To a cheering crowd he said: "It may be that our career will be ushered in the midst of a storm; it may be that as this morning opened with clouds, rain and mist, we shall have to encounter inconveniences at the beginning; but as the sun rose and lifted the mist it dispersed the clouds and left us the pure sunshine of heaven. So will progress the Southern Confederacy, and carry us safe into the harbor of constitutional liberty and political equality."[4]

The new nation had sprung so swiftly into being it as yet had no flag. When it did, another of the "matadors of the South," to use Marx's expression, again was to soar as he rendered the entire issue upon which the "Confederacy" was founded—calling the new banner "our holy flag—that symbol and sign of an adored trinity, cotton, niggers and chivalry."[5]

On January 20, 1861, Wendell Phillips spoke under the title "Argument for Disunion" from Theodore Parker's pulpit. To that date South Carolina and three states of the "Gulf Squadron" had passed ordinances of secession. Phillips began evenly: "Thirty years ago Southern leaders, sixteen years ago, Northern abolitionists, announced their purpose to seek dissolution of the American Union. Who dreamed that success would come so soon? South Carolina, bankrupt, alone, with a hundred thousand more slaves than whites, four blacks to three whites, within her borders, flings her gauntlet at the feet of twenty-five millions of people in defense of an idea, to maintain what she thinks her right. . . . This mistake of South Carolina is she fancies there is more chance of saving slavery outside the Union than inside."[6]

Toward the end of his exposition Phillips highlighted the actuality of the matter—"Disunion is abolition! That is all the value disunion has for me. . . . The music of disunion to me is that at its touch the slave breaks into voice, shouting his jubilee." At the conclusion of his speech, as before, Phillips was escorted home by his volunteer body-guard and by the police, who held back charges from would-be-assailants who again occupied the hall and crowded the approaches to it.

As the annual meeting of the Massachusetts Anti-Slavery Society convened at Tremont Temple, the hall again became the scene of fierce animosity as organized disrupters intent on closing the proceedings came tumbling in. The Reverend James Freeman Clarke led off from the speaker's rostrum. When Phillips rose to speak, the cry "All up!" sounded in the gallery. For fifteen minutes applause, catcalls, screeches, and yells deafened the hall, as a rain of cushioned seats fell from the galleries, and Phillips was unable to begin.

Whenever there was a lull Phillips's voice could be heard. At last he stooped to address reporters sitting in the first rows. Now someone in the gallery called out, "Speak louder. We want to hear what you're saying." It was a brilliant cuff against the ear of the mob. Phillips said:

> Abolitionists! Look here! Friends of the slave, look here! These pencils will do more to create opinion than a hundred thousand mobs. While I speak to these pencils, I speak to a million men. What, then, are these boys? (Applause.) We have got the press of the country in our hands. Whether they like us or not, they know that our speeches sell their papers. (Applause and laughter.) With five newspapers we may defy five hundred boys. Therefore, just allow me to make my speech to these gentlemen in front of me, and I can spike all those cannon. (Applause.) My voice is beaten by theirs but they cannot beat out types.

As the tumult rose again, adjournment came, and Phillips and a friend hastened to the State House to talk with Governor Andrew, urging him to protect the meeting from the rioters. Andrew refused to act, satisfied that he was not so empowered by Massachusetts statute. As the afternoon session began, the *Boston Traveller* related, in rushed "Breckinridgers, Negroes, Douglas men, Garrisonians, Bell men, North streeters, Beacon streeters, John Brown men, ministers of the Gospel, pickpockets, reporters, teamsters, dry goods jobbers, loafers, brokers, rum sellers, ladies, thieves, gentlemen, State officers, boys, policemen." With a mob in possession of the hall, the noise and distraction was even greater than that morning. Finally the mayor and the chief of police pushed toward the platform with a body of officers. As they reached the stage the mayor ordered quiet and declared that the trustees of the building had requested that he disperse the meeting. Challenged by trustees present in this construal of their order, the mayor refused discussion and urged all to respect him enough "to leave this place quietly and peaceably." Again the trustees and the officers of the Massachusetts Anti-Slavery Society declined to quit the meeting, and the mayor was forced to admit he'd only been asked "to quell the riot" and to protect property. Thus exposed, the mayor turned in confusion to the platform to ask the acting president what he should do. "Clear the galleries" and "Give us fifty policemen this evening to protect the meeting," he was told. Returning to City

Hall instead, the mayor issued an order to close the hall to "prevent any meeting being held there."[7] That evening a riotous procession converged through the streets to the home of Wendell Phillips, where, with a large police guard, the chief of police dispersed them.

In the winter of 1860–61, and on a number of occasions throughout the year, Phillips was to deliver one of his most famous discourses, a lecture on Toussaint L'Ouverture. First articulated in Watertown, New York, in December, Phillips's speech was intended not only to defend the black's abilities and accomplishments, but to prepare the largely white audiences for the coming bid for black equality and self-government. It has been the penchant of historians to enter upon this moment on the eve of the "War Between the States" to comment upon the "mute and luckless Negro," of the seeming mood of docility among the slaves. In contradistinction to this, Thomas Hamilton wrote, in the inaugural January 1859 issue of the *Anglo-African Magazine*, of which he was editor:

The wealth, the intellect, the Legislation (State and Federal,) the pulpit, and the science of America, have concentrated on no point so heartily as in the endeavor to write down the Negro as something less than a man: and yet at the very moment of the triumph of the effort, there runs through the marrow of those who make it, an unaccountable consciousness, an aching dread, that this *noir fainéant*, this great black sluggard, is somehow endowed with forces which are felt rather than seen which may in "some grim revel," "Shake the pillars of the commonweal!" And there is indeed reason for this "aching dread." The Negro is something more than mere endurance; he is a force. And when the energies which now imbrute him exhaust themselves—as they inevitably must—the force which he now expends in resistance will cause him to rise.

James T. Holly was also lecturing then on "The Negro Race, Self-Government, and the Haitian Revolution." Holly said, "Here in this black nationality of the New World . . . is the stand point that must be occupied, and the lever that must be exerted." The revolution in St. Domingo starting in 1791 had been "the noblest, grandest, and most justifiable outbursts . . . recorded in the pages of history," Holly noted, surpassing by an incomparable degree the American Revolution that had preceded it, because the black's tyrannical

oppressors had "not only imposed an absolute tax on their unrequited labor, but also usurped their very bodies." Before the slaves had
risen, however—as the slogan of the French Revolution, "Liberty,
Equality, Fraternity," had reached the island's shores—mulattoes,
occupying a station between white and black, rose to gain recognition
of their freedom. The blacks continued toiling with a seemingly
"careless reserve." This showed, said Holly, not only that they understood and appreciated the difficulties of their position, but also that
it was "one of the strongest traits of self-government." Silently waiting, they "had conscious faith in the ultimate designs of God," and
"in so doing," they had "given an evidence of their ability to govern
themselves that ought to silence all pro-slavery calumniators . . . once,
and forever."

But disappointed that justice would ever be done, they rose to fight
for eight bloody years, defeating at different times the armies of
France, Britain, and Spain; finally sacrificing to their freedom the
blood of the best troops of the Republic's First Counsel, who wanted
them out of the way while he crowned himself Emperor of France.
"There never was a slave rebellion successful but once, and that was
in St. Domingo. . . . The long toil of a century cries out, Eureka!—'I
have found it!'"— said Wendell Phillips—"the diamond of an immortal soul and an equal manhood under a black skin as truly as under a
white one."

In August 1859 a secretary of state of Haiti, F. E. Dubois, issued a
"Call for Emigration" to American blacks. The Haitian government
had opened a consulate in Philadelphia and offices in a few other
northern cities, and agents had been appointed. James Redpath
became the general agent of the bureau office established in Boston,
publishing in 1860 a *Guide to Hayti* he edited. As many as one hundred thousand emigrants were anticipated in a movement that had
gathered growing interest since Harper's Ferry. Advocating this emigration, Holly pronounced it "an important question for the Negro
race in America to well consider the weighty responsibility that the
present exigency devolves upon them, to contribute to the continued
advancement of this Negro nationality of the New World until its
glory and renown shall overspread and cover the whole earth and
redeem and regenerate by its influence in the future the benighted
fatherland of the race in Africa."

In spring 1861, after completing his travels abroad, Martin Delany
was back in the United States. In December 1859 he had signed a

treaty with eight kings of the Abbeokuta of the upper Niger Valley, providing unused land for the settlement of African Americans, stipulating they must possess skills, respect the laws and customs of the Egba people, and provide for the settlement's self-government. In his *Official Report of the Niger Valley Exploring Party*, published by Thomas Hamilton in New York that year, Delany wrote, "Our policy must be—and I hazard nothing in promulgating it; nay, without this design and feeling, there would be a great deficiency of self-respect, pride of race, and love of country, and we might never expect to challenge the respect of nations—Africa for the African race, and black men to rule them."

Once notified of the treaty, Henry Highland Garnet called a meeting of the African Civilization Society to further Delany's aims—and to develop the requisite alliances on the two continents. For a time however Delany would maintain his distance from Garnet's society because of the whites in its leadership, and he criticized Garnet anew for working with the Haitian Bureau, among whose principals was the "John Brownist" James Redpath. Garnet offered this sharp retort:

> You are indignant at the acknowledgement of the leadership of white men in any work that particularly concerns black men. . . . I see by the newspapers that in the convention held in 1848 [*sic*] in Chatham, C.W., one John Brown was appointed leader—commander-in-chief—of the Harper's Ferry invasion. There were several black men there, able and brave; and yet John Brown was appointed leader. The unfortunate Stevens moved for the appointment and one Dr. Martin Delany seconded the motion.
>
> Now, sir, tell me where I shall find your consistency, as John Brown was a very white man—his face and glorious hairs were all white. I am done with you on that point, Dr. Delany. You ought to have accepted the office of surgeon under that great white leader, as a surgeon's place is in the rear, out of harm's way.[8]

In January 1861, after a benediction at the wharf from the Reverend Garnet, the sailing vessel *Janet Kidson* embarked for Haiti with sixty-one black emigrants from New York City. Garnet was now on the staff of the one-and-a-half-year-old Haitian Bureau, as its agent for New York City and State. In April the Haitian Bureau chartered another vessel, a steamer, to carry emigrants and passengers from the United States to Haiti; among them, readers of *Douglass'*

Monthly were informed, would be Frederick Douglass himself, one of the chief critics of the emigration movement heretofore. The steamer was to embark from New Haven, Connecticut, about the 25th and reach Port-au-Prince by May 1. Douglass was anticipating a six- to eight-week stay, and reported in May 1861, in "A Trip to Haiti": "The intimation was accomplished with a generous offer of a free passage to ourself and daughter to and from Haiti, by Mr. Redpath, the Haitian Consulate at Philadelphia. We are not more thankful for this generous offer from the quarter whence it comes, than sensible of the kind considerations which it implies. We gratefully appreciate both, and shall promptly avail ourselves of the double favors."

In announcing the trip, Douglass wrote, free blacks "in all the States have been deeply exercised in relation to what may be our future in the United States." The apprehension was general, he observed, "that proscription, persecution and hardships are to wax more and more rigorous and more grievous with every year." For this reason, precisely at this crucial juncture, both for his people and for the life of the nation, Douglass was going to Haiti. The nation had "constantly been the victim of something like a down right conspiracy to rob her of the natural sympathy of the civilized world, and to shut her out of the fraternity of nations. No people have been compelled to meet and live down a prejudice so stubborn and so hatefully unjust." In undertaking the trip he could "paint her as she is, and . . . add the testimony of an honest witness to honest worth."

It was doubly significant too that Douglass was considering a trip to Haiti when the "Southern Confederacy" was gaining such latitude only weeks before the inauguration of Lincoln; a success predicated, in the first instance, upon the inaction of the Buchanan administration, and second, to the clamoring of the northern wing of the Democratic Party. Although Lincoln had made it clear he would adhere to the Chicago platform, Seward, acting as "premier" in the new government, declared in Lansing that everything was to be sacrificed to save the Union, whose paramount objects were "safety and security." In New York City Mayor Fernando Wood was proposing secession for that great center of shipping and merchant capital, while many conservatives were saying what must be backed down at all cost was the fearsome and uncompromising hostility to slavery. "Be it the business of the people everywhere to forget the Negro, and remember only the country," intoned Horace Greeley at the *Tribune*. Douglass was to write years later in his *Life and Times*, "Those who may wish to see to what depths of humility and self-abasement a noble people can be brought under the sentiment of fear will find no chapter of history more instructive than that which treats of events in official circles in Washington during the space between the months of November 1859, and March 1860." (The respective dates should appear, however, as November 1860 and March 1861.)

On February 11, 1861, Abraham Lincoln boarded the train that would carry him to Washington, with four officers of the United States Army, his son Robert, and his two secretaries. Known on the national scene only for two years, Lincoln had been out of elective office for ten years and had never held an executive position; yet before him lay the daunting task of preserving a Union he held to be perpetual. Although his features are justly described as homely, he was tall and made a striking figure, particularly as it was his custom to wear a high hat, often donning a cloak or shawl. Those who knew anything about him noted his sense of humor and fondness for story-telling, but above all he was distinguished for his personal and professional honesty. To these were added his strong commitment to realism, with powers of concentration mastered by self-discipline; mental facilities, it was said, that imparted to him a patience such as would wear down any opponent.

The train moving eastward from Springfield bearing the fifty-two-year-old president-elect also carried with it, through the press, the anxious attention of his countrymen; a journey that would take twelve days to arrive at the capital, three days before the inaugural. Crowds all along the route congregated trackside as Lincoln appeared on the rear platform to bow or wave, and at the major cities to give impromptu speeches and attend receptions. Many of these, it has been said, would have benefited from greater preparation as they began to impart the impression of a man inadequate to his office.

Indianapolis was the first major stop, where Lincoln's wife and younger son boarded, then on to Columbus where Lincoln addressed the Ohio legislature, then to Pittsburgh, doubling back to Cleveland. In Cleveland, after a parade in wet snow, Lincoln developed a theme he had broached in Pennsylvania—the suggestion that as tempers cooled "the troubles will come to an end." He said, "I think that there is no occasion for any excitement. The crisis, as it is called, is altogether an artificial crisis. . . . It has no foundation in facts. It was not argued up, as the saying is, and cannot, therefore, be argued down. Let it alone and it will go down of itself." Then the train headed through New York, to Buffalo and Rochester.[9]

On the day Jefferson Davis took the oath of office as president of the Confederate States of America, Lincoln addressed the New York legislature in Albany. Two days later he was in New York City, paying his respects to its mayor. Lincoln said, "there is nothing that can ever bring me willingly to consent to the destruction of the Union under which not only the commercial city of New York but the whole country has acquired its greatness." Newspapermen estimated Lincoln spoke fifty times in one week and appeared tired and dispirited, which only deepened as he reached Philadelphia and received reliable reports that an attempt at his assassination was being developed in Baltimore.

That Lincoln was scheduled to be in Baltimore on the afternoon of February 23, a Saturday, was public knowledge; however the city had not extended an official reception, nor had a police guard been contemplated. The plot called for secessionist sympathizers, in which the city abounded, to attack the train as it was switched at the relay station from the Philadelphia–Baltimore track to the Baltimore–Washington track, a transfer requiring the cars to be drawn a dozen blocks by horses. With commitments still to be fulfilled in Philadelphia and Harrisburg, Lincoln was urged to cut short his appearances and travel to Washington overnight secretly. He refused. But in Harrisburg the son of William H. Seward, Frederick Seward, brought further confirmation from his father and from the commander of the army, General Winfield Scott, of the plot's validity. The president-elect was now prevailed upon to take precautions, and slipped out of Harrisburg and away from his military escort in the company of Ward Lamon, a close friend who was armed with pistols and knives. In Philadelphia the pair changed to a Washington-bound sleeper and were joined by the detective Allen Pinkerton, while Lincoln assumed a nominal disguise consisting of a soft felt hat in place of his usual high hat. The party passed under the cover of night unmolested through Baltimore.

Arriving in Washington, Lincoln went to the Willard Hotel while Pinkerton telegraphed the presidential party given the slip in Harrisburg: "Plums delivered nuts safely." The story got around that Lincoln had entered Washington disguised in a Scottish cap and cloak, and as it appeared in the papers people believed it. It was, however, an inauspicious arrival upon his new office and altogether demeaning to the chief executive of the country. Frederick Douglass remarked in his *Monthly* in April 1861: "The manner in which Mr. Lincoln entered the Capital was in keeping with the menacing and troubled state of the times. He reached the Capital as a poor, hunted fugitive slave reaches the North, in disguise, seeking concealment, evading pursuers, by the underground railroad, between two days, not during the sunlight, but crawling and dodging under the sable wing of night."

As he settled into his rooms at the Willard three days before his inauguration, Lincoln was besieged on all sides with vexing questions. Washington was deluged, too, with office seekers such as no administration had ever experienced, because Buchanan had favored appointees from southern states and most of these were now leaving. In order to govern, the Republican Party, still largely a desperate coalition of factions, would now have to be knit together with the power of patronage. Lincoln's choices for his cabinet would reflect this, as his four main competitors for the nomination at Chicago were in it. Seward, still recognized as the most powerful politician of the party, would be secretary of state; his rival and also a power center in the party, Governor Salmon Chase, would take the treasury. Edward Bates, the favorite-son candidate of Missouri, would become attorney general; while to Simon Cameron, from the pivotal state of Pennsylvania, fell the spoils of the war department. Completing the cabinet, Montgomery Blair, son of a powerful father and key Lincoln advisor, became postmaster general; interior went to Caleb Smith; and Gideon Welles headed the Navy Department. All told, counting Lincoln, there were four former Whigs and four former Democrats, with two of these from slave states. Thus all elements comprising the party took a chair at the table; but with no helmsman as yet defined, there would be a period of disorder.

In sharp contrast was the regularity and order evidenced in Montgomery. The convention that produced a government for the Confederate states in February had continued meeting and begun drafting a permanent constitution designed to win the allegiance of the border slave states, with an eye too, to some of the states of the old Union. Davis's cabinet selections reflected this strategy, when he included none of the "fire-eaters" in an attempt to appeal to moderates in the border States. His chief rival for the helm, the former senator from Georgia Robert Toombs, was selected secretary of state; for the other positions, members from all other seceded states, excluding Mississippi, were selected. Some of these appointments reflected previous governmental experience; Stephen Mallory, former chairman of the Senate Committee on Naval Affairs, became secretary of the navy, while J. H. Regan, former chairman of the House Committee on the Post Office, became postmaster general. For attorney general he chose Judah Benjamin, Davis's confidant who would become his alter ego in the government; selected as secretary of treasury for his supposed expertise was a German-born South Carolinian, Christopher Memminger; and L. P. Walker was made secretary of war.

The South claimed, with much justification, to represent the true ground of the Constitution of 1787, which it insisted had been trampled upon by the North. But instead of the aversion of the founders to formally acknowledge slavery, the "confederate" version called a slave a slave. All in all, the Confederacy was thought by its proponents to offer a basis for reconstituting much of the old Union, minus New England, without the flaws of the original, on the basis of the universal recognition of African subordination. The two presidents now faced each other with diametrically opposing tasks. Davis's undertaking, with scarcely 10 percent of the country's white population and 5 percent of its industrial capacity, desperately needed the allegiance of the upper South; while Lincoln's chief concern was to prevent this enlargement. Thus Lincoln had before him the aim of uniting the North and dividing the South. Davis on the other hand needed the adherence of the slave states that had not seceded, and the sympathy of the northern wing of the Democratic Party. Compromise on the underlying issue of slavery was finished, yet peaceful separation was still possible. The choice lay with the North; with his inaugural address on March 4 Lincoln needed to declare himself.

In Montgomery, Alabama, as the day of Lincoln's inaugural dawned, the newly made banner of the Confederate States of America fluttered atop the capitol dome. The previous night had been stormy in Washington, with gale winds that left its mostly unpaved streets mired in mud. But as morning came, the skies had cleared. In the plaza east of the Capitol there was a large collection of construction implements and materials, castings and stone blocks for the uncompleted dome. In this unresolved setting a temporary platform had been constructed on the east portico of the building where Abraham Lincoln was to take the oath as the sixteenth president of the United

States. On buildings nearby and in the intersections of streets riflemen had been stationed by General Scott to prevent another rumored assassination attempt. There had also been reports that an armed body from Virginia might attempt to prevent the proceedings, and the city was filled with soldiers and parading artillery companies, and the air with the sounds of buglers and drummers.

On the day before, Lincoln's secretary of state designate had notified the president-elect he could not accept the post offered because he did not think he could serve effectively in the same cabinet with his rival Salmon Chase. Lincoln, having no doubt as to Seward's power and importance to the party, remarked to his private secretary: "I can't afford to let Seward take the first trick." Then he dictated a note asking Seward to reconsider: "The public interest, I think, demands that you should; my personal feelings are deeply enlisted in the same direction."[10] That afternoon Seward agreed to stay on. He had already had a major impact as a shaper of administration policy that would be reflected in the tone of Lincoln's inaugural speech; the government, he counseled, must be moderate, not meeting secession with threats, and make concessions to encourage Union men in the South and to hold the border states. True, the question of slavery had been answered in the election, and it was to be prevented from expanding beyond its current boundaries, but it would be an enormous blunder if dissolution were permitted to go through: a constitutional amendment to safeguard slavery may be required. Lincoln, for his part, seemed to misapprehend the earnestness of the South and was unworried, thinking that somehow the country could talk its way out of its dire predicament.

A week before, on February 25, Jefferson Davis named three commissioners to go to Washington in an amenable spirit to secure recognition for the Confederacy, treating with Lincoln, thought to be the nominal head, or with Seward, the ostensible head, as the case may be. Specifically they were to discuss the surrender of forts, arsenals, and other public property formerly belonging to the federal government, and arrive at an agreeable settlement of debts and other matters. The day before the inauguration Davis would also dispatch to Charleston a former captain in the United States Army who had resigned his commission when Louisiana seceded—General P. G. Toutant Beauregard. He was a soldier with a good professional reputation, a Creole, whose uniform was immaculate and manners faultless. Beauregard was ordered to take charge of the military preparations in the harbor, which he did with skill and tact, becoming in a short time the idol of Charleston.

On the morning of the inaugural the U.S. War Department received a telegram from Major Anderson that his garrison could only remain in Fort Sumter six more weeks without resupply. In the afternoon this message would be on the president's desk. That morning Lincoln rode to the swearing-in with the outgoing president, in a carriage driven within a hollow square of cavalry, three to four deep. As he mounted the stand before an audience of 25,000 and prepared to take the oath of office administered by Chief Justice Roger B. Taney, for an awkward moment Lincoln found he had nowhere to place his hat. Seeing his predicament Stephen Douglas took Lincoln's hat in hand and held it for the duration of the oath and the speech, thus perhaps establishing some small debt of obligation between the two.

In an extended analysis of Lincoln's carefully drafted brief in the April 1861 issue of *Douglass' Monthly*, which also published the entire speech, Frederick Douglass wrote of the "extraordinary and portentous" occasion: "Threats of riot, rebellion, violence and assassination had been freely, though darkly circulated, as among the probable events to occur on that memorable day. The life of Mr. Lincoln was believed, even by his least timid friends, to be in the most imminent danger. No mean courage was required to face the probabilities of the hour. He stood up before the pistol or dagger of the sworn assassin, to meet death from an unknown hand, while upon the very threshold of the office to which the suffrages of the nation had elected him."

Allowing for these circumstances, Douglass wrote, and for the atmosphere attendant in Washington that could be traced upon "the whole character of his performance. . . . we must declare the address to be but little better than our worst fears, and vastly below what we had fondly hoped it might be." Lincoln was, no doubt, counseled by his secretary of state, intent on bidding for time as passions cooled, the administration was organized, and a process of voluntary reconstruction begun. The ambiguity of the speech was intentional, but to it Douglass applied the apt invective "double-tongued." It concealed rather than declared a definite policy. The circumstance required lack of guile and resolution, but, wrote Douglass: "Overlooking the whole field of disturbing elements, he should have boldly rebuked them. He saw seven States in open rebellion, the Constitution set to naught, the national flag insulted, and his own life murderously sought by the slave-holding assassins. Does he expose and rebuke the enemies of the country, the men who are bent upon ruling or ruining the country? Not a bit of it. But at the very start he seeks to court their favor, to explain himself where nobody misunderstands him, and to deny intentions of which nobody had accused him. He turns away from his armed enemy and deals his blows on the head of an innocent bystander."

The address had opened with Lincoln declaring his complete fidelity to the slave states, affirming that the federal government had no lawful power to interfere with their species of property, and moreover that he had not the least "inclination" of doing so. Putting himself in line with Pierce and Buchanan, he said he regarded it as a constitutional duty to assist in the recapture of fugitive slaves. Snorted Douglass: "Whatever may be the honied phrases employed by Mr. Lincoln when confronted by actual disunion; however silvery and beautiful may be the subtle rhetoric of his longheaded Secretary of State . . . all know that the masses at the North (the power behind the throne) had determined to take and keep this Government out of the hands of the slave-holding oligarchy, and administer it hereafter to the advantage of free labor as against slave labor."

This denial was not only "discreditable to the head and heart of Mr. Lincoln . . . it could neither appease nor check the wild fury of the rebel Slave Power." Secession had not arisen from any misapprehensions of the policy of the Republican Party, Douglass pointed out, but from a clear perception of the meaning of the election, that the North was outstripping the South in population and that henceforth there was to be a change in the slaveholder position from ruler to ruled. Said Douglass: "They are not afraid that Lincoln will send out a proclamation to the slave States declaring all the slaves free, nor that Congress will pass a law to that effect. They are no such fools as to believe any such thing; but they do think . . . that the power of slavery is broken, and that its prestige is gone. . . . To those sagacious and crafty men, schooled into mastery over bondmen on the plantation, and thus the better able to assume the airs of superiority over Northern doughfaces, Mr. Lincoln's disclaimer of any power, right or inclination to interfere with slavery in the States, does not amount to more than a broken shoe-string!"

But the inaugural address did "not admit of entire and indiscriminate condemnation." In Douglass's view, Lincoln saved himself from the infamy of the Supreme Court's Dred Scott decision when, to his

admission of the right of slavery and slave-catching, he added: "In
any law on this subject, ought not all safeguards of liberty known in
humane and civilized jurisprudence be introduced, so that a free man
be not in any case surrendered as a slave." In providing this, was not
Lincoln suggesting that slavery was a condition from which a free
man ought by lawful means be saved, asked Douglass. And above
that, didn't this slight assertion mean that persons of African descent
had rights? This, and for a few other timid notions not in accord with
slaveholding gospel which no other American president had ever ventured to say—was something to be thankful for.

Now Lincoln proceeded to an exposition of his view on the perpe-
tuity of the federal union. While his argument was excellent, it came
too late, said Douglass:

> When men deliberately arm themselves with the avowed intention
> of breaking up the Government; when they openly insult its flag,
> seize its munitions of war, and organize a hostile Government, and
> boastfully declare that they will fight before they will submit, it
> would seem of little use to argue with them. If the argument was
> merely for the loyal citizen, it was unnecessary. If it was for those
> already in rebellion, it was casting pearls before swine. . . .
>
> It remains to be seen whether the Federal Government is really
> able to do more than hand over some John Brown to be hanged,
> suppress a slave insurrection, or catch a run-away slave—whether
> it is powerless for liberty, and only powerful for slavery.

In an earlier draft of his speech Lincoln had avowed his intention
to use "all the powers at my disposal" to "reclaim the public property and places which have fallen; to hold, occupy, and possess these,
and all other property and places belonging to the government" and
to "collect duties on imports." He promised that "the government
will not assail *you*, unless you *first* assail it." This was sure to be
regarded as "coercion," and both Seward and other confidants
advised this be dropped, and the finished speech vowed only to
"hold, occupy, and possess" its property, and "collect duties and
imposts." Lincoln now added a provision assuring the South whenever "in any interior locality" hostility to the federal government was
"so great and so universal, as to prevent competent resident citizens
from holding the Federal offices," government activities for "a time"
would be suspended.

In finishing his excoriation of Lincoln's inaugural speech, Douglass wrote, "A thousand things are less probable than that Mr. Lincoln and his Cabinet will be driven out of Washington, and made to go out, as they came in, by the Underground railroad. The game is completely in the hands of Mr. Jefferson Davis, and no doubt he will avail of every advantage."

Now there was to be half a month of indecision, where no regular meetings of the cabinet were held as Lincoln occupied himself, as he must, listening to office-seekers and dispensing the spoils of the Republican victory. But already the keystone of the administration's policy had been undermined with Major Anderson's dispatch stating he could not remain at Sumter without resupply. Upon this issue would run the entire question that had yet to be answered: how should the North look upon the seven "erring sisters"?

Two of the three commissioners sent by Jefferson Davis had arrived in Washington by March 4. Lincoln refused to acknowledge them, or even meet with them informally, while Seward, resisting the mediation of Senator Hunter of Virginia, kept them waiting. Communication between the parties was necessary, however, and a channel was opened through Justice John A. Campbell of the United States Supreme Court, although he was about to resign and return to his native Alabama. On March 15, Lincoln's cabinet voted 5 to 2 that relief of "the fort" was politically unwise, having been advised by General Scott that the navy could not relieve Sumter unless accompanied by twenty thousand soldiers. That day Seward replied to Campbell that "the evacuation of Sumter is as much as the administration can bear." This declaration was duly passed on to the Southern commissioners and was reaffirmed on March 21 in a communication by Campbell, stating: "As a result of my interviewing of today I have to say that I have still unabated confidence that Fort Sumter will be evacuated."[11]

Seward had made a commitment he had no authority to make, but Lincoln did nothing to check him. Lincoln knew, however, it was incumbent on him to make up his own mind, and while biding for time to bring unity to his cabinet and without committing an overt act that would spell war with the consequent loss of the border states, he sought to obtain facts on his own. Three agents were selected to travel to South Carolina and report back on the situation in

Charleston and at Fort Sumter. One of these was a retired naval officer and now a manufacturer in Massachusetts, Gustavus V. Fox, not
incidentally the brother-in-law of Montgomery Blair, who would later
become assistant secretary of the navy and a leading figure of the war.
Fox was ordered to confer with Major Anderson and privately assess
the possibilities for successful resupply of the fort. The other agents
were both Illinoisans and friends of the president: Stephen Hurlbut,
sent to probe the Union feeling of South Carolina, and Ward H.
Lamon, Lincoln's former law partner, sent to interview Governor
Pickens.

Fox reported that Sumter could be relieved; Hurlbut that he found
no Union sentiment; and Lamon merely reported his interview with
the governor, where he had assured him that the evacuation of the
fort would soon be ordered.

Although militarily it was known that Sumter could not be held if
there was war, to give it up was tantamount to acknowledging the
independence of the South. Such a move would rend the Republican
Party and lose for the administration the support of the Northwest.

Ben Wade, the fiery antislavery lecturer from Ohio and House
member, told Lincoln directly: "Give up fortress after fortress and Jeff
Davis will have you a prisoner of war in less than thirty days." From
the outset the crisis had involved not only Sumter, but also Fort
Pickens in Pensacola Bay, differing from the former in that it was in
Florida and not in South Carolina. It was in any case easily accessible
by sea and could be reinforced without firing a shot.

On March 28 Lincoln and his wife hosted their first official state
dinner, with the cabinet and foreign ambassadors and journalists
attending. General Scott appeared and before retiring handed the
president a memorandum informing him that it was his opinion that
not only must Fort Sumter be abandoned, but Fort Pickens as well.
After the dinner Lincoln asked his cabinet to stay while he read them
Scott's astonishing communication.

It was doubtful, the general wrote, "whether the voluntary evacuation of Fort Sumter alone would have a decisive effect upon the
States now wavering between adherence to the Union and secession.
It is known, indeed, that it would be charged to necessity, and the
holding of Fort Pickens would be adduced in support of that view."
Scott urged evacuation of both as it would "give confidence to the
eight remaining slaveholding states, and render their cordial adherence to the Union perpetual . . . [and] we should thereby recover the

From left to right, top to bottom: Jefferson Davis, president of the Confederate States; Judah Benjamin, Confederate attorney general; Robert Toombs, Confederate secretary of state; Abraham Lincoln, president of the United States; Salmon Chase, secretary of the treasury of the United States; William Seward, secretary of state of the United States.

States to which they geographically belong by the liberality of the act."[12]

Lincoln asked his cabinet to reassemble on the following day and submit in writing their thoughts on what ought to be done. Chase, who had voted with the minority in the previous polling, now spoke for the majority that he was definitely in favor of retaining Fort Pickens and "just as clearly in favor of provisioning Fort Sumter." Attorney General Bates concurred on Pickens but wavered as to the proper course in regard to Sumter, while Caleb Smith in alliance with Seward wanted the government to pull out of Sumter. In Seward's estimate Fort Sumter was too close to Washington, and he wanted the showdown to come at Fort Pickens and perhaps along the Texas coast; but he also had his reputation to protect as he had given assurances on his own authority that Sumter would be evacuated. The time had come for Seward to make his bid to establish his preeminence in shaping the administration's policy. Accordingly his famous memorandum, titled "Some Thoughts for the President's Consideration," was in Lincoln's hands on April 1. He began: "We are at the end of a month's administration, and yet without a policy, either domestic or foreign." The main thing, wrote the secretary, was to "change the question before the public from one upon slavery, or about slavery, for a question upon union or disunion. In other words, from what would be regarded as a party question to one of Patriotism or Union."13

In the meantime, Seward had a military officer, Captain Meigs, draw up a plan for the relief of Fort Pickens and had it submitted to the president by the afternoon of March 31. Lincoln liked the plan and told Meigs and the secretary of General Scott, who was present, to take it to the War Department and seek the approval of their superior. "I depend on you gentlemen," said Lincoln, "to push this thing through."14

Seward's memorandum had also detailed measures the administration should take in foreign relations. France and England had been pressing Mexico in regard to payment of debts, and Spain was interfering with Santo Domingo. The secretary recommended explanations should be demanded at once and emissaries sent throughout the hemisphere to awaken a "continental spirit of independence against European intervention." If explanations were not satisfactory, war should be declared. Seward calculated this would stay the hand of secession in the border slave states, and down the road invite voluntary reunification. Now the secretary was ready to propose that he assume power, writing: "But whatever policy we adopt, there must be energetic prosecution of it. For this purpose it must be somebody's business to pursue and direct it incessantly. Either the President must do it himself, and be all the while active in it, or devolve it on some member of his Cabinet. Once adopted, debates on it must end and all agree and abide." It was not his "especial province," Seward demurred, but he would "neither seek to evade nor assume responsibility."15

Although Lincoln was busy in his office, he lost no time in replying. The administration had a policy, he wrote, that had been put forward with the secretary's approval in the inaugural address. Furthermore, wrote Lincoln, he found it hard to follow the reasoning that reinforcement at Fort Sumter "would be done on a slavery, or party issue, while that of Fort Pickens would be on a more national or patriotic one." As to the rest on foreign policy, Lincoln remarked, "if this it to be done, *I* must do it." Then summing up, he wrote: "When a general line of policy is adopted, I apprehend there is no danger of its being changed without good reason, or continuing to be the subject of unnecessary debate; still, upon points arising in its progress, I wish, and I suppose I am entitled to have the advice of all the cabinet."[16]

At this point an inquiry came from South Carolina's governor Pickens via Justice Campbell to Secretary Seward as to what the latest word about Sumter might be. In the Confederate cabinet Robert Toombs was growing impatient, but had received reassurance from the southern commissioners in Washington: "If there is faith in man we may rely on the assurances we have here as to the status."[17] In answer to Governor Pickens, Seward told Campbell that Lincoln had no design to reinforce Sumter and would not undertake resupply without first giving notice to the governor. Although apparently satisfactory, the reply contained a scarcely concealed indication of what was to come. At New York's Governors Island two expeditions were being outfitted, one for Sumter and one for Pickens, and rumors of this were getting around.

In the days after April 1 Lincoln's mind was resolved: he would take a stand at Fort Sumter. On April 2 Toombs would write his president that all the incidents that would "inaugurate a civil war greater than any the world has yet seen," had fallen into place or by the way. Urging forbearance, Toombs wrote: "Mr. President, at this time it is suicide, murder, and will lose us every friend at the North. You will wantonly strike a hornet's nest which extends from mountains to ocean, and legions now quiet will swarm out and sting us to death. It is unnecessary; it puts us in the wrong; it is fatal."[18]

On April 4, Gustavus Fox, returning from New York, was summoned to the Executive Mansion and told that an effort to reprovision Sumter would be made and that he was the man delegated to do it. A messenger was dispatched to give due notice to South Carolina's governor that if the reprovisioning was unopposed no troops would be landed. If Jefferson Davis chose to shoot now, he, and South Carolina should stand before the civilized world as having fired upon bread.

On the morning of April 10 Confederate secretary of war Walker telegraphed Beauregard his instructions, the import of which were seen on April 12 as a mortar fired from Fort Johnson burning a red fuse arched high over Sumter, exploding squarely over its walls.

Among the first guns fired was that set off by the aged, white-headed Virginia secessionist Edmund Ruffin, having enlisted as a private in the South Carolina militia for the duration of the battle, who yanked its lanyard. The bombardment commencing at 4:30 in the morning lasted over thirty hours, consuming over 3,000 shells, shot, and mortar firings.

As the day dawned, off the coast outside the harbor Fox's relief expedition was unable to assist the beleaguered garrison. In the flash of battle, both North and South saw their reservations fade away, and as fast as the wires could carry the news, men from both sections sprang to arms. For the North, Sumter became the symbol of the Union, while for the South it became the emblem of all it could not tolerate if it was to be recognized as a sovereign nation. The attack on the stars and stripes, too, had stilled all talk of concessions and of letting the South "go in peace." It galvanized the Republican Party out of its moderation and converted the northern Democrats into "war Democrats."

Noting the fall of Sumter in the May issue of his *Monthly*, Frederick Douglass wrote: "To our thinking, the damage done to Fort Sumter is nothing in comparison with that done to the secession cause. The hail of fire of its terrible batteries has killed its friends and spared its enemies. Anderson lives, but where are the champions of concession at the North? Their traitor lips are pale and silent."

It was just as this was happening that Douglass was bringing to press his announcement of his Haitian trip. At the end of that article he inserted his resolve to forego it now: "The last ten days have made a tremendous revolution in all things pertaining to the possible future of the colored people of the United States. We shall stay here and watch the current of events, and serve the cause of freedom and humanity in any way that shall be open to us. . . . At any rate, this is no time for us to leave the country."

Two days after Fort Sumter was battered into submission, Lincoln issued a proclamation calling for 75,000 volunteers for three months' service, and in every city and village of the North war meetings were held to cheer the flag with fife and drum. A quarter of a million people gathered in New York City, and the *Tribune* announced that "the nations of Europe may rest assured that Jeff. Davis & Co. will be swinging from the battlements at Washington, at least, by the 4th of

The Confederate flag flying over Fort Sumter following its surrender on April 13, 1861.

July. We spit upon a later and longer deferred justice." In the Northwest where secession was reviled as no better than rebellion and treason, Michigan, Wisconsin, and Minnesota answered full-throated for war. In Illinois the *Chicago Tribune* reflected on the same day the Midwest's enthusiasm for the Union cause: "Let the East get out of the way; this is a war of the West. We can fight the battle, and successfully, within two or three months at the furthest. Illinois can whip the South by herself. We insist on the matter being turned over to us."

Ten times the number of men called would soon be mustering and drilling "to avenge the flag" throughout the North. After raising thirteen regiments in Ohio, the governor wired Washington, "I can hardly stop short of twenty." Massachusetts' governor Andrew wired Lincoln on April 17: "Two of our regiments will start this afternoon—one for Washington, the other for Fort Monroe; a third will be dispatched tomorrow, and the fourth before the end of the week." Garrison too abandoned his pacifism and declared for war: "When I said I would not sustain the Constitution because it was a covenant with death and an agreement with hell, I had no idea that I should live to see death and hell secede." Asked to occupy Theodore Parker's pulpit after the firing on Fort Sumter, and asked if the platform should bear the United States flag, Wendell Phillips replied, "Yes, deck the altar for the victim."[19]

Lincoln's proclamation called for "militia of the several States of the Union" to suppress "combinations too powerful to be suppressed by the ordinary course of judicial proceeding, or by powers vested in the Marshals by law." But on April 15 all calculations were shattered as Virginia's convention, which had been meeting throughout the "crisis," voted to secede, and the entire tier of border states appeared likely to go with it. Kentucky's governor Magoffin wired Lincoln that his state would "furnish no troops for the wicked purpose of subduing her sister Southern states." North Carolina's governor Ellis telegraphed the president that the proclamation was unconstitutional and a "gross usurpation of power" and "you can get no troops from North Carolina." From Tennessee governor Harris: "Tennessee will not furnish a single man for the purpose of coercion, but 50,000 if necessary, for the defense of our rights and those of our southern brethren." From Arkansas, Governor Rector wired he would send no troops, and warned the people of his state would "defend to the last extremity their honor, lives and property against Northern mendacity and usurpation." Governor Jackson of Missouri sent word that the call for troops was "illegal, unconstitutional, and revolutionary in its object, inhuman and diabolical, and cannot be complied with." Virginia's governor Letcher extended this argument when he wired: "In reply to this communication, I have only to say that the militia of Virginia will not be furnished to the powers at Washington for any such use or purpose . . . . Your object is to subjugate the Southern States, and a requisition made upon me for such an object—an object, in my judgment, not within the purview of the Constitution or the act of 1795—will not be complied with. You have chosen to inaugurate civil war, and having done so, we will meet it in a spirit as determined as the Administration has exhibited toward the South."[20]

As recruiting began across the North, blacks also offered their services to defend the government, recognizing their stake in the fight, and through public meetings, memorials, and resolutions pleaded for an equal chance. In Boston, Philadelphia, New York, Cleveland, Detroit, and Chicago, black men were turned away from recruiting stations. The editors of the *Boston Daily Atlas and Bee* received an anonymous letter dated April 19, 1861, stating: "The colored man will fight—not as a tool, but as an American patriot. He will fight most desperately, because he will be fighting against his enemy, slavery, and because he feels that among the leading claims he has to your feelings as fellow-countrymen, is that in the page of facts connected with the battles for liberty which his country has fought, his valor— the valor of black men—challenges comparison; and because he feels that those facts have weight in causing his countrymen to award to him all his rights as an American citizen."

But black men were rebuffed in all instances. The war was considered a white man's affair, it was said, a war for Union versus states' rights, and already far too much had been said about Negroes and slavery.

Lincoln's proclamation also called for both branches of Congress, not then in session, to convene at noon on Thursday on the 4th of July, "then and there to consider and determine such measures, as, in their wisdom, the public safety, and interest may seem to demand." Until then Lincoln would be the government, and he had substantially determined to fight for an enduring Union in which slavery would still remain where it was recognized by law. The support of Republicans in the conduct of the war was assured, but a public rapprochement was necessary between Lincoln and Senator Douglas to assure the support of northern Democrats. The two former rivals met the evening before the issuing of the proclamation, which Douglas found he could wholly endorse, but recommended against Lincoln's temerity that 200,000 troops be called. Drawing the president to a map of the United States, Douglas said: "You do not know the dishonest purposes of those men as well as I do." Then he pointed to four strategic points where it was necessary to concentrate the strength of the government: these were at Washington and Harper's Ferry, on the tip of the Virginia peninsula at Fort Monroe, and the junction of the Ohio and Mississippi rivers, at Cairo, Illinois. After the meeting Douglas gave a statement to the Associated Press saying that although he was a political opponent of the president, he would support him "in the exercise of all his Constitutional functions to preserve the union and maintain the Government and defend the Federal capital."[21]

Douglas then entrained for Chicago where he would meet with his political allies to urge they and the public support the war. The utility or perhaps debility of Stephen Douglas was to end, as he would die on June 3 at age forty-eight, succumbing to the strenuous efforts on the electoral canvas the previous year, and rheumatism and typhoid fever. Alexander Stephens would say Douglas's death had come both too late and too soon: too late to avoid the split in the Democratic Party, and too soon for the moderating influence he could have been presumed to exert on the subsequent course of the war.

On April 17 by a vote of 88 to 55 Virginia passed an ordinance of secession, with most of the pro-Union vote coming from the area beyond the mountains. The governor issued a proclamation reiterating what he had stated—that Lincoln's proclamation was unconstitutional and a dire threat to Virginia; then he called the state's volunteer regiments to active duty. Two days before this, Henry A. Wise convened a conference of militia officers in Richmond and reached consensus on a plan to seize the arsenal at Harper's Ferry and the Gosport Naval Yard at Norfolk. As a troop of one thousand drew within four miles of the town that John Brown had brought such prominence a year and a half earlier, they could smell smoke and taste ash. The company of regular United States infantry stationed there had fired it and were retreating on the road to Pennsylvania. Wise's troops promptly seized all arms still retrievable from the arsenal. Before the month was out 8,000 Confederate troops would be stationed in this town at the head of the Blue Ridge under General Thomas Jonathan Jackson who, like his predecessor at Harper's Ferry, saw the hand of God directing the scourge of war.

For the South, accordingly, the border slave states contained most of its available war resources—populations, industrial stock, livestock, and crops. Nearly half of all Confederate troops were to come from the border states. Although there was appreciable Union sentiment throughout, when the news flashed across the wires of the attack on Sumter, large crowds waving Confederate flags and cheering its victory and hurrahing Jeff Davis poured into the streets in Richmond, Raleigh, and Nashville, and in scores of other cities and towns. At the capitol in Richmond a hundred-gun salute was fired as the Stars and Stripes was run down and the Stars and Bars run up. But the tide that took the Gulf states out had not been enough to carry the border states entirely. Secession's "vital fire" burned only in those areas holding large masses of slaves, while there was a demonstrable lack of this sentiment in broad regions containing few slaves. This division was manifest most crucially in western Virginia and eastern Tennessee, but also in the west of North Carolina, the Cumberland region of Kentucky, and in the Allegheny range reaching into Maryland and the hill country around Frederick. A rift also affected the northern counties in Georgia and Alabama, and in

Missouri, where slavery held an importance economically only in the northwest portion of the state, and where secessionist sentiment was held in check by a large German population in St. Louis. Northern Arkansas, likewise, was tepid to secession's embrace.

Meanwhile Kentucky, with the Ohio River running along a nearly five-hundred-mile rolling border with Ohio, Indiana, and Illinois, presented a formidable frontier coveted by both sides. If possessed by the Confederacy the river presented a natural barrier that, if defended, could make invasion of the Deep South problematic and assure the retention of Tennessee. If possessed by the North, the Ohio River's two navigable tributaries, the Cumberland and Tennessee rivers, flowing into Tennessee and northern Alabama, offered an invasion route. In Tennessee, with rich cotton planting soil in the Nashville and Memphis districts, a military league with the Confederacy had been formed soon after Governor Harris refused his state's troops to Lincoln. But after Virginia's decision to secede the deepest quandary for the Lincoln administration lay in Maryland—if it went with the South, the federal government would have to abandon its seat and perhaps the war itself.

During the crisis in April 1861 Washington was underdefended. Stepping up to the exigency first was Jim Lane, just arriving to represent Kansas in the U.S. Senate, who promptly organized a Frontier Guard from among his followers and other Kansas men in Washington. Bivouacking in the East Room of the White House, at midnight on April 18 Lincoln and Secretary of War Cameron appeared at the door to inspect the guard armed with Sharps rifles and cutlasses. The first troops called up by the president's proclamation reaching the capital came on that day, 460 Pennsylvania volunteers and a company of regulars from posts in Minnesota.

Just before noon the next day, a full-strength militia regiment, the 6th Massachusetts, arrived in Baltimore, the only rail connection from the North to Washington. Travel on the line through Baltimore necessitated that the cars be hauled individually from the President Street to the Camden Street depot by horses, a break of a dozen blocks along the wharfs. The city was rife with secessionists, and many homes and businesses along the way were hung with Confederate flags. Nine cars carrying the 6th were hauled without overt trouble, but the way was swarming with hostile, jostling crowds, growing larger and more enraged as the transit of the "mercenary" troops went on. Finally, unfurling a Confederate flag, a jeering, stone-throwing mob put up a makeshift barricade of sand, stone, and anchors, barring the way for the remainder of the cars. Four companies were stranded, along with the regimental band and eight hundred unarmed volunteers from Philadelphia.

Shouldering arms the remaining companies of the 6th set out from President Street and were jostled and pelted with stones. When men along the way tried to wrest arms from the troops, their officers shouted for "double-quick." As the troops pressed through the streets, at times breaking into a run, confusion swirled all around—this became a signal for a full-scale riot. Shots were fired, killing four soldiers and wounding thirty-six, while scores went missing in the melee. Among Baltimoreans, eleven were dead and uncounted injured.

Now the police arrived, restoring enough order to allow the troops to reach Camden Street and board the train, as a dense angry swarm spread out along the tracks and pressed to the cars' windows, brandishing revolvers and knives. As the train pulled out amid hurrahs for Jeff Davis, answering volleys were fired inside the cars. One of Baltimore's leading citizens, who had been instigating the mob, was killed. The hapless Philadelphia volunteers, at the President Street depot, were set upon and beaten. The police had no alternative but to turn them back to the city from whence they had come.

That night a meeting in Monument Square demanded immediate secession for Maryland, as fiery speeches denouncing Lincoln's government were heard. Governor Hicks addressed the gathering, and affirmed: "I will suffer my right arm to be torn from my body before I will raise it to strike a sister State."[22] Any more government troops passing through the city, it was resolved, would meet there the direst consequences. The governor ordered the railroad bridges across the Susquehanna, the connecting links to Philadelphia and Harrisburg, be burned. Telegraph offices were seized and the wires cut at various junctures, isolating and leaving Washington seemingly stranded.

Just as Wise led troops to seize Harper's Ferry, Major-General Taliaferro also was dispatched to Norfolk with Virginia militia companies. The commandant at the Gosport Navy Yard, Commodore Charles McCauley, who had eight hundred sailors and marines in his command, was, like so many officials during the crisis, loathe to doing anything that could be construed as "coercion." The most important ship in the yard, which included two three-deckers, two sailing frigates, two sloops, and a dispatch boat, was the *Merrimack*, a steam frigate awaiting engine repairs. Secretary of the Navy Welles ordered temporary repairs to the powerful machine on April 11 so that it could be moved to safety in Philadelphia. But McCauley countermanded the order, and the vessel remained. Two days earlier, when Virginia passed its ordinance of secession, secessionists tried to block the harbor by sinking a few old hulls in the channel; now as Taliaferro arrived, he began preparing batteries to assail the yard, and McCauley ordered all the ships scuttled.

Just as this was happening a relief expedition sent from the Washington Navy Yard arrived, carrying a regiment of Massachusetts infantry whose officers had orders to secure the yard. But with the warships setting underwater, there was nothing to do but complete the destruction. The dry dock was mined, and the scuttled ships, with their upper decks and masts protruding, prepared to be burned. Cannon were spiked and dumped off the wharf, and the relief ships, the *Pawnee* and the *Cumberland* were loaded with everything of value. As floodtide came the next morning, the vessels hoisted anchors and began moving off. A single shot from the *Pawnee* turned the entire yard into "one vast sheet of flame." Though spectacular, the ruin was not total: the dry dock was found intact after the mine failed to explode; the *Merrimack* would be refloated, repaired, and fitted with iron plates; and nearly 1,200 heavy naval guns would be restored, and much of the machinery and equipment salvaged.

With 8,000 Virginia troops at Harper's Ferry, the Gosport Navy Yard in the hands of the same, and Jefferson Davis preparing to send an additional 16,000 soldiers into Virginia, it appeared the capital city might be captured together with the entire government. Washington was a city under siege, as public buildings were sandbagged and barricaded, and the Treasury building prepared as a redoubt for a last-ditch defense. While the prices of necessities soared and the city was rife with rumors, reports circulated of Lincoln's "trepidation." It seemed as if Lincoln was presiding over nothing so much as desolation. People began packing in the event flight became necessary, and barrels of provisions were secured in the General Post Office and in the basement of the Treasury building as General Scott estimated Washington could withstand an embargo for no more than ten days.

For its defense the government had only an unprepared and incomplete Pennsylvania regiment, the 6th Massachusetts, a handful of Army regulars, and an assortment of District of Columbia militia companies whose loyalty was doubtful. There were also Lane's Frontier Guard and the Strangers' Guard, organized by Cassius Clay, the tough Kentucky representative and abolitionist, who went about the capital armed with three pistols and three knives. Troops belonging to Rhode Island and New York were on the way, but their disposition was unknown; perhaps, it was conjectured, they were detained at Havre de Grace with 20,000 other Northern troops. Reviewing the arriving Massachusetts troops Lincoln could not conceal his stupefaction. He said: "I begin to believe there is no North. The 7th New York regiment is a myth. Rhode Island is not known to our geography any more. You are the only Northern realities."23

Relief came by way of an unlikely figure in a Massachusetts lawyer and politician named Benjamin F. Butler. Butler was a Democrat who had supported Jefferson Davis for the presidential nomination in Charleston, but when the party split, became a supporter of the Breckinridge proslavery ticket. Threadbare when it came to the martial arts, he was however skilled at argument, and after impressing the Massachusetts governor was appointed to lead state troops as brigadier-general. Stranded at the top of Chesapeake Bay in Havre de Grace, on April 20 he commandeered a steamer and, with the 8th Massachusetts Infantry, dropped anchor at Maryland's capital, Annapolis, only forty miles by rail from Washington. Maryland's governor, nominally a unionist, but fearing an incident that would give secessionists a pretext to wrest his state from the Union, refused Butler and his troops embarkation. By April 22, however, as another steamer carrying New York's 7th Regiment joined them, Butler was ready to have things his way, and all troops came ashore, Butler assuming the joint command.

From Annapolis the Baltimore & Ohio's rails ran twenty miles west to the link with the Baltimore-Washington line at Annapolis Junction; but its track, rolling stock, and bridges had all been damaged or disabled by secessionists. These were all promptly repaired by mechanics stepping forward from the ranks, and two companies were sent to guard against further interference. When communication with Washington was reestablished, Butler received a message from the War Department that he was to remain in Annapolis and keep the rail open, sending the bulk of the troops on to Washington. By April 25 these troops were marching down Pennsylvania Avenue toward the executive mansion where they would take a solemn oath of allegiance.

As secessionist fever flared on both sides of Maryland's Chesapeake Bay, it appeared likely Maryland would go with the South. But Ben Butler, with Massachusetts and New York troops, had opened the way from the north to Washington; now the task became that of making that road permanent by securing Maryland's adherence to the Union. The secessionists demanded that the state legislature assemble in Baltimore at once, issuing a proclamation calling for the same. To head off a precipitous flight into the arms of the Confederacy the governor called for a regular session of the legislature to convene on April 26 instead, specifying that, in as much as Annapolis was occupied by government troops, the meeting was to be held in Frederick, a town in the western part of the state more amenable to the Union. Lincoln now considered arresting the entire legislature, but decided against it since the assembly was completely legal, and "we cannot know in advance that their action will not be lawful and peaceful." The next day, however, Lincoln prepared to take extralegal action as he wrote to General Scott: "You are engaged in repressing an insurrection against the laws of the United States. If at any point on or in the vicinity of the military line, which is now (or which will be) used between the city of Philadelphia and the city of Washington, via Perryville, Annapolis city and Annapolis Junction, you find resistance which renders it necessary to suspend the writ of habeas corpus for the public safety, you, personally or through the officer in command at the point where the resistance occurs, are authorized to suspend that writ."[24]

A few days later Butler was ordered to occupy Relay House, an important railroad junction just southwest of Baltimore, and then march into the city proper, where he occupied and fortified Federal Hill overlooking the harbor and the business district. Arrests were made, arms confiscated, and the display of Confederate colors forbidden; while newspapers promulgating "disloyalty" were suspended and editors imprisoned. Butler was as vainglorious as a ham actor strutting the stage, adopting an air of high authority while favoring a uniform set off with gold braid—but he, Lincoln, and Maryland's governor Hicks, by their combined actions, saved Maryland for the Union.

Meanwhile one thousand miles to the west, the struggle to keep Missouri in the Union, where Claiborne Jackson, a former state senator and leader of the border ruffians, had attained the governor's seat in the 1860 election, was under way. By the time of his inauguration secession was in full flourish and he was determined that

Missouri would go out with the cotton states. Calling for a convention to deliberate on the matter, he found that a majority of delegates elected were Union supporters and so nothing could be done. The outcry over Fort Sumter presented a new opportunity, and on the day he spurned Lincoln's call for troops, Jackson called a special session of the legislature. The majority of the body and its political leaders took the position that the state should make common cause with "her sister slave-holding states," as the governor carefully weighed a plan to accomplish this. In a letter to Jefferson Davis, Jackson confided, "Missouri has been exceedingly slow and tardy in her movements hitherto, but I am now not without hope that she will promptly take her stand with her Southern sister states." He reported he had a plan for an assault against the federal arsenal at St. Louis—with some 60,000 muskets and other arms—that could be useful to their cause, or in the hands of government troops, be used against the secession movement in the state. To remove this obstacle in the path of their success, Davis arranged to ship two 12-pound howitzers and two 32-pound guns and the ammunition for them by steamer from the captured federal armory at Baton Rouge. They "will be effective, both against the garrison and to breach the inclosing walls of the place," once positioned on the hill commanding the arsenal, Davis pointed out in his return communication. He added: "We look anxiously and hopefully for the day when the star of Missouri shall be added to the constellation of the Confederate states of America."[25]

Although Jackson was making no overt statements that his state would secede, he was hopeful that it could be coordinated simultaneously with action in Tennessee and Kentucky. Meanwhile he had taken control of the St. Louis police, and secessionists seized a United States arsenal at Liberty, outside Kansas City. Frank Blair Jr., Republican congressman from Missouri, brother of Lincoln's postmaster general and son of one of his principal advisors, began taking steps to preclude Jackson's arrangements. First Blair saw to it that a reliable officer was placed in command at the St. Louis arsenal, Captain Nathaniel Lyon, formerly stationed at Fort Riley, Kansas, since 1855. Lyon had seen enough of border ruffian tactics to turn him into the staunchest Free-Soiler; he would not hesitate to meet those advancing the secessionist agenda with force. Blair had also seen to it that the U.S. Army's commander for the St. Louis district, General Harney, who was socially intimate with the local planters, was recalled to Washington for consultations.

The key to the situation in Missouri, however, lay with the German population of St. Louis. Emigrants by the thousand, many of them refugees from the unsuccessful European revolutions of 1830 and 1848, had been drawn west by the effective propaganda of Gottfried Duden. Known as "the gateway to the west," the city had doubled in size between censuses in 1850 and 1860 and had the largest foreign-born population of any city in the United States. By 1860 Germans were united into a solid voting block and had become a power in state politics, the basis of which was their gymnastic societies, which had assumed a semi-military status after antiforeigner riots in 1852. These were made the most of during Lincoln's campaign when Frank Blair called them to service as "wide-awakes." Two regiments would be formed, one commanded by Henry Boerstein, trained as an Austrian soldier and editor of the significant German language newspaper *Anzeiger des Westens*, and impresario of his own opera house; the other was headed by Franz Sigel, ex-commander of 30,000 in the European uprisings in 1848. It was Blair's intention to arm these regiments with government property in the arsenal at the proper time. Meanwhile to remove the excess arms Blair arranged with Illinois governor Yates, who had no desire to see a Confederate state on his western border, for a steamer to be sent to the St. Louis wharf at midnight on April 25. Accordingly 21,000 muskets were transferred across river, and thence to Springfield.

As Jackson waited for his shipment of heavy guns, on May 2 the state militia was mustered; but as Lyon stood atop the hill overlooking the arsenal, they were obliged to acquiesce, occupying a grove on the outskirts of St. Louis. The militia commander, West Pointer General D. M. Frost, who had not remained in the service, organized the encampment at Lindell Grove, promptly dubbed "Camp Jackson," with regimental streets taking the names "Davis," "Beauregard," and "Sumter." Drilling was carried out leisurely and included the sons of some prominent families of St. Louis, who, with undisguised satisfaction, came out to watch them; in the evenings, girlfriends and wives were out to stroll in the spring air on the arms of chivalrous lovers and husbands.

By May 7 Lyon was ready for action and called his volunteer officers to a conference: "We must take Camp Jackson and we must take it at once," he said.[26] The next night a steamer docked at St. Louis and large crates marked "marble" were quickly freighted out to the camp. While rumors were circulating of Lyon's impending attack,

General Frost directed a note to him asking clarification. As two regular army companies and two all-German volunteer companies marched out of the arsenal, the streets along the route to Lindell Grove began filling with people, as many stood in doorways and at windows. With his troops arriving at "Camp Jackson" and deploying in line of battle, Lyon wrote his reply to Frost.

When the note was received, three cheers went up at the general's tent from the militia gathered around it. Lyon's officers ordered troops to move their cartridge boxes to the front of their belts. Seeing he was outgunned, Frost requested more time to consider. He was promptly told he had ten minutes. Soon a note reached Lyon announcing "surrender." On the opposite sides of the street single files were formed, the militia disarmed, and Davis's guns seized. Now the entire parade, in one line, started for the arsenal.

As more and more people gathered, shouts of "Damn the Dutch" and "Hurrah for Jeff Davis" rent the air. One woman saw her lover among the prisoners and tried to reach him. She was turned aside by bayonet. Stones were thrown. A man said to be drunk tried to force his way through the line and was thrown down an embankment. He pulled out his pistol and fired, wounding an officer. Boerstein ordered fire returned. Suddenly there was firing on both sides. In the carnage fifteen people were killed on one corner alone, while in all over thirty died, two soldiers among them.

That night St. Louis was stricken by madness; thousands roamed the streets, brandishing the flag either of the Union or of the Confederacy, as groups of partisans marched and counter-marched. Weapons were fired in the darkness; several lone Germans were murdered. A firearms store was looted, as shouts were heard directing mobs to the offices of the *Democrat,* the Republican paper, and to the *Anzeiger des Westens.* The next day after Frost's troops were paroled, a German regiment marching out of the arsenal tangled with a mob at Fifth and Walnut streets, where a soldier was shot in the head. There was more firing, leaving more lifeless on the dumbstruck streets.

As the rival sections moved ever closer toward the precipice that would devour the lives of more than 600,000 men, it was clear neither side anticipated a war of the ferocity and length of the one that would be fought. The calculation by southerners in high government office would seem to have been to limit the possibility of a significant clash of arms as the crisis developed. John Floyd as secretary of war for Buchanan had seen to it that 115,000 muskets were transferred from arsenals in Springfield and Watertown to arsenals in the South, while the U.S. Navy had been dispatched to foreign ports as the regular army, too, with only 16,000 troops, were mostly dispersed on the frontier to guard against "Indians." Similarly Georgia's Howell Cobb, Buchanan's treasury secretary, left a largely depleted U.S. Treasury as his legacy to the new administration. "The crisis came in 1860," W. E. B. DuBois wrote in *Black Reconstruction*, "because the cotton crop of 1859 reached the phenomenal height of five million bales, compared with three million in 1850." On this the South made their bid to contest the challenges being raised, and shifted from a defense of slavery to an offensive strategy against the North. The party with Abraham Lincoln at its head would not be a government of Union that the Constitution made, and they would make war upon it.

When the so-called "seven sisters" seceded they did so over the issue of the national policy toward slavery; to be sure the question was framed not as one for or against the abolition of slavery, but whether the right to own slaves would be recognized in the United States territories. This was of importance to the South, not for economic reasons, as there was little chance slavery could become viable in the West; rather it was for constitutional rights and equality with a rival North. If slavery could be restricted in the territories, the next inevitable step, the thinking was, would be to restrict it in the states of the South. Secession, therefore, was promulgated to drive forward a change in policy as part of a political contest. It proceeded on a state-by-state basis, rather than by an all-Southern Convention, because then it could be presented to those states lagging behind as a fait accompli. The ground had been plowed and planted commencing with John Brown's Harper's Ferry raid, its leaders working ever since in a calculated way and warning under no event whatever would they submit to the rule of Abraham Lincoln.

If war were to result, the calculation was, much of the North would stand aside while secessionists and Washington maneuvered like duelists with raised rapiers. The contest that was foreseen, as a Republican electoral victory was widely expected, called for an "armed filibuster" by a united South to batter down the Washington government. It was expected that a concerted campaign could achieve victory in as little as three months.

As the "fort" became the contention, the entire issue bedeviling the nation became simplified, as the child of slavery, secession, grew to be more significant than its progenitor. When it suddenly became apparent after Fort Sumter how disastrously the enablers and abettors of this conspiracy had miscalculated, Lincoln and Seward were equally expectant of a brief contest after which the Union would be reconstructed. But with Lincoln's proclamation giving certain "combinations" twenty days to lay down arms and return to their homes, secessionist sentiment flamed in the border states, and the effort was made to unite the seceded states with the watchwords "tradition," "honor," and "sovereignty."

But as the second tier of slave states began to withdraw, there was to be a change in emphasis. Not only were all of the seceded states being claimed for the Confederacy, but all those states below the Mason-Dixon line and the territories to the west, and if they were successful, perhaps too, they might draw California and Oregon after them, and make the adherence of the Northwest and the Middle West impracticable to the Union. The argument for "states' rights," of course, had been heard as justification for secession in the lower South, but now as the two sections began to pull their masses onto the field of battle, the issue became one of rival sovereignties, with the North hoisting the banner "Union" and the South "Independence."

Slavery in the border states only had formal similarity to that in the Gulf. Except for the large landed estates in Virginia and in districts in western Tennessee and southern Arkansas, slaves were dispersed on small farms for the most part, cultivating staples like tobacco, hemp, and livestock. These states too had a thriving middle class comprised of merchants, industrialists, and professionals, making them in many respects closer to the states on their northern borders. But as they were drawn into the vortex of war, the issue became one of dominion. The reason Virginia, Tennessee, Arkansas, and North Carolina seceded, joining the "seven sisters," was the same. For Republicans to have had the temerity to suggest, as Lincoln did, that slavery was wrong, threatened nothing short of the direst catastrophe—outright revolution. Three of the border states that did not secede, Maryland, Kentucky, and Missouri—Delaware voted outright to remain in the Union—had powerful and vocal secessionist minorities, but had events proceeded in only a somewhat different way, they too would have been allied with the Confederacy, and its success for "independence" assured.

On May 5 the twenty days stipulated in Lincoln's proclamation expired, and the next day Jefferson Davis signed "an act recognizing the existence of war between the United States and the Confederate States." With the secession of Virginia, Richmond was selected as the capital of the Confederate States of America, lying across an exposed frontier only ninety miles from Washington. The blood spilled in Baltimore and in St. Louis at the end of the previous month had only been a token of that to come; but now there had been the necessary consecration . . . like the blood Nat Turner had seen in his master's fields on ears of corn "as though it were dew from heaven."[27]

Eleven

His Truth Is Marching On

I wish to say, furthermore, that you had better—all you people of the South—prepare yourselves for a settlement of this question, that must come up for settlement sooner than you are prepared for it. The sooner you are prepared the better. You may dispose of me very easily—I am nearly disposed of now; but this question is still to be settled—this Negro question I mean; the end of that is not yet.

—John Brown, interview, *New York Herald*

As Virginia passed its ordinance of secession, delegates from its western counties, meeting in Wheeling, created a new state government and drafted a constitution. Under the federal constitution, however, no state could divide itself but by general consent. Acting on the theory that "to the loyal people . . . belongs the government of the state," they chose F. H. Pierpont as governor and selected two new senators to Washington, declaring Virginia's secessionist government, sitting in Richmond, illegal and treasonous.

In the eastern districts of Tennessee this scenario was repeated when delegates with strong Union sympathies held a convention denouncing the governor and the state for their military alliance with the Confederacy. Kentucky was likewise of divided sentiment, Unionists still believing that their Senator Crittenden's compromise offered a way out of the crisis, as Governor Magoffin appealed to states on its northern and southern borders for a conference to mediate between the warring national factions. In mid-May the legislature resolved: "This state and the citizens thereof shall take no part in the Civil War now being waged," proffering a position of neutrality. Thousands of Kentucky citizens, however, were crossing the border to join units being organized for a Confederate army in Tennessee.

Meanwhile after the "Camp Jackson" incident, acting in haste, the governor and legislature in Jefferson City put Missouri on war readiness, appointing Sterling Price, a former governor and Mexican War general, commander of pro-Confederate militia. Returning from consultations in Washington, General Harney arrived in St. Louis the day after Lyon marched out to Lindell Grove. As Harney listened to the pleas of wealthy planters, Frank Blair hastened to Washington to inform Lincoln of the situation in his state, followed by a delegation sent by Harney intent on same. Both sides returned without resolution, but on May 20 Blair received a sealed letter from Lincoln relieving Harney of command and commissioning Lyon a brigadier-general. The next day, ignorant of Lincoln's decision, Harney announced he had concluded a truce with Jackson, whereby the state agreed not to invoke its military act in exchange for recognition of Missouri's neutrality. On May 30 when Harney began moving German regiments out of St. Louis, Blair sent him Lincoln's order removing him from command.

Jackson now called for a conference at the Planter House with the new federal commander in St. Louis, setting the date for June 11. At the appointed hour Jackson appeared with his secretary and Sterling Price, while Blair and Lyon represented the government. As Jackson presented the case for Missouri neutrality, Lyon became livid with anger; he thrust a finger at Jackson's breast: "Rather than concede to the State of Missouri for one single instant to dictate to my Government . . . I will see you, and you [Price], and you [Jackson's secretary], and every man, woman, and child in the State dead and buried. This means war." Taking out his pocket watch he advised, "In an hour one of my officers will call for you and conduct you through my lines."[1]

Early next morning, Jackson, Price, and his secretary were back in Jefferson City, stopping only to take on fuel or when cutting telegraph wires and burning bridges. Explaining the emergency to the legislature, Jackson prepared a proclamation to be distributed by daylight calling for fifty thousand volunteers, and ordered state documents packed for retreat. Three days later Lyon occupied the capital, and by June 17 caught Price and his army, the governor, and the legislature at Boonville. After a light skirmish the newly minted Union general sent them all reeling toward Arkansas.

In the spring and early summer of 1861 the popular enthusiasm sweeping North and South had not been requited in any decisive military confrontations. Both were mustering forces while their respective presses, clamoring for action, declared it high time they fought it out. General-in-chief Winfield Scott, however, offered a plan whereby he proposed to gradually envelop the adversary by sea blockade and a fleet of gunboats supported by troops along the Mississippi River, while mounting an amphibious expedition to capture New Orleans. Commerce would be restricted in this way and the secessionists brought to terms. Invasion from the north, he argued, would only produce "fifteen devastated provinces!" requiring that any conquered territory "be held for a generation, by heavy garrisons, at an expense quadruple the net duties or taxes which it would be possible to extort from them."[2]

Obviously this would take time; troops need be trained, ships and gunboats built. Scorning Scott's "Anaconda Plan," the northern press, led by the *New York Tribune*, began editorializing for invasion. Horace Greeley blazed forth: "Forward to Richmond! Forward to Richmond! The Rebel Congress must not be allowed to meet there on the 20th of July. By that date the place must be held by the national army."[3] Greeley's call was compatible with Lincoln's war aims and seemed to assure that the theater of the conflict could be limited to the area athwart the invasion routes from Washington.

Defending at Manassas, a railroad junction thirty miles southwest of Washington, were some 20,000 troops of the newly designated Army of the Potomac, commanded by the freshly laureled Gen. Beauregard. Another of the confederacy's generals, Joseph E. Johnston, the former quartermaster-general of the United States Army, commanded 11,000 troops at Harper's Ferry. Marching down from Hagerstown to oppose the latter were 15,000 Union troops under Gen. Robert Patterson—an elderly man who had accumulated wealth with a substantial stake in Pennsylvania railroads and in several Philadelphia steamship lines. On June 15 Johnston ordered the evacuation of Harper's Ferry and retreat to Winchester. In his *Southern History of the War*, Edward A. Pollard wrote, this brought one of those wild, fearful scenes which make the desolation that grows out of war. The splendid railroad bridge across the Potomac—one of the most superb structures of its kind on the Continent—was set on fire at its northern end, while about four hundred feet at its southern extremity was blown up, to prevent the flames from reaching other works which it was necessary to save. Many of the vast buildings were consigned to flames. Some of them were not only large, but very lofty, and crowned with tall towers and spires, and we may be able to fancy the sublimity of the scene, when more than a dozen of these huge fabrics, crowded into a small space, were blazing at once. So great was the heat and smoke, that many of the troops were forced out of the town, and the necessary labors of the removal were performed with the greatest difficulty.[4]

On June 29 Lincoln met with his general staff whereupon Scott again urged adoption of his plan. But with the enemy at Manassas and their congress scheduled to convene at Richmond, both the target and the timetable were set. A defeat at Manassas, moreover, might discredit the secessionists and leave Richmond open for capture. Scott's staff officer, Gen. Irwin McDowell, a soldier with no previous experience in field command, was ordered to prepare and begin an offensive by July 8. Suggesting that the ninety-day men were "green" and could scarcely be expected to maneuver effectively under fire, he urged postponement until the new three-year recruits began coming in and could be trained. Lincoln replied: "You are green, it is true, but they are green, also, you are all green alike."[5]

As July approached, and 35,000 troops of the newly designated Grand Army of the Union were assembling in Washington, the first session of the 37th Congress was set to meet under conditions dramatically different from the previous session. Republicans now had a commanding majority in both houses; out of a total of 48 senators, 32 were Republicans, and in the House the party held 106 of 176 seats. Kentucky's Senator Breckinridge was the lone proslavery voice, but by September he would exchange his seat for a generalship in the Confederacy. On July 5 Lincoln addressed the Congress, where he gave an account of events and what he had done in the emergency. The loyal citizens in western Virginia had proclaimed their adherence to the Union and would be recognized and protected "as being Virginia." The neutrality professed by some states had no more validity than did secession and was "treason in effect." Still promising peace to the South and continued life for slavery as "expressed in the inaugural address," "then as ever" Lincoln said he would be guided by the Constitution and the laws. Then he asserted, "This is essentially a People's contest," and thought it "worthy to note, that while in this, the government's hour of trial, large numbers of those in the Army and Navy, who have been favored with the offices, have resigned, and proved false to the hand which had pampered them, not one common soldier, or common sailor, is known to have deserted his flag." At the end of his address came acknowledgment of the length to which his government was prepared to go, indicating the potential scope of the war, as Lincoln asked Congress to arrange for a $400 million loan and authorize 400,000 new troops. Jefferson Davis, in his speech on July 20 before the Confederate Congress at Richmond, remarked that this stripped "the veil behind which the true policy and purpose of the Government of the United States had been previously concealed; their odious features now stand fully revealed; the message of their President and the action of their Congress during the present month, confess the intention of subjugating these States by a war whose folly is equaled by its wickedness."6

The army rising in Washington was the largest till then seen on the continent and was supplied with the best artillery and equipage known in the world. It would have 10,000 regular United States troops and upward of 25,000 three-month volunteers. But their assemblage proved unwieldy, and McDowell would delay his campaign due to a shortage of wagons and the necessity of integrating late-arriving troops. As McDowell got underway July 16, Beauregard was well aware of his plans, both from spies he employed and because the whole thing had been published in the *Tribune*, which he had been reading. But the troops marching across the Potomac on the Long Bridge, carrying shouldered muskets with bayonets and fifty-pound knapsacks, amid the jubilation and fanfare of the public, were an impressive sight. Few who saw them doubted they would be in Richmond within ten days to raise the Stars and Stripes over Capitol Square.

As the head of the Grand Army set out, its tail was still being composed, and as a sign of how the ninety-day men would bear up, several regiments whose terms were about to expire insisted on bringing up the rear. Columns soon found they could not march evenly, and as obstacles of felled trees and other debris were cleared, troops in the rear stood for hours in the sun. Only six miles were covered in the first day's march. The 20,000 Confederate troops awaiting them behind a muddy, tree-choked stream called Bull Run were deployed over eight miles between Manassas Junction and Stone Bridge. The country between was well intersected with roads and the stream had half a dozen fords; the main ford, Stone Bridge, crossed in the west on the turnpike from Centreville to Warrenton into rolling hills. It was along this road the Federals were advancing, while Beauregard had placed his main force to the southwest guarding the railroad. McDowell's plan was to attack these troops on their flank, while Beauregard planned to swing his army in a giant wheel at the advancing enemy. Essential to both was the status of Johnston's army. For the former this meant preventing them from reinforcing Manassas, and for the latter, the contrary.

Meanwhile, shuttling "Jeb" Stuart's cavalry in front of Patterson, Johnston gave him the slip, marching out of Winchester for the railroad at Piedmont, where his entire army entrained for Manassas over three days. As they began arriving, each army now was comprised of equal numbers. Both, however consisted of amateurs for the most part, led by men likewise without much experience. It is true there were veterans of the Mexican War, and even some who had seen service in 1812, as well as officers on both sides who were West Point graduates. But the mass were young men or mere boys. Both generals knew this and did what they could to surmount the weakness. For Beauregard the problem was least severe since his forces waited in defensive positions—they must simply hold on before the onslaught. Then too there was an expectation that southerners possessed a superior readiness for the field, in familiarity with arms and horsemanship, and by dint of having many officers bred for command. McDowell's troops had been organized into manageably sized brigades, each led by a regular army colonel with as many junior officers as were available. But with bare levels of efficiency the troops could scarcely be expected to maneuver under fire, so McDowell specified once the advance began there should be no step back.

On the morning of battle the whole concourse collecting at the rear of the Union army, wrote a *London Times* correspondent, "was like a holiday exhibitor on a race-course." Congress had adjourned to afford its members an opportunity to attend, and there were six senators and ten congressmen in the crowds of spectators, fashionably dressed ladies, sensation-seekers and editors jostling in every conceivable style of vehicle, with many parties spreading picnics and raising opera glasses to view the novel scenes. William Sprague, Rhode Island's wealthy new governor, was there with a consignment of champagne to toast the victory. Certainly none among the onlookers entertained any other idea.

The columns advancing on Manassas came down the Warrenton Turnpike, a straight road passing west from Washington through Fairfax Court House, climbing the rise around Centreville and then dipping into the valley of Bull Run. The turnpike forded the stream on a long arching stone bridge, and then was briefly girded on either side by hills welling up from the Virginia plain, when it curved southward to Warrenton. Stone Bridge, McDowell thought, was heavily guarded and perhaps mined for destruction. On July 18 Union artillery poured shot over the approaches at Mitchell's Ford but were held in check by answering guns. Then under the advantage of the steep slope of the northern bank of Bull Run a strong advance of three thousand infantry, supported with artillery and cavalry, were brought to bear on Blackburn's Ford. After a brief artillery barrage, infantry attempted to break open the crossing but were repulsed, and the encounter continued as an artillery duel.

McDowell now conceived a flank maneuver, whereby the opposite bank could be gained four miles upstream of the enemy's strong points. Unknown to him however, Johnston had made it through to Manassas accompanied by a brigade commanded by Thomas J. Jackson. McDowell ordered his troops prepare four days' rations on the prior evening, and begin to move out at 2 A.M. so as to be in place as the sun came up. These movements of the Grand Army of the Union were generally known in Washington and flashed over the wires throughout the North.

On that Sunday morning the scene framed against the Blue Ridge was of a glistening display of bayonets and streaming flags. Soon thousands of soldiers were filling the roads and clouds of dust obscured the view. Again the movement of the troops became a faltering march as those comprising the main strike force waited in the road for hours before they'd traveled three miles. Some of the soldiers saw the meadows of the bottomlands were filled with blackberry bushes; soon corpses would bear the stains of these saps on their fingers and mouths. In later years Beauregard would write: "There was much in this decisive conflict about to open not involved in any after battle, which pervaded the two armies and the people behind them and colored the responsibility of the respective commanders. The political hostilities of a generation were now face to face with weapons instead of words."[7]

As the cannonade began it was eleven o'clock before troops were over Bull Run, giving Confederate officers ample time to swing men and two artillery pieces across their path. Two hours into the contest, heavily pressed just north of the Warrenton Turnpike, the Confederates retreated up a hill to the south called Henry House Hill. As these troops began to break and flee, South Carolina's Brig.-Gen. Barnard Bee struggled to stiffen his brigade to hold the hill. Now McDowell called more troops into the engagement and it appeared the stronghold was to be overpowered. Reports began to drift to the rear that victory was imminent. By mid-afternoon news flashed across the North of a stirring triumph at Bull Run. During this time Jackson reached the hill's southern slope, posting five regiments below the crest so troops would have to gain the crown before firing on them. The whole fight was now for possession of the hill. Only Jackson's troops and those of Wade Hampton, the wealthy South Carolina planter who had financed his own "legion," remained to await the expected Union onslaught. General Bee rode down to confer with Jackson. "General, they are beating us back." Unmoved, Jackson replied, "Sir, we will give them the bayonet."

Bee rode back to the crest and into the smoke of battle, the sword presented him by his state extended. He cried, "Look! There is Jackson standing like a stone wall! Rally behind the Virginians!" Mortally wounded, Bee toppled from his horse and the sword from his hand. But when the Union assault surged forward, it lacked coordination and its effect was not nearly what it could have been. To stiffen it McDowell ordered two batteries pound the enemy at close range, but as infantry ordered to support them came up they ran headlong into Stuart's cavalry charging down Warrenton Turnpike.

These Union troops broke and fled, while answering close-range fire completely extinguished their battery. Reaching Henry House, McDowell climbed to the top floor to view the battlefield. For several hours possession of the bitterly contested ground see-sawed from side to side, with sustained charges coming in both directions. Finally a coordinated attack in front came from Beauregard's troops rising with one demonic shriek that was said to turn up a man's spine like a corkscrew—the rebel yell. Pollard wrote: "From the long-contested hill from which the enemy had been driven back, his retreating masses might be seen to break over the fields stretching beyond, as panic gathered in their rear. The rout had become general and confused; the fields were covered with black swarms of flying soldiers, while cheers and yells taken up along our line, for the distance of miles, rung in the ears of the panic-stricken fugitives."[8]

Just prior to this decisive moment Jefferson Davis arrived; riding toward the front with an aide he could only see signs of the disarray of wounded and beaten men. As Davis got to the field hospital he called out: "I am President Davis, follow me back to the field!" In the hospital, having a hand wound dressed, with his newly earned sobriquet, was "Stonewall" Jackson, who shouted, "We have whipped them! They ran like sheep. Give me five thousand men and I will be in Washington City tomorrow!"[9]

The fighting intensified with no fresh northern troops coming in, while Confederate troops were being added. Union troops began to fall back, stumbling a few dozen yards, then stopping to fire haphazardly. Soon whole companies were tumbling pell-mell down the hill, followed by caissons and cannon, ambulances and wagon trains. Officers began deserting their commands, passing the growing melee at a gallop. Soon exhausted soldiers, their lips blackened and cracked from powder, mouths agape with fear and confusion, were tumbling up the Warrenton Turnpike in a rising cloud of dust. Seeing this, dignitaries and politicians, correspondents and holiday seekers all ran for the nearest conveyances, adding to the growing welter. With the roads congested, cannon wrecked a wagon on the bridge as it reached mid-span, snarling traffic into a tangle. As the cry "cavalry is coming" rent the air, fear became panic; teamsters cut loose their horses and rode them bareback to safety; ambulances carrying the wounded were abandoned at the roadside.

McDowell wired Washington from Centreville that he had been driven from the battlefield. He posted reserve troops there and hoped to rally the routed and disintegrated mass behind them, but they would not be stopped even as the safety of Centreville was gained. Another attempt was made to halt the disintegrated army at Fairfax Court House, but to no avail. Nothing could now be done but fall back and prepare to defend the bridges at Washington. As the magnitude of the disaster began to unsettle Lincoln's government, authorities in Pennsylvania and New York were urged to send troops as soon as possible and the commander in Baltimore was alerted lest there be fresh trouble in the city. For what remained of the night and into the early morning, policy makers cast about for a solution. First George B. McClellan, a thirty-four-year-old army retiree who had been an observer a few years previous in the Crimean War and lately the manager of a railroad company, was ordered to advance into the Shenandoah Valley with his Ohio Volunteers bivouacked in western Virginia; a short time later he was told to remain where he was until he received new troops from Ohio. Finally the general was wired: "Circumstances make your presence here necessary. Charge Rosencrans or some other general with your present department and come hither without delay."[10] Davis and his generals likewise were taking stock. As the Grand Army began to stampede, Stuart's and other cavalry had been ordered in pursuit, and infantry commanders prepared to move up to Centreville. With such a thorough victory Davis had an unprecedented opportunity and was emboldened to think beyond mere survival. Envisaging pursuit to Washington, as he saw it, with the seat of the northern government occupied, its leaders and officials expelled or imprisoned, and Maryland free to ally itself with the South—the war could be decided at a stroke. In conference in Johnston's headquarters the president urged this position, in which Beauregard largely concurred, while Johnston stated that it was altogether beyond their capacity. They were lacking logistical support, and food stocks; the troops, too, were inexperienced and untrained, and had come close to breaking. An effective force could not be organized in a few hours, and the men, Beauregard reported, were as little disposed to fight after victory as was their counterpart in defeat.

Still the overriding factor in Davis's mind was the effect news of a crushing blow on Washington would have on northern morale; the defeated and fleeing rabble might be swept from the city at the mere approach of southern troops. Some enemy troops, Johnston pointed out, had been held in reserve; surely they would be sufficient to defend the bridges into Washington until reinforcements could reach the city. Then too Patterson's army could be brought over even before Confederate troops reached Washington. In the course of the conference Beauregard reported on the gains in material; twenty-eight artillery pieces, as many as five thousand muskets, sixty-four artillery horses with harnesses, twenty-six wagons, a garrison flag and ten colors, and tons of ammunition. Estimates were that the enemy suffered as much as a 15 percent casualty rate, with many hundreds dead on every part of the field, and 1,800 prisoners taken. In addition they had gained in blankets, hospital stores, and camp equipment.

As Davis and his officers continued to explore the issues, at 11
o'clock a report came that the Union troops were passing through
Centreville without stopping. Davis now dictated an order for immediate pursuit. But all the difficulties raised restrained his hand and his
order was not issued. The conference ended as Davis concurred with
his generals—at dawn infantry would be sent forward to make reconnaissance in force.

In the days following with no advance on Washington, Southern
editorialists and politicians expressed astonishment that an opportunity of such magnitude had been lost, and there would be recriminations between Davis and his two generals outlasting the war. Edmund
Ruffin, who appeared in Confederate ranks toting a musket, considered "this hard fought battle virtually the close of the war," confiding
in his diary that Beauregard's next move should be "a dash upon
Philadelphia, & laying it in ashes . . . as full settlement & acquittance
for the past northern outrages."[11]

Meanwhile the headlong flight from Bull Run began to precipitate
into Washington to the dismay of government supporters and the
delight of grinning secessionists, who thronged the streets. Only the
previous night it had been supposed the Grand Army had won a glorious victory; now there was nothing to distinguish the bedraggled
shuffling lot as soldiers, but their torn, dirty, and blood-stained uniforms. Subject to no authority, many rushed to the depot to continue
their flight to points north, while others continued into the country.
Many swarmed the streets, laying down in doorways and on pavements, or in open spaces for a much needed rest. Soon a guard was
stationed at the train depot to prevent men from boarding trains. By
noon there would be steady rain falling. One of many thousands who
came to Washington with the war was the Brooklyn newspaperman
and poet Walt Whitman, heralded for his *Leaves of Grass*. His observations, recorded as he began ministering to the wounded and dying
in the many hospital wards to spring up, is every bit the companion
of his better-known legacy.

Whitman confirmed Washington was "all over motley with these
defeated soldiers—queer looking objects, strange eyes and faces,
drench'd and fearfully worn, hungry, haggard, blister'd in the feet."
Their officers, he wrote, crowded into the barrooms, each reciting his
tale of woe, and shirking his responsibility for the calamity. In the
packed parlors of the Willard House they drank and ate, and
Whitman excoriated them: "There you are, shoulder-straps! But

where are your companies? Incompetents! Never tell of chances of battle, of getting strayed, and the like. I think this is your work, that retreat after all. Sneak, blow, put on airs, there in Willard's sumptuous parlors and barrooms, or anywhere—no explanation shall save you. Bull Run is your work; had you been half or one-tenth worth your men, this would have never happened."[12]

Gloom descended likewise on Horace Greeley, who had had so much to do with the popular clamor for a premature thrust on Richmond. After enduring a week of sleepless nights he wrote Lincoln on July 29, lamenting, "on every brow sits sullen, scorching, black despair." As a final point he concluded: "If it is best for the country and for mankind that we make peace with the rebels, and on their own terms, do not shrink even from that." The day after this letter, however, the *Tribune* editorialized: "It is not characteristic of Americans to sit down despondently after a defeat. . . . Reverses, though stunning at first, by their recoil stimulate and quicken to unwonted exertion. . . . Let us go to work, then, with a will."[13]

Although shaken, Lincoln and his War Department set about their task. Stations to feed and care for the scattered legions were established at specific locations in the city, and the troops began to be reconstituted as more new three-year recruits reached Washington. The day after the battle of Bull Run, or Manassas as the South called it, Lincoln signed a bill authorizing the enlistment of half a million three-year recruits, and signed another on July 25 for an equal number. The next day Gen. George B. McClellan arrived to begin organizing and training a new 100,000-man Army of the Potomac. Another commander for the Western Department arrived in St. Louis July 25, to replace the discredited Harney; this was John C. Frémont, renowned for mapping routes to California more than a decade before. Lincoln had been ruminating on the military problems facing his government. With no offensive developing on the capital and the city returning to normalcy for those extraordinary times, the president and his secretary of state crossed the Potomac to inspect the troop entrenchments and camps at Arlington. As they returned Lincoln wrote out an eight-point program, largely giving shape to the hard war ahead.

In the May 1861 issue of *Douglass' Monthly*, published under the emblem of the American eagle and the national flag with the slogan

"Freedom for all, or chains for all," the lead article bore the title "How to End the War." The source and center of the gigantic rebellion being waged against the American government was slavery, Douglass stated, and should therefore be stricken at its root, and the freedom of the slave proclaimed from the Capitol and "seen above the smoke and fire of every battle field, waving from every loyal flag." Reiterating the "John Brown Idea," Douglass declared: "This can be done at once by carrying the war into Africa. Let the slaves and the free colored people be called into service, and formed into a liberating army, to march into the South and raise the banner of Emancipation among the slaves." The Confederacy, after all, was availing itself of the slave; he was supplying its commissary department, building its forts, digging its entrenchments, and performing the other duties of camp. This left the rebel soldier free to fight while slave labor harvested corn and cotton "that made the rebellion sack stand on end." Destroy these, Douglass urged, "and you cripple and destroy the rebellion." By refusing to enlist the help of black men, the government fought "with their soft white hand, while they kept their black iron hand chained and helpless behind them."

Behind him now were the gloom and failure he had felt and seen gathering over the abolition cause in the Buchanan years. The humiliation of his hurried flight to England, the odium heaped upon him for seemingly abandoning John Brown, the personal tragedy he suffered at the death of his daughter, the declining influence and circulation of his weekly, his embrace of parties increasingly on the margins of the critical contest—all this had been carried away in the rapid march to war. "The Negro is the key of the situation—the pivot upon which the whole rebellion turns," Douglass proclaimed; and the prospect this brought put him in high spirits.

In the week that northern rail links were severed with Washington, Wendell Phillips too declared his support for a war of liberation under the banner of Union. In a speech at the Music Hall in Boston, Phillips proposed that abolitionists had imagined they would be able to lift four million human beings into liberty and justice by words alone, but they had made a great mistake. The nation indeed was made of clashing ideals—and Phillips avowed only arms could settle the controversy. As he closed his address on April 21 he said: "The war, then is not aggressive, but in self defence, and Washington has become the Thermopylae of Liberty and Justice. (Applause.) Rather than surrender that Capital, cover every square foot of it with a living body. (Loud cheers.) Crowd it with a million of men and empty every bank vault at the North to pay the cost. (Renewed cheering.) Teach the world once and for all, that North America belongs to the Stars and Stripes, and under them no man shall wear a chain. (Enthusiastic cheering.) In the whole of this conflict, I have looked only at Liberty—only at the slave."[14] After Boston's dailies failed to give notice to Phillips's speech, the *Liberator* printed it as an extra that sold twenty thousand copies.

On April 19, in a letter to Oliver Johnson, William Lloyd Garrison advised abolitionists to "'stand still, and see the salvation of God' rather than attempt to add anything to the general commotion." To this change in attitude Douglass confessed feeling a "moderate exultation." It was fully ten years since he articulated the position that the battle for freedom should be fought within the Union and not out of it, and had been branded an "apostate" and driven from fellowships he held dear, and the organization he venerated. In noting Garrison's change in regard to the American Union, Douglass wrote: "There are personal, as well as public reasons for our present joy; and though we might for the sake of appearances, attempt to conceal the personal reasons in the public ones, those who know anything of our humble history, during the last ten years, could not fail to see that Douglass feels a personal, if not a little malicious pleasure in finding Mr. Garrison and his friends in their present attitude."[15]

In indicating the prospectus of his journalistic work, Douglass wrote "The Real Peril of the Republic," in his *Monthly* in October 1861:

Speaking as we do, only once in each month, our communication ought to possess something of the quality of history. Indeed, a paper published monthly can be, in these fast times, a newspaper to but very few. The mission of our journal is, therefore, to be a faithful recorder, not of all events touching the great conflict going on between liberty and slavery in this country, whether in the field or in the councils of the nation, but enough of them, to enable all, whether near by or afar off, who may read our journal, to form an intelligent judgment in respect to the character of the whole controversy. In this capacity of recorder, it is our duty . . . to observe and criticize what is passing before us.

In formulating his own relation to Garrisonian abolition, Phillips would write in the following year: "I feel no wish to explain. . . . Besides *qui s'excuse, s'accuse*, has always been my motto. . . . Then I'm no longer exclusively a Garrisonian. I'm a citizen and prefer so to address my fellows just now." Actions and not words were the fitting duty of the hour. "Yet still," Wendell Phillips would pronounce in "The War for the Union" delivered in Boston and New York in December 1861, "cannon think in this day of ours, and it is only by putting thought behind arms that we render them worthy, in any degree, of the civilization of the nineteenth century."[16]

In clear-cut and trenchant speech, Douglass and Phillips sounded the refrain—rebellion and slavery were twin ogres and neither could be put down unless both were taken to the wall; it would not be possible to part freedom for the slave from the victory of the Union. These two articulated this historic imperative—and it was to their voices, and to similar influences, that the North began to respond.

It was the policy of Lincoln's government as the war opened that slavery should not be obstructed by United States arms, and that the constitutional obligation to return fugitive slaves and suppress insurrection would be observed. One of the first Union generals in the field, Benjamin Butler, shortly after landing in Annapolis tendered his services to the slaveholders of Anne Arundel County in putting down a rumored uprising among slaves. As McDowell's army itself had become "fugitive" it drew scores of slaves along with it. Some of these, it was reported, were persuaded to exchange their clothes for soldiers' uniforms, as some of the soldiers were intent on crossing the Susquehanna. In the days following Bull Run, Lincoln wrote a secret communication to McDowell asking "if it would be well to allow the armies to bring back those fugitive slaves which had crossed the Potomac with our troops." Not long after this, General Harney, while still commanding in Missouri, finding slaves seeking refuge in the camps of troops from the northern states, noted: "They were carefully sent back to their owners." McClellan crossing the Ohio River issued a circular "To the Union men of Western Virginia," promising all rights in regard to property "shall be religiously respected," and that they had not come in any way to interfere with their slaves. So as to leave no room for ambiguity he elaborated— "we shall . . . WITH AN IRON HAND, crush any attempt at insurrection on their part."[17]

But no sooner had this position been affirmed than it was to be modified. A few weeks after declaring martial law in Baltimore,

Butler was transferred to the tip of the Virginia peninsula at Fort Monroe. Across the water in Norfolk secessionists were busy erecting batteries to ensure their control of the harbor; when three fugitives came into Butler's lines he found they had been employed there as laborers. Soon one of their masters, Col. Charles Mallory, came to demand the return of his property. Already having thought the proposition over, Butler decided it was no advantage to him to let any property used by the Confederates—be it a shovel, a mule, arms, or whatever—fall out of his hands once it had been seized. He told Mallory he deemed them contraband of war and would hold them and use them as such.[18]

Passing from mouth to ear among the slaves in its remarkable way, word quickly spread and a new road to freedom was opened as blacks began to come of their own volition into Butler's lines. First there were a dozen, then a score, and finally many hundreds, and not only were men coming, but their wives and children came with them, and Butler suffered them to stay. He saw the men employed in the duties of camp life—chopping wood, tending roads, and building wharves—and the women as cooks and laundresses, while the children were simply allowed to exist.

Maj. J. B. Cary of the Virginia Volunteers artillery came to see Butler and remind him of his obligation—that he was required to return the slaves under the Fugitive Slave Act. Well, Butler replied, that act did not apply to a foreign country, which Virginia claimed to be, and now Virginia "must reckon it as one of the infelicities of her position that in so far at least she was taken at her word." Butler thought it prudent he write Washington for guidance and addressed a letter to Secretary of War Simon Cameron which soon found its way to wider circulation in northern newspapers. Butler wrote: "As a matter of property to the insurgents it will be very great moment, the number I now have amounting, as I am informed, to what in good times would be of the value of $60,000." Reveling in almost Shakespearean conceit, Butler asked: "Are these men, women, and children, slaves? Are they free? . . . If property, do they not become the property of the salvors? But we, their salvors, do not need and will not hold such property . . . has not, therefore, all proprietary relation ceased. . . . I confess that my own mind is compelled by this reason to look upon them as men and women. If not free born, yet free, manumitted, sent forth from the hand that held them never to be reclaimed."[19]

Cameron replied: "Your position in respect to the Negroes who came within your line from the service of the rebels is approved. The department is sensible of the embarrassments which must surround officers conducting military operations in a State by the laws of which slavery is sanctioned."

A policy in regard to these "contraband" had received a nod from Washington, but no sooner was it given than Lincoln was looking for ways to take it back. On July 22 and 25 the Senate and House respectively passed resolutions sponsored by Kentucky's John Crittenden and Tennessee's Andrew Johnson affirming that the government had no intention "of overthrowing or interfering with the rights or established institution of States" but only fought "to defend and maintain the supremacy of the Constitution and to preserve the Union with all the dignity, equality, and rights of the several States unimpaired."[20]

In his August *Monthly* under the title "The War and Slavery," Douglass remarked: "The impression which our Government seeks to make upon slaveholders seems to be that slavery is safer in, than out of the Union." But the government had very little to show in the way of progress in its side of the war; the rebellion, in fact, had only grown stronger and had gained a decided advantage. "Who shall explain to us why this is so?" asked Douglass. "Our solution of the whole matter is this: The South is in earnest, and the North is not. The South is whole, and the North is half. The South has one animus, the North another." United after the cannonade at Sumter, the North now risked being broken as the outline of a new bargain began to cast its shadow, namely of a government that would do nothing in regard to slavery that could embarrass its efforts to accomplish a speedy settlement. In a passing notice to West India emancipation in the same issue of his *Monthly*, Douglass wrote: "There have been many mistakes and blunders made by the Government during the war, but none greater than those which have been made by Abolitionists who have taken this war as a trump of jubilee, and a release of themselves from the toil of anti-slavery agitation."

The government, in some fashion, was engaged in a war against slavery, and although slavery may yet have an existence, once conquered the Slave Power would "part with its prestige and sink into weakness." The government had need to hear the appeals of abolitionists, Douglass reiterated, "and we believe from what we have been recently informed by a gentleman fully initiated, the Government is willing to do all that shall be demanded by the people."

After Bull Run Lincoln attempted to enlist the service of Garibaldi for the Union, and an envoy was sent to meet with the Italian patriot on his farm at Caprena, where he was offered command of troops as major-general. "If your war is for freedom," he replied, "I am with you with twenty thousand men." But he could only be useful, he felt, if he were given supreme command over all Union troops with the right to proclaim emancipation. This scotched Garibaldi's participation for Lincoln, and Douglass bitterly observed, "We might as well remove Mr. Lincoln out of the President's chair, and respectfully invite Jefferson Davis or some other slaveholding rebel to take his place." A new element must be infused into the government without delay. "Let it be known," Douglass wrote, in "Progress of the War," in September 1861, "that the American flag is the flag of freedom to all who will rally under it and defend it with their blood. Let colored troops from the North be enlisted and permitted to share the danger and honor of upholding the Government. Such a course would revive the languishing spirit of the North. . . . It would lift the war into the dignity of war for progress and civilization. . . . It would bring not only Garibaldi and his twenty thousand Italian braves to our side, but what is more important still, our own sense of right, and the sympathy of enlightened and humane men through the world."

Lincoln's policy of putting down a "slaveholder rebellion" while saving slavery, in Douglass's view, carried anarchy into the very underpinnings of the nation's ordeal and gave its enemies the advantage of seeming to be merely fighting for the right to govern themselves. The policy of slave-catching and suppression of insurrection awarded them the sympathy of international opinion which should have accrued naturally to the government, inviting interference by European powers in the coastal blockade. Lincoln thereby demonstrated, said Douglass, that he did not yet know slavery and did not know slaveholders.

But the reversal at Bull Run was also spurring the debate in the Congress. Just as the letter arrived from Butler at the War Department asking for clarification, Congress was considering a Confiscation Act. Passed on August 6, it allowed for the confiscation of property (including slaves) used directly to support the Confederate war effort. Both parties in Congress agreed that they had no constitutional right to legislate against slavery, whether in war or in peace, but the Republican proponents of the measure held that Congress could punish the treason of individuals, a consequence which could be levied without detriment to the institution of slavery itself.

With the reorganization of the Union army ordered by Lincoln, Nathaniel P. Banks, a former Massachusetts politician and Speaker of the House, assumed command of the army led by Patterson, just as Frémont replaced Harney in the Department of the West. Promptly seizing fifteen fugitive slaves in his camp, Banks returned them to be summarily dealt with by their "loyal" masters. Douglass weighed these contradictory vacillations of the government amid the reversals in the field and concluded what he had feared, that the government was not disposed to carry the war to the "abolition point." He wrote in a letter to Rev. Samuel May, Aug. 30, 1861:

> Who would have supposed, that General Banks would have signalized the first week of his campaign on the Potomac by capturing slaves and returning them to their masters? He has done less to punish the rebels than to punish their victims. Only think too, of Fremont with Edward M. Davis for quartermaster, cooping up two fugitive slaves in the arsenal of St. Louis and when the poor fellows succeeded in getting away from these their federal and abolition friends, their loyal owners were assured that they might expect to be duly paid for their runaway chattels. . . . Looking at the government in the light of these and similar examples, and the fact that the government consents only that Negroes shall smell powder in the character of cooks and body servants in the army, my anti-slavery confidence is blown to the winds. I wait and work relying more upon the stern logic of events than upon any disposition of the Federal army towards slavery.

Already the hopes raised at the beginning of the war were dissipating, but Douglass declared he would never accept anything "short of an open recognition of the Negro's manhood, his rights as such to have a country, to bear arms, and to defend that country equally with others." But others evidently were "preparing for another attempt to preserve the liberty of white men at the expense of that of the black."[21]

As John C. Frémont arrived in St. Louis he established his headquarters in the abandoned mansion of J. B. Brant on Chouteau Avenue.

Soon he had surrounded himself with scores of aides, both civilian and military, including officers of German, Hungarian, and Polish origin. His chief of staff, Alexander S. Asboth, was a Hungarian who had been a revolutionary with Kossuth. Soon Frémont reached out to suppress all manifestation of Confederate support in St. Louis. Marching columns appeared daily in the streets and a select group became the Jessie Frémont Guard, named for the general's wife, commanded by Charles Zagoni, another Hungarian.

John C. Frémont.

Jessie Frémont was the socially prominent daughter of ex-Missouri senator Thomas Hart Benton, who had been leader of the antislavery Democrats until his death in 1858. She wed her husband when he was a lieutenant, and with her father's backing made him the famous and popular "Pathfinder." Now with the help of Frank Blair, and because of his wife's position in St. Louis society, John C. Frémont had assumed command in the Western Department, including the vital states of Missouri, Kentucky, Illinois, and Kansas west of the mountain ranges.

As he came to St. Louis, Frémont found that Lyon and Blair had routed the Missouri governor and the legislature, pursuing them into the southwest corner of the state. Lyon was in Springfield with thirty-five hundred troops; Frank Blair was away in Washington, while Franz Sigel with a brigade of seventeen hundred Germans was in Neosho, near the Arkansas border, hopeful of meeting up with his quarry before they could be reinforced. When word came that a Confederate army of twenty thousand under former Texas Ranger Ben McCulloch had entered the state, Sigel began retreating and in an engagement heralded in the northern press for military skill and daring, evaded capture at Carthage. Separated from his supply line at Rolla and heavily outnumbered, Lyon called on the new commander at St. Louis for reinforcements.

Preoccupied with threatening formations near Cairo, Frémont felt he could spare no troops, and ordered Lyon to retreat or rely on his own resources. With full details of the disaster at Bull Run reaching the west, Lyon and his officers and men were anxious for a chance to redeem the Union, but the problem that confronted McDowell now harried him. Two-fifths of his enlistments were expiring; men were leaving daily and the whole campaign would dissolve just at the moment victory might be attained. What is more, war correspondents, representing papers in Chicago, St. Louis, San Francisco, and the leading New York dailies, had followed his columns; thus the scrutiny and hopes of the North followed his every action. Despite the odds, Lyon felt honor bound not to give up southwest Missouri without a fight.

On August 8 an officers' council was held where Lyon outlined his view of the situation: To entrench at Springfield was impracticable and supply lines precarious; to retreat outright in face of the enemy's superior force might result in their being excessively put upon. Lyon therefore proposed a plan put forward by Franz Sigel—their forces would divide and attack the enemy from two directions. Confounded and disorganized, the opponent might allow them to retreat safely. Meanwhile the combined forces of Sterling Price and McCulloch bearing on them had adopted a strategy of surrounding Lyon and cutting him off, with the expectation that Frémont would be lured to reinforce him. After defeating Lyon they would retake Jefferson City with one wing of the army and with the other circle Frémont's flank and take a weakened St. Louis. Thus would Missouri be gained for the Confederacy.

The battle of Wilson's Creek, as it was called in the North, or Oak Hill in the South, produced casualties three times in excess those recorded twenty days earlier in Virginia. Although Sigel survived, his brigade was decimated. Lyon was slain toward the end of the morning, shot through the heart. Had he survived he would have been hailed as a hero; his body was to lie in the rotunda in Washington in view of a grieving president and public. But as the broken ranks of the Union forces retreated to Rolla where they entrained for St. Louis, a tactical victory went to the Confederates. With his militia Price occupied Springfield and prepared to march up to Jefferson City, while McCulloch withdrew to Arkansas.

Joining the struggle for Missouri on its western border was James Lane, who had been commissioned a brigadier-general but could not accept it lest he lose his seat in the Senate. Straddling the issue by signing official orders "J. H. Lane, Commanding Army of the Western Border," he began urging enlistments for an expedition through Missouri to the Gulf of Mexico and to New Orleans. When enlistment began in Kansas, Lane hurried to Fort Scott, where he assumed command of regiments officered by Montgomery and Charles R. Jennison. Finding Fort Scott indefensible, Lane set men to work on another twelve miles up that he dubbed Fort Lincoln. When work was being completed at the end of the month, Price, with ten thousand veterans of the battle on August 10, began his bid to retake the state.

As this news reached Frémont he began drafting a proclamation to scuttle recruitment under Price's banner and put the state under martial law; when he finished it he read it to his wife, who gave her approval. Then he handed it to an adjunct for publication. The document read in part: "All persons who shall be taken with arms in their hands within these lines shall be tried by court martial, and, if found guilty, will be shot. The property, real and personal, of all persons in the State of Missouri, who shall take up arms against the United States, or who shall be directly proven to have taken active part with their enemies in the field, is declared to be confiscated to the public use, and their slaves, if any they have, are hereby declared free men."[22]

Frémont immediately began acting on this basis, issuing a deed of manumission against the property of the former secretary of Missouri's governor, then taking the field as the adjunct of Sterling Price. The title read in part:

> Whereas, Thomas L. Snead, of the city and county of St. Louis, State of Missouri, has been taking an active part with the enemies of the United States, in the present insurrectionary movement against the Government of the United States; now, therefore, I, John Charles Fremont, Major-General commanding the Western Department of the Army of the United States, by authority of law, and the power vested in me as such commanding general, declare Hiram Reed, heretofore held to service or labor by Thomas L. Snead, to be free, and forever discharged from the bonds of servitude, giving him full right and authority to have, use, and control his own labor or service as to him may seem proper, without any accountability whatever to said Thomas L. Snead, or any one to claim by, through, or under him. And this deed of manumission shall be respected and treated by all persons, and in all courts of justice, as the full and complete evidence of freedom of said Hiram Reed.

Frémont's stroke was widely hailed and published through the North. In "General Fremont's Proclamation," in his *Monthly* of

October 1861, Frederick Douglass declared it "the most important and salutary measure which has thus far emanated from any General during the whole progress of the war. . . . [It] declares the slaves of all duly convicted traitors in the State of Missouri, 'free men.' They are not only confiscated property, but liberated men." Frémont's proclamation was reviled in the South, and two days later guerrilla chieftain M. Jeff Thompson issued a counter-order. For each man executed he would "HANG, DRAW AND QUARTER a minion of said Abraham Lincoln."23

Soon a letter from Lincoln reached Frémont in which the president counseled privately not to execute anyone "without having my approbation," for "the Confederates would very certainly shoot our best men in their hands, in retaliation, and so, man for man, indefinitely." Furthermore, Lincoln wrote, the manumission measure would "alarm our Southern Union friends, and turn them against us—perhaps ruin our rather fair prospect for Kentucky." He asked if Frémont might not modify that part of his proclamation to conform to the Confiscation Act passed by the Congress, which was applicable only to slaves used directly in war activity.24

Fremont refused without a public order to follow the suggestions, and dispatched his wife to meet with Lincoln to explain the importance of his proclamation to the war effort. The day after Jessie Frémont's visit Lincoln made his order public, reasoning that if permanent rules on property could be made by a general or president by proclamation then they could no longer say the nation was ruled by a constitution and laws. Worried about the effect on so-called loyal slaveholders in the border states, particularly Kentucky, Lincoln confided in a letter to his Illinois friend Oliver Browning on September 22, 1861, that if the edict had not been modified by him, "the very arms we had furnished Kentucky would be turned against us. I think to lose Kentucky is nearly the same as to lose the whole game. Kentucky gone, we can not hold Missouri, nor, as I think, Maryland. These all against us, and the job on our hands is too large for us. We would as well consent to separation at once, including the surrender of this capitol."25

Lincoln's government had taken many stumbling steps, but this was the worst blunder yet, lamented Douglass, who wrote: "The Government should have thanked their wise and intrepid General for furnishing them an opportunity to convince the country and the world of their earnestness, that they have no terms for traitors; that with them the heaviest blow is the wisest and best blow; and that the rebels must be put down at all hazards, and in the most summary and exemplary way. But, poor souls! Instead of standing by the General, and approving his energetic conduct, they have humbled and crippled him in the presence of his enemies."

Evidently then, said Douglass, there was a stronger bond subsisting between the loyal slaveholders and rebel slaveholders than with the government; from the war's commencement these border slave states had "been the mill-stone about the neck of the Government, and their so-called loyalty has been the very best shield to the treason of the cotton States."[26]

Northern Missouri was divided north from south, east from west by two rail lines and two navigable rivers; the Missouri, flowing on the western boundary then traversing the state at its midsection, and on its eastern border the Mississippi, the two rivers having their confluence but a dozen miles upstream from St. Louis. The presence of a guerrilla force with allegiance to the Confederacy organized under Thomas A. Harris necessitated a large troop presence north of the Missouri as a counterbalance. These guerrillas, together with those in the swamps in southwest Missouri and Confederate pressure against Cairo, had rendered Frémont incapable of reinforcing Lyon in August. Lyon had to be sacrificed, and his loss had raised the ire of Frank Blair, whose efforts and influential family had placed Frémont in command of the Western Department. Now Blair found himself outranked and his voice, recently so vital in the cause of the government, diminished, just as the entire imbroglio over Frémont's proclamation was coming to a head.

On the western border Lane's 'command' skirmished with the left wing of Price's army in a fight at Dry Wood Creek on September 7, after which Lane withdrew. He had learned, however, from prisoners that Price was headed to Lexington, a wealthy town and an important center between St. Louis and Kansas City on the Missouri River. Lane sent a messenger requesting reinforcements to Fort Leavenworth, whose commander then wired Frémont of developments. Frémont ordered two Union commanders in northern Missouri, Pope and Sturgis, to converge on Lexington and prevent its capture. Lane was ordered to fall back to Leavenworth.

Before the Union columns could converge, Price was ready for action and began besieging the Lexington entrenchments commanded by an ambitious Chicago politician with a colonel's commission, James A. Mulligan. The next day as the besieged town's need for water became acute, Price's big guns churned up the breastworks as some skirmishing took place among hedges and intervening buildings. As another day dawned Mulligan and his Irish brigade were surprised to see a line of hemp bales moving in irregular motion toward them. In the night the staple had been dragged down to the river, and made impervious by the Missourians by soaking them, and men with levers were now prying them forward. With no relief column approaching and unable to penetrate the barrier, Mulligan's only alternative was surrender.

Officers and men were dutifully filed out under the magnanimous supervision of Price to give up their guns. Meanwhile when ordered back on Leavenworth, Lane had complied, but at a leisurely pace. Two days after Price occupied Lexington, Lane came upon Osceola, an important wholesale distribution center on the headwaters of the Osage River. Provisioning his army with flour, sugar, molasses, coffee, and bacon, along with some barrels of brandy and camp equipment, Lane also found part of Price's ammunition stores. Impressing teams and wagons, the supplies were loaded, while a drumhead court sentenced nine citizens to be shot under Frémont's martial law provisions. Lane also invoked the barred portion, freeing slaves, all told two hundred, along with confiscating three hundred fifty horses and four hundred cattle. Ordering the destruction of the courthouse with all its records, he set all but three houses afire. On his retreat dozens of blacks were now flocking to join Lane's van at every crossroad, some bringing horses and wagons. Hurrying toward Kansas, Lane rallied audiences, upholding his method as the best way to end "treason" while calling for enlistments to form an army to sweep across the South. A *New York Times* reporter described Lane as a "Joe Bagstock Nero fiddling and laughing over the burning of some Missouri Rome;" while in Chicago, the newspapers reported John Brown Jr. had passed through with sixty men eager to join Lane.[27]

Almost without ammunition and transportation, and confronted by seventy thousand Union troops positioned at various locations throughout the state, no sooner had Price attained Lexington than he found he must retreat. He was forced to disband most of his army, grown to twenty-three thousand, telling them to take care of their arms, cherish a determined spirit, and hold themselves in readiness for another opportunity to join his standard. On September 27 he began to retrace the roads whence he had recently come.

The high expectations in Washington and the North occasioned by the appointment of Frémont now plummeted. He failed to aid Lyon, and was letting Price traverse more than half the state and walk out unscathed. Blair's increasingly harsh criticism had been temporarily silenced with his arrest. Once released, he was again directing attention to Frémont's shortcomings and the *Times* and the *Herald* were taking his side. Besides his perceived military failings, Frémont had been heedlessly extravagant with contractors, while allowing people of questionable reputation into his orbit. The only thing that could save him at this point was a victory. Writing to General-in-chief Scott, Frémont announced, "I am taking the field myself, and hope to destroy the enemy either before or after the junction of forces under McCulloch. Please notify the president immediately."[28]

Dividing his army of forty thousand into five divisions, each deployed across mid-Missouri, Frémont saw to it that they were placed so as to be able to coordinate their activities and consolidate within twenty-four hours. In addition, another five thousand Kansas men under Lane and Sturgis were on the western border. As Frémont prepared to leave St. Louis for Jefferson City he ordered the rearrest of his nemesis Frank Blair. Secretary of War Cameron, together with Adjutant General Lorenzo Thomas, now arrived bearing a letter from the president for the general. Catching up with him at the terminus of the Pacific Railroad and the headquarters of Asboth's divisional command at Tipton, the general was shown the letter ordering his removal. Frémont begged for a chance to prove himself now that his army was trained and in the field. As he appeared master of the campaign, Cameron pocketed the letter and headed back to Washington.

By October 24 Frémont's columns were positioned within sixty miles of Springfield, after his engineering corps performed a Herculean task at the bridge at Osceola, reconstructing in thirty-six hours an eight–hundred-foot span that had been burned out the previous month. The quick advance created an expectation in his soldiery that they would be in Memphis by Thanksgiving and in New Orleans for Christmas. The next day, significant as the anniversary of the Charge of the Light Brigade, Major Zagoni led a mounted attack on the west of Springfield, routing the defenders and running his banner up at the courthouse. Seizing more than $4,000 in gold from a bank, he withdrew, to return two days later when Frémont's van entered Springfield.

Presented with an extensive military panoply, the reoccupation of the Missouri town was heralded in Frémont's dispatches as "atonement" for Bull Run. Parades demonstrated the army's power and Frémont rode out to the battlefield where Lyon lost his life. Satisfied that Frémont had retrieved all that had been lost since August, and now held the key to regaining the state, it seemed to many a turning point had been reached in the crusade to restore the Union. Even with the barring of Frémont's proclamation, in Lane's camp were hundreds of freed blacks. When John Brown Jr. arrived in Kansas four hundred men offered to serve under him for an expedition into the South similar to Lane's. Charles Jennison, an abolitionist with a colonel's commission, urged his men to show "no hesitation regarding the rebels and the man who sympathizes with them." In a speech to his company quoted by Marx in *Die Presse*, December 11, 1861, titled "The Crisis on the Question of Slavery," Jennison said:

I declared to General Fremont that I would not have taken up arms had I believed slavery would survive this war. Slaves who belonged to the rebels keep taking refuge in our camp, and we shall defend them to the last fighter, and to the last bullet. I want no men who are not Abolitionists; I have no place for them, and I hope there are no such people among us, since everyone knows that the question of slavery lies at the root of this accursed war, that it constitutes its essence and idea. . . . And, if the government does not approve of my course, then it can revoke the commission it gave me, but if this happens, I shall go ahead on my own hook, even though I have only half a dozen men in all to begin with.

It was Marx's contention that the Frémont affair marked a new phase in the North's grappling with the root question of the war, and Marx remarked in the article, "In the border slave states, especially in Maryland, and to a lesser degree in Kentucky, the question of slavery has already been decided in practice. An immense ebb and flow of slaves has been observed there."

Eighty miles away on the Arkansas–Missouri border at Neosho, although lacking a quorum, both houses of the Missouri legislature passed an ordinance of secession and elected a senator to represent the state in the Confederate Senate. Price had the honor of firing a hundred-gun salute to mark the occasion, to whose prospects he had contributed so much. McCulloch was on hand for a day with five thousand troops before retreating, quarreling with Price over whether it was better to confront Frémont on one or the other side of the state lines. Price remained in Neosho before departing to make preparations to receive Frémont at Pineville. Meanwhile Frémont ordered Sigel forward with ten thousand men. As everything seemed to be turning to his advantage, in only a few more days, on November 2 a courier arrived with a letter from the President. Frémont was writing at a long table in his tent when he learned his command was ended. The courier's instructions had been to withhold the letter under three conditions—if Frémont had fought a winning battle, was himself personally engaged in battle, or was at the immediate front of the enemy. With none of these conditions met, the letter was conveyed. The order specified one of Frémont's divisional commanders, David Hunter, would replace him as head of the Department of the West.

While Hunter had not yet reached Springfield, Frémont called his officer staff together and read them Lincoln's letter. Asboth and Sigel offered to resign as the news spread through the camps and men gathered around campfires. Whole companies threatened to throw down their arms. With signs of mutiny, 111 officers presented themselves at Frémont's tent urging he lead the army against the enemy at once, thus invalidating Lincoln's order. Frémont granted they wait another day; if Hunter did not appear, they would advance.

The plan drawn up the next day called for crushing blows to fall on Price from all sides. Sigel and Lane would circle in the west and attack from the rear, Asboth would hit from the east, while John McKinstry and Pope smashed down from the north. With the hour imminent, Hunter arrived to assume command. The next morning Frémont, his guard, and some of his staff rode toward Rolla. With no enemy in front of him and the army demoralized, General Hunter now ordered retreat. It looked as though Union authority was collapsing in Missouri. Sigel would pronounce this episode "an Outrage without parallel in history."[29] Union sympathizers would join Hunter's retreat in droves, flocking into St. Louis. When Frémont arrived, he met his wife and entrained with her and two aides for the east.

Lane's retreating columns meanwhile were blackening the sky. With more slaves joining him, Lane organized them into a "Black Brigade" under a council of three chaplains. Soon their train was a mile long, with men, women, and children packed in wagons together with their meager belongings. That night after foraging for corn and slaughtering a "traitors" steer, encamping six miles from the Kansas line at Dry Wood Creek, the jubilant throng settled down. Before dawn they clambered back on the road, and but a few hours later entered Kansas in the morning sun. The Reverend Hugh Fisher, standing in his stirrups, called to the freed people: "In the name of the Constitution of the United States, the Declaration of Independence and the authority of General James H. Lane, I proclaim you forever free."[30]

On November 19 Lincoln replaced Hunter with Gen. Henry M. Halleck, moving Hunter to Lane's post in a new Department of Kansas. Unsettled by the foreclosure of his military career, Lane returned to his seat in the Senate.

Giving utterance to the verity of the age, Douglass suggested a new stage in the struggle between North and South had been reached signifying the beginning of the end of slavery. Under the title "General Fremont's Proclamation to the Rebels of Missouri" in his *Monthly* in October 1861, Douglass wrote:

> Fremont's celebrated Proclamation is by far the most important and salutary measure, which has thus far emanated from any General during the whole tedious progress of the war. . . . For many days after its publication . . . the deepest anxiety existed throughout the country to learn whether that remarkable and startling document was the utterance of the Major-General, or that of the Cabinet at Washington—whether if only from the former, the President would approve it or condemn it? The suspense was truly painful, and attested the vast importance attached by the public to the measure. . . .
>
> The lawyer has prevailed over the warrior. The President, of whom [I] would gladly speak naught but [well], has interposed, most unseasonably.

Then under the title "The Real Peril of the Republic," Douglass counseled:

> If this great American Government . . . shall now, in this the first great trial of its strength, [go] down into the depths of social confusion, and into the midnight of wild Anarchy and chaos, the fact will not be explained by the power and ability arrayed against it—

for the rebels are notoriously a miserable, ill clad, ill fed, ill armed and poverty-stricken set. If [we fail, we shall fail] by moral causes, not by outward strength, but by internal weakness. . . . [The] Government is still in bondage to fear, not that which the battlefield inspires, but of the political power of slavery. [They] are allowing [their] contempt for the rights of man, and [their] old scrupulous regard for the interests of slaveholders to control all [their] movements towards the rebels, hoping to gain by conciliation instead of conquering by arms. The future historian will look at the facts of this war for the suppression of rebellion with astonishment.

Finally under the headline "The Duty of Abolitionists in the Present State of the Country," Douglass asked what ought men and women who believe in the brotherhood of man, in

the right of every person to his own body . . . who believe that slavery ought at once and forever and everywhere be abolished—What ought such men and women to do in the present crisis? Our first business is to save our Government from destruction. This we firmly hold. The dangerous and demoniacal character of slavery, which has brought the present distress upon us, we have been endeavoring to expose. . . . This is our task; and had the nation hearkened to our warnings . . . our land would not now have been plunged into all the troubles of civil war, and the nation rolling in fraternal blood. We now have the war upon us, and the question is what does duty require of us now? We can do no more to save our guilty country from destruction, than by doing all we can to make the Government and people an abolition Government and an abolition people. . . . The first duty of the American people is to put down and utterly abolish slavery.

Another ill-fated turn had been taken on October 21 when McClellan ordered a "slight demonstration" against the Confederates in the town of Leesburg, Virginia, midway between Harper's Ferry and Washington. While Union regiments marched upriver on the Virginia side to engage the Confederate flank, other troops attacked across the river from Maryland against a brigade posted in a wood atop a hundred-foot bank called Ball's Bluff. Attacking from a hopeless position, more than half of those engaged became casualties, as survivors were driven into the river where they were killed outright or by drowning. Command of these troops fell to Col. Edward Baker, a former Illinois senator and intimate of Lincoln. Baker was killed, and Lincoln was grief-stricken at the loss of his friend and the disastrous outcome of the battle. With the Potomac blockaded below the capital, while forty thousand Confederate troops held Centreville, the Army of the Potomac had been humiliatingly bloodied only forty miles from Washington. Adding to the gloom, it was at the end of the following week when Frémont learned of his untimely recall and there was near mutiny in the Union army in Missouri.

Douglass, who deserves to be consulted as fully as any historian of the time or since, wrote in his November *Monthly* under the title "Fremont and Freedom—Lincoln and Slavery": "There is a deep conviction in the public mind that the opposition to the rising young General, arises out of other than honorable and patriotic motives— motives which, if persisted in may lead to the complete and hopeless demoralization of the army, and pave the way for a civil war within a civil war." Douglass continued:

> Harsh and disagreeable as these words sound even to our own ear, truth consults no man's taste, and events enter without begging any man's permission. If Government shall humble merit, and exalt imbecility, displace Generals who are a terror to the rebels, and promote those who excite no alarm, and thus in fact allow the rebels to select only such Generals as they can whip, as well as choose the ground upon which they can whip them most easily— it would not be strange if the patience of the people should break down, and if some determined man of military genius should rise out of the social chaos, and displace the civil power altogether. Such things have taken place before, and what has been done may be done again. The voice of history and of human nature itself cries aloud with unsparing energy. "Have a care—have a care!"

Scheduled to lecture in Syracuse, Frederick Douglass arrived on November 14 to find a notice on every street corner beside those announcing his appearance. The bill read:

NIGGER FRED COMING
This reviler of the Constitution, and the author of "Death in the Pot!" and who once in this city called George Washington a Thief! Rascal!! And Traitor!!! Is advertised to lecture on "Slavery" again on Thursday and Friday evenings of this week at Wieting Hall!!!

Shall his vile sentiments again be tolerated in this community by a constitutional liberty-loving people? Or shall we give him a warm reception at this time for his insolence, as he deserves? Rally, then, one and all, and drive him from the city! Down on the arch fugitive to Europe, who is not only a coward but a traitor to this country!!!— Rally Freemen![31]

Before the appointed date the owner of the hall, Dr. Wieting, received repeated visits when it was put to him that he must close his doors to the occasion. To these impositions, he replied that "even if its walls were to be battered down by cannon" he would refuse to do any such thing. When reminded of Douglass's race, he said that "his principles of freedom applied to humanity, not to color."[32]

Anticipating mob action, the mayor appointed fifty special police to guard inside the hall and mustered a military guard to stand before the building. Gerrit Smith, passing through that afternoon and hearing of the proposed fracas, determined to stay on. At eight in the evening the Rev. Samuel May and Douglass entered the hall to a rousing ovation. Douglass's lecture was only the first of a series May announced to be held in Wieting Hall; the following lecturers were to be William Lloyd Garrison, Wendell Phillips, Gerrit Smith, and Parker Pillsbury. Douglass's lecture lasted ninety minutes, after which, sitting on the platform, Smith gave a short address.

The next day May wrote to Garrison of the "glorious triumph" in Syracuse, to which he received the reply, "Honor to your city." In Douglass's view, the aim of the threat "was to humble and mortify Frederick Douglass," but they had signally failed in silencing his "uninfluential Negro voice," and now had a whole series of abolition lectures to scuttle if they dared.[33] Noting the contrast between mobs that closed abolition meetings the previous year and this was satisfying, but Douglass particularly was heartened by coming again into the orbit of his estranged Garrisonian friends. He wrote: "Every man who is ready to work for the overthrow of slavery, whether a voter, or non-voter, a Garrisonian or a Gerrit Smith man, black or white is both clansman and kinsman of ours. We form a common league against slavery, and whatever political or personal differences, which have in other days divided and distracted us, a common object and a common emergency makes us for the time at least, forget those differences, and strike at the common foe—and to give victory to the common cause."[34]

That same week an Emancipation League was organized in Boston, which promptly selected Douglass to deliver the fourth lecture in its course. Framed by Samuel Gridley Howe, Franklin Sanborn, George Stearns, Wendell Phillips, and William Lloyd Garrison, among others, the League's function was one "of urging upon the people and the Government emancipation of the Slaves, as a measure of justice, and military necessity; it being the shortest, cheapest, and least bloody path to permanent peace, and the only method of maintaining the integrity of the nation." This opportunity of seeing the end of slavery, Douglass underlined, had been delivered to the nation by slaveholders themselves, and he had written to Gerrit Smith earlier that August: "I can't help feeling that we are not far from that event," adding that, "for the sake of Eternal Justice," his friend must "continue to urge upon the Government with all your powers of persuasion the duty of breaking the yoke. Now is the time. . . . It [was] weak as well as wicked not to strike now. Grandly Phillips comes up to the work. You and he are fortunate, and the cause is fortunate in that your thoughts now readily float to the millions through columns of the most widely circulating journals."[35]

The *New York Tribune* estimated that fifty thousand people heard Wendell Phillips's speeches and lectures in the winter of 1861–62, and no less than five million read the same as they were published in the dailies and journals. In New York and again in Boston in December 1861, Phillips offered his analysis of the war and its potential in a speech titled "The War for the Union." He began: "You and I come here tonight, not to criticize, not to find fault with the Cabinet. We come here to recognize the fact, that in moments like these the statesmanship of the Cabinet is but a pine shingle upon the rapids of Niagara, borne which way the great popular heart and the national purpose direct. It is in vain now, with these scenes about us, in this crisis, to endeavor to create public opinion; too late now to educate twenty millions of people. Our object now is to concentrate and to manifest, to make evident and to make intense, the matured purpose of the nation."

The South calculated that King Cotton held the stronger suit. But there never would have been this outbreak if, when "the thing first showed itself," Phillips continued,

> Jefferson Davis and Toombs and Keitt and Wise, and the rest, had been hanged for traitors at Washington, and a couple of frigates

anchored at Charleston, another couple in Savannah, and half a
dozen in New Orleans. . . . But you know we had nothing of the
kind, and the consequence is, what? Why, the amazed North has
been summoned by every defeat and every success, from its work-
shops and its factories, to gaze with wide-opened eyes at the lurid
heavens, until at last, divided, bewildered, confounded, as this
twenty millions were, we have all of us fused into one idea, that the
Union meant justice—shall mean justice—owns down to the Gulf,
and we will have it. What has taken place meanwhile at the South?
Why, the same thing. The divided, bewildered South has been summoned also out of her divisions by every success and every defeat
(and she has more of the first than we have), and the consequence
is that she too is fused into a swelling sea of State pride.[36]

Two equally problematical events for the war Lincoln's government was waging happened in the second week of November only
days apart. On November 8 a U.S. naval captain, Charles Wilkes,
stopped and boarded the *Trent*, a British steamer, in the Bermuda
channel. Two Confederate ministers, James Mason and John Slidell,
and their secretaries on board were seized and arrested as contraband, they being the "embodiment" of diplomatic dispatches. The
diplomats, headed for European capitals to represent the
Confederacy, were removed to Boston, while the British steamer was
allowed to proceed. The day before, a Union naval task force, after
encountering a gale off Cape Hatteras, steamed into South Carolina's
Port Royal Bay, confronting two defending Confederate forts.
Steaming in an oval pattern and pounding them broadside, both were
knocked out in only four hours. The defenders and all of the white
population fled inland, leaving the coastal islands and the connecting
waterways in Union hands, along with whole plantations embracing
as many as ten thousand slaves.

As news of the *Trent* affair flashed over the wires Captain Wilkes
was received in Boston with enthusiasm—here was a military commander willing to take action, a fitting rebuff to Britain's recognition
of the state of belligerency. Under international law, Britain's act,
coming after Lincoln ordered a blockade of the entire southern coast,
conferred legitimacy on the Confederacy, with the right to contract
loans, purchase arms, and commission cruisers with the power of
search and seizure. The House of Representatives passed a resolution
lauding Wilkes, while in England the Parliament expressed outrage

for the insult to the Union Jack. An ultimatum quickly arrived demanding an apology and release of the Confederate diplomats. Soon British troops were being sent to forts in Canada, its Atlantic fleet strengthened, and shipments of saltpeter, vital to the Union war effort, held in British ports. Now the public began having second thoughts; the government, Lincoln understood, had the capacity for only "one war at a time." Finally, to defuse the crisis without appearing to bow to the ultimatum, Seward acknowledged the U.S. had violated international law by failing to bring the *Trent* into port for adjudication before a prize court. The Southerners would be yielded on grounds that Wilkes acted without instructions. After a Christmas Day meeting between Lincoln and his cabinet, Mason and Slidell resumed their voyage to diplomatic posts in Europe.

Commenting on the *Trent* affair and the capture of Port Royal, Phillips, fully the equal of Douglass in assessing the fulcrum upon which the future was balanced, said:

The only danger of war with England is that, as soon as England declared war with us, she would recognize the Southern Confederacy immediately, just as she stands, slavery and all, as a military measure. As such, in the heat of passion the English people would allow such a recognition even of a slaveholding empire. War with England insures disunion. When England declares war, she gives slavery a fresh lease of fifty years. Even if we have no war with England, let another eight or ten months be as little successful as the last, and Europe will acknowledge the Southern Confederacy, slavery and all, as a matter of course. Further, any approach toward victory on our part, without freeing the slave, gives him free to Davis. So far, the South is sure to succeed, either by victory or defeat, unless we anticipate her. Indeed, the only way to break the Union, is to try to save it by protecting slavery. Unless we emancipate the slave, we shall never conquer the South without her trying emancipation. Every Southerner, from Toombs up to Fremont, has acknowledged it. Do you suppose that Davis and Beauregard mean to be exiles, wandering condemned in every great city of Europe, in order that they may maintain slavery and the Constitution of '89? They, like ourselves, will throw everything overboard before they will submit to defeat. I do not believe, therefore, that reconciliation is possible, nor do I believe the cabinet have any such hopes.

But the administration had provided only contradictory and vacillating indications that it intended to act in a decisive way to strike at slavery and remove it as a cornerstone in the nation's life. Phillips continued:

> If we look to the West, if we look to the Potomac, what is the policy? If, on the Potomac, with the aid of twenty governors, you assemble an army, and do nothing but return fugitive slaves, that proves you competent and efficient. If, on the banks of the Mississippi, unaided, the magic of your presence summons an army into existence, and you drive your enemy before you a hundred miles farther than your second in command thought it possible for you to advance, that proves you incompetent, and entitles your second in command to succeed you. Looking in another direction, you see the government announcing a policy in South Carolina. What is it? Well, Mr. Secretary Cameron says to the general in command there: "You are to welcome into your camp all comers; you are to organize them into squads and companies; use them any way you please—but there is to be no general arming." That is a very significant exception. The hint is broad enough for the dullest brain. But I suppose there is to be a very particular arming. But he goes on to add: "this is no greater interference with the institutions of South Carolina than is necessary—that war will cure." Does he mean he will give the slaves back when the war is over? I don't know. All I know is, that the Port Royal expedition proved one thing: it laid forever that ghost of an argument, that the blacks loved their master, it settled forever the question whether the blacks were with us or with the South. My opinion is, that the blacks are the key to our position. He that gets them wins, and he that loses them goes to the wall. Port Royal settled one thing: the blacks are with us, and not with the South. At present they are the only Unionists. I know nothing more touching in history, nothing that art will immortalize and poetry dwell upon more fondly—I know no tribute to the Stars and Stripes more impressive than that incident of the blacks coming to the waterside with their little bundles, in that simple faith which had endured through the long night of so many bitter years. They preferred to be shot rather than driven from the sight of that banner they had so long prayed to see.[37]

Twelve

Buffeted by Sleet
and by Storm

Help me to dodge the nigger—we want nothing to do with him. I am fighting to preserve the integrity of the Union. . . . To gain that end we cannot afford to mix up the negro question.
  —Gen. George B. McClellan to New York Democrat Samuel Barlow, November 8, 1861

A S THE YEAR 1861 DREW TO A CLOSE, THE SLAVE CONTINUED TO harry the Lincoln administration and its armies alike—while in the wings, George B. McClellan began to be mentioned as the next presidential candidate for the Democratic Party. Hailed as the savior of the Union and lionized in the press when he came to Washington after Bull Run, McClellan was said to carry "an indefinable air of success about him and something of the 'man of destiny.'" The adulation he received clearly went to his head. "I find myself in a strange position here," he had written to his wife; "president, Cabinet, Genl. Scott & all deferring to me. By some strange operation of magic I seem to have become *the* power of the land." He would, he informed his wife, "carry this thing 'en grand' & crush the rebels in one campaign."[1]

But McClellan was soon to manifest behavior that would mark him, in the words of Ulysses S. Grant, as "one of the mysteries of the war," displaying a chronic lack of temerity before the foe. Despite the proximity of Beauregard and Johnston to Washington, and the block-ade by Confederate forces of the Potomac, for seven months McClellan temporized, estimating the troop strength of the enemy near Manassas at 150,000, when in fact they had only 45,000 to oppose his 120,000. While Lincoln cautioned him that leading Republicans' call for action was "a reality, and must be taken into account," McClellan seethed with disgust "with this imbecile admin-istration." The president, he wrote, echoing the view of the South, "is nothing more than a well meaning baboon," and his cabinet "some of the greatest geese I have ever seen. . . . Seward the meanest of them all."[2]

Radical Republicans were now calling for the government to adopt an emancipation policy and for arming the freedmen. On December 1, without prior consultation with Lincoln, Secretary Cameron endorsed such moves in his annual report. His report stat-ed: "Those who make war against the Government justly forfeit all rights of property. . . . It is as clearly a right of the Government to arm slaves, when it may become necessary, as it is to use gunpowder taken from the enemy." An astonished Lincoln ordered the report with-drawn and the offending words deleted, but it had already reached print in some newspapers. Two days later, in his annual message to the Congress, Lincoln attempted to mollify welling sentiment against slavery, saying, "I have been anxious and careful" lest this war "degenerate into a violent and remorseless revolutionary struggle," advancing a recommendation of colonization for slaves and for free blacks to ameliorate the issue looming before the country. By a solid majority on December 4 the House voted not to reaffirm the Crittenden resolution, which disavowed an antislavery purpose in the war. Two days later Gerrit Smith issued a letter to Pennsylvania rep-resentative Thaddeus Stevens praising the congressman for calling for action on the question. Stevens had urged that the war become "a radical revolution" whereby the nation could "remodel [its] institu-tions," saying, "Free every slave—slay every traitor—burn every rebel mansion, if these things are necessary to preserve this temple of free-dom." Smith allowed that he "never contended that the nation could . . . act, in time of peace, directly on the great system of slavery, I have long contended that it could do so in time of war."[3]

Commending Smith for his timely letter, Douglass wrote on December 22: "It is thus far the noblest paper—and the most thrilling—I have met with on the war, either from your pen or that of any other statesman. As you ply the knife to our rotten Government, I shudder with a feeling of something like despair of finding any sound place upon which to build a hope of national salvation. I am bewildered by the spectacle of moral blindness, infatuation and help-less imbecility which the Government of Lincoln presents."[4]

What was the secret of Lincoln's subservience to slavery? Douglass asked.

The Northern and Western Members of Congress see and under-stand the cause of this rebellion, and know the true remedy; but the difficulty is, that the President and his Cabinet seem not to have ears for the anti-slavery voices that reach them.—The little finger of Virginia, Kentucky and Tennessee is greater than the loins of the whole loyal North. These border slaveholding States are just now playing the same game that distinguished them at the begin-ning of the rebellion. They act now as a grand breakwater against the anti-slavery tide which would speedily sweep away every ves-tige of the rebellion. These States, under the fair seeming garments of loyalty, are to-day, just as they were six months ago, the very best protection to the rebel cause.[5]

With the president and the Republican Congress locked in strug-gle on the issue of slavery, on January 11 Lincoln dismissed Secretary Cameron, appointing him ambassador to Russia, and brought in his stead Buchanan's last attorney general and a Douglas Democrat, Edwin Stanton. The Radicals in Congress, however, produced a new weapon in the just appointed Committee on the Conduct of the War, chaired by Ben Wade. The committee came into being to investigate Union losses at Bull Run and Ball's Bluff, but would hold hearings throughout the war's duration.

The general in command at Ball's Bluff, whom Lincoln trusted, was Charles P. Stone, a regular army officer who helped maintain order in Washington at the outbreak of war. Stone was appointed to lead regiments from New England and New York twenty miles above Washington when two fugitive slaves from a nearby Maryland farm came into his camp, finding refuge among soldiers of the 20th Massachusetts. Stone sent an officer with an armed squad to force out the fugitives and return them to their master. Infuriated soldiers wrote Governor Andrew, who immediately sent a letter of reprimand to Col. William Lee, their regimental commander. Lee passed the letter on to General Stone, who wrote Andrew reminding him of his government's policy and advising him to butt out of army affairs. Stone soon found himself denounced in a speech by Charles Sumner on the Senate floor. Then the catastrophe at Ball's Bluff occurred where Stone, at McClellan's direction, ordered Colonel Baker to attack across the Potomac.

The political importance of ex-senator Baker, the aggravating circumstance of the general's argument with two powerful Republicans, and the incidental capture of the 20th Massachusetts' Colonel Lee, who was lodged in prison in Richmond and threatened with hanging in reprisal for the expected execution of a Confederate officer of a privateer recently captured—cast a pall over Stone's military career. Then it was disclosed that secret messages had been exchanged across the Potomac possibly warning the enemy of the attack on Ball's Bluff.

The first subject before Wade's committee was Stone, whom it deemed disloyal, and it communicated its suspicions to War Secretary Stanton. At two in the morning one January night the general was awakened to hear he was under arrest for treason. McClellan first attempted to protect his subordinate, but realizing Stone was merely the committee's surrogate target, he supplied additional damning information to Stanton. The hapless Stone, locked up for six months in Fort Lafayette, was never formally charged and when released was restored to minor commands, his army career a shambles.

At National Hall in Philadelphia on January 14, 1862, Frederick Douglass delivered a speech titled "The Reasons for Our Troubles," telling his audience that thus far the government's efforts to put down the rebellion had struck "wide of the mark, and very feebly withal—that the temper of our steel has proved much better than the temper of our minds." He concluded his remarks:

> I am still hopeful that the Government will take direct and powerful abolition measures. That hope is founded on the fact that the Government has already traveled further in that direction than it promised. . . . No President, no Cabinet, no army can withstand the mighty current of events. . . . The first flash of rebel gunpowder, ten months ago, pouring shot and shell upon the starving handful of men at Sumter, instantly changed the whole policy of the nation. Until then, the ever hopeful North, of all parties, was still dreaming of compromise. The heavens were black, the thunder rattled, the air was heavy, and vivid lightning flashed all

around; but the sages were telling us there would be no rain. But all at once, down came the storm of hail and fire.

And now behold the change! Only one brief year ago, the great city of Boston, the Athens of America, was convulsed by a howling pro-slavery mob, madly trampling upon the great and sacred right of speech. It blocked the streets; it shut up the halls; it silenced and overawed the press, defied the Government, and clamored for the blood of Wendell Phillips, a name which will live and shine while Boston is remembered as the chief seat of American eloquence, philanthropy and learning. Where is that mob to-night? You must look for it on the sacred soil of old Virginia.

Nothing stands to-day where it stood yesterday. Humanity sweeps onward. Tonight with saints and angels, surrounded with the glorious army of martyrs and confessors, of whom our guilty world was not worthy, the brave spirit of old John Brown serenely looks down from his eternal rest, beholding his guilty murderers in torments of their own kindling, and the faith for which he nobly died steadily becoming the saving faith of the nation. He was "justly hanged," was the word from patriotic lips two years ago; but now, every loyal heart in the nation would gladly call him back again. Our armies now march by the inspiration of his name; and his son, young John Brown, from being hunted like a felon, is raised to a captaincy in the loyal army.[6]

In February Lincoln forced McClellan to commit the Army of the Potomac, but to the President's astonishment the general designed an oblique drive on Richmond by way of the Virginia Peninsula. McClellan's grand strategy, involving a transfer by flotilla—over 120,000 troops, 14,592 draft animals, 1,200 wagons, and 44 artillery batteries—was to march up the peninsula, while McDowell's corps came down from Fredericksburg to assist him. His rationale was twofold: first, he had no inclination to assault Confederate entrenchments at Manassas, and second, only two streams traversed the approach to Richmond while the more northern invasion route threw six streams across his path. It was also thirty miles shorter, while on the peninsular route too gunboats could lead on the James River, reducing Confederate defenses. Eager for action but not sharing his general's assessment, Lincoln assented to McClellan's plan on the promise that adequate troops stay behind for Washington's defense. Indeed, Lincoln had admonished that by going down the Chesapeake Bay "in search of a field, instead of fighting near Manassas" McClellan was "only shifting, and not surmounting, a difficulty"; he would "find the same enemy, and the same, or equal intrenchments, at either place."[7]

Just as the armada was getting under way, Union troops, after routing Confederates in Kentucky, were bearing down on Fort Henry in Tennessee. Success there was followed quickly by an assault on Fort Donelson where Brigadier General Grant spurned a proposition from the former Secretary of War John Floyd. Resigning his command lest he face arrest and imprisonment for actions during the Buchanan administration, Floyd ran off with 1,700 troops to Nashville as more than 13,000 of his command laid down their arms to Grant on February 16.

The Union victories on the Cumberland "had sent the echo back to Albemarle," Edward Pollard remarked in his *Southern History of the War*, as Burnside captured Roanoke Island in the east. With Grant now converging on Nashville, that metropolis was attended with "scenes of panic and distress . . . unparalleled in the annals of any American city. . . . An earthquake could not have shocked the city more." Pollard continued: "Buffeting sleet and storm, and by forced marches the enemy had seized Bowling Green, while Sigel fell suddenly upon Springfield; the enemy's gunboats threatened Savannah, and Gen. Butler hurried off his regiments and transports to the Gulf, for an attack *via* Ship Island upon New Orleans."

Throughout the month as the Union offensive beset the Confederacy "from Hatteras to Kansas," Frederick Douglass traveled over one thousand miles in an energetic campaign. On February 12 he delivered a speech in Boston's Tremont Temple before the Emancipation League titled "The Future of the Negro People of the Slave States," in which he predicted that the colonization scheme being advanced by Lincoln would become like the Fugitive Slave Law—dead upon the statute book, "having no other effect than to alarm the freed men of the South and disgrace the Congress by which it is passed."[8] Then he spoke in Milford, followed by an appearance in Providence. Six days after his Boston address he delivered the same speech at Cooper Institute presided over by Henry Highland Garnet. Then Douglass spoke in Jersey City, and subsequently at a string of engagements across New York State, in Naples, South Livonia, Hemlock Lake, Fowlerville, and Conesus. In his March *Monthly* he reported tens of thousands had heard him, and "the impression by all we have seen and heard is that the people are all ready to sweep slavery from the country would the Government lead off or stand out of the way."

When on March 6 Lincoln urged the Congress to pass a joint resolution affirming that the United States would "cooperate with any state which may adopt gradual abolishment of slavery, giving to such state pecuniary aid," Wendell Phillips hailed it as a sign of progress. Before the Emancipation League he said: "If the President has not entered Canaan, he has turned his face Zionward." Later in the speech, to the merriment of his audience, he added: "I observe that the cautious, and careful, and amiable, and good natured President, in his message to the Border States, did not speak of the 'abolition' of slavery—that is Garrison's phrase; he talked of 'abolishment.' Well, it is no matter, if he likes that way of spelling it better."[9]

In the meantime, just having finished its refitting into an iron-plated vessel with ten guns and a battering ram, the captured warship *Merrimack*, rechristened the *Virginia*, had steamed out of the former Gosport Naval Yard, quickly seizing supremacy in Hampton Roads, thereby threatening McClellan's peninsular campaign even before it began. First it rammed the twenty-four-gun *Cumberland*, sinking it, and then blew the fifty-gun *Congress* out of the water. But on March 9, as Navy Secretary Gideon Welles tried to calm the nerves of the cabinet, the Union ironclad *Monitor*, just completed at the Brooklyn Navy Yard—with only one gun but with greater maneuverability—dueled with the *Virginia* in Hampton Roads, fighting to a draw. The meeting of the two vessels revolutionized naval warfare, but of more urgent interest, McClellan was now able to put his grand scheme in motion.

Days after these events Congress spelled out the government's stance regarding "contrabands," prohibiting the armed services from returning fugitive slaves to any owner excepting those remaining "loyal," stipulating that any officer violating the law could be discharged and be forever ineligible for any appointment in the military. In Washington to deliver two lectures, Wendell Phillips was lauded on the Senate floor and introduced as a distinguished antislavery advocate. Vice president Hannibal Hamlin left his chair to greet him, and the speaker, Galusha Grow, hosted him at a dinner party. On successive evenings Phillips lectured to overflow audiences; in one of these he addressed the leaders directly:

> Gentlemen of Washington! You have spent for us two million dollars per day. You bury two regiments a month, two thousand men by disease without battle. You rob every laboring man of one-half of his pay for the next thirty years by your taxes. You place the curse of intolerable taxation on every cradle for the next generation. What do you give us in return? What is the other side of the balance sheet? The North has poured out its blood and money like water; it has leveled every fence of constitutional privilege, and Abraham Lincoln sits today a more unlimited despot than the world knows this side of China. What does he render the North for this unbounded confidence? Show us something; or I tell you that within two years the indignant reaction of the people will hurl the cabinet in contempt from their seats, and the devils that went out from yonder capital, for there has been no sweeping or garnishing, will come back seven times stronger, for I do not believe that Jefferson Davis, driven down to the Gulf, will go down to the waters and perish as certain brutes mentioned in the Gospel did.[10]

In an interview with Lincoln, who had come out one night to hear him, Phillips urged the dismissal of Seward for his conciliatory attitude, as it was Seward who was principally seen to be behind McClellan's inactivity. Then Phillips directly leveled with the president, telling him that if he started with the experiment of emancipation, and honestly devoted his energies to making it a fact, he would deserve to hold the helm until the experiment was finished—that "the people would not allow him to quit while it was trying." Even though this affirmation was not yet for Lincoln's ear, Phillips discerned he had moved. Previously, in Boston, regarding the president's March 6 message to the border states, he stated its implication was: "Now is your time. If you want your money, take it, and if, hereafter I should take your slaves without paying, don't say I did not offer to do it."[11]

Leaving Washington, Phillips took his message to the west, appearing before increasingly large and sympathetic audiences. Again, in an appearance at Pike's Opera House in Cincinnati, proslavery rowdies confronted him. As he began, cries of "Egg the nigger Phillips" and "Down with the traitor" went up. He was waiting for the disturbance to subside with his usual self-possession when a stone flung from the auditorium meanly skimmed Phillips's head, crashing among the footlights. Hard-fisted men now pressed down the aisles, yelling, "Put him out," "Tar and feather him," "Lynch the traitor." At this Phillips's bodyguard barred the way; the gaslights were turned off and he was guided safely onto the street. After several more engagements in other cities, Chicago was reached, where another attempt at disruption did not succeed. In his correspondence to the *Liberator* Phillips wrote: "You have no idea how the disturbance has stirred the West. I draw immense houses, and could stay here two months, talking every night, in large towns, to crowds."[12]

A bill for the abolition of slavery in Washington, D.C., had been introduced at the beginning of March by Senator Wilson, and throughout the month he and Sumner spurred on the effort in the Congress. The bill passed in a vote in the Senate on April 3 and was hailed by Sumner as "the first installment of the great debt which we all owe to an enslaved race, and will be recognized as one of the victories of humanity."[13] Exalting at his desk in Rochester, Douglass sent a missive off to the Massachusetts senator: "I want only a moment of your time to give you my thanks for your speech in the Senate on the Bill for the abolition of Slavery in the District of Columbia. I trust I am not dreaming but the events taking place seem like a dream. If Slavery is really dead in the District of Columbia, and merely waiting for the ceremony of 'Dust to dust,' by the president, to you, more than to any other American statesman, belongs the honor of this great triumph. . . . I rejoice for you. You have lived to strike down in Washington, the power which lifted the bludgeon against your own free voice."[14]

Only days after this emancipation act, armies of the United States met with armies of the Confederate States in a severe trial, with over 100,000 combined taking part. In an effort to replicate the rout at Bull Run, Albert Sidney Johnston and P. T. Beauregard surprised Union regiments under Grant near a church in southwestern Tennessee called Shiloh, assaulting them at daybreak in three parallel columns. The first day of the battle looked like a resounding southern victory as Grant's army all but disintegrated. But by nightfall, as Grant's troops fell back to Pittsburg Landing on the Tennessee River, another federal corps commanded by Don Carlos Buell crossed in reinforcement. Despite appalling losses, the combined commands readied for counterattack in the morning.

On the second day of battle the Southerners, facing unsustainable casualties, suddenly began withdrawing to their base at Corinth,

The Battle of Shiloh, April 6–7, 1862.

Mississippi. Grant lost 13,000 men and was superseded by Halleck; while Confederate casualties, with Albert Sidney Johnston killed in the first day of fighting, grew to over 10,000. On the first anniversary of war the fighting had reached unparalleled intensity; wave after wave of soldiery had clashed at Shiloh at point-blank range, while boys on both sides fled to the rear after "seeing the elephant" (in the parlance of the time) as commanders toiled feverishly to restructure shattered brigades.

As reports of the clash on the Mississippi and Tennessee border resounded, on April 11 the House passed the bill abolishing slavery in the capital city. The next day Lincoln's friend and confidant, Sen. Oliver Browning of Illinois, brought the bill to the executive mansion to lay it before him for his signature. Although dissatisfied with it, he would sign it, Lincoln allowed, but not until he could go talk with old Governor Wickliffe and give him warning to send two slaves, his personal servants, back to Kentucky. Lincoln favored gradual manumission in the District, terminating perhaps in twenty years, but did manage to have $100,000 included in the bill, to which the Congress added $500,000 to colonize the freed blacks. The bill stipulated $300 was payable to the owner of each freed slave. On April 16 Lincoln signed the bill outlawing slavery in the District of Columbia.

All through that month Henry David Thoreau lingered in bed in his mother's house till his death on May 6. An assessment of his life and death must tell of an utterly integral link to another individual that is altogether compelling. There is no doubt Thoreau found all his deepest requirements satisfied in John Brown, as no other person did, not even Emerson. Even before Brown had been tried and sentenced, the writer was profoundly moved to give voice to the martyr's apotheosis. Thoreau had written in his *Plea for Captain John Brown*: "This event advertises me that there is such a fact as death—the possibility of a man's dying. It seems as if no man had ever died in America before, for in order to die you must first have lived. I don't believe in the hearses, and palls, and funerals that they have had. There is no death in the case, because there had been no life; they merely rotted or sloughed off, pretty much as they had rotted or sloughed along. No temples wail was rent, only a hole dug somewhere."[15]

Ellery Channing reported he'd seen his friend's "hands involuntarily clenched together at the mention of Captain Brown"; but after Brown's execution Thoreau refused to accept that he had died at all. In his *Last Days of John Brown*—a compilation of the author's thoughts from his journal sent to the memorial service at Brown's graveside, which was read for him and published that December 1859 in the *Liberator*—Thoreau sounded his transcendental idea: "On the day of his translation, I heard, to be sure, that he was hung, but I did not know what that meant; I felt no sorrow on that account; but not for a day or two did I even hear that he was dead, and not after any number of days shall I believe it. Of all the men who were said to be my contemporaries, it seemed to me that John Brown was the only one who had not died."[16]

Thoreau was too clear-sighted not to have traced in his own lingering death the singular and meteorlike passing of the hero, even as the names and letters of other deaths were being spelled out in distant clashes. And so it is suggested, while perhaps he died from tuberculosis, in fact John Brown signified for him expiation, and he felt that he too must die.

Louisa May Alcott wrote of her Concord neighbor's death that spring in a poem titled "Thoreau's Flute":

> Spring comes to us in guise forlorn,
> The blue-bird chants a requiem,
> The willow-blossom waits for him,
> The genius of the wood is gone.[17]

Coming under increasing pressure from his party, and after consulting with the Committee on the Conduct of the War, Lincoln sought to gain a firmer hand with his generals. Appointing four new corps commanders to serve under McClellan, the president removed from the leader the coveted title "general in chief." This was followed by the announcement that a new military department would be created in western Virginia, an arrangement clearly meant to appeal to the rising sentiment to make the war aim one of emancipation, as Frémont was named its commander. By spring 1862 Union military fortunes seemed to be improving. In fact, before McClellan could set his plan in motion Johnston withdrew from Manassas, taking his position below the Rappahannock; with Confederate reverses in the west, Southern prospects appeared to have reached their nadir.

But then an unexpected jolt came from the Shenandoah Valley. Learning that three of Banks's divisions were to be sent to reinforce McClellan, a small Confederate force attacked a full federal division at Kernstown, just south of Winchester, supposing it to be merely a rear guard. Seriously mauled, the rebels commanded by Stonewall Jackson were compelled to retreat, but reasoning there must be a larger enemy force in the Valley than previously known, Lincoln abruptly cancelled the transfer of reinforcements to McClellan. He also ordered McDowell and his corps of 35,000 men to remain in northern Virginia for Washington's defense. Lincoln's decision to deprive McClellan of some 50,000 troops transformed a tactical defeat for Southern arms into a strategic victory.

The first week of April McClellan was ready to begin his first tentative movement with 55,000 troops. Only a few miles from his base at Fort Monroe, conjuring a ruse to beguile the overcautious general, the Confederate commander J. B Magruder, entrenched at Yorktown behind the Warwick River, marched his infantry in circles while shuffling his artillery, convincing McClellan the defenses could only be taken by siege. Hearing of his general's difficulties, an exasperated Lincoln wired: "I think you had better break the enemy's line . . . at once." Days later Lincoln tried again, indicating the extremity of McClellan's situation: "It is indispensable to you that you strike a blow. . . . I have never written you . . . in greater kindness of feeling than now, nor with a fuller purpose to sustain you. . . . But you must act."[18]

But the drift of the war in Washington's political circles was not to McClellan's liking, and he would be neither compelled nor cajoled. Complaining in a letter to his wife of his difficulties with "the rebels on one side, & the abolitionists & other scoundrels on the other," he obstinately remarked that if Lincoln wanted the lines broken, "he had better come & do it himself."[19] While McClellan continued grinding his heels, Jefferson Davis, in agreement with his military advisor Robert E. Lee, who had been recently transferred to Richmond from Savannah, ordered Johnston to move his army down the Virginia Peninsula to defend the line at Yorktown as long as possible. After nearly a month, just when McClellan felt ready to assail enemy defenses, Johnston withdrew, carrying his army back to Richmond. It looked as though the final hour of the Confederate government was nearing, as the state legislature and governmental offices began preparing to evacuate to Raleigh, North Carolina. But a strong rear guard action under James Longstreet blunted the Union pursuit at Williamsburg.

Contributing to the sense that the bid of the Confederate States of America for independence was about to be foreclosed was a victory with far-reaching consequences when the Union fleet under Farragut ran the gauntlet of enemy firepower below New Orleans. In anchorage before Jackson Square on April 29, the admiral sent a detachment of marines to raise the Stars and Stripes on the Crescent City's public buildings. At this arrival of the Yankees, crowds brandishing pistols and burning cotton bales swarmed the levees and the streets, venting their rage at the "invaders," while the navy leveled eleven-inch guns against them. Two days later, the ubiquitous Benjamin Butler entered New Orleans with fresh troops, and soon found the city was not to be subdued by civil means, as Pierre Soule, formerly U.S. ambassador to Spain, at the head of avowedly proslavery mobs decided to contest for control of the city. With General Butler's headquarters in the St. Charles Hotel surrounded, the commanding officer reported he was unable to maintain order. Butler ordered, "tell him, if he finds he cannot control the mob, to open on them with artillery."[20]

Now began Butler's "efficient but remorseless rule" that earned him the designation "outlaw" from the Confederate Congress and the popular sobriquet "the beast" in the South. Soon enough the hostile municipal authorities, whose cooperation was first sought, were confined to Fort Lafayette, and Butler abolished the whipping-house and prohibited all forms of corporal punishment, determining that Louisiana, when it left the Union, had "taken her black code with her." In the next two months Farragut would receive the surrender of Baton Rouge and then of Natchez, but would founder at Vicksburg, sitting atop a two-hundred-foot bluff on the Mississippi, and turn back.

Just previous to the loss of the lower Mississippi, on April 16 the Confederate Congress passed a conscription act calling into service men between ages thirty-five and forty-five, signaling a change of policy within the Davis government, with a consequent reorganization of the army. Pressed by events, Davis had been compelled to abandon the plan of a frontier defense, in favor of a concentration of forces in the interior, particularly in Virginia, for a decisive battle. Norfolk was now abandoned and everything of military value destroyed, including the fearsome ironclad that had so recently shown its prowess. With the James River open, the *Monitor* and a flotilla of five gunboats proceeded toward Richmond, while the Northern press and the boat's own officers speculated they might be able to emulate Farragut's victory at New Orleans.

Under his rule Butler at first prohibited fugitives from entering the army's camps and would seal New Orleans to their escape to await return to their masters. But with few loyalists among whites, this policy was rapidly undone. Shortly after entering the city Butler met with representatives of the large, educated, free black population, welcoming their cooperation, and in August would order that their military organization, the Native Guards, be accepted into the volunteer forces of the United States.

Adding to the measures then roiling the policy of Lincoln's government, Maj.-Gen. David Hunter, succeeding General T. W. Sherman in command of the Department of the South (South Carolina, Georgia, and Florida) would proclaim an emancipation even more far-reaching than Frémont's. With the fall of Port Royal the previous year and the capture of Beaufort and Hilton Head, the rebel defenders and white civilians fled the coastal Sea Islands and rich plantations along its connecting waterways, leaving them still under the care of slaves but occupied by federal forces. Despite the danger of being shot, blacks from the surrounding country began coming into Beaufort and Hilton Head, as a new Underground Railroad sprang into being. Butler had already articulated the dilemma facing Union forces, but with the extent of the occupied lands growing, it became evident to military men that something must be done about the growing slave population within their lines. They must be provisioned, provided with protection, instructed in self-governance—and they must be urged to continue their cultivation of the land.

These demands quickly attracted attention in the North, particularly in Boston, where freedmen's aid societies were forming as women and men volunteered for service among newly freed blacks to begin reconstruction on a new basis. Seeing the need, early that spring Harriet Tubman resolved to go south. In its May 1862 issue the *Liberator* announced a testimonial was held for her in Boston and donations taken. That same month with assistance from the Massachusetts governor, Tubman boarded the troop transport the *Atlantic*, embarking to Beaufort, South Carolina. Before leaving, looking to things yet to happen, she gave an interview with Lydia Maria Child, which the writer related in a letter to John Greenleaf Whittier, on January 21 of that year. Child wrote:

> She talks politics sometimes, and her uncouth utterance is wiser than the plans of politicians. She said the other day: "They may send the flower of their young men down south, to die of the fever in the summer and the ague in the winter. . . . They may send them one year, two year, three year, till they tire of sending or till they use up the young men. All of no use. God is ahead of Mister Lincoln. God won't let Mister Lincoln beat the South till he does the right thing. Mister Lincoln, he is a great man, and I'm a poor Negro; but this Negro can tell Mister Lincoln how to save the money and the young men. He can do it by setting the Negroes free. Suppose there was an awfully big snake down there on the floor. He bites you. You send for the doctor to cut the bite; but the snake, he rolls up there, and while the doctor is doing it, he bites you again. The doctor cuts down that bite, but while he's doing it the snake springs up and bites you again, and so he keeps doing till you kill him. That's what Mister Lincoln ought to know."

While Tubman may still have been in transit, in Beaufort Major-General Hunter issued his declaration of martial law on May 9 that abolished slavery in all the states under his purview. Like the stroke of Frémont, Hunter published his proclamation without informing the War Department or the president. The document read: "Slavery and martial law in a free country are altogether incompatible. The persons in these three States, Georgia, Florida, and South Carolina, heretofore held as slaves, are therefore declared forever free."

This thunderbolt made it into the newspapers and before he was even officially informed of it, Lincoln hastened to annul it. Secretary of the Treasury Salmon P. Chase was the only cabinet member urging the president's approval, but on May 19 Lincoln issued his revocation in these terms: "That neither Gen. Hunter nor any other commander or person has been authorized by the Government of the United States to make proclamation declaring the slaves of any State free; and that the supposed proclamation now in question, whether genuine or false, is altogether void, so far as respects such declaration." But the president followed with a caveat: "I further make known, that, whether it be competent for me, as Commander-in-Chief of the Army and Navy, to declare the slaves of any State or States free, and whether at any time or in any case it shall have become a necessity indispensable to the maintenance of the Government to exercise such a supposed power, are questions which, under my responsibility, I reserve to myself, and which I cannot feel justified in leaving to the decision of commanders in the field."[21]

Although revoked, the proclamations of Frémont and Hunter sounded deeply in the Northern public, the words "forever free" beaconing over the Union government's often stumbling war effort, hastening, rather than retarding adoption of the policy. But in 1862 Lincoln was looking through optics that did not seem to preclude a restoration of the "the Union as it was." Neither had he insisted on enforcement of Congress's Confiscation Act, nor forbidden his leading generals to force fugitives from their lines. McClellan even banned the gospel and antislavery Hutchinson Family Singers from performing in his camps, while the jails in Washington, D.C., were filling with fugitive slaves awaiting return to their masters.

Stalled before Yorktown for nearly a month, as Johnston consolidated in defense of Richmond, McClellan's 100,000 troops were slowed by heavy and frequent rains. But as McClellan came within earshot of Richmond's church bells, Lincoln ordered McDowell's corps at Fredericksburg to prepare to march south. Meanwhile—to preclude Union troops in the Shenandoah Valley from reinforcing McClellan, and perhaps to deflect McDowell—Johnston reinforced Jackson with a division commanded by Richard B. Ewell.

Still only half the strength of the opposing Union forces, Jackson now had 17,000 men, but had concluded he could make up for his numerical inferiority with speed. Earlier that spring at his behest Jackson's engineer had prepared new maps of the Valley with tables listing mileage from point to point, and he began relentlessly drilling and preparing his troops so they would be able to maintain combat effectiveness while covering maximal distances in the least time. Using its complex terrain, good roads along its corridors, and numerous gaps and trails, Jackson would attack and destroy isolated detachments piecemeal, while maneuvering to prevent combination against him. It would be difficult, he knew, even for an army of 100,000 to hold the country against him, and the Shenandoah Valley would be ideal for his purposes.

At the beginning of May, Jackson marched his force over a mountain pass and struck a detachment of Frémont's nascent army of 25,000 in western Virginia at the small hamlet of McDowell. Frémont's campaign, projected to push into eastern Tennessee and on to Knoxville, a favorite of Lincoln's, was spoiled before it could begin. Returning to the Valley, Jackson sent Turner Ashby's cavalry north on the road to Strasburg where Banks was then digging in. Ducking eastward, behind Massanutten Mountain, Jackson marched up the Luray Valley on a parallel course, overwhelming a small federal outpost at Front Royal. Jackson was now only ten miles from his enemy's flank with twice the force. Realizing his predicament, Banks hastily withdrew, heading northward, losing his supply train in the process. Reaching Winchester with the bulk of his troops, Banks turned and prepared to fight. On May 25, on the hills to the south and west of the town, the antagonists met in sharp battle where, bereft of ammunition and rations, Banks's divisions broke and fled for the safety of the Potomac.

With the Union army driven from the Shenandoah Valley, Jackson moved his force within a few miles of Harper's Ferry, sending dispatches to Richmond requesting permission to strike into the North. Before him was the possibility of destroying the crucial bridges over the Susquehanna, even the prospect of attacking and capturing Baltimore, allowing Maryland to join the Confederacy. But Jackson was ordered to remain in the Valley.

Ensconcing himself in a telegraph office, Lincoln sent a flurry of orders. Frémont must cross the mountains and attack the enemy's rear; McDowell must make haste and attack his flank; Banks, resupplied, must move down from the north. Cognizant of the danger to him, Jackson began racing south before Lincoln's pincers could close. On June 1 he cleared Strasburg; passing New Market, Jackson's force reached the southern extremity of the Valley at Port Republic. With Frémont converging along his line, while several of McDowell's divisions commanded by James Shield came down the eastern slope of Massanutten Mountain, Jackson held the only remaining bridge across the Shenandoah River. Facing two armies with a combined strength of over 30,000, Jackson, commanding half their number, determined to blunt Shield's advance, then turn to strike

Thomas "Stonewall" Jackson.

Frémont. But his plan was frustrated as his men were too weary and battered to carry it out. After a fierce engagement on June 9, Jackson withdrew to Brown's Gap in the Blue Ridge.

The mercurial commander had fought five battles in a campaign lasting over a month, outmaneuvered three armies, and disrupted two strategic movements. He had driven his men over 350 miles and emerged with an aura of invincibility. A correspondent who observed him during his signal passage said Jackson appeared to be "all fight," and had nostrils that flared as big as a horse's. Tinged with the same "fanaticism" that others had seen in John Brown, Jackson was withal a martinet so secretive that not even his officers would know his plans. Attired in his old Mexican War coat and a Virginia Military Institute cadet cap with a broken visor; he suffered from dyspepsia and sucked lemons for his condition.

McClellan considered the diversion of McDowell to the Valley to be a colossal blunder by Lincoln. Anticipating his aid in the campaign on Richmond, but also to protect his supplies, McClellan had placed more than half his troops north of the rain-swollen Chickahominy River, while the remainder of his army was isolated across several temporary bridges. On May 30 a heavy downpour washed these bridges out, and the next day, reinforced with troops from North Carolina, Johnston attacked. The fighting known as Seven Pines in the South and Fair Oaks in the North was inconclusive. But Johnston himself had been wounded and Davis immediately appointed Robert E. Lee to replace him. Recognizing the futility of continuing, Lee broke off the engagement.

A well-bred son of one of Virginia's most prominent families and a relative of "Light Horse" Harry Lee, Robert E. Lee displayed the exterior of a dour Episcopalian. Leaving the U.S. Army Engineering Corps to follow his state when it seceded, he had never commanded troops in combat, except to oversee the capture of John Brown at Harper's Ferry. Lee too had fared badly in western Virginia the previous year, and before being called to Richmond supervised the strengthening of defenses at Savannah. Pollard's *Richmond Examiner* upon his elevation called him "evacuating Lee, who has never risked a single battle with the invader," though Lee now put his soldiers to work strengthening the fortifications and digging trenches around Richmond, earning the untested warrior the appellation "the king of spades."

Between February and May, Union armies had conquered 50,000 square miles of territory; two state capitals, Nashville and Baton Rouge, had been occupied, while the Confederacy possessed the entirety of only three states—Texas, Alabama, and Georgia. On May 31, Corinth, Mississippi, would again be adjoined to the Union and Memphis on June 6. The Union, too, had regained one thousand miles of navigable river on the Cumberland, the Tennessee, and the Mississippi. Yet by July both sides would begin to realize that the limited war, such as McClellan envisioned, was being transformed into total war. Jackson and Lee, both men of military prowess, would now combine to offer the South hope of victory. Their very success, however, could only assure the destruction of everything the Old South stood for.

Richmond was pivotal to the Confederacy; if it fell then North Carolina, part of South Carolina, and eastern Tennessee would become untenable. As Lee hunkered down in his capital's defense, he was determined, as was Johnston before him, that McClellan's preparations not become a siege. Lee was in fact preparing a line that could be held by part of his forces; with the balance he would make an incisive thrust upon his enemy—the defensive-offensive conception of strategy that would become his special imprint. On June 12 Lee sent Jeb Stuart's cavalry on a four-day ride that started north of Richmond, then swung eastward, making a complete circuit around McClellan's army. In this storied spree Stuart discovered that McClellan had transferred most of his army to the south side of the Chickahominy, leaving Fitz John Porter's 5th Corps in an unprotected position on its north bank. On this exposed right flank Lee decided to focus his attack.

Lee's plan called for Jackson to stealthily march in to hit Porter's flank, while divisions from the Army of Northern Virginia, as Lee's army was now designated, crossed the Chickahominy and smashed its front. McClellan meanwhile continued to dicker with Washington about the timing of his own offensive. By June 24 he detected Jackson's approach and the next day he wired Stanton that he should "have to contend against vastly superior odds" and warned that if his army was destroyed, "the responsibility cannot be thrown on my shoulders; it must rest where it belongs."[22] This was an allusion to Lincoln's diversion of McDowell's corps, only a division of which had reached the Richmond front.

On June 26 the fighting north of the Chickahominy began, when, after vainly waiting for Jackson's assault on Porter's flank, A. P. Hill brought his division forward in the late afternoon near Mechanicsville. However, instead of going over to an offensive, McClellan ordered Porter to fall back to more defensible ground near Gaines' Mill, an adjustment caused by the mere appearance of Jackson near the battlefield.

Lee now called for a combined assault against Porter's new position, with Jackson again attacking on the right, Hill on the center, with Longstreet on the left. Again Jackson failed to engage the enemy and Hill's division bore the brunt of the fighting until sundown, when, in uncoordinated assaults, the other commands came into play. Finally John Bell Hood, leading a brigade of Texans, penetrated Porter's center and his line collapsed, but his surviving troops were able to cross the river during the night. The six hours of fighting at Gaines' Mill had produced more casualties than had two days of fighting at Shiloh. This had been Lee's baptism of fire.

McClellan wired Stanton on the evening of the 27th bitterly rebuking the government for not sustaining him. Saving McClellan's military career for several more months, an incredulous colonel in the telegraph office removed the last two sentences of his message: "If I save this army now, I tell you plainly that I owe no thanks to you or to any other persons in Washington. You have done your best to sacrifice this army." Apparently tossing aside his siege plan, McClellan began shifting his base and supplies to the James River. Lee now determined to harass McClellan's flank and ordered nine divisions to converge by six roads. Again a tardy Jackson and poor coordination brought the plan to grief. The next day's plan for a concentric attack

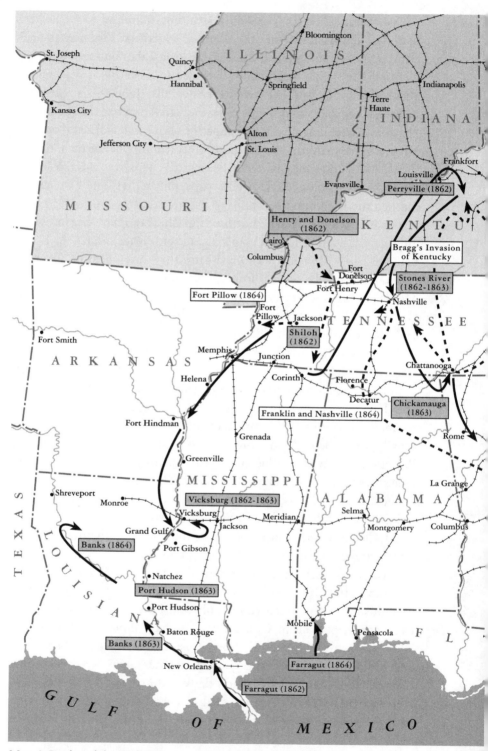

Map 4. Battles of the Civil War.

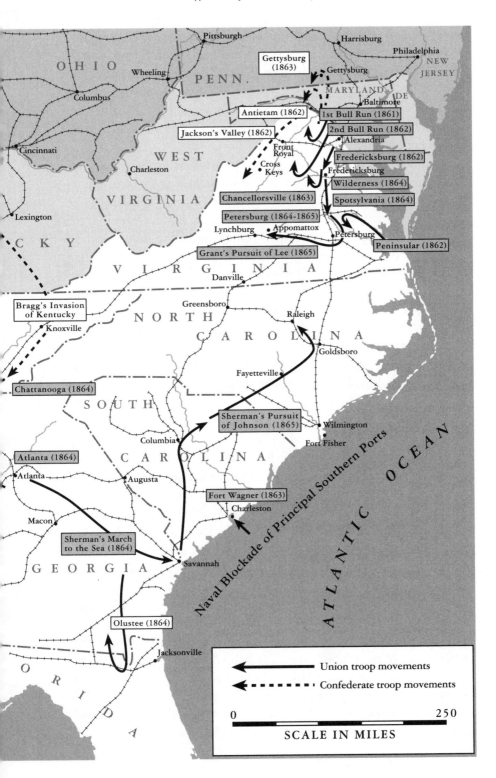

Union troop movements

Confederate troop movements

0   250
SCALE IN MILES

also produced an insignificant outcome. Finally, McClellan prepared for a decisive stand on Malvern Hill, a one-hundred-fifty-foot mound sided by deep ravines near the James River. The Union army now presented their strongest front yet, with 100 guns drawn axle to axle with four infantry divisions, and behind these another 150 guns and an additional four divisions. But at the back of this array, Lee knew, was an unnerved commander.

On July 1 Lee ordered an assault supported by artillery that produced double the casualties among Confederate soldiers to Union losses. Again Hill's division bore the brunt of this "murder," as he called it afterward, with a Union officer writing decades later that the view down the slopes on the next morning with its five thousand dead and wounded gave "the field a singular crawling effect."23 McClellan ordered another retreat to Harrison's Landing on the James River, under cover of gunboats. After a week of fighting that came to be called the Seven Days' Battle, the Confederates had come into possession of 30,000 small arms of all types, fifty cannon, as well as a vast array of equipment, and held 6,000 prisoners.

The despondency slackening the nerve of the North was as grim as that seen after Bull Run. Before the extent of the setback to Union fortunes could even be fully judged, however, Lincoln was preparing to intensify the contest by calling for 300,000 volunteers on July 2 to bring the war "to a speedy and satisfactory conclusion." This call-up was backdated by Seward to June 28, after consultation with Northern governors who were to supply the recruits on the basis of population, to defer any connection with McClellan's reversal. Lincoln then reached out to the West for a general, selecting John Pope to command a new army to be formed from the divisions of Banks, Frémont, and McDowell, to be designated the Army of Virginia. Now Frémont, stepped over for command, offered his resignation, which Lincoln immediately accepted. At the same time Halleck was called to Washington to be named general-in-chief of all armies. Then Lincoln hurried to Harrison's Landing for a meeting with McClellan.

In the previous year the consensus in the North had been that preservation of the Union was worth a three-month war, while largely anticipating the former relation of the states would be undisturbed. But at this moment it had to be asked: was it worth a year of war? By the summer of 1862 100,000 lives had been forfeited and one half billion dollars washed-out—and the war could already be called an old war. At this crucial juncture, Frederick Douglass declared, in an important speech delivered on the 4th of July at Himrods Corners, New York:

There are many men connected with the stupendous work of suppressing this slave-holding rebellion—but there are three men in the nation, from whose conduct the attention of the people should never be withdrawn: the first is President Lincoln. The single word of this man can set a million of armed men in motion: He can make and unmake generals, can lift up or cast down at will. The other two men are McClellan and Halleck. Between these two men nearly half a million of your brave and loyal sons are divided, the one on the Potomac and the other on the Mississippi. Are those two men loyal? Are they in earnest? Are they competent? Whatever may be said of the loyalty or competency of McClellan, I am fully persuaded by his whole course that he is not in earnest against the rebels, that he is to-day, as heretofore, in war as in peace a real proslavery Democrat. His whole course proves that his sympathies are with the rebels. . . . Bear in mind that slavery is the very soul and life of all the vigor which the rebels have thus far been able to throw into their daring attempt to overthrow and ruin this country. Bear in mind also that nothing could more directly and powerfully tend to break down the rebels . . . than the Insurrection or the running away of a large body of their slaves, and then, read General McClellan's proclamation, declaring that any attempt at a rising of the slaves against their rebel master would be put down, and put down with an iron hand. Bear in mind the fact that this General has in deference to the slave-holding rebels forbidden the singing of anti-slavery songs in his camp, and you will learn that this General's ideas of the demands of the hour are most miserable below the mark. . . . [He] is reported in the Richmond Dispatch to have said that he hated to war upon Virginia, and that he would far rather war against Massachusetts. This statement of the Richmond Dispatch in itself is not worth much, but if we find as I think we do find, in General McClellan's every movement an apparent reluctance to strike at Virginia rebels, we may well fear that his words have been no better than his deeds. . . . Unquestionably time is the mightiest ally that the rebels can rely on. Every month they can hold out against the Government gives them power at home, and prestige abroad, and increases the probabilities of final success. . . . Now I undertake to say that General McClellan has from the beginning so handled the army of the Potomac as to give the rebels the grand advantage of time.[24]

And what of the policy of the administration? Douglass asked. Lincoln was elected as the representative of the antislavery policy of the Republican Party. But thus far all the president had done in the war in reference to slavery had been "calculated in a marked and decided way to shield and protect slavery from the very blows which its horrible crimes have loudly and persistently invited." He had rejected the policy of arming slaves; he had refused to proclaim emancipation; he had arrested the antislavery policy of some of his most reliable generals, while he had assigned the most important position to notoriously proslavery ones, and so on. "It is from such action as this, that we must infer the policy of the Administration. That policy is simply and solely to reconstruct the union on the old basis of compromise, by which slavery shall retain all the power that it ever had."

"The question now arises," Douglass continued, "'Is such a reconstruction possible or desirable?' Mr. Lincoln can do many things, but Mr. Lincoln will never see the day when he can bring back or charm back, the scattered fragments of the Union into the shape and form they stood when they were shattered by this slaveholding rebellion."

What was the true course to be pursued in the war? was Douglass's next line of inquiry. The government and the people must recognize the fact that the only choice left to the nation was abolition or destruction; slavery must be abolished or the union abandoned. And how could slavery be abolished? One way would be a stringent confiscation bill by Congress; another would be a proclamation by the president at the head of the nation; another would be by the commanders of each division of the army. But the war had been wretchedly mismanaged, and Douglass confided to Gerrit Smith soon after this speech (in a letter dated September 8, 1862) that he shuddered "at what the future may still have in store for us. I think the nation was never more completely in the hands of the Slave power. . . . I think, in such hands, we shall do well if we at last succeed in buying a peace from our Southern masters, without fully indemnifying them for the entire expense to which they have been put in humbling us."[25]

When Lincoln arrived at Harrison's Landing on July 8, McClellan handed him a memorandum which Lincoln read without comment. In it the general instructed the president that the war "should not be

a war looking to the subjugation of the people," and that "neither confiscation of property, political executions of persons, territorial organization of states, nor forcible abolition of slavery should be contemplated for
the moment." If military power should be
used to interfere "with the relations of servitude," he cautioned, the northern armies
would "rapidly disintegrate." Hoping to
head off this gathering movement at the top,
McClellan, then being courted as a presiden-

George B. McClellan.

tial nominee for the Democrats (New York's
mayor Fernando Wood had been down to
see him), prepared his memorandum for the president and on the
same day he addressed an epistle to War Secretary Stanton warning
that the nation would not support a change of policy, nor would the
armies wage war under it. Furthermore, he stipulated he must have
an additional 50,000 troops to renew the campaign against
Richmond, the very number he had been refused in May.

Lincoln had not yet communicated his thoughts, but roiling in his
mind along with the shifting fortunes of the war was the question of
whether the organizational changes he was instituting in the military
should be accompanied by a change of policy. After all, there had
been a broad transformation of attitudes in the North and in
Washington on the question, where many who had opposed emancipation were now preparing to meet the issue. Earlier that June the
Congress had passed a bill which Lincoln promptly signed, declaring
that "there shall be neither slavery nor involuntary servitude in any
territories of the United States now existing, or which may at any
time hereafter be formed, or acquired." Then a bill was passed
authorizing the president to appoint diplomatic representatives to the
world's two black republics, Liberia and Haiti, previously unrecog-
nized, while the State Department stipulated that a "Negro" would
not be received as a foreign minister.

Just as Lee was thrashing McClellan, Union advances in the West,
where Stanton and Grant were advocating the capture of Vicksburg
as the next priority, began to stall. Mindful of Lincoln's desire to
regain eastern Tennessee, and faced with the daunting logistical and
organizational concerns of administering the conquered territory, as
well as the "contraband" camps in the upper Mississippi Valley,

Halleck decided that a move against the South's most important rail junction at Chattanooga was of prime importance. Ordering Grant's Army of West Tennessee posted for occupation and railroad repair duties, Halleck ordered Don Carlos Buell's Army of the Ohio to begin a campaign to take Chattanooga. Buell was a protégé of McClellan, sharing his conception of war by maneuver as well as his deferential regard for the social system of the South. His course through northern Alabama along the railroad from Corinth, considerate to the local population, consumed inordinate time, occupied as he was with the repair and protection of his rail link to his supply base. Just as Lincoln was conferring with McClellan, Buell's superior tried to impress upon him greater urgency, admonishing: "The President telegraphs that your progress is not satisfactory and that you should move more rapidly."[26] But Buell was still weeks from crossing the Tennessee River, when, five days later, Confederate cavalry under Nathan Bedford Forrest captured the Union garrison at Murfreesboro, thereby breaking Buell's critical link to Nashville.

As he returned to Washington, Lincoln had already made up his mind to be done with McClellan's manner of waging war—a game in which the government "stakes all, and its enemies stake nothing." Deference to loyal slaveholders, the cornerstone of government policy to date, had only stayed its hand like a "dead palsy." Summoning border state congressmen to the executive office on July 12, Lincoln urged action for the last time on the resolution he had advanced for a gradual, compensated manumission. He said: "I intend no reproach or complaint when I assure you that, in my opinion, if you had all voted for the resolution . . . of last March the war would now be substantially ended." If they did not act, he warned, "the institution in your states will be extinguished by mere friction and abrasion—by the mere incidents of the war . . . and you will have nothing valuable in lieu of it. How much better, to take this same money which else we sink in the War! How much better to do it while we can, lest the war ere long render us pecuniarily unable to do it! How much better for you as seller, and the nation as buyer, to sell out and buy out that without which the war could never have been, than to sink both the thing to be sold and the price of it in cutting one another's throats!"[27]

The next day Lincoln had his answer when he was handed a manifesto signed by two-thirds of the congressmen he had met with, detailing their objections. It was an unconstitutional intrusion in a state matter; it would draw the war out by compelling many Unionist slaveholders to join the rebellion; its price would be prohibitive; it was too radical a change. That same day Lincoln confided his thoughts on issuing an emancipation proclamation to two of his cabinet members, Welles and Seward. Welles recorded that the president said the question had "occupied his mind and thoughts day and night" for weeks, and he deemed it "a military necessity," and "absolutely essential to the preservation of the Union." "The slaves are undeniably an element of strength," he told the secretary, "to those who have their service, and we must decide whether that element should be with us or against us." If they wanted "the army to strike more vigorous blows," the administration "must set an example, and strike at the heart of the rebellion. We must free the slaves or be ourselves subdued."[28]

After five months of harsh debate on July 17 the Congress passed a second Confiscation Act. This provided "that every person who shall hereafter commit the crime of treason against the United States, and shall be adjudged guilty thereof, shall suffer death, and all his slaves, if any, shall be declared and made free." The act left to the discretion of courts whether imprisonment or fine would be levied in lieu of death and declared that all persons so convicted would be disqualified from holding office under the United States; and that all persons held by those in rebellion against the government would be free as they entered Union lines. Finally the act stated "that the president of the United States is authorized to employ as many persons of African descent as he may deem necessary and proper for the suppression of the rebellion, and for this purpose he may organize and use them in such manner as he may judge best for the public welfare."

The way was prepared for the momentous advance contemplated by Lincoln, a mandate no less necessitated by the tribulations of his government. Calling a meeting of his cabinet to lay before them a new policy, on July 22 the president said that he felt the plan of operations they had been pursuing had now "reached the end of the rope" and that he felt "that we had about played our last card, and must change our tactics, or lose the game."[29] He had prepared a draft of a proclamation and had called them together not to seek advice but to hear suggestions that would be in order after it was read.

While Lincoln had decided that it was a military necessity for his government to issue an emancipation proclamation, in none of his public statements or actions was he to give any indication that this was the case. In fact all of his utterances tended to discourage that expectation. He asked at one point rhetorically:

> What good would a proclamation of emancipation from me do, especially as we are now situated? I do not want to issue a document that the whole world see must necessarily be inoperative, like the Pope's Bull against the Comet! Would my word free the slaves, when I cannot even enforce the Constitution in the rebel states? Is there a single court or magistrate, or individual that would be influenced by it there? And what reason is there to think it would have any greater effect upon the slaves than the late law of Congress, which I approved and which offers protection and freedom to the slaves of rebel masters who come within our lines? Yet I cannot learn that that law has caused a single slave to come over to us.[30]

Until that point his policy had been dictated by expedience, of not risking an open breach with slaveholders in the border slave states, particularly in Kentucky, and that had failed. Now he was also under increasing pressure from prominent members in his own party and many outside it, which his government could not "afford to lose." He had given offense to this constituency when he revoked Hunter's proclamation two months earlier, and he could scarcely afford to dissatisfy them again. In the deliberations that followed in the cabinet meeting of July 22 only postmaster general Blair depreciated the change in policy on grounds that they would lose the support of a section of the Democrats to the war and hurt the administration in the fall elections. Nothing, however, was raised that Lincoln had not already anticipated, until Seward questioned "the expediency of its issue at this juncture." The secretary of state argued, "It may be viewed as the last measure of an exhausted government, a cry for help, the government stretching forth its hands to Ethiopia, instead of Ethiopia stretching forth her hands to the government. Now, while I approve the measure, I suggest, sir, that you postpone its issue, until you can give it to the country supported by military success, instead of issuing it, as would be the case now, upon the greatest disasters of the war."[31]

This observation struck the president as having the force of wisdom and he pocketed his paper until it could be determined that the government had the advantage on its side once again.

At that juncture another of the details retarding Buell and so vexing to Lincoln was one more Confederate horseman, a fastidious thirty-six-year-old Kentuckian named John Hunt Morgan. He had raised a brigade in his native state and in early July led them on a thousand-mile ride through Kentucky and Tennessee that netted 1,200 prisoners along with tons of stores. Then as Buell was finally poised to cross the Tennessee, Morgan would again cut off his rail link by rolling flaming boxcars into a tunnel, causing its collapse. These mobile successes caused Braxton Bragg, the new commander of the Confederate Army of the Mississippi, to see an opportunity for a thrust up through middle Tennessee into Kentucky, culminating with an invasion across the Ohio River and perhaps the capture of Cincinnati.

It was on July 23, the day following Lincoln's cabinet meeting, staying the hand, as it were, of righteous judgment, Bragg began to move his invasion force from Mississippi to Chattanooga. His solution was not to send them along the route which Buell had followed for six weeks, but to entrain them south to Mobile, then northeast to Atlanta and thence to Chattanooga, a roundabout journey of almost eight hundred miles in the largest troop movement of the war up to that time. Bragg could now make a conjunction with Kirby Smith in Knoxville and begin the great surge to recapture Tennessee, "free" Kentucky, and strike decisively into the North. Bragg wrote with bravura: "Van Dorn and Price will advance simultaneously with us from Mississippi on West Tennessee, and I trust we may all unite in Ohio."[32]

As McClellan remained inactive below Richmond, to counter Pope on July 13 Lee sent Jackson with 12,000 troops to the rail junction at Gordonsville northwest of Richmond, and then reinforced him with another 13,000 troops under A. P. Hill. The move magnified the threat to Pope, and Lincoln ordered McClellan's withdrawal and the abandonment of his peninsula campaign altogether. McClellan was to support the new effort against Richmond from the north. By now there was a mounting chorus urging McClellan's removal, but Lincoln was not yet ready to take that step. McClellan was revered by his troops and still had the support of a large number of officers, but heaping more indignity upon the North's former leading general, Pope issued a proclamation to his troops, saying in the West they had seen nothing of their enemy "but their backs." He told his army: "I desire you to dismiss from your minds certain phrases, which I am sorry to find much in vogue among you. I hear constantly of taking strong positions and holding them; of lines of retreat and bases of supplies. . . . Let us study the probable line of retreat of our opponents, and leave our own to take care of itself."[33]

Among the first orders Pope issued was pertaining to the seizure of "rebel property" without compensation, in line with the Congress's recent enactments. Captured guerrillas who had fired on Union troops would be shot, Pope decreed, and civilians refusing to take an oath of allegiance would be expelled from occupied territory. In Richmond, Lee fumed "this miscreant Pope" must be "suppressed."

With Jackson blocking Pope's advance, in early August Jackson's old adversary in the Valley, Nathaniel Banks, moved forward and the two clashed in a bloody battle at Cedar Mountain. Banks was stopped, but Jackson too was disappointed, as he had wished to destroy the Union armies piecemeal before they had opportunity to combine. Lee, however, was satisfied that the threat to the safety of Richmond lay from the north and not from the peninsula, from which McClellan's forces had not yet moved. On August 10, protesting bitterly the transfer of his army to assist Pope, McClellan wrote to his wife, evidently his most trusted confidant, predicting "Pope will be thrashed . . . & be disposed of. . . . Such a villain as he is, ought to bring defeat upon any cause that employs him."[34]

On August 30, 1862, *Die Presse* published "Abolitionist Demonstrations in America" by Karl Marx, in which he pointedly wrote that "in the present state of affairs" a speech by Wendell Phillips was "of greater importance than a battle bulletin." The speech, titled "The Cabinet," had been delivered by Phillips on August 1 in Abington, Massachusetts, on the anniversary of emancipation in Britain's West Indian colonies. Marx highlighted "striking passages" from it for his European readers:

> The government fights for the maintenance of slavery, and therefore it fights in vain. Lincoln wages a political war. Even at the present time he is more afraid of Kentucky than of the entire North. He believes in the South. The Negroes on the Southern battlefield, when asked whether the rain of cannon-balls and bombs that tore up the earth all round and split the trees asunder, did not terrify them, answered: "No, massa; we know that they are not meant for us!" The rebels could speak of McClellan's bombs in the same way. They know that they are not meant for them. . . . Have

no fear for Richmond; McClellan will not take it. If the war is continued in this fashion, without a rational aim, then it is a useless squandering of blood and gold. It would be better were the South independent today than to hazard one more human life for a war based on the present execrable policy. . . . But you cannot get rid of the South. . . . Recognize it tomorrow and you will have no peace. For eighty years it has lived with us, in fear of us the whole time, with hatred for us half the time, ever troubling and abusing us. Made presumptuous by conceding its present claims, it would not keep within an imaginary border line a year—nay, the moment that we speak of conditions of peace, it will cry victory! We shall never have peace until slavery is uprooted. So long as you retain the present tortoise at the head of our government, you make a hole with one hand in order to fill it with the other. . . . Had Jefferson Davis the power, he would not capture Washington. He knows that the bomb that fell in this Sodom would rouse the whole nation. The entire North would thunder with one voice: "Down with slavery, down with everything that stands in the way of saving the republic!" Jefferson Davis is quite satisfied with his successes. They are greater than he anticipated, far greater! If he can continue to swim on them till March 4, 1863, England will then, and this is in order, recognize the Southern Confederacy. . . . The President has not put the Confiscation Act into operation. He may be honest, but what has honesty to do with the matter? He has neither insight nor foresight. When I was in Washington, I ascertained that three months ago Lincoln had written the proclamation for a general emancipation of the slaves and that McClellan blustered him out of his decision and that the representative of Kentucky blustered him into the retention of McClellan . . . I know Lincoln. I have taken his measure in Washington. He is a first-rate second-rate man. He waits honestly, like another Vesenius, for the nation to take him in hand and sweep away slavery through him.[35]

A week later a *New York Herald* headline shouted, "Wendell Phillips Spouting Foul Treason." "Arrest him. He is a nuisance, a pest, and should be abated," chanted the *Boston Post*. Even the *New York Tribune* scourged Phillips for allegedly discouraging enlistments. This led Phillips to spell out again the basis of his criticism in the *Liberator*, on August 29, 1862: "I must educate, arouse, and mature a public opinion which shall compel the Administration to adopt and support it in pursuing the policy I can aid. This I do by frankly and candidly criticizing its present policy, civil and military. . . . My criticism is not, like that of the traitor presses, meant to paralyze the administration, but to goad it to more activity and vigor."

On August 6 Lincoln addressed a war meeting in Washington where he sought to quiet the growing controversy over the failure of the peninsula campaign, in which Stanton was being criticized for not reinforcing McClellan, while others were demanding McClellan's dismissal. Lincoln granted that "General McClellan is not to blame for asking for what he wanted and needed, and the Secretary of War is not to blame for not giving when he had none to give." In his September *Monthly* Douglass caustically remarked in "The President and His Speeches": "The President of the United States seems to possess an ever increasing passion for making himself appear silly and ridiculous, if nothing worse. Since the publication of our last number he has been unusually garrulous, characteristically foggy, remarkably illogical and untimely in his utterances, often saying that which nobody wanted to hear, and studiously leaving unsaid about the only things which the country and the times imperatively demand of him."

Douglass followed with an equally scathing description of the president meeting with a committee of Washington's most prominent blacks in the executive office on August 14, the first meeting ever between a president of the United States and members of the African American community: "Mr. Lincoln assumes the language and arguments of an itinerant Colonization lecturer, showing all his inconsistencies, his pride of race and blood, his contempt for Negroes and his canting hypocrisy. How an honest man could creep into such a character as that implied by this address we are not required to show." Lincoln was not capable of contemplating an edict like emancipation without accompanying it with the deportation of freed blacks from the country. He had found a suitable location for a colony in Central America, he said, and wanted his auditors to see what they could do to provide the colonists. As Lincoln mounted his "hobby horse" a reporter present took down his words that were widely disseminated in the northern press:

> Your race are suffering, in my judgment, the greatest wrong inflicted on any people. But even when you cease to be slaves, you are yet far removed from being placed on an equality with the white race. . . .

I do not propose to discuss this, but to present it as a fact with which we have to deal. I cannot alter it if I would. . . . See our present condition—the country engaged in war!—our white men cutting one another's throats . . . and then consider what we know to be the truth. But for your race among us there could not be war, although many men engaged on either side do not care for you one way or the other. Nevertheless, I repeat, without the institution of Slavery and the colored race as a basis, the war could not have an existence.

It is better for us both, therefore, to be separated. . . .

The practical thing I want to ascertain is, whether I can get a number of able-bodied men, with their wives and children, who are willing to go, when I present evidence of encouragement and protection. Could I get a hundred tolerably intelligent men, with their wives and children, and able to "cut their own fodder," so to speak? Can I have fifty? If I could find twenty-five able-bodied men, with a mixture of women and children, good things in the family relation, I think I could make a successful commencement.

I want you to let me know whether this can be done or not. This is the practical part of my wish to see you. . . .[36]

Taking up Lincoln's assertion that the presence of "Africans" in the country was "the real first cause of the war," Douglass wrote: "It does not require any great amount of skill to point out the fallacy and expose the unfairness of the assumption, for by this time every man who has an ounce of brain in his head, no matter to which party he may belong, and even Mr. Lincoln himself, must know quite well that the mere presence of the colored race never could have provoked this horrid and desolating rebellion. . . . A horse thief pleading that the existence of the horse is the apology for his theft or a highway man contending that the money in the traveler's pocket is the sole first cause of his robbery are about as much entitled to respect as is the President's reasoning at this point."

The "whole colonization scheme never appeared to us more detestable and wicked than at this moment," Douglass wrote in his *Monthly* in an article titled "The Spirit of Colonization." It "gives life and vigor to popular prejudice, gives it an air of philosophy, piety and respectability, and the violence of the mob, gives the facts to sustain their pious Negro-hating theories." Douglass's concerns were timely, for that summer there had been an attack on blacks in Brooklyn, where a tobacco warehouse with forty women and children inside was burned down. And in southern Illinois a mob attacked blacks from one of the contraband camps in Tennessee sent to help with the corn harvest; in Cincinnati and other cities there had also been riots against blacks.

At this crucial moment Horace Greeley published a letter titled "The Prayer of Twenty Millions" as an editorial in the *Tribune*, demanding Lincoln adopt a clear policy on emancipation. With masterly prevarication, so dear to generations of historians, Lincoln replied on August 22 in a letter to the editor: "My paramount object in this struggle is to save the Union, and is not either to save or to destroy slavery. If I could save the Union without freeing any slave, I would do it; if I could save it by freeing all the slaves, I would do it; and if I could save it by freeing some and leaving others alone, I would also do that."

With Pope waiting for McClellan to reinforce him before going over to an offensive, Jackson fell back on Gordonsville as Lee probed for an opening. Lee ordered Jackson to flank Pope on his right, only to see him step back from the Rapidan to the Rappahannock. Swinging farther north, Jackson again outflanked Pope, who once more gave up his position, withdrawing just south of Warrenton. Flanking his adversary again, Jackson was now twenty-five miles behind Pope, between the Union rear and Washington. In Alexandria, suggesting to the president that all available troops be held in reserve under his command to defend Washington, McClellan reported to his wife in a letter dated August 22, that if "Pope is beaten, they may want me to save Washington again. Nothing but their fears will induce them to give me any command of importance."[37]

Jackson's famished and ragged corps had marched fifty miles in two days when they fell unopposed on a large supply depot at Manassas, increasing their commissary at a stroke by fifty thousand pounds of bacon, one thousand barrels of beef, two thousand barrels of pork, and large quantities of flour, oats, and corn. Capturing several heavily laden trains, with ten locomotives, Jackson's ravished men fell like locusts upon the booty, burning what they could not eat or carry away in an immense bonfire. As the frantic Pope tried to locate and annihilate him before Lee could join him, Jackson secreted his men on Stony Ridge near the old Bull Run battlefield. As Longstreet approached on the same path Jackson had marched, one of Pope's divisions stumbled on Jackson's lair in an unfinished railroad bed. Pope began a fierce but uneven assault against Jackson's unyielding and determined troops, with Longstreet joining on his left flank; the contenders called the three days that followed the Battle of Second Manassas or Second Bull Run.

On August 29 there was terrible fighting with heavy Union losses. As Jackson and Longstreet adjusted their lines during the night, Pope mistook it for a retreat, declaring in his dispatch to Washington that the rebel army was whipped and driven from the field. The War Department permitted the communication to be published in the *Washington Star*, where it was also announced that Gen. Stonewall Jackson was cut off and captured. When the Union army advanced on the morning of August 30 they found Jackson's troops stuck fast in their roadbed trenches. During repeated assaults Longstreet began to enfilade the attackers with artillery. Pope's columns began to falter, even as Jackson's troops, wearied to the point of breaking, were running out of ammunition. Coming up like screaming banshees, Jackson's entire corps charged against scattered and leaderless troops. Within minutes the order to attack traveled down Longstreet's line and both armies were in motion. Now Pope's entire line was retreating, reassembling only to resume its retreat as the Confederates followed up at the double-quick using bayonets at close quarters.

As night fell the Union army was again being forced across Bull Run, their ranks reduced to a fleeing rabble. On reaching Centreville, officers rallied the debilitated troops into columns under the protection of 30,000 fresh troops. With a semblance of organization restored, Pope ordered a continued retreat to the entrenchments of Alexandria and Washington. At the War Department, Stanton appealed for volunteers to help care for the thousands of wounded, as Herman Haupt, a wizard of modern railroad engineering, organized trains to bring them in, and Clara Barton, leading scores of women nurses, made compresses and slings to bind wounds. Abetted by the cold and rainy weather, a pall of gloom fell over Washington and spread northward.

Thirteen

Lincoln's Emanicpation

[The slaves] cannot be neutral . . . as laborers or as soldiers they will be allies of the rebels or of the Union.
—Rep. George Julian of Indiana, speech on the floor of the House, January 14, 1862

T HE DAY POPE'S ARMY WAS DEMOLISHED AT MANASSAS, KIRBY Smith reached Richmond, Kentucky, twenty miles below Lexington and seventy-five miles south of Cincinnati, swamping a Union garrison of 6,500 raw recruits. In the west the Confederate plan called for Smith to divert Union forces by threatening Cincinnati, while Bragg, with 30,000 troops, captured Louisville, wresting control of the grain-growing and meat-producing region and expelling the Union troops from Kentucky. Meanwhile Van Dorn and Price had been left with 32,000 troops defending Vicksburg and central Mississippi. If Grant, with 67,000 dispersed troops, moved to aid Buell, these armies could move to regain western Tennessee.

On September 1 Kirby Smith took up the march to Lexington. Three days later he and his troops were given a boisterous welcome by thousands in that city, who were also thrilled as the dashing John Morgan and his cavalry joined them. Coming at the moment of Second Manassas, Jefferson Davis and General Lee saw this advance as their opportunity. Lee's army could not remain where it was and was not strong enough to attack Washington's defenses; the prudent thing would have been to withdraw for the defense of Richmond, but the prospect of recovering Maryland and Kentucky together was too tempting to pass by. In the worst event Lee might sever Washington's rail links with the west and destroy the railroad bridge over the Susquehanna River.

As the women of Lexington waved handkerchiefs at Kirby Smith in Kentucky, Lee, with some 55,000 men, crossed the Potomac River into Maryland thirty-five miles northwest of Washington. Although its forces were appallingly deficient in every regard compared to its adversary, September 1862 represented the high tide of the Confederacy. The South remained unconquered after a year and a half of war and the North had been humiliated. In England, Lord Palmerston's cabinet seemed likely to grant the Confederacy the recognition it sought; a consummation surely to be achieved if its armies advanced into Ohio and Pennsylvania.

Back in Washington Lincoln was moving fast to mend the disaster of Pope's defeat. With his generals dickering over who was to blame and cabinet members arguing against McClellan, who was seen to have let Pope fail, Lincoln merged Pope's army into the Army of the Potomac, putting McClellan back in command. McDowell, who had the misfortune again of being associated with defeat, was relieved of command, and Pope sent to watch Indians in Minnesota. Lincoln now became the scourge of Buell's slow-moving tack in Kentucky. Halleck was to warn: "The Government seems determined to apply the guillotine to all unsuccessful generals." Clearly not yet done with McClellan, but exasperated with him, the president ordered him to go after Lee, and "destroy the rebel army, if possible."[1]

To the strains of "Maryland, My Maryland," on September 6 Lee's army entered Frederick, forty-five miles west of Baltimore. His men were as bedraggled and starving a lot as any soldiers seen during the four years of war now known as the American Civil War. Nonetheless Lee gave strict orders that all troops were to respect the rights of civilians and to pay for everything they took, and issued a proclamation "To the People of Maryland," announcing in ringing tones: "We have come with the deepest sympathy [for] the wrongs and outrages that have been inflicted upon the citizens of a Commonwealth, allied to the States of the South by the strongest social, political, and commercial ties. . . . to aid you in throwing off this foreign yoke, to enable you again to enjoy the inalienable rights of free men."[2]

Lee's objective was spelled out in a letter to President Davis on September 8. With southern armies on northern soil, he wrote—a "proposal of peace would enable the people of the United States to determine at their coming elections whether they will support those who favor a prolongation of the war, or those who wish to bring it to a termination." He would occupy western Maryland, inviting Marylanders to his standard, seize Harper's Ferry (thereby preparing the way for a strike into Pennsylvania from Hagerstown), and if successful, emancipate Baltimore from Lincoln's tyranny. Lee now ordered Jackson's corps to proceed to capture Harper's Ferry, while Longstreet's and Hill's corps were put in position to cover his operations and to block McClellan's army, which was advancing in relief of the 11,000-man Union garrison there.

After his triumphal passage through Lexington, Kirby Smith had gone on to capture Paris and Cynthiana, establishing his forward line in sight of Cincinnati. If he had gone on to take the city with its valuable stores and yards for constructing gunboats, the result may well have precipitated the political crisis anticipated by Lee; but Smith's order had been merely to make a demonstration in that direction, then turn to support Bragg's advance into Kentucky. As Bragg crossed the Cumberland River east of Nashville, bypassing the slower-moving Buell, he issued a proclamation, declaring: "Kentuckians, I have entered your state . . . to restore to you the liberties of which you have been deprived by a cruel and relentless foe." Asking a helping hand "to secure you in your heritage of liberty," Bragg brought 15,000 extra rifles for those he expected to rally to his standard.[3]

In an article titled "The Situation in North America" in *Die Presse*, November 10, 1862, Marx highlighted Bragg's decree as showing the motive behind "the latest combined moves of the Confederacy." He wrote:

> Bragg's proclamation, addressed to the states of the Northwest, implies that his success in Kentucky is a matter of course, and obviously calculates on the contingency of a victorious advance into Ohio, the central state of the North. In the first place he declares the readiness of the Confederacy to guarantee free navigation on the Mississippi and the Ohio. This guarantee only acquires import from the time that the slaveholders find themselves in possession of the border states. At Richmond, therefore, it was implied that the simultaneous incursions of Lee into Maryland and Bragg into

Kentucky would secure possession of the border states at a blow. Bragg then goes on to prove the justification of the South, which only fights for its independence, but, for the rest, wants peace. The real, characteristic point of the proclamation, however, is the offer of a separate peace with the Northwestern states, the invitation to them to secede from the Union and join the Confederacy, since the economic interests of the Northwest and the South are just as harmonious as those of the Northwest and the Northeast are inimically opposed. We see: the South barely fancied itself safely in possession of the border states, when it officially blabbed out its ulterior object of the reconstruction of the Union, to the exclusion of the states of New England.

After occupying Martinsburg, Jackson then turned to come up behind Harper's Ferry. On the morning of September 14 the Union garrison there found itself deluged from all directions by guns which had been planted on the commanding heights. At 7 A.M., in a disgrace that was to reverberate across the North, they were compelled to hoist the white flag, raising the charge that the Union defenses had been treasonously compromised by another officer in the army with southern sympathies, Colonel Dixon Miles. Had he lived he would have surely faced court-martial, but at the moment of surrender an exploding shell carried away his left thigh. "My God, I am hit," he only had time to exclaim, before falling into the arms of his aide-de-camp. Jackson rode down through Harper's Ferry, leaving A. P. Hill to complete the surrender, leading his men in a severe fourteen-mile march to rejoin Lee at Sharpsburg.

Given McClellan's characteristic reticence, Lee calculated he had ample time to complete his movements. With Jackson securing Harper's Ferry and the supply line to the Shenandoah Valley, another of his corps advanced upon Hagerstown, while another, shielded by South Mountain, occupied the intermediate zone between. But in Frederick the paper bearing Lee's orders detailing these deployments had been discovered, supposedly lying in a field wrapped around three cigars, by one of McClellan's officers. Hastening to exploit the discovery—as it happened, a Marylander with southern sympathies who witnessed it hurried to warn Lee—McClellan began forcing passage through South Mountain simultaneous with Jackson's capture of the Ferry. Lee had no alternative but to abandon his invasion of Pennsylvania. He could have ordered retreat across the Potomac into northern Virginia, but he had come into Maryland to fight, so he ordered his armies to converge on Sharpsburg with the Potomac at their back. The ford there offered an exit in the event of defeat.

The next day McClellan's army began arriving at Antietam Creek, two miles east of Lee's position. On September 16, 60,000 Union troops, with another 15,000 nearby, were pressing on no more than 30,000 Confederates. Jackson's corps arrived to be deployed north of Sharpsburg on the Hagerstown Turnpike; on his right, along a sunken lane blocking Middle Bridge, were the divisions of D. H. Hill, and to their right, south of Sharpsburg, was Longstreet's Corps, facing Antietam Creek's Burnside Bridge. A brigade of Georgians covered this last position, which would become one of the keys to the battle known as Sharpsburg in the South and Antietam in the North. Here Brig.-Gen. Robert A. Toombs, thwarted in his political ambitions, would gain new laurels as a soldier.

After five hours battling on the left of Lee's lines, both sides began to break down—as 12,000 men lay dead or wounded. When Union troops began to prevail against the center and a sunken farm road used as a Confederate rifle pit, known as Bloody Lane, McClellan was too shaken to exploit the collapse. As the afternoon wore on the encounter shifted to the Union right, where two of Burnside's regiments had surged across the bridge over Antietam Creek after being held off for several hours by Georgia rifles. Just as Burnside was driving the Confederate line into Sharpsburg and threatening to cut off the escape across the Potomac, A. P. Hill arrived with his division from Harper's Ferry, smashing into the Union flank. In the surprise and confusion Burnside's troops began to retreat. As night fell both sides took stock of their appalling losses. Lee's divisional commanders reported casualties of 50 percent and more in several brigades.

The next morning McClellan had two fresh divisions on hand, and although Lee remained in Sharpsburg, he could not summon the nerve to force him out. Lee yielded on the night of the 18th, quietly ordering his army back to Virginia, taking with him the supplies gathered in Maryland, and the rich spoils of Harper's Ferry. McClellan wired Washington: "Maryland is entirely freed from the presence of the enemy, who has been driven across the Potomac. No fears need now be entertained for the safety of Pennsylvania."[4]

As the battle raged at Sharpsburg, Bragg captured a Union garrison of five thousand at Munfordville, Kentucky. Positioned to make an attack on Buell's flank, instead Bragg let him slip by, combining with Smith in Bardstown thirty-five miles from Louisville. While the two commanders prepared for the decisive battle for Kentucky, they took time out to witness the inauguration of the new Confederate governor at Frankfort; however, before the ceremony was completed the boom of cannon scuttled the proceedings. Buell, spurred by Lincoln, and after gaining 30,000 recruits in Louisville and an equal number from Cincinnati, had finally turned to fight. Three columns were sent to confront the enemy at Bardstown and one to capture the state capital; it was the latter that caused the newly inaugurated governor and dignitaries to take flight.

While Lee had been in Frederick, Lincoln confidentially committed himself to announce his proclamation as soon as Lee and his army were driven from Maryland. Neither the president nor anyone else believed he had the constitutional authority to decree an end to slavery, but in line with enactments already passed by the Congress he could confiscate enemy property in time of war. That war, Lincoln reiterated on September 22, when he issued a preliminary emancipation proclamation, was being conducted for the restoration of the constitutional relation between the United States, and he again repeated his offer that he would seek congressional aid for compensated "abolishment" in the border states, either immediate or gradual, coupled with a consensual emigration of the people set free. On January 1 in districts still "in rebellion against the United States," slaves "shall be thenceforward and forever free," Lincoln warned. Fugitives coming into northern military lines would likewise be free and military officers were ordered to assist in maintaining their freedom.

Lincoln's proclamation was of course a bold change, but there were obvious discrepancies. First, it did nothing to touch the real issue concerning slavery; it only pertained to persons, and the institution of slavery would remain in all the states where it had been before Lincoln spoke. Then too the proclamation would not become effective for another hundred days. Until then hope was held out that the seceded states would return to the Union, and it was not to apply to loyal slave states or to those portions of states where rebellion had been overcome. Many asked what manner of "military necessity" admitted to postponement. It is clear Lincoln was temporizing before the fall election and the certain and growing opposition from Democrats and border-state politicians. To many, Lincoln's proclamation also smacked of bad faith; they had supported a war "to restore the Union as it was," as the phrase went, and Lincoln himself had never gone into the war to put slavery down, only to put the flag back. Now he would be forced to rest his power exclusively with the Republican Party. On September 24 Lincoln issued another proclamation to gird his government, extending the application of martial law and the suspension of habeas corpus, aimed particularly at suppressing discouragement of enlistment.

To the South, Lincoln's decrees demonstrated that what he had failed to accomplish by military operations, he would now attempt "by the horrors of servile insurrection." The *Richmond Examiner* recalled events in Nat Turner's time and said that was the sort of work Lincoln wanted. Jefferson Davis addressed the Confederate Congress in Richmond, declaring there were "but one of three possible consequences" to result from the "impotent rage" of Lincoln's proclamations: "the extermination of the slaves, the exile of the whole white population of the Confederacy, or absolute and total separation of these states from the United States." Beauregard wired that after the period of one hundred days proscribed by Lincoln had expired, abolitionist prisoners should be executed "by garrote." Davis did ask congressional approval to hand captured Union officers over to the several states for punishment as "criminals engaged in inciting servile insurrection," once the proclamation became effective.[5]

In the North the *Tribune* expressed dissatisfaction because of the proclamation's tardiness, and for all that had been done to discourage the idea that freedom would ever be promulgated: "There was a time when this bit of paper could have brought the Negro to our side; but now slavery, the real rebel capital, has been surrounded by a Chickahominy swamp of blunders and outrages against the race which no paper spade can dig through." Massachusetts governor Andrew, however, wrote: "It is a poor document but a mighty act, slow, somewhat halting, wrong in its delay till January, but grand and sublime after all." Breaking through the despondency he had seen swirling around the Union cause, Lincoln's proclamation caused Frederick Douglass to shout with joy. To Garrison it was "a step in the right direction." "A step!" cried Wendell Phillips, "it's a stride." Having just finished delivering his address "American Civilization," Ralph Waldo Emerson inserted an addendum after September 22, observing: "This state-paper is the more interesting that it appears to be the President's individual act, done under a strong sense of duty. He speaks his own thought in his own style. All thanks and Honor to the

Head of State! The Message has been received throughout the country with praise, and, we doubt not, with more pleasure than has been spoken. If Congress accords with the President, it is not yet too late to begin the emancipation; but we think it will always be too late to make it gradual. All experience agrees that it should be immediate."[6]

In a speech titled "The State of the Country," Phillips provided a retort to Lincoln's notion of colonization: "Colonize the blacks! A man might as well colonize his hands; or when the robber enters his house, he might as well colonize his revolver. . . . We need the blacks even more than they need us. They know every inlet, the pathway of every wood, the whole country is mapped at night in their instinct. And they are inevitably on our side, ready as well as skilled to aid: the only element the South has which belongs to the nineteenth century."[7]

On October 8 the armies of Bragg and Buell collided in sanguinary fighting at Perryville, and like so many other engagements, this too was militarily indecisive. But the next day Bragg received news that Price and Van Dorn had been defeated five days earlier in the battle at Corinth, and the maintenance of his army in Kentucky was rendered untenable. On the night of October 12 Bragg's army commenced an exodus from Kentucky, carrying off an immense quantity of stores, together with some five thousand head of cattle, horses, and mules—as wagon after wagon joined the immense cavalcade. There were ammunition trains, trains of goods, trains of army stores, trains of captured muskets, batteries of artillery, escorts of cavalry, and then came private trains of families in every conceivable vehicle, together with their property in "negroes." In regard to this evacuation in Kentucky and elsewhere, Marx observed:

> Thus the Confederate campaign for the reconquest of the lost border slave states, which was undertaken on a large scale, with military skill and with the most favorable chances, has come utterly to grief. Apart from the immediate military results, these struggles contribute in another way to the removal of the main difficulty. The hold of the slave states proper on the border states naturally rests on the slave element of the latter, the same element that enforces diplomatic and constitutional considerations on the Union government in its struggle against slavery. In the border states, however, the principal theater of the Civil War, this element is in practice being reduced to nothing by the Civil War itself. A large section of the slaveholders, with its "black chattels" is con-

stantly migrating to the South, in order to bring its property to a place of safety. With each defeat of the Confederates this migration is renewed on a large scale. One of my friends, a German officer, who has fought under the star-spangled banner in Missouri, Arkansas, Kentucky and Tennessee in turn, writes to me that this migration is wholly reminiscent of the exodus from Ireland in 1847 and 1848.

Marx's acute assessment reached readers of the Vienna-based *Die Presse* in an article dated October 12, 1862, titled "On Events in North America," in which he called "Lincoln's manifesto on the abolition of slavery" of greater importance than the repulse of the Confederacy's two pronged invasion of Kentucky and Maryland. Marx wrote further:

An unshakeable conviction prevailed, not alone in the South, but in the North as well, that the appearance of the Confederates in Maryland would serve as the signal for a mass popular uprising against "Lincoln's satellites." The question here was not only one of war successes, but of a moral demonstration as well, which would electrify the supporters of the South in all the border states, and draw them with irresistible force into the maelstrom of events. . . . The simultaneous invasion of Kentucky, which is the most important of the border states . . . was, if it is examined in isolation, only a diversion. In conjunction with decisive successes in Maryland, however, it might have led to suppression of Union supporters in Tennessee, and a flank attack on Missouri; secured the safety of Arkansas and Texas; created a threat to New Orleans; and most important, carried the war into Ohio. . . . Following the defeat of the main rebel forces in Maryland, the invasion of Kentucky, carried out without the necessary vigor, and nowhere meeting with popular support, was reduced to a series of insignificant bushwhacker sorties. . . . In this manner, the Maryland invasion showed that the waves of secession did not possess a sufficiently powerful thrust to spread out towards the Potomac and reach Ohio. . . . Deprived of the border states . . . the South in this way has won nothing—except its own grave.

Lee's army had come close to being ruined at Sharpsburg, yet McClellan failed to seize the opportunity to finish it off. Positioning his forces in front of Winchester, scarcely twenty miles from Lee, McClellan did take a cautious step across the Potomac at Shepherdstown, but was bloodied and fell back over the river. Checked by supposed reserves Lee did not possess, McClellan was again becoming delusional. Soon after the battle at Antietam Creek, he wrote his wife: "Those in whose judgment I rely tell me that I fought the battle splendidly & that it was a masterpiece of art. . . . I feel that I have done all that can be asked in twice saving the country. . . . I feel some little pride in having, with a beaten & demoralized army, defeated Lee so utterly."[8]

In the following weeks McClellan had the chance to glower over Lincoln's proclamation, which he considered "infamous." To his wife he wrote that he "could not make up [his] mind to fight for such an accursed doctrine as that of a servile insurrection," and suggested he might tender his resignation. One of his corps commanders, Fitz John Porter, denounced it as an "absurd proclamation of a political coward"; while a staff officer, confiding to a colleague, said that "the game being played" required that neither army get the advantage of the other; the object was to keep both in the field to exhaustion, and then the political parties might "make a compromise and save slavery." This and other expressions began reaching the ear of Lincoln, and the offending officer, a major, was promptly cashiered as an example. Most soldiers, however, saw the value of denying the South the strength afforded by their slaves, and would sustain Lincoln. One private in the Army of the Potomac said, "I hold that nothing should stand in the way of the Union—niggers, nor anything else."[9]

Frustrated at McClellan's failure to destroy Lee and exasperated that he appeared to be making no preparations to follow him, in early October Lincoln journeyed to meet with his general at Antietam. A well-known photograph of their meeting, taken against a background of camp tents and the low roof of a country house in an open field, shows staff officers arrayed on either side of the two men. McClellan and Lincoln face each other no more than a yard apart, occupying the center of the portrait. Evidently a conference has just concluded in an opened tent to their right, and the men have obligingly congregated outside for the photograph. Both have unbuttoned coats. Lincoln, in top hat, rests his left hand without seeking support on the back of a chair as if holding himself in abeyance, like a man who has not been satisfied, but who has pushed as far as he is going to for now. McClellan is booted, as are most of the officers, reposing on his right leg, his left leg turned as if he were ready to swing into step. His cap visor shades eyes gazing up at the president in the attitude of a man sizing up a wearisome opponent. But Lincoln seems not to be answering McClellan's gaze, and is tolerating himself to be looked upon for the photographer's study. Without a doubt McClellan holds Lincoln in contempt; perhaps in the meeting he has insisted, as he did during this period, on the removal of Stanton, and that Halleck make way for himself as general-in-chief. Whatever Lincoln's thoughts, we know of his determination, as several days later he told Halleck unequivocally to order McClellan to "cross the Potomac and give battle."[10] McClellan would not move, instead making excuses that the army needed shoes, tents, and other supplies.

Halleck was livid with anger; Lincoln restrained. Could he not receive his supplies south as well as north of the Potomac? Finally Lincoln wrote McClellan on October 13, unable to conceal a parental tone: "You remember my speaking to you of what I called your over-cautiousness. Are you not over-cautious when you assume that you cannot do what the enemy is constantly doing?" When McClellan protested that the army's horses were fatigued, Lincoln queried with sarcasm: "Will you pardon me for asking what the horses of your army have done since the battle of Antietam that fatigues anything."[11]

McClellan's greatest and unforgivable cause for offense was delivered in his general order number 163 one day after receiving orders to attack Lee. In a testy response to Lincoln's proclamation, he forbade any demonstration against it in the army, but added, with a signal over the president's head and the coming November election: "The remedy for political errors, if any are committed, is to be found only in the action of the people at the polls." This entire imbroglio was finally capped when Jeb Stuart made a complete circuit around McClellan's army with two thousand cavalry, starting October 10 and ending two days later, where he seized Chambersburg, Pennsylvania, for a few hours. Aside from displaying Stuart's magnificent showmanship, however, the exhibition gained nothing of military significance.

The *New York Tribune*'s correspondent with the Army of the Potomac dispatched an article at this time that was to be excerpted in William Wells Brown's *The Negro in the American Rebellion*. In it the correspondent detailed an encounter on a road just beyond Charlestown with a fugitive named "John," the servant of Capt. A. Burnett Rhett, Light Artillery, Lee's Battalion, and brother of Barnwell Rhett, influential editor of the *Charleston Mercury*. It was related that John said he had been given a pass to search for butter and eggs between Winchester and Martinsburg, and was jogging along on Captain Rhett's horse when he came upon Union pickets and the correspondent. As he approached, he was ordered to halt. The *Tribune*'s account continues:

"Where are you from?"

"Southern Army, cap'n."

"Where are you going?"

"Coming to yous all."

"What do you want?"

"Protection, boss. You won't send me back, will you?"

They said they would not, and after learning who he was and his ostensible mission, questioned him further.

"Are there many Negroes in the rebel corps?"

"Heaps, boss."

"Would the most of them come to us if they could?"

"All of them, cap'n. There isn't a little pickaninny so high (waving his hand two feet from the ground) that wouldn't."

"Why did you expect protection?"

"Heard so in Maryland, before the Proclamation."

"Where did you hear about the Proclamation?"

"Read it, sir, in the Richmond paper."

"What is it?"

"That every slave is to be emancipated on and after the thirteenth day of January. I can't state it, boss."

John went on to relate he was from Charleston and the "property" of a northern woman who used to hire him out for the summer and have him wait on her every winter when she came south. It was she who had taught him to read. He had concealed the circumstance of his ownership because the Confederate government was seizing and selling such property, and he had slipped away into the service of its army, hiring to Captain Rhett for twenty-five dollars a month, although he'd never been paid. Twice, he said, he attempted to flee to northern lines but could not pass southern pickets. He was asked:

"Were you at Antietam?"

"Yes, boss. Mighty hard battle!"

"Who whipped?"

"Yous all, massa. They say you didn't; but I saw it, and know. If you had fought us that next day—Thursday—you would have captured our whole army. They say so themselves."

"Who?"

"Our officers, sir."

"Did you ever hear of old John Brown?"

"Hear of him? Lord bless you, yes, boss: I've read his life, and have it now in my trunk in Charleston; sent to New York by the steward of 'The James Adger,' and got it. I've read it to heaps of colored folks. Lord, they think John Brown was almost a god. Just say you was a friend of his, and any slave will almost kiss your feet, if you let him. They say, if he was only alive now, he would be king. How it did frighten the white folks when he raised the insurrection! It was Sunday when we hear of it. They wouldn't let a negro go into the streets. . . . I have a History of San Domingo, too, and a Life of Fred. Douglass, in my trunk, that I got in the same way."

At this point John was escorted to Gen. McClellan, who questioned him about the position, number, and organization of Lee's army, and found his knowledge to be full and valuable, and corroborated by facts learned from other sources. Toward the end of the interview, John asked apprehensively:

"General, you won't send me back, will you?"

"Yes, I believe I will."

"I hope you won't, general. If you say so, I know I will have to go; but I come to yous all for protection, and I hope you won't."

"Well, then, I suppose we will not. No, John, you are at liberty to go where you please. Stay with the army, if you like. No one can ever take you against your will."

The *Tribune* correspondent reported an hour later John was on duty as the servant of Captain Batchelor, Quartermaster of Couch's 2nd Division, "and I do not believe there was another heart in our corps so light as his in the unwonted joy of freedom," he wrote.[12]

Despite the prognostications of those who opposed the proclamation, it was not without results. At the same time the *Washington Republican* reported on the execution of seventeen slaves and free blacks in Culpeper County, Virginia, for possessing copies of Lincoln's proclamation cut from northern newspapers. In an article printed October 20, the *Republican* said its information was that there was "the greatest consternation imaginable among the whites in that section." Noting the same incident in his November *Monthly*, Frederick Douglass wrote: "The news today of insurrection in Virginia, though probably false, may at any moment become true. We should most deeply regret an insurrection now, for the slave's sake, but a formidable insurrection among the slaves would well nigh paralyze the arm now lifted against the Government and country. It would give Lee and (Stone)wall Jackson work to do outside of Maryland and Pennsylvania, and make Richmond an easy prey."

Democrats were quick to hold up Lincoln's proclamation as the main issue in the elections that November, attacking the president and the Republican Party for violating the Constitution, for subverting and usurping the government, and for the conduct of the war itself. The election became not only a referendum on the war, but also a vote between two policies—restoration of the Union as it was, "and the Niggers where they are," as the Ohio Democratic Party slogan had it, or a new union that would mean freedom for all. In New York, Horatio Seymour was nominated on the Democratic ticket that declared Lincoln's proclamation a strategy "of arson and murder," and the candidate asserted that if slavery needed to be abolished to save the Union, the South should be allowed to withdraw. Seymour won the governorship, with Democrats also winning in Pennsylvania and Ohio. In Illinois and Indiana both houses of the legislature fell under their control, while Republican governors were held over because there were no gubernatorial elections that year. Only in New Jersey were Democrats in full control. In an Illinois referendum "black suffrage" was overwhelmingly denied, and a law excluding black refugees from coming into the state was passed. While these results boded ill for the election in 1864, it was, however, far from ruinous. Although there was the possibility that a number of states would not cooperate in war measures, the Republican Party had retained its voting majority in Congress. The day after the election, Lincoln sacked McClellan for willful disobedience of a peremptory order, returning him to civilian life for the rest of the war.

Anxious to follow up with a military victory, Lincoln now chose Ambrose E. Burnside to lead the Army of the Potomac, and William S. Rosecrans to replace Buell as commander of the Army of Ohio. With Union forces out of Virginia, Lee's threadbare army had time to recover, as farmers in northern Virginia and the Shenandoah Valley harvested their crops, replenishing army stores. During the lull in the fighting Lee also formalized the command design being followed, promoting Jackson and Longstreet to corps commanders with the rank of lieutenant-general.

Although Burnside doubted his fitness to assume command in McClellan's stead, he began with dispatch, settling on a strategy that would bring him to Falmouth, opposite Fredericksburg on the Rappahannock River, where he could interpose his 110,000-man army between Lee and the city of Washington. By November 17 he had two corps in place, although the pontoons needed to bridge the river had not yet arrived. Lee arrived to begin digging in on the prominence called Marye's Heights, and on Prospect Hill on the southern bank of the river with his 75,000 men.

With the armies thus placed to renew battle, and with the northern press again hoisting the banner "On to Richmond" and prophesying the fall of the Confederate capital "within ten days," Lincoln addressed the Congress on December 1. Detailing a plan for compensated emancipation, he joined this with a plan of colonization, proposing that this be done by constitutional amendment. To be completed by January 1, 1900, with United States bonds, when the population was calculated to stand at 100 million, the plan, said Lincoln, would yield a great dividend. Confronting the concern that four million freed former slaves would injure and displace white laborers, the president asked, "Is it true, then that colored people can displace any more white labor, by being free, than by remaining slaves?" He answered his rhetorical question with this:

> If they stay in their old places, they jostle no white laborers; if they leave their old places, they leave them open to white laborers. Logically, there is neither more nor less of it. Emancipation, even without deportation, would probably enhance the wages of white labor, and very surely would not reduce them. Thus, the customary amount of labor would still have to be performed; the freed people would surely not do more than their old proportion of it and, very probably, for a time would do less, leaving an increased part to white laborers, bringing their labor into greater demand, and, consequently enhancing the wages of it. With deportation, even to a limited extent, enhanced wages to white labor is mathematically certain. . . . But it is dreaded that the freed people will swarm forth and cover the whole land! Are they not already in the

land? Will liberation make them any more numerous? Why should emancipation South, send the freed people North? People of any color seldom run, unless there be something to run from. Heretofore, colored people, to some extent, have fled North from bondage, and now, perhaps, from both bondage and destitution. But if gradual emancipation and deportation be adopted, they will have neither to flee from. Their old masters will give them wages at least until new laborers can be procured; and the freed men, in turn, will gladly give their labor for wages, till new homes can be found for them, in congenial climes, and with people of their own blood and race. This proposition can be trusted on the mutual interests involved. And, in any event, cannot the North decide for itself, whether to receive them?[13]

Ending on an inspirational note, Lincoln said: "In giving freedom to the slave, we assure freedom to the free—honorable alike in what we give and what we preserve. We shall nobly save, or meanly lose, the last, best hope of earth. Other means may succeed; this could not fail. The way is plain, peaceful, generous, just—a way which, if followed, the world forever applaud, and God must forever bless."

With expectations high, Burnside concluded that his offensive would gain an element of surprise by an audacious frontal crossing of the Rappahannock River. In the pre-dawn darkness on December 11 engineers began laying three pontoon bridges opposite Fredericksburg, and three more a few miles downstream opposite Prospect Hill. Singling out the engineers as soon as it became light, the rifles of a Mississippi brigade delayed the crossing at the town, while downstream under cover of artillery, construction was accomplished without hindrance. To silence the niggling foe in his front, Burnside sent three regiments across the river in boats to clear the town, already abandoned of civilians, in house-to-house fighting. The rest of the army then crossed, pillaging Fredericksburg as they occupied it.

The battle of Fredericksburg, beginning on the 13th, saw Burnside's army occupying the lowlands along the river and the town—100,000 men splendidly arrayed for five miles, behind which on the far bluff called Strafford Heights shown the Union artillery. Lee, with something over 70,000 men, waited atop an encircling amphitheater of low hills, his artillery partially concealed.

Longstreet's corps on the left, directly behind the town, occupied Marye's Heights—his line stretching some four miles and commanding an unrestricted fire over the ground the attackers would have to cross. On the right, on Prospect Hill, was Jackson's corps. Burnside's tactics called for a hard-hitting assault to begin against Prospect Hill while the defenses on Marye's Heights underwent probing.

A cover of fog lingered over the scene till mid-morning. As the fog lifted, Jackson's position came under attack and was penetrated along one of many ravines that scarred the terrain. But when Jackson threw his reserves into the breach, the attack was thrown back onto open field—a counterattack that was halted only by answering artillery on Strafford Heights. It was while watching this scene that Lee uttered his oft quoted remark to Longstreet: "It is well that war is so terrible—we should grow too fond of it."[14]

With the attack ceasing in this sector, wave after wave of brigade-sized Union detachments began marching out of Fredericksburg against Marye's Heights. To all evidence a hopeless assault, as behind a stone wall at the base of the heights were four rows of Confederate riflemen, shooting and reloading with maniacal rapidity. The surging blue lines got no closer than fifty yards of the abutment; New York's Irish brigade was so decimated many suspected it had been purposively sacrificed, hardening Irish hostility to the war, and in the next year to the draft.

Across the river Burnside sat weeping in his tent. The entire disastrous battle would be looked upon in the North as an act of stupidity and raise the question of criminal culpability among the general staff. There were nearly as many Union casualties at Fredericksburg as at Antietam—most of them in front of the stone wall. Learning of the calamity, Lincoln said, "If there is a worse place than Hell, I am in it." Again gloom settled upon Washington as rumors percolated: McClellan would soon be recalled to head a military government, it was said, and that Lincoln was about to step aside in favor of the vice president.[15]

On December 15 the House rejected a Democrat-sponsored resolution denouncing emancipation as "a high crime against the Constitution," and endorsed Lincoln's war measure by a party-line vote. In the days following, however, the Republican caucus in the Senate, with but one dissent, led behind the scenes by Secretary Chase, pressed for Seward's resignation. The secretary of state's undue influence with the president was impeding the war effort, they charged; he was behind the tardiness of the emancipation edict, and was blocking the appointment of antislavery generals and the recruitment of black troops. In these days, Senator Orville Browning, Lincoln's Illinois confidante, recorded in his diary that Lincoln felt his government was "now on the brink of destruction."[16]

On December 19 Lincoln met with a delegation of Republican senators in the executive mansion; listening to their concerns about his secretary of state, he invited them to return on the morrow. Unknown to the senators Seward had already tendered his resignation. When they returned they found the entire cabinet minus Seward assembled, while Lincoln went on to defend the secretary's actions in his absence. All of the cabinet, he asserted, had supported major policy decisions for which he alone was responsible. Looking to the cabinet for their assent, Lincoln caught his secretary of treasury in a way that he could only acknowledge the facts. Chase's "palace coup" was ended, and the next day he offered his resignation. Lincoln could now refuse both, and the squall threatening his presidency dissipated.

Even as Lincoln struggled with the helm, Davis had vexing problems of his own. The reaction to the preliminary emancipation proclamation had been deemed by Lincoln "not very satisfactory" for its effect in slowing recruitments, but there was something more worrying in November when Confederate raiders on a sea island in South Carolina seized four black soldiers in Union uniform. Immediately notified, Confederate secretary of war James Seddon and President Davis approved their summary execution. In December Davis would issue a general order requiring all former slaves and their officers captured in arms be delivered up to state officials for trial. Another event troubling Davis on the military front occurred during the night of the day after Fredericksburg, when the Union army was able to cross the Rappahannock and begin its reorganization, rendering the success of southern arms a negligible advantage. Lee's shortfall was similar to McClellan's and to Buell's when they had checked their opponents but not destroyed them.

Davis, equally with his northern counterpart, had a testy rapport with some of his generals, and with politicians in his own camp. His secretary of state, Robert Toombs, had resigned over disagreements on the direction of the war, and his vice president, Alexander Stephens, was largely absent from Richmond and had become his savage critic. In Stephens's assessment, Davis was a failed statesman who had "proved himself deficient in developing and directing the resources of the country, in finance and in diplomacy, as well as in military affairs." In November, Joseph E. Johnston, with whom Davis also had a strained relationship, reported he had recovered from his wounds and was ready for duty. He expected, or at least desired, to return to the command taken by Lee in his incapacity, but was appointed by Davis as "plenary commander" of the Department of the West. This assignment, embracing the territory west of the Appalachians and east of the Mississippi, placed Johnston in oversight of two strategic commands marked by dissatisfaction or dissension. These were the commands of John C. Pemberton in charge of the defenses at Vicksburg—an artillery expert with an undistinguished record, and unpopular with southerners because he was a Philadelphian who had married a Virginian—and of Braxton Bragg, who after his aborted campaign in Kentucky had occupied Murfreesboro, Tennessee, only twenty-five miles southwest of Nashville.

Johnston started west, but before he could reach Murfreesboro, Davis sent him a telegram that he himself would be coming west on a two-week inspection tour and he should meet him in Chattanooga. From the war's beginning Johnston had entertained doubts about the ability of Davis and of a government predicated on the theory of states' rights to create a sound command structure capable of meeting the challenges posed. Davis had been disposed to defend the entire frontier, thinking in terms of places, while Johnston thought in terms of battle position. In all his previous assignments he had felt obligated to express any uncertainties he felt and to ask for clarification of his mission; his grasp of strategy in the eyes of critics betrayed a reluctance to fight.

Johnston found the preparations at Vicksburg on inspection inadequate. Labor battalions had been assigned to get Vicksburg ready for a siege, but as a rambling fortification not as a compact stronghold, and the bulk of Pemberton's troops were digging in along an extended front 150 miles to the north, at Grenada. One hundred fifty miles south of Vicksburg was another fortification, Port Hudson; but Johnston knew if the first could not hold, then the second would fall. Davis made it clear that he expected a strong defense in both. To Johnston, this was just the situation the foe would want to find; the task as he saw it was to keep the army free to conduct a war of maneuver, get the invaders at a disadvantage, and annihilate them. Scrutiny of the situation in Tennessee suggested there would be no fighting for the foreseeable future. Bragg did face a numerically superior foe in Nashville, but it did not appear disposed to act, and Bragg's defenses with Stone's River at his front were strong. The dissension in his command did not invite intervention, but the visit did strengthen Johnston's conviction that he could not provide effective command in both Tennessee and Mississippi, and he requested the western district be split. Davis did not see it that way, but astonished Johnston when, without consulting him, he transferred a 7,500-man division from Bragg to Pemberton and ordered Vicksburg be sent more heavy guns.

After the fall elections, an Illinois war-Democrat, John A. McClernand, convinced Lincoln that he could fire the patriotism of Democrats in the Northwest if he was allowed to recruit his own army for a campaign against Vicksburg. McClernand had been a general on Grant's staff, but without informing his superior he had undertaken recruitment and began forwarding regiments to Memphis. Finding out about this, Grant wired Halleck for clarification. Assuring Grant full control of troops in his department, Halleck ordered McClernand's divisions be formed into two corps, one to be commanded by the politician-general and the other by William T. Sherman, Grant's most trusted subordinate.

By early December Grant had advanced in an overland campaign against Vicksburg with 40,000 troops as far as Oxford, Mississippi; while Sherman, in McClernand's absence, was ordered to prepare a river expedition. On December 20 Earl Van Dorn, relieved of his command by Davis after the loss of Corinth in October but given a cavalry command, wheeled northward from Grenada and wrecked Grant's supply base at Holly Springs. A week earlier Nathan Bedford Forrest's cavalry had been riding west from mid-Tennessee, tearing up railroad tracks and telegraph lines, while inflicting casualties and capturing or destroying great quantities of Union equipment and supplies. With his communication and supply line severed, Grant had no choice but to call off his campaign, while the unknowing Sherman proceeded.

Meanwhile in Murfreesboro Bragg's camp, despite the fact that the troops remaining to defend the Confederate bastion in Tennessee had been reduced, all was secure. Pollard remarked of a placid scene about to be interrupted:

> Balls, parties, and brilliant festivities relieved the ennui of the camp of the Confederates. On Christmas eve scenes of revelry enlivened

Murfreesboro, and officers and men alike gave themselves up to
the enjoyment of the hour, with an abandonment of all military
cares, indulging in a fancied security. The enemy's force at
Nashville, under command of Rosecrans, was not believed to have
been over forty thousand, and the opinion was confidently entertained that he would not attempt to advance until the Cumberland
should rise, to afford him the aid of his gunboats. Indeed, Morgan
had been sent to Kentucky to destroy the Nashville road and cut
off his supplies, so that he might force the enemy to come out and
meet us. Yet, that very night, when festivity prevailed, the enemy
was marching upon us![17]

With Lincoln prodding Rosecrans to action, he had marched out
of Nashville in three columns the day after Christmas. While Bragg's
cavalry under Joseph Wheeler wreaked havoc by capturing part of his
ammunition reserve, and Forrest and Morgan continued their depredations deep behind his lines, Rosecrans established his battle position within two miles of Murfreesboro. That night the two facing
armies were close enough to serenade one another. On the morrow
both commanders planned to attack on their enemy's right, to get into
the rear and cut off their base. The day before, Sherman had gotten
the majority of his force ready for an assault on the bluffs of
Vicksburg overlooking the Chickasaw Bayou, but the dug-in and outmanned defenders had little difficulty repelling the assault.

At Murfreesboro the rebels were the first to strike, hitting the
Yankees in a furious sweep as they ate breakfast. Driven three miles
into their rear, it was Philip Sheridan's division in the center, up and
under arms since 4 A.M., which prevented a complete rout. But four
hours' fighting cost Sheridan one third of his troops and all three of
his divisional commanders. Rosecrans was obliged to cancel his plan
and rush reserves to shore up his smashed right, his line having swung
around as if on a turnstile; yet now he held a commanding position
on an oval-shaped hill overlooking Bragg's center, protected by an
outcropping of rocks and dense cedar wood. Ignoring the field of carnage, he remarked of his adversary with an air of assurance: "I'll
show him a trick worth two of his."[18] Ordering his artillery be
massed, he prepared for Bragg's renewed assault.

The fighting at Murfreesboro, or Stone's River as the contenders
designated it, was said to be the fiercest of the war in proportion to
the numbers involved. At the end of a horrible year Pollard described

the aftermath: "The scene in the cedars was fearful and picturesque. A brilliant winter moon shed its luster amid the foliage of the forest of evergreens, and lighted up with silver sheen the ghastly battle-field. The dead lay stark and stiff at every step, with clenched hands and contracted limbs in the wild attitude in which they fell, congealed by the bitter cold. It was the eve of the new year. Moans of the neglected dying, mingled with the low peculiar shriek of artillery horses, chanted a *Miserere* of the dying year."[19]

Believing Rosecrans would retreat on New Year's Day, Bragg telegraphed Richmond of his victory, adding with his compliments, "God has granted us a happy new year."[20]

In Boston there was a fresh covering of snow on New Year's Day. As Frederick Douglass walked from his lodgings to Tremont Temple to join three thousand others to await the issuing of the Emancipation Proclamation, the way was lighted by thousands of lanterns and great candles presaging "the glorious morning of liberty about to dawn upon us," as he had said that day signified in an earlier meeting.[21] There had been abundant speculation throughout December as to whether Lincoln would issue his promised edict: Why, it was asked, had there been a delay of one hundred days if he was determined to carry it out; and why all the attention in his message to Congress to gradual emancipation, while he had made no mention of it in the proclamation of September 22? Would the rebuke of his party at the polls affect his resolve? What of the military reverses?

About 8 o'clock Douglass entered the packed hall and strode toward the platform amid greetings and hand clasping. Among other notables on the dais were the historian and orator William Wells Brown, Reverends J. Sella Martin and Leonard Grimes, pastor of Boston's Twelfth Baptist Church, and Anna M. Dickinson, woman's rights advocate and abolitionist, who would all address the auditorium that evening. In the first hour a restless audience listened to the eloquence of Reverend Martin and Miss Dickinson, while a line of messengers was established from the telegraph office to the platform to relay the good tidings when they arrived. As the moments of expectation lengthened, fears began to rise that the article would not be forthcoming. After 9 o'clock, through murmurs of discontent, Douglass and Brown sought to dispel uncertainty with brief, hopeful speeches. At 10 o'clock with the wires still silent, dejection deepened. Any postponement, Douglass had already declared in his *Monthly* addressing the subject of January 1, 1863, would have the effect of suppressing the "proclamation" altogether. What if the hour should come and the man were missing; if Lincoln were again to cower before "loyal" border state slavery, already nearly bankrupting his administration? What would become of his presidency then? What of the country? Could the North endure to be trifled with and betrayed? The dreadful thought of postponement would amount to the most stunning and disastrous blow yet received during the war, and the cold earth around Fredericksburg, still wet with the warm blood of northern men, would shrink back in vain sacrifice. These were doubts raised by Douglass and others, even as the conditions set by Lincoln for the proclamation's withdrawal—that states in rebellion return to the Union—had not been met.

On the morning of January 1 reports of the battle at Stone's River and the repulse at Vicksburg were undoubtedly coming into the War Department, but Lincoln did not let that deter him from greeting a long line of well-wishers at the executive mansion. In the hours the president was dedicating his hand to this duty of office, the official copy of the proclamation that was to be signed that day was at the State Department, where it had been embossed, and in the afternoon Secretary Seward had gone there to retrieve it. Having conveyed it to the presidential office, when Lincoln had finished with his visitors he went with Seward to his office and, dipping his pen in an inkwell, paused to remark, "I never in my life felt more certain that I was doing right, than I do in signing this paper. But I have been receiving calls and shaking hands since nine o'clock this morning, till my arm is stiff and numb. Now this signature is one that will be closely examined, and if they find my hand trembled they will say, 'he had some compunctions.' But anyway, it is going to be done."[22]

Then he carefully inscribed his name and, examining it, remarked, "That will do."

Nowhere discernable in the form and figure of the scribe was that of the "Great Emancipator" that would grace every schoolroom wall, along with that of the "Father of Our Country," in after years. The Emancipation Proclamation was "not the creation of individual design and calculation, but the grand result of stupendous, all controlling, wide sweeping national events," Douglass wrote of the edict. "Powerful as Mr. Lincoln is, he is but the hands of the clock."[23]

During the vigil in Boston—as 8 and 9 and 10 o'clock had come and passed—Douglass wrote, "when patience was well-nigh exhausted, and suspense was becoming agony, a man (I think it was Judge Russell) with hasty step advanced through the crowd, and face fairly illumined with the news he bore, exclaimed in tones that thrilled all hearts, 'It is coming! It is on the wires!!'" The hall erupted with shouts of joy; friends embraced, while many wept. Douglass continued: "My old friend Rue, a colored preacher, a man of wonderful vocal power, expressed the heartfelt emotion of the hour, when he led all voices in the anthem,

> Sound the loud timbre o'er Egypt's dark sea,
> Jehovah hath triumphed, his people are free.[24]

Soon the transmission of the entire text was completed and on hand, but it could not be read through without interruptions from the hearers.

In Boston's Music Hall that night another gathering had been held, among whose notables were Wendell Phillips and William Lloyd Garrison. Phillips declared: "Our rejoicing today is that at last the nation unsheathes its sword, and announces its purpose to be a nation. . . . But let me open for you the huts of three million of slaves, and what is that Proclamation there? It is the sunlight scattering the despair of centuries. . . . It is a word that makes the prayers of the poor and the victim the cornerstone of the Republic. Other nations since Greece have built their nationality on a Thermopylae or a great name—a victory or a knightly family. Our cornerstone, thank God, is the blessings of the poor."

That night also in the Music Hall Ralph Waldo Emerson read his poem "Boston Hymn," whose eighteenth stanza addressed Lincoln's idea for compensated emancipation:

> Pay ransom to the owner
> And fill the bag to the brim.
> Who is the owner? The slave is owner
> And ever was. Pay him.[25]

As midnight came, when the emotionally charged celebrants at Tremont Temple were required to vacate the hall, Reverend Grimes moved that the gathering adjourn to his church on Phillips Street. There they would hold forth till dawn, provided with refreshments, in a church packed from the pulpit to the doors. Douglass wrote: "It was one of the most affecting and thrilling occasions I ever witnessed, and a worthy celebration of the first step on the part of the nation in its departure from the thralldom of ages."[26]

As Douglass took his leave from his colleagues in the early morning and walked to his lodgings, his steps must have seemed borne on Boston's placid air. Could his meditations have returned to that meeting in Tremont Temple not long before, when a mob had shut down the commemoration for John Brown? It was evident to him that emancipation, even if as yet only an opening, had finally come to the nation and that a wondrous transformation had been set in motion by those now far-off events. Could these thoughts have brought others in tow—of his relation to the old man and how Brown had tried, desperately finally, to have him take a hand in helping the slaves to achieve their freedom? Now that that great moment in the nation's trial was nearer could he have contemplated his own relation to it— thoughts that could have only had reference to the sacrifice and loss that had been suffered on all sides? Had his faith that the just hand of the government under the Constitution could achieve emancipation been justified? All this had been arrived at obliquely, as "necessity." Perhaps Brown had been right; the setting free of the slaves could only come through the battle of arms, only by purging the land with blood? Whatever his resolve on this question, as he fell off to his rest Douglass would have been certain it was not yet the end; at best it was only the beginning of the end of slavery.

For the most part the opposing armies outside Murfreesboro had been quiet on New Year's Day. While Bragg expected Rosecrans to retreat, he had not; the Union general with a bulldog demeanor had merely adjusted his lines, withdrawing his troops and guns from Round Forest to a ridge along a sharp westward bend in Stone's River. Planting a strong battery of fifty-eight guns across the river, there he awaited the enemy's attack. With both sides waiting for the other to move, except for some brief cannonading, the morning of January 2 passed without incident.

Determined to assault the Union stronghold, Bragg had given the assignment to Breckinridge. After protesting the proposed attack, the Kentuckian formed his division into two lines. When the signal came the brigades moved rapidly forward out of the woods into an open field, their instructions being not to deliver their fire until close upon the enemy, then to charge with bayonets. Pollard related: "From the moment of gaining the field the enemy's artillery from the ridge opened a sweeping fire, and a whirlwind of Minnie balls from their infantry, with shot and shell, filled the air. Our men were ordered to lie down for a few minutes to let the fury of the storm pass. Then the cry from Breckinridge—'Up, my men and charge!'—rang out."[27]

Despite receiving reinforcements the Union lines were driven toward the river, overawed by this charge accompanied by the demonic "rebel yell." Soon the ridge was carried and guns brought up to hit the foe opposite. Pollard continues his account: "At this time the concentrated fire of the enemy became terrible and appalling. A sheet of flame was poured forth from their artillery on the hills on the opposite side of the river overlooking our left and front, and from their batteries on the river bank, while the opposite side also swarmed with their infantry, who poured in on us a most murderous fire. Still our men never quailed, but pressed forward and crossed the river, the enemy making frightful gaps in our ranks, but which were immediately closed up. Here it was that in less than half an hour over two thousand of our brave soldiers went down!"[28] In the face of utter hopelessness Breckinridge ordered his troops back across the river.

At dawn on January 3, with Rosecrans still holding and receiving reinforcements from Nashville, Bragg had no choice but to pull back. That night he ordered retreat to Tullahoma, a junction on the Nashville and Chattanooga Railroad twenty-five miles south. Lincoln telegraphed Rosecrans his heartfelt thanks, adding later in a letter to the general: "I can never forget, whilst I remember anything, you gave us a hard earned victory which, had there been a defeat instead, the nation could hardly have lived over."[29]

After the decimation of his division, Breckinridge challenged Bragg to a duel, while two other of Bragg's divisional commanders asked Davis to put Johnston in command, while another vowed never again to fight under Bragg's authority. Bragg in turn blamed Breckinridge for bungling the direction of his troops, court-martialed another commander for disobeying orders, and accused a second of drunkenness during the battle. After Bragg had written Davis, in effect requesting that someone relieve him, Johnston was asked to look into the situation. Reporting that there was indeed dissension among the officers but that the enlisted men had high morale, Johnston demurred from superseding his subordinate. If the War Department wanted otherwise, it was necessary for them to make the decision and send him orders.

With southern armies rife with desertion, in January Davis was obliged to leave Richmond on a tour of the states to urge the upholding of the conscription act that had raised a protest. Derided as "the twenty-slave law," it included among its allowances an exemption for any planter owning twenty or more slaves. To its most vocal critic, Gov. Joseph E. Brown of Georgia, national conscription was a usurpation of states' rights, and he tried to have the law nullified in the legislature and in the state's supreme court. Governors and legislators in North Carolina, Alabama, Mississippi, and Texas were following his lead.

Robert Toombs wrote while in command of troops at the front, "I shall be justified in any extremity to which the public interest would allow me to go in hostility to his [Davis's] illegal and unconstitutional course." Alexander Stephens denigrated the recent energy displayed by Davis with his touring, saying, it "seems to me like that of a turtle after a fire has been put upon it back."[30]

The other sector of Johnston's nominal command, Mississippi, where Pemberton had grown enamored of the idea of fortifying Vicksburg, Davis ordered held. To defeat Grant's spring thrust at Vicksburg, Johnston would insist Pemberton move east and consolidate with him at Jackson, but neither Davis nor his general could free their minds of the appeal of a Mississippi River stronghold.

The commanders in the Army of the Potomac were as dissatisfied with Burnside as Bragg's commanders were with him. William Franklin, who had refused to obey an order to continue the assault on Prospect Hill at Fredericksburg, had gone directly to Lincoln with his complaint. Joseph Hooker made no secret among fellow officers that he wanted the command for himself and obligingly commented in the press that what the country needed was a dictator. With the Democratic allies of McClellan clamoring for "little Mac's" return, this time "with unlimited and unfettered powers," Burnside ordered his army across the Rappahannock to flank Lee and force his army out of their trenches into a fight in the open. Misfortune struck immediately. On January 20, with rain falling in torrents, Burnside's entire army was mired in mud. Two days later the whole thing was called off. Soon the astonished general with distinguished whiskers found that Hooker was his successor. Lincoln admonished Hooker in a letter even as he appointed him: "I am not quite satisfied with you. . . . You have taken counsel of your ambition . . . in which you did a great wrong to the country, and to a most meritorious and honorable brother officer. I have heard, in such way as to believe it, of your recently saying that both the Army and the Government needed a Dictator. Of Course those generals who gain successes, can set up dictators. What I now ask of you is military success, and I will risk the dictatorship."[31]

Frederick Douglass retorted: "The Anglo-African well observes concerning this Potomac army—that McClellan has been removed but he has left his sting behind him. Pope might have won, but for Porter, and Burnside might have succeeded at Fredericksburg—but for Franklin. With disloyalty in the army, disaffection in the Cabinet, and a divided North against a united South, we might be hopeless for the result, but for the thought that the country has at last been placed upon ground making it deserve complete victory."[32]

The hesitation of the government to avail itself of "all its friends" was a marvel of folly and imbecility, wrote Douglass. "But," he noted, "the most hopeful sign of the times is the growing disposition to employ the black men of the country in the effort to save it from division and ruin."[33]

Fourteen

Men of Color,
to Arms!

Remember Denmark Vesey of Charleston; remember Nathaniel
Turner of Southampton; remember Shields Green and Copeland,
who followed noble John Brown, and fell as glorious martyrs for
the cause of the slave. . . . This is our golden opportunity. Let us
accept it, and forever wipe out the dark reproaches unsparingly
hurled against us by our enemies.

—Frederick Douglass, March 2, 1863

IT HAS BEEN ESTIMATED THAT IN THE FOUR CENTURIES OF THE
slave trade over ten million Africans were transported from Africa
to the Americas. Well over four million of their four and a half million descendants living in the United States in 1860 had been born in
the country; 11 percent of these were free, while 13 percent were visibly of mixed, with white (English, Irish, Scottish, French, Dutch, and
Spanish), and the various Indian nations and "Negro" blood.

As the war reached its midpoint, slavery was under severe strain.
Lincoln's Emancipation Proclamation of January 1 had signalized
that from that day forward the character of the war was to change: it
was no longer to be a war for the restoration of the old Union;
instead, as he had indicated to an official of the Interior Department,
T. J. Barnett, it would now be one of subjugation. The South, in

Barnett's paraphrase, was "to be destroyed and replaced by new propositions and ideas." To do this it had been necessary for Lincoln to pledge his government to a contest for the allegiance of the slave; and there is no doubt the blacks felt instinctively where their interests lay.[1]

It was in western Tennessee under sponsorship of Grant that the largest "contraband" camps existed. First appearing in Virginia under Butler at Fort Monroe, with the fall of Norfolk and the adjoining areas many thousands of blacks fell under the federal aegis, and Washington, Alexandria, and Georgetown eventually had large camps. With the fall of Port Royal, to these were added the densely populated areas of South Carolina's Sea Islands of St. Helena and Hilton Head, with their center in Beaufort; and following the success of Farragut and Butler, the slave-filled parishes in Louisiana around New Orleans, and the Crescent City itself, extending upriver to Baton Rouge—districts embracing in all half a million blacks who had transferred allegiance from their masters to the star-spangled banner. In all areas where blacks came in contact with the Union armies they were employed for the various camp tasks—as teamsters and laborers, in building fortifications and roads, as cooks and as servants and stretcher-bearers. With women and children coming into the camps the masses of black refugees were turning out to be as costly to feed and care for as a field army.

To these actual disruptions to slavery were added throughout the duration of the war the transport of upward of 150,000 slaves from districts in proximity to advancing Union lines into the Deep South and into Texas; a transfer amidst the privations of war in which thousands perished. Still this left over three million slaves under the control of slaveholders in the Confederate States; yet even among the most remote and isolated districts slaves were aware of the great stirring going on, and if they appeared passive and inarticulate to undiscerning observation, they had, in the words of one of their number, heard the good tidings "by keepin' still and mindin' things."[2]

In Massachusetts, with a share of government patronage and with the aid of abolition societies, missionaries had been enlisted to see what could be done by way of education and "uplift" for the masses of former slaves in South Carolina's coastal islands; schools would be set up, and social and family relations having no institutional existence under slavery would be established, the rights and obligations of citizens introduced. The Department of the South under Maj.-Gen.

David Hunter offered an opportunity for these early experiments in reconstruction. Many of the plantations abandoned by prominent slaveholders like Barnwell, Rhett, Cuthbert, and Phillips would again be cultivated using free black labor with white supervision, and hundreds of laborers would be brought in from the camps around Washington for the work.

When Harriet Tubman arrived in Beaufort and reported to Hunter, it soon became evident she could be a valuable liaison. With a high level of mistrust between locals, who knew whites only through servitude, and the government through its officers and soldiers—each was unfamiliar and incomprehensible to the other. Although Tubman was a proven leader—famed in the North for her exploits and for her association with John Brown—the people of the Sea Islands had no inkling of that legacy; moreover, their manners and language differed significantly from hers. Sarah Bradford was to relate Tubman saying: "Why, their language down there in the far south is just as different from ours in Maryland, as you can think. They laughed when they heard me talk, and I could not understand them." At first Tubman was allowed to draw rations as a soldier, but after encountering resentment and jealousy she relinquished these, instead supplying her wants by selling root beer and pies she made at night. Soon she was attached to a hospital established for the refugees whose medical director and surgeon was Dr. Henry K. Durrant. Earning a small stipend, Tubman used her money to build a bathhouse where the women could do laundry and earn money for their families.

In the summer of 1862 Hunter initiated an effort at forming black military companies, after the militia act of July 17; while in New Orleans Butler was accepting into the volunteer services a preexisting black military organization, the Native Guards, and in Kansas free blacks and "contrabands" would be organized for military service, in companies headed by Blunt, Hinton, and Jennison. In November 1862 Thomas Wentworth Higginson was on hand in South Carolina to take a colonelcy after the military governor, General Saxton, had written him of the intention of the department to raise a regiment from among local men. Higginson found himself at the head of the 1st South Carolina Volunteers, no doubt in "partial expiation," as he termed it in an interview with Katherine Mayo half a century later, for his failure to assist John Brown at the crucial moment; but this service was also deeply in line with his conviction and training. All commissioned officers of the black troops by stipulation were white, in which capacity James Montgomery would also appear on the scene. On December 10 after he had been at his new station several weeks, Higginson wrote his wife: "Who should drive out to see me today but Harriet Tubman who is living at Beaufort as a sort of nurse & general care taker; she sends her regards to you. All sorts of unexpected people turn up here."[3] Higginson's literary endeavors were to continue to be as important as his actual leadership of the first black regiment, he chronicled their activities in *Army Life in a Black Regiment* and in articles he wrote appearing in the *Atlantic Monthly*.

David Hunter.

Eager to get his regiment in the field as soon after Lincoln's proclamation as practicable, Higginson was to embark on January 23 on an expedition with 462 soldiers and officers aboard three steamers. Traveling down the Georgia coast, the principal action in the campaign was on the St. Mary's River, the boundary between Florida and Georgia, a narrow, winding stream bounded in many places by high bluffs. Higginson reported on the expedition in a letter to Saxton dated February 1, 1863, that the river was "the most dangerous in the department . . . left untraversed by our gunboats for many months." The bluffs blazed with rifle-shots, and many hundreds of mounted men dashed from point to point to meet them as they steamed forty miles into the interior. Although outgunned by the Union gunboat, the Confederates were daring in their fire, killing in one volley the captain of the vessel. Higginson wrote:

> Nobody knows any thing about these men who has not seen them in battle. I find that I myself knew nothing. There is a fiery energy about them beyond any thing of which I have ever read, unless it be the French Zouaves. It requires the strictest discipline to hold them in hand. During our first attack on the river, before I got them all penned below, they crowded at the open ends of the steamer, loading and firing with inconceivable rapidity, and shouting to each other, "Never give it up!" When collected into the hold, they actually fought each other for places at the few port-holes from which they could fire on the enemy. . . .

No officer in this regiment now doubts that the key to the successful prosecution of this war lies in the unlimited employment of
black troops. Their superiority lies simply in the fact that they
know the country, which white troops do not; and, moreover, that
they have peculiarities of temperament, position, and motive,
which belong to them alone. Instead of leaving their homes and
families to fight, they are fighting for their homes and families; and
they show the resolution and sagacity which a personal purpose
gives. It would have been madness to attempt with the bravest
white troops what I have successfully accomplished with black
ones.[4]

But the government's new policy immediately came into conflict
with entrenched biases and the practices long established in the
American caste system. On the evening of January 25 a delegation of
Bostonians, including Wendell Phillips, George Stearns, and Dr.
Samuel G. Howe, was ushered into the presidential office by
Massachusetts senator Wilson for a meeting with the president. The
appointment had been arranged principally to lodge a protest against
North Carolina's military governor Edwin Stanley, who had raised
indignation among abolitionists by prohibiting the schools established
in midsummer from teaching freedmen to read and write. Stanley took
as his guide the slave codes of North Carolina, saying any attempt at
educating former slaves was "premature." The delegation had
arranged the meeting with the president to urge his removal, and voice
concern at the emerging pattern of "reconstruction." Evidently Stanley
regarded his mission "to bind the Union to North Carolina, and not
North Carolina to the Union," Douglass snapped in "Closing Schools
for Contrabands in North Carolina" in his July *Monthly*.

Commanding the Department of the Gulf, replacing Butler in
December 1862, Nathaniel Banks was giving a similar tenor to reconstruction. Together with the administration of the military governor
of Louisiana, Brigadier-General Shelley, Banks had practically rein-
stated slavery by enforcing the old slave codes in New Orleans and
the outlying parishes. New Orleans chief of police Col. Jonas French
and his police were particularly zealous in persecuting blacks who
were on the city's streets and in places of "amusement" after 8:30 P.M.
without the required passes from their masters. The jails were nightly filled with men, women, and youths alike, along with free blacks
who did not reside in the city.

When Lincoln entered the room where the delegation was waiting he was enjoying a moment of amusement. His children, he related to his visitors, had told him that morning that the cat had had kittens, and as he was coming in he'd been told the dog had had pups. The house was in a prolific state, he said. As Senator Wilson began introductions Lincoln's face again showed the burdens he bore as he respectfully waved off the formality, saying he knew perfectly well who his visitors were.

Phillips began by expressing his approval of the Proclamation and asking how it seemed to be working. In a characteristic remark Lincoln replied that he had not expected much from it at first, and thus far had not been disappointed. But he hoped advantage would yet come of it. Then Lincoln said: "My own impression, Mr. Phillips, is that the masses of the country generally are only dissatisfied at our lack of military successes. Defeat and failure in the field make everything seem wrong. Most of us here present have been long working in minorities and may have got into a habit of being dissatisfied." At the sound of objection, the president put it before the delegation: "At any rate, it has been very rare that an opportunity of 'running' this administration has been lost."[5]

With disarming directness Phillips replied: "If we see this administration earnestly working to free the country from slavery and its rebellion, we will show you how we can run it into another four years of power."

"Oh, Mr. Phillips, I have ceased to have any personal feeling or expectation in that matter—I do not say I never had any—so abused and borne upon have I been."

"Nevertheless what I have said is true," said Phillips.

He now went on to detail the complaint against Stanley: he had shuttered the schools for contrabands, permitting only oral religious instruction; ordered the search of all vessels bound for the North to prevent freedmen from getting away from their masters; banished from the state the brother of Hinton Helper, author of *The Impending Crisis,* which was considered inflammatory in the South; and taken for his guidance the old slave code of North Carolina, saying he must do nothing which was likely to make him unwelcome to the people of the state—flouting the spirit of the Proclamation.

As Lincoln replied that Stanley could "stand" the Emancipation Proclamation, another of the delegation shot back, "Stand it! Might the nation not expect in such a place a man who can not merely stand its President's policy but rejoice in it?"

"Well, gentlemen, I have got the responsibility of this thing and must keep it."

Phillips again stepped in: "Yes, Mr. President, but you must be patient with us, for if the ship goes down, it doesn't carry down you alone. We are all in it."

"Well, gentlemen, whom would you put in Stanley's place?" Frémont was suggested. Lincoln replied: "I have great respect for General Frémont and his abilities, but the fact is that the pioneer of any movement is not generally the best man to carry that movement to the successful issue. It was so in old times—wasn't it? Moses began the emancipation of the Jews but didn't take Israel to the Promised Land after all. He had to make way for Joshua to complete the work. It looks as if the first reformer of a thing has to meet such a hard opposition and gets so battered and bespattered, that afterwards, when people find they have to accept this reform, they will accept it more easily from another man."

With both sides satisfied with the points made and the meeting nearing conclusion, Lincoln summed up his thinking when he remarked, "All I can say now is that I believe the Proclamation has knocked the bottom out of slavery though at no time have I expected any sudden results from it."

From the war's beginning Lincoln had prevaricated between the natural right of the slave to his liberty—in which he believed—and the right of the "loyal" slaveholder to his "property." After two years it was evident the North did not possess sufficient manpower or wealth to subdue the South without recourse to an emancipation policy. While only "military necessity" had been the stated reason compelling the shift in priorities, a crucial decision had nonetheless been reached. With emancipation anchored on the conflict's outcome, in Lincoln's view "the promise being made must be kept," and it followed that "the Negroes should cease helping the enemy, to that extent it weakened the enemy," and "whatever Negroes can be got to do in as soldiers, leaves just so much less for white soldiers to do in saving the Union."[6]

While the booming war economy had shrunk available manpower, thus reducing, along with the grim prospects of the war itself, the number of volunteers coming forward, early in 1863 65,000 Northern troops were also about to end their enlistments. The U.S. Congress was considering a conscription act, and the authorization of a provost marshal's bureau to enforce it. This act was passed in March, but the first enrollees would not be chosen until July. To ameliorate the nearing emergency, in early January the House passed a bill authorizing the president "to enroll, arm, equip and receive into the land and naval service of the United States" black volunteers; but the Senate returned the bill, refusing to act on it on the grounds that it was unnecessary, as the president had such power under previous acts of Congress. Pursuant the Senate's interpretation, Massachusetts governor Andrew requested permission from Secretary of War Stanton to raise two black regiments for three years' service. On January 20 he received a general order from the War Department authorizing "such volunteers to be enlisted . . . and may include persons of African descent, organized into separate corps." Andrew immediately announced his intention to form the first Northern black regiment, the 54th Massachusetts. But with a scarcity of black men among its general population from which to draw outside Boston and Springfield, recruiters found after six weeks they had only enrolled 100 volunteers.

Summoning George L. Stearns, a leading Republican, the chief behind the Kansas Aid Society movement, and as everyone knew an unapologetic supporter of John Brown, Governor Andrew speedily persuaded him to take complete charge of recruiting. Stearns lost no time setting up a committee that collected $5,000 and began advertising for enlistments, seeking to fill the regiment with men recruited throughout the North. With the help of such luminaries as William Wells Brown, Lewis Hayden, J. W. Loguen, J. Mercer Langston, Henry Highland Garnet, and Martin Delany, before long recruiting posts would be established from Boston to St. Louis. Gaining the support of Frederick Douglass, Stearns knew, would be crucial for this effort and he hastened to obtain his assistance, arriving in Rochester for a meeting with him on February 23.

Although he felt restrained by serious reservations, Douglass accepted the work to which Stearns invited him "with alacrity." In his *Life and Times* he wrote: "The nominal condition upon which colored men were asked to enlist were not satisfactory to me or to them, but assurances from Governor Andrew that they would in the end be made just and equal, together with my faith in the logic of events and my conviction that the wise thing for the colored man to do was to get into the army by any door open to him, no matter how narrow" compelled him to lay his qualms aside.[7]

Three days after Stearns met with him Douglass issued his clarion call—"Men of Color, to Arms!" Published in dailies throughout the North, and forthcoming in the columns of his March *Monthly*, also appearing as a pamphlet, he urged black men to "fly to arms, and smite with death the power that would bury the government and your liberty in the same hopeless grave." Applicants for the prospective regiment, he announced, should contact him in the next two weeks and he would supply them with the means to get to Boston. Douglass opened recruitment in Buffalo, where he made his headquarters, but his first enrollees had come from his own home, his son Charles, followed shortly thereafter by Lewis. Douglass obtained seven recruits in his first meeting in Buffalo, adding to these another thirteen from Rochester.

Just as Douglass had begun raising enlistments, Gerrit Smith, in a letter to Reverend May, had himself proposed a company be raised in New York, promising aid to assist in recruitment. Informing Smith of his activities in a letter dated March 6, and enclosing a copy of his call to arms, Douglass wrote:

> In your letter to Mr. May, you say that you will give two hundred dollars towards raising the proposed company. I have already been at the expense of two journeys to Buffalo and shall be at more before I get a hundred good names on my list. Mr. Stearns and I[,] talking over the matter, came to the conclusion to apply to you for my expenses in getting up the company within the limits of your promised two hundred dollars.
>
> I believe I can get up this company so as to hand it over to Governor Andrew's agents who will take them at the expense of Massachusetts, for less than the sum you promise to contribute. I shall be in Syracuse on Wednesday for the purpose of getting men for this company, and in Troy on Thursday, Friday in Albany. I shall go to New York and at the request of Mr. Stearns go to Philadelphia and stimulate enlistments there. . . . Mr. Stearns has promised to have my expenses paid and to allow me ten dollars per week for my services. I think my services ought to be worth a little more than this, and if you have the paying me, I shall get more. But more or less I am now fully bound to get up the company.[8]

Throughout March Douglass worked the northern tier of New York tirelessly, stopping in Auburn, Syracuse, and Ithaca, where enlistments totaled twenty-five. On the road to Troy and Albany, visits were made in Little Falls, Canajoharie, and Glen Falls, which contributed together twenty-five recruits. In Albany and Troy, aided by Stephen Myers, eighteen men came forward. By mid-April, with recruiting expenses running to $700, Douglass would forward more than a hundred men to Boston. In scarcely more than a month enough recruits had been raised to fill the regiment while recruitment continued for another, the 55th Massachusetts Regiment. In April men began drilling and training under the command of a young Harvard-educated volunteer, Robert Gould Shaw, a son of one of Boston's Brahmin families.

Lincoln's proclamation stated the duties of freed slaves admitted into Union service would be "to garrison forts, positions, stations, and other places, and to man vessels of all sorts." But already the president's thinking and that of the War Department had gone beyond the strictures of the Proclamation. On March 26, 1863, Lincoln wrote a letter to the military governor of Tennessee, Andrew Johnson—a staunch Unionist, but a recalcitrant Democrat on the "Negro Question"—words that reflected the extent to which United States policy had been revolutionized. "The bare sight of fifty thousand armed and drilled black soldiers on the banks of the Mississippi, would end the rebellion at once. And who doubts that we can present that sight, if we but take hold in earnest."[9] After a storm of criticism over the policy, Lincoln later would categorically assert had he not reverted to these measures, the government would have been "compelled to abandon the war in three weeks."

In early March Adj.-Gen. Lorenzo Thomas had been sent on an inspection tour in the valley of the Mississippi under Secretary of War Stanton's specific orders, which read:

The President desires that you should confer freely with Major-General Grant, and the officers with whom you may have communication, and explain to them the importance attached by the Government to the use of the colored population emancipated by the President's Proclamation, and particularly for the organization of their labor and military strength. . . .

You are authorized in this connection, to issue in the name of this department, letters of appointment for field and company officers, and to organize such troops for military service to the utmost extent to which they can be obtained in accordance with the rules and regulations of the service.[10]

At Cooper Institute in New York City, in a speech titled "The Proclamation and a Negro Army," Frederick Douglass's stirring oratory turned upon this great transformation taking place. He began:

> I congratulate you, upon what may be called the greatest event of our nation's history if not the greatest event of the century. . . . In the hurry and excitement of the moment, it is difficult to grasp the full and complete significance of President Lincoln's proclamation. The change in the attitude of the Government is vast and startling. For more than sixty years the Federal Government has been little better than a stupendous engine of slavery and oppression through which slavery has ruled us, as with a rod of iron. . . . The boast that Cotton is King was no empty boast. Assuming that our Government and people will sustain the President and his proclamation, we can scarcely conceive of a more complete revolution in the position of a nation. . . . I hail it as the doom of Slavery in all the States. I hail it as the end of all that miserable statesmanship, which for sixty years juggled and deceived the people, by professing to reconcile what is irreconcilable. . . . Color is no longer a crime or badge of bondage. At last the outspread wings of the American Eagle afford shelter and protection to men of all colors, all countries, and all climes, and the long oppressed black man may honorably fall or gloriously flourish under the star-spangled banner. (Applause) I stand here tonight not only as a colored man and an American, but, by the express decision of the Attorney-General of the United States, as a colored citizen, having, in common with all other citizens, a stake in the safety, prosperity, honor, and glory of a common country. (Cheering) We are all liberated by this proclamation. Everybody is liberated. The white man is liberated, the black man is liberated, the brave men now fighting the battles of their country against rebels and traitors are now liberated, and may strike with all their might, even if they do by thus manfully striking hurt the Rebels, at their most sensitive point. . . . (Applause) I congratulate you upon this amazing change—this amazing approximation toward the sacred truth of human liberty. . . . All the space between man's mind and God's mind, says Parker, is crowded with truths that wait to be discovered and organized into law for the better government of society.

Mr. Lincoln has not exactly discovered a new truth, but he has dared, in this dark hour of national peril, to apply an old truth,

Officers and men of the 1st Regiment of United States Colored Troops.

long ago acknowledged in theory by the nation—a truth which carried the American people safely through the war for independence, and one which will carry us, as I believe, safely through the present terrible and sanguinary conflict for national life.[11]

At the end of his speech Douglass came back to the touchstone from which he was never far—"The hope of the world—the progress of nations—the triumph of the truth and the reign of reason and righteousness among men are conditioned on free discussion. Good old John Brown (loud applause) was a madman at Harper's Ferry. Two years pass away, and the nation is as mad as he. (Great cheering) Every General and every soldier that now goes in good faith to Old Virginia, goes there for the very purpose that sent honest John Brown to Harper's Ferry."

The recruitment of black regiments was being dearly paid for by an exasperating backlash among many whites in the North, which in turn undermined the motivation for new enrollments. The black units, moreover, were to be segregated and commanded by white officers, some of whom would have little respect for their charges. It soon became apparent that the first battle would be for a chance for the newly enlisted regiments to prove themselves in combat. Would they fight, or would they throw down their arms rather than face the enemy? Douglass followed his exhortation to arms in the pages of his April *Monthly* with a personal statement titled "Another Word to Colored Men." As 1776 represented the hour of trial for white men, he wrote, 1863 was to try the souls of black men. The nation's gaze was now fixed upon them, and "They stand ready to applaud, or to hurl the bolt of condemnation—Which shall it be, my brave and strong hearted brothers?"

During the early months of 1863 the leader of the peace faction of the Democratic Party, the so-called Copperheads, defeated in his redrawn district in the election in 1862, set out his proposal for ending the war and his indictment of Lincoln's prosecution of it. In his farewell speech on January 14 before the House, subsequently published in pamphlet form for wider circulation, Clement L. Vallandigham conveyed what became the plank of his party for the remaining two years of Lincoln's term.

Republican fanaticism, he held, had provoked the war! And what had it achieved? The answer was to be heard in the stilled voices of the dead at Fredericksburg and Vicksburg; while the only trophies gained were "defeat," "debt," and "sepulchers." In the previous twenty months the country had been made into one of the worst despotisms on earth, but the South would never be conquered, he cautioned. The way out was to stop the fighting, make an armistice, and withdraw the northern armies from the seceded states. There was "more of barbarism and sin . . . in the continuance of this war . . . and the enslavement of the white race by debt and taxes and arbitrary power" than in black slavery. "In considering terms of settlement," Vallandigham laid out, "we look only to the welfare, and safety of the white race, without reference to the effect that settlement may have on the African." And he warned: "The people of the West demand peace, and they begin to more than suspect that New England is in the way."

In a mass meeting in New York, Democrats resolved that the war was unconstitutional and therefore illegal, and should not be sustained; while New York's Governor Seymour denounced Lincoln's emancipation as "bloody," "barbarous," and "revolutionary." In the Northwest, criticism of the government's course was growing with each defeat and was to become so acute that Lincoln remarked to Charles Sumner that he feared "the fire in the rear more than our military chances." As talk of a "northwest confederacy" gained currency, Ohio representative Samuel S. Cox foresaw "the erection of the states watered by the Mississippi and its tributaries into an independent Republic." Both lower houses of the legislatures in Indiana and Illinois, filled with newly elected Democrats, passed resolutions calling for an armistice to be followed by a peace conference, while Grant was compelled to disband two regiments of his army made up of fellow Illinoisans embittered over emancipation. Letter-writers urged soldiers to desert with the promise they would be protected by those back home; one letter stating, "the people are so enraged that you need not be alarmed if you hear of the whole of our Northwest killing off the abolitionists."[12]

A measure of the desperation gripping the North was seen in editorials in the *Tribune*; Horace Greeley, so recently urging Lincoln to adopt emancipation, began advocating the mediation of a European power to end the war, even addressing a letter to Vallandigham seeking his cooperation. In his *Monthly* in April 1863, Douglass wrote a stinging rebuke against the voice of the illustrious editor and the rising chorus for "peace" under the title "Do Not Forget Truth and Justice." He wrote:

> We know not where to look for a more marked and striking example of falling from principle than Mr. Greeley exhibits when he holds out to the rebels the idea that they may still preserve Slavery, by a return to the Union—and that the President's Emancipation Proclamation shall thereby be rendered inoperative and void. The suggestion is as mad as it is base—and that it should come from such a quarter, is all the proof needed of the sad prevalence of irreverence for the laws of truth, justice and humanity, even in the highest circles of American influence. Coming from such a quarter, the suggestion is a greater calamity than would be the loss of many battles. It is an attack on the soul of the nation. It proposes to break down the moral constitution of the nation as a means of securing a political Union.

At the end of 1862, after Fredericksburg and Stone's River, quiet fell upon the front lines that continued for three months. This calm was shattered on March 15 when Union gunboats under Farragut attempted to blast their way past the Confederate bastion at Port Hudson, Louisiana. Situated on a bend in the Mississippi some sixteen miles above Baton Rouge and three hundred miles below

Vicksburg, it gained strength by sitting atop a towering, nearly perpendicular cliff. Its heavy guns chastened Farragut's flotilla with plunging fire. Then on April 5 a large Union fleet, including eight ironclads, attempted to force entry into Charleston Harbor, and were battered back by Beauregard's heavy cannonry. But the great objective of the North under the oft-repeated headline "On to Richmond," abandoned after Burnside's ignominious "mud march," had been renewed under the leadership of "Fighting" Joseph Hooker. A brilliant organizer and strategic planner, though reviled in the South as an "abolitionist," Hooker had taken energetic measures to revitalize the morale and fighting capacity of his army. Rations for the troops were improved, camps and hospitals subjected to hygienic measures, shady quartermasters removed, and furloughs granted.

Hooker's army of 134,000 men was presented to the president upon his review as "the finest on the planet." The *Tribune* concurred: Hooker's was "an army of veterans, superior to that of the Peninsula," it noted. Had he, instead of McClellan, been at the head of the Army of the Potomac, "Fighting Joe" himself reported before the congressional Committee on the Conduct of the War, he could have marched into Richmond at any time. More than once disappointed by military suitors, Lincoln wryly remarked when he heard of Hooker's conceit, "The hen is the wisest of all the animal creation, because she never cackles until the egg is laid."[13]

Meanwhile, the problem for Grant was how to get sufficient quantities of men, guns, and supplies down the Mississippi for an attack on Vicksburg. To the west and north of the citadel was a lethal environment of streams and swamps presenting a formidable obstacle. The best approach was from the east; but how could he get troops and equipment there? Grant tried a number of schemes, the first involving digging a channel with "contraband" labor to allow passage of gunboats without going directly under the guns of Vicksburg. But rains and the rising river in February threatened the men with drowning. Grant's next project, known as the Lake Providence route, involved navigating the lake and Louisiana's rivers, swamps, and bayous, to gain firm ground below Vicksburg. This gargantuan task was likewise called off after the expenditure of futile effort in dredging and felling trees below water. The next approach was through the dense growth of the Yazoo Delta. Levees above Vicksburg were blown and the rivers flooded, but as the fleet ventured forth, overhanging cypress and cottonwood trees hampered their movement. As carping began to rise that Grant was getting nowhere and should be dismissed, Lincoln refused. "I think Grant has hardly a friend left, except myself," he said. "What I want . . . is generals who will fight battles and win victories. Grant has done this, and I propose to stand by him."[14]

With Grant's failure to circumvent Vicksburg, Pemberton became convinced that the Union general's design was now to withdraw from the Mississippi and join the fight against Bragg in Tennessee. On April 13 Pemberton telegraphed Johnston, elevated by Richmond to overall command in Mississippi: "I am satisfied Rosecrans will be reinforced from Grant's army. Shall I order troops to Tullahoma?" Realizing that "the problem for us was to move forward to a decisive victory, or our cause was lost," Grant determined instead upon an exploit that would mark him as one of the most dogged warriors of the war. He would accomplish this by marching an army of 23,000 down the west bank of the Mississippi through one hundred and fifty miles of hostile country. The arrangement also called for running a fleet of transports and gunboats past the batteries at Vicksburg to meet the troops in Louisiana, ferry them across the mile-wide river, then march through the heart of Mississippi by way of Jackson, coming up to the rear of Vicksburg.

In the last weeks of April, as a diversion in the east, a 1,700-man brigade of Union cavalry was sent slashing through Mississippi on a six-hundred-mile ride to Baton Rouge, tearing up railroad tracks, burning freight cars and depots, and cutting Vicksburg's supply lines. Then on the moonless night of April 16, as civilians and officers were treated to a gala ball in Vicksburg, the festive air of the Confederate "Gibraltar of the West" was rent by the sound of "Yankee gunboats" and transports running its batteries. With the loss of only one transport and one gunboat, a few nights later six more transports made the run past the guns, with five succeeding. At month's end Grant had his army down river for a rendezvous with the fleet, a crossing made unopposed on April 30.

Uniting with Sherman's corps on May 1, a Confederate force of 6,000 men at Port Gibson was swept aside. Now, rather than let a threat develop on his right where Johnston was raising an army, Grant decided to cut his army loose; before facing Vicksburg, he would reckon first with the state capital, Jackson. Stripping all plantations along the path of their advance, they marched over a small force at Raymond on May 12, and two days later attacked the entrenchments at Jackson head on. They wrecked all the buildings that could be used for war, burning railroad terminals, foundries, arsenals, factories, and machine shops, as whole neighborhoods went up in flames. With a pall of smoke covering the capitol, Pollard noted in his account, "The negroes were invited to assist and share in the pillage. Supposing that the year of jubilee had finally come, the blacks determined to enjoy it, with this end in view, they stole everything they could carry off."[15] The streets became a riot of incongruous articles—French mirrors held aloft, boots and shoes, pieces of calico, washstands and towels, hoopskirts, bags of tobacco, parasols.

Meanwhile, Johnston, considering that saving Pemberton's army was of greater value than holding Vicksburg, ordered him to march east to effect a conjunction of their armies. But wedded to the idea of the citadel on the Mississippi, Pemberton brought his force only midway between Vicksburg and Jackson.

In Virginia, without a new threat developing after Burnside's disaster, Lee sent Longstreet to confront Union thrusts from Norfolk and the North Carolina coast, and forage for supplies, leaving 60,000 troops in miles of trenches along the Rappahannock near Fredericksburg. For his part Hooker had no intention of repeating the past mistakes of others, instead proposing to maneuver Lee out of his defenses and into an open fight. At the end of April Hooker sent 10,000 newly reorganized Union cavalry across the river to cut enemy supply lines. Feigning an advance under Gen. John Sedgwick on Fredericksburg, he flanked Lee at a crossroad known as Chancellorsville, putting his army in a nearly trackless forest called the Wilderness. Hooker's order trumpeted: "Our enemy must ingloriously fly or come out from behind his defenses and give us battle on our own ground, where certain destruction awaits him."[16]

Marching the bulk of his army west to meet Hooker on May 1, Lee left only 10,000 troops to hold Fredericksburg's defenses. But even as the advantage of attack was on his side, Hooker ordered his army to take defensive positions around Chancellorsville, lessening the effect of his numbers and the effectiveness of his artillery, and thus evening the odds.

With the enemy too strong for frontal attack, Lee and Stonewall Jackson conferred that night. Hooker's left rested on the Rappahannock and could not be turned, while his right, three miles to the west, Stuart reported, was "in the air." The problem was to get a force around them unobserved. The solution came from a local man, who pointed out a path through the forest used to haul charcoal. The plan called for a coordinated attack on Hooker's right and front, and early on May 2 Jackson led 30,000 troops and artillery on a twelve-mile march to their attack position, while Lee remained with 15,000, leaving each segment of the Confederate army greatly outnumbered. But Hooker was convinced Lee had chosen retreat, and did nothing even when alerted of Jackson's movement.

By 5:15 Jackson's troops were deployed and the attack came in a yelling charge on a two-mile front three divisions deep. Hooker's right, facing south, was rolled up, while Lee attacked his front. After recoiling two miles, Hooker succeeded in bringing Jackson's attack to a halt, forming new lines as fighting continued for several hours into a moon-flooded night. But Jackson was not yet satisfied: if the enemy could be annihilated, Washington might be compelled to abandon the war. Determined to continue his attack Jackson rode forward with staff officers to reconnoiter. After an aide asked, "General, don't you think this is the wrong place for you?" Jackson replied: "The danger is all over—the enemy is routed! Go back and tell A. P. Hill to press right on!" As he returned to his lines Jackson and his aides were mistaken by their own troops for Union cavalry. Wounded with two shots in his left arm, Jackson fell. Stuart assumed command, while the battle-hardened gladiator was removed from the field. His arm would be amputated to save his life, but he would die of pneumonia eight days later.

During the night Hooker ordered his 6th Corps commanded by Sedgwick to attack the heights at Fredericksburg and press to Lee's rear. At dawn three divisions were thrown against Marye's Heights and twice thrown back. In a bayonet charge on the third try they broke through, capturing a thousand prisoners and putting the rest of the defenders to flight. But Hooker had remained inactive, allowing Lee and Stuart to reunite and mass their artillery in the one place it could be used effectively. With the attack falling on three corps at Chancellorsville, Hooker kept three other of his corps out of the fight, when by mid-morning a cannonball hit a column in front of his headquarters and knocked him unconscious. As fighting continued, Hooker recovered only to constrict his lines, his flanks touching upon both the Rapidan and the Rappahannock rivers. Both armies now broke off to rescue the wounded threatened by brush fires.

As Lee rode before the burning Chancellor mansion, wildly cheered by his troops, he received word of Sedgwick's advance. With Hooker corralled in front of him, Lee dispatched a division to stop Sedgwick midway between Fredericksburg and Chancellorsville. The next day Lee took yet another division—leaving Stuart with only 25,000 troops to hold Hooker's 75,000—renewed the attack on Sedgwick, but was repulsed. That night Hooker's corps commanders unanimously recommended they counterattack. But Hooker had had enough. Sedgwick ordered his corps across the river; in a rainstorm the following night Hooker would do the same. Receiving reports in the War Department telegraph office, Lincoln was stricken, exclaiming, "My God! My God! What will the country say?" Hearing the news Charles Sumner wailed, "Lost, lost, all is lost."[17]

Robert E. Lee had achieved his greatest victory of arms, yet Jackson's passing cost him his "right arm." While not being able to destroy his foe, he had defeated an opponent whose superiority in numbers was more than three to one. He now appeared invincible, and said of his own Army of Northern Virginia, "there never were such men in an army before. They will go anywhere and do anything if properly led."[18]

In the space of six months the Army of the Potomac had twice been turned back across the Rappahannock, but each time they retained a footing in the hills of Stafford Heights from which they could not easily be assailed. Lee and his army could not remain where they were, so Lee resolved the time was right to renew the experiment begun the previous year of carrying the war north of the Potomac.

In early May as the battle of Chancellorsville was concluding, with replenished stores from southeastern Virginia, James Longstreet was bringing his two divisions north. Stopping in Richmond on May 6 for a meeting with Confederate secretary of war James Seddon, he proposed his divisions move on to Tennessee, where he, Johnston, and Bragg would join forces and drive Rosecrans back across the Ohio, compelling Grant to break off at Vicksburg. Instead Seddon proposed that he reinforce Johnston and Pemberton to thrash Grant—a proposal favored by Davis—then turn to deal with Rosecrans, who as yet had scarcely stirred since the battle at Stone's River. The prospects of the fortress on the Mississippi at that moment, however, looked anything but dire, and as Lee arrived, he weighed in against both proposals. It would take too long to reach Mississippi with the damaged condition of southern railroads, and he forecast if Vicksburg could hold out "the climate in June will force the enemy to retire." The Army of the Potomac may in the meantime return to the offensive in Virginia, said Lee; "it becomes a question between Virginia and Mississippi."[19]

After losing command of the Army of the
Potomac, Burnside had been appointed commander of the Department of the Ohio,
embracing those states bordering the river. In
April he issued a general order intended to
suppress growing Copperhead dissent,
declaring that anyone committing offenses
that could be construed as disloyal or treasonous would be subject to trial by military
court and face punishment by imprisonment,
banishment, or death. As the first news was
coming over the wires of Hooker's collision
with Lee, Vallandigham, trying to ignite his

Robert E. Lee.

bid for the Democratic gubernatorial nomination in Ohio, was speaking at a rally in Mount Vernon in that state. With Burnside's staff officer on hand, Vallandigham denounced the government's "wicked,
cruel and unnecessary war," waged, he said, reiterating his standard
attack, "for the purpose of crushing out liberty and erecting a despotism . . . for the freedom of the blacks and the enslavement of the
whites."[20]

A few nights later, Vallandigham, his wife, and his sister-in-law
were startled, and the women frightened to hysterics, when soldiers
broke down the front door to their home in Dayton. Vallandigham
was seized and placed under arrest and hurried off to Cincinnati to
face a military court. Convicted of disloyalty on May 6, he was sentenced to imprisonment for the duration of the war. Soon afterward,
Lincoln, in a view he expressed to opponents, defended the arrest and
conviction because Vallandigham "was laboring, with some effect, to
prevent the raising of troops" and was encouraging desertions. His
speech was damaging to the army, Lincoln continued, "upon which
the existence and vigor of which the life of the nation depends" and
whereby, "under cover of 'liberty of speech,' 'liberty of the press,' and
'Habeas Corpus,' [the rebels] hoped to keep on foot . . . a most efficient corps of spies, informers, suppliers, and aiders and abettors of
their cause."[21]

A new direction was being signalized by Lincoln's government—
that the war had become national, extending throughout both North
and South. But in the swirl of events in spring 1863, Lincoln would
have to face the embarrassing predicament brought on by the arrest
and imprisonment of the former representative of Ohio. In an effort

to minimize the political backlash, Lincoln commuted Vallandigham's sentence to banishment, and on May 25 Union cavalry would escort him under flag of truce to Confederate lines south of Murfreesboro. He was to travel over a thousand miles through as yet unassailed territory, giving speeches and meeting with political leaders and army officers of the Confederacy. Reaching the port of Wilmington, North Carolina, he would board a blockade-runner for Canada, conducting his interrupted campaign for Ohio's governorship from Windsor, and win the nomination of his party later that year. When allowed to reenter the North that fall he would report that he had met no one in the South who was not prepared to die rather than yield to the pressure of arms. But after a career based on compromise, he had deluded himself that such as had occurred before was about to happen again. By this point the beleaguered leaders of the Confederate States of America had thrown aside all store in compromise, yet Vallandigham had given them hope that if they could hold out the year "the peace party of the North would sweep the Lincoln dynasty out of existence."22

Pursuant to Lincoln's policy on recruitment of black troops, in the spring of 1863 Adjutant-General Thomas made an appearance at Lake Providence before a large assembly of the army, composed of volunteers from all parts of the North. Thomas's voice rang out:

Fellow-Soldiers, I came from Washington clothed with the fullest power. . . . With this power, I can act as if the President of the United States were himself present. I am directed to refer nothing to Washington, but to act promptly—what I have to do to do at once; to strike down the unworthy and to elevate the deserving. Look along the river, and see the multitude of deserted plantations upon its banks. These are the places for these freedmen, where they can be self-sustaining and self-supporting. All of you will some day be on picket-duty; and I charge you all, if any of this unfortunate race come within your lines, that you do not turn them away, but receive them kindly and cordially. They are to be encouraged to come to us; they are to be received with open arms; they are to be fed and clothed; they are to be armed. This is the policy that has been fully determined upon. I am here to say that I am authorized to raise as many regiments of blacks as I can. I am authorized to give commissions, from the highest to the lowest; and I desire those persons who are earnest in this work to take hold of it. I desire only those whose hearts are in it, and to them alone will I give commissions.[23]

Now Thomas explained the shift in policy to the officer corps. The rebels had been able to send every available fighting man into the field because they were able to keep their slaves at home to raise subsistence for their armies; they had been able to bring their strength against the government, while the North was able to send only a portion of its eligible fighting men, being compelled to leave behind a portion to support the wants of an immense army. He told the officers: "The administration has determined to take from the rebels this source of supply—to take their Negroes and compel them to send back a portion of their whites to cultivate their deserted plantations—and very poor persons they would be to fill the place of the dark-hued laborer. They must do this, or their armies will starve."

On May 4 Major-General Hunter addressed a letter to Governor Andrew expressing his "eminent satisfaction with the results of the organization of Negro regiments" in his department. While the freedmen were energetic and brave, he wrote, familiar with the country, and possessing great natural aptitude for arms, they were also "deeply imbued with that religious sentiment—call it fanaticism, such as like, which made the soldiers of Cromwell invincible. They believe that now is the time appointed by God for their deliverance; and, under the heroic incitement of this faith, I believe them capable of showing a courage, and persistency of purpose, which must, in the end, extort both victory and admiration."[24]

Hunter now looked to future and expanded operations with the addition of the Massachusetts 54th, and he concluded: "With a brigade of liberated slaves already in the field, a few more regiments of intelligent colored men from the North would soon place this force in a condition to make extensive incursion upon the main land, through the most densely populated slave regions; and, from expeditions of this character, I make no doubt the most beneficial results would arise."

The forceful laying out of Lincoln's new policy by Thomas and Hunter's initiative, predicated on the difficulties besetting the Union armies, had astonishing implications. On surveying the situation of the "colored" population in America before the thirteenth annual meeting of the American Anti-Slavery Society, held in New York City on May 12, Robert Purvis was to declare, "the slave power no longer rules at Washington," and he offered three actions of the cabinet to bolster his argument. First, Attorney General Bates had pronounced blacks citizens, overturning "the damnable doctrine of the detestable Taney." Second, Secretary Seward had recognized the right of blacks to "take out a passport and travel to the uttermost parts of the earth protected by the broad aegis of the government" as he had done in the case of Henry Highland Garnet on August 26, 1861, so that his voice would carry to England to advocate for the cause of the slave and the Union. Third, the secretary of war and the entire administration had recognized the black as "a citizen soldier." Blacks had thus been invested with the prerogative of which they had been basely robbed, said Purvis, and he affirmed, "I am proud to be an American citizen." He continued: "You know, Mr. Chairman, how bitterly I used to denounce the United States as the basest despotism the sun ever shone upon; and I take nothing back that I ever said. When this government was, as it used to be a slaveholding oligarchy . . . I hated it with a wrath which words could not express, and I denounced it with all the bitterness of my indignant soul. . . . But now I forget the past; joy fills my soul at the prospects of the future. . . . The good time which has so long been coming is at hand. I feel it, I see it in the air."25

On May 19 and again on the 22nd at Vicksburg, Grant would be repulsed in bloody frontal assaults on his defenses, and soon afterward Banks would likewise be bloodied at Port Hudson. Just prior to any of this, however, Lee was back in Richmond for a meeting with Davis with a plan for an invasion of Pennsylvania intended to inflict a devastating setback on the Yankees. Lee would remove the threat on the Rappahannock by maneuvering Hooker out of Virginia, clear the Shenandoah of Union troops, invade the enemy's country where he could supply his troops while relieving war-ravaged Virginia, and prepare a blow against the train center at Harrisburg and the bridges across the Susquehanna. The capitol at Harrisburg itself could be captured, and the campaign continued through the rich farming districts of Lancaster County, with the possibility of striking Philadelphia and Washington, and even New York. With Lee's fame spreading worldwide, the general and his army possessed immense credibility and Davis had little choice but to acquiesce. But Lee's success, he saw, might result in Grant's strategic withdrawal from his home state.

Admirers of Lee marvel at the daring gambler that lay concealed beneath a taciturn demeanor. His most remarkable characteristics, it is often said, were his devotion to duty and to Virginia, with little intimation of any political views. Yet undeniably his military strategies were predicated upon them, as they were informed and based upon his faith in "the rising peace party of the North" which he saw "as a means of dividing and weakening our enemies." True as it was, Lee wrote Davis on June 10, that the Copperheads favored reunion as the object of peace negotiations, whereas the South aimed for independence, but it would not hurt to play along to weaken northern support for the war and discredit the Republicans. Lee wrote, that effect "after all is what we are interested in bringing about. When peace is proposed to us it will be time enough to discuss its terms, and it is not part of prudence to spurn the proposition in advance, merely because those who made it believe, or affect to believe, that it will result in bringing us back to the Union."[26] A successful strike into the North, too, in the view of all in the Confederate cabinet with only a few exceptions, might reopen the question of foreign recognition. It was just the right time for an overture for peace—and Alexander Stephens offered to be the one to approach his old friend Lincoln under a flag of truce.

As Grant was tearing through Mississippi, Banks had been ordered to come upriver to assault Port Hudson. Situated in the sugar- and cotton-growing region of Louisiana along Bayou Teche, Port Hudson possessed formidable natural defenses, further strengthened by man-made military projections. Between it and Vicksburg the Confederacy controlled nearly three hundred miles of the Mississippi, blocking all commerce from the northwest and allowing communication between Richmond with the slave states west of the river. The Native Guard, renamed the 1st Louisiana Regiment, and another comprised of freedmen, the 3rd Louisiana Regiment, were part of Banks's force. Commanded by Gen. William Dwight, the 3rd's officers, in line with government policy, were white, but those of the 1st were black, the only regiment to be so officered during the war. While the regiments were in Baton Rouge white officers demanded that the 1st's officers be dismissed, and a number, amidst jeers and taunts, were compelled to tear up their commissions. William Wells Brown in *The Negro in the American Rebellion: His Heroism and His Fidelity,* wrote:

Among these were First Lieut. Joseph Howard of Company I, and Second Lieut. Joseph G. Parker of Company C. These gentlemen were both possessed of ample wealth, and had entered the army, not as a matter of speculation . . . but from a love of military life. Lieut. Howard was a man of more than ordinary ability in military tactics; and a braver or more daring officer could not be found in the Valley of the Mississippi. He was well educated, speaking the English, French, and Spanish languages fluently, and was considered a scholar of rare literary attainments. He, with his friend Parker, felt sorely the humiliation attending their dismissal from the army, and seldom showed themselves on the streets of their native city, to which they had returned.

On May 26 Banks arrived and ordered that an advance begin on the rifle pits and gun emplacements of Port Hudson. It was a lovely southern night, William Wells Brown wrote, "with its silvery moonshine on the gleaming waters of the Mississippi . . . while the fresh soft breeze was bearing such sweet scents from the odoriferous trees and plants." As a portent the day would be hot; in the morning, it was reported, the sun was deep red. As Banks pushed his infantry within a mile of the works, clouds of dust rose from the parched earth. The assault, lasting from early morning to mid-afternoon, would fall along the line, but the decision to use black troops on the gun emplacements, where the most difficult fighting occurred, had come down to the fact, Brown noted, that to General Dwight, the officer in their command, they were "only niggars."

Beneath these breastworks was a bayou too deep for man to ford; on the left was a massed battery whose guns could rake the entire field, while commanding the front were three or four guns, and on the right, six heavy pieces that could enfilade the rear as the charge was made. It was clear the guns could not be taken. Nevertheless the line officers of the 1st Louisiana, foremost of who was Capt. André Callioux, began to prepare their troops, while the white officers and men looked on, asking whether these blacks could withstand so severe a test, even as the enemy, feeling their power, offered defiance. When the attack came, exploding shells covered entire companies in felled branches while at every pace the ranks were reaped in a terrible harvest. But the advancing troops closed up as others fell. When up close they fired a last volley and went in with bayonets. A *New York Herald* correspondent reported "one negro . . . with a rebel soldier in his grasp, tearing the flesh from his face with his teeth, other weapons having failed him."[27] Soon the ground was littered with

The 1st and 3rd Louisiana Regiments assaulting the Confederate breastworks at Port Hudson on May 27, 1863.

maimed and mangled bodies, yet, wrote Brown, "charge after charge was ordered and carried out under all these disasters with Spartan firmness."

A scene that became emblematic concerned the 1st's Color Sergeant Anselmas Planciancoise. As he received the regimental flag before the charge he vowed, "Colonel, I will bring back the colors with honor or report to God the reason why." That morning, it would be reported, pierced by five shots, he clung to the flag as he fell, while two corporals struggled to raise the banner. One of these also succumbed to the intense musketry, the other charging ahead to lead the advance. Captain Callioux—who, Brown wrote, prided himself in having the darkest hued skin in New Orleans—with his left arm broken and dangling, his right holding up his unsheathed sword, was seen, his hoarse voice faintly cheering his men, before he went down. Brown added that his demise "so exasperated his men, that they appeared to be filled with new enthusiasm; and they rushed forward with a recklessness that probably has never been surpassed."

After this assault, the 3rd Louisiana was ordered to follow up. These charges attained the perimeter of the defenses, but fell fifty paces short of the guns. A week later the *Herald* reported: "The first Regiment Louisiana . . . went on the advance, and, when they came out, six hundred out of nine hundred men could not be accounted for. It is said on every side that they fought with the desperation of tigers." Two days later the *New York Tribune* sounded a similar refrain: "That heap of six hundred corpses, lying there dark and grim and silent before and within the rebel works, is a better proclamation of freedom than even President Lincoln's."[28]

The day after the assault on Port Hudson, with orders received to proceed to the Department of the South, the 54th Regiment of Massachusetts Volunteer Infantry broke camp and entrained for Boston. Reaching the depot the regiment marched through the city to the Common as people came out cheering in the streets. The *Boston Transcript* reported: "Since Massachusetts first began to send her brave sons into the field, no single regiment has attracted larger crowds into the street than the 54th." After only three months of camp life the troops marched, a reporter noted, with a "general precision attending their evolutions."[29] At the Common the regiment formed a hollow square, while the stand of dignitaries and officials and invited guests occupied the middle, among whom stood Frederick Douglass, who had an especial relation with the regiment, not only for the fact that two of his sons were of its rank. After prayer offered by the Reverend Grimes, Governor Andrew and Colonel Shaw would address the assembly.

In his speech the governor presented the regimental colors, the national and state flags, an emblematic banner bearing the figure of the Goddess of Justice on one side and the words "Liberty, Loyalty, and Unity" on the other, and a banner with a gold cross with a star against a blue field with the motto "In hoc signo vinces" written beneath. As he handed over the emblem, in his remarks the governor said: "The cross which represents the passion of our Lord, I dare to pass into your soldier hands; for we are fighting now a battle not merely for country, not merely for humanity, not only for civilization, but for the religion of our Lord itself. When this cause shall ultimately fall, if ever failure at the last shall be possible, it will only fail when the last patriot, the last philanthropist, and the last Christian shall have tasted death, and left no descendants behind them upon the soil of Massachusetts."[30]

Remarking on this solemn and dignified picture, William Wells Brown wrote: "By a strange, and yet not strange, providence, God has made this despised race the bearers of his standard. They are thus the real leaders of the nation."

After the remarks of the governor, Colonel Shaw delivered words whose youthful eloquence its speaker thought "small potatoes" compared to those preceding. Now the colors were carried to their place in line by the guard and the regiment reviewed by the governor. Brown wrote: "Thence they marched out of the Common, down Tremont Street, down Court Street, by the Court House, chained hardly a decade ago to save slavery and the Union. Thence down State Street, trampling on the very pavement over which Sims and Burns marched to their fate, encompassed by soldiers of the United States."

When they passed Wendell Phillips's residence some officers of the 54th raised their hats in regard, as Phillips looked down from the room of his invalid wife. On the balcony of Phillips's study below stood William Lloyd Garrison with his daughter, her hands resting on the sculptor Edwin Brackett's bust of John Brown, placed there symbolically to oversee the procession. As the regiment had wheeled onto State Street the band struck up "John Brown's Body," to the enthusiastic cheers of the throng. Shortly after 1 o'clock the regiment reached Battery Wharf and boarded the *De Molay*, getting on also its guns and horses. Douglass followed the regiment from the train station into the Common, standing as an honored guest on the reviewing platform, although he delivered no remarks, and then walked beside them on their march to the wharf. There he boarded the vessel with the troops and officers, remaining even as it eased out of its moorings, bound for Port Royal, South Carolina. Only when they were well out into the harbor did he bid farewell and Godspeed to the regiment, returning on a tug.

Far down the seascape into which Douglass and others may have gazed, before the terrible wounds of battle could be staunched, the account of the 1st and 3rd Louisiana Volunteer regiments was descending the Mississippi to New Orleans, reaching those whose sympathies lay with the troops, who implicitly understood the significance of their contribution. It was not at all important they had not been the victors—only that they had died.

When the news reached Howard and Parker, the disgraced officers of the 1st Louisiana at once determined to travel upriver and rejoin their regiment, as privates if need be. When they arrived, with the troops still in proximity of the works, the two were hailed in a joyous demonstration, and reassigned to command a company each. Only days after the onslaught on May 27 the white officers who had taken their places had renounced their commissions. In a renewed assault on June 5 Howard was to distinguish himself by leading his company in a charge against the enemy's rifle pits, capturing and holding them for three hours before retreating under withering artillery fire. That same night General Banks came to see him, shaking his hand and congratulating him on his decision to return, remarking to his aides that "the man" was worthy of an elevated position.

In spring 1863, in a development wholly out of its unique situation, the Department of the South began a more ambitious plan of operations. Headquartered in the environs of Hilton Head, Port Royal, and Beaufort, the command had the task of projecting federal power on the southern Atlantic coastal region and on the various islands and coastal rivers. With limited resources General Hunter began to call upon the abundant local African-descended inhabitants for his soldiery. Higginson arrived to command the first regiment organized, while James Montgomery headed another. With Gen. Rufus Saxton in charge of their training and the Massachusetts governor supplying two northern regiments for the department, it differed greatly from any other command in the war. Given her previous experience of piloting fugitives on Maryland's Eastern Shore, it is not surprising that Harriet Tubman soon made the transition to work as a scout and a spy in her new surroundings, having in her band nine men who had grown up in the area. With this company Tubman began to probe the condition and deployment of the enemy in the coastal area, and develop an active Underground Railroad into Union lines. Soon she became quite a recognized figure, carrying a rifle and a satchel filled with first aid necessities, many officers saluting her respectfully as she moved about. She reported to Generals Hunter, Saxton, or Gilmore. With the arrival of Montgomery, she was presented with another opportunity of broadening her activity.

Since receiving his baptism of fire in Kansas, like John Brown, Montgomery felt no obligation to observe the usual amenities of war being stipulated elsewhere, holding that the most expeditious way to attack a slaveholding enemy was to ruthlessly undermine its economic viability—burning buildings, seizing or despoiling property, and forcefully emancipating slaves. Early in the spring he had led an expedition in a raid on Palatka, Florida, attaining a notoriety that marked him personally for destruction by the Confederacy.

Now Tubman and the new commander would jointly conduct a noteworthy campaign on the Combahee River, a narrow stream running about thirty miles from South Carolina's interior, emptying into St. Helena Sound. Only ten or twelve miles upriver were a few great estates, each with lordly homes bearing such names of slaveholding families as Middleton, Blake, Lowndes, and Heyward. To protect these plantations the rebels had established picket lines from Combahee Ferry to Green Pond, and placed a number of torpedoes in the river to prevent navigation by Yankee gunboats. Tubman's spies had penetrated these lines, ascertained the location of the torpedoes, and determined that the area was ripe for invasion.

With the objective of taking up the torpedoes, destroying railroads and bridges, and cutting off supplies to rebel troops, General Hunter asked Harriet to accompany an expedition of three gunboats. She replied she would if Colonel Montgomery were appointed commander. Set to begin on June 2, the whole thing took on such an air that it became noised about in the *New York Tribune*, which announced "that the Negro troops at Hilton Head, S.C., will soon start upon an expedition, under the command of Colonel Montgomery."

Discussing the movement afterward the Confederate investigating officer wrote: "The enemy seems to have been well posted as to the character and capacity of our troops and their small chance of encountering opposition, and to have been well guided by persons thoroughly acquainted with the river and country."[31]

The party, including one hundred fifty troops, set off from Port Royal in three gunboats. In the early hours of June 2, navigating the water between Beaufort and St. Helena Island, the force came into the sound. At the mouth of the river enemy pickets at Field's Point dispatched word of this to the commander at Green Pond. As the abolitionist crafts came upriver, other pickets stationed at Combahee Ferry and Chisholmville did the same but, being poorly armed and outnumbered, began to withdraw. Now the steamers began to land parties of ten to twelve men on both sides of the river to rouse the plantations. The Confederate officer entered into his report that the boats "passed safely the point where the torpedoes were placed and finally reached the bridge at the ferry, which they immediately commenced cutting way, landed to all appearances a group at Mr. Middleton's and in a few minutes his buildings were in flames."[32] It was not long before several buildings on other large plantations were also set on fire, the slaves fleeing toward "Lincoln's gunboats" and the plantation owners, their families, and overseers fleeing inland.

As this was happening a company of rebels went into action and, trying to stop the flight of slaves, shot and killed a young girl. It was about 6:30 in the morning when the steamboats blew their whistles, and hundreds of slaves who had been hiding in the woods or had been cowed before overseers' whips, began to flood to the riverbank. Tubman later related to Sarah Bradford:

> I never saw such a sight. We laughed and laughed and laughed. Here you'd see a woman with a pail on her head, rice a-smoking in it just as she'd taken it from the fire, young one hanging on behind, one hand around her forehead to hold on, the other hand digging into the rice pot, eating with all its might; a-hold of her dress two or three more; down her back a bag with a pig in it. One woman brought two pigs, a white and a black one; we took them all on board; named the white pig Beauregard and the black pig Jeff Davis. Sometimes the women would come with twins hanging around their necks; it appears I never saw so many twins in my life; bags on their shoulders, baskets on their heads, and young ones tagging behind, all loaded; pigs squealing, chickens screaming, young ones squealing.[33]

The rush became so frantic that from the upper deck of the steamer Montgomery called down, "Moses, you'll have to give them a song." As Tubman sang a verse, those clinging to the boats were filled with such feeling that they raised their hands and shouted "Glory!" Only then were the boats able to cast off. In the struggle to get the woman with the pigs aboard Tubman's foot pinned the hem of her dress and it was nearly torn off. After this she vowed to get and wear bloomers in the style of the women's suffragists, and did obtain a pair from Boston.

Seven hundred and fifty-six slaves were brought away to freedom in this campaign. All of the males of military age were enrolled in the new black regiments and the women with their children took up lodgings at a place called Old Fort Plantation in Beaufort, afterward known as Montgomery Hill. This was a row of more than a dozen boxlike buildings divided into four compartments, each with a fireplace and a single window with a board shutter. Every compartment included a double row of berths, with benches and tables filling out the accruements. In her *First Days Amongst the Contrabands*, Elizabeth Hyde Botume wrote: "It was rough and crude living, and compact and hasty, but the Negroes were free and they preferred this to the slightly larger cabins of the plantations they had quit."

Blacks captured in arms were to be treated as felons and insurrectionists, President Davis had declared, as would the white officers leading them. Confederate resolve was demonstrated late in 1862 with the execution of four captured blacks wearing blue from South Carolina's 1st Colored Regiment, and again in the murder of twenty black teamsters near Murfreesboro in early spring 1863. The first wholesale demonstration of this policy came on June 7 at Milliken's Bend, Louisiana, where the ferocity with which black soldiers fought back, after seeing their overtaken comrades murdered, marked a turning point in the thinking of many. William Wells Brown noted this circumstance when he wrote: "This battle satisfied the slave-master of the South that their charm was gone; and that the Negro, as a slave, was lost forever. Yet there was one fact connected with the battle . . . which will descend to posterity, as testimony against the humanity of the slaveholders; and that is, that no Negro was ever found alive that was taken a prisoner by the rebels in this fight."

Milliken's Bend is the name borne by the wayward meandering of the Mississippi only miles above its confluence with the Yazoo, and about twenty miles above another bend in the river on whose bluffs Vicksburg sits. Here were stationed two newly organized, and hence untrained and poorly armed, black regiments, the 9th and 11th Louisiana, comprising together about five hundred men, with another company of one hundred white soldiers from the 23rd Iowa. During the early morning of June 7 the Texas 16th, one of three brigades sent by Gen. Richard Taylor (son of Zachary Taylor) from Port Hudson to assist Vicksburg, attacked in an attempt to cut Grant's supply line. The first intimation that they were about to be overrun came at 3 o'clock in the morning when one of the "contraband" recruits entered his colonel's tent—"Massa, the secesh are in camp." The officer told him to have the men load their muskets at once, whereupon the messenger reported—"We have done did dat now, massa."[34]

Over three-fifths of the combined black soldiery became casualties; but they and the gunboats stopped the enemy's attempt to give relief to Vicksburg from the west. Now Vicksburg would have to anticipate help from Johnston in the east. Grant agreeably remarked that the troops had "behaved well," particularly for raw recruits. The commander of the 9th Louisiana, Captain Miller from Galena, Illinois, wrote in his report of the battle: "I never felt more grieved and sick at heart, than when I saw how my brave soldiers had been slaughtered— one with six wounds, all the rest with two or three, none less than two wounds. Two of my colored sergeants were killed; both brave, noble men, always prompt, vigilant, and ready for the fray. I never more wish to hear the expression, 'The niggers won't fight.'"[35]

The fierce counter-charge of "contraband" troops at Milliken's Bend, the operations of the Department of the South, the experiment in reconstruction on Hilton Head each have received little account in the wider context of the war. The places and events concerned were far from the great battles, but the relationship between the various theaters stands out once the focus becomes Lincoln's strategy in the spring and summer of 1863: to win the allegiance and strength of the black masses for the Union. One of the keenest chroniclers of the era, Thomas Wentworth Higginson, observed that Sherman's pivotal march to the sea, begun in winter 1864 with its end point at Port Royal, was predicated on the success of what had appeared merely subsidiary and attendant. In his *Army Life in a Black Regiment*, Higginson wrote: "Next to the merit of those who made the march was that of those who held open the door. That service will always remain among the laurels of the black regiments."[36]

In preparing for his second invasion of Pennsylvania, Lee set as his first task the reorganization of his army. Dividing Jackson's corps with some new additions between R. S. Ewell and A. P. Hill, Longstreet completed the triumvir. Longstreet's corps decamped from Fredericksburg on June 3 in the direction of Culpeper Court House, with the other two corps departing in succession between June 5 and 8. On the former date Hooker made a demonstration-in-force, crossing the Rappahannock as if he intended to move on Lee's extended lines. Instead he recrossed the river and began packing stores and readying troops on Stafford Heights. On June 9 he sent a large force of cavalry back across the Rappahannock, surprising Jeb Stuart at Brandy Station. While screening Lee's movements, Stuart recovered sufficiently to repulse the attack in the largest cavalry clash of the war.

Evidently anticipating that Lee was withdrawing to the Shenandoah Valley, Hooker wired the War Department that he could be expected to come down upon Washington through an interior route, and since Lee and the Army of Northern Virginia were moving north, he should move south and march the Army of the Potomac into Richmond. Sensing that his general was again showing lack of resolve, Lincoln retorted: "I think Lee's Army, and not Richmond, is your true objective point. If he comes toward the Upper Potomac, follow on his flank, and on the inside track. . . . Fight him when opportunity offers."[37]

In the succeeding week Confederate troops crossed the Shenandoah at Front Royal, battering the Union garrison there, capturing Winchester and Martinsburg, then pushing the defenders out of Harper's Ferry, gaining a large amount of military stores, twenty-nine pieces of artillery, and two hundred horses, while netting more than 4,000 prisoners. When Ewell's corps crossed the Potomac at Williamsport, sending a division eastward from Chambersburg across South Mountain to York, and proceeding with the remainder toward Carlisle, Hooker offered no challenge. On June 24 the whole of A. P. Hill's corps crossed the Potomac at Shepherdstown, uniting with Longstreet at Hagerstown, and arrived at Chambersburg on the 27th. All of south-central Pennsylvania was now invested by Confederate troops, with forward deployments within artillery range of Harrisburg. The initiative lay with Lee, and with his communication firmly established on the other side of the Potomac, he was "in a position," wrote Pollard in his *Southern History of the War*, "to hurl his forces wherever he might desire." Pollard further noted, "the North was thrown into paroxysms of terror. At the first news of the invasion, Lincoln had called for a hundred thousand men to defend Washington. Governor Andrew offered the whole military strength of Massachusetts in the terrible crisis. Governor Seymour of New York, summoned McClellan to grave consultations respecting the defences of Pennsylvania. The bells were set ringing in Brooklyn. Regiment after regiment was sent off from New York to Philadelphia. The famous Seventh regiment took the field and proceeded to Harrisburg. The Dutch farmers in the valley drove their cattle to the mountains, and the archives were removed from Harrisburg."[38]

Unlike the Yankees who had been wrecking the South, Lee gave orders against pillaging, stipulating that to conciliate the people of the North and opinion in Europe, troops should show forbearance. But forced requisitions were levied, payment offered in Confederate script or IOUs, and railroad stock destroyed, shoes, clothing, horses, and cattle seized. One woman in Chambersburg recorded in her diary seeing Confederate troops seize several black women with their children, who were then driven like cattle. An iron furnace owned by Thaddeus Stevens was destroyed at Caledonia; and remembering the fugitive slave resistance in Christiana as an "outrage against Southern honor," as the saying went, Lee inquired about its location, for he wished to burn it.

Not accepting that this was as grave a crisis as any he'd been faced with, Lincoln hailed it as another opportunity to destroy Lee. Although Hooker was still in Maryland, this was proximity enough for Lee, and via couriers he ordered his commanders to concentrate the army in the Cashtown area, just east of South Mountain and west of the prosperous market town of Gettysburg. With no discernible movement by Hooker, a drawn and weary president gathered his cabinet on June 28, telling them that to his mortification Hooker had turned out to be another McClellan. Relieved of command, Hooker was replaced by Gen. George Gordon Meade.

On July 1, A. P. Hill authorized a division to move in the direction of Gettysburg. (There is a persistent myth that Hill ordered the division to claim a warehouse filled with shoes in Gettysburg, but that is not the case.) Marching out of the west that morning, the division found two brigades of Union cavalry waiting for them armed with breech-loading carbines, commanded by a Kentuckian named John Buford. With Confederate divisions hastening along the Chambersburg, Mummasburg, and Harrisburg pikes, by 10 o'clock Buford's dismounted cavalry was reinforced by Pennsylvania's 1st Infantry Corps. Gen. John Reynolds rode forward to inspect the ground and select a position for their line of battle when he was struck in the neck by several balls and was dead before his body touched ground. With Confederate troops arriving from the northeast, as the battle wore on Union troops were compelled to retreat through the town, rallying finally on Cemetery Hill.

Lee had been victorious in that day's fight, but with his foe dug in he was unable to continue the attack. Through the night from the east along the Baltimore Pike, Union reinforcements were marching at the double-quick under a full moon. By daybreak, with supply wagons and reserve artillery positioned behind, Meade had extended the defensive line along Cemetery Ridge, from Culp's Hill to Little Round Top.

West of Gettysburg on a north-south axis, where the first day's fighting had taken place, is McPherson's Ridge, crossed by the Chambersburg Pike. Just south of the town and west of Cemetery Ridge, and running parallel to it, is Seminary Ridge. The Emmitsburg Road crosses diagonally between these ridges, and in the interval was a peach orchard and wheat field, and opposite the hill called Little

Round Top a jumbled labyrinth of stones called Devil's Den. As the Emmitsburg Road skirted Cemetery Ridge before terminating at Gettysburg, there was a stone wall whose tangents formed an angle beside which stood a clump of trees. This was terrain to be translated in three days by the clash of two armies—July 1, 2, and 3.

Lee expected to win and could well have won a decisive victory; as it was, he made a fatal mistake when he let his foe seize the high ground and fight on the defensive. Meade had been commanding the Army of the Potomac for only six days, and for three of those days he'd been fighting to save it and his government from catastrophe. Perhaps he could barely believe when he'd won a victory over a man and army whose prestige had loomed so large. When urged to mount a counterattack with 20,000 fresh reserves, immediately after Lee's debacle with Pickett's Charge, Meade felt constrained both by uncertainty as to Lee's remaining resources and in deference to the immense numbers of dead and wounded. The winning general later explained he did not want to ruin his army as Lee had done in "attacking a strong position."

The news coming out of Gettysburg on July 4 had an electrifying effect on the public mood in the North, followed two days later by news of Grant's success at Vicksburg. Lost to many of the celebrants may have been the detail that Rosecrans, on the move in Tennessee, had maneuvered Bragg out of the state, compelling him to fall back to Chattanooga. The Confederacy was now halved; and as Port Hudson, unable to stand alone, fell on July 9, the Mississippi was largely freed of impediments to the commerce of the Northwest. The victory at Vicksburg, although overshadowed by the three days in Pennsylvania, was of greater strategic importance. Lincoln described Grant's campaign as "one of the most brilliant in the world"; and Grant himself in his *Personal Memoirs* dated the demise of the Confederacy from Vicksburg's fall. Meade's success was devalued in Lincoln's eyes too because he had not pursued Lee, letting him cross the Potomac unmolested at Williamsport on July 14. On July 21 the president was to make clear his disappointment, saying: "I was deeply mortified by the escape of Lee across the Potomac, because the substantial destruction of his army would have ended the war, and because I believed such destruction was perfectly easy—believed that General Meade and his noble Army had expended all the skill and toil and blood, up to ripe harvest, and then let the crop go to waste."[39]

Summer 1863 now marks the midpoint in the maelstrom called the American Civil War, as two torturous years were now to be followed by two more years of war. But if the recruitment of black soldiery, as Du Bois maintained, was the deciding factor in determining its outcome, then the key moment of the war occurred, as it were, behind the swathe of battles viewed as paramount in the conflict's history. Already the disturbing news that the black dead at Port Hudson had not been accorded burial, while the cry "no quarter" filled the air at Milliken's Bend as black troops were murdered, had gained wide circulation; even as the city of Philadelphia, on June 17, received permission to raise its own black regiment. Prior to this, black recruits had been promised pay and conditions on an equal footing with all other recruits: $13 per month plus an allowance of $3.50 for clothing. But now Congress decided to bring the pay of black soldiers under the enactment of July 1862 regarding the employment of "contrabands." With the 54th already in the field in South Carolina, the United States Paymaster declared black soldiers would receive the same pay as laborers, $7 per month plus $3 allowance for clothing; nor would they, it was decided, receive an enlistment bounty as white recruits did, and it was reiterated, no black could receive a commission.

In consequence black recruitments in the North were beginning to slow. After a few weeks with only paltry results to show in Philadelphia, Maj. George L. Stearns, now head of all black recruitment, called on Frederick Douglass to address the situation and arranged a mass meeting for July 6 at Philadelphia's National Hall. Douglass was heard as the last of three speakers, following the lengthier talks urging enlistment of black men in Philadelphia by Judge Kelly and Anna Dickinson.

Perhaps some of those in the audience had been fugitive slaves, but most came from Philadelphia's large free black population, men who were working in the city's catering businesses or as barbers and waiters, laborers and teamsters, dock workers and warehouse men, with a few tradesmen. Many in the audience could, no doubt, read and write and were active members of churches, some from abolition societies and the vigilance committee—and undoubtedly among them stood some of the men whom John Brown had hoped to recruit. To them, Douglass spoke in a thoroughly plain and commonsense manner. He would not oppose the view that held that black men should enter the service on the same terms as white men. Here was the son who wanted his mother cared for in his absence, here a husband who wanted his wife provided for, a brother who wanted a sister secured. "I honor you for you solicitude. Your mothers, your wives, and your sisters, ought to be cared for, and an association of gentlemen, composed of responsible white and colored men, is now being organized in this city for this purpose," said Douglass. Do you say you want the same pay that white men get? he asked rhetorically.

> I believe that the justice and magnanimity of your country will speedily grant it. But will you be overnice about this matter? Do you get as good wages now as white men get by staying out of the service? Don't you work for less every day than white men get?
> You know you do. Do I hear you say you want black officers? Very well, and I have not the slightest doubt that in the progress of this war, we shall see black officers, black colonels, and generals even. But is it not ridiculous in us in all at once refusing to be commanded by white men in time of war, when, we are everywhere commanded by white men in time of peace.[40]

Two governments were struggling for possession of and endeavoring to bear rule over the United States, Douglass told the throng. One had its capital in Richmond and was represented by Mr. Jefferson Davis, and the other was in Washington and was represented by "Honest Old Abe." These two rival powers were "now confronting each other with vast armies upon many a bloody field, north and south, on the banks of the Mississippi, and under the shadows of the Alleghenies. Now, the question . . . is what attitude is assumed by these respective governments and armies towards the rights and liberties of the colored race in this country; which is for us, and which against us!"

If the Davis government triumphs in this contest, Douglass warned, "woe, woe, ten thousand woes, to the black man!" And what is the attitude of the Washington government toward us? Douglass queried.

> Mind, I do not ask what was its attitude . . . before this bloody rebellion broke out. I do not ask what was its disposition when it was controlled by the very men who are now fighting to destroy it when they could no longer control it. I do not even ask what it was two years ago, when McClellan shamelessly gave out that in a war between loyal slaves and disloyal master, he would take the side of the masters, against the slaves—when he openly proclaimed his

purpose to put down slave insurrections with an iron hand—when glorious Ben. Butler, now stunned into a conversion to anti-slavery principles . . . proffered his services to the Government of Maryland, to suppress a slave insurrection, while treason ran riot in that State, and the warm, red blood of Massachusetts soldiers still stained the pavements of Baltimore. . . .

I do not ask you about the dead past. I bring you to the living present. Events more mighty than men, eternal Providence, all-wise and all-controlling, have placed us in new relations to the Government and the Government to us. . . .

. . . and now, so far from there being any opposition, so far from excluding us from the army as soldiers, the President at Washington, the Cabinet and the Congress, the generals commanding and the whole army of the nation unite in giving us one thunderous welcome to share with them in the honor and glory of suppressing treason and upholding the star-spangled banner. The revolution is tremendous."

Then Douglass cautioned his audience: "Do not flatter yourselves . . . that you are more important to the government than the government is to you. You stand but as the plank to the ship. This rebellion can be put down without your help. Slavery can be abolished by white men; but liberty so won for the black man, while it may leave him an object of pity, can never make him an object of respect."

In conclusion Douglass called out, "Young men of Philadelphia, you are without excuse. The hour has arrived, and your place is in the Union army. Remember that the musket—the United States musket with its bayonet of steel—is better than all mere parchment guarantees of liberty. In your hands that musket means liberty; and should your constitutional right at the close of this war be denied, which, in the nature of things, it cannot be, your brethren are safe while you have a Constitution which proclaims your right to keep and bear arms."

Although the contending sides had allowed for a cessation for the recovery of white remains, the corpses of the six hundred black men who fell in the initial assault on Port Hudson would linger on the field into July when that stronghold fell. And while the memory of their heroic assault was extant, General Banks authorized "Port Hudson" be inscribed on every banner represented in the fight, but those regiments. Yet those who fell at Port Hudson were heard. Capt. Andre Cailloux had fallen leading his men in a charge on the enemy works on May 27, and had lain together with his comrades under the vigilant eye of sharpshooters until the surrender of the place on July 8. As soon as the body was recovered it was conveyed down river to his native city, where the judgment of black citizens was given vent at a funeral on July 11. A correspondent for a *New York Journal* cited by William Wells Brown wrote: "The arrival of the body developed to the white population here that the colored people had powerful organizations in the form of civic societies; as the Friends of Order, of which Captain Cailloux was a prominent member, received the body, and had the coffin containing it, draped with the American flag, exposed in state in the commodious hall. Around the coffin, flowers were strewn in the greatest profusion, and candles were kept continually burning. All the rights of the Catholic Church were strictly complied with. The guard paced silently to and fro, and altogether it presented as solemn a scene as was ever witnessed."[41]

The streets around the building were impassable, as men and women from thirty-seven societies lined Esplanade Street for more than a mile, joined by thousands who had recently been slaves. The coffin, borne from the hall by eight soldiers, six members of the Friends of Order, and six black officers, was joined in procession by two companies of the 6th Louisiana Regiment, the 42nd Massachusetts Regimental band, and about a hundred convalescing wounded from the black regiments represented at Port Hudson, with six coffins bearing other fallen soldiers and the carriages of Cailloux's family and military officers following. But for the band's dirge, the procession wended its way to the Bienville Street cemetery through the downtown streets in complete silence. The New York correspondent wrote in his dispatch dated August 1, 1863: "The long pageant has passed away, but there is left deeply impressed on the minds of those who witnessed this extraordinary sight the fact that thousands of people born in slavery had, by the events of the Rebellion, been disenthralled enough to appear in the streets of New Orleans, bearing to the tomb a man of their own color, who had fallen gallantly fighting for the flag and his country—a man who had sealed with his blood the inspiration he received from Mr. Lincoln's Emancipation Proclamation. The thousands of the unfortunates who followed his remains had the flag of the Union in miniature form waving in their hands, or pinned tastefully on their persons."

After exhorting black men in Philadelphia, Douglass went to Camp William Penn to assist in filling up enlistments for the "colored" regiments. In that week, following the Conscription Act of the previous March, draft officers were preparing to draw names all across the North, and in largely Democratic districts Copperhead politicians had been fanning resentment. Reviling the enactment and Lincoln's emancipation alike as unconstitutional, they goaded that conscription was only a vile attempt to make white working men fight for the freedom of blacks who would then inundate the North and take away their jobs. They would not support the "abolition crusade" and would "resist to the death all attempts to draft any of our citizens into the army." In a 4th of July oration in New York City the state's governor cautioned Republicans: "Remember this—that the bloody and treasonable doctrine of public necessity can be proclaimed by a mob as well as by a government."[42]

The process was begun without incident in New York City on July 11, a Saturday, but in many of the bars where Irish laborers congregated, men vowed to put an end to it by breaking up the draft offices when they opened again. Good to their word, on Monday morning draft offices were besieged, with a number of them burned. Marching through the city, men called at factories, yards, and docks where Irish were employed, and the protest swelled to thousands. Stores were broken into, hotels and saloons looted. Soon whole neighborhoods were swept by mob violence and a hunt was begun for blacks—seeking them out in their places of employment, chasing them through the streets, and driving them from their homes. A black woman was murdered and her house set aflame, while at least half a dozen black men were seized and lynched. With attacks continuing blacks fled to the wharfs, boarding steamboats for New Jersey or to Long Island, seeking refuge in the west in the woods in back of North Bergen and in Hoboken, and in the east on farms. At 4 o'clock, rioters—mostly women and children—inundated the Orphan Asylum for Colored Children located on Fifth Avenue between 43rd and 44th streets. Housing upward of eight hundred children, the building was ransacked from top to bottom, and threatened with destruction by fire, while the chief engineer of the institution showed a flag of truce and pleaded it not be burned. Soon, however, the building was in flames, and became a smoldering ruin and collapsed.

Frederick Douglass happened to be on his way to Rochester and wrote in his *Life and Times*:

I was met by a friend at Newark, who informed me of the condition of things. I, however, pressed on my way to the Chambers Street Station of the Hudson River Railroad in safety, the mob, fortunately for me, being in the upper part of the city, for not only my color, but my known activity in procuring enlistments would have made me especially obnoxious to its murderous spirit. This was not the first time I had been in imminent peril in New York City. My first arrival there, after my escape from slavery, was full of danger. My passage through its borders after the attack of John Brown on Harper's Ferry was scarcely less safe. . . . Such men as Franklin Pierce and Horatio Seymour had done much in their utterances to encourage resistance to the drafts.[43]

Mobs would continue to hold sway in New York for nearly a week, as most of the troops that had been stationed in the city were away in response to the recent emergency, and the police were no match for their numbers or ferocity. The homes of prominent Republicans and abolitionists were sacked; the offices of the *Tribune* were attacked as rioters sought Horace Greeley's blood, burning out the ground floor. At the *Times* Henry Raymond had three Gatling guns mounted on the roof to dissuade aggression. By July 15 the War Department rushed several regiments from Gettysburg to New York, who poured lead into the ranks of the city's Irish, leaving many scores dead. With the metropolis returning to order, troop strength would rise to 20,000 by August when drafting resumed.

Just prior to the carnage in New York with its signature attack on the black population answering specifically the question of black freedom and black soldiery, news of the raid on the Combahee in South Carolina reached Boston via a dispatch from a correspondent of the *Wisconsin State Journal*, and became known to the public in the first week of July in the pages of the *Commonwealth*. The following week the Franklin Sanborn–edited paper carried a biographical sketch of Harriet Tubman on its front page. As this article was circulating, Tubman was on the move—she relied on a dozen or so certificates from officers and officials of the government to move about freely— following the Massachusetts 54th up the Sea Islands to the mouth of Charleston Harbor, where Brigadier-General Gilmore was preparing a campaign against the citadel of secession.

The base for Gilmore's operation against Charleston was Folly Island—but the key to its success was possession of Morris Island, a strip of sand some three and a half miles long lying directly on the ocean commanding the channel leading to the city. On the northern end of the island bearing on Fort Sumter was Cumming's Point Battery. A large Union fleet had assembled in Stono Inlet awaiting an advance on Sumter, pending possession of Fort Wagner, a strong earthwork midway on Morris Island protecting the battery. A landing on James Island to conceal the fleet's true object and the troop movements on Folly Island had been effected on the evening of July 9.

On July 16 the Massachusetts 54th was attacked on James Island by a larger Confederate force in which the black troops gave good account, driving the enemy from the field.

That same day Colonel Shaw received orders to evacuate the island for Cole Island, a movement across marshes, streams, and dikes, which would be accomplished at night during a storm, part of the way single file on narrow footbridges. The next day the 54th rested on the beach opposite the south end of Folly Island. At ten that night Colonel Shaw received orders to report with his command to General Strong at Morris Island, and the entire regiment was transferred fifty at a time via the steamboat *General Hunter*. On Morris Island, William Wells Brown wrote, the regiment "breakfasted on the same fare, and had no other food before entering into the assault on Fort Wagner in the evening." Hildegarde Hoyt Swift, author of *The Railroad to Freedom*, stated that Tubman served Colonel Shaw his last meal before the battle, indicating her presence on Cole Island, and without doubt she was on Morris Island when the troops assumed order of battle. While Fort Wagner could have been taken by siege, it was to be assaulted that evening and the 54th had been assigned "the post where the most severe work was to be done and the highest honor was to be won."

A rain of fire had been falling on the fort since daylight, from a fleet of four ironclads, a frigate, and four gunboats mounted with mortars, which increased as the morning wore on. Pollard wrote:

> Until six o'clock in the evening the firing was incessant. There was scarcely an interval that did not contain a reverberation of the heavy guns, and the shock of the rapid discharges trembling through the city called hundreds of citizens to the battery, wharves, steeples, and various look-outs, where, with an interest never felt before, they looked on a contest that might decide the fate of their fair city. Above Battery Wagner, bursting high in air, striking the

sides of the work or plunging into the beach, and throwing up pillars of earth, were to be seen the quickly succeeding shells and round shots of the enemy's guns. Battery Gregg at Cumming's Point and Fort Sumter took part in the thundering chorus. As the shades of evening fell upon the scene the entire horizon appeared to be lighted up with the fitful flashings of the lurid flames that shot out from monster guns on land and sea.[44]

Inside the works, rounded by a sloping earthwork parapet and covered by a massive roofing of felled trees, were two thousand men commanded by General Taliaferro. An eyewitness related: "Wagner loomed, black, grim and silent."[45]

The line of battle was formed within six hundred yards of the fort; the 54th—which had no training in assaulting fortifications and who had seen its first battle only two days previous—was forwardmost, backed by the 6th Connecticut and the 9th from Maine, followed by regiments from New York, Pennsylvania, and New Hampshire: two full brigades in all. General Strong addressed the 54th, then their colonel, of whom an eyewitness wrote that all could see that the twenty-seven-year-old "had counted the cost of the undertaking before him; for his words were spoken ominously, his lips were compressed, and now and then there was visible a slight twitching of the corners of the mouth, like one bent on accomplishing or dying."[46]

Shaw told the men—among whom were some of the most deliberate and committed of northern blacks, including two of Frederick Douglass's sons, "how the eyes of thousands would look upon the night's work they were about to enter on." Then he walked along the line speaking words of encouragement, as one soldier struck by his solemnity, exclaimed, "Colonel, I will stay by you till I die."

The order to charge was given at quarter to eight, when the 54th began to move forward at quick time in complete silence. Although General Taliaferro had ordered his entire garrison to the parapet, there was no musketry, only solid shot falling between the battalions and to their right. At one hundred yards the fort's batteries opened with grape and canister in "a blinding sheet of vivid light," while the 54th stumbled to close gaps in their ranks. As the first line hesitated, Colonel Shaw sprang forward, waving his sword and shouting, "forward my brave boys!" Pollard engraved the scene in his volume: "Barely waiting for the Yankees to get within a destructive range our infantry opened their fusillade, and from a fringe of fire that lined the parapet leaped forth a thousand messengers of death. Staggering under the shock, the first line seemed for a moment checked, but, pushed on by those in the rear, the whole now commenced a charge at a 'double-quick.' Our men could not charge back, but they gave a Southern yell in response to the Yankee cheer, and awaited the attack."

Robert G. Shaw.

Tripping and faltering over hills and pits of sand, on came the 54th. Reaching the ditch of the battery, they took but a moment to clamber up the sloped wall against fearful fire. Shaw was among the first to reach the parapet, where he stood urging the men on. The first wave's color-sergeant was wounded and Sgt. William H. Carney took up the colors before they could touch the ground. Mounting the parapet, he received several wounds, but did not fall. Sgt.-Maj. Lewis H. Douglass, close behind, cried out, "Come, boys, come, let's fight for God and Governor Andrew." As soldiers scrambled up, Colonel Shaw was shot dead and fell into the fort where he would be discovered with twenty of his men around him, two having fallen over his body.

Pollard continued, "The antagonists were breast to breast, and Southern rifles and Southern bayonets made short of human life. We could stop to take no prisoners then. The parapet was lined with dead bodies, white and black, and every second was adding to the number."

In a quarter hour a second column had formed to the left of the defeated attempt, and the experiment was repeated. In deepening darkness this assault succeeded in attaining the chambers occupied by two of Fort Wagner's guns. Most of these men were taken prisoner, and by midnight the assault was ended.

Over 1,550 soldiers were sacrificed; the killed or wounded of the 54th alone totaled 261. Six hundred of the dead Union soldiers were buried by Confederate troops in mass graves beneath the fort's outer barrier. When inquiry was made under flag of truce for the body of Robert Shaw, the answer was given: "We have buried him with his niggers!"[47]

Sergeant-Major Douglass had been the last soldier of the 54th to retreat when the charge was repulsed. Days later he wrote his sweetheart, Amelia Loguen, the daughter of J. W. Loguen:

I have been in two fights, and am unhurt. I am about to go in another I believe tonight. Our men fought well on both occasions. The last was desperate. We charged that terrible battery on Morris Island known as Fort Wagner, and were repulsed. . . . De Forest of your city is wounded. George Washington is missing, Jacob Carter is missing, Charles Reason wounded, Charles Whiting, Charles Creamer all wounded.

I escaped unhurt from amidst that perfect hail of shot and shell. It was terrible. I need not particularize, the papers will give a better [account] than I have time to give. My thoughts are with you often, you are as dear as ever."[48]

The apotheosis of the 54th and Robert Gould Shaw, taking place just after New York's antidraft riots, was a potent rebuke to Copperhead sentiment. Over a year later, on September 8, 1864, the *Tribune* editorialized the battle had "made Fort Wagner such a name to the colored race as Bunker Hill had been for ninety years to white Yankees." Harriet Tubman thereafter spoke of the young Shaw with the same reverence with which she spoke of John Brown. The historian Albert Bushnell Hart, who knew Tubman, chronicled "her extraordinary power of statement" in *Slavery and Abolition*, when, undoubtedly of the events of July 18 she had witnessed, she narrated: "And then we saw the lightning, and that was the guns; and then we heard the thunder, and that was the big guns; and then we heard the rain falling, and that was the drops of blood falling; and when we came to get in the crops, it was dead men that we reaped."[49]

Fifteen

Battles for Liberty, Battles for Union

FAIL, NOW, & OUR RACE IS DOOMED,
SILENCE THE TONGUE OF CALUMNY
Of prejudice and hate, let us rise now and fly to arms.
We have seen what VALOR AND HEROISM
our brothers displayed at
PORT HUDSON AND MILLIKEN'S BEND,
ARE FREEMEN LESS BRAVE THAN SLAVES?
OUR LAST OPPORTUNITY HAS COME,
Men of Color, Brothers and Fathers!
We Appeal to you!
STRIKE NOW!
—Recruitment broadside, U.S. Colored Troops, 1863

IN A SPEECH AT THE CHURCH OF THE PURITANS IN NEW YORK
titled "The Present and Future of the Colored Race in America" in
May 1863, Frederick Douglass asked: What would the effect be on
"colored men" of fighting and winning battles for the republic? What
would the effect be on the country? He concluded, "We are opening
a new account with the American people and with the whole human
family."[1]

The answers to Douglass's question came rapidly in the valor and in the hard-won glory in the succeeding months. But the "curse of the Negro" was also at once palpable—that his life and liberty were held in "the utterest indifference and contempt." As he concluded his effort to facilitate black enlistments in Philadelphia, Douglass was invited by George Stearns to continue the work in Pittsburgh, and as he prepared to do so he denounced the abuse of black troops and placed the onus for its redress squarely on the commander-in-chief. In his August *Monthly* of 1863, Douglass wrote:

> Whatever else may be said of President Lincoln, the most malig-nant Copperhead in the country cannot reproach him with any undue solicitude for the lives and liberties of the brave black men, who are now giving their arms and hearts to the support of his Government. . . . Unhappily the same indifference and contempt for the lives of colored men is found wherever slavery has an advo-cate or treason an apologist. . . . Such has been our national edu-cation on the subject, and that it still has power over Mr. Lincoln seems evident from the fact, that no measures have been openly taken by him to cause the laws of civilized warfare to be observed toward his colored soldiers. . . . More than six months ago Mr. Jefferson Davis told Mr. Lincoln and the world that he meant to treat blacks not as soldiers but as felons. The threat was openly made, and has been faithfully executed by the rebel chief. . . . Thousands of Negroes are now being enrolled in the service of the Federal Government. The Government calls them, and they come. They freely and joyously rally around the flag of the Union, and take all the risks, ordinary and extraordinary, involved in this war. . . . If any class of men in this war can claim the honor of fighting for principle, and not from passion, for ideas, not from brutal mal-ice, the colored soldier can make that claim preeminently. He strikes for manhood and freedom. . . . Yet he is now openly threat-ened with slavery and assassination by the rebel Government—and the threat has been savagely executed. What has Mr. Lincoln to say about this slavery and murder? What has he said?—Not one word. In the hearing of the nation he is as silent as an oyster on the whole subject.

As a leading agent in the North for recruitment of black troops, Douglass felt it his obligation to bring his concerns directly to the top of the government, to the War Department and to the president. Gen. Lorenzo Thomas's efforts at Memphis were being stepped up, and Stearns was now charged with overseeing the entire office under the title Bureau of Colored Recruitment. With Stearns's encouragement, and with the help of Kansas senator Samuel Pomeroy, an interview was promptly arranged between Douglass and the president and his secretary of war. The only firsthand accounts of the meetings are found in Douglass's *Life and Times*, and in a speech he gave at the 13th annual meeting of the American Anti-Slavery Society held in Philadelphia on December 3, 1863; in these is ample material to draw out a compelling impression of that remarkable encounter. Although the date of these appointments is uncertain, they occurred between July 18 and 30, but surely nearer to the end of the month as Douglass had written to Stearns on August 1 abjuring from further recruitment because of his reservations. On August 30, too, Lincoln issued an executive proclamation ordering "that for every soldier of the United States killed in violation of the laws of war, a rebel soldier shall be executed."2 Douglass was urging such upon him, but was not yet aware of its consideration.

Douglass, or any other black man or woman of prominence, would have been hesitant to travel openly before mid-1863 over the borders of Delaware and Maryland. But Douglass was now to set out through the last address of his bondage, Baltimore, on his way to the national capital; touching this precinct was like "rubbing against my prison wall." He could not be certain how a black man would be received at the presidential mansion; a year earlier, in his first official audience with blacks, Lincoln impetuously declared them to be the "cause" of the war. Had his thinking evolved appreciably, or were Douglass's recent criticisms agonizingly on the mark? When he arrived Douglass found the stairway outside Lincoln's office packed with applicants, and his face "the only dark spot among them." He expected to wait half a day at least, and had heard of men waiting a week; however, Douglass was surprised when within two minutes of presenting his card, a messenger returned inviting in "Mr. Douglass." Pressing his way to the front of the line, Douglass reported that a despairing voice remarked, "a Peace Democrat," he supposed. "Yes, damn it, I knew they would let the nigger through."

Entering the presidential office, Douglass related that Lincoln was surrounded by a large number of documents and busy secretaries, "sitting in his usual position . . . with his feet in different parts of the room, taking it easy." He appeared overworked and tired, with "long lines of care . . . deeply written" on his forehead. As Douglass approached and Lincoln began to rise, Senator Pomeroy began his introduction, but Lincoln interrupted, saying, "Mr. Douglass, I know you; I have read about you, and Mr. Seward has told me about you." Showing that he too was aware of Douglass's censure of his policies, Lincoln brought out that in one of his published speeches he read he stated the saddest and most disheartening feature of the present political and military situation, was not the various disasters experienced by the armies, "but the tardy, hesitating, vacillating policy of the President of the United States." Lincoln said, "Mr. Douglass, I have been charged with being tardy, and the like, and partly admit that I might seem slow. But I am charged with vacillating. Mr. Douglass, I do not think that charge can be sustained. I think it can be shown that when I once take a position, I have never retreated from it."

Douglass regarded this point as considerable and proceeded to the issues that brought him to the interview. He had, he told the president, been assisting in the raising of "colored troops," and for several months had been very successful, but now found it harder to induce men into service because of the feeling that the government did not deal with them fairly. Asked for particulars, Douglass brought forth three: The government had been tardy in proclaiming equal protection for "colored soldiers and prisoners." They ought to be "exchanged as readily and on the same terms as any other prisoners, and if Jefferson Davis should shoot or hang colored soldiers in cold blood the United States should, without delay, retaliate in kind and degree upon Confederate prisoners in its hands." Continuing, Douglass said that "colored soldiers" ought to receive the same pay as white soldiers, and that they ought to "be rewarded by distinction and promotion precisely as white soldiers are."

For his part Lincoln remarked that the employ of these troops at all was a great gain for the colored people, and that it could not have been successfully adopted at the beginning of the war. In respect to the issue of equal pay the wisdom of making them soldiers was still doubted, that their enlistment was a serious offense to popular prejudice, and that they ought to be willing to enter the service upon any condition. The fact that they were not receiving the same pay as white soldiers was a necessary concession to smooth the way for their employment at all. Ultimately, Lincoln said he did not doubt they would receive the same. In respect to the equal protection of colored prisoners, the case was more difficult. If he issued a proclamation of retaliation, which he was considering, for treating captured black troops as felons, Lincoln said: "All the hatred which is poured on the head of the Negro race will be visited on my administration. Retaliation is a terrible remedy, and one which is very difficult to apply—if once begun there is no telling where it will end. If I could get hold of the Confederate soldiers who are guilty of treating colored soldiers as felons I could easily retaliate, but the thought of hanging men for a crime perpetrated by others is revolting to my feelings." "Remember this, Mr. Douglass; remember that Milliken's Bend, Port Hudson and Fort Wagner are recent events; and that these were necessary to prepare the way for this proclamation of mine." He thought the rebels themselves would soon stop such barbarous warfare, and that he already had information that colored soldiers were being treated as prisoners of war. On Douglass's last point, that the nation's black soldiers receive commissions, while Lincoln did not commit himself, he did say that he would sign any commission to black soldiers whom his secretary of war should commend to him.

Although far from being satisfied with Lincoln's response to his concerns, Douglass was to come away from the interview with a favorable view of the man. He had been given a careful hearing and the president made his replies with thoughtful gravity, impressing Douglass as to his core honesty. Here was a man few as yet discerned; a man Douglass saw had none of the base ill feeling universally prevalent among white Americans against his race. For Lincoln the connections between victory for the Union and emancipation was now one of obligation, and despite his high office, Douglass felt, he could easily have placed a hand on Lincoln's shoulder during their exchange.

The success of his audience with the president in a district that was more than a novelty to him, and at a time of unmatched danger to the republic, undoubtedly impressed even greater urgency upon Douglass as he continued his mission on behalf of black soldiery in an audience with Secretary Stanton. Hastening to the War Department, Douglass arrived at the office of Assistant Secretary Charles A. Dana, whom he'd known since his Brook Farm days and then as managing editor at the *Tribune*. Dana would accompany Douglass to his superior and make their introduction.

Coming into Stanton's office, Douglass was immediately struck by the contrast to his previous meeting. Douglass's account in his *Life and Times* reads: "Every line in Mr. Stanton's face told me that my

From left to right: Frederick Douglass, Edwin Stanton, and Lorenzo Thomas.

communication with him must be brief, clear, and to the point—that he might turn his back upon me as a bore at any moment." The offices of the two were alike busy, and Douglass quickly reiterated the issues he had just gone over with the president. He wrote: "As I ended, I was surprised by seeing a changed man before me. Contempt and suspicion and brusqueness had all disappeared from his face and manner, and for a few minutes he made the best defense that I had then heard from anybody of the treatment of colored soldiers by the government." Douglass at last felt it was proper to propose a radical measure, announcing he himself was willing to take a commission to further the government's ends. Stanton said he was prepared to confer upon him authority as "assistant adjutant" to General Thomas. Asked how soon he could be ready, Douglass replied that he could have his affairs in order within two weeks; the commission, he informed the secretary, could be sent to him at Rochester.

Hastening to his home, Douglass would prepare to publish the last issue of his paper that would reflect all the trying circumstances and vicissitudes of the time. *Douglass' Monthly* in August 1863 included his speech in Philadelphia urging black enlistments, his letter to Stearns declining his request to promote enlistment in Pittsburgh, a speech castigating Lincoln for his silence on the atrocities against black soldiers, and an article critical of Secretary Seward. In his last effort Douglass would also publish his reply to Parker T. Smith, who had asserted in the pages of the *Anglo-African* that men in Philadelphia only awaited his ascent to a captaincy to follow him into battle. Then Douglass penned his valedictory to his "Respected Readers," dated August 16. He wrote: "I am going South to assist

Adjutant General Thomas, in the organization of colored troops, who shall win for the millions in bondage the inestimable blessings of liberty and country." While he was ceasing publication under his own name after sixteen years, there were, Douglass told his readers, other publications that were now prepared, as had not been the case when he began, where blacks could be published. These included three New York papers, the *Tribune,* the *Anglo-African,* and the *Independent.* The last had especially invited him into its pages, through its editor Theodore Tilton.

The final paragraph prepared for his sheets played across an eventful life, but perhaps above all upon his lasting debt to John Brown, who had urged a similar role on him. Douglass wrote:

> Slavery has chosen to submit her claims to the decision of the God of battles. She has deliberately taken the sword and it is meant that she should perish by the sword. Let the oppressor fall by the hand of the oppressed, and the guilty slaveholder, whom the voice of truth and reason could not reach, let him fall by the hand of the slave. . . . Eternal justice can thunder forth no higher vindication of her majesty nor proclaim a warning more salutary to a world steeped in cruelty and wickedness, than such a termination of our system of slavery. . . . The oppressor has hardened his heart, blinded his mind and deliberately rushed upon merited destruction. Let his blood be upon his own head. That I should take some humble part in the physical as well as the moral struggle against slavery and urge my long enslaved people to vindicate their manhood by bravely striking for their liberty and country is natural and consistent. I have indicated my course. . . . With a heart full and warm with gratitude to you for all you have done in furtherance of the cause of these to whom I have devoted my life, I bid you an affectionate farewell.

A letter from Assistant Adj. Gen. C. W. Foster dated August 13 arrived at Douglass's home directing him to report to General Thomas at Vicksburg. The envelope enclosed a pass testifying to the character and status of Frederick Douglass signed by A. Lincoln, the secretary of the interior, and Senator Pomeroy, as well as Postmaster-General Montgomery Blair, allowing him to travel through military lines. However no commission accompanied the communiqué. Writing for clarification, Douglass asked on what conditions he was "expected to enter upon the recruiting service in the South West." The reply he received dated August 21 still contained no commission, but did establish that the recruiting commissioner, Major Stearns, would pay him. The only added point of explanation stated: "It is of course expected that you go to aid General Thomas in any way that your influence with the colored race can be made available to advance the object in view."[3]

Still retaining his wish to begin his new course, considering it one of importance and opportunity, on August 19 Douglass sent a letter to Thomas Webster of Philadelphia, chairman of the recruitment committee for black soldiers. He wrote: "I can hardly hope to see you before leaving or while en route to Vicksburg whither it seems I am to go. I do not yet know by what route my transportation will require me to go nor upon what conditions, rank, pay or duty. It pleases his excellency the Sec of War to keep me in the dark on all essential points. He only commands me to go. Like King Lewis, he thinks a citizen a person having duties but no rights. I shall obey, however, hoping that all will be well in the end."[4]

In the next week a letter arrived from Stearns informing Douglass that his salary would be one hundred dollars per month plus subsistence, and likely transportation; but still there was no word on his commission. Douglass concluded that in offering the position in the first place, Stanton was demonstrating he was fully aware the government's hand in recruiting black regiments would be greatly strengthened by having a figure of prominence with visible marks of rank, but finally neither he (nor Lincoln) was ready to take that step, calculating they would face a renewed chorus of Copperhead criticism. Because of this, Douglass decided he must withdraw from his commitment to assist General Thomas in Vicksburg.

The war that had fallen fitfully upon the nation after the presidential contest in 1860 was essentially a political war, where fighting men on both sides, prior to departing for the front, assembled before cheering, flag-waving crowds, and where battle, many thought, would quickly emblazon their standards with emblems of success. The accomplishments of southern arms in the first half of 1863 gave way in the latter part of the year to a situation that was critical to its cause. But this at last allowed its leaders to consider a more suitable strategy—not that of conquering political power to implement a national slaveholder policy, as had been their aim at war's commencement, nor of a hoped-for military victory to usher in a diplomatic coup to achieve independence. Rather, based upon the strength of their resistance and the growing stress upon the North's capacity to prosecute a war of subjugation, the Confederacy sought to compell Lincoln's government, through an alliance with northern Peace Democrats, toward a cessation of hostilities.

With the fall of Vicksburg, and an important victory soon afterward in Arkansas won by two columns of Kansas regiments led by Gen. James G. Blunt and Col. William A. Phillips, the greater portion of the southwest (with Mississippi and Louisiana in part, and Missouri and Kentucky wholly) was occupied by armies of the North. Jefferson Davis assigned Edmund Kirby Smith to command what remained of the Confederate Trans-Mississippi Department, including portions of Arkansas, Louisiana, and Texas, which, cut off from Richmond, became known as Kirby-Smithdom. Meanwhile, as Bragg fell back on Chattanooga, Knoxville and eastern Tennessee had been delivered into the hands of the Army of the Ohio, commanded by Burnside on September 3. Then, as Bragg retreated farther into northern Georgia in preparation to reduce any Union drive into the center of the Confederacy, Chattanooga was delivered up to the Army of the Cumberland commanded by Rosecrans. In the Southeast, parts of North Carolina, South Carolina, and Alabama were occupied by Union armies, while in Charleston, Beauregard mounted a vigorous defense, and Lee, despite his rebuff in Pennsylvania, held sway in northern Virginia.

Seeing the reversals of July–September as "the darkest hour of our political existence," Jefferson Davis was determined to turn the Confederacy's fortunes around. Hadn't Jackson and Lee done just that against McClellan in the previous year? Reinforcing Bragg with two divisions from Joseph Johnston's army in Mississippi, making it the equal of Rosecrans's, Davis urged Lee to take command. Lee, however, wanted to take the offensive against Meade on the Rappahannock, perhaps maneuvering strategically between the Army of the Potomac and Washington. Longstreet, in contrast, proposed that he be sent with two of his divisions (Pickett's hadn't been reconstituted yet) to join Bragg, arguing a check on the Yankee advance at Chattanooga would allow the reoccupation of Knoxville and open up the way for a countermarch clear to the Ohio. Seizing on this, Davis overruled Lee and ordered Longstreet to Georgia.

On September 9, as veteran troops of the Army of Northern Virginia entrained on a circuitous 900-mile route through the Carolinas and Georgia, Bragg began his offensive with a ploy that signaled to Rosecrans that he was again retreating. With the Union army pushing three separate columns into the valley south of Chattanooga between Lookout Mountain on the west and Pigeon Mountain on the east, from September 10 to 13 Bragg bypassed three opportunities to assail Union armies piecemeal. Now alerted, Rosecrans ordered his army to concentrate in the valley of West Chickamauga Creek—but he was to pay dearly for his misjudgment of Bragg's intentions.

As the opposing armies met head-on across Chickamauga Creek on September 19, the initiative lay with Bragg. His plan called for turning his enemy's left to cut him off from Chattanooga, then driving him southward into a narrow, dead-end valley. All through the day division-sized assaults were futilely made against the Union left commanded by Gen. George H. Thomas, whose corps fought from the cover of thick underbrush in dense woods, while Rosecrans shuttled in reinforcements. That evening Longstreet arrived to take personal command of troops that had been disembarking all day and marching for the front. The next morning Bragg decided on an echelon attack from right to left, with Leonidas Polk commanding on the right and Longstreet on the left. But Polk started late against positions protected by breastworks thrown up during the night. Bragg now canceled the coordinated strike and ordered Longstreet forward with everything he had.

Beginning at 11:30 on the morning of September 20, Longstreet's assault had the luck to drive straight into a gap in Rosecrans's line, collapsing the entire Union right and sending the stunned general, with a third of his army, fleeing headlong to Chattanooga. Longstreet now saw that the enemy's whole force might be captured, and that no obstacle could be thrown in his way before reaching the Ohio. But the Union left had formed a new line and stood firm, earning Thomas the sobriquet "the Rock of Chickamauga." Bragg was so appalled by the wastage of men and material strewn along the creek that he countermanded Longstreet, directing that arms and stragglers be picked up instead, netting over eight thousand prisoners, and bringing in a haul of 15,000 stand of small arms and fifty-one pieces of artillery, in addition to large quantities of ammunition, with a great number of wagons, ambulances, teams, and hospital stores. Although victory had lain before him, Bragg had to contend with the loss of nearly a third of his army, while Rosecrans could not admit the full ignominy of his defeat, pointing out that in falling back on Chattanooga he had prevented its capture. After Chickamauga both generals lost considerable capital with their respective political leaders, and both in due course would lose their commands.

During the night of September 20, Thomas, too, fell back to Chattanooga, and Rosecrans began preparing a fortified line along the southeastern and eastern approaches to the railroad city. Bragg's devise now became one of investment, placing artillery on the summit of Lookout Mountain to the south, and infantry to the east along Missionary Ridge, with infantry blocking all the river roads to the west. Rosecrans's only supply route was over the Cumberland range from the north, subject to cavalry raids, and not nearly able to sustain the flow of provisions needed. Soon his horses were dying and his troops reduced to half rations or less. But Rosecrans allowed the retention of eastern Tennessee, rendering Bragg's victory tactically barren.

If Lincoln had learned anything during the long and torturous war it was that he must display forbearance, in public and private. It seemed in Washington that general-in-chief Halleck was making no preparation for the relief of Chattanooga—Burnside was a hundred fifty miles distant guarding Knoxville, and Grant's effective force had just been sent across the Mississippi where it was not needed. Meanwhile Meade was doing nothing, and so far as it could be learned did not want to do anything; what's more, Sherman had been ordered to the relief of Chattanooga from Vicksburg, but would have to rebuild the railroad on the way, requiring weeks of hard toil. Immediately after Chickamauga, Secretary of the Navy Gideon Welles preserved an interview in his diary he had with Lincoln.

Lincoln remarked: "It is the same old story of this Army of the Potomac. Imbecility, inefficiency—don't want to do—is defending the capital. I inquired of Meade, what force was in front? Meade replied he thought there were 40,000 infantry. I replied he might have said 50,000, and if Lee with 50,000 could defend their capital against our 90,000—and if defense is all our armies are to do—we might, I thought, detach 50,000 from his command, and thus leave him with 40,000 to defend us. Oh, it is terrible, terrible, this weakness, this indifference of our Potomac generals, with such armies of good and brave men."[5]

The escape of Lee across the Potomac had distressed him "almost beyond any occurrence of the war," Welles remarked. Welles asked why the president did not rid himself of Meade, who had "not breadth or strength," and although a good man and officer, was not a great general, and "certainly [was] not the man for the position he occupies"? With all this Lincoln concurred, but asked: "What can I do with such generals as we have? Who among them is any better than Meade? To sweep away the whole of them from the chief command and substitute a new man would cause a shock, and be likely to lead to combinations and troubles greater than we now have. I see all the difficulties as you do. They oppress me."

It was during the depths of his disquiet over military reversals that there came to Lincoln a delegation of radicals from Missouri and Kansas led by Senator Lane, demanding the removal of another of the generals—John Schofield of the Department of the Border. In August the Missouri guerrilla Quantrill had sought redress for Lee's defeat at Gettysburg by sacking Lawrence, Kansas, where he slaughtered many scores of men. In reply to the atrocity one of Schofield's subordinate officers gave an order for the expulsion of all people from Jackson, Cass, Bates, and Vernon counties, in Missouri, excepting those near large towns. Meanwhile the general himself was retaining troops in his command who had been paroled as Confederate prisoners, and was accused of not doing enough to stop the sort of activity presented by the likes of Jo Shelby, who just then, starting with six hundred raiders turned into a column of fifteen hundred, began a fifteen-hundred-mile bushwhacking raid into Missouri. Lincoln stood firm on retaining Schofield, but the next day, persuaded by Stanton, he agreed to the extraordinary transfer of two corps—or more than 20,000 men—from the Army of the Potomac to Rosecrans, under command of Joseph Hooker. Departing from Culpeper for the more than thousand-mile journey, men, material, artillery, and horses began arriving after eleven days at the railhead near Chattanooga.

On October 6 Jefferson Davis left Richmond for a meeting with Bragg, where Davis would take the unprecedented step of soliciting criticism of Bragg from his subordinates in his presence, offering Bragg's command to Longstreet. But Longstreet would not hear of it, recommending instead Joseph Johnston. Davis in turn would not accept this, nor consider Beauregard, and so Bragg was retained for the moment. Lincoln, however, with Hooker's force arriving, and Sherman's to arrive in mid-November, saw no choice but to search out new leadership. In a bold move on October 18, Ulysses S. Grant was appointed commander of a newly created Military Division of

the Mississippi "with his headquarters in the field," comprising the
Departments of the Ohio, the Cumberland, and the Tennessee,
embracing the whole region between the Mississippi River and the
Appalachian Range. Grant relieved Rosecrans of command, replacing
him with Thomas.

With Grant's arrival in Chattanooga officers and men immediately began "to see things move." In only a week the stranglehold was
loosened as the roads to the west and the Tennessee River were
opened. When, in early November, as Davis had counseled, Bragg
detached Longstreet's corps for a campaign to recapture Knoxville,
Grant, with Sherman arriving, began a drive to open the gateway to
Georgia by turning the Confederate entrenchments on Missionary
Ridge. Hooker's divisions went up the northern slope of Lookout
Mountain; on the other end of the line, Sherman pressed his divisions
forward against considerable resistance. To prevent Bragg from reinforcing either flank Grant now ordered Thomas forward in an assault
on the first line of trenches on Missionary Ridge. Along a two-mile
front 23,000 men streamed against what seemed impregnable defenses. To the astonishment of all they swept through the first trench with
ease, sending the defenders scrambling to the top of the ridge and into
the second and third line. Thomas's divisions, whom Grant thought
too demoralized to assume the offensive, without orders continued to
the next line of the enemy's defenses. His teeth gripping a cigar, Grant
peered at the sight of sixty regimental flags streaming to the top of
Missionary Ridge, angrily snapping, "Thomas, who ordered those
men up the ridge?" Grant's subordinate replied, "I don't know. I did
not." Intoxicated by the prospect of avenging their reputation, the
attackers broke upon the enemy trenches yelling "Chickamauga!
Chickamauga!" as Bragg's lines unexpectedly disintegrated. The fleeing Confederates would not regroup until their retreat had yielded
thirty miles on the road to Atlanta.

As Longstreet was repulsed at Knoxville, Bragg tendered his resignation in a letter to Davis. Joseph Johnston belatedly took his command, and Bragg moved on to Richmond to serve as military advisor
to Davis. Belatedly, too, after the defeat of Democrats in the fall elections across the North, and most important in the key contests for the
governorships in Ohio and Pennsylvania, Lincoln felt able to address
his problems on the border, replacing Schofield with the hapless
Rosecrans.

In the words of Francis Williams, a student at Avery College (Negro), Allegheny City, a shift in the rationale of the "war against the rebellion of slaveholders" and former "statesmen, Congressmen, and Cabinet officers, the once petted favorites of the government," had been heralded on January 1, 1863. But the fusing of the promise of Liberty and Union into the bedrock of a single purpose—which Lincoln would enunciate in his memorial address at Gettysburg on November 19 as a "new birth of freedom"—could only have been accomplished with the successful introduction of black regiments into the northern armies. By August 1863 there were 70,000 blacks employed in these armies, with 22,000 bearing arms; numbers that would leap to 100,000 army workers by December of that year as the total wearing the blue uniform as soldiers rose to 50,000. In all departments by war's end there were over 200,000 ex-slaves employed in camps as laborers, servants, cooks, and teamsters, with official figures listing 186,017 black soldiers serving. At least half a million former slaves at one time or another were involved directly in the war effort for the Union, and possibly as many as 700,000.

But, inaccessible on plantations, where the Emancipation Proclamation could not be enforced, many hundreds of thousands were left untouched by the war. That the Southerner preened about the absence of violence and the supposed docility on his plantations, was because, as DuBois noted, they "found an easier, more effective and more decent way to freedom." It took a superior wisdom, when, despite the fact that nine-tenths were illiterate, they began to decide where their best interest lay. To abandon plantations meant, moreover, they were leaving the site where they had always been assured food and shelter, and despite their condition, a modicum of safety. Many chose to wait on the plantations to which they belonged, their liberation a mere contingency, waiting "quietly, for the development of things."[6]

In the winter of 1863–64 Louisiana's Confederate governor reported that more blacks had perished than the combined war casualties of both armies in the state. In the Mississippi Valley, opened with the fall of Vicksburg and Port Hudson, recruitment of black regiments was being vigorously pursued; in practice that meant separation of families. Between Memphis and Natchez camps were established which congregated as many as 50,000 persons, where all able-bodied men had been culled for military service. A report in a northern paper told of 35,000, mostly women and children and the aged, in camps between Helena and Natchez that were in "extreme destitution," surviving on poor food with tattered shelters, many without beds or bedding, and living amidst filth. One fourth of them, it was reported, had "but one single worn and scanty garment . . . and are afflicted with all fatal diseases." An officer detailed to supervise the camps confided, "It is my conviction that, unrelieved, the half of them will perish before the spring. Last winter, during the months of February, March and April, I buried, at Memphis alone, out of an average of about four thousand, twelve hundred of these people, or twelve a day."[7]

In the North many black males were eager for military service, but their population was not great, and the overwhelming number of new recruits came from the South. Unscrupulous recruiters and substitute brokers were soon swarming around "contraband" camps in Nashville and Memphis as the price for a substitute had soared to $1,000, only a fraction of which would be paid to the enrollee. Provost marshals sent to each congressional district had their work cut out for them in canvassing door to door to find conscripts, and encountered a notable problem in Irish neighborhoods in New York, in the Pennsylvania coal-mining regions, and in the "Butternut" counties of Ohio, Indiana, and Illinois—areas settled by people from the upper South and Pennsylvania, with an economy of corn, hogs, and whiskey, wearing plain homespun clothes dyed in the oil of butternut trees or walnut, which lent them this generic name. Enrollees, however, could be exempted as physically or mentally unfit or for being the sole support of dependent relatives, hire a substitute, or pay a commutation fee. If a man wanted to, he found it easy enough to avoid conscription, hiring substitutes among immigrant communities who could be lured by the pay. All of this was an invitation for fraud. This reality, along with the success of the black regiments in the field, caused many whites to reconsider their initial opposition to black recruitment. After all, each black who joined the army kept one white at home.

Martin Delany, who had been appointed recruiting agent for western Ohio, after fulfilling his appointed duties as a state recruiting agent hastened to Nashville. But after making only a few speeches urging enlistment, revolted by the corruption he saw, and the sight of freedmen being strong-armed and marched away like cattle, resigned. Delany wrote: "I became convinced that the business of recruiting had reached such a state of demoralization that no honorable men, except a U.S. commissioned officer, could continue it successfully without jeopardizing his own reputation."[8]

The pay differential between white and black troops did not appreciably hinder the government's rush to enlist an overwhelming "African" presence in its armed forces, but it became a matter of serious contention. Governor Andrew felt honor bound, pledging to the Massachusetts 54th and 55th Volunteers: "I shall not rest until you shall have secured all of your rights." When the Massachusetts legislature appropriated money to equalize the compensation, the 54th rejected it out of hand, deciding as a body not to accept the stipulation of Congress that black soldiers be paid as "laborers." One soldier of the 54th wrote: "Sir, the Fifty-fourth and Fifty-fifth Regiments would sooner consent to fight for the whole three years, gratis, than to be put upon the footing of contrabands. It is not that we think ourselves any better than they; for we are not . . . but we have enlisted as Massachusetts Volunteers, and we will not surrender that proud position, come what may."[9]

Working as an agent for enlistments in Ohio, John M. Langston demanded his audience heed the importance of the opportunity before them, saying, "Pay or no pay, let us volunteer. The good results of such a course are manifold. But this one alone is all that needs to be mentioned in this connection. I refer to thorough organization. This is the great need of the colored American."[10]

The moment for Frederick Douglass was one for reflection. His son Lewis was home recuperating from a wound received in the battling around Charleston, and for three weeks Douglass was constantly at his bedside. The time was an important one for him, not for anything he had achieved or failed to achieve, but for the chance it afforded him "to look facts sternly in the face." Almost within his reach had been an important position in the army that could have elevated a leader before a people in their trial for deliverance. So significant had it been to him, originating as it did from his interviews with the highest officers in the land, that Douglass abjured all other duties, even his journalistic enterprise, to grasp the opportunity. But it was not to be. He had toiled for fifteen years with his pen, and now that significant contribution was largely curtailed. In the previous year as the Emancipation Proclamation was announced, surprisingly Douglass had thought of retiring. Informing his old friend and coadjudicator Julia Griffiths of his intention to give up his editorship and buy a farm, she wrote from England on December 5, 1862:

Now a word, my dear friend, about your personal matters & give attention to what I say. Even if all goes as you wish it on the 1st of January 63 you must not give up your paper. . . . The paper was started from this side the water and the ground of obtaining material aid for your branch of the cause is the paper. Surely, the more free colored people are in the North, the more they will need a paper—to assist in elevating them & educating them. No, my dear friend, do not be led astray, or make a mistake by giving up the paper. You know nothing about farming yourself and would be like a fish out of water without mental labors & public work! I wish I could fly over the water & have a consultation with you.[11]

With the first shot fired at Fort Sumter, many among the abolitionists asserted that slavery expired. But Douglass had no such confidence. At the year's end Douglass had given up his notion of retirement in order to grasp a mission he thought incomparably greater. As he contemplated the issues and the way forward at his son's bedside, of his own loss and the losses to the country, temperamentally he could not fall into despondency, but began formulating his thoughts for an important speech that he would repeat week after week across the North in the winter of 1863–64, titled "The Mission of the War." The blow the loyal people of this country strike against the slaveholding rebels, he would pronounce, "is not merely to free a country or continent—but the whole world from Slavery—for when Slavery fails here—it will fall everywhere. We have no business to mourn over our mission. We are writing the statutes of eternal justice and liberty in the blood of the worst of tyrants as a warning to all after-comers. We should rejoice that there was normal life and health enough in us to stand in our appointed place, and do this great service for mankind."[12]

There was much, however, that could sweep back and defeat this sacred mission, destroying its potential, and Douglass was to warn his audiences:

The saying that revolutions never go backward, must be taken with limitations. The revolution of 1848 was one of the grandest that ever dazzled a gazing world. It overturned the French throne, sent Louis Philippe into exile, shook every throne in Europe, and inaugurated a glorious Republic. Looking from a distance, the friends of democratic liberty saw in the convulsion the death of kingcraft in Europe and throughout the world. Great was their disappointment. Almost in the twinkling of an eye, the latent forces of despotism rallied. The Republic disappeared. Her noblest defenders were sent into exile, and the hopes of democratic liberty were blasted in the moment of their bloom. Politics and perfidy proved too strong for the principles of liberty and justice in that contest. I wish I could say that no such liabilities darken the horizon around us. But the same elements are plainly involved here as there.[13]

On February 17, 1864, Douglass wrote to a colleague: "I am, this winter, doing more with my voice than with my pen. I am heard with more than usual attention and hope I am doing some good in my day and generation." The week before, he had spoken in Boston to sustained applause. On February 13 the address was delivered in New York's Cooper Institute, an event sponsored by the Women's Loyalty League, filling the auditorium to capacity, the only time that winter besides a speech by Wendell Phillips that it was done. The next day substantial portions of the text were published in the *Tribune*.

The war had been a long one, Douglass acknowledged, developing in its initial stage from what had been expected to be a ninety-day conflict into one that was seemingly interminable. But this very slow progress was an essential element of its effectiveness. The nation had been in a very low condition before the outbreak of war due to the proslavery compromises endemic to its politics. A radical change was needed in the whole system, and nothing, it appeared to Douglass, was better calculated to effect the desired change than the slow, steady, and certain progress of the war. The charge was now leveled by the so-called Peace Democrats that the war had become diverted from its original aim into an abolition war, into a war for the subjugation of the South. Although abolition was now a vast power, the nation had bitterly hated it, and because it was still loathed the opponents of the war now sought to rely upon this to serve their ends. This was Douglass's counterargument:

For one, I am not careful to deny this charge. But it is instructive to observe how this charge is brought and how it is met. . . . The charge in a comprehensive sense is most true, and it is a pity that it is true, but it would be a vast pity if it were not true. Would that it were more true than it is. When our Government and people shall bravely avow this to be an Abolition war, then the country will be safe. Then our work will be fairly mapped out. . . . Had slavery been abolished at the very beginning of the war, as it ought to have been—had it been abolished in Missouri, as it would have been but for Presidential interference—there would now be no Rebellion in the Southern States—for instead of having to watch these Border States, as they have done, our armies would have marched in overpowering numbers directly upon the Rebels and overwhelmed them. I now hold that a sacred regard for truth, as well as sound policy, makes it our duty to own and avow . . . that this war is, and of right ought to be an Abolition war.[14]

The 30th anniversary meeting of the American Anti-Slavery Society had been held on December 3 and 4, 1863, in Philadelphia, where many illustrious names were present. Garrison and Phillips were there, as were Henry C. Wright and Samuel J. May, Robert Purvis, Lucretia Mott, J. Miller McKim and the two Fosters, Abby Kelly and her husband Stephen; Henry Ward Beecher and Senator Wilson attended some of the sessions, and Douglass was there throughout. As the meeting opened, an honor delegation of black soldiers sat on the platform, while a slave auction block served the speakers as a stand. The tenor of the gathering was one of celebration and of reminiscence. Samuel May, Lucretia Mott, and Miller McKim spoke of the founding convention in 1833. When Robert Purvis spoke, drawing a more urgent notice, he called attention to the existence throughout the country of anti-Negro prejudice, saying this should imply caution that the society's work was not yet done. After all, he would say, they had organized with two connected objectives—the emancipation of the slave and of his elevation.

Finally, acknowledging his presence and his importance to the last twenty years of their work, after several calls for him from the floor, Frederick Douglass came to the platform. Mounting the auction block, Douglass told the assembly the tone of reminiscence he was hearing disturbed him. Although he had many memories of his own in connection with the society and of his experience as an American slave, he would offer none except to say that when he first heard the words of "the honorable President of this association," "words so clear, so simple and truthful, and adapted to the human heart," he felt only five years would be required for slavery to be abolished.[15] But he soon realized he had been mistaken. Referring to Purvis's speech, Douglass said while it was evident that the mission of the war was to remove the chains from the slave, with the other purpose of the society—the African's elevation—they would find a harder resistance. Douglass confided: "I am afraid of this powerful element of prejudice against color. While it exists, I want the voice of the American Anti-Slavery Society to be continually protesting, continually exposing it." Because of the "large minority called Democratic, in every State of the North," there was a danger of compromise and there existed "a powerful nucleus for the most infernal reaction in favor of slavery," he warned. Thus far the best safeguard against reaction had been the ferocity with which the rebel armies had fought, compelling Lincoln to wage what he was still loath to call an "Abolition war." Abolitionists must now insist upon an "Abolition Peace!" Douglass would say. "The day that shall see Jeff. Davis fling down his Montgomery constitution, and call home his Generals, will be the most trying day to the virtue of this people that this country has ever seen. When the slaveholders shall give up the contest, and ask for readmission into the Union, then, as Mr. Wilson has told us, we shall see the trying time in this country. Your Democracy will clamor for peace, and for restoring the old order of things, because that old order of things was the life of the Democratic party."

The only way to stop this was to admit the black "as a full member in good and regular standing in the American body politic"; and the only way to reestablish and reorganize republican institutions in the South was the immediate enfranchisement of the slave. The old Union had been outlived long before the rebellion that came to announce it, Douglass held. The people of the North—the Unionists and loyalists, "had already become utterly disgusted with the idea of playing the part of bloodhounds for the slave-masters, watch-dogs for the plantations. They had come to detest the principle upon which the Slave States had a larger representation in Congress than the Free States. They had already come to think that the little finger of dear old John Brown was worth more to the world than all the slaveholders in Virginia put together."

Concluding his speech, Douglass stood forth amidst the applause for his last statement, saying to even greater applause, "What business, then, have we to fight for the old Union? We are not fighting for it. We are fighting for unity; unity of idea, unity of sentiment, unity of object, unity of institutions, in which there shall be no North, no South, no East, no West, no black, no white, but a solidarity of the nation, making every slave free, and every free man a voter."[16]

The conditions upon which "peace" would be made between the United States and those in "rebellion" against it were officially announced as Lincoln addressed the Congress on December 8, 1863, and sent out an accompanying proclamation. Given under the auspicious view that the crisis that threatened the Union in the previous year was largely past, and the moment was a good one to try to undermine the morale of the South before the ongoing war weakened the resolve of the North, the cardinal points of this policy were the Confiscation Act of 1862 and the Emancipation Proclamation.

Lincoln revealed, moreover, that the key to reconstruction would become amnesty and the pardoning power of the president. On condition of swearing an oath of allegiance to the U.S. government, a southerner was to be granted a pardon, excepting all important ranks in the army and in the political arena. In addition, one-tenth of the voters in any of the seceded states, declaring for these terms of allegiance, "should be recognized as the true government of the State."

These declarations were designed to influence political developments, particularly in Louisiana, where two representatives had already been admitted to Congress, and in Arkansas, where Little Rock had just been occupied by Union troops, but also in North Carolina and Georgia, where dissatisfaction with the course of the war and the Davis administration was being encouraged. As Lincoln's proclamation put it: "Whereas, it is now desired by some persons heretofore engaged in said rebellion, to assume their allegiance to the United States, and to re-inaugurate loyal state Governments within and for their respective States. . . . And I do further proclaim, declare, and make known that any provision which may be adopted by such state government in relation to the freed people of said state which shall recognize and declare their permanent freedom, provide for their education, and which may yet be consistent as a temporary arrangement with their present condition as a laboring, landless and homeless class, will not be objected to by the National Executive."[17]

While Lincoln's plan was positively, even enthusiastically received by most of the antislavery bloc in Congress, it was evident to his critics that he contemplated leaving the political power of the planter untouched and the old political structure intact. Wendell Phillips vented his objection in a letter to General Butler, on December 13. The president's amnesty, with restoration of property and the franchise, he wrote, "leaves the large landed proprietors of the South still to domineer over its politics, and makes the Negro's freedom a mere sham." When these men again secure power, Phillips warned, "The Revolution may be easily checked with the aid of the Administration, which is willing that the Negro should be free but seeks nothing else for him. . . . What McClellan was on the battlefield—'Do as little hurt as possible!'—Lincoln is in civil affairs—'Make as little change as possible!'" In the following week Phillips addressed a packed auditorium at Cooper Institute where he chastised the "lickspittle" administration. The president was no leader, but rather the agent of public opinion; it was true that he had learned in the course of the war that he had his face "Zionward," but he had only moved because he had been pushed.

Yet this criticism was no injustice to the president, said Phillips; "Mr. Lincoln has done such service in this rebellion, has carved for himself a niche so high in the world's history, that he can well afford to have his faults told." Compromise was inevitable in conflicts such as this, and no one could tell where that line lay; the question was, at what point in the war would the North draw back and compromise? To an approving audience Phillips confirmed: "Today the helm is in our hands, and you and I, if faithful, can say this to the nation, and the future. You may compromise when and where you please, with one exception, and that is, that the tap root of slavery shall be cut."[18] After being saved by the Negro the nation owed him, said Phillips, who went on to advocate confiscation of the large estates in the South, and that they be put in the hands of the black and white men who had fought for Union. Most of all the African was owed, not technical freedom, but protection in all his rights; he was owed land and education, and it was Phillips's emphatic point that the debt must be paid.

To prolonged applause Phillips then hit the drum that had been struck in Philadelphia by the American Anti-Slavery Society: "What I ask of Mr. Lincoln in his behalf is, an amendment of the Constitution which his advice to Congress would pass in sixty days, that hereafter there shall be neither slavery nor involuntary servitude in any State of this Union. Mr. Seward wants the Mississippi chairs—the Senate chamber filled. So do I. He is for having them filled as they are. I am for making them so hot that a slaveholder cannot sit in them."

"Emancipation, Abolition, Confiscation, Southern lands for landless Negroes! This is the programme—" ran a column in the *New York World* December 23. While it reprinted Phillips's speech in its entirety, it depreciated its substance. But the paper also realized that Phillips was in the vanguard of a revolution. Its editors spluttered: "The *Tribune* will, as usual, wait six months, and then follow Wendell Phillips' lead face foremost. The *Times* will wait about ten months, and then follow, as usual, back foremost."

At this moment the very pertinent question as to just how the freedom of blacks was to be maintained came to the fore. Louisiana had 300,000 persons held in slavery before the war with about 40,000 free blacks; while in the city of New Orleans, there were upwards of 20,000 free blacks, nearly all of whom could read and write. Taxed for the support of public-school education and on their property, but barred from sending their children to the common schools and compelled to contribute toward the general expense of sustaining the state, they had always been prohibited from the elective franchise. Refusing to accept this exclusion, blacks in New Orleans drew up a petition on January 5, 1864, addressed to both President Lincoln and the Congress with over a thousand signatures. Two of the endorsees, Jean Baptiste Roudanez and Arnold Bertonneau, were selected to travel to Washington where they presented the petition to Lincoln on March 12, while Charles Sumner presented it to the Senate March 15. A day after he met with the delegates, Lincoln addressed a private letter to governor-elect of Louisiana Michael Hahn. He wrote: "I congratulate you on having fixed your name in history as the first Free state Governor of Louisiana. Now you are about to have a convention, which, among other things, will probably define the elective franchise. I barely suggest, for your private consideration whether some of the colored people may not be let in—as, for instance, the very intelligent, and especially those who have fought gallantly in our ranks. They would probably help in some trying time in the future to keep the jewel of Liberty in the family of freedom. But this is only a suggestion, not to the public, but to you alone."[19]

Although crouched in his customary terminology, for the first time the far-reaching proposition had been broached on an authoritative level. Working with General Banks, the governor was able to eliminate a clause forbidding black suffrage in the state constitution, but only to secure another sanctioning the franchise for blacks in the future merely if the legislature "saw fit."

By March Lincoln recognized "loyal" governments in Louisiana and Arkansas, in Tennessee and Virginia. In Lincoln's view the states of the South that formed the "Confederacy" had been usurped, and an insurrection existed in those areas that only awaited the reestablishment of federal authority for the genuine Unionists to prevail—the states themselves were indestructible. In the week after Lincoln's proclamation, the House appointed a committee to look into the subject, chaired by Maryland representative Henry Winter Davis. Many in Congress felt reconstruction should not be limited in its consequences to the war emergency, as did Lincoln, and it belonged rightly to the legislature rather than the executive. In the House, Thaddeus Stevens set forth the view as early as the debate on West Virginia, propounding that, although secession was illegal, those following that course had indeed left the Union; in being subjugated, they became mere property of the United States. Then there was the view first offered by Charles Sumner in a resolution in the Senate in February 1862. It stated that the failure of any state to perform its obligations to the national government created an abrogation of its rights; so it fell under the exclusive jurisdiction of Congress as other territory. In Stevens's outlook there was no constitutional limitation on the conquering power's right to punishment or to control the offending former states; while Sumner stipulated the same constitutional restrictions placed upon Congress in relation to the territories. The Davis committee's bill would be reported in May 1864, and after debate and amendment, passed in the House by a twenty-six-vote majority. In the Senate the bill was referred to Senator Wade's Committee on Territories, and the result in conference was known as the Wade-Davis Bill.

The talk of Washington, so it went in the winter and spring of 1864, was that Lincoln was serving out the last of his presidency. But Lincoln was by no means without resources. The restored states of the South with their "loyal electorates" would surely offset a possible surge of Democratic votes in the North. One of those states that appeared ready for the picking was Florida. At that point in the war Florida had been invaded three times by troops of the Department of the South; first in the spring of 1862, then again that October, and finally in March 1863, as part of the policy aimed at the recruitment of black regiments. On each occasion Jacksonville was occupied. Deeming the state only sparingly defended by Confederate arms on the line from Jacksonville through Lake City to Tallahassee, the Lincoln government decided the time was auspicious for taking hold of the state, assembling a new state government, and reincorporating it into the Union.

With the campaign against Charleston at a standstill, a force of seven thousand Union troops embarked in eighteen transports, including the Massachusetts 54th, under command of Maj.-Gen. T. Seymour. These forces, it was proposed, would occupy Jacksonville, march along the rail-line to capture Lake City and thereby quickly advance on Tallahassee, then fall back on St. Mark's, securing a base by water on the west of the peninsula. Florida was in Beauregard's command and he immediately began counter moves, dispatching brigade level detachments from South Carolina and Georgia, including the Chatham Artillery of Savannah that had borne the brunt at Fort Wagner for many weeks.

Colonel Higginson received orders from General Gilmore that the 1st South Carolina Volunteers should strike their camp near Beaufort, and begin loading men, equipment, and material aboard the steamer *Delaware* to join the expedition. With the vessel loaded and his men ready to give battle to their enemies in Florida, on February 19 Higginson's orders were countermanded. Too many of his officers and men were suffering from the effect of malarial fever. Already extant were reports of a court-martial among black troops garrisoned in Jacksonville, where Sgt. William Walker had been shot for leading the men of Company A of the 3rd South Carolina Volunteers to stack arms before their captain's tent in protest. His motive, a *Tribune* correspondent reported, was that "he could no longer remain a soldier for seven dollars per month." The correspondent wrote: "He met his death unflinchingly. He was a smart soldier and an able man, dangerous as a leader in a revolt."[20]

Higginson commented on the incident in his *Army Life in a Black Regiment*: "The fear of such tragedies spread a cloud of solicitude over every camp of colored soldiers for more than a year. Many of them had families to provide for, and between the actual distress, the sense of wrong, the taunts of those who had refused to enlist from the fear of being cheated, and the doubt how much farther the cheat might be carried, the poor fellows were goaded to the utmost."

Pressing the cause in the pages of the *Tribune* with a letter dated January 22, 1864, Higginson concluded: "The mere delay in the fulfillment of [the government's] contract has already inflicted untold suffering, has impaired discipline, has relaxed loyalty, and has begun to implant a feeling of sullen distrust in the very regiments whose early career solved the problem of the nation, created a new army, and made peaceful emancipation possible."

Earlier that month Massachusetts senator Wilson had introduced legislation to redress the pay differential between white and black troops, but two of his fellow Republicans— senators Fressenden and Doolittle—promptly defeated the measure. Fessenden retorted astonishingly that the government was not bound by unauthorized promises of recruiting officers, and Doolittle said that white soldiers should receive higher pay than black ones, because the families of the latter were often supported by the government. These tidings were the precursors of the Battle of Olustee, as the North called it, or the Battle of Ocean Pond in the South. Thirty-five miles west of Jacksonville three black units—the Massachusetts 54th, the 1st North Carolina, and the 8th United States Battery—fought in combination with three white units—the 7th New Hampshire, the 7th Connecticut, and the 3rd United States Artillery.

By early afternoon on February 20, 1864, the expeditionary force had proceeded to Sanderson, advancing on the railroad bed, with Union cavalry approaching within three miles of Lake City. Concentrated a few miles beyond, in numbers greater than expected, the Confederates were fortifying an area traversing the rail line from north to south, between Ocean Pond—one of Florida's inland lakes— and a swamp behind the village bearing an Indian name, Olustee. A formidable barrier of rifle pits and redoubts were a niggling proposition as the 7th Connecticut came into the enemy, deployed as skirmishers armed with Spencer rifles. Soon the line of battle extended a mile, with the 8th United States Battery, which had only been formed weeks earlier and had never been under fire, came in on the right. Ordered to advance upon the railroad bed, they rushed in, dropping like grass before a sickle, as the saying goes. Closing ranks, they maintained their cohesion within two hundred yards of the fortification for nearly three hours before yielding. As evening settled in, with the Union left and right threatened, the 54th Massachusetts and the 1st North Carolina, which had remained in the rear, were ordered up at the double-quick. As the Confederates prepared a counterattack toward the center, the 54th rushed forward with a cheer. William Wells Brown related: "It was the irrepressible Negro humor, with something of sarcasm, that prompted the battle-cry, 'Three cheers for Old Massachusetts, and seven dollars a month!'" The enemy's charge was blunted by the 54th, followed by the 1st—both regiments inflicting and receiving heavy casualties, before the 1st fell back against overwhelming numbers. Firing with such rapidity that they exhaust-

ed their ammunition, the 54th stood, Brown wrote, "with fixed bayonets till the broken columns had time to retreat, and though once
entirely outflanked, the enemy getting sixty yards in their rear, their
undaunted front and loud cheering caused the enemy to pause, and
allowed them time to change front."[21]

The battle had been a disaster for the North. As many as two thousand men were lost, five hundred becoming prisoners, together with
five pieces of artillery, two stand of colors, and two thousand small
arms. The retreat back to Jacksonville was lined for miles with
wounded and stragglers. General Seymour was reported as looking
"haggard and pale" as he passed by. Denouncing the entire affair as
Lincoln's attempt to secure the three electoral votes of Florida, the
New York Herald nevertheless had emphatic praise for some of the
combatants. Their correspondent wrote: "The First North Carolina
and the Fifty-fourth Massachusetts, of the colored troops, did
admirably. The First North Carolina held the positions it was placed
in with the greatest tenacity, and inflicted heavy loss on the enemy. It
was cool and steady, and never flinched for a moment. The Fifty-
fourth sustained a reputation they had gained at Wagner, and bore
themselves like soldiers throughout the battle."[22]

In the retreat the 54th protected the rear. When all seemed safely
out of danger it was learned a train bearing wounded soldiers had
broken down a few miles outside of town, and the 54th was sent to
render assistance. With the engine disabled, the men were the only
motive power available, and ropes were attached to the cars for the
worn-out troops to haul back to Jacksonville late into the night.

The first months of 1864 offered a preview of military movements to
come that fell like shadows before the flickering flames of later battles. These were Sherman's march through middle Mississippi and
Banks's expedition up the Red River. Grant, the mastermind behind
Sherman's campaign, determined that possession of Chattanooga was
insufficient to support a drive into north Georgia on account of the
great distance between the front and the base of supplies, and the ten-
uous communications. Sherman would pass through Jackson and
Meridian in Mississippi, with the evident objective of capturing
Selma, Alabama, the far corner of a triangle formed by the
Tombigbee and Alabama rivers and the railroad from Meridian to
Montgomery. Once in possession of this salient it was thought it

would be impossible for Sherman to be dislodged, enabling him to begin a campaign overland, supported by columns converging from Pensacola and New Orleans and an assault by the Union fleet, on Mobile. The campaign would rule out reinforcements being sent to Georgia once the campaign began against Atlanta from the north, and would open a water base to support a combined convergence from the southwest. Grant's plan contemplated the removal of the line of the Mississippi from New Orleans to Vicksburg with another from Mobile to Montgomery along the Tombigbee and Alabama, as a flank movement upon Johnston's army.

Banks's expedition through Alexandria and up the Red River, with the end objective of Shreveport, was undertaken for economic as well as political reasons. Passing through the rich cotton region of Louisiana where many hundreds of thousands of bales sat warehoused and ready for shipment, the campaign also offered an important opportunity to expand Lincoln's experiment in reconstruction. Grant, however, was adamant in opposing Banks's Louisiana campaign. Troops would be diverted from Vicksburg, he objected, and until it was successfully completed, it would prevent a combined movement on Mobile. But those higher in the line of command supported it, and as Lincoln favored it, columns proceeded from New Orleans on March 1, taking up a line of advance along the Bayou Teche; while on March 10 troops embarked on twenty transports from Vicksburg to the mouth of the Red River, to join twenty gunboats assembled there for the expedition.

Sherman had departed Vicksburg at the beginning of February at the head of a column of 35,000 infantry, 3,000 cavalry and eighty artillery pieces. Simultaneously nearly 10,000 cavalry and mounted infantry commanded by Benjamin Grierson left Memphis for the ride through northern Mississippi into the central part of the state. Passing quickly through Jackson, Sherman cautiously approached Meridian where he found the Confederates, under the command of Leonidas Polk, had successfully evacuated all their supplies and the rolling stock belonging to the railroad. Reposing in the perfect solitude of Pine Barrens, Polk prepared to meet him across the border in Demopolis, Alabama. Grierson's column, passing some of the richest plantations dotting Mississippi's countryside, was followed by a numerically inferior cavalry newly recruited in west Tennessee under Nathan Bedford Forrest.

Sherman's ambitions rested on his conjunction with Grierson's columns, but Forrest engaged him near West Point on the Tibbee River before that could happen. With Grierson completely outflanking and outmanning Forrest's horsemen three to one, Forrest ordered them to dismount and scatter in the brush. As a fierce Union charge broke upon them, they let fly volley after volley until the ranks of the oncoming cavalry began to falter in confusion. After another unhappy engagement on February 21, Grierson's entire column was compelled to retreat to Memphis. A few days later, Gen. George H. Thomas, in immediate command of the "On to Atlanta" drive in its northern wing, was compelled to fall back after hard fighting. His supplies and communication disrupted, Sherman was likewise compelled to return to Vicksburg.

In expectation of a more auspicious future, Lincoln now conferred upon Grant the grade of commanding general of the armies of the United States. This commission, last held by Washington, had been revived by the Congress and, bearing the date March 2, 1864, was presented to Grant in person on March 9. Halleck, who had been general-in-chief, was relieved of duty, to become chief of staff of the army. Grant immediately transferred his personal presence to the Army of the Potomac, although Meade retained that command. Sherman assumed Grant's former position as head of the Department of the West, and Thomas, losing his post at Chickamauga, was transferred to Nashville.

For the armies of the South there was to be a two-month reprieve for repair and replenishment before the expected supreme struggle to come—the campaigns in Virginia and Georgia—that most historians, weighted by all that is now known, mark as the demise of the Confederacy. But such chronicles could not then be conceived. It was during this time, along with the disaster suffered by Banks, that pages written in blood and fire came at the hand of Forrest, who, after thwarting Sherman, made a rapid advance as far north as the Ohio River.

Banks's flotilla easily reached and occupied Alexandria on March 16, but his forces only managed to secure four thousand bales of cotton before as much as $60 million of the fiber went up, fired by the torches of "unreconstructed" planters. In the coming weeks Banks was to endure inglorious defeats at Sabine Cross Roads and Pleasant Hill by only 15,000 troops commanded by Richard Taylor. Banks managed to return with a mere remnant of his force toward the end of May,

when his flotilla, harassed by sharpshooters
and foundering in the falls above Alexandria,
was rescued by a regiment comprised of
Wisconsin woodsmen, constructing a six-
hundred-foot dam allowing the boats to float
free. Relieved of his command, Banks
retained his important administrative assignment overseeing Louisiana's reconstruction.

Nathan Bedford Forrest is regarded as
one of the war's natural chieftains; he was
indeed a ferocious fighter, unflinching in
inflicting brutality on his opponents, who
often prevailed by mere shock effect and sur-

Ulysses S. Grant.

prise. He was also the owner of several large plantations near
Memphis, who was reputedly "far more humane" to his Negroes
than was the usual practice, keeping families together and providing
"new clothes and decent food." When he returned to his native west
Tennessee in mid-March his name had already gained legendary status and he easily attracted new recruits. In a series of raids ending at
Paducah, Kentucky, on April 25, Forrest gained even more recruits,
arming them with Union arms he had seized. To the south of
Paducah, situated on a high bluff at a bend in the Mississippi River
forty miles above Memphis, was a Union garrison with less than six
hundred men. More than half were white Tennesseans of the 1st
Battalion, 13th Tennessee Cavalry; the rest were black from the 1st
Battalion, 6th U.S. Heavy Artillery, and Company D, U.S. Light
Artillery. There were also a reported hundred noncombatants in the
fort, the wives and children of the black troops, and some whites who
sought sanctuary from Confederate conscription. Forrest took the
opportunity on his ride from Paducah to "attend to" Fort Pillow.

The fort had been built by secessionists early in the war to stop
gunboats and was taken over by Union troops. Consisting of a stock-
ade protected by a dirt parapet 125 yards long and 8 feet high by 4
1/2 feet thick, it was further protected by three crescent-shaped
trenches. The ground in front of the fort sloped steeply toward the
river, and was sided forty to sixty yards above and below by two deep
tree-and brush-choked gorges.

The approach from the rear was cleared, sloping gently down, and
was exposed to raking fire from the fort on two sides. Docked in the
river was a Union gunboat, *New Era*. In the mind of Forrest and his

men, the Tennessee soldiers in the fort were "home grown Yankees" and "renegades," whereas black men in arms in the Union army were an anathema. Forrest meant to discourage and take revenge on both.

Forrest's assault began at sunrise on April 12, with the driving in of the blue-clad pickets. Attacking the outer defenses, several attempts to charge were repulsed, but with constant reinforcements, Forrest's lines soon extended to the river on both sides of the fort. Throughout the fight, guided by signals, the *New Era* kept up its fire, without much effect, however. At noon the Union troops withdrew from the outer trenches, for a more determined fight within the fort, whereupon Forrest ordered his men to shelter in the gorges. It was now clear on both sides, given the preponderance of numbers, that the garrison was to be overwhelmed. Sending out a flag of truce, Forrest demanded surrender. The ranking officer of the fort had been killed early in the fight, and his second asked an hour to consider. Forrest allowed twenty minutes.

The defending officers reached a unanimous decision—they would not capitulate, and a reply stating such was sent bearing the name of the deceased commanding officer. Forrest's men, meanwhile, had positioned themselves in the most advantageous way for their final charge. About four in the afternoon the rush was made and the parapet quickly stormed, just as a Union transport, the *Olive Branch,* was gliding down river to dock with fresh troops. As men were slain all around, finding themselves brutally overpowered, others threw down their arms in surrender. With the killing continuing, terrified men ran for the river, as cries of "no quarter" and "kill the damn niggers, shoot them!" rent the air. Many of the fleeing soldiers were shot as they ran; some who remained dropped to the ground feigning death and were dispatched where they lay. All the officers in the fort, some especially being singled out for retribution and wantonly shot and killed, were euphemistically reported as "while trying to escape." Many captives were shot after being lined up, and their riddled bodies rolled into the trenches.

Over one hundred and thirty white troops were killed and nearly one hundred and ninety black. According to Forrest's report, among those surviving were "164 Federals and 75 negro troops, and about 40 negro women and children." The captured blacks were transported to slave labor on Confederate defenses. There has always been a divergence of opinion as to what happened in the massacre at Fort Pillow. Forrest denied that any atrocity had been planned in advance and his apologists point out the Union troops made no attempt to surrender prior to the storming of the fort; that no white flag had been raised, nor had the U.S. flag been lowered until it was hauled down and cut loose and handed to Forrest. Just at that moment, too, a Union transport was beginning to dock and it appeared that the fort would be reinforced; it was also stated that Forrest ordered his men to stop firing once Fort Pillow had been taken, for witnesses recounted that he shot one man who disobeyed his order and cut another with a saber. But the accounts of witnesses are also ample that men, particularly if they were black, and even women and children, were murdered by the score after they surrendered. The tally of those killed shows that this was manifestly the case.

Soon afterward, the Joint Select Committee on the Conduct of the War sent to the site some of its members, who reported that two weeks later despite heavy rains, matted human vestiges—of hair, brains, and blood—were still unmistakable. Since many of the bodies had been carelessly thrown into trenches and only partially covered, the report to the committee read:

Portions of heads and faces were found protruding through the earth in every direction. . . .
It will appear from the testimony that was taken, that the atrocities committed at Fort Pillow were not the result of passion elicited by the heat of conflict, but were the results of a policy deliberately decided upon, and unhesitatingly announced. Even if the uncertainty of the fate of those officers and men belonging to colored regiments, who have heretofore been taken prisoners by the rebels, has failed to convince the authorities of our Government of this fact, the testimony herewith submitted must convince even the most skeptical, that it is the intention of the rebel authorities not to recognize the officers and men of our colored regiments as entitled to the treatment accorded by all civilized nations to prisoners of war.[23]

Although he always maintained his honor in this affair, it is clear that Forrest and other Confederate commanders considered of no consequence the lives of blacks who refused acceptance of their status as "property." In one of several reports on the combat at Fort Pillow, Forrest affirmed, "I regard captured negroes as I do other captured property and not as captured soldiers." In another he announced, "It is hoped that these facts will demonstrate to the Northern people that negro soldiers cannot cope with Southerners."[24]

Although this was the largest atrocity committed against black United States soldiery by Confederate arms, there were many instances of the black flag being unfurled, signifying that no quarter would be given. Less than a week later another little-heralded killing occurred in a southern Arkansas town called Poison Springs, where a forage train guarded by heavily outmanned Union troops was mauled by Confederate cavalry. The 1st Kansas Colored Volunteer Infantry suffered the same fate that their counterpart had in Tennessee. The voice of Confederate Arkansas, the *Washington Telegraph*, justified the merciless killing of black prisoners and wounded, writing: "We cannot treat negroes . . . as prisoners of war without the destruction of the social system for which we contend. . . . We must claim the full control of all negroes who may fall into our hands, to punish with death, or any other penalty."

In late spring 1864, in a letter to an English correspondent, Frederick Douglass conceded that his patience with the course being pursued by Lincoln had been worn threadbare. The government had repeatedly promised to protect the lives of its black soldiers, but had failed to do so when faced with slaughter or exchange when captured. And although promised it when they enlisted, black men had been refused equal compensation, and were prohibited from holding commissions. Most disturbing was Lincoln's approval of Banks's order No. 12 issued on January 29, 1863, prescribing for blacks in Louisiana conditions of labor little different from their old situation. Lincoln's proclamation on reconstruction had insisted that as a condition for returning to the Union, "rebels" recognize the permanence of black freedom, suggesting it would be expedient to introduce a system of apprenticeship in which the young were provided with education. But he had also laid out that temporary arrangements would "not be objected to by the National Executive . . . Consistent . . . with their present condition as a laboring, landless, and homeless class."[25] How temporary? How long would this status persist? were the questions that arose. To Douglass, particularly because Lincoln was pursuing reconstruction on the votes of 10 percent of the voters in 1860, this meant emancipation was "a mockery and delusion."

By mid-1864 the recruitment of black troops was approaching 100,000, placing it equal in size with any other Union army. The administrative burden on the War Department had become so onerous that a Bureau of Colored Troops had been created under the aegis of the Adjutant General's Office. As the bureau's work progressed, the black regiments, formerly using state designations, became United States Colored Troops (USCT); only the Massachusetts 54th and 55th, and its 5th Cavalry, as well as the 29th Connecticut Infantry, retained their original titles. With already noted exceptions, black troops had been used largely to do the work white troops found demeaning, from fatigue duty to garrisoning forts. But they had also performed substantial labor in building fortifications, and the numerous manual tasks that arose, all tasks performed with equal effectiveness by blacks for the Confederate armies. These activities prohibited many from being properly drilled and trained for combat, and those that had been were often issued dated smoothbore muskets or defective rifled muskets. But by 1864 the Union army began to assign these units as integral parts of divisions, corps, and brigades.

Before the war the new general-in-chief had appeared a man of ordinary attainment. Ulysses S. Grant attended West Point but graduated at the bottom of his class. After seeing war in Mexico and duty on the frontier, he had been cashiered for drunkenness, afterward working in his family's tannery. As war began he was initially refused a commission, then accidentally selected to lead a regiment of Illinois recruits. Lionized in the press for his victories at Fort Donelson and at Vicksburg, while his near disaster at Shiloh, though tactically not a defeat, was nearly forgotten—his star had risen. In battle, his contemporaries observed, Grant was unperturbed by the wastage of troops, as he was unconcerned of the disposition of his enemy "out of his sight." His generalship was not one of maneuver but of deliberate and heavy strokes. Pollard remarks that his "Scotch pertinacity of character was a constant and valuable assistance in his military campaigns."[26]

When Grant assumed high command, war had been ravaging and bleeding the country for three years. The fate of Lincoln's presidency and of his emancipation strategy—whether they were to go down altogether or were to triumph—tottered in the balance. Each now depended, as Lincoln himself was to say, upon the advance of arms. Within eight weeks of receiving his commission Grant had devised a plan and had his armies ready. His strategic design called for parallel and simultaneous operations in Virginia and in Georgia, while Union forces in the Mississippi Valley exerted and extended control, and a new thrust was developed against Mobile from land and sea.

Sherman with 112,000 veteran soldiers was ordered to go after Johnston's 60,000-man army and break it up, then to get as far into the interior of Georgia as possible, inflicting all the damage he could against Confederate war resources. To Meade, who would be subject to his strategic orders, Grant instructed: "Lee's Army will be your objective point. Wherever Lee goes, there will you go also." While his 118,700 troops would advance across the Rapidan against Lee's 70,000, Benjamin Butler with 36,000 troops was to advance up the James River from Fort Monroe to cut the railroad between Petersburg and Richmond, and then threaten the Confederate capital from the south. Franz Sigel's 6,000-troop command was to move down the Shenandoah Valley pinning down its defenders, then cut the railroad connecting Lee's army with his principal base of supplies at Lynchburg. Banks, still in command of his troops, mistakenly project-ed to be finished with his Red River expedition by the start of the campaign, was to begin a drive against Mobile together with Farragut's fleet, and then push northward to prevent Alabama rein-forcements from joining Johnston. Presented with the particulars of Grant's plan to finish off the Confederacy, Lincoln, using the collo-quialism of the frontier of which he was fond, remarked agreeably, "Those not skinning can hold a leg."

No black regiments had been assigned to the Army of the Potomac, and its successive generals, a few of them implacably, had been opposed to Lincoln's adoption of these policies after 1862. When he was elevated to strategic command Grant transferred two divisions of black troops from the western theater to northern Virginia. In addition, black troops were transferred from Gilmore's command to Butler's, and orders were issued by the adjutant general forbidding the exclusive employment of black troops for fatigue labor, a practice encouraged by both Meade and Sherman. For every combat soldier four more were required in support, securing lines of communication, guarding train lines, bridges, supply depots, and gar-risoning forts, and for holding a far-flung and hostile territory. That these troops had become the bulwark of the Union was indubitable; an all-out-effort to crush the Confederacy would not have been prac-ticable without this preponderance of black force, both as labor and as arms. It was their crusade for freedom; as one black noncommis-sioned officer put it: "They had been to the armory of God, and had received weapons of the heart, that made them daring and dangerous foes—men to be really reckoned with."27

Sixteen

War for the Total
Expiation of Slavery

Fondly do we hope—fervently do we pray—that this mighty scourge of war may speedily pass away. Yet, if God will that it continue, until all the wealth piled by the bondman's two hundred and fifty years of unrequited toil shall be sunk, and until every drop of blood drawn with the lash, shall be paid by another drawn with the sword, as was said three thousand years ago, so still it must be said "the judgments of the Lord, are true and righteous altogether."
—Abraham Lincoln, Second Inaugural Address

A T DAWN ON MAY 5, 1864, IN A BRILLIANT FLOTILLA OF gunboats and transports, Butler's Army of the James steamed up river for a landing between Petersburg and Richmond. Meanwhile, Meade's Army of the Potomac crossed the Rapidan followed closely by a separate army corps commanded by Burnside, while simultaneously William Tecumseh Sherman's army converged in three columns on Johnston's position astride the Atlanta–Chattanooga railroad at Dalton, Georgia. It was Sherman's intent to extend his right beyond Johnston's capacity to resist while moving down the line to Atlanta; while Grant's was to move by right flank to force Lee out of his trenches and into an open fight.

In the afternoon when Butler's flotilla began to land at Bermuda Hundred there were only 5,000 Confederate troops, along with some hurriedly mobilized government clerks, in the defenses of Petersburg and Richmond. On May 7 the connecting link between the two bastions was severed when a bridge was blown, and Butler could have marched into the Confederate capital with overwhelming force; instead, he began entrenching close to the enemy's defenses on Drury's Bluff. By the next week when he felt ready to push on, Beauregard—newly transferred from Charleston with reinforcements from the Carolinas—had moved in more than enough troops to meet his challenge.

In northern Virginia, as Union columns marched on roads crossing the Wilderness, an area of thick forests and dense undergrowth west of Fredericksburg, they found themselves all but trapped near where Hooker had been caught the year before. In the first day of a two-day struggle, fierce fighting surged from side to side as the underbrush and forest exploded into flame; but by dusk Grant had gained a position to attack Lee's right. The next morning the attack sent the enemy reeling through the woods clear into Lee's headquarters. However, by mid-morning the lost ground had been recovered; by noon, attacking straight ahead and by flank from the overgrowth of an unmarked roadbed, they had rolled up an entire Union regiment. But the vigor of this counter-thrust was lost when Longstreet was wounded in the neck and was out of the battle. In the spreading fire the breastworks were ignited, and the conflagration cremated the dead. Then a surprise assault at dusk by Confederates on an exposed Union right smashed through the line, capturing two generals.

With his army seemingly on the verge of rout, Grant reassessed his position. Where previous northern generals had withdrawn to the river's safety, Grant sidestepped Lee, forcing him into another fight, or else retreat toward Richmond. He would head east using the road to Fredericksburg, clearing Lee's right, and then turn south to seize Spotsylvania a dozen miles away. When the Union columns reached the juncture where they again would turn south, the ranks began to sing as if a burden had been lifted—it seemed as if a turning point in the war had been reached. In the next few days the Northern press heralded a victory with banner headlines that read "A Waterloo Defeat of the Confederates."[1] But Grant had suffered over 17,000 casualties versus 10,000 for his opponent—what is more, Lee had perfectly understood his intention and accepted the challenge.

Taking up a line along the Po River, Lee would send Jeb Stuart to block the Union advance, while Longstreet's corps (commanded by Gen. Robert H. Anderson) hurried to seize the new battle area. Winning the race, the Confederates quickly began work on an arrangement of fortifications, with networks of trenches and earthen breastworks studded with abatis, redoubts, and artillery emplacements. Spotsylvania was to see the strongest field works up to that point in the war.

Grant now sent Gen. Philip Sheridan with 10,000 cavalry to Lee's rear to cut his communications and destroy his supplies. Countering, Lee sent the South's "plumed cavalier" harrying at his heels. Six miles north of Richmond on May 11, outmanned and outgunned, Stuart made his last stand at Yellow Tavern. Joining Stuart's cavalry on that fateful day was the 5th Virginia Cavalry commanded by Colonel Henry Clay Pate. With Sheridan's men swarming with rapid-fire carbines, Stuart asked Pate how long he could hold out. Pate answered, "Until I die." Minutes later while leading a charge he was interrupted with finality; a short time later near the same ground, Stuart too was mortally wounded. As Sheridan crossed the James to unite with Butler, and as Sigel came down the Shenandoah Valley, Grant confronted Lee.

At Spotsylvania there were two alternatives for Grant: On May 9 he tried to turn Lee's left but was countered when Lee shifted two divisions; believing the center was now vulnerable Grant tried to smash through on May 10, but found it reinforced from Lee's right. But along a mile of the front there had been a notable success—twelve regiments, forming in four lines commanded by Col. Emory Upton, made a dash without firing across 200 yards of open ground. Overwhelming a salient in the defenses, they captured a thousand prisoners; but as supporting regiments came under artillery fire, Upton's storming party was forced back in the gathering darkness. Grant, however, was persuaded to try the same tactic massively. On the eve of the great clash he telegraphed Washington, "I propose to fight it out on this line if it takes all summer."[2]

The next day 15,000 troops advanced on Confederate defenses in rain and fog. With his army cleaved nearly in two, Lee personally organized a counterattack, as he said, "to drive those people from our works." Fighting through the jumbled confusion the Confederates forced incoherent attacks back into the trenches, even as the flags of the antagonists waved over the breastworks. From dawn till midnight

Soldiers of the 1st Massachusetts Heavy Artillery burying dead after the battles around Spotsylvania, May 1864.

the fighting raged, producing another 18,000 casualties for Grant and casualties proportionally as terrible for Lee. The fiercest struggle occurred at the Bloody Angle, as the salient was known, where men, in the fight of their lives, trampled on the dead in an oozy tangle of blood and matter. All told Grant had lost 32,000 men in the week since his drive began, and was now known among his rank and file as "the butcher." Yet so far the deaths had been as unproductive militarily as those at Fredericksburg.

As news of the carnage began reaching the North, on May 15 Confederate forces near New Market in the Valley commanded by Breckinridge met Sigel's columns. Adding to the mood of despondency already gripping Washington came further report of heavy losses in men, arms, and supplies. The next day Beauregard struck Butler's position, inflicting another 5,000 losses, turning Butler's flank and crushing him backward toward the Bermuda Hundred, with Beauregard entrenching across his front. When Grant learned of it he observed with disgust that Butler's army was "as completely shut off from further operations . . . as if it had been in a bottle strongly corked."[3] It was at this time, too, the misfortune of Bank's reversal on the Red River was becoming known—there would be no movement on Mobile.

Forced out of Dalton, Johnston, in order to meet Sherman's advance, had fallen back on Resaca, and there the two armies contested on May 14–15, yielding 5,500 casualties combined. Again

Sherman's flanking movement forced Johnston to retreat, with skirmishing following at Adairsville, Kingston, and Cassville. Meanwhile Grant, after determining that neither a head-on assault nor short flanking maneuvers would shake Lee from his defenses, decided to circumvent Lee's right again. Arriving just beyond the North Anna River, there too Grant found the same entrenched enemy. After several days' skirmishing, Grant again sidestepped Lee, while keeping his left open to secure supply lines, crossing the Parmunkey River. Now Grant was on the roads leading directly to Richmond.

The numerical advantage Grant enjoyed over Lee was nearly equaled by that of Sherman over Johnston. But as Sherman advanced into Georgia and Johnston yielded, he encountered an increasing logistical difficulty. His supply and communications stretched along a single line that ran some three hundred miles in his rear to Nashville; if broken for an extended period it could be disastrous. This was precisely where Johnston's best prospect lay, and to facilitate it he urged "the immediate movement of Forrest into Middle Tennessee." A slashing attack by Forrest was such a concern that Sherman offered a major-general's commission to anyone able to stop him. A contender was found in Gen. S. D. Sturgis, who led a column of cavalry and infantry out of Memphis, with twenty-two cannon and a train of 250 supply wagons and ambulances on June 1. Sturgis's orders were to destroy all property, foodstuffs, and industrial and rolling stock that could be used by the enemy. A part of his forces were black regiments flourishing the moniker "Avengers of Fort Pillow."

The crisis facing Lincoln's government in summer 1864 was every bit as severe as the crisis in 1861. The president ran the gauntlet of criticism: he was unmethodical in his office and deficient in nerve. Some argued he was an honest and good man, but where persons were concerned he was oversentimental. His aim was wrong—in his stumbling way he endangered the precious load he carried, and had failed to come to grips with the significance of the age. His hope of reconciling the South was futile, and he had been reluctant to champion freedom and equality when they should have been paramount.

These were issues roiling the Massachusetts Anti-Slavery Society earlier that year, when a split between Wendell Phillips and William Lloyd Garrison became public. To Phillips's motion that the society declare "that the government is ready to sacrifice the honor and interest of the North to secure a sham peace and have the freedmen under the control of the late slave-holders," Garrison offered the amendment that "the government was only in danger of doing so." The two quarreled again at the meeting of the New England Anti-Slavery Society when Parker Pillsbury urged adoption of a resolution condemning the administration. Garrison, and at least half of the old-line abolitionists, felt they had a moral obligation to support Lincoln in the coming electoral season, and that, as Lydia Child said, "We can have no better president than 'honest Abe.'" Phillips however would be at his most outspoken: "My charge against the administration . . . is that it seeks to adjourn the battle from cannon shot to the forum, from Grant to the Senate House, and to leave the poisoned remnants of the slave system for a quarter of a century to come."[4]

Opposition to Lincoln had been rife among the so-called Radicals too. Within his own party other names had been floated and would fall before the nominating convention was held June 8. First was Secretary of the Treasury Salmon P. Chase, with an important patronage at his disposal and an outsized ambition, whose name had emerged as Lincoln's replacement in a circular by Kansas senator Pomeroy in January. Then Butler's name had floated, only to be punctured by his failure in his latest military campaign, as Grant's name too had been mentioned as his stock had risen. Lincoln was not the man to carry the war to its conclusion and set the nation on a new footing, many suggested. Horace Greeley offered in the *Tribune* that the nominating convention be postponed until September in hope such a man would surface.

Another challenge to Lincoln would be mounted by a third party nominating John C. Frémont. A call for such had come from three separate quarters: a committee of Radical Republicans headed by Missouri's B. Gratz Brown, a second from a committee in New York headed by Lucius Robinson, and issuing a third was a committee of abolitionists. This movement attracted broad support in German-American communities, particularly in St. Louis and Chicago, and was led in Boston by Wendell Phillips's protector Karl Heinzen. Douglass replied to the call in the pages of the *New York Times* on May 27, and laid out the platform to be followed as he saw it:

> I mean the complete abolition of every vestige, form and modification of Slavery in every part of the United States, perfect equality for the black man in every State before the law, in the jury box, at

the ballot-box and on the battlefield; ample and salutary retaliation for every instance of enslavement or slaughter of prisoners of color. I mean that in the distribution of offices and honors under this Government no discrimination shall be made in favor of or against any class of citizens, whether black or white, of native or foreign birth. And supposing the convention which is to meet at Cleveland means the same thing, I cheerfully give my name as one of the signers of the call.

As forthright as Douglass's declaration was, it was tepid in comparison to the stance taken by Phillips, who wrote in his letter to the Cleveland convention, published in the *Liberator,* June 3, 1864:

We had three tools with which to crush the rebellion—men, money, and the emancipation of the Negro. We were warned to be quick and sharp in the use of these, because every year the war lasted hardened the South from a Rebellion into a nation, and doubled the danger of foreign interference. For three years the administration has lavished money without stint, and drenched the land in blood, . . . Meanwhile slavery was too sacred to be used; that was saved lest the feelings of rebels should be hurt. . . . Mr. Lincoln's model of reconstruction puts all power into the hands of the unchanged white race, soured by defeat, hating the laboring class, plotting constantly for aristocratic institutions. . . . To reconstruct the rebel States on that model is only continuing the war in the Senate chamber after we have closed it in the field. . . . Such reconstruction makes the freedom of the Negro a sham, and perpetuates slavery under a softer name. . . . There is no plan of reconstruction possible within twenty years unless we admit black citizenship and the ballot, and use him with the white, as the basis of States. . . . The administration I regard as a civil and military failure, and its avowed policy ruinous to the North. Mr. Lincoln may wish the end—peace and freedom—but he is wholly unwilling to use the means which can secure that end. Frémont is my choice.

On May 31, meeting in Cleveland's Cosmopolitan Hall, four hundred delegates in a tightly orchestrated proceeding nominated Frémont for president and Gen. John Cochrane, a Democrat and nephew of Gerrit Smith, for vice president. The convention's platform called for uncompromising prosecution of the war, a constitutional amendment prohibiting slavery, and reconstruction exclusively under the aegis of Congress. In its most radical plank the convention called for confiscation of rebel land to be divided among U.S. soldiers and actual settlers. The convention also called for a one-term presidency and offered criticism of Lincoln's restriction on habeas corpus, free speech, and the press, to attract Democrats.

After abandoning Resaca, Georgia, Johnston fell back to Cassville, a railroad junction twenty-five miles to the south. Then again Johnston deftly withdrew, winning a stronger lodgment at Allatoona Pass. Seeing the futility of assaulting him there, Sherman swung west, cutting loose from the railroad and his communication. Finally on May 25–28 the antagonists engaged in savage battle near Dallas at New Hope Church. Amid heavy rains as the fighting dwindled to skirmishing in the early weeks of June, Johnston fell back to Kennesaw Mountain in front of Marietta, his strongest lodgment yet. Although he had yielded mile after mile on the road to Atlanta, he had at last attained a position that threatened to send Sherman reeling back. But, as he had given up ground, almost the whole of the white population, with a large number of slaves, peeled off the land, heading for the Georgia interior.

The day before Sturgis left Memphis, in Virginia Sheridan's cavalry had seized the crossroad village named Cold Harbor, east of the Chickahominy River and eight miles from Richmond, precipitating a severe fight with his counterpart, Fitzhugh Lee. The following day, as he received reinforcements in infantry, Sheridan's force withstood a counterattack. That night, facing each other along a seven-mile front stretching from Totopotomy Creek to the Chickahominy, both Lee's and Grant's armies feverishly entrenched. In New York City, anticipating a great victory, a mass meeting drawing 25,000 was called, as editorialists "discounted" Richmond's capture. The true presumption, however, as Pollard remarked in his *Southern History of the War*, predicated on Grant's repeated disappointments, was not victory but defeat.

While Robert E. Lee's army had been badly wasted in more than twenty days' fighting and entrenching, Grant's too was suffering terribly. Scarcely a month before, Grant had embarked on an impressive arrangement; now that that design had been thwarted Grant found himself occupying ground he could have had, without as great loss, had he chosen another route. But victory at Cold Harbor might end the war, yet another flanking maneuver would only ensnare Grant in the same bottomlands McClellan had staggered in and allow Lee to step back into the formidable defenses of Richmond. If he could drive Lee back on the Chickahominy, his army might well be annihilated. Accordingly Grant ordered an assault against imposing Confederate trenches and field fortifications to begin at dawn on June 3. For his part Lee recognized that he had to do everything he could to destroy Grant's army before he reached the James, at which point the war would become a siege operation, and its inevitable denouement would only be a matter of time. When the blue lines surged forward on that foggy morning, score upon score of the men had pinned their names and addresses on the outside of their uniforms in hope their graves would be properly marked and their kin notified. Many of the 7,000 Union dead in the battle expired in a single half hour. By evening, a visibly stunned Grant commented, "I regret this assault more than any one I have ever ordered."[5]

Meanwhile Sturgis's advance into northern Mississippi was being accompanied by a wasting that was now characteristic of the war. At Brice's Crossroads he and Forrest converged warily on June 10. Reaching there first, Forrest intended to whip Sturgis's cavalry, then lash into the worn-out infantry as they came up. "It is going to be hot as hell," he predicted. A correspondent to the *Richmond Examiner* wrote:

> Before the battle fugitives from the counties through which Sturgis and his troops were advancing, came into camp detailing incidents which made men shudder, who are accustomed to scenes of violence and bloodshed. I cannot relate the stories of these poor frightened people. Robbery, rapine, and the assassination of men and women, were the least crimes committed, while the "Avengers of Fort Pillow" overran and desolated the country. Rude unlettered men, who had fought at Shiloh, and in many subsequent battles, wept like children when they heard of the enormities to which their mothers, sisters, and wives had been subjected by the negro mercenaries of Sturgis.[6]

Forrest, to preclude being struck first, ordered a spoiling attack, and then at the optimal moment led a full charge into the Union cavalry. By noon Forrest's opposite had arrived with the forward detach-

ments of his infantry, drawing them up in a
taut semicircle in front of his guns with a
large reserve in the rear. Pressing into
Sturgis's right, Forrest simultaneously sent a
unit to strike him in the rear. Soon confusion
beset his foe and Forrest gave the signal for
an all-out assault. Hit in both flanks, Sturgis
was forced to flee with his personal escort,
leaving a third of his men prisoners, losing
nearly all of his cannon and his entire train
of supplies. When news of the disaster

Nathan B. Forrest.

reached Sherman he ordered another, bigger
expedition out of Memphis to "follow
Forrest to the death, if it cost 10,000 lives
and breaks the Treasury." He would "have the matter of Sturgis critically examined," Sherman wired the secretary of war, "and if he
should be at fault he shall have no mercy at my hands. . . . Forrest is
the devil, and I think he has got some of our troops under cover."[7]

After the bitter defeat at Cold Harbor, Grant ordered David
Hunter, now replacing Sigel in the Shenandoah Valley, to move southward, tearing up the railroad tracks as he went, while he ordered two
cavalry divisions under Sheridan to ride northward and wreck the
railroads on the other end. Hunter was to reach across the Blue
Ridge, demolishing Lee's supply depots at Lynchburg; then Sheridan
and Hunter were to make a conjunction and advance east toward
Richmond, wrecking the railroad and the James River canal. At the
same time, Grant proposed to cross the James and seize Petersburg
with its hub of railroads, forcing Lee into the open and into a decisive fight.

By June 5, Hunter seized the town of Staunton, killing its
Confederate commander and capturing over one thousand prisoners.
On June 7, Sheridan crossed the Pamunkey and was moving toward
Gordonsville. On the 9th, with nine regiments, Butler began probing
the outer defenses of Petersburg. Wreaking general devastation
through the country, Hunter reached Lexington on June 12, where
the Virginia Military Institute and the home of the ex-governor
Letcher were burned and Washington College was sacked. By June 13
Hunter was reported to be moving against Lynchburg. The previous
night, Grant had begun moving the main army across the James. But
already a crucial movement of the campaign had gone awry. Wade

Hampton's cavalry had intercepted Sheridan at Trevillian Station, compelling him to withdraw across the North Anna, while, to meet the threat at his rear, Lee dispatched Jubal Early with 10,000 veterans of Jackson's old command to counter Hunter. Attacked on the south side of Lynchburg on June 17–18, Hunter like Sigel was compelled to retreat to western Virginia.

Grant's movement by contrast went smoothly, his engineers building a 2,100-foot pontoon bridge for the crossing. The defenses at Petersburg presented him with the obstacle of a twenty-foot-thick breastwork ten miles long, brimming with artillery-studded salients, and fronted with a ditch fifteen feet deep. With two of Grant's corps approaching its undermanned garrison, Gen. Henry Wise gave an address to rouse the troops of his command, saying, "Petersburg is to be, and shall be defended on her outer walls, on her inner lines, at her corporation bounds, in every street, and around every temple of God, and altar of man."[8]

On June 15, sending troops commanded by William "Baldy" Smith forth for a sundown assault, a portion of the Petersburg line was captured. Not realizing the citadel lay at his mercy, Smith made little headway for three days. With Davis and Bragg vacillating as Beauregard called for reinforcements, Lee ordered 48,000 of his troops to invest the lines south of Richmond. Only now did Davis transfer overall command in the district to Lee. At this point a syndrome seemed to afflict both commanders, and troops of Grant's army, as regiments balked in promptly obeying orders, as if waiting for other units to move first. Meade's field telegraph pounded out his orders: "Finding it impossible to effect cooperation by appointing an hour for attack, I have sent an order to each corps commander to attack at all hazards and without reference to each other."[9]

At four in the morning on June 18 an assault was made and repulsed, and again at noon and four in the afternoon, with both being repulsed. Meade called off the assaults as Grant concurred, saying, "We will rest the men and use the spade for their protection until a new vein has been struck."[10]

Grant's next move was against the Weldon Railroad southwest of Petersburg, where again he found misfortune as the assaulting party was pierced at its center, resulting in the loss of four guns and the capture of an entire brigade and part of another. An expedition six days later also ended in disaster, with the loss of more artillery and a complete wagon train.

As this phase of Grant's campaign concluded, in Georgia to demonstrate and to stiffen his armies, on June 27 Sherman directed that Johnston's position at Kennesaw Mountain be attacked. After feinting against his enemy's flanks, in sweltering heat a charge rushed into Johnston's center. Protected by defenses that rivaled those at Cold Harbor, Sherman sustained an estimated 6,000 casualties against a tenth of that for his foe. But Sherman approached within twenty miles of Atlanta as Johnston fell back upon Peachtree Creek.

After seven weeks of hard fighting, Grant's army had lost 65,000 men killed, wounded, or missing—more than Lee had in his entire army. But Lee's losses had been proportionally as great. Sherman's losses approached 17,000, while Johnston sustained 14,000. On all fronts in May and June the North had losses of 90,000 men, while the South's stood at just over half that number. But Lee's latitude for offensive operations had been smothered, and he had been severed from part of his communications with the rest of the Confederacy. Johnston, in conceding the road to Atlanta, had given up a great portion of Georgia, including one of the best wheat-growing regions of the South, and lost two iron-rolling mills of considerable importance, sorely perplexing President Davis. By this strategy, however, Johnston had kept his army intact and was able to fall back on the defensive works surrounding Atlanta, nearly the rival of those at Petersburg. He had also drawn out Sherman's logistical vulnerability, with Forrest operating in his rear, to the maximal extent.

As a successful termination of the war for the Union seemed more remote than ever, the question loomed in the North: What purpose could justify the cost; had the wastage, both human and material, resolved anything? At a fund-raising speech before the philanthropic Sanitary Commission on June 16 in Philadelphia, Lincoln hoped the war would end when the "worthy object . . . of restoring the national authority over the whole national domain . . . is attained." To the cheers of his audience he recalled Grant's avowal to fight it out "on this line if it takes all summer," adding, "I say, we are going through on this line if it takes three years more."[11] But at this point his prospects were strongly laid on showing substantial progress in the proximate two months.

The helm ought to pass to firmer hands, was the categorical assessment of Horace Greeley: "Mr. Lincoln is already beaten. He cannot be elected." The *Evening Post*'s William Cullen Bryant wrote he was "so disgusted with Lincoln's behavior that he could not muster enough courage to write to him." The *New York World*, embarking on a period of attacks that became so vociferous that the publication was temporarily suspended, wrote: "Who shall revive the withered hopes that bloomed at the opening of Grant's campaign?"[12]

The National Union Convention meeting in Baltimore on June 8 had nominated Lincoln for a second term amid ongoing mutterings and silent opposition. Sen. James Lane put Lincoln's name before the convention and concluded by saying, "If we nominate any other than Abraham Lincoln, we nominate ruin." That was to be the unanimous consensus; and with the need to shore up support, the convention jettisoned Vice President Hannibal Hamlin, selecting Andrew Johnson, a War Democrat and governor of Tennessee who had led reconstruction in his state. The convention also culled out a platform calling for uncompromising prosecution of the war and a constitutional amendment abolishing slavery. When the latter item on the plank was presented, William Lloyd Garrison, an observer at the convention, wrote: "The whole body of delegates sprang to their feet . . . in prolonged cheering. Was not a spectacle like that rich compensation for more than thirty years of personal opprobrium?"[13]

After the convention Garrison endorsed Lincoln in the pages of his paper and was welcomed into the executive mansion, as Phillips jeered: "A million dollars would have been a cheap purchase for the administration for the *Liberator*'s article on the Presidency." The next day, addressing the National Union League, Lincoln acknowledged the widespread dissatisfaction with his continuance in the executive chair, supposing the convention had concluded that "It is not best to swap horses while crossing the river, and have further concluded that I am not so poor a horse that they might not make a botch of it in trying to swap."[14]

On the battlefield, the superiority in numbers so carefully counted on by Grant was about to diminish. Many of the veterans in the Union armies, who had already suffered a large proportion of the casualties on 1864, were due to muster out as their three-year enlistments expired. Even with inducements—a thirty-day furlough, a $400 bounty, and a special chevron to wear on their sleeves—the Army of the Potomac had a reenlistment rate of only 50 percent, and more than 100,000 men were set to leave the service. Lincoln was now faced with having to resort to another unpopular draft, while his political and military rivals grasped for every wedge to exploit fault lines into full-fledged ruptures. The struggle though was not as simple as it would have been had it merely been between Democrat and Republican Unionists. Lincoln's difficulty with the radicals in the Congress, too, was to come into sharper focus. On May 4 Henry Davis's bill on "reconstruction" passed the House, and referred to Wade's Committee on Territories, passed in the Senate July 2. The resultant Wade-Davis Bill contrasted stridently with Lincoln's December proclamation: it would postpone reconstruction until the war was won and put the process entirely in the hands of the Congress. Since the bill had passed at the end of the session Lincoln had only to use his pocket veto, but he also issued a statement, warning that Congress had no right to abolish slavery by statute. It must be done by constitutional amendment, and he would not be "inflexibly committed to any single plan of restoration." To do so, he held, would put "the free-state constitutions and governments, already adopted and installed in Arkansas and Louisiana," in jeopardy.[15]

In the Republican Party a division between radicals, whose power was growing, and conservatives intensified the strife; while in the Democratic Party there was division between those who would restore the Union "as it was" and those who would stop the fighting altogether. Accordingly, the Democratic nominating convention, scheduled for July 4, was postponed until the end of the summer to await developments. Clement Vallandigham, the chief representative of the Peace Democrats, had returned to Ohio from his Niagara exile in June. Not wishing to create a martyr, Lincoln did nothing, but his homecoming had ignited a blaze of editorialists clamoring for an end to the fighting.

Soon after his return Vallandigham was elected grand commander of the Sons of Liberty, a secretive pro-Southern organization embracing thousands in the southern regions of Ohio, Indiana, and Illinois. Funded in February from a $5 million appropriation by the Confederate Congress, clandestine political and sabotage operations were set afoot across the North. With Peace Democrats operating on the "inside," Davis sent two agents to Canada—Jacob Thompson and Clement Clay, to St. Catherines and Niagara, respectively—to coordinate matters on the "outside."

On the military front, after stopping Hunter at Lynchburg, Early was authorized by Lee to use the Shenandoah Valley as Jackson had and to strike if possible into the North. On July 7 a distraught Horace

Greeley wrote to Lincoln: "Our bleeding, bankrupt, almost dying country longs for peace—shudders at the prospect of fresh conscription, of further wholesale devastations, and of new rivers of human blood. I entreat you to submit overtures for pacification to Southern insurgents." Although given to temperamental nervousness, Horace Greeley had every reason to shudder, as the news of Jubal Early's arrival just above Harper's Ferry on July 3 was circulating with consternation in the North. In the next days, he routed his opponents at Monocacy Bridge and opened the road to Washington, as his army of 15,000 gathered up large herds of cattle and horses. Burning the private residence of Maryland's governor and Postmaster General Blair's home in Silver Spring, on July 11 Early appeared before Washington's defenses, only five miles from the executive mansion. Pollard wrote: "The Yankee capital was in imminent peril, and the whisper ran through the North that it was already lost or surely doomed."[16]

In fact behind Washington's formidable defensive works there were only a few army units along with some militia and convalescents, its garrison having been depleted to support Grant's campaign in Virginia. The situation appeared so dire that the navy prepared a ship on the Potomac to take the president away. But Early was unable to come to a decision and Grant had time to express an entire corps— the 6th—for Washington's relief.

As Early sent forces out to reconnoiter, intercepting trains and destroying telegraph lines, Lincoln himself, amid the buzz of sharpshooter bullets, appeared on the walls of Fort Stevens to survey the situation, wearing his stovetop hat and bobbing up and down to look at the action. A captain of the 6th Corps assigned to the 20th Regiment of Massachusetts Volunteers—Oliver Wendell Holmes, Jr.—shouted, not knowing who he was, "Get down, you damn fool, before you get shot!"[17]

With the 6th Corps within Washington's defenses and his rear threatened, Early had no choice but to withdraw, having missed an opportunity that would surely have given him the lasting fame to rival that achieved by "Stonewall" Jackson. Early had, however, supplied his cavalry and artillery with fresh horses, carrying away a reported five thousand, and twenty-five hundred beef cattle. Inflicting further losses at Kernstown before occupying Martinsburg, several hundred of Early's cavalry made a raid on Chambersburg, demanding a requisition of half a million dollars. As this was refused, an extensive portion of the town was burned.

Removed by several days from impacting opinion, and not directly connected to fighting in the east, the fighting in Georgia stirred
equal consternation in Richmond in the Davis government. Despite
his assurances to the contrary, Johnston had been forced to yield
ground again by the flanking maneuvers of Sherman, and on July 4
he found himself with his back directly on the Chattahoochee River.
Abandoning his customary contrivance around his enemy's left,
Sherman now directed a cavalry division and an entire infantry corps
around Johnston's right. With a part of his army across the river, the
Confederate Army of Tennessee was compelled to withdraw to another fortified position behind Peachtree Creek, four miles outside
Atlanta. As the Atlanta populace clamored for space aboard south-
bound trains, Davis sent Bragg to assess the situation at the front. On
July 17 Jefferson Davis dismissed Joseph Johnston in a telegraphed
message stating: "You are hereby relieved from command of the
Army and the Department of Tennessee, which you will immediately
turn over to General Hood."[18]

On the next day Horace Greeley and John Hay, one of Lincoln's
private secretaries, met at Lincoln's direction with Confederate agents
in Niagara Falls, Canada. The agents, it was reported, were soliciting
safe conduct to Washington, and that "terms of peace were already
passing over the wires." In fact Greeley and Hay had arrived with a
letter from Lincoln simply addressed "To Whom It May Concern," so
as not to acknowledge any official capacity. The message did offer
safe conduct to Washington but only to discuss "any proposition
which embraces the restoration of peace, the integrity of the whole
Union, and the abandonment of slavery." Davis's agents, however,
were not authorized to negotiate any terms, especially as those
phrased would have been tantamount to surrender, and the negotiation came to naught. In a message to the Associated Press, published
by the *Tribune*, the two southern "gentlemen" sought a propaganda
victory by saying that for their brethren with hopes for peace,
Lincoln's terms "will strip from their eyes the last film of such delusion." For northern "patriots or Christians," they wrote, "who
shrink appalled from the illimitable vistas of private misery and public calamity," they must "recall the abused authority and vindicate the
outraged civilization of their country."[19]

Across the North, people had been stunned on July 18, when,
instituting a new round of the draft, Lincoln issued a call for half a
million men to fill the Union armies. One of Lincoln's secretaries

notated in his diary: "Everything now is darkness and doubt and discouragement."[20]

With Grant stymied at Petersburg and searching for a new "vein," an officer who had been an engineer in the Pennsylvania coal mines suggested one to his incredulous superiors. A shaft could be dug under the fortification and packed with gunpowder and blown up, he said, stunning the "rebels" and creating a breach through which to take the city. A regiment from the Pennsylvania coalfields was soon put to work, and after solving the difficulty of forcing air through a shaft so long, at the end of July nearly 511 feet had been excavated, with a lateral tunnel of 40 feet at the end to house the powder.

After Sturgis's debacle with Forrest, Sherman had ordered out a second, stronger expedition to put an end to "the wizard of the saddle." On July 14 Forrest was checked near Tupelo, Mississippi, when he was wounded in the foot, and rumor spread that he had been killed. To rally his men he rode before unit after unit on horseback, till forced by pain to quit for the more comfortable transport of a buggy. Forrest had been set back, but he was far from out of the war. Meanwhile in the east, angered by the ineffectuality of Hunter in coming to terms with Early, Grant reached out for another horseman, Phil Sheridan, to command the newly created Army of the Shenandoah, ordering him to follow Early "to the death." Lincoln read Grant's directive and wrote that it was good, but given his own past experience with reluctant generals he recommended Grant follow up to make sure that the job was done. Sheridan was to carry out to the letter Grant's original order to Hunter, to move back into the Valley and "to eat out Virginia clear and clean as far as they go, so that crows flying over it for the balance of the season will have to carry their provender with them."[21]

In Georgia, with Sherman pressing on Atlanta, Hood had gone on the offensive. John Bell Hood was a thirty-three-year-old Texan who had already lost an arm and a leg in the fighting. Although Lee conceded he was a "lion" in combat, he lamented Hood did not have enough of the "fox" in him. After lashing out at Sherman's divisions crossing Peachtree Creek, he was driven back into Atlanta's defenses. The next day Hood struck again, this time at the exposed flank commanded by James B. McPherson, killing the general. In one day Hood had lost more men than Johnston had in ten weeks; and in three battles over eight days he would use up 15,000 men, as opposed to 6,000 expended by Sherman. Despite his losses the fighting spirit shown by the Army of Tennessee cheered the South. "Sherman's army is *doomed*," one observer exalted in Richmond.[22] Although his situation was critical, Sherman persisted at Atlanta's front, raining shells on its streets and buildings and on the works without.

At Petersburg the work of excavating a shaft under a Confederate gun battery only 150 yards from the Union trenches had been completed by July 20. The project under Burnside's jurisdiction had gotten the reluctant support of Meade and of Grant, who was willing to try anything to breach Petersburg's defenses. One of four divisions that comprised Burnside's corps was a black division; the three white divisions had seen action in the Wilderness campaign, but the blacks were without combat experience, having been delegated to guard supply depots. Because of the effect they were sure to excite in the enemy, this division was designated to lead in the assault once the mine was exploded, and they had begun a rigorous tactical training lasting weeks. There were thirty-three black regiments in the trenches before Petersburg at the time, some of whom had already performed creditably in combat, and the regiment chosen was anxious to prove its mettle.

In the pre-dawn hours of July 30 the eight tons of gunpowder heaped at the end of the burrow in lateral chambers were scheduled to explode, simultaneous with a massive artillery barrage. The plan was for assaulting units to storm the first line of entrenchments, and to follow up immediately with new divisions to capture the second line 400 yards behind it. Called Cemetery Hill, this was a long row of earthworks, as Pollard described it, "pitted with redoubts and redans, and ridged with serried salients and curtains."[23] Its fall would break the Confederate armor and open the gate to Richmond. To draw the attention of the defenders away from the front, divisions had been arranged north of the James, and an empty supply train sent across the river as a decoy. The plan began to unravel, however, as Meade had second thoughts about making the initial assault with black troops only hours before the charge was set to go off. It was decided the black troops would be part of the second wave, and white troops would go in first.

When the explosion went off—leaving a crater 170 feet long, 60 feet wide, and 30 feet deep—an entire Confederate regiment and battery were destroyed, with the defenders on either side fleeing their posts. But the assaulting party, chosen by lots, stumbled forward several hours after the blast had occurred, awestruck at the sight of the destruction, and instead of skirting the edges of the cavity charged directly into it and were slowed by debris. Two following assaults only added to the confusion, and soon Confederate batteries were raining shot and mortar on an assault with no cohesion. When the black troops initially chosen charged in, they collided pell-mell with retreating white troops just at the moment the enemy command had organized a counterattack. The black troops recoiled after briefly attaining the opposite parapet, retreating to the assumed safety of the ruins of the exploded fort where they were shot like fish in a barrel. Many who tried to surrender were murdered. Burnside, who lost control of the operation after Meade's intervention, later remarked that had the black division led the assault as planned it would have been a success. As it was, the Battle of the Crater cost 5,000 Union casualties against 1,200 for the defenders. Grant characterized it as a "stupendous failure." He wired Halleck: "It was the saddest affair I have witnessed in the war. Such opportunity for carrying fortifications I have never seen and do not expect again to have."[24]

On August 22 the Republican National Committee met in New York and through its chairman Henry Raymond, editor of the *New York Times* and a friend of Lincoln, wrote the president that "the tide is setting strongly against us." Raymond stipulated in his letter: "Two special causes are assigned to this great reaction in public sentiment—the want of military success, and the impression . . . that we can have peace with Union if we would . . . [but that you are] fighting not for Union but for the abolition of slavery." The national committee went on to urge Lincoln to send a commissioner to Richmond "to make distinct proffers of peace of Davis . . . on the sole condition of acknowledging the supremacy of the constitution—all the other questions to be settled in a convention of the people of all the States."[25] On the next day, with this matter in the balance, Lincoln entered a meeting with his cabinet, placing before them his famous "blind memorandum" that he had put in an envelope and sealed. Passing it around, Lincoln asked each to sign his name on the back, and when this was done he returned it to his pocket. The memorandum read: "This morning, as for some days past, it seems exceedingly probable that this Administration will not be re-elected, then it will be my duty to so co-operate with the President elect, as to save the Union between the election and the inauguration; as he will have secured his election on such ground that he can not possibly save it afterwards."[26]

The president had also authorized that Raymond go to Richmond on behalf of the government to propose "that upon the restoration of the Union and the national authority, the war shall cease at once, all remaining questions to be left for adjustment by peaceful modes." But such was the turmoil of Lincoln's predicament that when Raymond called on him two days later, he urged him to abandon the idea for a "conference for peace." Such an offer, he said, in words recorded by his private secretary, "would be worse than losing the Presidential contest—it would be ignominiously surrendering it in advance."[27]

It was in these weeks, with resolution in abeyance, that Frederick Douglass met Rev. John Eaton in Toledo, Ohio. Grant had appointed Eaton—who had a colonel's commission—superintendent of "contrabands" for the Mississippi Valley in November 1862. He had proved an effective and sympathetic organizer-advocate for the freedmen, regularizing their employment and providing protection for the camps where they lived, and Douglass found in him a solicitous ear for his bitterest denunciations of the government.

Said Douglass, the respect which Lincoln's administration asked from the world for abolishing slavery was a "swindle."[28] In the session of Congress just ended Thaddeus Stevens had been a leader in the fight for equalizing the pay of black and white soldiers; but instead of salving that wound he had thrown salt on it. The bill that was passed provided that only those black soldiers who had been free prior to April 18, 1861, would receive the full disbursement of thirteen dollars per month; those not free before that day would continue to receive the same pay as "contraband" laborers. An exasperated Col. Thomas Wentworth Higginson wrote the *Tribune* on August 12, 1864: "In other words a freeman (since April 19, 1861), has no rights which a white man is bound to respect." Said Douglass, the government defrauded its "colored" soldiers in refusing them equal pay, after it had been promised them, and it withheld them its active protection as captured combatants. They were good enough to fight for the country, but not good enough for the franchise, and thus they were being denied the only sound basis for citizenship. In his present course, Douglass argued, Lincoln was doing nothing less than preparing to hand the "Negro" back to those who were fighting the government. No rebuke therefore, not even loss of the election, was too strong for this dishonorable conduct of the presidential office.

Eaton was so impressed by these forceful utterances that a short time later when he met with the president he conveyed some of Douglass's criticisms to him. Lincoln too was impressed, remarking that Douglass, in his opinion, was one of the most, "if not the most meritorious man in the United States"; and he must make some arrangement to see him and hear his thoughts himself. Eaton agreed to arrange the interview at the nearest opportunity.

With these questions looming, on August 5 Senator Wade and Representative Davis jointly published a "manifesto" in the *New York Tribune* addressed to the "Supporters of the Government." Aimed at the president's policy on reconstruction and his pocket veto of their bill in July, Wade and Davis brazenly challenged Lincoln's authority. The manifesto declared: "This rash and fatal act of the President [was] a blow at the friends of his Administration, at the rights of humanity, and at the principles of Republican Government." They assailed the usurpation of the executive and accused Lincoln of craving power and trying to secure his reelection on the votes of the states he was reconstructing.

Unlike his policy, the congressional bill, they stated, sought to protect "the loyal men of the nation [against the] great dangers [of a] return to power of the guilty leaders of the rebellion [and] the continuance of slavery." The manifesto ended with an appeal to the supporters of the government, saying that if Lincoln wanted Republican support for his reelection, "he must confine himself to his Executive duties—to obey and execute, not make the laws—to suppress by arms armed rebellion, and leave political reorganization to Congress." With Lincoln's National Union Party going into the election with no platform on reconstruction, and with the radical Republicans having lost at the nominating convention, they could win now by forcing Lincoln to withdraw, or run on their platform.

Thurlow Weed was convinced Lincoln's reelection was impossible, and he and other Republicans began urging Lincoln to step aside for a stronger candidate. Lincoln was also feeling considerable heat from Democrats and moderate Republicans. Holding up his "To Whom It May Concern" letter as a case in point, they accused him of shifting his war aim from restoring the Union to fighting for the abolition of slavery. Lincoln was holding out for the war to go on even if the former was attained, they said, and ignoring the fact that people all across the North were expressing a growing weariness with the war and were calling for peace. In a letter written to a War Democrat on

August 17—a letter that he did not send—Lincoln concluded: "If
Jefferson Davis . . . wished to know what I would do if he was to offer
peace and reunion, saying nothing about slavery, let him try me." Yet
on the same day, to another correspondent in Wisconsin he wrote
that to cast off emancipation "would ruin the Union cause itself. All
recruiting of colored men would instantly cease, and all colored men
now in our service would instantly desert us. And rightly too. Why
should they give their lives for us, with full notice of our purpose to
betray them?"[29]

To another Wisconsin Republican in a meeting on August 19,
Lincoln said, "There have been men who have proposed to me to
return to slavery the black warriors of Port Hudson & Olustee. I
should be damned in time & eternity for so doing. The world shall
know that I will keep my faith to friends & enemies, come what
will." When asked if he had read the "manifesto" or any of Wendell
Phillips's speeches, Lincoln replied: "I have not seen them, nor do I
care to see them. I have seen enough to satisfy me that I am a failure,
not only in the opinion of the people in rebellion, but of many distinguished politicians of my own party."[30]

On the morning of the publication of the Wade-Davis manifesto,
with everyone focused on the dire predicament of the war, they may
have been too hard pressed to appreciate the breakthrough Farragut's
fleet achieved in Mobile Bay. During a terrific duel between
Confederate shore batteries and the Union fleet, Farragut had himself
lashed to the mast while he peered above the smoke of battle, rising
to lasting fame with the phrase "Damn the torpedoes! Full speed
ahead." The deepening savagery of the war and the dimming
prospects of his own reelection were staggering burdens daily to
Lincoln—as was the vexing issue of black freedom. The summer
months of 1864 were probably among the most crucial of Lincoln's
presidency. Walt Whitman, who often saw him passing on the streets
of Washington, offers a discerning portrait:

> He was in his plain two-horse barouche, and look'd very much
> worn and tired; the time, indeed, of vast responsibilities, intricate
> questions, and demands of life and death, cut deeper than ever
> upon his dark brown eyes; yet all the old goodness, tenderness,
> sadness, and canny shrewdness, underneath the furrows. (I never
> see that man without feeling that he is one to become personally
> attach'd to, for his combination of purest, heartiest tenderness,

and native western form of manliness.) By his side sat his little boy, of ten years. There were no soldiers, only a lot of civilians on horseback with huge yellow scarfs over their shoulders, riding around the carriage. . . . They pass'd me once very close, and I saw the President in the face fully, as they were moving slowly, and his look, though abstracted, happen'd to be directed steadily in my eye. He bow'd and smiled, but far beneath his smile I noticed well the expression I have alluded to. None of the artists or pictures has caught the deep, though subtle and indirect expression of this man's face. There is something else there. One of the great portrait painters of two or three centuries ago is needed.[31]

Along with the deluge in war deaths, and with all else that was directed against Lincoln it is no wonder his appearance began to take on the aspect it did. Two of the epithets making the rounds at the time, indicative of the cross-purposes of the opposition to him, were the mock honorific "Abraham Africanus the first," and the dire "Abe the Widow-maker." One Democratic newspaper editorialized that tens of thousands of white men had bitten the dust to allay "the Negro mania of the President"; while a song parody, to the tune of "When Johnny Comes Marching Home," set out

> The widow-maker soon must cave
> Hurrah, Hurrah
> We'll plant him in some nigger's grave.

Visiting the executive mansion during this period was the German-American Carl Schurz, then a major-general, who captured in his *Reminiscences* Lincoln's words in regard to the pressure being exerted by radical Republicans: "They urge me with violent language to withdraw from the contest, although I have been unanimously nominated, in order to make room for a better man. I wish I could. Perhaps some other man might do this business better than I, that is possible. I do not deny it. But I am here, and that better man is not here, and if I should step aside to make room for him, it is not at all sure—perhaps not even probable—that he can get here. It is much more likely that the factions opposed to me would fall to fighting among themselves, and that those who want me to make room for a better man would get a man whom most of them would not want at all."

It is likely that it was on August 25 that Frederick Douglass "most gladly" responded to Lincoln's request to consult with him on the national situation. In the vast trove of the ever-growing Lincoln bibliography one wonders if there are more than a few references to this important and illuminating occurrence, yet in Douglass's own hand there is basis for fixing the time and subject of this audience with Abraham Lincoln. In a passage in his *Life and Times* he places it without specifying during Grant's Wilderness campaign but prior to McClellan's nomination; and in a letter to Theodore Tilton dated October 15 which elaborates details beyond what is in his autobiography, Douglass refers to the meeting as occurring six weeks past. Finally in a letter to the president dated August 29, Douglass gives reference to their meeting of a few days before. In his autobiography Douglass remarked:

> It is due perhaps to myself to say here that I did not take Mr. Lincoln's attention as due to my merits or personal qualities. While I have no doubt that Messrs. Seward and Chase had spoken well of me to him, and that the fact of my having been a slave and gained my freedom and of having picked up some sort of an education, and being in some sense a "self-made-man," and having made myself useful as an advocate of the claims of my people, gave me favor in his eyes, yet I am quite sure that the main thing which gave me consideration with him was my well-known relation to the colored people of the Republic, and especially the help which that relation enabled me to give to the work of suppressing the rebellion and of placing the Union on a firmer basis than it ever had or could have sustained in the days of slavery.

Then in his letter to Tilton, Douglass wrote: "The President's 'To Whom It May Concern' frightened his party and his party in return frightened the President. I found him in this alarmed condition when I called upon him six weeks ago, and it is well to note the time. The country was struck with one of those bewilderments which dethrone reason for the moment. Everybody was thinking and dreaming of peace, and the impression had gone abroad that the President's antislavery policy was about the only thing which prevented a peaceful settlement with the Rebels."[32]

Arriving in one of the anterior rooms to the presidential office, Douglass sent in his calling card. As he entered, in addition to one of Lincoln's personal secretaries, War Secretary Stanton was there, and Lincoln had at hand the draft of a letter he was considering for publication. It was a letter he hoped might placate the swirling fervor raised after Greeley's failed peace mission, and Douglass detailed the points it made in his communication with Tilton, writing:

> The first . . . was the important fact that no man or set of men authorized to speak for the Confederate Government had ever submitted a proposition for peace to him. Hence the charge that he had in some way stood in the way of peace fell to the ground. He had always stood ready to listen to any such propositions. The next point referred to was the charge that he had in his Niagara letter committed himself and the country to an abolition war rather than a war for the union, so that even if the latter could be attained by negotiation, the war would go on for Abolition. . . .
>
> The president did not propose to take back what he had said in his Niagara letter, but wished to relieve the fears of his peace friends, by making it appear that the thing which they feared could not happen, and was wholly beyond his power.

Lincoln continued, "Even if I would, I could not carry on the war for the abolition of slavery. The country would not sustain such a war, and I could do nothing without the support of Congress. I could not make the abolition of slavery an absolute prior condition to the reestablishment of the Union." "Now," the president invited Douglass, "the question I want to put to you, shall I send forth this letter?" Douglass's reply was emphatic and worthy of Lincoln's best advisor. He said: "Certainly not. It would be given broader meaning than you intend to convey; it would be taken as a complete surrender of your anti-slavery policy, and do you serious damage. In answer to your Copperhead accusers, your friends can make the argument of your want of power, but you cannot wisely say a word on that point."

Surely this represents another of the crucial moments of his presidency, equal to Lincoln's decision to make a stand on the issue of Sumter and his resolution on emancipation. Through his war powers he had taken the issue about as far as he could constitutionally, he now found he was forced into another fateful choice—not only because of a moral imperative, but also by the fact that the Union armies and navy had 130,000 black men in them, and they "must be prompted by the strongest motive—even the promise of freedom. And the promise being made, must be kept." In the summer of 1864 this was the dilemma Abraham Lincoln had to come to grips with as

he faced reelection and a lagging war effort. The answer he gave
translated him into a figure of transcendent historic greatness.

The president made an extraordinary proposition to Douglass.
"The slaves are not coming so rapidly and so numerously as I
hoped," Lincoln said in a "regretful tone." In his letter to Tilton
Douglass elaborated on the suggestion that "The President said he
wanted some plan devised by which he could get more of the slaves
within our lines. He thought now was their time—and that such only
of them as succeeded in getting within our lines would be free after
the war was over." This was an alarming statement to Douglass, for
in it he could see that the President believed his proclamation only
had effect during the war, and that its operation ceased once it ended.
In his *Life and Times* he relates he replied that the slaveholders knew
well now to keep such things from their slaves. "Probably very few
know of [your] proclamation."

Lincoln proposed: "Well, I want you to set about devising some
means of making them acquainted with it, and bringing them into our
lines." Douglass wrote that Lincoln "saw the danger of premature
peace" and "wished to provide means to render such consummation
as harmless as possible. I was the more impressed by this benevolent
consideration because he before said, in answer to the peace clamor,
that his object was to save the Union, and to do so with or without
slavery. What he said on this day showed a deeper moral conviction
against slavery than I had ever seen before in anything spoken or
written by him."

Douglass quickly agreed to undertake the project, and after the
necessary consultation with coadjutors whose "hearts were in the
cause," he would report on a plan in writing to the president.
Concluding his account, Douglass wrote in his autobiography: "An
incident occurred during this interview which illustrates the character
of this great man, though the mention of it may savor a little of vanity on my part. While in conversation with him his Secretary twice
announced 'Governor Buckingham of Connecticut,'" one of the
noblest and most patriotic of the loyal governors. Mr. Lincoln said,
'Tell Governor Buckingham to wait, for I want to have a long talk
with my friend Frederick Douglass.'"

By August 29 Douglass was back in Rochester and had ready a
draft of a plan. What consultation he did hold—for he wrote Lincoln,
"I have freely conversed with several trustworthy and Patriotic
Colored men"—must have taken him to hurriedly arranged meetings

along his route home, perhaps in Philadelphia, New York City, Albany, and Syracuse. The guiding idea, he wrote, would be "to inform the Slaves in the Rebel States of the true state of affairs in relation to them and to warn them as to what will be their probable condition should peace be concluded while they remain within the Rebel lines and more especially to urge them the necessity of making their escape." Recommending a general agent be appointed by the president with the power to employ from twenty to twenty-five men, Douglass wrote that these agents should be given permission to travel and "visit such points at the front as are most accessible to large bodies of Slaves in the Rebel States." "The general agent should also have a kind of roving commission within our lines," given "direct and effective oversight of the whole work," ensuring "activity and faithfulness on the part of his agents," and place himself "where he can most readily receive communications from and send them to his agents."[33]

Douglass continued: each agent should have the power to appoint subagents who would know and be known in the locality they were required to operate in. These men would be well instructed in the representation they were to make to the slaves, while they conducted them in squads "as . . . may be able to collect safely within the Loyal lines." Finally, Douglass proposed, so that these agents shall not be arrested or hindered in their work, "let them be properly ordered to report to the commanding generals in the departments they may visit," and receive permission to pursue their vocation. The slaves brought within the Union lines should receive subsistence until such as are fit to enter the service of the country or be otherwise employed." Douglass further recommended that each agent be paid "a sum not exceeding two dollars per day while upon active duty," and then concluded: "This is but an imperfect outline of the plan, but I think it enough to give your Excellency an idea of how the desirable work shall be executed."

In his autobiography Douglass remarked that this "business should be somewhat after the original plan of John Brown." The admiration he felt, and the reflexive identification between the two men he most revered, perhaps caused him, as he sat down in after years to write his *Life and Times*, to rhetorically counterbalance John Brown's "original" plan with his "Harper's Ferry" plan, as he remained unaware he had misplaced his recollection of the former by a decade. As for his own proposal, nothing further would be heard of it, for in only a few days—perhaps even before Douglass's letter reached the president—the wires would be crackling with lightning that would change all political calculations.

When Vallandigham returned to Ohio in June, ostensibly it had been to attend the state's Democratic Party convention. Denouncing the war as "unnecessary," one of the resolutions adopted by delegates to the convention called for the "immediate cessation of hostilities." Finding a ready echo for this platform throughout the North, Vallandigham vaulted to the top of the Copperhead opposition. By the early summer it was clear the peace faction would control the national Democratic convention, first scheduled to convene in Chicago on July 4 but postponed to August 29.

To ferment resistance to "Lincoln's war" into a veritable fifth column, Jefferson Davis had dispatched an agent, Capt. Thomas Hines—he had been a scout with Morgan's cavalry in Kentucky during 1862–63—with instructions to "confer with the leading persons friendly or attached to the cause of the Confederacy, or who may be advocates of peace." Hines's orders were to carry out "appropriate enterprises of war against our enemies," together with other agents headquartered at St. Catharines.[34] Two of the most prominent men involved in these activities—Jacob Thompson and Clement Clay— had been high officials in former United States administrations and had numerous and cordial contacts with northern Democrats whom they could count on to aid in organizing uprisings. Although none was to become full-blown, just short of half a dozen are known to have been contemplated and were in various stages of readiness.

One was to have coincided with Vallandigham's return in anticipation of his re-arrest; another was to occur on July 4 to counter expected attacks by vigilantes on the Democratic convention in Chicago; yet another was to concur with the announcement of a resumption of the draft. Caches of arms and incendiary devices had been hidden in various locations to aid in "insurrections," notably looking to the "expulsion or death of the abolitionists and free negroes." Finally Hines, along with former Confederate soldiers and sympathizers from the large clandestine organization known as the Sons of Liberty, planned an attack to free Confederate prisoners at Camp Douglas, to coincide with the Democratic convention at its rescheduled date. But this plan too would collapse, due to the operation of the Pinkerton organization and the timely warning given by John Brown Jr., who lived in Put-in-Bay at that time.

Operations set afoot also included the appropriation of funds to support peace candidates for elections, as well as subsidies paid to several Democratic newspapers. Sabotage operations included the burning of half a dozen Union gunboats at St. Louis, incendiary attacks against several hotels in New York City and an army warehouse at Mattoon, Illinois, while armed brigands robbed banks in St. Albans, Vermont. But

Clement Vallandigham.

this bellicose activity was secondary to the real objective of Davis: that war be stopped and talks on reconciliation started, leading to an independent Confederate government—a consummation to follow once a candidate emerged from the Democratic convention on that platform.

When delegates assembled in Chicago they were nearly evenly divided between War and Peace Democrats. It was the backroom consensus that the former would have their candidate in McClellan, while the latter would formulate the platform he would run on. When the convention was permanently organized, New York's governor Seymour was appointed chairman. In his speech to the convention he reviled the Lincoln administration and the Republicans, saying: "This Administration cannot now save the country if it would. It has placed obstacles in its pathway which it cannot overcome. It has hampered its own freedom of action by un-constitutionalities. This Administration cannot save the Union. We can. We demand no conditions for the restoration of the Union. We are for fraternal relations with the people of the South. We demand for them what we demand for ourselves, the full recognition of the rights of the States."[35]

Vallandigham sat on the platform committee and headed its subcommittee on resolutions. The most important of those adopted read:

> Resolved, That this Convention does explicitly declare, as the sense of the American people, that after four years of failure to restore the Union by the experiment of war, during which, under the pretence of a military necessity . . . the Constitution itself has been disregarded in every part, and public liberty and private right

alike trodden down, and the material prosperity of the country essentially impaired, justice, humanity, liberty, and the public welfare demand that immediate efforts be made for the cessation of hostilities, with a view to an ultimate convention of all the States, or other peaceable means, to the end that, at the earliest practicable moment, peace may be restored on the basis of the federal Union of the States.[36]

Would the former general-in-chief stand on Vallandigham's plank? For him to be professedly of the Copperhead camp would have been as good as betrayal of the soldiers he had led; but McClellan was by far the most popular and widely known Democrat and an outspoken critic of Lincoln's war policies, particularly emancipation. His candidacy would be a potent symbol for the opposition to the war. McClellan's advisors assured party insiders that if elected he would "recommend an immediate armistice," then "call for a convention of all the states and insist upon exhausting all and every means to secure peace without further bloodshed." This was enough for most peace delegates, but for added assurance, "Gentleman" George Pendleton, an ally of Vallandigham, was advanced for vice president on the ticket. From an old Virginia family, Pendleton was sympathetic to the South and had opposed the war from its commencement. When balloting began on August 31, Horatio Seymour refused to stand for nomination. The only candidate besides McClellan receiving substantial support was Thomas H. Seymour who had run for governor in Connecticut as a peace candidate and lost. On motion of Vallandigham the vote for McClellan was made unanimous, and the peace plank adopted.

Just as the convention was ending, General Sherman telegraphed Washington: "Atlanta is ours, and fairly won." The question immediately arose, could McClellan, in light of this Union victory, honorably stand on the peace platform now?

At his home in New Jersey, even as he sat composing his acceptance letter, Vallandigham sent McClellan a warning: "Do not listen to your Eastern friends, who, in an evil hour, may advise you to insinuate even a little war into your letter of acceptance. . . . If anything implying war is presented, two hundred thousand men in the West will withhold support." Laboring through three drafts, McClellan tried to reconcile with the convention's plank, writing he was in "cordial concurrence" with its call for a "cessation of hostilities," and stating, "We have fought enough to satisfy the military honor of the two sections." But in a fourth draft of the letter, August Belmont, of the Rothschilds' banking house and chairman of the Democratic National Committee, convinced him to delete even these allowances. When McClellan's letter of acceptance was released on September 8 it had only qualified support for the peace plank—when the South was ready for peace on the basis of Union, negotiation could begin. McClellan wrote: "I could not look in the faces of gallant comrades of the army and navy . . . and tell them that the labor and sacrifice of our slain wounded brethren had been in vain. . . . The Union is the one condition of peace—we ask no more."[37]

Mulling over the situation from his post in Canada, Clement Clay concluded, in a letter to Judah Benjamin, former senator from Louisiana and secretary of war and state successively in the Confederate government, on September 12: "McClellan will be under the control of the true peace men. . . . At all events, he is committed by the platform to cease hostilities and to try negotiations. An armistice will inevitably result in peace. The war cannot be renewed if once stopped, even for a short time."[38]

Many Peace Democrats were revolted, even going so far as holding another convention on October 18 in Cincinnati. But they nominated no candidates and the movement ended without further political effect.

By that time too, the radicals opposed to Lincoln could see that their hope of electing a more robust antislavery candidate was waning, and any ideas of seeking a new convention to supersede the president were abandoned. On his own behalf Lincoln was quick to exploit this sea change. To placate Chase and the radicals, the president removed Montgomery Blair, a voice of moderation on the war and slavery. His post was offered to Horace Greeley, who declined it; but he produced a two-column editorial in his paper: "Henceforth we fly the banner of Abraham Lincoln for the next President." The editor of the *Herald*, James Gordon Bennett, was offered the ambassadorship to France, which he also declined, but his pages too pronounced in favor of Lincoln. An overture was made to Sen. Henry Davis to take the vice presidency on the ticket, which was refused. With the tide moving in his favor Lincoln hurried to shore up support among German Americans, conferring the rank of major-general to officers with such names as Osterhaus and Hovey. Frémont himself was offered an active and important new command, but after deliberating, he turned it down on grounds that it was too late in the war. Finally after intense consultation with his backers, Frémont withdrew from the presidential contest altogether, writing on September 22: "I consider that [Lincoln's] administration has been politically, militarily, and financially a failure, and that its necessary continuance is a cause of regret for the country." It was his patriotic duty however, he held, to stand aside to assure McClellan's defeat. After Chicago, Frederick Douglass also concluded, he would be backing "Mr. Lincoln."[39]

Determined to rally the southern cause in a way that would recoil favorably in the North and galvanize the "peace vote," President Davis traveled to Georgia. Arriving at Hood's headquarters on September 18, he, his staff, and Hood devised a plan to drive Sherman from the state. The next day in a review of the Confederate Army of Tennessee, Davis addressed the soldiery in ringing tones: "Be of good cheer, for within a short while your faces will be turned homeward, and your feet pressing Tennessee soil."[40]

Hood's army was to march over the roads Sherman and Johnston had parried, cut off the enemy from his supplies and communications, and draw him into a battle of desperation. In this scenario Sherman's army would be broken apart and destroyed piecemeal, middle Tennessee recovered, and perhaps Knoxville and the eastern part of the state. To accomplish this Hood knew his army needed replenishment, but Davis had already consulted with Lee and knew no reinforcements would be forthcoming. Help, however, could be expected from Richard Taylor, who would leave Louisiana to join him with his army in northern Alabama.

Toward the end of the month Davis was in Macon, Georgia, and in a speech subsequently repeated before cheering crowds in several towns in South Carolina, he gave a vivid rendition of things to come: "I see no chance for Sherman to escape from a defeat or a disgraceful retreat. The fate that befell the army of the French Empire in its retreat from Moscow will be re-enacted. Our cavalry and our people will harass and destroy his army, as did the Cossacks that of Napoleon, and the Yankee general, like him, will escape with only body guard." The Army of Tennessee would return to the state whose name it bore and would "draw from twenty thousand to thirty thousand to our standard, and . . . push the enemy back to the banks of the Ohio and thus give the peace party of the North an accretion no puny editorial can give."[41]

When Grant read these inflated words he retorted, "Who is to furnish the snow for this Moscow retreat?"[42]

Aware of the record of Union disaster during the course of the war in the Shenandoah Valley, Sheridan had begun his campaign cautiously. While Early's forces occupied Winchester, reinforced with two divisions from Lee, Sheridan established his headquarters at Harper's Ferry. Late in August when Grant cut the Weldon Railroad severing Richmond's link to the port of Wilmington, Lee recalled one of these divisions. Learning of this, Sheridan struck in overwhelming force on September 19, destroying a quarter of Early's army, and sending the survivors whirling some twenty miles to the south. Sheridan hit again on September 22 on Fisher's Hill near Strasburg; this time the blow threw Early sixty miles further into the Valley, landing in a pass in the Blue Ridge below Port Republic.

The entire Valley was now at Sheridan's disposal, and a vital food-producing region lost for Richmond, with Lynchburg again threatened. Sheridan sent his cavalry out to capture Staunton and destroy all storehouses, machine shops, and Confederate government buildings, as well as all military material. Then, moving on to Waynesboro, seven miles of the Virginia Central Railroad was torn up and the depot destroyed along with an iron bridge over the Shenandoah River. In the next several weeks two thousand barns were burned, seventy mills, numerous tanneries, blast furnaces, foundries, distilleries, and storehouses, while Sheridan reserved for the supply of his troops herds of cattle, sheep, and horses. "The people must be left nothing but their eyes to weep with over the war," Sheridan wrote.[43]

Hood began his march north on September 29, while Sherman, leaving an entire army corps to guard Atlanta, moved in pursuit on October 3. On the same day Grant struck at Chaffin's Farm, an important component of the Richmond defenses. Before the war one of the largest and most prosperous farms in Virginia, it now had within its boundary, situated on New Market Heights and on Laurel Hill, three closely linked and mutually supportive forts manned by some of Lee's best troops. Grant had designed this thrust as groundwork to gaining position for the final assault on Richmond—preparations that included consolidating three all-black army corps into one large corps, the 25th under the command of General Weitzel. The fighting at Chaffin's Farm would involve more black soldiers than at any other place in the war, with 3,000 troops from General Paine's 3rd Division of the USCT taking part, leading assaults on New Market Heights and Fort Harrison. By October 6 Grant had 40,000 troops on the north side of the James River, southeast of Richmond.

Christian Fleetwood, awarded the Medal of Honor during the action at Chaffin's Farm, September 29, 1864.

On October 16, leaving his army encamped near Cedar Creek fifteen miles south of Winchester, Sheridan entrained for Washington. Reinforced with an infantry division and a cavalry brigade, Early decided to make a surprise predawn attack, striking a severe blow on unsuspecting Union troops on October 18. Many men were slain in their camps, fifteen hundred were captured, along with eighteen pieces of artillery, numerous wagons with small arms, and everything on the ground. But instead of following up Sheridan's routed army, Early's famished and ill-clad troops fell upon their plunder. During a lull of four or five hours, Sheridan, having returned the previous evening and alerted by the rumbling guns to the south, saddled up for his ride, as Thomas Read's much recited poem in the decades following told it, down the "road from Winchester town."

When Sheridan came up at full gallop he met stragglers and the wounded from the battle. As they began to cheer, he shouted: "God damn you don't cheer me! If you love your country, come up to the front. . . . There's lots of fight in you men yet! Come up, God damn you! Come up!"[44] Ordering a new line of battle, by 3 o'clock Sheridan was on the offensive. Soon the rout of his army was turning into a stunning defeat for the Confederates, as they in turn would become panic-stricken fugitives, losing more ground than they had gained.

In Georgia, on October 5 Hood would assault the garrison at Alatoona, while Sherman signaled from Kennesaw Mountain to hold out till he got there. By means of swift marches, by October 12 Hood had captured Dalton, and by the 15th, moving through the gap of Pigeon Mountain, he occupied Lafayette. It now appeared that the battleground at Chickamauga would be the site of significant new fighting, but instead Hood moved into Alabama at Gadsden, while Sherman extended his army as far as Gaylesville, on the north shore of Weiss Lake. By October 23 Hood was moving far into his rear.

The rapidity of Hood's movement startled Sherman; he had not even been aware he'd left Gadsden. But as Hood reached Decatur, meeting up with Taylor's army, Sherman announced he would follow him no longer. "If he will go to the river, I will give him his rations," he said. Sherman would divide his army, sending part under Thomas to cover Nashville, and part under Schofield to cover Knoxville, while the balance prepared to return to Atlanta. In his communication with Grant, Sherman argued that it would be impossible to protect the railroads "now that Hood, Forrest, and Wheeler, and the whole batch of devils, are turned loose, without home or habitation."[45] To hold the roads he would need thousands of men and there would be no appreciable gain from it.

On the electoral front, speaking at Tremont Temple on October 20 Wendell Phillips lashed out at Lincoln for his "halting, half-way course, neither hot nor cold."

> Observe how tender the President had been towards the South, how unduly and dangerously reluctant he has been to approach the Negro and use his aid. . . .
>
> Let me allow that he is the only candidate in the field. As for that Confederate gunboat which anchored off Chicago, August 29, and invited G. McClellan to be Captain, my only wish is may she soon meet her Kearsarge, and join her sister pirate Alabama in the ocean's depths.
>
> Reconstruction will be a matter of bargain. In a bargain, neither party ever gets all he sets out with asking. We must expect, therefore, that when the bargain is made, one or the other of the two claims made at Niagara (Union and Abolition) will be wholly or in part surrendered. This is inevitable. Which is it like to be?[46]

As to his stance on Lincoln's reelection Phillips declared: "I mean to agitate till I bayonet him and his party into justice."

This too was effectively where Douglass stood. With Democrats brandishing banners associating the Republican Party with emancipation and the arming of the blacks, and suggesting that the "final fruit" of Republican prosecution of the war would be a "blending of the white and black"—*miscegenation* was a word coined for the presidential election of 1864—Lincoln's campaign was eager to counter the shibboleth "Black Republican." Douglass was to remark in his letter to Theodore Tilton: "To all appearance we have been more

ashamed of the Negro during this canvass than those of '56 and'60."
He also confided, "I am not doing much in this Presidential canvass
for the reason that Republican committees do not wish to expose
themselves to the charge of being the 'N—r' party. The Negro is the
deformed child, which is put out of the room when company
comes."47

The Democrats, Douglass had remarked on October 4 at the
Colored National Convention assembling in Syracuse, were under the
power of slavery, while Republicans were under the sway of its shadow—color prejudice. That gathering in Syracuse, ten days after the
withdrawal of Frémont from the presidential race, was the first meeting of its kind since 1854. When the attendees were called to order by
Rev. Henry Highland Garnet, there were 144 delegates from eighteen
states. John Mercer Langston was elected temporary chairman and
Douglass was named president.

Outside the hall on their way to the assembly as the men had
passed through the streets the remark had been heard around
Syracuse, "Where are the damned niggers going?" Now Douglass
rose to declare that they were meeting to deal with the attitude of the
country toward the "Negro people," and he observed appreciatively
the set of younger men on the platform who "had come up in this
time of whirlwind and storm." Said Douglass: "In what is to be done,
we shall give offense to none but the mean and sordid haters of our
race." After hearing from several speakers the convention proceeded
to organize the National Equal Rights League. In its declaration the
league petitioned Congress to remove "invidious distinction, based
upon color, as to pay, labor, and promotion" still affecting black soldiers, and due note was given the advances already attained during
four years of war; viz., the abolition of slavery in the District of
Columbia, recognition of Liberia and Haiti, and the retaliatory military order of August 30, 1863. The delegates also drew up a resolution commending Sen. Charles Sumner and General Butler for their
activities in behalf of the "Negro people."48

But the timeliness of the convention, and its basic dilemma—
whither the nation and the war?—were as Phillips framed them: A
bargain to be made, which was it likely to be? While the Republican
platform as well as the president were strongly committed on the matter of the abolition of slavery, was there not yet room for equivoca-
tion? Was the government prepared to make the abolition of slavery
a precedent to reestablishment of the Union? The president's "To

Whom It May Concern" letter seemed to rest on it, but many attempts had been made since then to explain it away. On September 3 Secretary of State Seward, in an address at Auburn, New York, assured his audience that the abolition of slavery was not part of the plan of the government at all. Seward elaborated: "When the insurgents shall have disbanded their armies, and laid down their arms, the war will instantly cease; and all the war measures then existing, including those which affect slavery, will cease also; and all the moral, economical, and political questions, as well affecting slavery as others, which shall then be existing between individuals and States and the Federal Government, whether they arose before the civil war began, or whether they grew out of it, will, by force of the Constitution, pass over to the arbitrament of courts of law, and the counsels of legislation."[49]

When Douglass rose at the convention in Syracuse to give the keynote address, "The Cause of the Negro People," this was the issue he had to tackle head on. He announced directly: "The fear of continued war, and the hope of speedy peace, alike mark this as the time for America to choose her destiny. Another such opportunity as is now furnished in the state of the country, and in the state of the national heart, may not come again in a century. Come, then, and let us reason together." But as he continued he reduced the issue to the proposition that it was "not upon the disposition of the Republican party, not upon the disposition of President Lincoln; but upon the slender thread of Rebel power, pride, and persistence" that "the hope of the speedy and complete abolition of slavery" hung. "Only this, is our surest and best ground of hope, namely, that the Rebels, in their madness, will continue to make war upon the Government, until they shall not only become destitute of men, money, and the munitions of war, but utterly divested of their slaves also."[50]

There were powerful forces in the country working to "reverse the entire order and tendency of the events of the last three years," said Douglass. These forces had even caused fissures to appear within the friends of abolition. Gerrit Smith in his circular on McClellan's acceptance speech cautioned abolitionists against a "disposition to pervert the war to abolitionism." Whereupon Douglass sent his old mentor this rejoinder: "You and I know that the natural use of this war is to abolish slavery, and that that only is a perversion of it which would divert it from this its natural work." Then Garrison, in defending Lincoln's policy on reconstruction in Louisiana, denied the efficacy of granting the franchise to freedmen. He said, "Chattels personal may be instantly translated from the auction-block into freemen, but when were they ever taken at the same time to the ballot-box, and invested with all political rights and immunities? According to the laws of development and progress it is not practicable." The *Anti-slavery Standard*, too, had published an editorial denying that the American Anti-Slavery Society advocated black suffrage. Douglass lamented that these remarks "injure us more vitally than all the ribald jests of the whole proslavery press." Taking a stand for the "absolute overthrow of color caste in America," it only followed that prejudice in matters of government "should be allowed no voice whatever."[51]

In a republic complete equality could only have a basis in the franchise, and therefore Douglass demanded as the voice of the convention, that for "the peace and welfare of our common country. . . . We want the elective franchise in all the States now in the Union, and the same in all such States as may come into the Union hereafter." Again, after marshaling all the compelling augments for establishing one law for white and black alike, the final appeal came down to the contribution of blacks as soldiers:

> If the arguments addressed to your sense of honor, in these pages, in favor of extending the elective franchise to the colored people of the whole country, be strong, that which we are prepared to present to you in behalf of the colored people of rebellious States can be made tenfold stronger. By calling them to take part with you in the war to subdue their rebellious masters, and the fact that thousands of them have done so, and thousands more would gladly do so, you have exposed them to special resentment and wrath; which, without the elective franchise, will descend upon them in unmitigated fury. To break with your friends, and make peace with your enemies; to weaken your friends, and strengthen your enemies; to abase your friends, and exalt your enemies; to disarm your friends, and arm your enemies; to disfranchise your loyal friends, and enfranchise your disloyal enemies—is not the policy of honor, but of infamy."[52]

Meanwhile on October 27 to electrify the North and carry the presidential election for Lincoln, Grant had begun an "On to Richmond" drive. Commencing with an assault on Richmond's lines with artillery and heavy skirmishing, then advancing simultaneously with the Army of the Potomac and the Army of the James, Grant attempted to find and turn Lee's left flank. He was met and thwarted at every point. At Nine Mile Road, where some of the heaviest fighting occurred, three black brigades carried the Confederate fortifications, but were subsequently overrun, suffering terrible retribution at the point of the blade by the 24th Virginia and Wade Hampton's Legion.

Grant's ambitious movement resulted in the loss of 1,200 men as prisoners, more than a thousand killed, with many thousands wounded. Yet he had forced Lee to extend his defenses over thirty-five miles between Petersburg and Richmond. "I fear a great calamity will befall us," if more troops could not be gotten, Lee informed Davis.[53]

On November 4 a message from Sherman reached Washington: "Hood has crossed the Tennessee. Thomas will take care of him and Nashville, while Schofield will not let him into Chattanooga or Knoxville. Georgia and South Carolina are at my mercy—and I shall strike." Sherman confidently told Grant he would "cut a swath through to the sea, divide the Confederacy in two, and come up on the rear of Lee." Although doubtful as to the tractability of this course, Grant gave his consent, and as Lincoln too entertained misgivings, he nevertheless deferred to his soldiers. Yet if Hood reconquered Nashville, Knoxville and Chattanooga might also fall, and Lee's army could be reinforced from the west. Just prior to the election, the Confederate Congress met in Richmond and was addressed by President Davis. In his review of the year he would point to some notable successes: Texas was completely free of enemy troops, and in Louisiana Banks's expedition had met serious rebuff; Arkansas, Mississippi, and Alabama were mostly in Confederate possession, while Tennessee was the site of active operations. In the east Grant's army had been dealt a series of defeats and become bogged down for months before Petersburg. Sherman had captured Atlanta, but this had given the Yankees no real advantage. "The Confederacy," Davis concluded, "had no vital points." Even if its seaports and major centers were all captured, he said, "The Confederacy would remain as defiant as ever, and no peace would be made which did not recognize its independence."[54]

By Election Day—November 8, 1864—Lincoln's prospects for reelection were considerably improved. But Lincoln and his campaign were not willing to rest content. In some states where the vote would be close it was suggested that granting soldiers furloughs so that they might appear at their polling places would be an important influence for a Union vote. Might not a word dropped to the general-in-chief be helpful, Lincoln wondered? He did not know if Grant would be amenable; perhaps Meade or Sheridan might be asked. On hearing Sheridan's name Lincoln's countenance lit up, "Oh, I can trust Phil. He's all right." To Sherman, Lincoln wrote: "The loss of [Indiana] to the friends of the Government would go far towards losing the Union cause."[55]

As the polling day neared the New York *Journal of Commerce* reported: "Several thousand more soldiers arrived from Washington Friday. They were pouring through some of the streets from morning to night, making their way to the railroad, depots, and steamboat landings, where they could take passage home to vote. When asked how many are on the route, they laugh and say that 'It's only begun to sprinkle yet, but a smart shower may be looked for on Saturday, Sunday or Monday." One sergeant on the way through Albany was reported to have "boasted he had brought on sixty-nine soldiers—all Republicans—on their way to Utica to vote and had left every damned Democrat behind to take charge of the battery and horses."[56]

When the votes were cast and counted Lincoln received a majority in all but three states, Delaware, Kentucky, and New Jersey, giving him 213 electoral votes to McClellan's 21. The popular vote totaled 2,213,665 for Lincoln to 1,802,237 for McClellan.

Ralph Waldo Emerson wrote to a friend on November 16: "I give you joy of the election. Seldom in history was so much staked on a popular vote, I suppose never in history."[57]

Seventeen

One More River to Cross

O, Jordan bank was a great old bank,
Dere ain't but one more river to cross.
We have some valiant soldier here,
Dere ain't but one more river to cross.
—from a spiritual sung by ex-slaves

ON NOVEMBER 15, WITH NO SUBSTANTIAL ENEMY TO IMPEDE him, Sherman started south, his destination—Port Royal, South Carolina—285 miles away. Before leaving Atlanta the Union general ordered the burning of nearly all inhabitable buildings, some four or five thousand in all, including thirty-two hundred homes. Georgia's governor later wrote of the result in his official report: "The suburbs present to the eye one vast, naked, ruined, deserted camp."[1]

The Union armies moving out of Atlanta were deployed as two vast columns on a front of varying breadth from twenty-five to sixty miles, with cavalry covering both flanks. The pace set would allow for a dozen miles per day, and with scant discipline the march became a huge "frolic." In the army's wake came a growing train of liberated slaves and Georgia Unionists, called "bummers," who, along with foraging soldiers, took everything they wanted from plantations and farms, carrying off anything they could use that couldn't be destroyed or burned. With Sherman easily passing through Georgia, demonstrating Grant's contention that the South was a hollow shell, the paucity of the military strategy of Davis and Hood also became evident.

On November 20, Hood took up a line of march into Tennessee, from Florence to Waynesboro, with Schofield simultaneously abandoning northern Alabama and concentrating at Pulaski. On reaching Waynesboro and joined by Forrest's cavalry, Hood advanced toward Columbia, forcing a Union retreat from Pulaski. By forced march Schofield avoided being trapped, but was again flanked by Hood, as he sent Forrest to cut railroad communication from Nashville. On the night of the 26th with Forrest's cavalry in his rear, Schofield again fell back, reaching Franklin as Confederate infantry filed across the Duck River. On the evening of November 30 Hood's army approached Union entrenchments being hastily thrown up only fifteen miles south of Nashville before the crossings of the Harpeth River.

Annoyed at not being able to catch his enemy earlier, Hood resolved to attack at once, even though his artillery was not yet at the front. The defensive works to be overcome, he reasoned, would only be greater if he waited until morning, and a successful attack might decide the Battle of Nashville. With major-generals, brigadiers, and colonels riding in front waving them on, the Confederate army surged across level, unbroken ground, Hood himself riding along the lines urging the men on. Schofield's left was hit at 5 o'clock; soon 20,000 rifles and muskets let loose, at some points men engaging in hand-to-hand fighting as fierce as at Spotsylvania. Twilight disgorged hundreds, then thousands of dead—as numerous as those seen at Cold Harbor; many of Hood's top command, including a dozen generals and over fifty regimental colonels were killed—the attackers sustaining casualties three times that of the defenders. But Hood had knocked through the first line of defenses, and now the assaulting troops swept on to the second. At midnight Schofield broke off and retreated to Nashville, while Hood advanced, laying siege to the "Athens of the South" on December 2.

By November 20 one wing of Sherman's marchers had occupied Milledgeville, then functioning as the state capital of Georgia, while the other wing made a feint toward Macon. When Sherman left Milledgeville on December 2 he did so with enough food for forty days' rations for each man, and ammunition to supply every rifle with eighty rounds, with wagons packed with every necessary good.

Georgia's principal rail lines had been thoroughly destroyed, along with many of its bridges and industrial works—the rail lines heated in great bonfires and twisted around trees to make "Sherman neckties." Sherman now intended to crown his campaign with the capture of Savannah. Marching with deliberate speed in a dozen columns over as many roads, with cavalry as vanguard and as rearguard, his troops reached the outer defenses of the city by December 10.

While chroniclers still awaited the finale of this March to the Sea, it appeared that Thomas was temporizing at Nashville. Had Hood chosen to cross the Cumberland, as some in Richmond were urging, he might have cut Thomas's communications and compelled a further Union retreat. Stanton seethed at the War Department that he was repeating the "do nothing" strategy of McClellan and Rosecrans, as Grant barraged him with telegraphed missives exhorting "action." Finally with no movement evident, Grant himself started for Nashville with the intention of relieving Thomas of command. But as Hood invested the southern extremity of Tennessee's capital, Thomas had already begun pulling in his army from disparate locations so as to have an overwhelming force available.

On the morning of December 15 Thomas hurled one division (including two black brigades) at Hood's left, while three infantry corps and a cavalry corps swung around, smashing him in the right flank. In bitter fighting, the already weakened Army of Tennessee held positions throughout the afternoon, but by nightfall Hood ordered a retreat two miles to the south to a stronger lodgment anchored by two hills on either end. The next day Thomas renewed battle with another overwhelming onslaught. By slender means the Confederates held until dusk, when, in pouring rain, their lines completely collapsed. Thousands of tired, hungry, and ragged men surrendered, as the remainder, dropping weapons and discarding equipment, fled south. In two days' fighting at the Battle of Nashville, Hood had lost nearly half his army, effectively putting it out of existence. On January 23, at Tupelo, Mississippi, he would relinquish his command.

Meanwhile siege operations at Savannah had commenced with barrages of heavy artillery. On December 16 Sherman demanded the surrender of the city from its commander, Hardee, who refused. But as Sherman hurried more heavy guns to the lines, on December 20 Hardee evacuated Savannah's defenses. The next day Union troops marched in to receive the city's surrender from its mayor. On

December 22 in a telegram to Lincoln the general jauntily wrote: "I beg to present you, as a Christmas gift, the city of Savannah, with 150 heavy guns and . . . about 25,000 bales of cotton."[2]

Sherman's feat has since been elevated into one of the signature actions of the war, but he had a more tempered view. He wrote: "I am now a great favorite because I have been successful; but if Thomas had not whipped Hood at Nashville, six hundred miles away, my plans would have failed, and I would have been denounced the world over."[3]

The capture of Atlanta had played a crucial role in Lincoln's reelection, now the President sent his personal acknowledgment to the general:

> Many, many, thanks for your Christmas-gift—the capture of Savannah.
>
> When you were about leaving Atlanta for the Atlantic coast, I was anxious, if not fearful; but feeling that you were the better judge, and remembering that "nothing risked, nothing gained" I did not interfere. Now, the undertaking being a success, the honor is all yours; for I believe none of us went farther than to acquiesce. And, taking the work of Gen. Thomas into the count, as it should be taken, it is indeed a great success. Not only does it afford the obvious and immediate military advantages; but, in showing to the world that your army could be divided, putting the stronger part to an important new service, and yet leaving enough to vanquish the old opposing force of the whole—Hood's army—it brings those who sat in darkness, to see a great light.[4]

Reelection for Lincoln was a mandate to press Congress to pass an amendment abolishing slavery. The Thirteenth Amendment had already won approval in the Senate on April 8, 1864, by a vote of 36 to 6, but the vote in the House on June 15, although with a solid majority, found it lacking the necessary two-thirds margin. Intending to call a special session of the 39th Congress in March for a new vote, Lincoln decided to do it before then. Most Democrats overtly opposed the measure, but after the election many lame duck Congressmen understood it was time they, as was said, "cut loose from the dead carcass of Negro slavery."[5] Accordingly Seward headed the effort of jaw-boning, swaying some votes with favors and moving others with flattery, to assure passage at the end of the second session of the 38th Congress. The vote came on January 31, 1865, when the Thirteenth Amendment to the United States Constitution was reported to the chamber:

> SECTION I. Neither slavery nor involuntary servitude, except as a punishment for crime whereof the party shall have been duly convicted, shall exist within the United States, or any place subject to their jurisdiction.
> SECTION II. Congress shall have power to enforce this article by appropriate legislation.

Silence settled onto the floor of the House and into the spectators' gallery, filled to capacity with soldiers and reporters, delegations of black citizens, men and women, as the clerk called roll. When a Democratic representative from Connecticut was called and he responded "Aye," the hush was broken by murmurs of approval. Soon another of his colleagues answered with an affirmative vote, and the chamber responded with applause. Sixteen of eighty Democrats were to join Republicans in voting for the amendment, while eight Democrats abstained allowing it to pass 119 to 56, with two votes to spare for a two-thirds margin. As the result was announced members leapt to their feet, stood atop chairs, reached across desks to shake hands of colleagues or clasp a shoulder. Some waved their hats or threw them triumphantly into the air. The gallery too erupted in celebration; women waved handkerchiefs, many persons embraced and wept as the shouting and cheering reached tumult. When order was restored a member rose to express the sentiment of the floor: "Mr. Speaker, in honor of this immortal and sublime event, I move that this House do now adjourn."[6]

Outside Washington batteries boomed salutes to a *new* country; a throng accompanied by a band hurried to the executive mansion intent on serenading the president. When Lincoln appeared and the noise subsided, he said: "The great job is ended. The occasion is one of congratulation, and I cannot but congratulate all present, myself, the country, and the whole world upon this great moral victory!" He added that his proclamation might have been objected to as unconstitutional and partial in its operation, "but this amendment is a King's cure-all for all the evils. It winds the whole thing up."[7]

The day after the historic vote, and no doubt flush with the urgency of the just consummated revolution in the American body politic, Sen. Charles Sumner accompanied Boston lawyer John S. Rock for a presentation before newly confirmed Chief Justice of the Supreme Court Salmon P. Chase. Justice Taney had died in October 1864, and to conciliate radicals in the Republican Party in December Lincoln appointed his former secretary of the treasury and rival to the Supreme Court. Rock was accredited to practice in all the courts within the jurisdiction of the United States. Two weeks later the Rev. Henry Highland Garnet, then serving as pastor of the 15th Street Presbyterian Church in Washington, D.C., was to preach a sermon in the Hall of the United States House of Representatives to commemorate the passage of the Thirteenth Amendment. Garnet and Rock were the first African Americans to rise before these highest of the nation's forums, and *The Anglo-African* exulted: "Who shall say we are not on the onward march?"[8]

The full measure of the advance marked in the winter of 1865 was articulated by Garnet on that February 12, when well into a lengthy discourse he said: "With all the moral attributes of God on our side, cheered as we are by the voices of universal human nature—in view of the best interests of the present and future generations—animated with the noble desire to furnish the nations of the earth with a worthy example, let the verdict of death which has been brought in against slavery by the Thirty-eighth Congress be affirmed and executed by the people. Let the gigantic monster perish. Yes, perish now and perish forever!"[9]

While the policies of emancipation and the arming of freed slaves had allowed the government to continue the war at a crucial moment, it was the war itself that finally broke slavery. Now the question loomed—what to do with the freedmen? Douglass had already answered the question in 1862 in "What Shall Be Done with the four million slaves if they are emancipated?" Do nothing with them, he advised; leave them alone to rise by their own work and industry; let them have land, and education, and the franchise. "Let the American people, who have thus far only kept the colored race staggering between partial philanthropy and cruel force, be induced to try what virtue there is in justice."[10]

The whole vexing question was raised again as "contrabands" had attached themselves to Sherman's army on their march through Georgia. With the capture of Savannah, additional tens of thousands came under his military jurisdiction, from a vast area from the Sea Islands of South Carolina, down the Georgia coast to Jacksonville, Florida. Conservative on this question about the freedmen and preoccupied with military affairs, Sherman appeared indifferent to the plight of blacks, and reports began to reach Washington of their ill treatment by the officers and soldiers of his corps. To sort out the matter and put Sherman and his staff on the footing now desired by the administration, Secretary of War Stanton journeyed to Savannah to meet with the freshly laureled general. He immediately won Sherman's cooperation and that of his staff, and on January 12, 1865, a meeting to ascertain the needs and expectations of the newly freed people was convened at Sherman's headquarters at Savannah. Twenty leaders of the black community, ministers, and church elders, were invited to a special interview with the general, the secretary of war, and the assistant adjutant-general of the army, E. D. Townsend, and other officials. Fifteen of the invited interviewees had been slaves, four were freeborn, and their spokesman, sixty-seven-year-old Rev. Garrison Frazier, had been a slave in North Carolina until purchasing his freedom in 1857. The minutes of the interview were to be published in the *Liberator*, February 24, 1865, as authenticated by Townsend. That transcript records a series of eleven questions put by the chair, with the answers of Reverend Frazier, three of which follow:

"State what your understanding is in regard to the Acts of Congress, and President Lincoln's Proclamation, touching the condition of the colored people in the rebel states."

Answer: "So far as I understand President Lincoln's Proclamation to the rebellious States, it is, that if they would lay down their arms and submit to the laws of the United States before the 1st of January, 1863, all should be well, but if they did not, then all the slaves in the rebel States should be free, henceforth and forever; that is what I understood."

"State what you understand by slavery, and the freedom that was to be given by the President's Proclamation."

Answer: "Slavery is receiving by irresistible power the work of another man, and not by his consent. The freedom, as I understand it, promised by the Proclamation, is taking us from under the yoke of bondage, and placing us where we could reap the fruit of our own labor, and take care of ourselves, and assist the Government in maintaining our freedom."

"State in what manner you think you can take care of yourselves, and how you can best assist the Government in maintaining your freedom."

Answer: "The way we can best take care of ourselves is to have land, and turn in and till it by our labor—that is, by the labor of the women, and children, and old men—and we can soon maintain ourselves and have something to spare; and to assist the Government, the young men should enlist in the service of the Government, and serve in such manner as they may be wanted. . . . We want to be placed on land until we are able to buy it, and make it our own."

Acting with a boldness characteristic of his military campaigns, on January 16 Sherman issued Special Field Order No. 15. That order designated the Sea Islands between Charleston and Jacksonville and thirty miles inland for settlement by freedmen, giving each head of a family "possessor title" to forty acres of land until Congress "shall regulate their title." At this stroke the possibility of a true revolution in Southern affairs was thrown open, with the proviso that it was subject to future legislative action and or adjudication. In the next several months General Saxton supervised the settlement of some of the richest formerly planter-owned lands to 40,000 freedmen.

During the winter of 1864–65 Lee lost as much as half of his army to desertion. To be sure, the defenses at Petersburg and Richmond still held; but on January 15 Fort Fisher, twenty miles upriver from Wilmington, North Carolina, fell, and thereby Lee's principal source of supply was terminated. The desertions and the loss of the Confederacy's last coastal outlet were stunning blows, and as dissention with and criticism of Davis mounted, the Confederate Congress created the post of general-in-chief as a rebuke to his incessant intermeddling in military affairs. Davis promptly appointed Lee to the position, and at Lee's prompting Joseph Johnston was given a new command. This time it would be in the Carolinas to oppose Sherman's advance from Savannah and prevent his conjunction with Grant.

During its winter session the Confederate Congress, to the incredulity of posterity, began to consider measures to extend conscription to slaves. Estimating there were as many as 600,000 males from which to draw, it was proposed in the initial draft that blacks employed in the trenches in defense of the capital would free up to 15,000 white soldiers who could be used on the front. Now the question was asked, was it not better that the South use the blacks for its defense rather than they were used against them? It was said the North had demonstrated that blacks could be serviceable soldiers; with the proper inducements—like making him a freemen in his own home and led by officers that "better understood his nature"— the South would find the black soldier would be even better in its ranks. Lee and his officer staff notably promoted this view. Remarking on this ironic turn, Douglass later wrote in his autobiography: "It [the South] had reached that verge of madness when it called upon the Negro for help to fight against the freedom which he so longed to find, for the bondage he would escape."[11]

On March 4 and 6 the Virginia legislature issued four joint resolutions authorizing a limited call-up in the armed service of the Confederacy of blacks. But large slaveholders were as obdurate as ever, saying it would open a wedge to abolition. The *Charleston Mercury* declared if the measure was pursued, South Carolina would have no further interest in prosecuting the war. In the end, a bill authorizing Davis to ask for and to accept from the owners of slaves as many able-bodied men as he might think expedient was passed on March 7. Then on March 15 the Confederate War Department authorized officers "to raise a company or companies of Negro soldiers," and recruiting immediately began, men joining singly or several at a time. Pollard wrote: "The entire results of this ridiculously small and visionary legislation, which proposed to obtain negro soldiers from such volunteers as their masters might patriotically dedicate to the Confederate service, and was ominously silent on the subject of their freedom, were two fancy companies raised in the city of Richmond, who were allowed to give balls at the Libby, and to parade in Capitol Square, and were scarcely intended to be more than decoys to obtain sable recruits. But they served not even this purpose."[12]

This issue was already extant at the time of the interview in Savannah of the United States Army and government with black ministers and leaders, where it was asked:

"If the rebel leaders were to arm the slaves, what would be the effect?"

Answer: "I think they would fight as long as they were before the bayonet, and just as soon as they could get away they would desert, in my opinion."[13]

The sentiment for a cessation of the fighting that nearly derailed Lincoln's candidacy now began to disturb the resolve of the South. As its prospects for military victory began to wane, a victory of a sort

could be achieved by negotiations, it was thought, by dint of representing the "indomitable will of the Southern people." In early
January an old Jacksonian Democrat, Francis P. Blair, proposed to
Lincoln that he seek a meeting with Confederate commissioners,
holding out the possibility of a joint campaign to expel France from
Mexico as a basis to reunite North and South. Lincoln of course
wanted nothing to do with the scheme, but did provide a pass for
Blair to go to Richmond to meet with Davis. After the meeting had
taken place, Davis gave Blair a message stating he was ready to send
commissioners to confer with the "Northern President," if he could
be assured they would be received to "enter into conference with a
view to secure peace to the two countries." Lincoln promptly
addressed a note to Blair acknowledging he had seen Davis's communication, saying he would be willing to receive an agent of Mr. Davis,
or any influential person now actually resisting the authority of the
Government, "with the view of securing peace to the people of our
one common country."[14] Davis took this as an opportunity to galvanize support for the war until independence was acknowledged, feeling that a meeting would only demonstrate Lincoln sought the complete subjugation of the South.

Davis appointed three commissioners—his vice president,
Alexander Stephens, the president *pro tem* of the Confederate Senate,
Robert Hunter, and the former justice of the U.S. Supreme Court and
assistant secretary of war John Campbell—while Lincoln dispatched
his secretary of state. The meeting outside Fort Monroe began amid
recriminations between the "two countries" and our "common country" agendas. Seward had been instructed to adhere to three fundamental suppositions: "The restoration of the National authority
throughout all the States. No receding by the Executive of the United
States on the Slavery question. No cessation of hostilities short of an
end of the war, and the disbanding of all forces hostile to the government."[15]

As the parties seemed irreconcilable at the outset, and convinced
of the sincerity of their desire for peace after he had talked with
Stephens and Hunter, Grant wired Washington suggesting that sending the commissioners home without a meeting would leave a bad
impression. With the newly passed Thirteenth Amendment in his
pocket Lincoln jumped at the opportunity, determining to join his
secretary of state in a face-to-face meeting with the Confederate commissioners.

The meeting, lasting four hours, took place February 3 aboard the steamer *River Queen* off Hampton Roads. While no notes were taken, the discussion on the part of both parties was known to be "full and explicit," as the three Confederate commissioners would report to President Davis on February 6. Alexander Stephens would later provide the only extended firsthand account in his two-volume memoir *Constitutional View of the Late War Between the States*, while both Hunter and Campbell furnished commentary, and a brief of the exchange was subsequently provided in testimony given to Congress by Seward. Coming after four years of war, and as it did so close to Lincoln's unanticipated departure from the scene, the encounter had an air of high drama, as well as being elucidatory of Lincoln's views—and where he was willing to bargain and where he was not.

Lincoln began by suggesting that in sending commissioners, Davis wanted to discredit the peace movement in the South by identifying it with the surrender terms he had laid before Congress in the previous December. For their part both Stephens and Hunter tried to ply Lincoln away from his inflexible union position, the former by averring to Blair's Mexico scheme and the latter with a proposal for an armistice and a convention of states. As Seward summarized: "What the insurgent party seemed chiefly to favor was a postponement of the question of separation upon which the war was waged, and a mutual direction of efforts of the Government, as well as those of the insurgents, to some extraneous policy or scheme for a season, during which passions might be expected to subside, and the armies be reduced, and trade and intercourse between the people of both sections be resumed."[16]

Lincoln replied he would treat no proposals because that would be as good as recognition of their existence as a separate power, conditioning a cessation of hostilities only on the basis of the disbandment of Confederate forces. In his remarks Hunter wrote that in addressing the Confederate vice president, Lincoln said: "Stephens, let me write 'union' at the top of this page and you may write below it whatever you please." According to Campbell Lincoln advocated that the leaders return to their respective states and convene their legislatures to expedite peace. Since that is where secession had been initiated, he said, let their last act be reunion! Hunter pointed out that during the English Civil War even Charles I had entered into agreements with rebels against the government. Lincoln replied: "I do not profess to be posted in history. All I distinctly recollect about the case of Charles I, is, that he lost his head."[17]

There has ever been uncertainty as to Lincoln's course on reconstruction, had he lived, and how that may have differed from that undertaken. Particularly, what was the meaning of "no receding . . . on the Slavery question." Emancipation had been promulgated as a "military necessity," and he considered its effect void as the war ended, hence the urgency of a constitutional amendment abolishing slavery. Surely Lincoln held that no slaves freed by the war would be reenslaved, but in the interim, what of the rest? And what would the status of the freed people be? At the Hampton Roads conference it is clear that Lincoln alluded to a figure he had in mind of compensating slave owners for their loss of property to the amount of $400 million, only about 15 percent of the 1860 value of slave property. Stephens recalled that Lincoln urged him to go home and recommend that Georgia be taken out of the war, saying the legislature could ratify the Thirteenth Amendment "prospectively." While this was constitutionally impossible, it undoubtedly indicates the course Lincoln favored for blacks acquiring full rights of citizenship; he thought their assumption to equal status should be mitigated over time. In all events, both slavery and the rebellion were condemned to pass away, said Lincoln, and only the courts could decide on the many attenuating and extenuating circumstances. As for the remaining question of punishment of rebel leaders and confiscation of their property, Lincoln assured openhanded treatment based on the president's constitutional power of pardon. But without authority to negotiate, the commissioners returned to Richmond with only a reiteration of Lincoln's stipulation of unconditional surrender.

Two days later Lincoln called his cabinet to an evening meeting where he read to them the draft of a resolution he intended as a peace proposal to the states in rebellion. On condition of the cessation of war-like acts by April 1, 1865, and ratification of the Thirteenth Amendment the president would be empowered to pay $400 million to the states for the loss of property in slaves. The cabinet unanimously opposed the proposal; thereupon Lincoln folded the paper and put it away.

The next day President Davis addressed a message to the Confederate Congress accompanying the letter of his commissioners about the "negotiations" at Hampton Roads, where they had only learned the conditions set for their "humiliating surrender."

With all business suspended in Richmond at noon on February 7, a mass meeting was called in front of the African Church where orators addressed appeals "to stir the heart and nerve the resolution of a people fighting for liberty." Davis was one of those who denounced the 'Northern president' as "his Majesty Abraham the First"; Seward and Lincoln would yet find, he predicted, "they had been speaking to their masters." Carried away by grandiloquence, Davis predicted great victories were in preparation. Military affairs were in excellent condition; before the summer solstice fell upon the country it would be the Yankees who would be asking for terms of peace and the grace of conferences in which the Confederates might make known their demands; Sherman's march through Georgia would be his last. Edward Pollard noted that in his unfortunate address,

> there was one just and remarkable sentiment. He referred to the judgment of history upon Kossuth, who had been so weak as to abandon the cause of Hungary with an army of thirty thousand men in the field; and spoke of the disgrace of surrender, if the Confederates should abandon their cause with an army on our side and actually in the field more numerous than those which had made the most brilliant pages in European history; an army more numerous than that with which Napoleon achieved his reputation; an army standing among its homesteads. . . .
>
> It was very clear that the Confederacy was very far from the historical necessity of subjugation. But it was at any time near the catastrophe of a panic.[18]

As Davis was uttering his words of "unconquerable defiance," Sherman was setting out on a march that, although less heralded than the March to the Sea, was of equal significance. After winning Grant to the idea, his forces were in motion on February 1—one wing by water up to Beaufort from which it penetrated up to the Pocotaligo, convincing the Confederates of his intention of moving on Charleston; while a second wing marched up the Savannah River on both the Georgian and Carolina sides, toward Augusta. With the first wing approaching the juncture of the Branchville and Charleston Railroad, and the second approaching within thirty miles of Augusta, Sherman directed a large force at the middle of the state. In a letter to Halleck detailing his plans he wrote: "The truth is the whole army is burning with an insatiable desire to wreak vengeance upon South

Carolina. I almost tremble at her fate, but feel that she deserves all that seems to be in store for her." In his two-volume *Memoirs of W. T. Sherman*, he later elaborated: "My aim then was to whip the rebels, to humble their pride, to follow them to their inmost recesses, and make them fear and dread us."[19]

Strategically Sherman's aim was to bring his powerful army up to Goldsboro, North Carolina, where they would be reinforced by troops coming up from Wilmington, when they would join Grant in crushing the Army of Northern Virginia, or in Grant's phrase, "wipe out Lee." The obstacles in front of Sherman were formidable. For one thing, his objective was 425 miles away and required the fording of nine rivers with extensive swamps and bottomlands and the crossing of scores of tributaries. He did not yet know it, but he would face one of the wettest winters in twenty years, submerging many of the roads and making it nearly impossible to march an army across the lower portion of the state. But with Sherman's army were thousands of freedmen working with soldiers in specially organized "pioneer battalions," felling trees to corduroy the roads, building bridges, and laying causeways. The enemy military forces too were not inconsiderable; Beauregard was in command of 10,000 troops at Columbia, while Hardee, after his retreat from Savannah, had reinforced an equal number at Charleston with 10,000 of his own troops. In defense of Augusta a corps of Hood's shattered army was hastening, while Lee sent a brigade of cavalry under Wade Hampton to rally his threatened state. But these forces were scattered and kept in check by the development of Sherman's feints, while his main force penetrated to the center of South Carolina.

Despite heavy rains Sherman's marchers were striding eighteen miles a day, soon reaching the swollen Salkehatchie River, fifty miles north of the Georgia border. "The Salk is impassable," declared General Hardee; Sherman's pioneers built bridges and his army crossed it. Gen. Joseph Johnston, who would assume command on February 22, said, "When I learned that Sherman's army was marching through the Salk swamps, making its own corduroy roads at the rate of a dozen miles a day, I made up my mind that there had been no such army in existence since the days of Julius Caesar."[20]

Sherman's unexpected progress soon brought him in contact with the railroad linking Branchville and Charleston, compelling the Confederate force to evacuate the former on February 11. Now Sherman's pioneers fell to tearing up the railroad, preventing

Charleston's reinforcement from the west. Continuing the feint toward Augusta, Sherman dexterously entered Orangeburg on the 16th, threatening all the dispersed Confederate armies arrayed against him; forces, pending Johnston's advent, commanded by Cheatham, Beauregard, and Hardee, who resolved to unite to block the Union army's path into North Carolina.

Withdrawing from Charleston, Hardee left detachments in the city until the morning of the 18th with instructions to burn every building holding cotton. Meanwhile, with Beauregard hastily evacuating Columbia, and a white flag fluttering from city hall, Sherman's troops marched smartly down Main Street to Capitol Square with bands playing and banners flying. In Charleston a large shed at the Savannah Railroad wharf was torched with several others on Lucas Street, both fires rapidly communicating to adjoining buildings. Soon the western district of the city was in flames; the bridge over the Ashley was burned, as well as the arsenals, the quartermaster's stores, two ironclads, and several vessels in the shipyard, which exploded in a tremendous shower of debris. While the conflagration raged engines were brought out as the fire department, joined by many blacks still in the city, fought to keep the surrounding buildings from igniting. Compounding the scope of the catastrophe, a large stockpile of ordnance had been left at the depot of the Northwestern Railroad, and the building exploded, shaking every foundation in Charleston, sending corpses up in a whirling mass of ruin. United States troops of the Southern Department were already entering the city from James Island, lending a hand to the stricken city, while the Confederate flags still flying on the various outlying forts gave way to the Stars and Stripes. One of the first regiment's occupying Charleston was the 54th Massachusetts. In his history of the regiment Luis Emilio wrote:

> We could not but be exultant, for by day and night, in sunshine and storm, through close combat and far-reaching cannonade, the city and its defences were the special objects of our endeavour for many months. Moving up Meeting and King Streets, through the margin of the "burnt district," we saw all those fearful evidences of fire and shell. Many colored people were there to welcome the regiment, as the one whose prisoners were so long confined in their midst. Passing the Mills House, Charleston Hotel, and the citadel, the Fifty-fourth proceeded over the plank road one and a half miles to the Neck, where the Confederate intrenchments extended clear

across the peninsula. Turning to the right, we entered Magnolia Cemetery, through which the line of works ran, and camped along it among the graves.[21]

Devastated as Charleston was, it was more fortunate than Columbia, beset with thousands of bales of cotton, many of them smoldering. Union troops entered the city with a large contingency of extraneous hangers-on. As the streets began to fill, congregations formed before public buildings and on street corners, and acts of robbery and pillage began to occur. Watches were especially in demand, but women's personal jewelry was also taken, silver being prized; and there was an extensive search for provisions of every kind. The large quantity of liquor shipped from Charleston for safekeeping introduced a growing inebriation into the scene of triumphalism and revenge. With cotton aflame, in the evening some buildings began to burn, while in some places fires were deliberately set. High winds quickly spread the fires up and down Main Street and into the eastern section of the city. While the ultimate responsibility for the destruction may not have lain with Sherman, there is no doubt that it occurred within the overall frame of his design. Even as he and his officers and troops battled to contain the scourge, they had no apology for it.

In the next days Sherman continued northward, moving in two columns on Winnsboro and rendering the Columbia and Charlotte Railroad valueless. Adding to his entourage was a growing train of refugees who followed the army under the charge of officers escaped from Confederate prisons, and who were supplying themselves liberally with horses and mules and wagons of every type, and were completely self-sufficient by their foraging. Whole villages were burned, as were many plantation houses and even slave quarters. Sherman's march through the Carolinas could be tracked, it was said, by columns of smoke, and stragglers never found difficulty in rejoining it.

The revolution of January 1, 1863, through January 31, 1865, in the American Republic—years Walt Whitman would call "the real parturition years (more than 1776–'83) of this hence forth homogeneous Union"[22]—had been ushered in with the force of black arms. The dawn of a new era too was discernible when the Congress passed legislation March 3 establishing the Bureau of Refugees, Freedmen and Abandoned Lands, shortened popularly to Freedmen's Bureau.

Following the pledge of his government to maintain black freedom Lincoln immediately signed the bill. Congress stipulated that the bureau was to last "during the present War of Rebellion, and for one year thereafter" as an arm of the War Department, with a commissioner appointed by the president with the consent of the Senate, and with the possibility of army officers being named as assistant commissioners for each of the states in rebellion. It was Sherman's order writ large, leasing not more than forty acres to freedmen and white refugees for a term of three years, at the end of which the land could be purchased at an appraised value. Not only had attitudes been altered by the course of the war; the standing of blacks had been transformed. As they had risen to a higher humanity with the demand for black equality and the rights of citizenship and participation in full self-government, the extent these were to flourish, became the question.

Now that the Confederate government itself was considering emulating the Union experiment in conscripting black troops, the creation of a black-officered Corps d'Afrique took on urgency in the minds of many. The sight of blacks fighting blacks would be a horrendous setback; the world would ridicule them in their unwitting disorder. The way forward had been set out in the previous year in the pages of the *Liberator*, which carried a letter from a sergeant of the 54th, then engaged in the as yet unsuccessful campaign to capture Charleston. Published anonymously, and dated October 7, the letter writer referred to the black men serving in the Union armies. The sergeant wrote:

> Somehow it is a religious or superstitious belief with me that this country will be saved by us yet. I say I believe this; but it is not a mere blind belief. I know that we shall have to labor hard, and put up with a great deal before we are allowed to participate in the government of this country. I am aware that we in the army have done about all we can do, and that to you civilians at home falls the duty of speaking out for all—as we have done the fighting and marching, and suffered cold, heat, and hunger, for you and all of us. My friend, we want black commissioned officers; and only because we want men we can understand, and who can understand us. We want men whose hearts are truly loyal to the rights of man. We want to be represented in courts martial, where so many of us are liable to be tried and sentenced. We want to demonstrate our

ability to rule, as we have demonstrated out willingness to obey. In short, we want simple justice.

Two proponents of this transformation, Martin Delany and Frederick Douglass, who had assisted in the recruitment of black regiments and had withdrawn after seeing it perverted by abuses that withheld authentic equality with whites, have come to represent a divergence of views. Toward the close of 1861, only months after returning from his trip abroad, and continuing into the summer in 1862, Delany toured the East and Midwest giving speeches to mostly white audiences under the titles "The Moral and Social Relations of Africans in Africa" and "The Commercial Advantages of Africa." The *Chicago Tribune* covered one of these, where it reported the doctor wore a long dark robe called a dashiki, "with curious scrolls upon the neck as a collar." Douglass attended three of Delany's lectures in Rochester and wrote that he "is one of the very best arguments that Africa has to offer. Fine looking, broad chested, full of life and energy, shinning like polished black Italian marble, and possessing a voice which when exerted to full capacity might cause a whole troop of African Tigers to stand and tremble. . . . He is the intensest embodiment of black Nationality to be met outside the valley of the Niger."[23]

In November 1864, deeming it offered the kind of environment he sought for his children, Delany moved his family from Canada to Wilberforce, Ohio. Originally a summer resort southeast of Dayton near Xenia, Wilberforce had been purchased by the African Methodist Episcopal Church to set up a college to train ex-slaves to become teachers and ministers. No sooner had his family settled in a cottage on seven acres he had purchased from the school's president, the Rev. Daniel A. Payne, than Delany hastened to Washington. He would be in the gallery at the House of Representatives when Garnet delivered his historic sermon, but had come to the capital especially to see what could be done to prevent the "horrible iniquity" represented by black conscription in the South.

Delany now desired an audience with Lincoln, who he had reason to believe might be sympathetic to this view. Although forewarned by Garnet that obtaining an appointment would be difficult, on a brisk morning he joined the queue at the gate of the presidential residence on Pennsylvania Avenue. After sending in his card, Delany was surprised and encouraged when he was told to return promptly on the following day. Four years later Delany offered this recollection to his biographer:

"On entering the executive chamber, and being introduced to his excellency, a generous grasp and shake of the hand brought me to a seat in front of him. No one could mistake the fact that an able and master spirit was before me. Serious without sadness, and pleasant withal, he was soon seated, placing himself at ease, the better to give me a patient audience. He opened the conversation first."

Lincoln: "What can I do for you sir?'

Delany: "Nothing, Mr. President. I've come to propose something to you which I think will be beneficial to the nation."

Encouraged, Delany brought to the president's attention the intent of the Confederate government to arm slaves; he had, he told the president, a proposal to thwart it.

Delany: "I purpose, sir, an army of blacks, commanded by black officers. This army to penetrate the heart of the South, with the banner of Emancipation unfurled, proclaiming freedom as they go. By arming the emancipated, taking them as fresh troops, we could soon have an army of 40,000 blacks in motion. It would be an irresistible force."

With Lincoln clearly interested, seeing the advantage of operating in the enemy's interior, Delany continued, "They could subsist on the country as they went along." Lincoln mused that "cavalry and a few light artillery would be all that was necessary." As these troops went into the interior, the white troops could remain at the front doing the siege work and holding the enemy armies in check. Thus black freedom, said Delany, as was not now the case, would be sustained and protected in the South as the emancipating army passed on, "leaving a few veterans among the new freedmen when occasion requires, keeping the banner unfurled until every slave is free, according to the letter of your proclamation. . . . An army of blacks commanded by blacks would win every slave for the Union and speedily bring the war to a close."

Lincoln confided that when he had issued the preliminary emancipation proclamation he had this in mind when he had forbidden the army to interfere with any efforts the slaves "may make for their actual freedom."

Delany: "But Mr. President, these poor people couldn't read your proclamation. Many still don't know anything about it."

Lincoln said he had "hoped and prayed" for such a course to develop but did not wish to initiate it unless it came as a policy suggested by others.

"Will you take command?" he asked suddenly.

Delany: "If there be none better qualified than I, sir, I will. As black men we have had no experience in the service as officers."

Lincoln: "That matters but little. Some of the finest officers we have never studied tactics till they entered the army. What we require most are men of executive ability."

As Delany started to present letters in introduction, Lincoln dismissed them, saying, "Not now. I know all about you. There's nothing to be done but to give you a line of introduction to the Secretary of War."

As he reached for a pen on his desk and began to write the rumble of cannon shook the room. "Stanton is firing! Listen! He's in his glory!" remarked the president with delight.

"What's the firing about, sir?" queried Delany. "Haven't you heard the news? Charleston is ours." The president handed Delany a note addressed to his secretary of war, dated February 18, 1865, bearing his signature, *A. Lincoln*: "Do not fail to have an interview with this most extraordinary and intelligent black man."[24]

Delany's audience with the president had taken forty-five minutes, but had resulted in what he called "the crowning act of the noble President's life." As he left the executive mansion, no doubt elated at the prospect of his new opportunity, Delany walked to the Seventeenth Street address of the War Department, where he found Stanton behind a high desk in a crowded reception room where he received the general public for an hour daily. After glancing at the card on which Lincoln had written, Stanton briskly nodded his assent. "Come back on Monday," he said.

That Monday when he met with Stanton, Delany found the secretary had already been briefed about his talk with Lincoln, and made up his mind to send him South at once. He would be assigned to Charleston, South Carolina, under the command of Major-General Saxton. After a handshake, Stanton turned Delany over to Colonel Foster, chief of the Bureau of Colored Troops, for examination. By February 26 both men were back in Stanton's office, where the secretary signed Delany's commission as major in the Army of the United States, and administered the oath. Stanton then said:

"Major Delany, I take great pleasure in handing you this commission. You are the first of your race who has been thus honored by the government. Therefore much depends and will be expected of you. But I feel assured it is safe in your hands."

Well-wishers flocked to Garnet's home that night to offer the new major congratulations and wish him Godspeed. The next edition of the *Anglo-African* editorialized: "We rejoice at this appointment of our brother. Having devoted the whole of a long life to an unselfish advocacy of our people, he becomes a fit instrument through which to do them honor."

Portrait of Maj. Martin Robison Delany.

After a brief trip to Wilberforce to arrange things with his family, Delany journeyed to New York City before transport to Charleston from Washington. In Wilberforce and New York Delany held consultations with other black leaders and gave speeches to both the committed and the curious. Under the titles "The Progress of the Government" and "The Capacity of the African Race to the Highest Civilization," Delany's speeches struck both new and familiar themes: The government was intent on dealing justly with black troops, he said. It would now commission qualified men as officers, would treat them well, and intended to make no distinction between soldiers. But he also urged his black audience to have "a higher opinion of themselves. We must declare ourselves to be the equals of white men, if not their superiors. In no other way can we attain to our proper position in the body politic."[25] An article appeared on the front page of the *New York Times* under the heading "A Black Major"' and soon the celebrated portrait of Delany, by the artist Bogardus of New York's Broadway, went on sale for 25 cents a copy. There in full uniform, one sees, standing erect before a field tent, with campaign hat, shoulder straps, sash, dress sword and spurs, sword in his right hand with its tip resting on the ground—Major Martin Robison Delany.

Because of ill health but also to look after her parents, in May 1864 Harriet Tubman had returned to her home in Auburn. It was during this time that she gave extended interviews and formed a friendship with a woman from Geneva, New York, Sarah Hopkins Bradford, and on this basis there appeared the book *Scenes in the Life of Harriet Tubman*, published in 1868. In the late summer and fall Tubman was in Boston staying in the home of John S. Rock, much of her time taken with seeking avenues to get just compensation from the government for her war services. A notice of her attendance in the city appeared in the *Commonwealth*, and the paper informed its public: "Her services to her people and to the army seem to have been inadequately recompensed by the military authorities, and such money as she has received, she has expended for others as her custom is. Any contribution of money or clothing sent to her at this office will be received by her, and the givers may be assured that she will use them with fidelity for the good of the colored race."26

In late fall she was staying with Gerrit Smith in Peterboro, and by early 1865, feeling well enough to return to her work in the South, she was in Washington. Before embarkation Major Delany held an interview with Tubman in Washington, where the two happened to cross paths, and Delany secured her cooperation.

A number of others who were "leading spirits of the Underground Railroad" had also promised to join Delany, among them William Howard Day. Delany wrote to Colonel Foster concerning two recruits who had come with him from New York. One of these men, he reported, would prove "an excellent Scout who was born and raised in Charleston and belonged to Major Rhett, rebel officer. The other is an intelligent young man who has been an officer and drill-master of a colored volunteer company in Detroit, Michigan for several years. I wish to present them to Brev. Maj. Gen. Saxton, as I know them to be just such persons as we shall want. Would you be pleased to present these facts to the honorable Secretary of War, and obtain the necessary transportation, sending it on immediately?"27

It is known that Tubman's authorization to use government transportation was issued by order of the secretary of war, but Delany's resourcefulness in regard to others was discouraged. Colonel Foster replied: "It is not considered expedient to furnish such transportation as the men could not be paid for services as 'Scouts.' It is only by entering the military service that they can be employed and paid for services in connection with the recruitment of colored troops."28

March 4 was the second inaugural for Abraham Lincoln. The day began with steady rain, and in its earliest hours strong winds lashed the city. As traffic and pedestrians began to fill the streets, mud rose

ankle deep making the journey from the special trains to the east portico of the Capitol unpleasant. With 30,000 citizens expected at the
ceremony, Washington was rife with rumors of an assassination
attempt on the president and other members of his government. By
order of the secretary of war sharpshooters had been posted atop
buildings around the Capitol, and all the roads and bridges into
Washington had been picketed days before with extra troops, while
Pinkerton detectives plied the streets. In addition, at the front of the
city, at Fairfax Court House, the 8th Illinois Cavalry had been posted to be on the lookout for "suspicious characters."

Taking his place in the crowd in the company of Mrs. Dorsey, the
wife of Thomas Dorsey, a well-to-do caterer and leading figure of
Philadelphia's black community, Douglass later reflected in his *Life
and Times* on the "dark and lowering" air of Washington:

> The friends of the Confederate cause here were neither few nor
> insignificant. They were among the rich and influential. A wink or
> a nod from such men might unchain the hand of violence and set
> order and law at defiance. To those who saw beneath the surface
> it was clearly perceived that there was danger abroad, and as the
> procession passed down Pennsylvania Avenue I for one felt an
> instinctive apprehension that at any moment a shot from some
> assassin in the crowd might end the glittering pageant and throw
> the country into the depths of anarchy. I did not then know, what
> has since become history, that the plot was already formed and its
> execution which, through several weeks delayed, at last accom-
> plished its deadly work as contemplated for that very day.

As the official procession reached the Capitol and people took
their seats, Lincoln, with his visibly drunken vice president at his side,
spied Douglass standing within hailing distance of the platform and
pointed him out. Douglass wrote:

> There are moments in the lives of most men when the doors of
> their souls are open, and, unconsciously to themselves, their true
> characters may be read by the observant eye. It was at such an
> instant that I caught a glimpse of the real nature of this man, which
> all subsequent developments proved true. I was standing in the
> crowd by the side of Mrs. Thomas J. Dorsey, when Mr. Lincoln
> touched Mr. Johnson and pointed me out to him. The first expres-

sion which came to his face, and which I think was the true index
of his heart, was one of bitter contempt and aversion. Seeing that
I observed him, he tried to assume a more friendly appearance, but
it was too late; it is useless to close the door when all within has
been seen. His first glance was the frown of the man; the second
was the bland and sickly smile of the demagogue. I turned to Mrs.
Dorsey and said, "Whatever Andrew Johnson may be, he certainly is no friend of our race."[29]

By the time the ceremony started and Chief Justice Chase administered the oath of office, the rain had subsided, and as the president
began his address the sky cleared somewhat and for a moment a shaft
of light illuminated the spot where he stood. Although much
remarked upon in the press, it was the words themselves that remain
as daylight, as Lincoln drew out in a few paragraphs the entire shad-
owy and sorrowful journey both he and the country had engaged in
to finally realize the war's true meaning. One witness, Charles Francis
Adams Jr., was to remark "in its grand simplicity and directness" the
address was "for all time the historical keynote of this war."[30]
Douglass observed, it "sounded more like a sermon than like a state
paper."[31]

The South had referred to its system of labor and racial subordination as a "peculiar institution," and Lincoln now laid the war's
coming to the behest of that "powerful interest" whose raison d'être
had been to transmute the labor of black skins into gold. Lincoln continued:

> All knew that this interest was, somehow, the cause of the war. To
> strengthen, perpetuate and extend this interest was the object for
> which the insurgents would rend the Union, even by war; while the
> government claimed no right to do more than to restrict the terri-
> torial enlargement of it. Neither anticipated that the cause of the
> conflict might cease with, or even before, the conflict itself should
> cease. Each looked for an easier triumph, and a result less funda-
> mental and astounding. Both read the same Bible, and pray to the
> same God; and each invokes His aid against the other. It may seem
> strange that any men should dare to ask a just God's assistance in
> wringing their bread from the sweat of other men's faces; but let us
> judge not that we be not judged.

Douglass remarked: "I know not how many times and before how many people I have quoted these solemn words of our martyred President. They struck me at the time, and have seemed to me ever since to contain more vital substance than I have ever seen compressed in a space so narrow; yet on this memorable occasion, when I clapped my hands in gladness and thanksgiving at their utterance, I saw in the faces of many about me expressions of widely different emotion."[32]

That evening Douglass desired to stand in a place and on an occasion no less undreamt, and resolved to be among the thousands that would be thronging to the presidential residence. Not able to convince any of his brethren to summon the nerve for such an exercise, again Douglass took Mrs. Dorsey on his arm to join the procession. However, when they reached the entrance two policemen roughly grabbed Douglass, ordering him to stand back; "no person of color can be admitted," had been their orders. Douglass immediately responded: "Surely you are mistaken, for no such order could have emanated from President Lincoln. If he knew I was at the door he would desire my admission."[33]

To put an end to the parley and, he supposed, to remove an obstruction from the doorway, Douglass relates the policemen assumed an air of politeness, offering to conduct him in. Following their lead the duo soon found themselves walking up planks arranged as a temporary exit for visitors through a window. Douglass and his companion halted, as he said: "You have deceived me. I shall not go out of this building till I see President Lincoln."

Douglass continued:

At this moment a gentleman who was passing in recognized me, and I said to him: "Be so kind as to say to Mr. Lincoln that Frederick Douglass is detained by officers at the door." It was not long before Mrs. Dorsey and I walked into the spacious East Room, amid a scene of elegance such as in this country I had never before witnessed. Like a mountain pine high above all others, Mr. Lincoln stood, in his grand simplicity, and homely beauty. Recognizing me, even before I reached him, he exclaimed, so that all around could hear him, "Here comes my friend Douglass." Taking me by the hand, he said, "I am glad to see you. I saw you in the crowd today, listening to my inaugural address; how did you like it?" I said, "Mr. Lincoln, I must not detain you with my poor opinion, when there are thousands waiting to shake hands with you." "No, no," he said, "you must stop a little Douglass; there is no man in the country whose opinion I value more than yours. I want to know what you think of it?" I replied, "Mr. Lincoln, that was a sacred effort." "I am glad you liked it!" he said, and I passed on, feeling that any man, however distinguished, might well regard himself honored by such expressions, from such a man.

With the Carolinas being exercised by fire and sword, the only other extensive area of the South untouched by the war, outside of Texas, was Alabama below its northern tier and beyond Mobile Bay. Simultaneous to Sherman's campaign, troops from Pensacola commanded by General Canby were to invade Alabama attacking Confederate fortifications across the bay from Mobile, while Thomas's 13,000-man cavalry, armed with repeating carbines, invaded the state from Tennessee. The objectives of the horsemen were the state capital at Montgomery and the munitions-producing center at Selma. While the latter completely outranged Forrest, they burned large stores of cotton, destroying bridges and factories, railroads, and rolling stock; Mobile only fell to Canby on April 9—as Grant would remark, a year too late to be of any strategic value. These operations in the spring of 1865 were part of a strategy to destroy the war-making capacity of the Confederacy, and just as surely to bring the scourge of war to a recalcitrant population.

By the time the Alabama operations were under way, Sheridan's cavalry divisions had completely devastated the remnants of Early's army at Waynesboro and started east to assist Grant at Petersburg, destroying the Virginia Central Railroad as he came. Meanwhile, leaving Columbia, Sherman's ravaging army and ever-growing train of "bummers" soon reached Winnsboro, burning and pillaging plantation houses along the route, as well as every corncrib, smokehouse, and cotton gin they came upon. Leaving the carcasses of animals littering roadsides, crops and stores that could not be carried off were burned, houses and personal belongings ransacked; while blankets, sheets, and quilts were carried away, farm and garden implements, fences, everything of use, broken up or burned; wagons and carriages harnessed to horses and driven off. Edward Pollard wrote that a succession of "tall chimneys standing solitary and alone, and blackened embers, as it were, laying at their feet," pointed out due north on the road leading to Blackstock.

Directing the march at Charlotte, Sherman misled Johnston into uncovering Fayetteville, beyond which he intended to combine with another powerful column coming up from the coast. After Blackstock, Sherman veered east to Cheraw on the Great Pee Dee River; three days later, fording the river, Sherman crossed into North Carolina in four columns, engaging Wade Hampton's brigade in the vicinity of Rockingham. Fayetteville was gained March 11, and Sherman communicated with the column advancing from the coast to consummate their conjunction at Goldsboro. Half of Hardee's divisions, entrenched near the confluence of the Cape Fear and the Black rivers, engaged with this force in a bloody clash before falling back halfway between Fayetteville and Raleigh. It was Johnston's purpose now to cripple Sherman before the junction could be effected and a supply established from the coast.

By forced marches to Bentonville, Johnston soon gained a position from which to strike. After achieving initial success with an ambush on a strung-out wing of Sherman's advance on March 19, Johnston was repulsed and withdrew toward Raleigh. Calling off a counterattack on the 21st, Sherman let him slip by. Crossing a pontoon bridge laid across the Neuse River, Sherman was in Goldsboro March 22, less than two months after leaving Savannah. Only a hundred and fifty miles now separated Sherman's army from Grant.

Deciding he needed a respite from Washington, Lincoln now arrived at City Point, Virginia, as Grant's guest, while Sherman hastened for an interview with the two men. Some observers in Richmond conjectured this meant that "peace negotiations" were to be proffered; soon they were to find it meant something else altogether.

As the railroad line would have to be relaid, Sherman could not be expected to be in Grant's vicinity with his army until late April. But Grant already outnumbered Lee by more than two to one, and with Sherman's conquest of the Carolinas, Lee was threatened with complete encirclement. Grant, moreover, wanted the Army of the Potomac to earn the laurels of vanquishing "their old enemy." "I mean to end the business here," Grant told the hard-driving Phil Sheridan after his conference with President Lincoln and General Sherman. At the same time Lee knew he must abandon his positions or face encirclement; withdrawal would mean the fall of Richmond, but if he could keep his army together the Confederacy might still survive. The government could remove to Danville and he might be able to unite with Johnston for a retributive strike on Sherman.

At daylight on March 25 Lee's breakout attempt came east of Petersburg with a surprise assault on Fort Stedman. Swarming over the fort, tattered, wizened troops seized several batteries and a half-mile of Union trenches. While Lee hoped to follow up this success by taking hold of the neighboring works and achieving a commanding position over the Union military railroad, cutting off Grant's left from its base and with the Army of the Potomac north of the James, roused Union troops counterattacked. Opening with a barrage of heavy artillery, Lee lost nearly 5,000 men versus 2,000 for Grant.

On March 29 Grant retaliated again with a heavy movement toward Lee's line on the Southside Railroad near Five Forks. The fighting was desperate, and when Grant ordered an all-out assault at dawn on April 2 along the Petersburg front, Lee found his army nearly shorn in two.

News of the military situation had been tightly controlled by the Confederate War Department, and as not a shot had been fired on the defenses of the city, Richmond remained unperturbed as the battle raged to its south. Only as the day wore on did the extent of the disaster begin to trickle in. At the end of the day, a Sunday, Lee withdrew his military within the inner defenses of Petersburg, so they could get away under cover of night. That morning Lee had sent a message to Davis that arrived while he was attending service at St. Paul's Church in Richmond: The defenses of Petersburg had been broken in three places, Lee's dispatch stated, and if the lines could not be reestablished preparations should be made to evacuate Richmond by 8 o'clock that evening. As Davis read the note, his complexion turned waxen and he rose from his pew and quickly left without uttering a word. As the parishioners were dismissed uneasy whispers pulsed through the air, and soon rumor ran to the ends of the city: the army and the government were to withdraw.

With wagons loading in front of various government departments for transport to the Danville depot, many in the general population decided they had better emulate the soon-to-be fugitive government. There was a run on deposits at the banks, and the price of rental conveyance shot up. Pollard remarked: "Suddenly, as if by magic, the streets became filled with men, walking as though for a wager, and behind them excited negroes with trunks, bundles, and luggage of every description. All over the city it was the same—wagons, trunks, bandboxes, and their owners, a mass of hurrying fugitives, filling the streets."[34]

As the battle for Petersburg developed, half the Union army investing Richmond had been withdrawn to assist to the south. General Weitzel, left in command of the Richmond trenches, had one division of the 24th Corps and two divisions of the 25th, with orders to make as great a show as possible, and whenever satisfied of his ability to enter the city, he should push on. Accordingly, as night came all his bands began sending strains up into the night air. This bravura performance ended at midnight, replaced by boding silence, hovering before every wakened eye. That evening the Richmond city counsel met and resolved that, to avoid disorder, all liquor in the city should be destroyed. This work was begun about midnight when hundreds of barrels were rolled into the streets and their heads smashed in, while fine cases of brandy were flung from open windows. Even so, some of it did get imbibed and the sound of breaking glass and riot began to fill the air.

But an even greater peril appeared in the order of Confederate military authorities to burn the four principal tobacco warehouses in the city and the numerous buildings housing cotton, and to burn or blow up all military property. The mayor sent a committee to warn that such action would threaten the entire business district of Richmond, but their remonstrations fell on deaf ears. About 3 in the morning the quiet was rent by tremendous blasts on the James River, as the rams the *Richmond* and *Virginia* were blown skyward, and lurid lights were seen as warehouses and the bridges of Richmond went up in flames. As General Weitzel put his troops at the ready, the desolation awaiting them would be even greater than that seen in Charleston or Columbia. Thomas Morris Chester, reporting for the *Philadelphia Press*, wrote: "The soldiers along the line gathered upon the breastworks to witness the scene and exchange congratulations." Pollard wrote another of his acerbic passages: "Morning broke upon a scene such as those who witnessed it can never forget. The roar of an immense conflagration sounded in the ears; tongues of flame leaped from street to street; and in this baleful glare were to be seen, as of demons, the figures of busy plunderers, moving, pushing, rioting, though the black smoke and into the open street, bearing away every conceivable sot of plunder."[35]

General Weitzel's orders had specified that no offensive operations would be undertaken at night, as the ground in front of the trenches and the fortifications had been sown with torpedoes. But while it was still dark, with the help of prisoners who knew where the infernal machines were laid, he had his troops in motion. About forty horsemen from the 4th Massachusetts Cavalry were sent to discover the condition of affairs and found they could ride steadily into the burning city. Reaching Capitol Square at a gallop, they hauled down the Confederate colors and ran up the Stars and Stripes. Weitzel's troops of the 25th Corps and those of the 24th under General Kautz were now filing and jostling along the roads leading into Richmond from the southwest—along Osborne Turnpike and New Market Road, the Darbytown and the Charles City roads, and the Williamsburg Road. On the outskirts of the capital all of these merged into a single avenue, and so here the race to be the first Union troops into the fallen bastion was to be resolved. General Kautz wrote: "The colored men gave the whites the road but as long as there was room they kept the head of their column abreast of the other, until they reached the narrow streets where there was room for one column only." Reporter Chester noted: "Along the road which the troops marched, or rather double quicked, batches of Negroes were gathered together testifying by unmistakable signs their delight at our coming."[36]

Hoisting a fanfare of regimental flags and banners, the marchers strode to fife and drum, as a long line of black cavalry, the 5th Massachusetts, swept in, brandishing swords and "uttering severe oaths." It is often remarked that one company of black soldiery ended its march singing "John Brown's Body"—and to be sure the service record of Company B reads: "The Company was part of the 9th Regt. USCT entered the city of Richmond on the morning of April 3rd 1865. Drums beating, colors flying and men singing the John Brown hymn—Gloria in excelsis!"[37]

Without doubt this was an astonishing and wonderful moment; Richmond had long been the object of their waking and fighting hours, of their idle moments and of their dreams, and of the death of many. Above and beyond flag and union, they had been fighting for home, for wife and children—fighting to unloose the confounding knot of bondage. They had endured long years of oppression with nothing less than their extraordinary faculty of patience, and now in this moment as they strode into the capital of "Sech" a fiery ecstasy must have risen in their invocation.

There was more than one version of the song, one declaring Jeff Davis would be hung from a sour apple tree, but the song sung by

these soldiers was surely not that. Their exaltation likely began with
the verse:

> Old John Brown's body lies a mouldering in the grave,
> While weep the sons of bondage, whom he ventured to save;
> But though he lost his life in struggling for the slave,
> His soul is marching on.
> Glory, glory, Hallelujah!
> Glory, glory, Hallelujah!
> His soul is marching on!

As they smartly stepped on Richmond's streets, the still raging fire
was consuming some of the city's most prominent buildings. Beneath
trees on the sward of Main Street, people had dragged their belongings from their burning homes. Piles of furniture and personal effects,
carpets, blankets, and basins, were laid amid people standing in family groupings gasping for fresh air. The strains of Company B's marching song rose like an oath against the sky.

> John Brown was a hero, undaunted, true, and brave,
> And Kansas knew his valor, when fought her rights to save;
> And now, though the grass grows green above his grave,
> For his soul is marching on.

West on Main, the fire was stayed at 9th Street, sweeping toward
the river. On Richmond's northern side the flames halted between
13th and 14th streets; then arching back to 8th, consuming Bank
Street—where the Bank of Richmond, Trader's Bank, Bank of the
Commonwealth, Bank of Virginia, and Farmer's Bank were all smoldering ruins.

The fire had spread wings to encompass twenty square blocks of
the business and governmental quarters of Richmond. Down to the
river the destruction had been complete, leaving only blackened walls
and solitary chimneys protruding through smoking ruins. And the
chanting black warriors continued:

> John Brown was John the Baptist, of the Christ we are to
> see—
> Christ, who of the bondman shall the Liberator be;
> And soon throughout the sunny south the slaves shall all be
> free,
> For his soul is marching on.

Steadily treading through the hellish circle, past the ruins of the American Hotel and the Columbian Hotel, past the smoking *Enquirer* building, past the remains of the post office to the corner of Franklin Street where the burning rubble of the State Court House stood,wheeling left they may have entered the Capitol Square.

> The conflict that he heralded, he looks from heaven to view,
> On the army of the Union, with its flag, red, white, and blue;
> And heaven shall ring with anthems o'er the deed they mean to do,
> For his soul is marching on.
> Ye soldiers of freedom then strike, while strike ye may,
> The death-blow of oppression in a better time and way;
> For the dawn of old John Brown has brightened into day.

By afternoon, with the help of these soldiers, the fire was spent. General Weitzel, establishing his headquarters in the State Capitol—formerly occupied by the Virginia House of Delegates—instituted measures to restore order. Only hours after Confederate forces had withdrawn, President Lincoln and General Grant had entered Petersburg. As Grant organized his army for pursuit of Lee, Lincoln had returned to City Point on the James River. Meeting with Admiral Porter, he said: "Thank God I have lived to see this. It seems to me that I have been dreaming a horrid dream for four years, and now the nightmare is gone. I want to see Richmond."[38]

Soon Porter had an escort arranged—twelve sailors with carbines and an assortment of officers, both navy and army. The president and his son were rowed up the James to the wharfs where they alighted about a mile from Capitol Square. No carriages were available so the entourage began to walk, six sailors proceeding and six in the rear, the president and his son flanked on either side by officers.

Walking at Lincoln's side, Porter peered warily into every doorway and window on the route, but Lincoln, tall with his distinctive accoutrement, was immediately spotted, first by a contingent of black laborers, who abandoned their supervising officers to flock around the president. Straight away the exhilaration spread when a reporter for the *Boston Journal*, walking with the entourage, said to a black woman nearby: "There is the man who made you free." Breaking into shouts of joy, "Glory, Glory, Glory!" her voice was drowned in the rising cheer of a clapping, dancing, shouting, whirling crowd. The correspondent wrote: "What a spectacle it was! Such a hurly-burly, such wild, indescribable, ecstatic joy I never witnessed. A colored man acted as guide."[39]

A woman standing in a doorway shouted, "Thank you dear Jesus, for this! Thank you, Jesus!" Another standing on the sidewalk clapped her hands, shouting, "Bless de Lord." Women and men, boys and girls were waving bonnets and hats, while the President walked in silence acknowledging the cries, and the salutes of soldiers and officers who had been drawn to the scene. One man shouted: "Bless de Lord, dere is de great Messiah! I know'd him as soon as I seed him. He's been in my heart for long years, an is come at las' to free his children from der bondage. Glory hallelujah!" Several people approached Lincoln just to touch him. Visibly moved, as one man fell on his knees, Lincoln extended a hand: "Don't kneel to me. That is not right. You must kneel to God only, and thank Him for the liberty you will enjoy hereafter."[40]

J. J. Hill, who witnessed these events as a soldier of the 29th Regiment of Connecticut Colored Troops, wrote in his *Sketch*: "It was a man of the people among the people. It was a great deliverer among the delivered. No wonder tears came to his eyes when he looked on the poor colored people who were once slaves, and heard the blessings uttered from thankful hearts and thanksgiving to God and Jesus. They were earnest and heartfelt expressions of gratitude to Almighty God, and thousands of colored men in Richmond would have laid down their lives for President Lincoln."[41]

When Lincoln had made his way to the executive mansion, vacated by Jefferson Davis just forty hours before, he sat at his desk; for a few moments, as it was described, the look on his face was "a serious, dreamy expression." Then he moved on to the State Capitol where he delivered a short impromptu speech to the crowd gathered there. "In reference to you, colored people, let me say God has made you free. Although you have been deprived of your God-given rights by your so-called masters, you are now as free as I am, and if those that claim to be your superiors do not know that you are free, take the sword and bayonet and teach them that you are—for God created all men free, giving to each rights of life, liberty and the pursuit of happiness."[42]

Thomas Morris Chester of the *Philadelphia Press* wrote: "Richmond has never before presented such a spectacle of jubilee. What a wonderful change has come over the spirit of Southern dreams."

Robert E. Lee was the scion of one of Virginia's first families. Five years before being driven from the Petersburg defenses with the remnants of his once proud but now desperate army, he had delivered the verdict on John Brown at Harper's Ferry as "the attempt of a fanatic or madman which could only end in failure."[43] Now the irony was that the work John Brown set in motion had come to fruition, while Lee faced a breakdown that was to bring in its wake the astonishing collapse of the Confederacy. With only 35,000 men left from the Petersburg and Richmond garrisons, Lee's retreating divisions streamed westward along the north bank of the Appomattox River in the direction of Amelia Court House. There they expected to find rations to allay their near starvation, but instead found a train loaded with ammunition, necessitating a delay while troops foraged the countryside. As Sheridan was in direct pursuit to cut off retreating columns and harass Lee's rear, Grant moved along the interior lines to head him off. But Lee's army was now running through his fingers; in only three days many thousands deserted, leaving in squads for their homes. All along the road were strewn the arms and accoutrements of this disintegration, and nothing could be done to reassert discipline.

Sheridan telegraphed Grant on the evening of April 5: "I feel confident of capturing the entire Army of Northern Virginia, if we exert ourselves. I see no escape for Lee."[44]

As Union armies maneuvered to cut off any avenue of escape without resort to battle, Lee and Grant exchanged a flurry of notes over the next two days, Grant stating the terms of surrender and Lee angling for "the restoration of peace" without surrender. But the end had come, and Lee, though reviling it, at last was made to face that he must accept defeat. On the morning of April 9 near Appomattox Court House Lee sent Grant a letter embracing that proposition.

That afternoon, as surrender formalities were being discussed in the parlor of a private home near Appomattox, six hundred miles away the last major infantry combat of the war was taking place on the western shore of the Tensaw River across from the city of Mobile—the assault on Fort Blakeley. The battle was begun with nine black regiments of General Hawkins's 1st Division leading on the left, while white regiments attacked on the right. Confederate gunners in the fort were incensed at seeing the massed black troops and directed a galling fire into their midst. They were bearing up under the ferocity of the barrage when at last the cry was heard—"Remember Fort Pillow!" These words had an instantaneous effect, jumping like an electric arc among the ranks, cascading down the line to become an overwhelming impulse. Without orders, to a man they surged forward, their cries hanging in the air. Seeing they were being overwhelmed by blacks, the defenders began to flee, some to the opposite side of the fort so as to be able to surrender to white troops. A lieutenant wrote: "The rebs were panic-struck. Numbers of them jumped into the river and were drowned attempting to cross, or were shot while swimming. Still others threw down their arms and ran for their lives over to the white troops on our left, to give themselves up, to save being butchered by our niggers. The niggers did not take a prisoner, they killed all they took to a man."[45]

The Rev. Henry McNeal Turner, chaplain of the 1st USCT, was to publicly rebuke the division for this atrocity.

On April 3, as Martin Delany had reached Charleston, he immediately embarked on his duties enlisting freedmen in the service of the United States. Black Charlestonians flocked into his office on St. Philip Street at the corner of Calhoun, to both seek advice and offer congratulations to the black man wearing the shoulder straps of a United States Army officer. Adding measurably to the strides blacks were beginning to make as a free community, in May, Bishop Daniel Payne arrived to organize a congregation of the AME Church in the city. Before long Delany recruited enough men to fill two regiments, and they began drilling at the racetrack, while the ranks of a third regiment began to fill up.

At Lee's surrender there had been a five-hundred-gun cannonade from the battlements at Washington, with great rejoicing in that city and in New York and at numerous other points. For his part Grant forbade all demonstrations by the Army of the Potomac and sent three days' rations for 25,000 men across the lines. Lincoln was forward looking, contemplating the future peace and restoration, and on April 11 from the executive mansion's balcony spoke to a celebratory gathering of some of his intimations of what was to come: "There is no authorized organ for us to treat with. . . . We must simply begin with and mould from disorganized and discordant elements."[46] The way, Lincoln continued, had been pointed out in the experience of Louisiana, Arkansas, and Tennessee, all of which had formed new governments under direction of the executive during the course of the war. He would have preferred, he conceded, referring to Louisiana, that its reconstructed government had conferred the franchise to literate blacks and to those who had served as soldiers. He would have an announcement for a new policy for restoration soon.

Among Lincoln's auditors was the dapper stage idol and veteran Confederate agent John Wilkes Booth, who snarled to a cohort, "That means nigger citizenship. Now, by God, I'll put him through. That is the last speech he will ever make."[47]

Lincoln had designated April 14, 1865—the fourth anniversary of the fall of Fort Sumter—as a day of national celebration to mark the victory of the Union, and its centerpiece and symbol was the raising of the flag over the fortification's ruins in Charleston Harbor. Shredded and tattered, the flag would be the very one hauled down four years before, and the officiating officer would be Major-General Anderson, who had been the officer in command when the fort was surrendered in 1861. Politicians and army officers and assorted dignitaries from Washington and New York and as far away as Boston were among invited guests, including William Lloyd Garrison and the Scottish abolitionist George Thompson, making a gathering of several hundred. Some observing the ceremony were ferried from the wharf out into the harbor on *The Planter*, the steamer commandeered by Robert Smalls in 1862 and of which he had been given charge. When Smalls had returned to Charleston, crowds of freedmen hurried to greet the captain of his vessel, one old man crying, "Too sweet to think of."[48] Among the passengers on the quarterdeck with Captain Smalls were Major Delany, his son Toussaint—a soldier with the 54th Massachusetts who had been delegated to assist his father—and another son, that of Denmark Vesey, the mastermind of the slave rebellion in Charleston forty years earlier.

As Garrison had arrived in Charleston he went to the graveside of John C. Calhoun to pronounce, "Down to a deeper grave than this, slavery has gone and for it there is no resurrection."[49] Afterward standing outside St. Michael's Church, when he heard a regimental band marching to the strains of "John Brown's Body," he wept uncontrollably.

Also that morning Garrison would stand together with Major Delany in Citadel Square as a parade of black children, boys and girls newly admitted to schools under the superintendence of James Redpath, passed through. Later that afternoon Delany and Garrison, together with General Saxton, addressed an overflow audience of six thousand freedmen at the Zion Church. When Garrison was pointed out, he was swept up and carried to the dais on the shoulders of ex-slaves, where he told the assembly: "Once I could not feel any gladness at the sight of the American flag, because it was stained with your blood. Now it floats purged of its gory stains. It symbolizes freedom for all, without distinction of race or color." In his remarks, Delany proposed establishing a newspaper "to advocate the interest of the colored population" in the city. When he spoke, having been advised by some of the leading black churchmen of the necessity of gaining the franchise, Saxton recommended they petition the president. "I want you to elect a committee to draft this petition and get every colored man to sign it." He added: "I can get 3000 in Beaufort to sign it, but I want it started here in the City of Charleston, the leading city of the rebellion." An observer at the meeting wrote, "The freed people could not leave until they made a target of the major's head, by aiming at it bouquets, and grasping his hand until it was sore."[50]

On that seemingly auspicious day the president was feeling elated, relieved of the harsh burden of many years. On that day in April for decades thereafter, Walt Whitman would recount, the afternoon paper of Washington, the *Evening Star*—"had splattered all over its third page, divided among the advertisements in a sensational manner, in a hundred different places: The President and his Lady will be at the theatre this evening." It would be during the playing of the comedy *Our American Cousin* that there came in the second act the murder of Abraham Lincoln.

By April 12 Union forces had occupied the city of Mobile, and on the same day Montgomery was peaceably surrendered. In the week following Sherman and Johnston concluded a "military convention" to disband all Confederate forces in North and South Carolina, Georgia, and Florida, which President Johnson and his cabinet later promptly rejected because it suggested that the state governments continued to function pending a ruling of the U.S. Supreme Court. Grant joined Sherman and Johnston in a new conference where the terms of surrender accepted by Lee were imposed. Dick Taylor surrendered to General Canby all "the forces, munitions of war, etc., in the Department of Alabama, Mississippi, and East Louisiana" on May 4.

On May 23 Kirby Smith sent officers from his headquarters at Shreveport to negotiate terms at Baton Rouge with Canby, and the surrender of Confederate forces in western Louisiana and in Texas was conceded May 26. Jefferson Davis had been captured on May 10 in Georgia and conveyed in shackles to Fort Monroe.

By the end of the spring, the War Department called a halt to Delany's recruiting; the 54th was returned to Boston to be mustered out, and Delany himself went to General Saxton's headquarters in Beaufort for assignment to the Freedmen's Bureau. In July Delany was to be on St. Helena Island, South Carolina, where slaves had toiled in isolation for generations on some of the wealthiest plantations of the South, producing long-staple cotton that brought premium prices in New York and Liverpool. Turbaned women, parcels or baskets balanced on their heads, walked roads vaulted on either side by massive live oaks, their branches draped with Spanish moss; while oarsmen chanted gentle, plaintive dirges as they plied the river. People here spoke a rapid-fire patois, retaining some African words, and a number of the men and women bore tribal markings on their faces. The Sea Islands for many miles inland bore the striking suggestion of Africa, but as Delany saw, an Africa completely superintended by white Northerners. Known in its early years as the "Port Royal Experiment," missionaries from Boston and New York had come to set up schools and churches with the help of Freedmen's Aid societies, but by 1865 they had shed some of their philanthropic motives, many of which were compiling, through payment of low wages and discipline reminiscent of slavery, substantial profit. Delany was indignant at much of what he saw; delivering a speech to freedmen at the Brick Church one Sunday in July on how they could best sustain their gains from the war. Addressing between five and six hundred former slaves, a report of Delany's speech was made by Lt. Edward M. Stoeber in a letter to the assistant commissioner of the Freedmen's Bureau, Brevet Major S. M. Taylor, which subsequently became one of the items in the National Archive. The lieutenant reported:

> As introduction Maj. Delany made them acquainted with the fact, that slavery is absolutely abolished, throwing thunders of damnations and malediction on all the former slave owners and people of the South, and almost condemned their souls to hell. He says "It was only a War policy of the Government, to declare the slaves of the South free, knowing that the whole power of the South, laid in

the possession of the slaves. But I want you to understand, that we would not have become free, had we not armed ourselves and fought out our independence." (He repeated this twice.)

The lieutenant concluded his report:

My opinion of the whole affair is that Major Delany is a thorough hater of the white race and excites the colored people unnecessarily. He even tries to injure the magnanimous conduct of the Government towards them, either intentionally or through want of knowledge. He tells them to remember "that they would not have become free, had they not armed themselves and fought for their independence." This is a falsehood and misrepresentation.—Our President Abraham Lincoln declared the colored race free, before there was even an idea of arming colored men. This is decidedly calculated to create bad feeling against the Government. . . . the mention of having two hundred thousand men well in arms:—does he not hint to them what to do?"[51]

EPILOGUE

Though more than twenty years have rolled between us and the Harper's Ferry raid, though since then the armies of the nation have found it necessary to do on a large scale what John Brown attempted to do on a small one, and the great captain who fought his way through slavery has filled with honor the Presidential chair, we yet stand too near the days of slavery, and the life and times of John Brown, to see clearly the true martyr and hero that he was and rightly to estimate the value of the man and his works.
—Frederick Douglass at Harper's Ferry,
West Virginia, May 30, 1881

TO THE LAST DAYS OF HIS LIFE LINCOLN ENTERTAINED ambiguous notions on the future of the people freed in and by the United States as a result of the war of Union/Secession: Would they become citizens on an equal basis with whites or should the issue be resolved with resort to colonization? Only days before his murder, the president met with Benjamin Butler seeking his advice on the matter. For the important role he played in four years of war, Butler had an unlikely appearance. A man of wide girth with large sagging moustache and a great domed forehead and big popping eyes, his drooping eyelids gave him a permanently tired look. Yet he was a man of considerable energy and innate humanity. Lincoln once said of him that he reminded him of "Jim Jett's Brother": "Jim used to say that his brother was the damndest scoundrel that ever lived, but by the infinite mercy of Providence he was also the damndest fool."[1] Whether rogue or fool, it could also plausibly be said that he was the man who actually freed the slaves, or at any rate those freed directly by the war.

Kicked out of the Democratic Party for his growing radicalism, Butler joined the Republicans only to be thrown out for exposing its rot in the Reconstruction Era. Toward the end of the war, too, he had lost his command when Grant sacked him for his muddled leadership at the start of the campaign to capture Fort Fisher.

Calling on the former general for his advice, Lincoln said to him:

> What shall we do with the Negroes after they are free? I can hardly believe that the South and North can live in peace unless we get rid of the Negroes whom we have armed and disciplined and who have fought with us, to the amount, I believe, of some 150,000 men. I believe that it would be better to export them all to some fertile country with a good climate, which they could have to themselves. You have been a staunch friend of the race from the time you first advised me to enlist them at New Orleans. You have had a great deal of experience in moving bodies of men by water— your movement up the James was a magnificent one. Now we shall have no use for our very large navy. What then are our difficulties in sending the blacks away . . . ? I wish you would examine the question and give me your views upon it and go into the figures as you did before in some degree so as to show whether the Negroes can be exported.[2]

Butler replied, "I will go over this matter with all diligence and tell you my conclusions as soon as I can."

After two days, Butler called on the president early in the morning. He reported: "Mr. President, I have gone very carefully over my calculations as to the power of the country to export the Negroes of the South and I assure you that, using all your naval vessels and all the merchant marine fit to cross the seas with safety, it will be impossible for you to transport to the nearest place that can be found fit for them—and that is the Island of San Domingo, half as fast as Negro children will be born here."

After his assassination Lincoln's successor, Andrew Johnson—who had been an inveterate despiser of the slaveholding aristocracy that had ruled the South—seemed intent on ignoring the freedmen and overturning all of the gains made by the war. To the champions of the "Negro's Cause"—Charles Sumner and Thaddeus Stevens in the Senate and House respectively, Wendell Phillips as the conscience and beacon of the "revolution of 1863–65," and Frederick Douglass as tribune of his people—the freedmen would need protection from the political machinations of the old master class. For this they advocated immediate franchise, the granting of land, and the bestowal of education, without which, they argued, the freedmen would inevitably be drawn back toward slavery. This was the minimum policy required and the debt owed by the nation for the crime of slavery.

At first Johnson let it appear that he was solicitous of their goals, letting Sumner think he had his ear, and sending Carl Schurz, a German émigré and social progressive, on an extended tour to report on the condition of the freedmen in the South. But Sumner was to be frustrated when Johnson let it be known that suffrage could wait so as not to antagonize influential Southerners who were flocking to see him, and to whom he was granting pardons; and he at first refused to meet with Schurz upon the completion of his tour, and then to ever read his important report. Schurz's report succinctly summed up the matter, writing, "The people boast that when they get freedmen's affairs in their own hands, to use their own expression, 'the niggers will catch hell.'"[3]

At the turning of the year Johnson refused to hear an appeal for the franchise for black men from a delegation consisting of Frederick Douglass and his son Lewis, George T. Downing, John Jones, William Whipper, and others; reciting his own set piece for three-quarters of an hour on the matter. Knowing that Johnson's lecture to them would be published across the nation in the next day's papers, the delegation decided to draft a reply, choosing the elder Douglass for the task. The reply began: "In consideration of a delicate sense of propriety as well as of your own repeated intimations of indisposition to discuss or listen to a reply to the views and opinions you were pleased to express to us in your elaborate speech today, the undersigned would respectfully take this method of replying thereto. Believing as we do that the views and opinions you expressed are entirely unsound and prejudicial to the highest interests of our race as well as to our country at large, we cannot do other than expose the same and, as far as may be in our power, arrest their dangerous influence."[4]

But Johnson had never been solicitous of abolitionist views. Speaking in the Senate in 1860, as secession began to gather steam a month after his predecessor's election, Johnson had turned to the Southern leaders, saying: "I voted against him [Lincoln], I spoke against him, I spent my money to defeat him, but still I love my country; I love the Constitution; I intend to insist upon its guarantees." And in the previous year as the crisis broke, hastening fitfully toward its denouement, he had spoken against John Brown: "John Brown stands before the country as a murderer. . . . The time has arrived when these things ought to be stopped, when encroachments on the institution of the South ought to cease . . . when you must preserve the Constitution or you must destroy the Union."[5]

Murder had ushered "the Rebellion" into the presidential chair, was the ringing denunciation of Wendell Phillips in 1866. The armies had left the field of battle after Lee's surrender, but the war went on "more secretly, more spasmodically, and yet as truly as before the peace," in the assessment of DuBois years afterward.[6]

With the failure of the Lincoln–Johnson approach to Reconstruction commenced ten years of congressional Reconstruction, beginning with the establishment of a Joint Committee of Fifteen and the organization of five military districts for the rule of the South. Forthcoming deliberative action resulted in the passage of the first Civil Rights Act over Johnson's veto and the granting of the continuance of the Freedmen's Bureau, and in 1868, the impeachment of Andrew Johnson. Johnson was able to avoid the disgrace of removal from office by the margin of only one vote. In the years following Grant's election in 1868, the passage of the Fourteenth and Fifteenth Amendments to the Constitution mandated the end of slavery in law, nationalizing citizenship and freedom, and removing the color bar in voting rights.

With the growing power of cartels in the new era of finance and industrial capital given impetus by the war, Reconstruction's downfall and the withdrawal of black franchise were stipulations of the disputed presidential election of 1876. That election went by the narrowest margin to Republican Rutherford B. Hayes, with the understanding he would withdraw federal troops from Columbia, South Carolina, and from New Orleans. Whereupon, with a growing reign of white terror that included the lynching of thousands of blacks over the next several decades lasting well into the twentieth century, came the system of "crime peonage," backed by stringent laws on vagrancy, guardianship, and labor contracts enacted across the South. This new slavery was doubled by the widespread use of a crop-lien system and the growing indebtedness of the toiling black masses, and finally as the decades passed, the institutionalization of Jim Crow. While the period known as "Reconstruction" was long vilified and ignored in its true significance, it is now seen that on the bedrock of the new legislation, a process of real democratic reform was begun in the South, hampered only by a lack of a thoroughgoing commitment to the education and to the economic development of the freedmen. "History does not furnish an example of emancipation less friendly to the emancipated class than this American example," was the bitter judgment of Frederick Douglass.[7]

At the war's end, after twenty-five years as a writer and editor and antislavery lecturer, Douglass found ready employment on the lecture circuit, carrying on for some years making as much as a hundred dollars an appearance. After his home burned down in 1872, and with no further ties in Rochester, he moved to a small brick dwelling in Washington, D.C., where he took on the editorship of the *National Era*. When the paper closed down, he was briefly president of the Freedman's Bank before its tragic bankruptcy. Then he began to look to Republican Party patronage for employment, taking the post of marshal under the Hayes administration, recorder of deeds under Garfield's brief tenure and that of his successor, Arthur, and finally minister to Haiti under Harrison's aegis. With the abolition struggle of thirty-five years consigned to bygone days and regarded as a "controversial" topic, Douglass continued to fulfill the position as tribune and spokesman for "his people." He had lived to see the completion of his work, seen the destruction of slavery; yet "the country," as he would say, "had not quite survived the effects and influence of its great war of existence." The long and appalling shadow of slavery still fell "broad and large over the face of the whole country." "There is no disguising the fact," he wrote in *Life and Times*, "that the American people are much interested and mystified about the mere matter of color as connected with manhood."[8]

In 1878 Douglass purchased a substantial home with twenty rooms sitting on a hill amid fifteen acres just across the bridge spanning the Anacostia branch of the Potomac River in Washington, D.C.

Numerous cedar trees, as well as oaks and hickories surrounded the house, and after the tree most in profusion Douglass named his estate Cedar Hill. In the downstairs parlor were hung large portraits of Benjamin Lundy and William Lloyd Garrison, Gerrit Smith and Wendell Phillips, and of Abraham Lincoln; and flanking either side of the entrance hallway were Susan B. Anthony and Elizabeth Cady Stanton. Looking back on that momentous earlier period—on the numerous personages he had known, and the countless friendships he had had in the struggle, the various organizations and differing points of view, on the draconic laws that were enacted proscribing blacks, the hunting of the fugitive slave and the refuge he found, the Kansas War and his association with John Brown, Lincoln's election and the secession crisis, the coming of the war and the emancipation and its aftermath—he narrated it all with his own story. Written in his book-lined study, his *Life and Times of Frederick Douglass* was first published in 1881, and then in a completed and revised edition in 1892. By then, one of his biographers, Benjamin Quarles, has written, "His emotionalized memory played tricks on his sense of perspective," and his reminiscences took on the aspect "of faded clover leaves in the family Bible." Douglass himself loved to take up his fiddle and play, but confided to a friend, "I sometimes try my old violin; but after all, the music of the past is sweeter than any my unpracticed and unskilled bow can produce. So I lay my dear, old fiddle aside, and listen to the soft, silent, distant music of other days."[9]

Two decades after he parted with Brown in the stone-quarry, on May 30, 1881, at the celebration of the fourteenth anniversary of Storer College at Harper's Ferry, Douglass was allowed the last word on his "old friend," where he not only defended him, but extolled "him as a hero and martyr to the cause of liberty." He wrote in *Life and Times*: "I confess that as I looked out upon the scene before me and the towering heights around me, and remembered the bloody drama there enacted—as I saw the log house in the distance where John Brown collected his men and saw the little engine-house where the brave old Puritan fortified himself against a dozen of Virginia Militia, and the place where he was captured by the United States troops under Col. Robert E. Lee—I was a little shocked at my own boldness in attempting to deliver in such presence an address of the character advertised in advance of my coming."[10]

Sitting on the platform with him was Andrew Hunter, the district attorney who had prosecuted Brown and secured his execution.

Douglass asked, at the conclusion of his memorable address, of those other notables who stood over him after United States Marines had been ordered by Lee to batter Brown out of his "fort"—"Did John Brown fail?"

> Ask Henry A. Wise in whose house less than two years after, a school for the emancipated slaves was taught. Did John Brown fail? Ask James M. Mason, the author of the inhuman fugitive slave bill, who was cooped up in Fort Warren, as a traitor less than two years from the time that he stood over the prostrate body of John Brown. Did John Brown fail? Ask Clement C. Vallandigham, one other of the inquisitorial party; for he too went down in the tremendous whirlpool created by the powerful hand of this bold invader. If John Brown did not end the war that ended slavery, he did at least begin the war that ended slavery. If we look over the dates, places and men, for which this honor is claimed, we shall find that not Carolina, but Virginia—not Fort Sumter, but Harper's Ferry and the arsenal—not Col. Anderson, but John Brown, began the war that ended American slavery and made this a free Republic. Until this blow was struck, the prospect for freedom was dim, shadowy and uncertain. The irrepressible conflict was one of words, votes and compromises. When John Brown stretched forth his arm the sky was cleared. The time for compromises was gone—the armed hosts of freedom stood face to face over the chasm of a broken Union—and the clash of arms was at hand. The South staked all upon getting possession of the Federal Government, and failing to do that, drew the sword of rebellion and thus made her own, and not Brown's the lost cause of the century.[11]

After this speech Douglass returned to his home with a memento— a single red brick from the enginehouse which had been dismantled and used for some years by a farmer for a shed, and was then purchased by the college and reassembled and was standing on the grounds of the school a half mile from its original site. This clod of earthwork, its color reminiscent of blood, found an honored place amid Douglass's two thousand-volume study, along with a table and a desk from the effects of Charles Sumner, who had died in 1874, and Douglass would show this item to special visitors. Might he ever have wondered what would have been the outcome if he had joined John

Brown at Harper's Ferry? What if they had come up together and marched across Virginia freeing slaves to reach the Blue Ridge with a large cache of arms and an army of fifteen hundred men; would the standing of blacks in the United States have been less or more inclusive than it then was? Was it, perhaps, that John Brown's attempt to realize black freedom had been the better opening than was Lincoln's emancipation, the better opportunity to achieve an overt aspect of nationality within the American polity for Africa, without which resolution of the "Negro problem" would be elusive?

It had been a matter of "anxious solicitude" for Douglass as to how he could best honor his friend; he simply paid tribute to him with these words: "His zeal in the cause of my race was far greater than mine—it was the burning sun to my taper light—mine was bounded by time, his stretched away to the boundless shores of eternity. I could live for the slave, but he could die for him."[12]

NOTES

PROLOGUE

1. Stampp, *The Peculiar Institution*; DuBois, *Black Reconstruction*.
2. Speech of Robert Toombs at the first Confederate Congress in 1861, in Freehling and Simpson, ed., *Secession Debated*.
3. "Mysterious spiritual telegraph" is a phrase quoted in W. E. B. DuBois's *Black Reconstruction*, 63; also used by James Redpath in *The Roving Editor*.

CHAPTER ONE: LIFE AND TIMES

1. Dillon, *The Abolitionists*, 93–97; Wendell Phillips, "On the Murder of Lovejoy," speech at Faneuil Hall, December 8, 1837.
2. DuBois, *John Brown*, chap. 1, "Africa and America," 15–20.
3. Ruchames, ed., *John Brown*, 86–87.
4. *Frederick Douglass' Paper*, Feb. 25, 1852, "Letter to Kossuth, Concerning Freedom and Slavery in the United States," in Foner, ed., *Life and Writings of Frederick Douglass*, vol. 2, 170–172.
5. William Addison Phillips, "Three Interviews with Old John Brown," in Ruchames, ed., *John Brown*, 216–226.
6. Sanborn, *Life and Letters of John Brown*, 97.
7. Higginson, "A Visit to John Brown's Household in 1859," in *Contemporaries*, 225.
8. DuBois, *John Brown*, 110.
9. Letter dated Jan. 23, 1852; in Ruchames, ed., *John Brown*, 87.
10. John Brown Jr. in 1885, in ibid., 189–190.
11. Buckmaster, *Let My People Go*, 191.
12. Speech made by Loguen, Oct. 4, 1850, in Aptheker, ed., *Documentary History of the Negro People in the United States*, vol. 1, 306–308.
13. Ruchames, ed., *John Brown*, 80, 83.
14. Sherwin, *Prophet of Liberty*, 325–326.
15. *Life and Times of Frederick Douglass*, 275.
16. Martin, *Mind of Frederick Douglass*, 46.
17. Letter from Delany dated March 23, 1853, in Foner, *Life and Writings of Frederick Douglass*, vol. 5, 274–275.

18. Speech of H. Ford Douglass before the Emigration Convention in Cleveland, Aug. 27, 1854, in Aptheker, ed., *Documentary History of the Negro People in the United States*, vol. 1, 366–368.

CHAPTER TWO: BLEEDING KANSAS

1. Sanborn, *The Life and Letters of John Brown*, 188–190.
2. Ibid., 110–111.
3. Letter of John Brown Jr. dated May 20, 1854, quoted in Abels, *Man on Fire*, 40–41.
4. Brown to his wife, June 28, 1855, Sanborn, *The Life and Letters of John Brown*, 193–194.
5. See Brown, letter to the Editor of *Summit Beacon*, Dec. 20, 1855, in Ruchames, ed., *John Brown*, 97–100.
6. See Phillips, *The Conquest of Kansas by Missouri and Her Allies*; Monaghan, *Civil War on the Western Border*, 52–59.
7. John Brown Jr. letter quoted in Abels, *Man on Fire*, 56–57.
8. John Brown's views on confronting the slave power in battle are discussed in Richard Hinton's *John Brown and His Men*.
9. Atchison's speech quoted in Monaghan, *Civil War on the Western Border*, 58.
10. Reminiscences of Salmon Brown, Ruchames, ed., *John Brown*, 197–205.
11. Franklin Sanborn quoted in Abels, *Man on Fire*, 80.
12. See Phillips, *The Conquest of Kansas by Missouri and Her Allies*.
13. Sanborn, *The Life and Letters of John Brown*, 236–241; also see Redpath, *The Public Life of Captain John Brown*.
14. R.C. Elliot quoted in Hinton, *John Brown and His Men*.
15. Quoted in Abels, *Man on Fire*, 92.
16. Monaghan, *Civil War on the Western Border*, 68.
17. Hinton, *John Brown and His Men*, 201–204.
18. Monaghan, *Civil War on the Western Border*, 73.
19. "The Republican Party—Our Position," *Frederick Douglass' Paper*, Dec. 1855; "What Is My Duty as an Anti-Slavery Voter?" *Frederick Douglass' Paper*, April 1856, in Foner, *Life and Writings of Frederick Douglass*, vol. 2, 379–383, 390–391.
20. "The Danger of the Republican Movement," Frederick Douglass before the Radical Abolition party nominating convention, in Foner, *Life and Writings of Frederick Douglass*, vol. 5, 385–390.
21. Letter to Gerrit Smith, May 23, 1856, in Foner, *Life and Writings of Frederick Douglass*, vol. 2, 395–396.
22. *Frederick Douglass' Paper*, Aug. 1856, Frémont and Dayton, in Foner, *Life and Writings of Frederick Douglass*, vol. 2, 396–401.
23. Letter to Gerrit Smith, Sept. 6, 1856, in Foner, *Life and Writings of Frederick Douglass*, vol. 2, 402–403.
24. Ibid., 405.
25. Monaghan, *Civil War on the Western Border*, 78–80.
26. Abels, *Man on Fire*, 104.

27. Abels, *Man on Fire*, 104.

28. Missouri press quoted in Sanborn, *The Life and Letters of John Brown*, 321.

29. Monaghan, *Civil War on the Western Border*, 82–84.

30. See *Transactions of the Kansas State Historical Society*, Governor Geary, 739.

31. Speech in Redpath, *The Public Life of Captain John Brown*, 163–164.

32. Ibid., 164–165.

33. Sanborn, *The Life and Letters of John Brown*, 332–333.

34. Monaghan, *Civil War on the Western Border*, 90–91.

35. Quoted in Abels, *Man on Fire*, 110; Cooke's report dated Oct. 7, 1856.

36. Sanborn, *The Life and Letters of John Brown*, 333.

CHAPTER THREE: RAISING AN ARMY OF ONE HUNDRED VOLUN-TEERS

1. Nelson, ed., *Documents of Upheaval: Selections from William Lloyd Garrison's The Liberator, 1831–1865*, Annual meeting of the Massachusetts Anti-Slavery Society, Feb. 6, 1857, 219–235.

2. Aptheker, *American Negro Slave Revolts*, 340–358.

3. Reminiscences of Mrs. Mary E. Stearns, in Hinton, *John Brown and His Men*, 719–727; Emerson, *Complete Works*, 1202–1206, quoted in Hinton, *John Brown and His Men*, 720; *The Writings of Henry David Thoreau*, 197–236; Sanborn, *Recollections of Seventy Years*.

4. Abels, *Man on Fire*, 130.

5. Sanborn, *Recollections of Seventy Years*.

6. Sherwin, *Prophet of Liberty*, 163–169; Ables, *Man on Fire*, 119.

7. Frederick Douglass's speech before the American Anti-Slavery Society, May 11, 1857, "The Dred Scott Decision," in Foner, *Life and Writings of Frederick Douglass*, vol. 2, 407–424.

8. Bearden and Butler, *The Life and Times of Mary Shadd Cary*, Part 3, *The Provincial Years*, 128–208.

9. Reminiscences of Dr. Wayland, in Sanborn, *The Life and Letters of John Brown*, 381; see also Eli Thayer, *A History of the Kansas Crusade: Its Friends and Its Foes* (New York: Harper and Brothers, 1889).

10. Ables, *Man on Fire*, 144–145.

11. Abels, *Man on Fire*, 136–137; Sanborn, *Recollections of Seventy Years*; Thoreau, *A Plea for Captain John Brown*.

12. Sanborn, *Recollections of Seventy Years*.

13. Ibid.

14. *New York Evening Post*, Oct. 23, 1909, Brown in hiding and in jail; Katherine Mayo's interview with Mrs. Judge Russell, in Ruchames, ed., *John Brown*, 242–248; Hinton, *John Brown and His Men*, reminiscences of Mrs. Mary E. Stearns, 719–727.

15. Sanborn, *The Life and Letters of John Brown*, letter to son John dated April 15, 1857.

16. Abels, *Man on Fire*, 146–147.

17. Brown's meeting with Thoreau narrated in Sanborn, *Recollections of Seventy Years*; Thoreau, *A Plea for Captain John Brown*.

18. Ruchames, ed., *John Brown*, 111; letter to Dear Wife, March 31, 1857.

19. Sanborn, *The Life and Letters of John Brown*.

20. Abels, *Man on Fire*, 152–157.

21. Sanborn, *The Life and Letters of John Brown*, 393, letter to Augustus Wattles.

22. Ibid.

23. Abels, *Man on Fire*, letter of James H. Holmes, 149.

24. *New York Herald*, Oct. 29, 1859, letter from Hugh Forbes to Dr. Samuel G. Howe, May 14, 1858.

25. Monaghan, *Civil War on the Western Border*, chapter 8, "Buchanan Tries His Hand."

26. Sanborn, *The Life and Letters of John Brown*, correspondence of Lane and Brown, 401–402; Abels, *Man on Fire*, 152–157.

27. Ibid.

28. Ibid.

29. Ibid.

30. Monaghan, *Civil War on the Western Border*, chapter 8.

31. Sanborn, *The Life and Letters of John Brown*.

32. Ibid.

33. Hinton, *John Brown and His Men*, confession of John E. Cook, 700–701.

34. Ruchames, ed., *John Brown*, reminiscences of Salmon Brown.

35. Hinton, *John Brown and His Men*, 156–157.

36. Cook in Hinton, *John Brown and His Men*.

37. Abels, *Man on Fire*, 162.

38. McPherson, *Battle Cry of Freedom*, 168.

39. Richman, *John Brown Among the Quakers*.

40. Hinton, *John Brown and His Men*, 156–157.

CHAPTER FOUR: CARRYING THE WAR INTO AFRICA

Note to epigraph: Letter to My Dear Wife and Children, Everyone, dated January 30, 1858, in Ruchames, ed., *John Brown*, 117–118.

1. *Life and Times of Frederick Douglass*, 315.

2. Interview with John Brown, Richard Hinton, *John Brown and His Men*, 672–675.

3. *Life and Times of Frederick Douglass*, 314–315.

4. Frederick Douglass's speech at Storer College at Harper's Ferry on May 30, 1881, in Louis Ruchames, ed., *John Brown*, 294–295; *Life and Times of Frederick Douglass*, 274–275.

5. *Life and Times of Frederick Douglass*, 316.

6. Sanborn, *The Life and Letters of John Brown*, 434–435.

7. Warch and Fanton, ed., *John Brown*, 37.

8. Ibid., 36.

9. Sanborn, *The Life and Letters of John Brown*, 436.

10. Ibid., 438–440.

11. Ibid.

12. Ibid., 440–441.

13. Ibid., 444–445.

14. Ibid., 442–443.

15. Abels, *Man on Fire*, 78.

16. Sanborn, *The Life and Letters of John Brown*, 450–451.

17. Earl Ofari, *"Let Your Motto Be Resistance": The Life and Thought of Henry Highland Garnet*, 41, 105–108.

18. Letter to My Dear Wife and Children, Everyone, dated January 30, 1858, in Ruchames, ed., *John Brown*, 117–118.

19. Earl Conrad, *Harriet Tubman*, 113.

20. Thomas Wentworth Higginson, *Cheerful Yesterdays*.

21. Sanborn, *The Life and Letters of John Brown*, 452.

22. James Hamilton, *John Brown in Canada*, reminiscences of J. M. Jones, 14–15.

23. Frank Rollin, *Life and Public Services of Martin R. Delany*, 85–90.

24. Richman, *John Brown Among the Quakers*.

25. Dorothy Sterling, *The Making of an Afro-American, Martin Robison Delany*, 226–227.

26. Reports of Senate Committees, 36th Congress, 1st Session, No. 278; Testimony of Richard Realf.

27. Osborne Anderson, *A Voice from Harper's Ferry*, chapter II.

28. James Hamilton, *John Brown in Canada*.

29. Anderson, *A Voice from Harper's Ferry*, chapter III.

30. Ibid., chapter II.

31. Letter dated May 7, 1858, Warch and Fanton, ed., *John Brown*, 49.

32. *The Life and Letters of John Brown*, 448–461; issue treated in W. E. B. DuBois, *John Brown*, 226–270.

33. Abels, *Man on Fire*, 197.

34. Richman, *John Brown Among the Quakers*, 40–41.

35. Hinton, *John Brown and His Men*.

36. Abels, *Man on Fire*, 192.

37. Sanborn, *The Life and Letters of John Brown*, 463–464.

38. Reports of Senate Committees, 36th Congress, 1st Session, No. 278; Testimony of Richard Realf.

CHAPTER FIVE: AN AMERICAN SPARTACUS

1. Monaghan, *Civil War on the Western Border*, 99.

2. Ibid., 103–104.

3. Sanborn, *The Life and Letters of John Brown*, 470–477.

4. Abels, *Man on Fire*, 208.

5. Redpath, *The Public Life of Captain John Brown*, 199–206.

6. Abels, *Man on Fire*, 203.

7. Secretary Cass quoted in Abels, *Man on Fire*; Brown's letter in Sanborn, *The Life and Letters of John Brown*.

8. Abels, *Man on Fire*, 211.

9. Hinton, *John Brown and His Men*, 672–675.

10. Reports of Senate Committees, 36th Congress, 1st Session, No. 278; Testimony of Richard Realf.

11. Tragle, ed., *The Southampton Slave Revolt of 1831*; Section IV, Thomas Hamilton, editor of the *Anglo-African Magazine*, 1859, 325–326.

12. Tragle, ed., *The Southampton Slave Revolt of 1831*; extracts from the diary and correspondence of Governor John Floyd of Virginia, letter 14, 275–276.

13. Monaghan, *Civil War on the Western Border*, 112.

14. Abels, *Man on Fire*, 222.

15. Ibid.

16. Monaghan, *Civil War on the Western Border*, 111.

17. Ruchames, ed., *John Brown*; William Addison Phillips, "Three Interviews with Old John Brown," *Atlantic Monthly* (December 1879): 216–226.

18. Abels, *Man on Fire*, 228.

19. Ibid.

20. Sanborn, *The Life and Letters of John Brown*, 491.

21. Quarles, *Allies for Freedom*, 76.

22. Abels, *Man on Fire*, 234–235.

CHAPTER SIX: A SETTING POSSESSED OF IMPOSING GRANDEUR

1. *Life and Times of Frederick Douglass*, 317.

2. "A Plan of Anti-slavery Action," *Frederick Douglass' Paper*, July 8, 1859, in Foner, ed., *The Life and Writings of Frederick Douglass*, vol. 5, 453–454.

3. Conrad, *Harriet Tubman*, 122; Sanborn, *The Life and Letters of John Brown*, letter dated June 1 to Sanborn.

4. Sanborn, *The Life and Letters of John Brown*, diary of A. Bronson Alcott, 504–505.

5. Villard, *John Brown, 1800–1850*, 396.

6. Sanborn, *The Life and Letters of John Brown*.

7. Ables, *Man on Fire*, 239–240.

8. Sanborn, *The Life and Letters of John Brown*.

9. *Letters and Journals of Thomas Wentworth Higginson, 1846–1906*

10. Sherwin, *Prophet of Liberty*, 239–240.

11. This version of the story cited by an author is apocryphal; these words were spoken to James Redpath in a similar spirit by Brown in another setting.

12. Ables, *Man on Fire*, 240.

13. Abels, *Man on Fire*, 255.

14. Letter of John Brown to Wife and Children All, dated July 27, 1859.

15. Ables, *Man on Fire*, 256.

16. Hinton, *John Brown and His Men*, 259.

17. John Brown Jr. to Kagi, Aug. 11, 1859, in Sanborn. *The Life and Letters of John Brown*.

18. Ibid.

19. Ibid.

20. Ibid.

21. Ibid.

22. Ibid.

23. Villard, *John Brown*; Salmon Brown interviewed by Katherine Mayo.

24. Reports of Senate Committees, 36th Congress, 1st Session, No. 278; Testimony of Richard Realf.

25. *Life and Times of Frederick Douglass*, 319.

26. In Douglass's speech at Storer College in 1881; in Ruchames, ed., *John Brown*, 294.

27. Abels, *Man on Fire*, 242.

28. Abels, *Man on Fire*, Sanborn cited on 248.

29. *Life and Times of Frederick Douglass*, 275.

30. See "The Prospect for the Future," in *Douglass' Monthly*, Aug. 1860— "The Future of the Abolition Cause," April 1861—"Nemesis," June 1861— "Danger to the Abolition Cause," July 1861—"Notes on the War," etc., in Foner, ed., *The Life and Writings of Frederick Douglass*, vol. 3.

31. *The Life and Times of Frederick Douglass*, 320.

CHAPTER SEVEN: THE BATTLE OF HARPER'S FERRY

1. Katz, *Resistance at Christiana*, The Freedman's Story quoted 27.

2. Katz, *Resistance at Christiana*, John Brown Jr.'s letter cited 281.

3. Hinton, *John Brown and His Men*.

4. Letters of April 11 and 23, quoted in Quarles, *Allies for Freedom*, 86–87.

5. Hinton, *John Brown and His Men*.

6. Abels, *Man on Fire*, 266; letter dated Sept. 8, 1859.

7. Anderson, *A Voice from Harper's Ferry*, chapter V; letter dated Sept. 14, 1859.

8. Conrad, *Harriet Tubman*, 124–125.

9. Hinton, *John Brown and His Men*, 570.

10. Ibid.

11. Quarles, *Allies for Freedom*, 80; Harris to Kagi dated Aug. 22, 1859.

12. Hinton, *John Brown and His Men*.

13. Cohen, *John Brown: A Pictorial Heritage*, 64; letter no. 8 dated Sept. 27, 1859, facsimile of a sheet from the *Baltimore Sun*.

14. Osborne, *A Voice from Harper's Ferry*, chapter V.

15. Ibid., chapter VII.

16. Ibid., chapter III.

17. Abels, *Man on Fire*, 257.

18. Villard, *John Brown*, 422–423.

19. Ruchames, ed., *John Brown*, 281; Douglass's speech at Storer College.

20. Anderson, *A Voice from Harper's Ferry*, chapter VIII.

21. Ibid.; also cited with slightly different wording in John Cook's confession.

22. Reports of Senate Committees, 36th Congress, 1st Session, No. 278; Testimony of Daniel Wheeler.

23. Anderson, *A Voice from Harper's Ferry*, chapter X.

24. Ibid., chapter IX—order number 11.

25. Ibid., chapter X.

26. Abels, *Man on Fire*, 283.

27. Ibid.

28. Ibid.

29. Keller, *Thunder at Harper's Ferry*; an hour-by-hour account of John Brown's raid.

30. Anderson, *A Voice from Harper's Ferry*, chapter XI.

31. Ibid.

32. Robinson, *The Kansas Conflict*.

33. Anderson, *A Voice from Harper's Ferry*, chapter XI.

34. Ruchames, ed., *John Brown*, 286; Douglass's speech at Storer College.

35. Abels, *Man on Fire*, 284.

36. Anderson, *A Voice from Harper's Ferry*, chapter XII.

37. Keller, *Thunder at Harper's Ferry*, 66.

38. Anderson, *A Voice from Harper's Ferry*, chapter XI.

39. Ibid., chapter XII.

40. Ibid.

41. This story related by Anderson is told *Life and Times of Frederick Douglass*, 321.

42. John E. P. Daingerfield, "John Brown at Harper's Ferry," *Century* magazine, June 1885; quoted in DuBois, *John Brown*, 326.

43. Abels, *Man on Fire*, 290.

44. *The Public Life of Captain John Brown*, 320–321.

45. Leary's death as reported by the *Baltimore Sun* reporters given in Keller, *Thunder at Harper's Ferry*.

46. Ruchames, ed., *John Brown*, 126; interview in the *New York Herald*, Oct. 21, 1859.

47. Ibid.

48. *Life and Times of Frederick Douglass*, 310–312.

49. Quarles, *Allies for Freedom*, 105; letter published in the *Evening Star*, Nov. 9, 1859.

50. Sanborn, *The Life and Letters of John Brown*, 589–591; letter of Brown to Rev. H. L. Vaill, Nov. 15, 1859.

CHAPTER EIGHT: YEAR OF METEORS

1. Ross, *Memoirs of a Reformer*, 91–92.

2. John E. P. Daingerfield, "John Brown at Harper's Ferry," *Century* magazine, June 1885.

3. Brown's interview with Mason, Vallandigham, Governor Wise and others reported in the *New York Herald*, in Sanborn, *The Life and Letters of John Brown*, 562–571.

4. *Life and Times of Frederick Douglass*, 308.

5. From Governor Wise's stump speeches in front of the Wager House and inside the Armory grounds on the evening of October 18, quoted in John S. Wise, *The End of an Era* (Boston: Houghton Mifflin, 1901), 132.

6. Quoted from Wise's speech of December 26, 1859, "The Question of the Day."

7. Higginson quoted in Abels, *Man on Fire*, 340–341.

8. *Life and Times of Frederick Douglass*, 321.

9. Wise's entire speech is found in Richard Scheidenhelm, *The Response to John Brown* (Belmont, Calif.: Wadsworth, 1972), 154–166.

10. Quoted in Abels, *Man on Fire*, 324.

11. Ibid., 324–325.

12. Ibid., 320.

13. Ibid., 326; newspaper report quoted in Redpath, *Public Life of Captain John Brown*, 337.

14. Thoreau quoted in Abels, *Man on Fire*, 388.

15. Speech in Scheidenhelm, *The Response to John Brown*, 50–51.

16. This speech is among the most quoted in American letters; it appears in Ruchames, ed., *John Brown*, 133–135.

17. Quoted in Abels, *Man on Fire*, 353.

18. Journalist quoted in Quarles, *Allies for Freedom*, 107; letter to My Dear Wife, Nov. 21, 1859.

19. Keller, *Thunder at Harper's Ferry*.

20. Speech on Dec. 5, 1859.

21. Anderson, *A Voice from Harper's Ferry*, chapter XIX.

22. To Rev. James W. McFarland of Wooster, Ohio, Nov. 23, 1859.

23. Ross, *Memoirs of a Reformer*.

24. Katherine Mayo, "Brown in Hiding and in Jail": Interview with Mrs. Russell, Oct. 23, 1909, in Ruchames, ed., *John Brown*, 242–248.

25. See Prison Letters in Ruchames, ed., *John Brown*, Oct. 31, Nov. 8, Nov. 16, Nov. 21, Nov. 27; and two letters to Higginson on the subject of his wife coming to Charlestown, dated Nov. 4 and Nov. 22.

26. This scene narrated in Abels, *Man on Fire*, 360.

27. Letter dated Nov. 29, 1859, in Sanborn, *Life and Letters of John Brown*, 610–611.

28. DuBois, *John Brown*, 365.

29. Quarles, *Allies for Freedom*, 127.

30. For these Martyr Day observances see Quarles, *Black Abolitionists*, 241–244.

31. Speech of William Lloyd Garrison in Nelson, ed., *Documents of Upheaval*, 263–267.

32. See Brown, *The Heroes of Insurrection*.

33. Quarles, *Allies for Freedom*, 130.

34. Wendell Phillips, "Burial of John Brown," in Ruchames, ed., *John Brown*, 266–269.

35. Henry David Thoreau, "The Last Days of John Brown," in Bode, ed., *The Portable Thoreau*, 676–682.

CHAPTER NINE: 1860

1. *Life and Times of Frederick Douglass*, 322.

2. Ibid.

3. Letter to Helen Boucaster, Dec. 7, 1859, in Foner, ed., *Life and Writings of Frederick Douglass*, vol. 5, 459.

4. Speech in Glasgow, Scotland, March 26, 1860: "The Constitution of the United States: Is it Pro-Slavery or Anti-Slavery?" in Foner, ed., *Life and Writings of Frederick Douglass*, vol. 2, 467–480.

5. "The Mason Report: Harper's Ferry Invasion," June 15, 1860, in Scheidenhelm, ed., *The Response to John Brown*, 9–36.

6. Abels, *Man on Fire*, 374–376.

7. Ibid.

8. "The Mason Report: Harper's Ferry Invasion."

9. McPherson, *Battle Cry of Freedom*, 214; Catton, *The Coming Fury*, 18–21.

10. Ibid.

11. Sherwin, *Prophet of Liberty*, 405–406.

12. Channing, *Crisis of Fear*, chap. 1, "The Fear."

13. Sanborn, *The Life and Letters of John Brown*; see Higginson's correspondence with Spooner.

14. Higginson, *Contemporaries*.

15. Karl Marx and Frederick Engels, "The Civil War in the U.S.," in *Correspondence*, 221–222.

16. Ibid.

17. Aptheker, *American Negro Slave Revolts*, 237; see also "A Secret Organization for Freedom" in Aptheker, *Documentary History*, vol. 1, 378–380.

18. Wendell Phillips, "Civil Rights and Freedom," in Filler, ed., *John Brown and Harper's Ferry*, 95–113.

19. Ibid., cited in note on xiii.

20. Speech of William L. Yancey, quoted by McPherson, *Battle Cry of Freedom*, 215.

21. *Life and Times of Frederick Douglass*, 324.

22. Conrad, *Harriet Tubman*, 143, quoting *Boston Commonwealth*, July 17, 1863.

23. Sherwin, *Prophet of Liberty*, contains a good account of Phillips and the women's rights movement; for Hinton see his book and Sanborn.

24. "Attitude of Parties in the United States," *New York Times*, April 10, 1860, from the *London Times*, March 22.

25. See Fish, *The American Civil War*; this paragraph is a synopsis of his chapter 1.

26. Sherwin, *Prophet of Liberty*, chap. 34, "Fearless Lips," chap. 35, "Election of 1860."

27. *Douglass' Monthly*, June, 1860, in Foner, ed., *Life and Writings of Frederick Douglass*, vol. 2, 480–483.

28. Letter to James Redpath, June 29, 1860, in Foner, ed., *Life and Writings of Frederick Douglass*, vol. 2, 487.

29. Letter to William Still, July 2, 1860, in Foner, ed., *Life and Writings of Frederick Douglass*, vol. 2, 488.

30 Letter to Redpath.

31. Sherwin, *Prophet of Liberty*, 356; Charles Sumner, *His Complete Works*.

32. "The Republican Party," *Douglass' Monthly*, August, 1860. Foner, ed., *Life and Writings of Frederick Douglass*, vol. 2, 490–493.

33. Letter to Elizabeth Cady Stanton, Aug. 25, 1860, in Foner, ed., *Life and Writings of Frederick Douglass*, vol. 2, 497–498.

34. Channing, *Crisis of Fear*, chap. 4, "The Radical Mind on the Eve of the Conventions."

35. Speech cited in McPherson, *Battle Cry of Freedom*, 230.

36. McPherson, *Battle Cry of Freedom*.

37. McPherson, *Battle Cry of Freedom*, 250–251; Fish, *The American Civil War*, 67.

38. *Life and Times of Frederick Douglass*, 331.

39. In Sanborn, *The Life and Letters of John Brown*.

40. Cited in Foner, ed., *The Life and Writings of Frederick Douglass*, 100, from *Boston Evening Transcript*, Dec. 3, 1860.

41. Speech on John Brown, in Foner, ed., *The Life and Writings of Frederick Douglass*, vol. 2, 533–538.

42. Ibid.

43. "Plea for Free Speech in Boston," Dec. 10, 1860, in Foner. ed., *The Life and Writings of Frederick Douglass*, vol. 2, 538–540.

44. Cited in Sherwin, *Prophet of Liberty*, 418–420.

45. Conrad, *Harriet Tubman*, 150–151.

46. Bradford, *Scenes in the Life of Harriet Tubman*.

CHAPTER TEN: TERRIBLE SWIFT SWORD

1. Sherwin, *Prophet of Liberty*, 422.

2. Catton, *The Coming Fury*, 185–186.

3. Ibid., 215.

4. Ibid.

5. Quoted in DuBois, *Black Reconstruction*, 72.

6. "The Argument for Disunion," in Filler, ed., *Wendell Phillips on Civil Rights and Freedom*, 116.

7. Sherwin, *Prophet of Liberty*, 435–436.

8. Ofari, *"Let Your Motto Be Resistance,"* 100.

9. Catton, *The Coming Fury*, 221.

10. Lincoln to Seward, March 4, 1861, *Abraham Lincoln; Complete Works, Comprising His Speeches, Letters*, ed. John G. Nicolay and John Hay, vol. 2 (New York: Century Co., 1894), 8.

11. Fish, *The American Civil War*, 118.

12. Catton, *The Coming Fury*, 285–286.

13. Ibid., 290–292.

14. Ibid., 289.

15. Ibid., 291.

16. Ibid., 292.

17. Ibid., 295.

18. Anders, *Fighting Confederates*, 51–52.

19. Sherwin, *Prophet of Liberty*, 444.

20. Catton, *The Coming Fury*, 334.
21. Ibid., 332.
22. Pollard, *Southern History of the War*, 70.
23. McPherson, *Battle Cry of Freedom*, 286.
24. Catton, *The Coming Fury*, 357.
25. Ibid., 375.
26. Monaghan, *Civil War on the Western Border*, 127–131.
27. Aptheker, *Nat Turner's Slave Rebellion*, including the full text of Nat Turner's 1831 "Confession," 137.

CHAPTER ELEVEN: HIS TRUTH IS MARCHING ON

1. Monaghan, *Civil War on the Western Border*, 135.
2. McPherson, *Battle Cry of Freedom*, 333.
3. Ibid., 334.
4. Pollard, *Southern History of the War*, 88.
5. McPherson, *Battle Cry of Freedom*, 336.
6. Catton, *The Coming Fury*, 434.
7. Catton, *The Coming Fury*, 452.
8. Pollard, *Southern History of the War*, 118.
9. Catton, *The Coming Fury*, 463.
10. Ibid., 467.
11. McPherson, *Battle Cry of Freedom*, 347.
12. Whitman, "The Battle of Bull Run, July, 1861," *Specimen Days*, in *The Complete Poems and Prose of Walt Whitman*.
13. McPherson, *Battle Cry of Freedom*, 348.
14. Sherwin, *Prophet of Liberty*, 445.
15. "A Change of Attitude," *Douglass' Monthly*, June 1861, in Foner, ed., *Life and Writings of Frederick Douglass*, 137.
16. "The War for the Union," in Filler, ed., *Wendell Phillips on Civil Rights and Freedom*, 137.
17. "Progress of the War," *Douglass' Monthly*, Sept. 1861; in Foner, *Life and Writings of Frederick Douglass*, vol. 3, 145–149.
18. Donovan, *Mr. Lincoln's Proclamation*, 94–96.
19. McPherson, *Battle Cry of Freedom*, 355.
20. McPherson, *Battle Cry of Freedom*, 312.
21. To Rev. Samuel J. May, Aug. 30, 1861, in Foner, ed., *The Life and Writings of Frederick Douglass*, vol. 3, 158–159.
22. Donovan, *Mr. Lincoln's Proclamation*, 97–98.
23. Monaghan, *Civil War on the Western Border*, 185.
24. McPherson, *Battle Cry of Freedom*, 352–353.
25. Ibid., 357.
26. "General Fremont's Proclamation," *Douglass' Monthly*, Oct. 1861, in Foner, *Life and Writings of Frederick Douglass*, vol. 3, 159–162.
27. Monaghan, *Civil War on the Western Border*, 197.
28. Ibid., 199.
29. Monaghan, *Civil War on the Western Border*, 206.

30. Ibid., 207.

31. *Douglass' Monthly*, Dec. 1861, handbill is in "The Would-Be Democrats of Syracuse," in Foner, *Life and Writings of Frederick Douglass*, vol. 3, 180.

32. Foner, *Frederick Douglass*, 19.

33. *Douglass' Monthly*, Dec. 1861, "The Would-Be Democrats of Syracuse," in Foner, *Life and Writings of Frederick Douglass*, vol. 3, 182.

34. Foner, *Frederick Douglass*, 19–20.

35. To Hon. Gerrit Smith, Aug. 12, 1861, in Foner, *Life and Writings of Frederick Douglass*, vol. 3, 142.

36. "The War for the Union," in Filler, ed., *Wendell Phillips on Civil Rights and Freedom*, 145.

37. Phillips, "The War for the Union," 151–152.

CHAPTER TWELVE: BUFFETED BY SLEET AND BY STORM

Note to epigraph: Quoted in McPherson, *Battle Cry of Freedom*, 361.

1. McPherson, *Battle Cry of Freedom*, 361; McClellan to Ellen Marcy McClellan, Aug. 2, 1861.

2. McPherson, *Battle Cry of Freedom*, 364.

3. Frederick Douglass to Hon. Gerrit Smith, Dec. 22, 1861, in Foner, ed., *Life and Writings of Frederick Douglass*, vol. 3, 184; for Smith's statement see note 44.

4. Ibid.

5. "The Slave Power at Washington," *Douglass' Monthly*, Jan. 1862, in Foner, ed., *Life and Writings of Frederick Douglass*, vol. 3, 186.

6. "The Reasons for Our Troubles," *Douglass' Monthly*, Feb. 1862, in Foner, ed., *Life and Writings of Frederick Douglass*, vol. 3, 207–208.

7. McPherson, *Battle Cry of Freedom*, 423.

8. "The Future of the Negro People of the Slave States," *Douglass' Monthly*, March 1862, in Foner, ed., *Life and Writings of Frederick Douglass*, vol. 3, 224.

9. Sherwin, *Prophet of Liberty*, 458.

10. DuBois, *Black Reconstruction*, 84–85.

11. Sherwin, *Prophet of Liberty*, 458.

12. Letter dated March 27, 1862, ibid., 460.

13. Foner, ed., *The Life and Writings of Frederick Douglass*, vol. 3, 21–22.

14. Frederick Douglass to Hon. Charles Sumner, April 8, 1862, in Foner, ed., *Life and Writings of Frederick Douglass*, vol. 3, 233.

15. Scheidenhelm, ed., *The Response to John Brown*; Thoreau, *A Plea for Captain John Brown*, 55.

16. Bode, ed., *The Portable Thoreau*, 682.

17. Louisa Alcott's poem in Edward Waldo Emerson, *Henry Thoreau as Remembered by a Young Friend*, 63, note 34.

18. McPherson, *Battle Cry of Freedom*, 426; Lincoln to McClellan, April 6, 1862.

19. McPherson, *Battle Cry of Freedom*, 426; McClellan to Ellen Marcy McClellan, April 8, 1862.

20. Brown, *The Negro in the American Rebellion*, 87.

21. Ibid., 72.

22. McPherson, *Battle Cry of Freedom*, 464.

23. Ibid., 477.

24. Frederick Douglass, "The Slaveholders Rebellion," speech delivered on July 4, 1862, at Himrods Corners, Yates County, New York; in Foner, ed., *Life and Writings of Frederick Douglass*, vol. 3, 242–259.

25. Ibid., 260.

26. McPherson, *Battle Cry of Freedom*, 513.

27. Donovan, *Mr. Lincoln's Proclamation*, 101–102.

28. McPherson, *Battle Cry of Freedom*, 504.

29. Donovan, *Mr. Lincoln's Proclamation*, 107.

30. Ibid., 105.

31. Ibid., 108.

32. McPherson, *Battle Cry of Freedom*, 516.

33. Ibid., 524.

34. McClellan to Ellen Marcy McClellan, Aug. 10, 1862, in McPherson, *Battle Cry of Freedom*, 525.

35. Marx and Engels, *The Civil War in the United States*, 201–206.

36. Donovan, *Mr. Lincoln's Proclamation*, 83–84; DuBois, *Black Reconstruction*, 147–148.

37. McClellan to Ellen Marcy McClellan, Aug. 22, 1862, in McPherson, *Battle Cry of Freedom*, 528.

CHAPTER THIRTEEN: LINCOLN'S EMANCIPATION

Note to epigraph: George W. Julian, *Speeches: Political Questions*, with an introduction by L. Maria Child (New York: Hurd and Houghton, 1872), 168.

1. Lincoln to McClellan, Sept. 15, 1862, in McPherson, *Battle Cry of Freedom*, 534.

2. Ibid., 536.

3. Ibid., 517.

4. Ibid., 545.

5. Donovan, *Mr. Lincoln's Proclamation*, 112–113.

6. *The Complete Writings of Ralph Waldo Emerson*, 1208–1212.

7. Sherwin, *Prophet of Liberty*, 447.

8. McClellan to Ellen Marcy McClellan, Sept. 18, 1862, in McPherson, *Battle Cry of Freedom*, 545.

9. McPherson, *Battle Cry of Freedom*, 559.

10. Halleck to McClellan, Oct. 6, 1862; in McPherson, *Battle Cry of Freedom*, 568.

11. Ibid., 569.

12. *New York Tribune* article published in the weeks after the battle at Antietam, quoted by Brown in *The Negro in the American Rebellion*, 64–68.

13. Donovan, *Mr. Lincoln's Proclamation*, 114–115.

14. McPherson, *Battle Cry of Freedom*, 572.

15. Ibid., 574.

16. Ibid., 575.

17. Pollard, *Southern History of the War*, vol. 1, 555.

18. Ibid., 559.

19. Ibid., 560.

20. Ibid.

21. "A Day for Poetry and Song," remarks at Zion Church, Dec. 28, 1862, *Douglass' Monthly*, Jan. 1863, in Foner, ed., *Life and Writings of Frederick Douglass*, vol. 3, 310.

22. Sherwin, *Prophet of Liberty*, 471–472; Donovan, *Mr. Lincoln's Proclamation*, 118–119.

23. "January First, 1863," *Douglass' Monthly*, Jan. 1863, in Foner, ed., *Life and Writings of Frederick Douglass*, vol. 3, 306.

24. *The Life and Times of Frederick Douglass*, 353.

25. Sherwin, *Prophet of Liberty*, 458.

26. *The Life and Times of Frederick Douglass*, 354.

27. Pollard, *Southern History of the War*, vol. 1, 563.

28. Ibid.

29. McPherson, *Battle Cry of Freedom*, 582.

30. Pollard, *Southern History of the War*, vol. 1, 636–641.

31. McPherson, *Battle Cry of Freedom*, 585.

32. "Condition of the Country," *Douglass' Monthly*, Feb. 1863, in Foner, ed., *Life and Writings of Frederick Douglass*, vol. 3, 314–317.

33. Ibid.

CHAPTER FOURTEEN: MEN OF COLOR, TO ARMS!

Note to epigraph: Frederick Douglass's Broadside "Men of Color, to Arms!" issued March 21, 1863; in Foner, ed., *Life and Writings of Frederick Douglass*, vol. 3, 317–319.

1. T. J. Barnett to Samuel Barlow, Sept. 25, 1862, in McPherson, *Battle Cry of Freedom*, 558.

2. Harriet Martineau, *Retrospect of Western Travel*, vol. 1 (1838; New York: Haskell House, 1969), 246.

3. Conrad, *Harriet Tubman*, 165.

4. Brown, *The Negro in the American Rebellion*, 159–162; also published in the *Tribune*, Feb. 11, 1863.

5. Sherwin, *Prophet of Liberty*, 474–476.

6. Donovan, *Mr. Lincoln's Proclamation*, 126.

7. *The Life and Times of Frederick Douglass*, 342.

8. Foner, ed., *The Life and Writings of Frederick Douglass*, vol. 3, 320–321.

9. Lincoln to Johnson, March 26, 1863, in McPherson, *Battle Cry of Freedom*, 565.

10. DuBois, *Black Reconstruction*, 98.

11. "The Proclamation and a Negro Army," *Douglass' Monthly*, March 1863, in Foner, ed., *Life and Writings of Frederick Douglass*, vol. 3, 321–337.

12. McPherson, *Battle Cry of Freedom*, 595.

13. Ibid., 585–586.

14. Ibid., 588.

15. Pollard, *Southern History of the War*, vol. 2, 55.

16. McPherson, *Battle Cry of Freedom*, 639.

17. Ibid., 645.

18. Ibid., 647.

19. Ibid., 646–647.

20. Ibid., 596.

21. Ibid., 598.

22. Ibid.

23. Brown, *The Negro in the American Rebellion*, 124–126.

24. Ibid., 131–132.

25. Foner, *The Voice of Black America*, vol. 1, 293–294.

26. McPherson, *Battle Cry of Freedom*, 650.

27. Brown, *The Negro in the American Rebellion*, 175.

28. Ibid., 175–176.

29. Reporter for *Boston Transcript*, quoted in Quarles, *The Negro in the Civil War*, 11.

30. Brown, *The Negro in the American Rebellion*, 153.

31. Report of Captain John F. Lay investigating the Combahee raid in Conrad, *Harriet Tubman*, 171.

32. Ibid., 172–173.

33. Bradford, *Scenes in the Life of Harriet Tubman*, 39–40.

34. Brown, *The Negro in the American Rebellion*, 138.

35. Ibid., 139.

36. Higginson *Army Life in a Black Regiment*, 263.

37. McPherson, *Battle Cry of Freedom*, 651.

38. Pollard, *Southern History of the War*, vol. 2, 277.

39. Lincoln to Howard, July 21, 1863, in *Life and Letters of General Meade*, vol. 2, 138.

40. "Address for the Promotion of Colored Enlistments," *Douglass' Monthly*, Aug. 1863, in Foner, ed., *Life and Writings of Frederick Douglass*, vol. 3, 361–366.

41. Brown, *The Negro in the American Rebellion*, 187.

42. McPherson, *Battle Cry of Freedom*, 609.

43. *Life and Times of Frederick Douglass*, 433.

44. Pollard, *Southern History of the War*, vol. 1, 95.

45. DuBois, *Black Reconstruction*, 109.

46. Brown, *The Negro in the American Rebellion*, 205–206.

47. Ibid., 201.

48. Lewis Douglass to My Dear Amelia, in Aptheker, *A Documentary History of the Negro in the United States*, vol. 1, 482.

49. Hart, *Slavery and Abolition*, 209.

CHAPTER FIFTEEN: BATTLES FOR LIBERTY, BATTLES FOR UNION

1. *Douglass' Monthly*, June 1863, in Foner, ed., *Life and Writings of Frederick Douglass*, vol. 3, 347–359.

2. McPherson, *Battle Cry of Freedom*, 794.

3. Frederick Douglass, "Emancipation to Appomattox," in Foner, ed., *Life and Writings of Frederick Douglass*, vol. 3, 39.

4. To Thomas Webster, Esq., Aug. 19, 1863, in Foner, ed., *Life and Writings of Frederick Douglass*, vol. 3, 377.

5. Gideon Welles's diary, quoted in Rhodes, *History of the Civil War*, 291.

6. DuBois, *Black Reconstruction*, 104.

7. Forrest quoted in Grant, *Personal Memoirs*, chap. 47.

8. Sterling, *Martin Robison Delany*, 238.

9. Brown, *The Negro in the American Rebellion*, 251.

10. DuBois, *Black Reconstruction*, 99.

11. Frederick Douglass, "Emancipation Proclamation to Appomattox," in Foner, ed., *Life and Writings of Frederick Douglass*, vol. 3, 37–38.

12. *New York Tribune*, Jan. 14, 1864, in Foner, ed., *Life and Writings of Frederick Douglass*, vol. 3, 386–403.

13. Ibid., 387.

14. Ibid., 390–391.

15. Frederick Douglass, *Selected Speeches and Writings*, 547.

16. Frederick Douglass "Our Work Is Not Done," speech to American Anti-Slavery Society annual meeting, Philadelphia, Dec. 3–4, 1863, in Foner, ed., *Life and Writings of Frederick Douglass*, vol. 3, 378–386.

17. DuBois, *Black Reconstruction*, 151–152; McPherson, *Battle Cry of Freedom*, 699–700.

18. Sherwin, *Prophet of Liberty*, 486.

19. DuBois, *Black Reconstruction*, 157.

20. Brown, *The Negro in the American Rebellion*, 252.

21. Ibid., 223.

22. Ibid., 224.

23. Report of the Committee on the Conduct of the War on the Fort Pillow Massacre, extract in Brown, *The Negro in the American Rebellion*, 241–247.

24. Glatthar, *Forged in Battle*, 145.

25. DuBois, *Black Reconstruction*, 152; Frederick Douglass, "To an English Correspondent," in Foner, ed., *Life and Writings of Frederick Douglass*, vol. 3, 404.

26. Pollard *Southern History of the War*, vol. 2, 312.

27. Quoted in Glatthar, *Forged in Battle*, 145.

CHAPTER SIXTEEN: WAR FOR THE TOTAL EXPIATION OF SLAVERY

1. Pollard, *Southern History of the War*, vol. 2, 270.

2. McPherson, *Battle Cry of Freedom*, 731.

3. Quoted in Anders, *Henry Halleck's War*, 564.

4. Sherwin, *Prophet of Liberty*, 490.

5. McPherson, *Battle Cry of Freedom*, 735.

6. Pollard, *Southern History of the War*, vol. 2, 350.

7. Anders, *Fighting Confederates*, 144.

8. Pollard, *Southern History of the War*, vol. 2, 330.

9. McPherson, *Battle Cry of Freedom*, 741.

10. Ibid.

11. Ibid., 742.

12. Sherwin, *Prophet of Liberty*, 495.

13. McPherson, *Battle Cry of Freedom*, 716.

14. Sherwin, *Prophet of Liberty*, 494.

15. McPherson, *Battle Cry of Freedom*, 713.

16. Pollard, *Southern History of the War*, vol. 2, 355.

17. McPherson, *Battle Cry of Freedom*, 757.

18. Anders, *Fighting Confederates*, 103.

19. McPherson, *Battle Cry of Freedom*, 767.

20. Quoted in Rhodes, *History of the Civil War*, 335.

21. McPherson, *Battle Cry of Freedom*, 778.

22. Quoted in McPherson, *Battle Cry of Freedom*, 755.

23. Pollard, *Southern History of the War*, vol. 2, 334.

24. McPherson, *Battle Cry of Freedom*, 760.

25. Ibid., 770.

26. Sherwin, *Prophet of Liberty*, 499.

27. McPherson, *Battle Cry of Freedom*, 770.

28. Frederick Douglass, "To an English Correspondent," June 1864, in Foner, ed., *Life and Writings of Frederick Douglass*, vol. 3, 404.

29. McPherson, *Battle Cry of Freedom*, 769.

30. Sherwin, *Prophet of Liberty*, 498.

31. Whitman, *Complete Prose Works*, 38.

32. Frederick Douglass to Theodore Tilton, Oct. 15, 1864, in Foner, ed., *Life and Writings of Frederick Douglass*, vol. 3, 422–424.

33. Frederick Douglass to Hon. Abraham Lincoln, President of the United States, Aug. 29, 1864, in Foner, ed., *Life and Writings of Frederick Douglass*, vol. 3, 405–406.

34. McPherson, *Battle Cry of Freedom*, 763.

35. Pollard, *Southern History of the War*, vol. 2, 366.

36. Ibid., 365–366.

37. McPherson, *Battle Cry of Freedom*, 776.

38. Ibid., 803.

39. McPherson, *Battle Cry of Freedom*, 776; Sherwin, *Prophet of Liberty*, 505–506; Frederick Douglass to William Lloyd Garrison, Sept. 17, 1864, in Foner, ed., *Life and Writings of Frederick Douglass*, vol. 3, 406–407.

40. Pollard, *Southern History of the War*, vol. 2, 388.

41. McPherson, *Battle Cry of Freedom*, 807.

42. Ibid., 807–808.

43. Ibid., 778.

44. Ibid., 779.

45. Ibid., 808.

46. Sherwin, *Prophet of Liberty*, 505.

47. Douglass to Tilton, Oct. 15, 1864, in Foner, ed., *Life and Writings of Frederick Douglass*, vol. 3, 422–424.

48. Frederick Douglass, "The Cause of the Negro People," address at the Colored National Convention to the People of the United States, in Foner, ed., *Life and Writings of Frederick Douglass*, vol. 3, 408–422.

49. Seward quoted in "The Cause of the Negro People," 415.

50. Ibid.

51. Frederick Douglass, "Emancipation Proclamation to Appomattox," in Foner, ed., *Life and Writings of Frederick Douglass*, vol. 3, 49.

52. "The Cause of the Negro People," 422.

53. Freeman, *Lee*, 442.

54. Pollard, *Southern History of the War*, vol. 2, 413.

55. McPherson, *Battle Cry of Freedom*, 804.

56. Sherwin, *Prophet of Liberty*, 507.

57. Ibid.

CHAPTER SEVENTEEN: ONE MORE RIVER TO CROSS

1. Pollard, *Southern History of the War*, vol. 2, 427.

2. Ibid., 432.

3. Ibid.

4. Abraham Lincoln in Shapiro, ed., *Mystic Chords of Memory*, 62.

5. McPherson, *Battle Cry of Freedom*, 839.

6. Sherwin, *Prophet of Liberty*, 509.

7. Ibid.

8. Sterling, *The Making of an Afro-American*, 245.

9. Henry Highland Garnet, "Let the Monster Perish," Feb. 12, 1865; in Foner, ed., *The Voice of Black America*, vol. 1, 335–344.

10. "The Future of the Negro People of the Slave States," *Douglass' Monthly*, March 1862, in Foner, ed., *The Life and Writings of Frederick Douglass*, vol. 3, 210–225.

11. *Life and Times of Frederick Douglass*, 440.

12. Pollard, *Southern History of the War*, vol. 2, 473.

13. "The Interview," *Liberator*, Feb. 24, 1865, in Aptheker, *A Documentary History of the Negro People*, vol. 1, 496–498.

14. Pollard, *Southern History of the War*, vol. 2, 467–468.

15. McPherson, *Battle Cry of Freedom*, 822.

16. Pollard, *Southern History of the War*, vol. 2, 469.

17. McPherson, *Battle Cry of Freedom*, 822–823.

18. Pollard, *Southern History of the War*, vol. 2, 470–471.

19. McPherson, *Battle Cry of Freedom*, 826, 827.

20. Ibid., 828.

21. Emilio, *History of the Fifty-Fourth Regiment of the Massachusetts Volunteer Infantry, 1863–1865*.

22. *The Complete Poems and Prose of Walt Whitman*.

23. Sterling, *The Making of an Afro-American*, 226–227.

24. Ibid., 243.

25. Ibid., 246–247.

26. Conrad, *Harriet Tubman*, 183.

27. Sterling, *The Making of an Afro-American*, 248.

28. Ibid.

29. *Life and Times of Frederick Douglass*, 364.

30. Quoted in Foner, *The Fiery Trial*, 328.

31. *Life and Times of Frederick Douglass*, 363.

32. Ibid., 364.

33. Ibid., 365–366.

34. Pollard, *Southern History of the War*, vol. 2, 492.

35. Ibid., 493.

36. Dispatch of T. Morris Chester to the *Philadelphia Press*, April 11, 1865.

37. Record Group 94, National Archives.

38. McPherson, *Battle Cry of Freedom*, 846.

39. Brown, *The Negro in the American Rebellion*, 324.

40. McPherson, *Battle Cry of Freedom*, 847.

41. J. J. Hill, "A Sketch of the 29th Regiment of Connecticut Colored Troops," in Aptheker, *A Documentary History of the Negro People in the United States*, vol. 1, 488–490.

42. Ibid.

43. Craven, *An Historian and the Civil War*, 118.

44. Pollard, *Southern History of the War*, vol. 2, 505.

45. Glatthaar, *Forged in Battle*, 158.

46. McPherson, *Battle Cry of Freedom*, 851–852.

47. Ibid.

48. Brown, *The Negro in the American Rebellion*, 293.

49. Sherwin, *Prophet of Liberty*, 510.

50. Sterling, *The Making of an Afro-American*, 249.

51. Delany's speech as reported by Lieutenant Edward Stoeber in Foner, ed., *The Voice of Black America*, vol. 1, 348–353.

Epilogue

1. Donovan, *Mr. Lincoln's Proclamation*, 94.

2. DuBois, *Black Reconstruction*, 149.

3. Ibid., 136.

4. *Life and Times of Frederick Douglass*, 383–385.

5. Bowers, *The Tragic Era*, 32.

6. DuBois, *Black Reconstruction*.

7. *Life and Times of Frederick Douglass*, 503.

8. Ibid.

9. Quarles, *Frederick Douglass*, 271, 272.

10. *Life and Times of Frederick Douglass*, 451.

11. Ruchames, ed., *John Brown*, 298–299.

12. Ibid., 282.

BIBLIOGRAPHY

A MONG THE FIRST TEXTS PRODUCED BY THE CIVIL WAR COULD BE said to be the Scot émigré James Redpath's *Public Life of Captain John Brown*, published in 1860, followed in 1861 by Osborne Perry Anderson's pamphlet-sized *A Voice from Harper's Ferry*. Harbingers of a vast literature to come include works by the leaders of the Confederacy: Alexander Stephens's *Constitutional View of the Late War Between the States* (1868–70), Jefferson Davis's *Rise and Fall of the Confederate Government* (1881), and Edward Pollard's *Southern History of the War* (vol. 1, 1862; vols. 2–3, 1864), all three maintaining that slavery merely provided an occasion for conflict. What was essential, they held, was that it had been a contest between opposing ideas as to the nature of government, viz. the Hamiltonian and the Jeffersonian concepts; the former holding the government to be thoroughly national, and the latter strictly federal. The union adopted among independent American states was mutual, and could be dissolved by mutual consent—a notion that is being reasserted even today.

At last came studies authored by names like Rhodes (1917), Milton (1941), Channing (1925), Beard (1927), Cole (1919), among others, alternately contending the war was an "irrepressible" and a "repressible" conflict, but invariably with a tepid and merely peripheral view of slavery. The conclusion, however, was finally reached that the war amounted to "a second American revolution," although its main feature was largely underappreciated—that it was based upon the precedent of the 1850s, a fact, however, that was not lost on keen observers at the time.

In addition to the sources above, the New York Public Library has an extensive collection of what had been catalogued as "Controversial Literature," that is, the abolitionist tracts. These include William Lloyd Garrison's *A Letter to Louis Kossuth* (1852), William A. Phillips's *The Conquest of Kansas by Missouri and Her Allies* (1856), and others mentioned in this text. During the course of my research, I consulted libraries and special collections, including the public library in Chatham, Ontario, Canada, where John Brown held his convention; the special collection of diaries kept by slave-holding women at libraries in Birmingham and Eufaula, Alabama; the library in Galveston, Texas; and the library at the English Cemetery in Florence where Theodore Parker is buried. This book has also benefited from the burgeoning publishing activity in the so-called Negro, then Black, and subsequently African American history.

A reader will readily discern many of the primary sources that are cited as they are used in the text. But these especially include: *Life and Times of Frederick Douglass* (1881), as well as the numerous writings, speeches, and letters of the same, as contained in the five-volume series edited and introduced by Philip S. Foner: *The Life and Writings of Frederick Douglass* (1950–1975). The reading on John Brown, in addition to Redpath's and Anderson's works, principally includes: W. E. B. DuBois's, *John Brown: A Biography* (1909), Franklin Sanborn's *The Life and Letters of John Brown* (1891), Richard Hinton's *John Brown and His Men* (1894), and Oswald Villard's *John Brown, 1800–1859: A Biography Fifty Years After* (1910), as well as numerous others having less direct utility in the narrative, but whose authors include Warren (1929), Wilson (1913), Malin (1942), Featherstonhaugh (1897), Woodward (1968), Abels (1971), Oates (1970), Reynolds (2005), and Horwitz (2011).

Other readings that have opened a course for this book are as follows: Wendell Phillips's speeches, particularly "The Lesson of the Hour" (1859); Henry David Thoreau's *A Plea for Captain John Brown* (1859); *The Confessions of Nat Turner—as made to Thomas R. Gray* (1831); David Walker's *Appeal to the Coloured Citizens of the World* (1829); William Wells Brown, *The Negro in the American Rebellion* (1867); Benjamin Quarles's *Frederick Douglass* (1968); *Black Abolitionists* (1970); and *Allies for Freedom: Blacks and John Brown* (1974). Also consulted are Earl Ofari's *"Let Your Motto Be Resistance": The Life and Thought of Henry Garnet Highland* (1972); Oscar Sherwin's *Prophet of Liberty: The Life and Times of*

Wendell Phillips (1958); Earl Conrad's *Harriet Tubman* (1942); Sterling Stuckey's *The Ideological Origins of Black Nationalism* (1972); Herbert Aptheker's *American Negro Slave Revolts* (1963) and *To be Free* (1992); James M. McPherson's *The Negro's Civil War* (1965); Jay Monaghan's *Civil War on the Western Border, 1854–1865* (1985); Jonathan Katz's *Resistance at Christiana* (1974); Bearden and Butler's *The Life and Time of Mary Shadd Cary* (1977); and numerous writings of W. E. B. DuBois, including *Black Reconstruction in America, 1860-1880* (1935) and *The Souls of Black Folk* (1903).

The following edited works have all been indispensable: *A Documentary History of the Negro People in the United States, Vol. I* (1951), edited by Herbert Aptheker; *The Southampton Slave Revolt* (1971), edited by Henry Irving Tragle; *The Voices of Black America* (1972), edited by Philip Foner; *John Brown: The Making of a Revolutionary* (1969), edited by Louis Ruchames; *Documents of Upheaval* (1966), edited by Truman Nelson; and *The Portable Thoreau* (1947), edited by Carl Bode, whose epilogue contains the connection to be inferred between John Brown's death and that of Thoreau.

Other sources include *AntiSlavery* (1961) by Dwight Lowell Dumond; *The Slave Catchers, Enforcement of the Fugitive Slave Law, 1850–1860* (1970) by Stanley Campbell; *Black Jacobins* (1963) by C. L. R. James; and *William Styron's Nat Turner: Ten Black Writers Respond* (1968), edited by John Henrik Clark.

For direct quotations, whether of John Brown, Frederick Douglass, Abraham Lincoln, or any other persons, other works I have particularly relied upon include: James M. McPherson, *Battle Cry of Freedom* (1988); Steven Channing, *Crisis of Fear: Secession in South Carolina* (1970); Bruce Catton, *The Coming Fury* (1961); Curt Anders, *Fighting Confederates* (1968); and Carl Russell Fish, *The American Civil War* (1937).

BOOKS

Abels, Jules. *Man on Fire: John Brown and the Cause of Liberty*. New York: Macmillan, 1971.

Anders, Curt. *Fighting Confederates*. New York: Putnam, 1968.

———. *Henry Halleck's War: A Fresh Look at Lincoln's Controversial General-In-Chief*. Carmel: Guild Press of Indiana, 1999.

Anderson, Osborne Perry. *A Voice from Harper's Ferry: A Narrative of Events at Harper's Ferry; with Incidents Prior and Subsequent to Its Capture by Captain Brown and His Men.* Boston: Privately printed, 1861.

Aptheker, Herbert. *American Negro Slave Revolts.* New York: International Publishers, 1963.

_____. *A Documentary History of the Negro People in the United States, Vol. 1: Colonial Times to the Civil War.* New York: Citadel Press, 1951.

_____. *To Be Free: Pioneering Studies in Afro-American History.* New York: Citadel Press, 1992.

Avey, Elijah. *The Capture and Execution of John Brown: A Tale of Martyrdom.* Elgin, Ill.: Brethren Publishing House, 1905.

Barbour, Floyd B. *The Black Power Revolt.* Boston: Extending Horizons Books, 1968.

Bayliss, John F., ed. *Black Slave Narratives.* New York: Collier Books, 1970.

Beard, Charles Austin. *History of the United States.* New York: Macmillan, 1927.

Beard, John R. *The Life of Toussaint L'ouverture: The Negro Patriot of Hayti.* Westport, Conn.: Negro Universities Press, 1970.

Bearden, Jim and Linda Jean Butler. *Shadd: The Life and Time of Mary Shadd Cary.* Toronto: NC Press, 1977.

Berlin, Ira. *Slaves without Masters: The Free Negro in the Antebellum South.* New York: New Press, 1974.

Botume, Elizabeth Hyde. *First Days Amongst the Contrabands.* (1893) Reprint. New York: Arno Press, 1968.

Bowers, Claude C. *The Tragic Era: The Revolution After Lincoln.* Boston: Houghton Mifflin, 1929.

Bradford, Sarah. *Scenes in the Life of Harriet Tubman.* Auburn, New York. 1869. Reprint. New York: Books for Libraries, 1971.

Brown, Henry Box. *Narrative of the Life of Henry Box Brown.* Edited by John Ernest. Chapel Hill: University of North Carolina Press, 2008.

Brown, William Wells. *The Negro in the American Rebellion: His Heroism and His Fidelity* (1867). Reprint. New York: Johnson Reprint Company, 1968.

Buckmaster, Henrietta. *Let My People Go: The Story of the Underground Railroad and the Growth of the Abolition Movement.* (1941) Reprint. Columbia: University of South Carolina Press, 1992.

Campbell, Stanley. *The Slave Catchers: Enforcement of the Fugitive Slave Law, 1850–1860.* Chapel Hill: University of North Carolina Press, 1970.

Catton, Bruce. *America Goes to War*. New York: Hill and Wang, 1958.

_____. *The Coming Fury*. Garden City, N.Y.: Doubleday, 1961.

_____. *Glory Road*. Garden City, N.Y.: Doubleday, 1952.

_____. *Mr. Lincoln's Army*. Garden City, N.Y.: Doubleday, 1951.

_____. *Never Call Retreat*. Garden City, N.Y.: Doubleday, 1965.

_____. *A Stillness at Appomattox*. Garden City, N.Y.: Doubleday, 1953.

_____. *Terrible Swift Sword*. Garden City, N.Y.: Doubleday, 1963.

Channing, Edward. *The War for Southern Independence, 1849–1865*. New York: Macmillan, 1926.

Channing, Steven A. *Crisis of Fear: Secession in South Carolina*. New York: Simon and Schuster, 1970.

Channing, William Henry. *The Civil War in America, or, The Slaveholders' Conspiracy*. Ithaca, N.Y.: Cornell University Press, 1861.

Clarke, John Henrik, ed. *William Styron's Nat Turner, Ten Black Writers Respond*. Boston: Beacon Press, 1968.

Cole, Arthur Charles. *The Era of the Civil War, 1848–1870*. Springfield: Illinois Centennial Commission, 1919.

Conrad, Earl. *Harriet Tubman: Negro Soldier and Abolitionist*. New York: International Publishers, 1942.

Cornish, Dudley T. *The Sable Arm: Negro Troops in the Union Army, 1861–1865*. New York: W. W. Norton, 1966.

Craton, Michael. *Testing the Chains! Resistance to Slavery in the British West Indies*. Ithaca, N.Y.: Cornell University Press, 1982.

Craven, Avery O. *An Historian and the Civil War*. Chicago: University of Chicago Press, 1964.

Cruse, Harold. *The Crisis of the Negro Intellectual, From Its Origins to the Present*. New York: William Morrow & Company, Inc., 1967.

Davis, Jefferson. *Rise and Fall of the Confederate Government*. New York: Appleton, 1881.

Delany, Martin R. *Blake, or the Huts of America*. Boston: Beacon Press. 1970.

De Witt, R. M. *The Life, Trial and Execution of Captain John Brown, Known as "Old Brown of Ossawatomie," with a Full Account of the Attempted Insurrection at Harper's Ferry*. New York: R. M. De Witt, 1859.

Dillon, Merton Lynn. *The Abolitionists: The Growth of a Dissenting Minority*. New York: W. W. Norton, 1979.

Donald, David Herbert, ed. *Why the North Won the Civil War*. New York: Collier Books, 1972.

Donovan, Frank. *Mr. Lincoln's Proclamation: The Story of the Emancipation Proclamation*. New York: Dodd, Mead, 1964.

Douglass, Frederick. *Life and Times of Frederick Douglass* (1881, revised 1892). Reprint. New York: Thomas Crowell, 1966.

———. *My Bondage and My Freedom* (1855). Reprint. New York: Washington Square Press, 2003.

———. *Narrative of the Life of Frederick Douglass, an American Slave* (1845). Reprint. New Haven: Yale University Press, 2001.

DuBois, W. E. B., *Black Reconstruction in America: An Essay Toward a History of the Part Which Black Folk Played in the Attempt to Reconstruct Democracy in America, 1860–1880.* (1935) Reprint. New York: Oxford University Press, 2007.

———. *John Brown: A Biography.* (1909) Reprint. Introduction by David R. Roediger. New York: Modern Library, 2001.

———. *The Negro.* (1915) Reprint. Afterword by Robert Gregg. Philadelphia: University of Pennsylvania Press, 2001.

———. *The Souls of Black Folk.* (1903) Reprint. New York: New American Library, 1969.

———. "Suppression of the African Slave-Trade to the United States, 1638–1870." Ph.D. dissertation (1896). Reprint. New York: Russell & Russell, 1965.

Dumond, Dwight Lowell. *AntiSlavery: The Crusade for Freedom in America.* Ann Arbor: University of Michigan Press, 1961.

Emerson, Ralph Waldo, *The Complete Works of Ralph Waldo Emerson.* New York: William H. Wise & Co., 1929.

Emilio, Luis F. *History of the Fifty-Fourth Regiment of the Massachusetts Volunteer Infantry, 1863–1865.* Boston: Boston Book Company, 1894.

Featherstonhaugh, Thomas. *A Bibliography of John Brown.* Baltimore: Friedenwald Company, 1897.

Filler, Louis. *The Crusade Against Slavery, 1830–1960.* New York: Harper & Brothers, 1961.

Fish, Carl Russell. *The American Civil War: An Interpretation.* London: Longmans, Green and Company, 1937.

Foner, Philip S. *The Fiery Trial: Abraham Lincoln and American Slavery.* New York: W. W. Norton, 2010.

———, ed. *Frederick Douglass: Selected Speeches and Writings.* Chicago: Chicago Review Press, 2000.

———, ed. *Frederick Douglass on Women's Rights.* Westport, Conn.: Greenwood Press, 1976.

———, ed. *The Life and Writings of Frederick Douglass.* 5 vols. New York: International Publishers, 1950–1975.

———, ed. *The Voices of Black America: Major Speeches by Negroes in the United States, 1797–1971.* New York: Simon and Schuster, 1972.

Franklin, John Hope. *Reconstruction: After the Civil War*. Chicago: University of Chicago Press, 1961.

Fredrickson, George M., ed. *William Lloyd Garrison, Great Lives Observed*. Englewood Cliffs, N.J.: Prentice-Hall, Inc., 1968.

Freehling, William W. and Craig M. Simpson. *Secession Debated: Georgia's Showdown in 1860*. New York: Oxford University Press, 1992.

Freeman, Douglas Southall. *Lee*. Abridgement by Richard Harwell. New York: Simon and Schuster, 1991.

Garrison, William Lloyd. *A Letter to Louis Kossuth; Concerning Freedom and Slavery in the United States*. Boston: R. F. Wallcut, 1852.

Genovese, Eugene D. *From Rebellion to Revolution, Afro-American slave Revolts in the Making of the New World*. New York: Random House, 1979.

_____. *In Red and Black: Marxian Explorations in Southern and Afro-American History*. Knoxville: University of Tennessee Press, 1984.

_____. *The Political Economy of Slavery, Studies in the Economy and Society of the Slave South*. New York: Random House, 1967.

_____. *The World the Slaveholders Made*. New York: Random House, 1971.

Glatthar, Joseph T. *Forged in Battle: The Civil War Alliance of Black Soldiers and White Officers*. New York: Free Press, 1990.

Grant, Ulysses S. *Personal Memoirs of U. S. Grant*, New York: Charles L. Webster and Company, 1885.

Grodzins, Dean. *American Heretic, Theodore Parker & Transcendentalism*. Chapel Hill: University of North Carolina Press, 2002.

Halstead, Murat. *Caucuses of 1860*. Columbus, Ohio: Follett, Foster and Company, 1860.

Hart, Albert Bushnell. *Slavery and Abolition, 1831–1841*. New York: Harper & Bros., 1906.

Helper, Hinton Rowan. *The Impending Crisis: How to Meet It*. Reprint. Cambridge, Mass.: Belknap Press of Harvard University Press, 1968.

Higginson, Thomas Wentworth. *Army Life in a Black Regiment*. Boston: Fields, Osgood, and Company, 1870.

_____. *Contemporaries*. Boston: Houghton Mifflin, 1900.

Hinton, Richard J. *John Brown and His Men*. (1894) Reprint. New York: Ayer, 1968.

Horwitz, Tony. *Midnight Rising: John Brown and the Raid That Sparked the Civil War*. New York: Henry Holt, 2011.

James, C. L. R. *Black Jacobins: Toussaint L'Ouverture and the San Domingo Revolution*. New York: Vintage Books, 1963.

Katz, Jonathan. *Resistance at Christiana: The Fugitive Slave Rebellion, Christiana, Pennsylvania, September 11, 1851, a Documentary Account*. New York: Thomas Crowell, 1974.

Kettell, Thomas Prentice. *History of the Great Rebellion*. Hartford, Conn.: L. Stebbins, 1865.

Leech, Samuel Vanderlip. *The Raid of John Brown at Harpers's Ferry as I Saw It*. Washington, D.C.: Privately printed, 1909.

Malin, James Claude. *John Brown and the Legend of Fifty-Six*. (1942) Reprint. Brooklyn, N.Y.: Haskell House, 1970.

Martin, Waldo E., Jr. *The Mind of Frederick Douglass*. Chapel Hill: University of North Carolina Press, 1984.

Marx, Karl, and Frederick Engels. *The Civil War in the United States*. New York: International Publishers, 1961.

McPherson, James M. *Battle Cry of Freedom: The Civil War Era*. New York: Oxford University Press, 1988.

_____. *The Negro's Civil War: How American Negroes Felt and Acted During the War for the Union*. New York: Pantheon Books, 1965.

_____. *The Struggle for Equality: Abolitionists and the Negro in the Civil War and Reconstruction*. Princeton, N.J.: Princeton University Press, 1964.

Meade, George Gordon. *Life and Letters of General George Gordon Meade*. 2 vols. New York: Charles Scribner's Sons, 1913.

Meltzer, Milton. *In Their Own Words, A History of the American Negro*. New York: Thomas Crowell, 1964.

_____. *Tongue of Flame, The Life of Lydia Maria Child*. New York: Thomas Crowell, 1965.

Milton, George Fort. *Conflict: The American Civil War*. New York: Coward McCann, 1941.

Monaghan, Jay. *Civil War on the Western Border, 1854–1865*. Lincoln: University of Nebraska Press, 1985.

Nelson, Truman, ed. *Documents of Upheaval: Selections from William Lloyd Garrison's The Liberator, 1831–1865*. New York: Hill and Wang, 1966.

Oakes, James. *The Radical and the Republican, Frederick Douglass, Abraham Lincoln, and the Triumph of Antislavery Politics*. New York: W. W. Norton, 2007.

Oates, Stephen B. *To Purge This Land with Blood: A Biography of John Brown*. New York: HarperCollins, 1970.

Ofari, Earl. *"Let Your Motto Be Resistance": The Life and Thought of Henry Garnet Highland*. Boston: Beacon Press, 1972.

Olmsted, Frederick Law. *A Journey through Texas or, a saddle-trip on the Southwestern Frontier*. Reprint. Austin: University of Texas Press, 1978.

_____. *The Slave States before the Civil War*. New York: Capricorn Books, 1959.

Phillips, Wendell. *Speeches, Lectures, and Letters*. Boston: Lee and Shepard, 1884.

Phillips, William A. *The Conquest of Kansas by Missouri and Her Allies: A History of the Troubles in Kansas from the Passage of the Organic Act Until the Close of July 1856*. Boston: Phillips, Sampson and Company, 1856.

Pollard, Edward. *Southern History of the War*. 3 vols. (1862, 1864) Reprint. Freeport, N.Y.: Books for Libraries Press, 1969.

Quarles, Benjamin. *Allies for Freedom: Blacks and John Brown*. New York: Oxford University Press, 1974.

_____. *Black Abolitionists*. New York: Oxford University Press, 1970.

_____. *Frederick Douglass*. New York: Macmillan, 1968.

_____, ed. *Frederick Douglass, Great Lives Observed*. Prentice-Hall. 1968

_____. *Lincoln and the Negro*. New York: Oxford University Press, 1962.

_____. *The Negro in the Civil War*. Boston: Little, Brown, 1969.

Redpath, James. *Echoes of Harper's Ferry*. 1860.

_____. *Public Life of Captain John Brown*. Boston: Thayer & Eldridge, 1860.

_____. *The Roving Editor*. New York. 1859.

Reynolds, David S. *John Brown, Abolitionist: The Man Who Killed Slavery, Sparked the Civil War, and Seeded Civil Rights*. New York: Knopf, 2005.

Rhodes, James Ford. *History of the Civil War, 1861–1865*. New York: Macmillan, 1917.

Roberts, Timothy Mason. *Distant Revolutions: 1848 and the Challenge to American Exceptionalism*. Charlottesville: University of Virginia Press, 2009.

Rollin, Frank A. [Frances Rollin Whipper]. *Life and Public Service of Major Martin R. Delany*. Boston. (1868) Reprint. New York: Arno Press & New York Times, 1969.

Ruchames, Louis, ed. *John Brown: The Making of a Revolutionary*. New York: Grosset & Dunlap, 1969.

_____. ed. *A John Brown Reader: The Story of John Brown in His Own Words, in the Words of those Who Knew Him, and in the Poetry and Prose of the Literary Heritage*. London: Abelard–Schuman, 1959.

Sanborn, Franklin B. *The Life and Letters of John Brown: Liberator of Kansas and Martyr of Virginia*. Reprint. Boston: Roberts, 1891.

Scheidenhelm, Richard, ed. *The Response to John Brown*. Belmont, Calif.: Wadsworth Publishing Company, 1972.

Sernett, Milton C. *Harriet Tubman: Myth, Memory, and History*. Durham, N.C.: Duke University Press, 2007.

Shapiro, Larry, ed. *Abraham Lincoln: Mystic Chords of Memory*. New York: Book-of-the-Month Club, 1984.

Sherwin, Oscar. *Prophet of Liberty: The Life and Times of Wendell Phillips*. New York: Bookman Associates, 1958.

Simone, Timothy Maliqalim. *About Face, Race in Postmodern America*. New York: Automedia, 1998.

Stampp, Kenneth M. *The Peculiar Institution: Slavery in the Ante-Bellum South*. New York: Knopf, 1956.

Starobin, Robert S., ed. *Denmark Vesey, The Slave Conspiracy of 1822*. Englewood Cliffs, N.J.: Prentice-Hall, 1970.

Stephens, Alexander Hamilton. *Constitutional View of the Late War Between the States: Its Causes, Character, Conduct and Results*. Philadelphia: National Publishing Company, 1868–70.

Sterling, Dorothy. *The Making of an Afro-American: Martin Robison Delany*. Garden City, N.Y.: Doubleday, 1971.

Still, William. *Underground Railroad*. Philadelphia. (1872) Reprint. New York: Arno Press & New York Times, 1968.

Stoddard, William Osborn. *Inside the White House in War Times*. (1890) Reprint. Ed. Michael Burlingame. Lincoln: University of Nebraska Press, 2000.

Stuckey, Sterling. *The Ideological Origins of Black Nationalism*. Boston: Beacon Press, 1972.

Thoreau, Henry David. *A Plea for Captain John Brown: Read to the Citizens of Concord, Massachusetts on Sunday Evening, October Thirtieth, Eighteen Fifty-Nine*. (1859) Reprint. Boston: David R. Godine, 1969.

———. *The Portable Thoreau*. Carl Bode, ed. New York: Viking Press, 1947.

Tragle, Henry Irving. *The Southampton Slave Revolt of 1831: A Compilation of Source Material*. Amherst: University of Massachusetts Press, 1971.

Turner, Nat and Thomas R. Gray. *The Confessions of Nat Turner . . . made to Thomas R. Gray* (1831). Reprint. Las Vegas: Classic Americana Publishing, 2000.

Villard, Oswald G. *John Brown, 1800–1859: A Biography Fifty Years After*. (1910) Reprint. New York: Knopf, 1943.

Walker, David. *Walker's Appeal, in Four Articles Together With a Preamble, to the Coloured Citizens of the World, but in Particular, and Very Expressly, to Those of the United States of America.* (1829) Reprint. Chapel Hill: University of North Carolina Press, 2011.

Warch, Richard. *John Brown.* Englewood Cliffs, N.J.: Prentice-Hall, 1973.

Warren, Robert Penn. *John Brown: The Making of a Martyr.* (1929) Reprint. Nashville: J. S. Sanders Books, 1993.

White, Ronald C. Jr. *Lincoln's Greatest Speech, the Second Inaugural.* New York: Simon & Schuster, 2002.

Whitman, Walt. *Complete Prose Works: Specimen Days and Collect, November Boughs and Good Bye My Fancy.* Boston: Small, Maynard, 1901.

Wills, Gary. *Lincoln at Gettysburg, The Words that Remade America.* New York: Simon & Schuster, 1992.

Wilson, Hill Peebles. *John Brown, Soldier of Fortune: A Critique.* Lawrence, Kans.: H. P. Wilson, 1913.

Wilson, Joseph T. *Black Phalanx: A History of Negro Soldiers of the United States.* Hartford, Conn.: American Publishing Company, 1888.

Winkley, Jonathan. *John Brown, the Hero: Personal Reminiscences.* Boston: James H. West, 1905.

Wish, Harvey, ed. *Ante Bellum.* New York: Capricorn Books, 1960.

Woodward, C. Vann. *The Burden of Southern History.* Baton Rouge: Louisiana State University Press, 1968.

NEWSPAPERS AND PERIODICALS

Anglo-African
Anglo-African Magazine
Appeal
Argus (Weston, Mo.)
Atlantic Monthly
Austin State Gazette
Baltimore American
Boston Courier
Boston Journal
Boston Post
Boston Transcript
Century (magazine)
Charleston Mercury
Chicago Tribune
Cincinnati Commercial
Cleveland Leader

Columbus (Ga.) *Sun*
Commonwealth (Boston)
Constitutional Whig
Daily Atlas and Bee (Boston)
Democrat and American (Rochester, N.Y.)
Des Moines Register
Detroit Free Press
Die Presse (Vienna, Austria)
Douglass's Monthly
Enquirer (Richmond, Va.)
Evening Journal (Albany, N.Y.)
Evening Post (New York)
Evening Star
Herald (Syracuse, N.Y.)
Herald of Freedom (Lawrence, Kans.)
Journal of Commerce
Journal of Negro History
Kansas Republican (Lawrence, Kans.)
Liberator
Liberty Bell
Missouri Democrat
Montgomery (Ala.) *Advertiser*
New Orleans Crescent
New York Herald
New York Independent
New York Times
New York Tribune
New York Weekly Tribune
New York World
North Star
Philadelphia Press
Provincial Freeman (Chatham, Ontario)
Republican (Savannah, Ga.)
Richmond Dispatch
Richmond Examiner
Springfield (Ill.) *State Journal*
Standard
State Journal (Columbus, Ohio)
Summit Beacon (Akron)
Times (London)
Transactions of the Kansas State Historical Society
Traveller (Boston)
Troy (N.Y.) *Whig*

Washington Republican
Washington Star
Washington Telegraph (Ark.)
Weekly Anglo African

Illustration Credits

All images are from the Library of Congress except:

p. 5. National Portrait Gallery, Smithsonian Institution.

p. 9. Massachusetts Historical Society.

p. 11. Frontispiece to *A Memorial Discourse*, 1865.

p. 41. Boyd B. Stutler Collection, West Virginia State Archives.

p. 71. George Stearns and Samuel Gridley Howe, Boyd B. Stutler Collection, West Virginia State Archives; Benjamin Franklin Sanborn, Concord Free Public Library.

p. 123. Special Collections of Alderman Library, University of Virginia.

p. 205. Boyd B. Stutler Collection, West Virginia State Archives.

p. 505. Frederick Douglass, New-York Historical Society.

p. 595. National Portrait Gallery, Smithsonian Institution.

ACKNOWLEDGMENTS

I HAVE BENEFITED FROM THE RESEARCH OF THE MANY AUTHORS listed in the bibliography, and from the collections of the libraries and archives, notably the New York Public Library, I consulted, but this book is entirely my responsibility.

After I had finished this research and writing of many years, I embarked on a campaign to find a suitable publisher, drawing on a list of university and small presses that might be amenable to a first-time writer. Nearly exhausting these, I began, as I must, to think I might have to roll up my manuscript and leave it for one or another of my children long after I was gone. At last I found among the stacks at a local bookstore a title, *The Caning: The Assault that Drove America to Civil War*. Sending a query to its publisher, I promptly received an invitation to send the manuscript. And in almost the time it took to receive, open and read, I received an acceptance letter. This was from Bruce H. Franklin, the sole proprietor of Westholme Publishing. It is he who has lifted this book out of its presumed destiny and into its possible destiny, and he who I must acknowledge, and all those allied with Westholme—copyeditor Noreen O'Connor-Abel, designer Trudi Gershenov, cartographer Tracy Dungan, proofreader Mike Kopf, indexer Kendra Millis—who have made a public presentation of this book a reality.

I thank my wife, Patricia, for her support throughout this entire process. She says she misses the clack of the typewriter as this manuscript was being prepared. I pounded two or three into oblivion before getting my first computer.

I chose the title from an anecdote recorded by historian Henry Tragle in 1969 while interviewing a seventy-year-old life-long resident of Boykins, Virginia, named Percy Claud regarding the "folk" memory of the Nat Turner insurrection. Asked whether he felt the future would be better for blacks, Claud replied, "God's going to destroy this here wicked race and he goin' to raise up a nation, a race that's goin' to be here, just like he did in Noah's day. . . . He's goin' to raise a nation, obey him, and do as his command, and love one another." It is to the voices of Americans like Claud and his forebears still abiding with and in us as part of our national memory that I have given my ear.

INDEX

Abbeokuta, 323
Abbott, J. B., 149
Abolition Bill (District of Columbia), 399
Adams, Charles Francis, 308, 598
Adams, John Quincy, 180
African Civilization Society, 158, 323
Alabama, 317–318, 342–343, 600–601
Albany Evening Journal, 308
Alcott, A. Bronson, 68, 165–166
Alcott, Louisa May, 400
Allstadt, Bryne, 217, 228
Allstadt, John, 217–220, 228
American and Foreign Anti-Slavery Society, 78
American Anti-Slavery Society, 79, 476, 502, 518–519, 572
"American Civil War, The" (Marx), 184–185
American Negro Slave Revolts (Aptheker), 285
Amistad mutiny, 2
Anderson, Jeremiah, 144, 168–169, 172, 205, 213, 228
Anderson, Osborne Perry: arrival in Virginia of, 204; Brown's appeal to black leaders and, 112; Brown's Chatham Convention and, 115–120, 122; escape of, 232–233; flight of to Canada, 248–249; in Harper's Ferry, 202, 205–208, 211, 216–218, 220–221, 224, 226–228, 235; portrait of, 205ph
Anderson, Robert H., 316, 320, 330, 334, 537, 610
Andrew, John, 68, 320, 432, 515

Anglo-African, 161, 236, 321, 505, 580, 592
Anthony, Henry B., 308
Anthony, Susan B., 295
Anthony Burns case, 22–25
antidraft riots, 494–495
Antietam, Battle of, 429–430, 435
Anti-Slavery Standard, 4, 6, 572
Anzeiger des Westens, 349–350
Appeal to the Coloured Citizens of the World (Walker), 2, 11, 142
Aptheker, Herbert, 135, 284–285
"Argument for Disunion, The" (Phillips), 184
Arkansas, 317, 340, 342–343, 352, 522–523
Army Life in a Black Regiment (Higginson), 457, 486, 524
Army of Northern Virginia, 409, 486, 588
Army of Tennessee, 550, 552, 566
Army of the James, 535, 573
Army of the Mississippi, 419
Army of the Ohio, 416, 439
Army of the Potomac, 365, 427, 435–436, 439, 452–453, 489, 534, 547–548, 573
Army of the Shenandoah, 551
Army of West Tennessee, 416
Asboth, Alexander S., 373, 381
Ashby, Turner, 406
Assing, Ottila, 99, 123, 246
Atchison, David R., 32, 41–42, 56, 59–62
Atlanta, 564, 575–576
Atlantic Monthly, 47, 150, 197, 238, 457
Avey, Elijah, 148
Avis, John, 258, 267, 268

Bailey, Frederick. *See* Douglass, Frederick
Baker, Edward, 384, 393
Ball's Bluff, Battle of, 383–384, 392–393
Baltimore American, 257
Banks, Nathaniel P: blacks in Louisiana and, 522; emancipated slaves and, 458; 1st Louisiana and, 482; Port Huron and, 476–477; Red River expedition of, 526–529, 538; Shenandoah Valley battles, 406–407; slavery in Civil War and, 372
Barnett, T. J., 454–455
Barry, Joseph, 221
Barton, Clara, 425
Bates, Edward, 327, 335, 476
"Battle Hymn of the Republic" (Howe), 70
Beauregard, P. G. Toutant, 337; Battle of Bull Run and, 358–361; Battle of Shiloh and, 398; at Bermuda Hundred, 538; in Carolinas, 588–589; at Charleston Harbor, 468; defense of Richmond and, 545; Florida and, 524; pre-war negotiations and, 329–330; in South Carolina, 508
Beckham, Fontaine, 230–231
Bee, Barnard, 361
Beecher, Henry Ward, 48
Bell, James Madison, 112, 115–116, 118, 122, 196, 203–204, 271, 297, 306
Belmont, August, 565
Benjamin, Judah, 328, 335*ph*
Bennett, James Gordon, 565
Benton, Thomas Hart, 77, 373
Bertonneau, Arnold, 522
Black, Jeremiah S., 316
Black Law (Illinois), 28, 157
Black Law (Kansas), 37
black officers, 591–592
Black Reconstruction, 351
black suffrage, 439, 571–572
black volunteers: Army of the Potomac and, 534; Battle at Chaffin's Farm and, 567–568; Fort Pillow Massacre, 529–532, 539; Milliken's Bend and, 490; at

Nine Mile Road, 573; pay for, 525; Poison Springs and, 523; Port Hudson and, 490; taking of Richmond and, 607
Blair, Charles, 80, 169
Blair, Francis P., 584
Blair, Frank, Jr., 347–348, 355, 373, 377
Blair, Montgomery, 327, 565
"blind memorandum", 553–554
Blunt, J. G., 147, 508
Boerly, Thomas, 220
Bondi, August, 52, 55
Booth, John Wilkes, 269, 610
Border Journal, 606
Border Post, 421
Border Star, 41
Boston Commonwealth, 596
Boston Courier, 309
Boston Daily Atlas and Bee, 340
Boston Transcript, 480
Boston Traveller, 320
Boston Vigilance Committee, 176
Boteler, Alexander, 232, 265
Botume, Elizabeth Hyde, 484
Boucaster, Helen, 275
Boyle, William A., 237
Bradford, Sarah, 174, 293, 314, 456, 484, 596
Bragg, Braxton: Battle of Chickamauga Creek and, 509; Battle of Murfreesboro and, 445–447, 450–451; Battle of Perryville and, 433; conference with Seddon and, 472; defense of Richmond and, 545; J. Johnston and, 444; in Kentucky, 428, 430–431; press to Chattanooga of, 419; withdrawal from Tennessee, 489
Brandy Station, Battle of, 486
Breckinridge, John C., 298, 306, 450–451, 538
Brewer, Joseph, 222–223, 230
Brice's Crossroads, Battle of, 543–544
British and Foreign Anti-Slavery Society, 5
Brochett, W. B., 130
Brown, Anne, 165, 171–172, 202, 206
Brown, B. Gratz, 540

Brown, Edward, 9

Brown, Esau. *See* Green, Shields

Brown, Frederick, 15, 34, 42, 56–57, 133

Brown, Henry "Box", 141–142

Brown, Jason, 15, 34, 43, 46, 57–58

Brown, John: anti-slavery movement and, 3–4; appeal to black leaders, 107–109; burial arrangements of, 267–268; Burns case and, 23; business of, 13–14, 16; Chatham Convention of, 115–123; as a commander, 52; continued fighting in Kansas and, 131–133, 143, 143–149; correspondence with wife, 83; correspondence with Douglass, 10, 22; decision to settle in Kansas, 33–36; defeat of at Harper's Ferry, 226–230, 232; deferral of return to Kansas, 87; Delany and, 113–114; departure from Kansas, 62; Detroit meeting of, 157–158; establishment of Harper's Ferry base and, 170–173; execution of, 265, 268–272; "Farewell" of, 81–82; first meeting with Douglass, 3, 7–10; Forbes and, 99, 123–125, 127; free-state legislature and, 46–47; Fugitive Slave Law and, 21–22; funeral of, 271–273; on government and Kansas, 63; Harper's Ferry raid and, 210, 212–213, 219–225, 234–238, 242–245; influence on Douglass of, 26; initial trip to Canada, 110–113; insurrectionary movements and, 135–137, 184–186; invocation of by abolitionist leaders, 310–311, 323, 366, 394, 465; A. Johnson on, 617; in Kansas Territory, 36–41; in Kansas war, 45–46, 48–51, 55–60; Lincoln inauguration and, 332; Marais des Cygnes Massacre, 131; meetings with Boston leaders, 103–105; meetings with Douglass in 1859, 163; meetings with Douglass in Chambersburg, 177–184, 190–195; meeting with Lane, 48–49; Mount Alto meeting of, 202–203; New England Anti-Slavery Conference (1860) and, 295; Oberlin-Wellington rescue trial and, 159–162; "Parallels" of, 147–148; Phillips interview, 150–151; plans for armed revolt, 8–10; plans for South and, 186, 188–189; portrait of, 9*ph*; in prison, 247, 261–265; Radical Abolition Party and, 36; raising of funds for army, 72–73, 168; recruiting for Virginia and, 89–91; on recruitment for army, 200; religion and, 1–2, 16; rescue attempts and, 258; retaliation for sacking of Lawrence and, 42–43; revolutionary influences of, 10–12; Smith settlements and, 14–16; Spooner and, 284; standoff with Pate and, 43–45; stay with Quakers, 95–96; steps to provision army, 65–66, 82–83; story of Harper's Ferry rifle, 65; Sumner and, 106; in Tabor, 84–86; travels east of, 67–73, 79–81; trial of, 250–252, 254–256; trip to Canada with fugitive slaves of, 149, 153–156; N. Turner and, 136; U.S. Arsenal at Harper's Ferry seizure and, xiii–xiv; United States League of Gileadites and, 21–22; on victory over Pate, 45; Virginia plans of, 100–101, 133–135, 179–180, 181–182; in W. E. B. DuBois biography, xii–xiv

Brown, John Jr.: advance trip South of, 106; attack of abolitionist meetings and, 309–311; attempted rescue of Brown and, 258; on Brown and religion, 16; on Brown's business, 13; Brown's Mount Alto meeting and, 203; Confederate sabotage plans and, 562–563; correspondence with Brown, 132; correspondence with Kagi, 202, 205, 209–210; in Kansas, 33–36, 42–43, 81; as liaison for father, 169, 171–172, 175–177, 196, 205–206; on *Life and Times* account of Brown meeting, 10, 98; Mason Committee and, 277; portrait of, 41*ph*; Radical Abolition Party

and, 36; on C. Robinson, 52
Brown, Joseph E., 452
Brown, Martha, 171–172
Brown, Mary, 83, 95, 109, 263–264, 266–267
Brown, Oliver, 36, 42, 172–173, 207, 213*ph*, 223, 228, 232, 240
Brown, Owen: in Cleveland, 125–126; decision to settle in Kansas, 34; in Harper's Ferry, 172–173; Harper's Ferry raid and, 212, 222, 232; journey to Kennedy Farm of, 199–200; in Kansas, 42, 48; portrait of, 91*ph*; return to Kansas, 83, 85; at Tabor, 90–91; on trip to Ohio with Brown, 91
Brown, Salmon, 34, 42, 48, 67, 89, 179, 190, 224
Brown, Watson, 36, 67, 172, 205, 212, 216, 228–230, 242, 247
Brown, William Wells, 111, 271, 436–438, 447–448, 477–480, 485, 525–526
Browning, Oliver, 399, 443
Brown Liberty Songsters, 271
Bryant, William Cullen, 547
Buchanan, James, 63; appointment of R. Walker, 86; civil war and, 181; Douglass and, 249; Dred Scot case and, 73; 1860 election and, 307–308; Harper's Ferry raid and, 226, 247; Kansas Territory and, 66, 88, 94, 129, 133; Oberlin-Wellington rescue trial, 159; presidential campaign of 1860 and, 304; proslavery party and, 185; secession of South Carolina and, 316
Buell, Don Carlos, 398, 416, 419, 427, 430–431, 433
Buford, Jefferson, 39, 41
Buford, John, 488
Bull Run, Battle of, 357–364, 392–393
Bureau of Colored Recruitment, 502
Bureau of Colored Troops, 532–533
Burns, Anthony, 22–25
Burnside, Ambrose E., 395, 439, 441–442, 452, 473, 552–553
Bushnell, Simeon, 159, 161
Butler, Benjamin F.: attack on New

Orleans and, 395; Battle of Spotsylvania and, 537; at Bermuda Hundred, 535–536, 538; blacks in Civil War and, 403; capture of New Orleans, 402–403; contraband camps and, 455; defense of Washington and, 346–347; entry in Maryland, 347; National Equal Rights League and, 570; Republican convention of 1864 and, 540; San Domingo plan and, 614–615; slavery in Civil War and, 368–369
Byrne, Terrance, 222, 225, 235

Calhoun, John C., 610
California, 352
Callioux, André, 478–479, 493
Cameron, Simon, 327, 343, 369–370, 379, 391–392
Campbell, Alexander, 56
Campbell, John A., 333, 337, 586
Canada, 110, 259
Canby, Edward, 600, 611–612
Cannibals All! (Fitzhugh), 77
Capture and Execution of John Brown, The (Avey), 148
Cary, J. B., 369
Cary, Mary Shadd, 3, 79, 112, 115
Cary, Thomas, 112, 118, 196
Cass, Lewis, 133
Cato, Sterling, 41, 61
Caucuses of 1860, 298
Cedar Creek, Battle of, 568
Cedar Mountain, Battle of, 420
Chaffin's Farm, Battle of, 567
Chambersburg, 487–488
Chancellorsville, Battle of, 470–472
Chapman, Maria Weston, 6
Charleston, 589–590
Charleston Harbor, 468, 610–611
Charleston Mercury, 305, 437, 583
Chase, Salmon P., 160; Fort Sumter and, 335; Hunter's proclamation and, 405; Lincoln's cabinet and, 327–329; Lincoln's second inaugural and, 598; portrait of, 335*ph*; Republican convention of 1864 and, 540; Seward and, 442–443; Thirteenth Amendment and, 580
Chattanooga, fall of, 508
Chatham Convention, 115–123,

134, 188

Cheerful Yesterdays (Higginson), 166, 173–174

Chester, Thomas Morris, 603–604, 607

Chestnut, James, Jr., 307

Chicago Tribune, 339

Chickamauga Creek, Battle of, 509–510

Child, Lydia Maria, 6, 261, 293, 295, 540

Christiana Fugitive Slave Rescue, 3

Cincinnati Commercial, 298

Cinqué, Joseph, 2

Civil Rights Act, 617

Clarke, George W., 41, 56, 130, 143

Clarke, James Freeman, 106, 270–271, 319–321

Clay, Cassius, 346

Clay, Clement, 548, 562, 565

Cleveland Leader, 160–161

Cline, James B., 55, 65

Cochrane, John, 541

Cold Harbor, Battle of, 542–543

Coleman, Lucy, 164

Colored Citizens of the State of New York, 36–37

Colored National Convention, 570

Columbia, 590

Columbus State Journal, 161

Columbus Sun, 285

Committee on the Conduct of the War, 392–393, 531

Committee on the Territories, 548

Compromise of 1850, 74

Condition, Elevation, Emigration and Destiny of the Colored People of the United States, Politically Considered, The, 113

Confederate Army of Tennessee, 550, 582–583

Confederate States of America, 318, 328, 342–343, 387–389, 403, 427, 452, 573

Confederate War Department, 583, 602

"Confessions of Nat Turner, The" (Hamilton), 136–137

Confiscation Acts, 371–372, 376, 405, 417, 520

Conquest of Kansas by Missouri and Her Allies, The, 44

Constitutional Union Party, 297

Constitutional Whig (Richmond), 140–141

Cook, Joe. *See* Lane, James

Cook, John: on Brown's meeting with Douglass, 98; in Cleveland, 125–126; on Douglass, 234–235, 248; execution of, 276; in Harper's Ferry, 127, 173, 207; Harper's Ferry raid and, 212, 216, 218, 222, 232, 235; in Kansas, 40; portrait of, 91*ph*; in prison, 268; recruitment of, 89; rescue attempts and, 258; stay with Quakers, 95; trial of, 250–252

Cooke, Phillips St. George, 58, 62

Cooper, Ezikial, 112

Copeland, John: arrival of in Virginia, 208–209; execution of, 276; Harper's Ferry raid and, 213, 221, 229, 240; Oberlin-Wellington rescue trial and, 161–162; in prison, 247, 268; trial of, 252

Copperheads, 466, 473, 477, 559, 562

Coppoc, Barclay, 91*ph*, 95, 172, 212, 222

Coppoc, Edwin: capture of at Harper's Ferry, 242; execution of, 276; in Harper's Ferry, 172; Harper's Ferry raid and, 213, 220, 228, 230; in prison, 247, 268; recruitment of, 95; trial of, 252

Corinth, Battle of, 433

cotton gin, xi

Cotton Kingdom, The (Olmsted), 75–76

cotton production: economics of, 75–77; expansion of, xi-xii; Harper's Ferry and, 284; onset of Civil War and, 351

Cox, Samuel S, 466–467

Crater, Battle of the, 552–553

Crimean War, 94, 363

Crittenden, John J., 305, 308, 317, 370

Crittenden Compromise, 308, 317, 354, 391

Cromwell, John A., 143

Cuba, 307

Cumberland, 345, 396

Curry, John Steuart, 33

Cushing, Caleb, 288–289, 298

Daingerfield, J. E. P., 220, 240–242
Dana, Charles A., 209
Daniels, Jim, 146–147
Darrell, James, 219–220
Davis, Henry, 555, 565
Davis, Jefferson, 187; Battle of Bull
    Run and, 362–364; Battle of
    Yorktown and, 402; border states
    and, 342; Bragg and, 451; on
    Brown, 281; cabinet of, 327–328;
    Confederate Congress of 1864
    and, 573; Confederate generals
    and, 443–444, 451–452, 511;
    Confederate States of America
    and, 318–319, 326; criticism of,
    582; Crittenden Compromise and,
    308; defense of Richmond and,
    545; Democratic convention of
    1864 and, 563; First Confederate
    Congress and, 358; Fort Sumter
    and, 333, 337; in Georgia, 566;
    Hampton Road negotiations and,
    584–587; Kansas and, 44, 59;
    Lee's entry in to Maryland and,
    426–427; Lincoln's preliminary
    emancipation proclamation and,
    432; Mason Committee and, 276;
    military strategy of, 403; Missouri
    secession fight and, 348; onset of
    Civil War and, 353; portrait of,
    335*ph*; pre-war negotiations with
    Union, 329; resignation from
    Senate, 317; secession of Southern
    States and, 316–317; on slavery,
    282–283; slavery in Civil War
    and, 443; on straight slave code,
    280; treatment of blacks in Union
    army, 484–485
Davis slave code, 280, 288
Day, William H., 29, 110, 158, 596
Dayton, William L., 54
De Baptiste, George, 157, 277
Deitzler, George, 129
Delamater, George B., 9
Delany, Martin: in anti-slavery
    movement, 3; black officers and,
    592–594; Brown's appeal to black
    leaders and, 109, 113–114,
    158–159; Brown's Chatham
    Convention and, 115–123;
    Charleston Harbor celebration
    and, 610–611; commission of as
    officer, 594–595; Douglass and,
    114; as emigrationist, 322–323,
    592; Freedmen's Bureau and,
    612–613; meeting with Lincoln of,
    592–594; National Emigration
    Convention and, 30; portrait of,
    123*ph*, 595*ph*; on recruiting black
    volunteers, 514–515; recruiting of
    halted, 612; Stowe proposal and,
    27–29; Tubman and, 596
Delaware, 352
Democratic Party: Buchanan and,
    66; divisions of in 1864, 548; Fort
    Sumter and, 338; Free Soil party
    and, 26; Kansas and, 47; Lincoln's
    preliminary emancipation procla-
    mation and, 439; McClellan and,
    390–391, 415; nominating con-
    vention of 1864, 562–565; nomi-
    nating conventions of 1860,
    287–289, 298–299; platform to
    end war, 466–467; presidential
    campaign of 1860 and, 303–306;
    slave state compromise of, xii;
    Thirteenth Amendment and,
    578–580; 36th Congress and, 276,
    280; war proclamation and, 341
De Molay, 481
Denver, James, 129, 133
Department of Kansas, 382
Department of the Border, 511
Department of the South, 403,
    455–456, 480, 482
Department of the West, 365,
    381–382
Detroit Free Press, 157
Detroit Vigilance Committee, 157
Dickinson, Anna M., 447, 490
Dickinson, Emily, 70
District of Columbia, 398–399
Disunion Convention, 79
Dorsey, Thomas, 207, 597
Douglas, Stephen, 23; Democratic
    convention of 1860 and, 298;
    1860 election and, 306–307;
    Kansas and, 86, 88, 94, 133;
    Lincoln and, 282; Lincoln inaugu-
    ration and, 330–333; presidential
    campaign of 1860 and, 305–306;
    slavery as political issue and,

78–79; 36th Congress and, 276; war proclamation and, 341

Douglass, Charles, 462

Douglass, Frederick: after the Civil War, 618–619; American Anti-Slavery Society meeting of 1863 and, 518–519; O. Anderson and, 233; attack of abolitionist meetings and, 309–312; black officers and, 592; on black suffrage, 572; on black volunteers, 464–466, 500–501; on Brown in Kansas, 52–53, 67; on Brown's appeal to black leaders, 97, 100, 103, 107–109, 202; Brown's Chatham Convention and, 115; Brown's stay with, 99–101; Brown's Virginia plans and, 98–101, 175; on Civil War, 371; on Compromise of 1850, 74, 78; on Confederate slave conscription, 583; correspondence with Brown, 10, 22, 34, 83; correspondence with Still, 299; "council of war" meeting with John Brown, xii–xiv; on J. Davis, 333; death of daughter and, 290; Delany and, 114; on Democratic conventions of 1860, 299; on Democratic Party, 569–571; on District of Columbia abolition bill, 398; on disunionists, 79; on Dred Scot case, 74–75, 79, 249; J. Eaton and, 554–555; Emancipation Proclamation and, 447–450; emigrationists and, 30–31, 79, 157; endorsement of Republican Party and, 53–55; on escaped slaves, 297; in Europe, 275–276, 289; evolution of anti-slavery ideas, 25–26; on failure of Brown, 620–621; 54th Regiment and, 481; first meeting with Brown, 3, 7–10; flight of to Canada, 248–249; Forbes and, 123; Fort Sumter and, 338; on Frémont nomination, 540–541; on Frémont's manumission proclamation, 375–376; on Fugitive Slave Law, 17, 20; on Garrison, 27, 367; on Greeley, 467; Harper's Ferry raid and, 211, 225, 234–237, 239, 246, 248, 260,

299–300; informing slaves of emancipation and, 560–561; Johnson administration and, 616–617; on Kansas Territory, 36–37; letter of summons to Harper's Ferry, 203; on letter to Kossuth, 12–13; on Lincoln, 327, 414, 422–423, 532; Lincoln inauguration and, 330–331; on Lincoln nomination, 566; Lincoln's preliminary emancipation proclamation and, 432; Lincoln's second inaugural and, 597; Mason Committee and, 277, 297; on McClellan, 413–414; meeting at Brown's grave and, 299; meetings with Brown, 157–158, 165, 177–184, 190–195; meetings with Lincoln, 502–504, 557–560; meeting with Stanton, 504–505; Midwest speaking tour of, 156–157; *North Star* and, 6; offer to assist in Vicksburg and, 505–507; W. Parker and, 197–199; portrait of, *5ph*, *505ph*; presidential campaign of 1860 and, 306; purchase of freedom of, 6; Radical Abolition Party and, 35–36, 302–303; recruiting of black volunteers, 461–463, 490–492, 494; on Republican convention of 1860, 296, 301; on Republican Party, 569–571; rise to prominence of, 2–5, 25; on secession of Southern states, 308, 324–325; on slavery and Civil War, 370–372, 376–377, 382–385, 391–394, 414, 516–518; on slavery as political goal, 77–78; speaking tour of 1861 and, 395–396; speech in Chambersburg of, 189–190; E. Stanley and, 458; Stowe and, 27–30; Sumner's 1860 Senate speech and, 301; Syracuse appearance of, 385; on travel to Haiti, 323–324; on U.S. Constitution, 275–276; on Union generals, 453; on violence against blacks, 423–424; on Virginia insurrections, 439; on war and abolition, 571; women's rights movement

and, 6–7
Douglass, H. Ford, 30–31, 156–157
Douglass, Lewis, 175, 462,
  497–499, 515
Douglass' Monthly, 25, 156, 236,
  297, 299–301, 303, 323–324,
  365–366, 465
Downing, George T., 12
Doy, John, 201–202
Drewry, William S., 138
Dubois, F. E., 322
DuBois, W. E. B: on blacks of anti-
  slavery movement, 3; on black
  volunteers, 490; on Brown in
  Kansas, 45–46; on Brown's appeal
  to black leaders, 97, 107, 119; on
  Brown's business endeavors, 13;
  on Brown's preparations for raid,
  207; on Canada, 110; on "council
  of war" meeting between Brown
  and Douglass, xiii–xiv; on
  Harper's Ferry raid, 233, 235,
  261; onset of Civil War and, 351;
  on slaves after Emancipation
  Proclamation, 513; on Tubman,
  174
Duden, Gottfried, 349
Durrant, Henry K., 456

Early, Jubal, 545, 549, 567–568,
  568, 600
Eaton, John, 554
Elliot, R. C., 46
Ellsworth, A. M., 122
Emancipation League, 386, 396
Emancipation Proclamation, 190,
  447–450, 454–455, 520, 522
Emerson, Ralph Waldo, 68–69,
  80–81, 186–187, 432–433, 449,
  574
emigrationists, 30–31, 79, 114
Engels, Frederick, 284
England, 387–389, 427
Everett, Edward, 297
Ewell, Richard B., 405
*Exposition of the Constitutional
  Duty of the Federal Government
  to Abolish Slavery* (McCune
  Smith), 36

Fair Oaks, Battle of, 407
Farragut, David, 402–403, 467–468,
556
Fayetteville, taking of, 601
Fifteenth Amendment, 617
Fillmore, Millard, 12
financial panic of 1857-1858, 94–95
First Days Amongst the Contraband
  (Botume), 484
Fisher's Hill, Battle of, 567
Fitzhugh, George, 77–78
Five Forks, Battle of, 602
Fleetwood, Christian, 568*ph*
Florida, 317, 523–526
Floyd, John, 140, 225, 278, 351,
  395
Forbes, Hugh, 82, 84–86, 123–127,
  277
Forbes, John Murray, 167
Forrest, Nathan Beford: Battle of
  Brice's Crossroads and, 543–544;
  Battle of Murfreesboro and, 416,
  446–447; Battle of Nashville and,
  576; Fort Pillow Massacre and,
  529–532; in Georgia, 539; in
  Mississippi, 551; portrait of,
  544*ph*; Sherman's march through
  Mississippi and, 528; Vicksburg
  and, 445
Fort Blakeley massacre, 608–609
Fort Donelson, Battle of, 395
Fort Fisher, 582
Fort Henry, Battle of, 395
Fort Moultrie, 315–316
Fort Pickens, 334
Fort Pillow Massacre, 529, 539
Fort Stedman, Battle of, 602
Fort Sumter, 315–317, 330,
  333–338, 352
Fort Wagner, Battle of, 495–499
Foster, Abbey Kelly, 293
Foster, Stephen S., 302–303
Fourteenth Amendment, 617
Fox, Gustavus, 334, 337–338
Franklin, William, 452
Frazier, Garrison, 581
Frederick Douglass' Paper, 25
Fredericksburg, Battle of, 441–442
Freedmen's Bureau, 70, 590–591,
  612, 617
Freedom's Journal, 2
Free Soil party, 26
Free State Democrats, 37, 39, 46–47
Free State Hotel, 38, 41–42

Frémont, Jessie, 373
Frémont, John C., 54; Department
of the West and, 365; fight for
Missouri and, 372–373, 379–381;
Lincoln and, 565–566; manumission proclamation of, 375–377;
portrait of, 373*ph*; presidential
nomination in 1864 and,
540–541; removal of, 381; resignation of, 412; Shenandoah Valley
battles, 406–407; withdrawal
from 1864 election of, 566
Friends of Order, 493
Front Royal, Battle of, 487
Frost, D. M., 349–350
Fugitive Slave Law, 17, 20–22, 71,
110, 161, 244, 307, 309, 369
Fuller, Abram, 153

Gaines' Mill, Battle of, 408–409
Garibaldi, Giuseppe, 300, 371
Garnet, Henry Highland: in anti-
slavery movement, 3; Brown's
appeal to black leaders and, 100,
102–103, 107–109, 158–159;
Delany and, 323; portrait of,
11*ph*; revolutionary ideas of,
10–12; Thirteenth Amendment
and, 580; Tubman and, 313; on
unrest after Harper's Ferry, 257
Garrett, Thomas, 312–313
Garrison, William Lloyd: anti-slavery campaigns of, 3; on black suffrage, 571–572; Brown execution
and, 270–271; Brown's travels
east and, 68–69; Charleston
Harbor celebration and, 610–611;
Civil War and, 367–368; Disunion
Convention and, 79; Douglass
and, 4, 26–27; Emancipation
League and, 386; Emancipation
Proclamation and, 449; 54th
Regiment and, 481; introduction
of the *Liberator* and, 2; Lincoln's
preliminary emancipation procla-
mation and, 432; New England
Anti-Slavery Conference (1860)
and, 295; W. Parker and, 198; on
Republican convention of 1864,
547; split with Phillips, 539–540
Geary, John W., 59–63
Georgia, 317–318, 342–343

Gettysburg, Battle of, 488–489
Gettysburg Address, 513
Giddings, Joshua R., 160–161, 246,
277–279, 295, 308
Gilead League. *See* United States
League of Gileadites
Gill, George: on Brown, 145–147;
recruitment of, 95; reluctance to
go to Virginia, 172; on Reynolds
visit, 125–126; on road to
Harper's Ferry, 209; trip to
Canada with fugitive slaves and,
153–155
Gilmore, James R. *See* Kirke,
Edmund
Gloucester, Elizabeth, 178
Gloucester, James, 103–105, 107,
158–159
Gorsuch, Edward, 198–199
Gosport Naval Yard, 342, 344–345,
396
Grant, Ulysses S.: Battle of Chaffin's
Farm and, 567–568; Battle of
Cold Harbor and, 542–543; Battle
of Fort Donelson and, 395; Battle
of Fort Stedman and, 602; Battle
of Milliken's Bend and, 485;
Battle of Nashville and, 577;
Battle of Shiloh and, 398; Battle
of Spotsylvania and, 537–538;
Battle of the Crater and, 552–553;
Battle of the Wilderness and,
535–536; conference with Lincoln
in Virginia, 601–602; "contraband
camps" and, 455; drive to
Richmond of, 572–573; at
Jackson, 469–470; Lee's surrender
and, 608; McClellan and, 390;
McClernand and, 445; at
Petersburg, 545; portrait of,
529*ph*; promotions of, 511–512,
528; Republican convention of
1864 and, 540; in Shenandoah
Valley, 544–545; Sherman's
Carolinas march and, 573, 588;
Sherman's Mississippi march and,
526–527; strategy as general-in-
chief, 533–534, 546; surrender
negotiations and, 611; on taking
of Mobile, 600; taking of
Richmond and, 606; transfer of
black troops to Army of the

Potomac, 534; Vicksburg and, 415, 445, 468–469, 476, 489; in Virginia, 549, 567

Greeley, Horace, 123; Battle of Bull Run and, 365; on Brown, 148; on ending the war, 467; Forbes and, 84; *The Impending Crisis* and, 276; on invasion of Confederacy, 356; Kansas Territory and, 143; letters to Lincoln of, 424, 548–549; on Lincoln and war, 546–547; on maintaining the union, 308, 324; on nomination of Lincoln, 565; peace negotiations and, 550; on Republican convention of 1864, 540; Republican Osawatomie meeting and, 201

Green, Shields: Brown's Mount Alto meeting and, 202–203; capture of at Harper's Ferry, 242; execution of, 276; in Harper's Ferry, 178–179, 189, 193, 195, 206–207; Harper's Ferry raid and, 216, 227–228; journey to Kennedy Farm of, 199–200; in prison, 247, 268; trial of, 252

Grierson, Benjamin, 527–528

Griffiths, Julia, 7, 198, 515–516

Grimes, Leonard, 271, 447, 449

Grinnell, Josiah Bushnell, 155

Guesses at the Beautiful (Realf), 89

Guide to Hayti, 322

Hahn, Michael, 522

Hairgrove, William, 130–131

Haiti, 322, 415

Haitian Bureau, 323

Halleck, Henry M.: Battle of Shiloh and, 399; Buell and, 427; Department of the West and, 382; McClellan and, 436; McClernand and, 445; press to Chattanooga of, 415–416; removal of general-in-chief duties, 528

Hall's Rifle Works, 170, 181, 213

Halstead, Murat, 298

Hambleton, Charles A., 130

Hamilton, James, Jr., 142

Hamilton, James C., 116–117

Hamilton, Thomas, 136–137, 321

Hamlin, Hannibal, 547

Hammond, James H., 307

Hampton, Wade, 168, 361, 544–545, 573, 588, 601

Hampton Road negotiations, 584–586

Harper, Samuel, 153

Harper's Ferry, Battle of, 429

Harper's Ferry: described, 169–170; place in Brown's Virginia plans, 101; strategic importance of, 179–181; surrender to Confederate Army of, 429

Harper's Ferry Raid: capture of train station, 218–220; eve of, 211–212; militia response to, 226–230; onset of Civil War and, 351; reasons for failure of, 233–238; slaves and, 224–225, 260; taking of Armory in, 212–213; taking of estates in, 216–217; timing of, 210; U.S. Marine suppression of, 240–242; waiting period in, 223–225

Harriet, the Moses of Her People (Bradford), 174, 314

Harris, James H., 118, 205, 271

Harris, Thomas A., 377

Hart, Albert Bushnell, 499

Harvey, James, 48, 57–59

Haupt, Herman, 425

Hay, John, 550

Hayden, Lewis, 3, 23–24; Brown's Virginia plans and, 176; letter to Brown party in Virginia, 204; Mason Committee and, 277; Merriam and, 204, 207; Tubman and, 209

Hayes, Rutherford B., 617

Hayward, Sherwood, 218

Hazlett, Albert, 144, 172; escape of, 232–233; execution of, 278; Harper's Ferry raid and, 213, 228; portrait of, 213*ph*; rescue attempts and, 258; trial of, 250–252

Heinzen, Karl, 312, 540

Helper, Hinton Rowan, 276, 459

Henry, William, 291

Herald of Freedom, 38, 42

Hicks, Governor, 344, 347

Higginson, Thomas Wentworth, 23–24; black regiment of,

456–458, 482; on black volunteers, 486, 524; Brown's appeal to black leaders and, 100, 102; on Brown's search for "coadjutors", 15; Brown's travels east and, 68, 70, 72, 166, 168; on Brown's Virginia plans, 102, 105; on denial of secret six members, 249–250; Disunion Convention and, 79; escort of Brown's wife to prison of, 266; Forbes delay and, 124, 127; on Free State faction in Kansas, 43; on Lane, 48; Merriam and, 207–208; on Nebraska City, 58–59; on W. Phillips, 287; portrait of, 71*ph*; on Spooner circular, 283–284; on Tubman, 167–168, 173–174

Hill, A. P., 409, 412, 429, 430, 488

Hill, D. H., 430

Hill, J. J., 607

Hines, Thomas, 562–563

Hinton, Richard J.: on Brown in Kansas, 48–49; on Brown's business endeavors, 13; Brown's Mount Alto meeting and, 203; Brown's recruiting and, 205; on Brown's Virginia plans, 100–101, 134–135; G. Gill and, 145–147; Harper's Ferry Raid and, 237–238; journey to Harper's Ferry and, 209; in Kansas, 40, 48, 62; on *Life and Times* account of Brown meeting, 10, 98; Mason Committee and, 277; on J. Montgomery, 132; recruitment of for Virginia, 89; on slave insurrection, 201; on stay with Quakers, 95–96; on Stevens, 90; Tenth National Women's Rights Convention, 293

Holden, Isaac, 118–119, 158

Holley, Sallie, 293

Holly, James T., 321–322

Holmes, James H., 55, 62, 84–85

Holmes, Oliver Wendell, 549

Homestead Act, 295

Hood, John Bell: Army of Tennessee and, 550; Battle of Nashville and, 576; in Carolinas, 588; in Georgia, 551, 566–567, 568–569; resignation of command, 577

Hooker, Joseph, 452, 468, 470–472, 486

Howe, Julia Ward, 70

Howe, Samuel Gridley: Brown's travels east and, 68, 70, 72, 167–169; Brown's Virginia plans and, 105, 176; Burns case and, 23; Emancipation League and, 386; evolution of ideas on slavery and, 168; Forbes and, 124; on Harper's Ferry raid, 249–250; Mason Committee and, 277; meeting with Lincoln and, 458–460; portrait of, 71*ph*

Hoyt, George H., 258, 265

Hunter, Andrew, 247–248, 252–253, 256–257, 265, 268–269, 619

Hunter, David: abolition proclamation of, 404–405; black volunteers and, 475, 482–483; Department of Kansas and, 382; Department of the West and, 381; emancipated slaves and, 456; portrait of, 457*ph*; in Shenandoah Valley, 544–545; slavery in Civil War and, 403–404

Hunter, Robert, 78

Hurd, H. B., 72

Hutchinson Family Singers, 405

Hyatt, Thaddeus, 55, 263, 277

Hyndale, Hector, 266

Illinois, black suffrage referendum and, 439

Impending Crisis, The (Helper), 276, 459

Ingalls, John J., 56

insurrectionary movements, 74–75, 85–86, 135–137; after Harper's Ferry, 284–286; Brown trial and, 256–257; 1856-1857, 64–65; Spooner and, 283–284; Turner and, 137–143

International Workingmen's Association, 238

Jackson, Claiborne, 347–350, 355–356

Jackson, Thomas J. "Stonewall": Battle of Antietam and, 429–430; Battle of Bull Run and, 360–362; Battle of Cedar Mountain and,

420; Battle of Chancellorsville and, 470–472; Battle of Fredericksburg and, 442; Battle of Gaines' Mill and, 409; Battle of Kernstown and, 401; Brown execution and, 269; Confederate Army and, 408; portrait of, 407*ph*; promotion of, 440; reinforcement of Richmond and, 419; Second Bull Run and, 424–425; Shenandoah Valley battles, 405–407; Virginia secession and, 342

Janet Kidson, 323

Jefferson Guard, 227

Jennison, Charles R., 375

Jim (Washington's coachman), 217–218, 260

John (slave of A. B. Rhett), 437

John Brown, A Biography Fifty Years After (Villard), 51

John Brown (DuBois), xii–xiv, 1

John Brown Among the Quakers (Richman), 96

John Brown and His Men (Hinton), 48, 90, 125–126, 134, 146

John Brown hymn, 604–606

John Brown in Canada (Hamilton), 116–117, 153

"John Brown Song", 70

John Brown the Hero (Winkley), 51

Johnson, Andrew, 370, 547, 597–598, 611, 616–617

Johnston, Albert Sidney, 398–399

Johnston, Joseph E.: Army of the Potomac and, 356; Battle of Bull Run and, 359, 363–364; Bragg and, 451; conference with Seddon and, 472; at Dalton, 535; Davis and, 444–445; in Georgia, 539, 542, 546, 550, 566; position in Carolinas and, 582; relief of command of, 550; on Sherman, 588; surrender negotiations and, 611; Vicksburg and, 470, 472; withdrawal from Manassas, 401

Jones, James Munroe, 116–119, 158

Jones, John, 29, 156–157

Journal of Negro History, 143

Kagi, John Henry: Brown and Douglas "council of war" meeting and, xiii–xiv; Brown's Chatham Convention and, 117–119, 122; Brown's Mount Alto meeting and, 202; Brown's plans for South and, 186; correspondence with John Brown, Jr., 196–197, 199, 205, 209–210; in Harper's Ferry, 171–173, 177–179, 205; Harper's Ferry raid and, 213, 221, 224–225, 228–229, 238; in Kansas, 40, 144–145; Oberlin-Wellington rescue trial and, 160; Phillips interview and, 150; portrait of, 91*ph*; reconnaissance in Missouri and, 133; recruitment of for Virginia, 89–90; stay with Quakers, 95; trip to Canada with fugitive slaves and, 154–155

Kaiser, Charles, 57

Kansas Free State, 42

Kansas-Nebraska Bill, 22, 32, 74

Kansas Republican, 253

Kansas Territory: creation of, 32–33; elections and, 34, 86, 129; fighting in, 41–47, 50–51, 55–61; first elections in, 34–35; Free State elections, 39; free-state legislature and, 46–47; Lecompton Constitution and, 88, 91, 94, 129; Lecompton legislature and, 86–87; Marais des Cygnes Massacre, 130–131; rising tensions in, 37–41; sacking of Lawrence, 41–42; settlement of, 33–34

Keitt, Lawrence M., 303–304

Kelly, Abby, 6, 518

Kelly, Stephen, 518

Kemp, Walter, 219–220

Kentucky: Frémont's manumission proclamation and, 376; proclamation of emancipation and, 418; secession and, 317, 342–343, 352; statement of neutrality and, 354–355; war proclamation and, 340

Kernstown, Battle of, 401

Kinnard, Thomas, 118, 120–122

Kirke, Edmund, 286

Kitzmiller, A. M., 219, 230, 257

Know-Nothing Party, 276

Kossuth, Louis, 12–13

Lambert, William, 3, 30, 118, 157, 196, 271

Lamon, Ward, 326

Lane, James: fight for Missouri, 374–375, 377; Free State Hotel negotiations, 38–39; Frémont's manumission proclamation and, 378, 380; Frontier Guard and, 343, 346; Kansas elections and, 129; in Kansas war, 48, 50, 56–59, 61–62; on kidnapping, 149; on Lincoln nomination of 1864, 547; Montgomery's "Self-Protection Company" and, 130; removal of, 382; request for Brown's return and arms, 87–88; Schofield and, 511

Langston, Charles H., 159, 161, 205, 271

Langston, John Mercer, 159, 515

Last Days of John Brown (Thoreau), 400

Law and Order Party, 37

Lawrence, Amos, 68–69, 83, 166

Lawrence, Kansas, sacking of, 41

League of Liberty, 196

Leary, Lewis Sheridan: arrival of in Virginia, 208–209; Harper's Ferry raid and, 216, 221, 225, 229, 231; Oberlin-Wellington rescue trial and, 161–162

Leaves of Grass (Whitman), 364–365

Lecompton Constitution, 88, 91, 94, 129

Lee, Fitzhugh, 542

Lee, Robert E.: Battle of Antietam and, 429–430; Battle of Chancellorsville and, 470–472; Battle of Cold Harbor and, 542–543; Battle of Fort Stedman and, 602; Battle of Fredericksburg and, 441–442; Battle of Gaines' Mill and, 408–409; Battle of Gettysburg and, 488–489; Battle of Malvern Hill and, 412; Battle of Shepherdstown and, 435; Battle of Spotsylvania and, 537–538; Battle of the Wilderness and, 535–536; Battle of Trevillian Station and, 545; Battle of Yorktown and, 402; Brown execu-

tion and, 266; Confederate Army and, 408; defense of Richmond and, 545; entry in to Maryland, 426–428; general-in-chief position and, 582; Grant's strategy and, 546; Harper's Ferry raid and, 232, 240–242; invasion of Pennsylvania and, 476, 487–488; on Pope, 420; portrait of, 473*ph*; reinforcement of Richmond and, 419; reorganization of the army and, 286; Second Bull Run and, 424; strategic goals of, 477; surrender at Appomattox, 608; surrender negotiations and, 611

Lee, William, 392–393

Leeman, William, 172, 218, 220, 222, 225, 228, 230, 235

Lenhart, Charles, 44, 258

"Letter to Louis Kossuth, A", 12–13

Liberator, 2, 4, 6, 11, 25, 142, 301–302, 398, 400, 591–592

Liberia, 415

Liberty Party, 35–36

Life and Letters of John Brown, Liberator of Kansas, and Martyr of Virginia, The (Sanborn), 52, 203

Life and Public Service of Martin R. Delany (Rollin), 114

Life and Times of Frederick Douglass (Douglass): on Brown's stay with Douglass, 99; on "council of war" meeting with John Brown, xii–xiv; on first meeting with Brown, 9–10; on letter of summons to Harper's Ferry, 203; Mason Committee and, 297; meetings with Brown in Chambersburg, 178; meeting with Lincoln and, 560; on reception of *North Star* plans, 6; on W. Parker, 198; recruiting of black volunteers and, 461–463; on Stowe, 29–30; on travels in British Isles, 5

Lincoln, Abraham: arrival in Washington, 325; assassination of, 611; Bank's Red River expedition and, 527; Battle of Ball's Bluff and, 384; Battle of Bull Run and, 357, 363–365; Battle of Chancellorsville and, 472; Battle of Fredericksburg and, 442; Battle of Murfreesboro and, 451; Battle of Vicksburg and, 489; black volunteers and, 474; "blind memorandum" of, 553–554; cabinet of, 327; call for volunteers, 550; calls for emancipation proclamation and, 391–392; colonization and, 440–441; conscription and, 460–461; Cooper Institute speech of, 282; on Douglass's criticisms, 55; 1860 election and, 306–307; 1864 election and, 573–574; election of 1864 and, 578; emancipation and, 559–560; Emancipation Proclamation and, 190, 418, 447–450; at Fort Stevens, 549; Fort Sumter and, 333–338, 352; Frémont's manumission proclamation and, 376; Frontier Guard and, 343; Gettysburg Address of, 513; gradual abolition proposal, 396, 416–417; on Grant, 469, 534; Greeley's letter and, 424; Hampton Road negotiations and, 584–586; on Hooker, 468; Hooker and, 486–487; Hunter's proclamation and, 405; inauguration of, 324, 328–329, 330–333; Lee's surrender and, 609–610; on length of war, 546; Maryland legislature and, 347; McClellan and, 391, 394–395, 401, 412, 414, 427, 435–436, 439; on Meade, 489; meeting with abolitionists, 458–460; meeting with Delany, 592–594; meeting with Douglass, 502–504, 557–560; peace conditions of, 520; on Peninsula campaign, 422; Phillips and, 397; portrait of, 335*ph*; preliminary emancipation proclamation, 431–433; presidential campaign of 1860 of, 303–306; presidential nomination of 1864 and, 547; proclamation of retaliation, 502–504; reconstruction and, 523, 532; relief of Chattanooga, 511; Republican convention of 1860 and, 295; Republican Party and, 555–557; second inaugural of, 596–600; senatorial campaign of, 78–79;

Seward's memorandum and, 336–337; Seward's resignation and, 442–443; Shenandoah Valley battles, 405–406; slavery in Civil War and, 368, 370–371; taking of Richmond and, 606–607; Thirteenth Amendment and, 578–590; 37th Congress and, 357–358; Union generals and, 401, 427, 452–453, 510–511, 573–574; on Vallandigham, 473; war proclamation and, 338–341, 353

Loguen, Amelia, 498

Loguen, Jermain W.: in anti-slavery movement, 3; Brown's appeal to black leaders and, 100, 103, 107, 109, 158–159, 205; Brown's Chatham Convention and, 115, 122; Brown's Virginia plans and, 175–176; in Canada, 110–112; on Fugitive Slave Law, 21; recruiting for Brown's army of, 196; Tubman and, 111, 174

London Times, 359–360

Longstreet, James, 424–425, 440, 472, 509–510

Louisiana, 317, 522

Louisiana Purchase, xi

Lovejoy, Elijah, 2–3, 8

Lundy, Benjamin, 2

Lyon, Nathanial, 348–350, 355–356, 373–374

Magoffin, Gov., 354–355

Magruder, J. B., 401

Mallory, Charles, 369

Mallory, Stephen, 328

Malvern Hill, Battle of, 412

Manassas, Battle of, 357–365

Manual for the Patriotic Volunteer (Forbes), 82

Marais des Cygnes Massacre, 130–131

Martin, J. Sella, 447

Marx, Karl, 47, 184–185, 238, 284, 319, 380, 420–421, 428–429, 433–434

Maryland, secession and, 342, 344, 347, 352

Mason, James, 234, 242–245, 276, 387–388

Mason Committee, 180, 253, 276–279, 297–298

Massachusetts, 339, 480–481

Massachusetts Anti-Slavery Society, 4, 64, 173, 201, 319–321, 539–540

Massachusetts Arms Company, 82

Massachusetts Emigrant Aid Society, 38

Massachusetts Kansas Committee, 48, 69–72, 124

May, Samuel J., 295, 309–310, 385

Mayo, Katherine, 265, 456

McCauley, Charles, 344–345

McClellan, George B.: Army of the Potomac and, 365; Battle of Antietam and, 429–430, 435; Battle of Ball's Bluff and, 383; Battle of Bull Run and, 363; Battle of Gaines' Mill and, 409; Battle of Shepherdstown and, 435; Battle of Yorktown and, 401–402; Committee on the Conduct of the War and, 393; Democratic convention of 1864 and, 563–565; Douglass on, 413–414; Lincoln and, 435–436; Lincoln's preliminary emancipation proclamation and, 435; military hesitations of, 390–391; Peninsula Campaign and, 394–395, 401, 419; on Pope, 420; portrait of, 415*ph*; reinforcement of Richmond and, 419; removal of, 439; on Seward, 391; Shenandoah Valley battles, 405, 407; slavery in Civil War and, 368, 405, 414–415

McClernand, John A., 445

McCulloch, Ben, 373–374, 380–381

McCune Smith, James, 34, 35–36, 99, 124

McDowell, Irwin, 357–362, 368, 401, 405–407, 427

McFarland, Samuel, 53

McGuire, Horace, 163

McKinstry, John, 381

McPherson, James B., 551

Meade, George, 489, 534, 545, 552–553

Meigs, Captain, 336

Memminger, Christopher, 328

Merriam, Francis J., 204, 207, 212, 222

Merrimack, 344–345, 396

Mexican War, 74

Michigan, 339

Miles, Dixon, 429

Military Division of the Mississippi, 511–512

Milliken's Bend, Battle of, 485–486

Minnesota, 339

miscegenation, 569

Mississippi, 316–317

Missouri: Dred Scott case and, 73; elections in, 66; fight over secession, 347–350, 355–356; Frémont's manumission proclamation, 375–377; Kansas War and, 69; secession and, 317, 342–343, 352; war proclamation and, 340

Missouri Democrat, 284

Mobile, taking of, 600

Mobile Bay, Battle of, 556

Moffet, Charles, 209

Monitor, 396, 403

Montgomery, James, 56; black regiment of, 482; break with Brown, 149; Civil War activities of, 482–484; fight for Missouri and, 375; fighting in Kansas and, 130, 143; Gill's account of, 145–147; Harper's Ferry and, 258; meeting with Brown and, 132–133

Montgomery Advertiser, 285

Morgan, John Hunt, 419, 426, 446–447

Morton, Edwin, 102, 165, 175–176

Mott, Lucretia, 269–270

Mulligan, James A., 377–378

Munroe, William C., 118, 157, 158

Murfreesboro, Battle of, 445–448, 450–451

Musgrove, Thomas M., 262

My Bondage and My Freedom (Smith), 25

Myers, Stephen, 463

Mystic Red, 285

Nalle, Charles, 291–292

Napoleon, xi

Narrative of Henry Box Brown, 141–142

Narrative of the Life of Frederick Douglass (Douglass), 5

Nashville, Battle of, 576
National Emigration Convention, 30–31
National Equal Rights League, 570
National Kansas Committee, 47, 50, 66, 72, 84
National Women's Rights Convention, 293
Native Guards, 403
Nat Turner's Slave Rebellion (Aptheker), 135
Negro in the American Rebellion, The, 436–438, 477–480
Newby, Dangerfield, 200, 219–220, 223, 227–228, 230
New England Anti-Slavery Conference (1860), 295–296
New England Anti-Slavery Society, 540
New England Asylum for the Blind, 70
New England Colored Citizens, 174
New England Emigrant Aid Company, 33, 47
New Era, 529–530
New Orleans Crescent, 305, 455
New York Herald, 245–246, 248, 252, 270, 285, 421, 478–479, 526
New York Independent, 261
New York State, 303, 324
New York State Asylum for the Insane, 249
New York Times, 59, 149, 378, 595
New York Tribune, 14, 38, 45, 73, 82, 90, 147–148, 161, 209, 245, 308, 338–339, 356, 386, 421, 479, 483, 555
New York Weekly Tribune, 65
North Carolina, 317–318, 340, 342, 352
North Star, 6, 11, 25

Oak Hill, Battle of, 373–374
Oberlin-Wellington rescue trial, 159–162, 244
Ocean Pond, Battle of, 525–526
O'Connell, Daniel, 5
Official Report of the Niger Valley Exploring Party (Delany), 323
Ohio, 339
Ohio Anti-Slavery Society, 159
Old John Brown Liberty League, 271
Olive Branch, 530
Olmsted, Frederick Law, 75–76
Olustee, Battle of, 525, 526
Oregon, 352
Osawatomie, Battle of, 56–57
Our American Cousin, 611

Painter, John, 95
"Parallels" (Brown), 147–148
Parker, Theodore, 23, 62; Brown's appeal to black leaders and, 100–102, 103; Brown's travel's east and, 67–71, 81–82; Brown's Virginia plans and, 105; death of, 168, 295; Forbes and, 126–127; Forbes manual and, 105–106; on Harper's Ferry raid, 261; Mason Committee and, 277; portrait of, 71*ph*; Spooner and, 283
Parker, William, 3, 21, 112, 197–199
Parsons, L. F., 89
Parsons, Luke, 209
Pate, Henry Clay, 41, 43–46, 59, 261, 537
Patterson, Robert, 356, 358, 363
Pawnee, 345
Peachtree Creek, Battle of, 546
Pemberton, John C., 444, 469, 470, 472
Pendleton, George, 564
Pennington, J. W. C., 29
Pennington, William, 280
Pennsylvania Anti-Slavery Society, 108
Perryville, Battle of, 433
"Personal Liberty Laws", 308–309
Phil (Allstadt's slave), 217, 228, 260
Philadelphia Press, 603, 607
Phillips, Andrew, 218–220
Phillips, Wendell, 23–25; attack of abolitionist meetings and, 310–312; on black race, 322; Brown funeral and, 272–273; on Brown's Harper's Ferry pronouncements, 245; on Brown's plans for South, 187; Brown's travels east and, 68–69; "Cabinet" speech of, 420–422; Civil War and, 366–368, 386–387; on 1860 election, 306–307; Emancipation

League and, 386; Emancipation Proclamation and, 449; England, on *Trent* affair, 388–389; 54th Regiment and, 480–481; on Frémont nomination, 541; on Garrison's endorsement of Lincoln, 547; gradual abolition proposal, 396; on insurrection, 259–260, 286–287; Johnson administration and, 616–617; on Lincoln, 301–302, 433, 569; Lincoln's preliminary emancipation proclamation and, 432; on meeting with Brown and Tubman, 167; meeting with Lincoln and, 458–460; New England Anti-Slavery Conference (1860) and, 295–296; on peace condition proclamation, 520–522; on secession, 319; on Seward, 294, 397; on slavery, 184; split with Garrison, 539–540; Tenth National Women's Rights Convention, 293; war proclamation and, 340; in Washington, 396–398

Phillips, William Addison, 14, 44, 47, 84–85, 130, 150–153, 209, 508

Pierce, Franklin: Burns case and, 24; Kansas Territory and, 39, 44–45, 47, 59; proslavery party and, 185

Pierpoint, John, 302–303

Pierpont, F. H., 354

Pillsbury, Parker, 540

Pinkerton, Allen, 156, 326, 562–563

Pionier, Der, 312

Plain Dealer (Cleveland), 160

Planciancoise, Anselmas, 479

Planter, The, 610

Plea for Captain John Brown (Thoreau), 21, 253–254, 400

Plumb, Preston, 130

Plumb, Ralph, 209

Plumb, Samuel, 209

Poison Springs, 532

Polk, Leonidas, 509, 527

Pollard, Edward A.: on Battle of Bull Run, 356–357, 361–362; on Battle of Cold Harbor, 542; Battle of Murfreesboro and, 445–447, 450–451; on Battle of Richmond, 602–603; on Cemetery Hill, 552; on Davis' Hampton Road address, 587; Fort Wagner and, 496–497; on Jackson, 470; on Roanoke Island, 395; on Sherman, 600–601; on Washington, 549

Pomeroy, Samuel C., 261, 502, 540

Pope, John, 381, 412, 419–420, 424–425

Porter, Fitz John, 408–409, 435, 606

Port Huron, assault of, 477–482, 489, 492

Pottawatomie Massacre, 43

Presse, Die, 47, 380, 420, 428, 434

Preston, William, 298

Price, John, 159, 161

Price, Sterling, 355–356, 374, 377, 380–381, 433

Provincial Freeman, 112, 116, 204

Pugh, George A., 288

Purvis, Robert, 270, 295, 476

Quakers, Brown and, 95–96

Radical Abolition Party, 35–36, 53, 53–55, 302–303

Radical Abolition Society, 164–165

Radical Republicans, 540–542

Randolph, Thomas Jefferson, 142

Raymond, Henry, 553–554

Realf, Richard: on Brown and insurrectionary movements, 136; Brown's Chatham Convention and, 117–119, 122; in Cleveland, 126; desertion of, 127–128; Forbes and, 125; in Kansas, 40, 84; Mason Committee and, 180, 277; portrait of, 91*ph*; recruitment of for Virginia, 89; stay with Quakers, 95

reconstruction: after the Civil War, 616–618; Bank's Red River expedition and, 527; Douglass on, 532; Freedmen's Bureau and, 590–591; A. Johnson and, 547; meeting in Savannah with "contrabands", 580–584; Special Field Order No. 15, 582; Thirteenth Amendment and, 580–581; Wade-Davis Bill and, 548; Wade-Davis manifesto and, 555

Redpath, James: African Civilization

Society and, 323; on Brown in Kansas, 60; in Kansas, 40, 62; Mason Committee and, 277; meeting at Brown's grave and, 299; on Sumner visit, 106
Reeder, Andrew, 81
Regan, J. H., 328
Reid, John W., 56–57, 61
Reminiscences (Schurz), 557
Remond, Charles Lenox, 79, 115, 295
Remond, Sarah Parker, 275
Republican National Committee peace convention, 553–554
Republican Party: Brown on, 192; convention of 1860 of, 294–295; convention of 1864 and, 540, 547; divisions of in 1864, 548; Douglass and, 53–55; elections of 1864 and, 439; Fort Sumter and, 334, 338; founding of, 35; Kansas and, 47; Lincoln administration and, 327; Lincoln and, 282, 555–557; Lincoln inauguration and, 331; Lincoln's preliminary emancipation proclamation and, 432; in Osawatomie, 201–202; platform of 1856 and, 53; presidential campaign of 1860 and, 303–306; slavery issue of, 78–79; Thirteenth Amendment and, 578–580; 36th Congress and, 276, 280; war proclamation and, 341
Reynolds, G. J., 118, 121, 125–126
Reynolds, John, 488
Rhett, A. B., 437
Rhett, R. B., Jr., 288
Richardson, Richard, 90–91, 95, 118
Richmond, 603
Richmond: capture of, 603–607; defense of, 545; evacuation of, 602–603; John Brown hymn and, 604–606
Richmond Dispatch, 289
Richmond Enquirer, 141
Richmond Examiner, 432, 543
Roanoke Island, Battle of, 395
Robinson, Charles, 38–39, 41, 52, 56, 59, 61, 223
Robinson, Lucius, 540
Rochester Democrat and American, 235, 248
Rock, John S., 159, 301, 580, 596
Rollin, Frank A. *See* Whipper, Frances Rollin
Root, J. P., 66–67
Rosencrans, William, 439, 446–447, 450–451, 489, 509–510
Ross, Alexander M., 239–240, 262
Roudanez, Jean Baptiste, 522
Ruffin, Edwin, 256, 280–281, 338, 364
Russell, Thomas, 68, 81, 262, 265

Sanborn, Franklin: attack of abolitionist meetings and, 309; on Brown in Kansas, 43, 60–61; Brown's appeal to black leaders and, 100; on Brown's business endeavors, 13; Brown's travels east and, 68–69, 72, 80–81; Brown's Virginia plans and, 103–105, 176–177; correspondence with Brown, 85; Emancipation League and, 386; flight to Maine of, 249; Forbes and, 124, 126–127; letter from John Brown Jr., 52; on *Life and Times* account of Brown meeting, 10, 98; Mason Committee and, 277–278; meetings with Brown in 1859, 166; Merriam and, 207–208; on J. Montgomery, 132–133; portrait of, 71*ph*; Tubman and, 167, 200, 293, 313, 495
Savannah, 577–578
Savannah Republican, 281
Saxton, Rufus, 482, 594, 596, 611
Scenes in the Life of Harriet Tubman (Bradford), 596
Schofield, John, 511–512, 569, 573, 576
Schurz, Carl, 557, 616
Scott, Dred, 73–74
Scott, Winfield: "Anaconda" plan of, 356–357; defense of Washington and, 345; Fort Sumter and, 333–336; plot against Lincoln of 1860 and, 326, 329; secession of border states and, 347
secession movements, 304

Second Bull Run, 424–425

Second Manassas, 424

secret six, 249

Seddon, James, 472

Sedgwick, John, 470–472

Sepoy War, 194

Seven Days' Battle, 412

Seven Pines, Battle of, 407

Seward, Frederick, 326

Seward, William H.: on abolition, 571; bid for 1860 nomination and, 293–295; blacks and, 476; Crittenden Compromise and, 308; Emancipation Proclamation and, 448; England, *Trent* affair and, 388; Forbes and, 124; Fort Sumter and, 333–337, 352; Hampton Road negotiations and, 584–585; Lincoln's abolition proposal and, 417; Lincoln's cabinet and, 327–329; on maintaining the Union, 324; Mason Committee and, 277; memorandum to Lincoln, 336–337; portrait of, 335*ph*; presidential campaign of 1860 and, 305–306; proclamation of emancipation and, 418; Republican convention of 1860 and, 294–295; resignation of, 442–443; secession of Southern states and, 308; Thirteenth Amendment and, 578–579

Seymour, Horatio, 439, 564

Seymour, Thomas H., 564

Shadd, Abraham, Brown's appeal to black leaders and, 112–113

Shadd, I. D., 112, 118, 122, 196, 204

Sharpsburg, Battle of, 429–430, 435

Shaw, Robert Gould, 463, 497, 498*ph*, 499

Sheffield Anti-Slavery Association, 275

Shenandoah Valley battles, 405–407

Shepherdstown, Battle of, 435

Sheridan, Philip: Battle of Cedar Creek and, 568; Battle of Cold Harbor and, 542; Battle of Murfreesboro and, 446; Battle of Spotsylvania and, 537; Battle of Trevillian Station and, 544–545; Lee's surrender and, 608; in

Shenandoah Valley, 567–568; in Virginia, 551; at Waynesboro, 600

Sherman, John: at Dalton, 535; 36th Congress and, 276, 280

Sherman, William T.: blacks and, 580–582; black troops and, 534; burning of Atlanta, 575–576; capture of Savannah and, 577–578; Charleston and, 589–590; Columbia and, 590; conference with Lincoln in Virginia, 601; at Dalton, 535; Department of the South and, 403; Department of the West and, 528; on Forrest, 544; in Georgia, 538–539, 542, 546, 550–552, 566–567, 568–569; at Jackson, 469–470; march through Carolinas, 587–590, 600–601; march through Mississippi of, 526–528; march to the sea of, 576–577; McClernand and, 445; plan to march on to South Carolina, 573; Special Field Order No. 15, 582; surrender negotiations and, 611; taking of Atlanta, 564; Vicksburg and, 446

Shield, James, 407

Shiloh, Battle of, 398–399

Sigel, Franz, 349, 373–374, 381, 395, 537–538

slavery: Civil War and, 368–372; conditions in Virginia, 187; cotton production and, 76–77; J. Davis on, 282–283; distribution of, 18–19*m*; insurrectionary movements and, 64–65, 74–75, 85–86, 135–137; numerical power and, 184–186; onset of Civil War and, 351; presidential campaign of 1860 and, 303–306; Thirteenth Amendment and, 578–580; U.S. expansion of, xi-xii

Slavery and Abolition (Hart), 499

slaves, 224; after Emancipation Proclamation, 513; communication among, 286; Confederate conscription, 582–583; distribution of, 18–19*m*; Eaton and, 554; at Harper's Ferry, 260; reconstruction efforts for emancipated, 455–456; as soldiers, 366, 455;

Thirteenth Amendment and, 578–580; Union position towards, 368–372; vigilante action and, 281

slave uprisings, 64–65, 74–75, 85–86, 135–137, 256–257

Slidell, John, 387–388

Smalls, Robert, 610

Smith, Caleb, 327, 335

Smith, Gerrit, 14–15; arrangements for support of Brown's family, 83; Brown's appeal to black leaders and, 100, 102; Brown's decision to go to Kansas and, 34; on Brown's raid in Missouri, 148; Brown's travels east and, 68; Brown's Virginia plans and, 103–104; calls for emancipation proclamation and, 391; denial of Brown, 249; Douglass and Radical Abolition Party and, 54–55; Douglass's Syracuse appearance and, 385; Forbes and, 126–127; implication of in Harper's Ferry raid, 246; influence of Douglass, 26; on legislation of slavery, 180; Mason Committee and, 277; meetings with Brown in 1859, 165; portrait of, 71*ph*; Radical Abolition Party and, 35–36, 53, 302; recruiting of black volunteers, 462; on slavery, 201; Tubman and, 111, 209, 290; on war and abolition, 571

Smith, Kirby, 426–427, 428, 612

Smith, Parker T., 236

Smith, Stephan, 103, 107

Smith, William "Baldy", 545

Snead, Thomas L., 375

Snyder, Eli, 131, 133

Sociology of the South (Fitzhugh), 77

Sons of Liberty, 548, 562–563

Soule, Pierre, 402

Southampton Insurrection, The (Drewry), 138

South Carolina, 281, 307, 316–319, 583

Southern History of the War (Pollard), 356–357, 395, 542

Special Field Order No., 15, 582

Spooner, Lysander, 283–284

Spotsylvania, Battle of, 537–538

Sprague, William, 360

Spring, Rebecca, 262, 266

Springfield, Battle of, 395

Springfield State Journal, 303

Squatter Sovereign, 41

"Squatter Sovereignty Bill". *See* Kansas-Nebraska Bill

St. Domingo revolution, xi

St. Louis Democrat, 350

St. Louis Observer, 2

Stanley, Edwin, 458–460

Stanton, Edwin: Battle of Gaines' Mill and, 409; black volunteers and, 463–464; commission of Delany and, 594–595; Fort Sumter and, 316; Lincoln's cabinet and, 392; meeting with Douglass, 504–505; Second Bull Run and, 425; Sherman and, 581; Vicksburg and, 415

Stanton, Elizabeth Cady, 6–7, 293, 303

Starry, John, 218–221

Stearns, George L.: arrangements for support of Brown's family, 83; Brown's appeal to black leaders and, 100, 103; Brown's requests for funds, 82; Brown's travel's east and, 67–70, 72, 167–169; correspondence with Brown and, 87–89; Emancipation League and, 386; flight to Canada of, 249; Forbes and, 124–125, 126–127; Mason Committee and, 277–278; meeting with Lincoln and, 458–460; portrait of, 71*ph*; recruiting of black volunteers, 461, 490, 502; Tubman and, 209

Stephens, Alexander, 94, 318, 341–342, 443–444, 452, 584–586

Stevens, Aaron D. "Colonel Whipple": in Cleveland, 125–126; continued fighting in Kansas and, 144–145; execution of, 278; first meeting with Brown, 49–50; in Harper's Ferry, 172–173, 206; Harper's Ferry raid and, 212–213, 216–217, 220, 225, 227–228, 230, 231, 240; Kansas War and, 59; portrait of, 91*ph*; in prison, 247, 261–262, 268; recruitment of for Virginia, 89–90; taken in to custody, 242; trial of, 252–253; trip to Canada with fugitive slaves and, 153

Stevens, Thaddeus, 391, 616

Still, William: in anti-slavery movement, 3; Brown's appeal to black leaders and, 103, 107–109; correspondence with Douglass, 299; Henry "Box" Brown and, 141; on Sumner's 1860 Senate speech, 300; on Tubman, 111

Stoeber, Edward M., 612–613

Stone, Charles P., 392–393

Stone, Lucy, 293

Stone's River, Battle of, 445–448, 450–451

Stowe, Harriet Beecher, 27–30

Strangers' Guard, 346

Stringer, Thomas, 112, 118

Stringfellow, J. H., 41

Stuart, J. E. B., 45; Battle of Brandy Station and, 486; Battle of Bull Run and, 359, 361; Battle of Chancellorsville and, 470, 472; Battle of Spotsylvania and, 537; at Chambersburg, 436; Harper's Ferry raid and, 240–241; reconnaissance of McClellan, 408

Sturges, Joseph, 5

Sturgis, S. D, 539, 542, 543–544

Summit Beacon, 39

Sumner, Charles, 42, 53, 106; Battle of Chancellorsville and, 472; District of Columbia abolition bill and, 398; Johnson administration and, 616; National Equal Rights League and, 570; on rights of Confederate states, 523; on slavery, 300–301; Thirteenth Amendment and, 580

Sumner, Edwin, 44–45, 46, 47

Syracuse Herald, 201

Taney, Roger, 73–74, 330

Taylor, Richard, 485, 528, 566, 569

Taylor, S. M., 612

Taylor, Stewart, 95, 172, 212, 228

Tennessee: anti-secessionist movement and, 354; secession and, 317–318, 342, 352; Union recognition of loyal government in,

522–523; war proclamation and, 340

Texas, 74, 317

Thayer, Eli, 47, 68, 79, 81

Thirteenth Amendment, 578–580

36th Congress, 276, 280

Thomas, George H., 509–510, 528, 577

Thomas, Lorenzo, 379, 463–464, 474–476, 502, 505

Thomas, Thomas, 36

Thompson, Adolphus, 15, 213

Thompson, Anne, 15

Thompson, Dauphin, 172

Thompson, George, 5, 275

Thompson, Henry, 15, 36, 42, 48, 109

Thompson, Jacob, 548, 562

Thompson, Mary, 206

Thompson, Ruth Brown, 15, 109, 272

Thompson, William, 15, 48, 172, 213*ph*, 223, 225–226, 229–231, 235

Thoreau, Henry David, 21; on Brown, 264; Brown and, 400; on Brown execution, 273; Brown's travels east and, 68, 80, 82; death of, 400; on Harper's Ferry raid, 253–254, 261

Tidd, Charles Plummer, 88; in Harper's Ferry, 172–173; on Harper's Ferry raid, 233; Harper's Ferry raid and, 212, 218, 222, 232, 235; in Kansas, 144–145; Oberlin-Wellington rescue trial and, 161; reconnaissance in Missouri and, 133

Tilton, Theodore, 506

Titus, Harry, 41, 50–51, 61–63

Todd, John, 67, 84

Toombs, Robert: Battle of Antietam and, 430; Crittenden Compromise and, 308; Davis and, 452; Davis's cabinet and, 327–328, 443; Fort Sumter and, 337; portrait of, 335*ph*

Toronto Enquirer, 240

Townsend, E. D., 581

Tragic Prelude (Curry), 33

Transactions (Kansas State Historical Society), 51–52

Transcendentalists, 68
Tremont Temple, 309–311
Trent, 387–389
Trevillian Station, Battle of, 544–545
Troy Whig, 291–292
Truth, Sojourner, 26
Tubman, Harriet: anti-abolitionist sentiment and, 312–314; in anti-slavery movement, 3; bond with Brown of, 174–175; Bradford and, 595–596; Brown's appeal to black leaders and, 110–112, 158–159, 202; Brown's Chatham Convention and, 115; Brown's Virginia plans and, 175–176; Civil War activities of, 404, 482–484, 495–496; Delany and, 596; emancipated slaves and, 456; meetings with Brown in 1859, 165–166; Nalle and, 290–292; New England Anti-Slavery Conference (1860) and, 295; portrait of, 113*ph*; Shaw and, 499; trips to Boston of, 167–168, 173–174, 293
Turner, Benjamin, 137
Turner, George W., 227, 256
Turner, Henry McNeal, 609
Turner, Nat, 3, 135–136, 137–141
Turn Verein, 312

U.S. Armory at Harper's Ferry: establishment of, 170; seizing of by Virginia secessionists, 342; taking of, 212–213
U.S. Marines, 232, 240
U.S. Supreme Court, Dred Scot case and, 66, 73–75
Uncle Tom's Cabin (Stowe), 27
Underground Railroad: Fugitive Slave Law and, 20–21; in Harper's Ferry, 177; importance of, 2; in Missouri, 201–202; Oberlin-Wellington rescue trial, 159–162; as source of "coadjutors" for Brown, 15
Underground Railroad (Still), 108
United States League of Gileadites, 21–22, 68
Upton, Emory, 537

Vaill, H. L., 263
Vallandigham, Clement C., 242–245; arrest of, 473–474; Democratic Party and, 466, 562–564; portrait of, 563*ph*; Sons of Liberty and, 548
Van Dorn, Earl, 433, 445
Vashon, George, 271
Vicksburg, Battle of, 403, 448, 489
Vigilance Committee, 3, 21, 23, 71
Villard, Oswald Garrison, 51, 265
Virginia, 396, 603
Virginia: aftermath of Turner revolt and, 141–143; anti-secessionist movement and, 354; conditions for slaves and, 187; recognition of neutral Virginia, 357–358; secession and, 317–318, 340, 342, 352–354; Turner revolt and, 137–141; Union recognition of loyal government in, 522–523; war proclamation and, 340
Voice from Harper's Ferry, A (Anderson), 204
von Holst, Hermann E., 223
voting laws, 14

Wade, Benjamin, 308, 334, 392, 548, 555
Wade-Davis Bill, 523, 548
Wade-Davis manifesto, 555
Wagoner, Henry O., 28, 300–301
Walker, David, 2
Walker, George, 68–69
Walker, L. P., 328, 337
Walker, Robert, 86–87, 88
Walker, Samuel, 49–51, 56–58, 63
Walker, William, 62
Ward, Artemus, 160
Washington, D.C., 343, 345–346, 364, 549
Washington, Lewis W., 216–220, 228
Washington Republican, 438
Washington Star, 425
Watkins, Frances Ellen, 263
Watkins, William, 271
Watson, Henry, 171, 177, 190, 209
Wattles, Augustus, 84–85, 146–147
Webb, William, 157–158
Webster, Thomas, 507
Weed, Thurlow, 294, 308, 555

Weekly Anglo African, 115
Weitzel, Gen., 567–568, 603–604, 606
Welles, Gideon, 327, 345, 396, 417, 510–511
Weston Argus, 56
Wheeler, Joseph, 446
Whipper, Frances Rollin, 114
Whipple, Alfred, 118, 122
White, Martin, 133
Whitman, Edmund, 84, 88, 149
Whitman, Walt, 364–365, 556–557, 590, 611
Whitney, Eli, 170
Wickliffe, Governor, 399
Wide-Awakes, 303, 349
Wigfall, Louis T., 285
Wilberforce, Ohio, 592
Wilderness, Battle of the, 535–536
Wilkes, Charles, 387–389
Williams, "Fiddling", 130, 143–145
Wilson, Henry, 167, 277–278, 398
Wilson's Creek, Battle of, 373–374
Winkley, Jonathan, 51
Winters, Joseph, 177, 200
Wisconsin, 339
Wise, Henry A.: Brown execution and, 266; on Canada and Douglass, 259; Douglass and, 249–250; on Harper's Ferry raid, 247; insurrectionary movements and, 256–258, 258–259; interrogation of Brown and, 242–245; militia response to Harper's Ferry raid and, 226, 232, 237, 240; at Petersburg, 545; on response to Harper's Ferry, 245–246; Ross and, 262; secession and, 342; Stevens' trial and, 252–253; visit to Brown in prison of, 262, 264
Women's Loyalty League, 517
women's rights movement, 6–7, 293
Wood, Fernando, 324, 415
Wright, Henry C., 64
Wright, Theodore, 108

Yancey, William Lowndes, 288–289, 318
Yorktown, Battle of, 401–402
Young, Joshua, 272

Zagoni, Major, 379